PACIFIC SYMPOSIUM ON BIOCOMPUTING 2008

PACIFIC SYMPOSIUM ON BIOCOMPUTING 2008

Kohala Coast, Hawaii, USA
4–8 January 2008

Edited by

Russ B. Altman
Stanford University, USA

A. Keith Dunker
Indiana University, USA

Lawrence Hunter
University of Colorado Health Sciences Center, USA

Tiffany Murray
Stanford University, USA

Teri E. Klein
Stanford University, USA

World Scientific

NEW JERSEY • LONDON • SINGAPORE • BEIJING • SHANGHAI • HONG KONG • TAIPEI • CHENNAI

Published by

World Scientific Publishing Co. Pte. Ltd.
5 Toh Tuck Link, Singapore 596224
USA office: 27 Warren Street, Suite 401-402, Hackensack, NJ 07601
UK office: 57 Shelton Street, Covent Garden, London WC2H 9HE

British Library Cataloguing-in-Publication Data
A catalogue record for this book is available from the British Library.

BIOCOMPUTING 2008
Proceedings of the Pacific Symposium

ISBN-13 978-981-277-608-2
ISBN-10 981-277-608-7

Printed in Singapore by Mainland Press Pte Ltd

PACIFIC SYMPOSIUM ON BIOCOMPUTING 2008

This year, PSB returns to its most common venue, the Fairmont Orchid on the Big Island of Hawaii. We are now in our thirteenth year, and had a record number of both proposed sessions (we accepted 9) as well as submissions to the conference this year (150). Many sessions at PSB have a lifetime of approximately three years. The first year is a test of the interest in the field, and the ability to attract a critical mass of papers. With success, the second year is usually a larger, more competitive session. The third (and rarely fourth years) usually come as the subdiscipline is recognized at the general biocomputing meetings such as ISMB, RECOMB and others, and often this is when the PSB organizers conclude that their work is done, and the session can be "retired." PSB aficionados will notice new sessions this year in next-generation sequencing, tiling array analysis, multiscale modeling, small regulatory RNAs, and other areas. The richness of exciting new areas has led us to have more total sessions, and thus the average session size is smaller. We consider this an experiment, and look forward to seeing how it goes.

We would like to thank our keynote speakers. Dr. Andrew McCulloch, Professor and Chair, Department of Bioengineering, University of California, San Diego, will talk about "Systems biology and multi-scale modeling of the heart." Our keynote in the area of Ethical, Legal and Social implications of technology will be John Dupre, Director of Egenis (ESRC Centre for Genomics in Society) and Professor of Philosophy of Science, University of Exeter.

PSB provides sessions focusing on hot new areas in biomedical computation. These sessions are usually conceived during earlier PSB meetings, as emerging fields are identified and targeted. The new sessions are lead by relatively junior faculty members trying to define a scientific niche and bring together leaders in these exciting new areas. Many areas in biocomputing have first been highlighted at PSB. If you have an idea for a new session, contact the organizers at the meeting or by e-mail.

Again, the diligence and efforts of a dedicated group of researchers has led to an outstanding set of sessions, with associated introductory tutorials. These organizers provide the scientific core of PSB, and their sessions are as follows:

Michael Brudno, Randy Linder, Bernard Moret, and Tandy Warnow. *Beyond Gap Models: Reconstructing Alignments and Phylogenies Under Genomic-Scale Events*

Doron Betel, Christina Leslie, and Nikolaus Rajewsky. *Computational Challenges in the Study of Small Regulatory RNAs*

Francisco De La Vega, Gabor Marth, and Granger Sutton. *Computational tools for next-generation sequencing applications*

Michael Ochs, John Quackenbush, and Ramana Davuluri. *Knowledge-Driven Analysis and Data Integration for High-Throughput Biological Data*

Atul Butte, Maricel Kann, Yves Lussier, Yanay Ofran, Marco Punta, and Predrag Radivojac. *Molecular Bioinformatics for Diseases: Protein Interactions and Phenomics*

Jung-Chi Liao, Peter Arzberger, Roy Kerckhoffs, Anushka Michailova, and Jeff Reinbolt. *Multiscale Modeling and Simulation: from Molecules to Cells to Organisms*

Martha Bulyk, Ernest Fraenkel, Alexander Hartemink, and Yael Mandel-Gutfreund. *Protein-Nucleic Acid Interactions: Integrating Structure, Sequence, and Function*

Antonio Piccolboni and Srinka Ghosh. *Tiling Microarray Data Analysis Methods and Algorithms*

Kevin Bretonnel Cohen, Philip Bourne, Lynette Hirschman, and Hong Yu. *Translating Biology: Text Mining Tools that Work*

Tiffany Murray and Susan Kaiser provided outstanding support of the peer review process, assembly of the proceedings, and meeting logistics. We thank the National Institutes of Health, National Science Foundation, Applied Biosystems, and the International Society for Computational Biology for travel grant support. We also acknowledge the many busy researchers who reviewed the submitted manuscripts on a very tight schedule. The partial list following this preface does not include many who wished to remain anonymous, and of course we apologize to any who may have been left out by mistake.

Aloha! Pacific Symposium on Biocomputing Co-Chairs,
September 28, 2007

Russ B. Altman
Departments of Bioengineering, Genetics & Medicine, Stanford University
A. Keith Dunker
Department of Biochemistry and Molecular Biology, Indiana University School of Medicine
Lawrence Hunter
Department of Pharmacology, University of Colorado Health Sciences Center
Teri E. Klein
Department of Genetics, Stanford University

Thanks to the reviewers...

Finally, we wish to thank the scores of reviewers. PSB requires that every paper in this volume be reviewed by at least three independent referees. Since there is a large volume of submitted papers, paper reviews require a great deal of work from many people. We are grateful to all of you listed below and to anyone whose name we may have accidentally omitted or who wished to remain anonymous.

Joshua Adelman
Baltazar Aguda
Eric Alm
Rommie Amaro
Sophia Ananiadou
Alan Aronson
Joel Bader
Chris Baker
Kim Baldridge
Brad Barbazuk
Ziv Bar-Joseph
James Bassingthwaighte
Serafim Batzoglou
William Baumgartner Jr.
Daniel Beard
Takis Benos
Bonnie Berger
Ghislain Bidaut
Judith Blake
Christian Blaschke
Guillaume Blin
Olivier Bodenreider
Benjamin Bolstad
James Bonfield
Rich Bourgon
Phil Bourne
Guillaume Bourque
Karl Broman
Mike Brownstein
Christopher Bruns
Philipp Bucher
Martha L. Bulyk
Herman Bussemaker
Diego Calzolari
Stuart Campbell
Amancio Carnero
Bob Carpenter
David Case
Mark Chaisson
Wendy Chapman
Gal Chechik
Asif Chinwalla
Kuo Ping Chui
Melissa Cline
Aaron Cohen
Sarah Cohen-Boulakia
Carlo Colantuoni
Francois Collin
Leslie Cope
Tony Cox
Mark Craven
Aaron Darling
Debo Das
Christopher Davies
Francisco De La Vega
Dennis Dean
Arthur Delcher
Dina Demner-Fushman
Ye Ding
Katharina Dittmar de la Cruz
Marko Djordjevic
Ian Donaldson
Joaquin Dopazo
Eran Eden
Daniel Einstein
Ahmet Erdemir
Jason Ernst
Steven Eschrich
Alexander Favorov
Hakan Ferhatosmanoglu
Carlos Ferreira
David Finkelstein
Russell Finley
Juliane Fluck
Lynne Fox
Ernest Fraenkel
Christine Gaspin
Curtis Gehman
Debashis Ghosh
Fabian Glaser
Margy Glasner
Jarret Glasscock
Harley Gorrell
Ivo Grosse
Trent Guess
Donna Harman
David Haussler
William Hayes
Marti Hearst

Bill Hersh
Jana Hertel
Lynette Hirschman
Mark Holder
Jeffrey Holmes
Masahiko Hoshijma
Jim Huang
Alex Hudek
Timothy R. Hughes
Ela Hunt
Ilya Ioschikhes
Lakshmanan Iyer
Leighton Izu
Elling Jacobsen
Jeffrey Jacot
Saleet Jafri
Anil Jegga
Lars Juhl Jensen
Susan Jones
John Kececioglu
Roy Kerckhoffs
Abbas Khalili
Seungchen Kim
Sun Kim
Judith Klein-Seetharaman
Jim Knight
Andrew Kossenkov
Martin Krallinger
Ellen Kuhl
Martin Kurtev
Alain Laederach
Jens Lagergren
Juan Pablo Lewinger
Ming Li
James Liao
Jung-Chi Liao
Jimmy Lin
Ross Lippert
Guoying Liu
Yunlong Liu
Sandya Liyanarachchi
Kenzie MacIsaac
Yael Mandel-Gutfreund
Luigi Marchionni
Hanah Margalit
Debora Marks
Gabor Marth
Satoshi Matsuoka
Andrew McCulloch
Anushka Michailova
Julie Mitchell
Edwin Moore
Alex Morgan
Burkhard Morgenstern
David Morrison
Salvatore Mungal
Luay Nakhleh
Shu-Kay Ng
Bill Noble
Cedric Notredame
Uwe Ohler
Sean O'Rourke
David Parker
Suraj Peri
Helene Perreault
Graziano Pesole
Steve Piazza
George Plopper
Mihai Pop
Dustin Potter
Amol Prakash
Jose Puglisi
Huaxia Qin
Randy Radmer
Marco Ramoni
Ronald Ranauro
Wouter-Jan Rappel
John Rasmussen
John Jeremy Rice
Thomas Rindflesch
Phocbc Roberts
Carlos Rodriguez
Antonios Rokas
Michael Rosenberg
Mikhail Roytberg
Andrey Rzhetsky
Ravi Sachidanandam
Frank Sachse
Akinori Sarai
I. Neil Sarkar
Jeffrey Saucerman
Rob Scharpf
Ariel Schwartz
Paola Sebastiani
Ilya Serebriiskii
Sohrab Shah
Harris Shapiro
Changyu Shen
Robert Sheridan
Michael Sherman
Asim Siddiqui
Jonathan Silva
Gregory Singer
Saurabh Sinha
Steve Skiena
Barry Smith
Doug Smith
Larry Smith
Nicholas Socci
Melissa St. Hilaire
David States
David Steffen
Tim Stockwell
Chris Stoeckert
Gary Stormo
Jens Stoye
Krishna Subramanian

Chuck Sugnet
Hao Sun
Granger Sutton
Merryn Tawhai
Sarah Teichmann
Alun Thomas
Nicki Tiffin
Jun'ichi Tsujii
David Tuck
Simon Twigger
Rajanikanth
Vadigepalli
Alfonso Valencia
Giorgio Valle
Karin Verspoor
Todd Vision
Bonnie Webber
Zasha Weinberg
W. John Wilbur
Derek Wilson
Zhohar Yakhini
Yuzhen Ye
Zeyun Yu
Aleksey Zimin
Pierre Zweigenbaum
Derrick Zwickle

CONTENTS

PROTEIN-NUCLEIC ACID INTERACTIONS: INTEGRATING STRUCTURE, SEQUENCE, AND FUNCTION

TILING MICROARRAY DATA ANALYSIS METHODS AND ALGORITHMS

TRANSLATING BIOLOGY: TEXT MINING TOOLS THAT WORK

BEYOND GAP MODELS: RECONSTRUCTING ALIGNMENTS AND PHYLOGENIES UNDER GENOMIC-SCALE EVENTS

MICHAEL BRUDNO
University of Toronto

BERNARD MORET
EPFL, Switzerland

RANDY LINDER
The University of Texas at Austin

TANDY WARNOW
The University of Texas at Austin

Multiple sequence alignment (MSA) has long been a mainstay of bioinformatics, particularly in the alignment of well conserved protein and DNA sequences and in phylogenetic reconstruction for such data. Sequence datasets with low percentage identity, on the other hand, typically yield poor alignments. Now that researchers want to produce alignments among widely divergent genomes, including both coding and noncoding sequences it is necessary to revisit sequence alignment and phylogenetic reconstruction under more ambitious models of sequence evolution that take into account the plethora of genomic events that have been observed.

Most current methods postulate only two types of events: substitutions (modeled with a transition matrix, such as PAM or BLOSUM matrices for protein data) and insertions/deletions or *indels* (rarely modelled beyond a simple affine cost function of the size of the gap). While these two events can indeed transform any sequence into any other, this model of genomic events is far too simplistic: substitutions are not location- or neighbor-independent, and indels can be caused by a variety of complex events, such as uneven recombination, insertion of transposable elements, gene duplication/loss, lateral transfer, etc. Moreover, genomic rearrangement events can completely mislead procedures based on most current models, resulting in a total loss of alignment when a homologous element has undergone an inversion or a duplication.

The aim of our session is to bring together researchers in multiple sequence alignment, phylogenetic reconstruction, comparative genomics, DNA sequence analysis, and genetics to examine the state of the art in multiple sequence alignment, discuss how methods can be improved, and whether current projects will suffice for the emerging applications in various biological fields. The four papers in our session, while centering around the topic of sequence comparison, represent the breadth of interests of scientists in the field: algorithms to generate and analyze alignments, the estimation of phylogenetic trees and how these phylogenies affect alignment algorithms, and analyzing the frequencies of genome rearrangements in various locations of the genome.

Our session has four papers addressing different aspects of the general problem. The paper by Dalca and Brudno presents a unifying view of many sequence alignment algorithms. The authors propose the rectangular scoring scheme framework and demonstrate algorithms to speed up comparison of sequences with arbitrary rectangular scoring. While the resulting program is too slow for whole-genome applications, it can allow for easy prototyping of complex scoring schemes for alignments.

The paper by Landan and Graur addresses the problem of finding the regions of high reliability within a multiple alignment. The authors present an elegant algorithm that determines if the alignments are changed when the sequences are reversed – an indication of a region where the alignment is less reliable. This work has implications for phylogeny estimation, since low-confidence regions within the alignment can then be down-weighted (or even eliminated) during a phylogeny estimation, and thus potentially lead to more accurate phylogenetic estimates.

The paper by Nelesen and colleagues addresses the impact of the choice of guide tree on multiple alignment methods, and on the phylogenetic estimations obtained using the resultant multiple alignments. Their simulation study shows that some methods (for example, ProbCons) are highly responsive to the particular guide tree, while others (for example, Muscle) are less responsive. In addition, they provide a particular technique for producing the guide tree that results in much better estimates of phylogenies than the current gold standard.

The fourth paper in the session by Sinha and Meller addresses the use of genome rearrangements in the estimation of evolutionary relationships between genomes. The potential for genome rearrangements to reveal evolutionary histories is great, but accurate reconstructions require better understandings of the frequencies of the various events, such as inversions, transpositions, and duplications. Sinha and Meller make important inroads on this problem by analyzing how varying definitions of a synteny block affect the observed inversion and breakpoint rates. One of the most interesting conclusions is that the definition of a synteny block has little effect on the estimation of the reuse of breakpoints, shedding additional light on an ongoing academic controversy in the field.

We are excited by the breadth of research taking place in the fields of MSA and phylogeny estimation, and are hopeful that our session will help bring together researchers in these areas. The four papers presented at our session were selected with the help of several reviewers, whose help we gratefully acknowledge.

FRESCO: FLEXIBLE ALIGNMENT WITH RECTANGLE SCORING SCHEMES *

A.V. DALCA[1] AND M. BRUDNO[1,2]

1. Department of Computer Science, and 2. Donnelly Center for Cellular and Biomolecular Research, University of Toronto, Toronto, Canada
{dalcaadr, brudno}@cs.toronto.edu

While the popular DNA sequence alignment tools incorporate powerful heuristics to allow for fast and accurate alignment of DNA, most of them still optimize the classical Needleman Wunsch scoring scheme. The development of novel scoring schemes is often hampered by the difficulty of finding an optimizing algorithm for each non-trivial scheme. In this paper we define the broad class of *rectangle scoring schemes*, and describe an algorithm and tool that can align two sequences with an arbitrary rectangle scoring scheme in polynomial time. Rectangle scoring schemes encompass some of the popular alignment scoring metrics currently in use, as well as many other functions. We investigate a novel scoring function based on minimizing the expected number of random diagonals observed with the given scores and show that it rivals the LAGAN and Clustal-W aligners, without using any biological or evolutionary parameters. The FRESCO program, freely available at http://compbio.cs.toronto.edu/fresco, gives bioinformatics researchers the ability to quickly compare the performance of other complex scoring formulas without having to implement new algorithms to optimize them.

1. Introduction

Sequence alignment is one of the most successful applications of computer science to biology, with classical sequence alignment programs, such as BLAST[1] and Clustal-W[2], having become standard tools used by all biologists. These tools, developed while the majority of the available biological sequences were of coding regions, are not as effective at aligning DNA[3]. Consequently, the last ten years have seen the development of a large number of tools for fast and accurate alignment of DNA sequences. These alignments are typically not an end in themselves - they are further used as input to tools that do phylogeny inference, gene prediction, search for tran-

*This work is supported by an NSERC USRA and Discovery grants scription factor binding sites, highlight areas of conservation, or produce another biological result.

Within the field of sequence alignment, several authors[4,5,6,7] have noted the distinction between the function that is used to score the alignment, and the algorithm that finds the best alignment for a given function. Given an arbitrary scoring scheme, it is normally easy to assign a score to an already-aligned pair of sequences, but potentially more complicated to yield a maximal-scoring alignment if the sequences are not aligned to begin with. While for some scoring schemes, such as edit distance or Needleman-Wunsch, the optimizing algorithm is simple to write once the scoring function is defined, in other cases, such as the DIALIGN scoring metric, it is trivial to score a given alignment, but the algorithm which one could use to compute the optimal alignment under the given metric may be difficult to devise. Because of this complexity, sequence alignment programs concentrate on a single scoring scheme, allowing the user to vary a range of parameters, but not the scheme itself.

Among the many DNA alignment programs developed over the last few years, most have attempted to use various heuristics to quickly optimize the Needleman-Wunsch metric. In this paper we propose algorithms and software to enable bioinformatics researchers to explore a plethora of richer scoring schemes for sequence alignments. First, we define the class of *rectangle scoring schemes*, which encompass a large number of scoring metrics, including Needleman-Wunsch, Karlin-Altschul E-value, DIALIGN, and many others. Secondly we demonstrate an efficient polynomial-time algorithm to compute the optimal alignment for an arbitrary rectangle scoring scheme, and present both provably optimal and non-optimal heuristics to speed up this search. Our algorithms are implemented in a tool, FRESCO, which can be used to investigate the efficacy of various rectangle scoring schemes. Finally, we illustrate two examples of scoring functions that produce accurate alignments without any prior biological knowledge.

2. Scoring Schemes

2.1. *Previous work*

The work on scoring an alignment is closely tied to the problem of defining a distance between two strings. Classical work on formulating such distances are due to Hamming[8] for ungapped similarities and Levenshtein[9] for similarity of sequences with gaps. The Needleman-Wunsch algorithm[10] expanded on Levenshtein's approach to allow for varying match scores, and mismatch and gap penalties. Notably, the Needleman-Wunsch algorithm, as described by the original paper, supports arbitrary gap functions and runs in $O(n^3)$ time. The special case of affine gaps being computable in quadratic time was demonstrated by Gotoh[11] in 1982. Most of the widely used DNA sequence alignment programs such as BLASTZ[12], AVID[13] and LAGAN[14] use the Needleman-Wunsch scoring scheme with affine gaps. The DIALIGN scoring scheme[4,5] is notable because it was one of the first scoring schemes that allowed for scoring not on a per-letter, but on a per-region (diagonal) basis. The score of a single diagonal was defined as the probability that the observed number of matches within a diagonal of a given length would occur by chance, and the algorithm sought to minimize the product of all these probabilities.

The Karlin-Altschul (KA) E-value[15], which estimates the expected number of alignments with a certain score or higher between two random strings, can be formulated not only as a confidence, but also as a scoring scheme, as was heuristically done in the OWEN program[16]. After the single best local alignment between the two sequences is found and fixed, the algorithm begins a search for the second best alignments, but restricts the location to be either before the first alignment in both sequences, or after in both. The KA E-value depends on the lengths of sequences being aligned, and because the effective sequence lengths are reduced by the first local alignment, the KA E-value of the second alignment depends on the choice of the first. The OWEN program, which uses the greedy heuristic, does not always return the optimal alignment under the KA E-value scoring scheme. Other alignment scoring schemes include scoring metrics to find an alignment which most closely matches a given evolutionary model. These can be heuristically optimized using a Markov Chain Montecarlo (MCMC) algorithm, for example the MCAlign program[17]. Alignment scoring schemes which are based on various interpretations of probabilistic models, e.g. the ProbCons alignment program that finds the alignment with the maximum expected number of correct matches, are another example. Within the context of alignments based on probabilistic models there has been work on methods to effectively learn the optimal values for the various parameters of the common alignment schemes using Expectation Maximization or other unsupervised learning algorithms[18].

2.2. *Rectangle Scoring Schemes*

In this section we will define the concept of a rectangle scoring scheme, and illustrate how some of the classic alignment algorithms are all special cases of such schemes. Consider a 2D matrix M defined under the dynamic programming paradigm, on whose axes we set the two sequences being aligned. We define a diagonal in an alignment as referring to a sequence of matched letters between two gaps. We define a diagonal's bounding rectangle as the sub-rectangle in M delimited by the previous diagonal's last match and the next diagonal's first match (Fig. 1a). Thus, a diagonal's bounding rectangle includes the diagonal itself as well as the preceding and subsequent gaps. A *rectangle scoring scheme* is one that makes use of gap and diagonal information from within this rectangle (such as the number of matches, area of the rectangle, lengths of the dimensions, etc), while the scores for all rectangles can be computed independently and then combined[a]. For example, Needleman-Wunsch[10] is one such scheme: the score of a rectangle is defined as the sum of match and mismatch scores for the diagonal, minus half of the gap penalties for the two gaps before and after the diagonal. The Karlin-Altschul E-value[15] ($E = -Kmne^{-\lambda s}$) is another example, as the E-value depends on m and n, the entire lengths of the two strings being compared. The DIALIGN scoring function is another example of a rectangle scoring scheme.

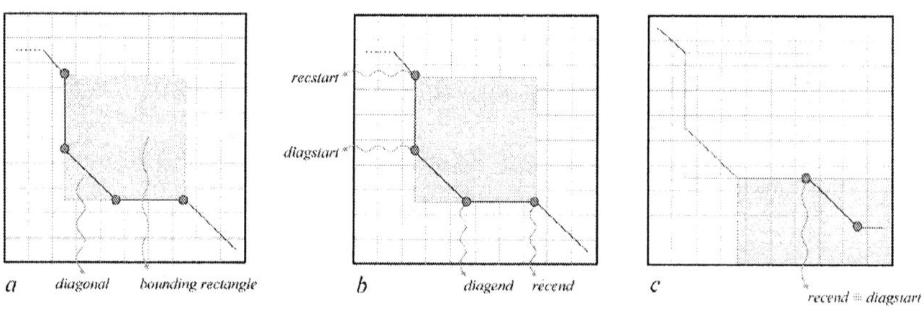

Figure 1. (a) Definition of a bounding rectangle of a diagonal - the rectangle in M delimited by the previous diagonal's last match and the next diagonal's first match. (b) Shows the 4 important points within a rectangle: *recstart* & *recend* - the start and end (top left and bottom right, respectively) points of a rectangle, and *diagstart* & *diagend* - the starting and ending points of a diagonal. (c) Note how a *recend* is equivalent to the next diagonal's *diagstart*.

[a]Currently FRESCO assumes the operation combining rectangle scores is addition or multiplication, as this is most often the case, but can be trivially modified to allow for any operation/formula.

3. Algorithm

In this section we will present an overview of the algorithm that we use to find the best alignment under an arbitrary rectangle scoring scheme. We will again make use of the 2D dynamic programming matrix M defined above, on whose axes we set the two sequences being aligned.

3.1. *Basic FRESCO Algorithm*

Given any rectangle scoring scheme, FRESCO computes an optimal alignment between two sequences. For clarity, we define *recstart* as the starting point of a rectangle and *recend* as the endpoint of a rectangle, and, similarly, *diagstart* and *diagend* points to be the starting and ending points of a diagonal (Fig. 1b). By definition of a rectangle of a given diagonal, a *recstart* is equivalent to the previous diagonal's *diagend* and a *recend* is equivalent to the next diagonal's *diagstart* (Fig. 1c).

The FRESCO algorithm can be explained within a slightly modified dynamic programming algorithm paradigm.

1. *Matrix.* First we create the dynamic programming matrix M with the two sequences on the axes.

2. *Recursion.* Here we describe the recursion relation. We iterate through the matrix M row-wise.

- **Terminology**: a *diagend* cell C can form a gap with several possible *recend* cells D (cells that come after C on the same row or column), as shown in Figure 2a. Note that this {C, D} pair can thus be part of a number of rectangles {A, B, C, D}, where A is the *recstart* and B is the *diagstart.* To view all of these, one would consider all the possible *diagstarts*, and for each, all the possible *recstarts*, as shown in Figure 2(b-d). We use this notion of a {C, D} pair and {A, B, C, D} rectangle below.
- **Invariant**: (true at the end of each step (i,j)). Let $C = M[i, j]$ and consider this cell as a possible *diagend.* We have computed, for each possible pair {C, D} described above, a rectangle representing the best alignment up to {C, D}.
- **Recursion**: Assume cell C is a *diagend.* For every cell D as described above:
  - Find all the possible rectangles through {C,D}: {A, B, C, D} as described above (for every possible *diagstart* consider every possible *recstart*)
  - For each rectangle R = {A, B, C, D}, we will have computed the best alignment & associated score S_B up to {A, B} (via the invariant) and we can compute the score S_R of R alone via the current rectangle scoring scheme. Adding $S_T = S_B + S_R$ will give us the score of the best alignment through {A, B, C, D}.

- ◦ After computing all the S_T (total) scores for each R, we take the maximum, giving us the optimal alignment & score up to {C, D}. This completes the recursion. For the purposes of recreating the alignment, we hold, for each {C, D} pair, a pointer to the optimal {A, B} choice.

3. *Computing the alignment.*

- Let the final cell be denoted by F, $F \equiv M[m, n]$. We will have $m + n - 1$ pairs {C, F}, (where C is on the rightmost column of bottommost row) that will hold the best alignment and score up to {C, F}. Taking the maximum of these will give us the best alignment up to F. Having stored pointers from each {C, D} pair to its optimal {A, B} pair, we simply follow the pointers back through each rectangle up to $M[0, 0]$, thus recreating the alignment.

The proof of correctness is by induction and follows very similarly. The algorithm can be trivially modified to allow for unaligned regions by setting the diagonal score to the score of the maximum contiguous subsequence.

3.1.1. *Running Time and Resources*

Let the larger of the sequences be of length n. The algorithm iterates over all the points of the matrix M — $O(n^2)$ iterations. In the recursion, we look ahead at most $2n$ *recends* D and look back at no more than n *diagstarts* B. For each of these {B, C, D} sets, we search through at most $2n$ *recstarts* A. Thus we have $O(n^3)$ computation and $O(n)$ storage. Consequently we have an overall running time of $O(n^5)$ and storage of $O(n^3)$.

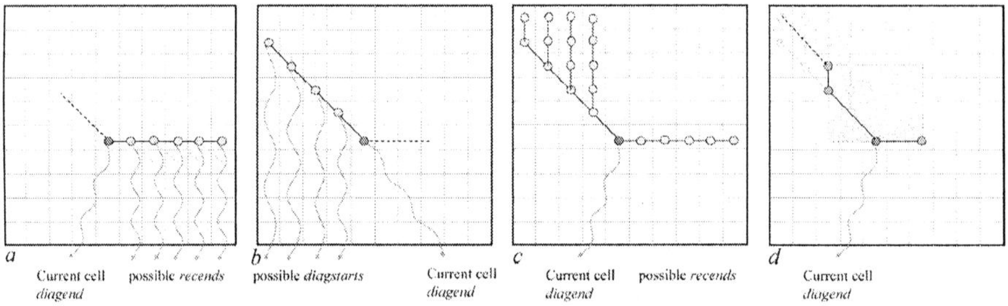

Figure 2. The figure illustrates the search, described in section 3.1, for the best rectangle assuming the current point acts as a *diagend*. For the current cell being considered (dark gray), referred to as C, (a) shows possible *recends* D; and hence pairings {C, D}. (D could also be on the same column as C). (b) illustrates the possible *diagstarts* (B) considered for each of these {C, D}. For each {B, C, D} set we have possibilities such as those shown in (c), all of which form rectangles {A, B, C, D} that go through the diagend C we begin with. We choose the optimal of these rectangles, as shown in (d).

3.2. *FRESCO Speed Ups*

Under most scoring schemes, a large portion of the calculations above become redundant. We have built into FRESCO several optional features that take advantage of the properties of possible scoring functions with the aim of lowering the time and storage requirements for the algorithm. These can be separated into two categories: optimal (the resulting alignment is still optimal) and heuristic (without optimality guarantees).

Optimal

- *Pareto Efficiency.* Most relevant scoring schemes will score a specific diagonal's rectangle lower if its length or width are larger than another rectangle with the same diagonal (and if the other parameter is the same). Given this likely property, we have implemented an optional feature in FRESCO whereby, for each set {B, C, D}, we will have eliminated any points A (*recstarts*) where we have another closer A with a better overall score. This defines a pareto-efficient set[19]. While it is difficult to predict the exact size of this reduction, empirically, we observed that about order *logn recstarts* of the originally available $O(n)$ are retained, allowing us to reduce the running time and space requirements by almost a factor of n. This holds for both unrelated and highly similar sequences.

- *SMAWK.* Given that the scoring function has the same concavity with respect to the rectangle area throughout (i.e. the function is always concave, or always convex), we can further speed up the alignment using the SMAWK[20] algorithm. In the recursion, we can reduce the number of rectangles we look at if we change the order of the iterations: first we consider pairs of *diagbegin* and *diagend* points {B, C}, and then compute the total scores at all relevant *recends* (Ds) and *recbegins* (As). When the computation is done in this manner, we can view this as the search for all of the column minima of a matrix $D[NxN]$, where each row corresponds to a particular *recbegin* point, each column corresponds to a *recend* point, and the cell $D[i, j]$ is the score of the path that enters the given diagonal through *recbegin* point i and exists it through recend point j. This matrix has been previously used in literature, and is commonly known as the DIST matrix[21]. If the scoring function is either concave or convex, the DIST matrix is totally monotone, and all of its column minima can be found in time linear in the number of columns and rows using the SMAWK algorithm. This optimization decreases the computation time for each possible *diagend* to $O(n^2)$, speeding up the overall alignment by $O(n)$.

Because the user may be interested in exploring non-uniform scoring schemes we have made both SMAWK and Pareto-efficency optional features in FRESCO, which can be turned on or off using compile-time options. However, with both the Pareto-efficency and the SMAWK speedups, the overall running time, originally $O(n^5)$, is observed to grow as $n^3 logn$ when both speed-ups are enabled. The observed running times are summarized in Figure 3.

Heuristic (Non-Optimal)

We also introduce two speed ups that, while not guaranteeing an optimal overall score, have been observed to work well in practice.

- *Maximum diagonal length.* Since one key parameter that limits the running time of our algorithm is having to compute diagonals of all possible lengths, we have added an optional limit on the length of the diagonal, forcing each long diagonal to be scored as several shorter ones. For many scoring schemes this does not greatly affect the final alignment, while the running time is reduced by O(n). This improvement was also employed in the DIALIGN program[5].

- *Banded Alignment.* We have also added an option to FRESCO which forces the rectangle scoring scheme to act only within a band in the matrix M around an already-computed alignment. Because most genome sequence alignment tools are going to agree overall on strong areas of similarity, banded alignment heuristics have commonly been used to improve on an existing alignment. Since FRESCO allows the testing of abilities of various scoring schemes, this improvment technique may be of particular interest when used with FRESCO. We have performed empirical tests by running FRESCO within a band around the optimal alignment to investigate the running time, and empirically observed a running time linear in n.

Figure 3 displays the running time of FRESCO using various optimization techniques for sequences of length 100 to 1000 nucleotides.

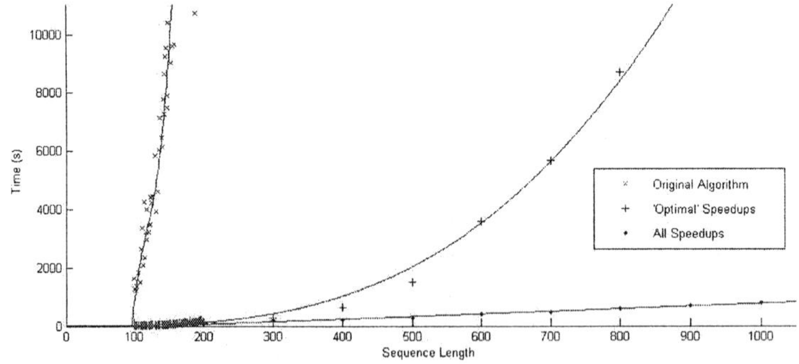

Figure 3. We show the improvements in running time from the original FRESCO algorithm, indicated by (x) and modeled by a polynomial, to the running time of FRESCO with the Pareto and Ranges (SMAWK) utilities on, indicated by (+) and modeled by a polynomial, and finally applying all speedups described in the text (including band size of 20 bp, maximum diagonal length of 30 bp), and resulting in linear running time (•).

4. Results

4.1. *Functions allowed*

The main power of the FRESCO tool is its ability to create alignments dictated by any rectangle scoring scheme. This will allow researchers to test schemes based on any motivations, such as evolution-based or statistical models. Since the creation of a new algorithm is not required for each of these schemes, we now have the ability to quickly compare the performance of complex scoring schemes. We have investigated traditional scoring schemes and aligners Clustal-W and LAGAN, against two novel scoring functions based on a parameter-less global E-value, described below.

4.2. *Example function & performance*

Given a diagonal and its bounding rectangle, the global E-value is the expected number of diagonals within this rectangle with equal or higher score. We calculate this by computing, for every possible diagonal in the rectangle, the probability that it has a score higher than the one in our diagonal, and summing these indicator variables. Note that our global E-value is different from the Karlin-Altschul statistic. To compute the global E-value we first define a random variable corresponding to the score of matching two random (non-homologous) letters. The expected value of this random variable (referred to as R below) is determined by computing the frequency of all nucleotides in the input strings, and for all 16 possible pairings multiplying the score of a match by the product of the frequencies. The variance (V) of the variable is the sum of the squared differences from the expectation. We model a diagonal of length d as a set of repeated, independent samplings of this random variable. The probability $g(s, d)$ that the sum of these d trials has a *score* $> s$ can be approximated as the integral of the tail of a Gaussian, with mean R_d and variance V_d:

$$g(s,d) = \int_s^{\infty} \frac{1}{V_d\sqrt{2\pi}} e^{-\frac{(x-R_d)^2}{2V_d^2}}\, dx \tag{1}$$

Note that $g(s, d)$ is also the expected value of the variable which indicates whether or not a particular diagonal of length d has a score higher than s. The expected number of diagonals within a rectangle with a given score or higher is equal (by linearity of expectation) to the sum of expectations of indicator variables corresponding to individual diagonals, yelding the formula

$$E = \sum_{i=1}^{min(m,n)} g(s,i) + |m-n| g(s,d) \tag{2}$$

The E-values for the individual rectangles can be combined in a variety of ways, leading to various alignment qualities. Below we will demonstrate results for two ways of combining the functions:

$$E - Value\ I : \sum_{i=1}^{N} \log(\frac{2}{E_i} + \epsilon) = \prod_{i=1}^{N} (\frac{2}{E_i} + \epsilon) = \epsilon^N + 2 \prod_{i=1}^{N} E_i^{-1} \tag{3}$$

$$E - Value\ II : \sum_{i=1}^{N} \log(\log(\frac{1}{E_i} + \epsilon)) = \prod_{i=1}^{N} \log(\frac{1}{E_i} + \epsilon) \tag{4}$$

Where ϵ is used to avoid asymptotic behaviour. We used $\epsilon = 0.1$.

Performance

The evaluation of DNA alignment accuracy is a difficult problem, without a clear solution. In this paper we have chosen to simulate the evolution of DNA sequence, and compare the alignments generated by each program with the "gold standard" produced by the program that evolved the sequences. We used ROSE[22] to generate sequences of length 100-200 nucleotides from a wide range of evolutionary distances and ratios of insertions/deletions (indels) to substitution, using a Jukes-Cantor[23] model and equal nucleotide frequency probability (See Table 1). The evolved sequences were aligned with FRESCO using several E-value based scoring functions (described above), as well as with the Clustal-W and LAGAN aligners, with default parameters. The accuracy of each alignment was evaluated both on a per-nucleotide basis with the program described in Pollard et al, 2004[24], as well as based on how closely the number of indels in the generated alignments matched the number of indels in correct alignments. The results are summarized in Figure 4. While the per nucleotide accuracy of the LAGAN aligner is best, the E-value II function we have defined manages to top the ClustalW aligner in accuracy and estimate the indel ratio better than both LAGAN and ClustalW in most tests, without using any biological or evolutionary knowledge. It is important to note that the improvement of the global E-value over ClustalW becomes more pronounced with greater evolutionary distance.

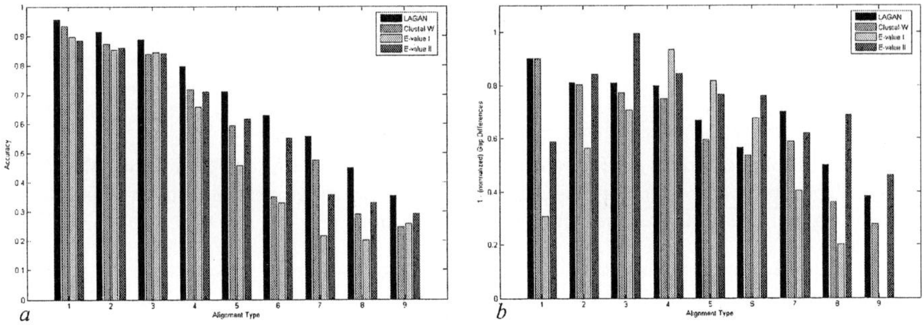

Figure 4. We evaluated the E-value scoring functions on a set of ROSE-generated alignments based on the accuracy (a) and 1 - the gap frequency difference (b) between the observed and evolved alignment, and compared with results from the LAGAN and ClustalW aligners. For alignment types 1-9, evolutionary distance 0.25, 0.50 & 0.75 subs/site, from left to right, we tried three indel per substitution ratios 0.06, 0.09, and 0.12 each. While the accuracy of the E-value II scheme fell between LAGAN and ClustalW, the indel ratio is in general better (than both aligners) with the E-value II function. The details of the analysis are included in the appendices.

Table 1. Summary of evolutionary parameters used to generate test data. Sequences were evolved using three different evolutionary distances (substitutions per site), each with three different indel to substitution ratios.

Type	1	2	3	4	5	6	7	8	9
Subs Per Site	0.25	0.25	0.50	0.50	0.50	0.75	0.75	0.75	0.75
Indel/Subs	0.06	0.09	0.12	0.06	0.09	0.12	0.06	0.09	0.12

5. Discussion

In this paper we generalize several schemes that have been previously used to align genomes into a single, more general class of *rectangle scoring schemes*. We have developed FRESCO, a tool that can find the optimal alignment for two sequences under any scoring scheme from this large class. While the tool we have built only allows for alignment of short sequences, and is not usable for whole genomes (it is many-fold slower than anchored aligners such as LAGAN and AVID), we believe that it should enable bioinformaticians to explore a large set of schemas, and once they find one that fits their needs, it will be possible to write a faster, specialized program for that scoring scheme. In this paper we provide an example of a rectange scoring function that incorporates no biological knowledge but performs on par with popular alignment algorithms, and we believe that even more accurate schemas can be found using the FRESCO tool.

6. Implementation and Supplementary Information

FRESCO was developed solely in C. The scoring scheme is supplied as a '.c' file, in which we allow a definition of the scoring function (in C code) as well as any pre-computations and global variables necessary for the scheme. A script to test the FRESCO results against other aligners or the *true* alignment is written to aid in comparing scoring schemes, implemented in a combination of perl and shell scripts. All are available at http://compbio.cs.toronto.edu/fresco. At this same address one can find an appendix and the generated datasets used in the results section.

References

1. S. F. Altschul, T. L. Madden, A. A. Schffer, J. Zhang, Z. Zhang, W. Miller and D. J. Lipman, *Nucleic Acids Res* **25**, 3389 (1997).
2. J. D. Thompson, D. G. Higgins and T. J. Gibson, *Nucleic Acids Res* **22**, 4673 (1994).
3. C. M. Bergman and M. Kreitman, *Genome Res* **11**, 1335 (2001).
4. B. Morgenstern, A. Dress and T. Werner, *Proc Natl Acad Sci* **93**, 12098 (1996).
5. B. Morgenstern, *Bioinformatics* **16**, 948 (2000).
6. C. Notredame, D. G. Higgins and J. Heringa, *J Mol Biol* **302**, 205 (2000).
7. C. Do, M. Brudno and S. Batzoglou, *Nineteenth National Conference on Artificial Intelligence AAAI* (2004).
8. R. Hamming, *Bell System Technical Journal* **26**, 147 (1950).
9. V. I. Levenshtein, *Soviet Physics Doklady* **10**, p. 707 (1966).
10. S. B. Needleman and C. D. Wunsch, *J Mol Biol* **48**, 443 (1970).
11. O. Gotoh, *J Mol Biol* **162**, 705 (1982).
12. S. Schwartz, Z. Zhang, K. A. Frazer, A. Smit, C. Riemer, J. Bouck, R. Gibbs, R. Hardison and W. Miller, *Genome Res.* **10**, 577 (2000).
13. N. Bray, I. Dubchak and L. Pachter, *Genome Res.* **13**, 97 (2003).
14. M. Brudno, C. B. Do, G. M. Cooper, M. F. Kim, E. Davydov, E. D. Green, A. Sidow and S. Batzoglou, *Genome Res* **13**, 721 (2003).
15. S. Karlin and S. F. Altschul, *Proc Natl Acad Sci* **87**, 2264 (1990).
16. M. A. Roytberg, A. Y. Ogurtsov, S. A. Shabalina and A. S. Kondrashov, *Bioinformatics* **18**, 1673 (2002).
17. P. D. Keightley and T. Johnson, *Genome Res.* **14**, 442 (2004).
18. C. Do, S. Gross and S. Batzoglou, *Tenth Annual International Conference on Computational Molecular Biology (RECOMB)* (2006).
19. M. J. Osborne and A. Rubinstein, *A Course in Game Theory* 1994.
20. A. Aggarwal, M. M. Klawe, S. Moran, P. Shor and R. Wilber, *Algorithmica* **2**, 195 (1987).
21. J. P. Schmidt, *SIAM J. Comput.* **27**, 972 (1998).
22. J. Stoye, D. Evers and F. Meyer, *Bioinformatics* **14**, 157 (1998).
23. T. H. Jukes and C. R. Cantor, *Evolution of Protein Molecules* 1969.
24. D. A. Pollard, C. M. Bergman, J. Stoye, S. E. Celniker and M. B. Eisen, *BMC Bioinformatics* **5** (2004).

LOCAL RELIABILITY MEASURES FROM SETS OF CO-OPTIMAL MULTIPLE SEQUENCE ALIGNMENTS

GIDDY LANDAN
DAN GRAUR
Department of Biology & Biochemistry, University of Houston, Houston, TX 77204

The question of multiple sequence alignment quality has received much attention from developers of alignment methods. Less forthcoming, however, are practical measures for quantifying alignment reliability in real life settings. Here, we present a method to identify and quantify uncertainties in multiple sequence alignments. The proposed method is based upon the observation that under any objective function or evolutionary model, some portions of reconstructed alignments are uniquely optimal, while other parts constitute an arbitrary choice from a set of co-optimal alternatives. The co-optimal portions of reconstructed alignments are, thus, at most half as reliable as the uniquely optimal portions. For pairwise alignments, this irreducible uncertainty can be quantified by the comparison of the high-road and low-road alignments, which form the co-optimality envelope for the two sequences. We extend this approach for the case of progressive multiple sequence alignment by forming a large set of equally likely co-optimal alignments that bracket the co-optimality space. This set can, then, be used to derive a series of local reliability measures for any candidate alignment. The resulting reliability measures can be used as predictors and classifiers of alignment errors. We report a simulation study that demonstrates the superior power of the proposed local reliability measures.

1. Introduction

Multiple sequence alignment (MSA) is the first step in comparative molecular biology. It is the foundation of a multitude of subsequent biological analyses, such as motif discovery, calculation of genetic distances, identification of homologous strings, phylogenetic reconstruction, identification of functional domains, three-dimensional structure prediction by homology modeling, functional genome annotation, and primer design [1]. The fundamental role of multiple sequence alignment is best demonstrated by noting that a paper describing a popular multiple-alignment reconstruction method, ClustalW [2], has been cited close to 25,000 times since its publication (i.e., an average of five times a day). Being a fundamental ingredient in a wide variety of analyses, the reliability and accuracy of multiple sequence alignment is an issue of utmost importance; analyses based on erroneously reconstructed alignments are bound to be severely handicapped [e.g., 3-9]. The question of multiple sequence alignment quality has received much attention from developers of alignment methods [10-15]. Unfortunately, practical measures for addressing alignment-quality issues in real life settings are sorely missing.

Multiple sequence alignment is frequently treated as a "black box"; the possibility that it may yield artifactual results is usually ignored. Moreover, in a manner reminiscent of basic laboratory disposables, the vast majority of multiple sequence alignments are produced robotically and discarded unthinkingly on the road to some other goal, such as a phylogenetic tree or a 3D structure. We speculate that more than 99% of all multiple sequence alignments that ultimately yield publishable results are never even looked at by a human being. Yet, when an occasional alignment is actually inspected, it is usually found wanting. Multiple sequence alignments are so notoriously inadequate, that the literature is littered with phrases such as "the alignment was subsequently corrected by hand" [e.g., 16-22]. Unfortunately, "hand correction" is neither objective nor reproducible, and as such we should strive to replace it by a scientifically legitimate method.

Errors in reconstructed alignments are typically attributed to the inadequacy of the evolutionary model and its parameters. Understandably, then, the recent proliferation of new reconstruction methods is mainly concerned with developing new optimality criteria and optimization heuristics. Unfortunately, the second source of reconstruction errors, i.e., the fact that the objective function usually possesses multiple optima even when the evolutionary model is adequate, is rarely addressed. Moreover, the full co-optimal solution set is often far too large to enumerate explicitly [23], and current MSA programs arbitrarily report only one of these co-optimal solutions. Reporting only one alternative from among the multitude of equally optimal or co-optimal alignments obscures the fact that the entire set of co-optimal alignments possesses valuable information; some portions of the alignments are uniquely optimal and are reproduced in every solution, while other portions differ among the solutions. Since the choice between such co-optimal alternatives is necessarily arbitrary, these portions of the alignments represent inherent irreducible uncertainty.

When dealing with pairwise alignments, we can capture this information by considering two extreme cases, termed the high-road and the low-road [24-25], which bracket the set of all co-optimal alignments. Alignment programs usually report either the high-road or the low-road as the final alignment. In such cases the other extreme alignment can be easily obtained by reversing the sequence residue order in the input [26]. Reversing the sequences amounts to inverting the direction of the two axes of the alignment dot matrix, thereby converting the high road to the low road and the low road to the high road. Columns that are identical in the two alignments define parts of the alignment where a single optimum of the objective function exists, whereas columns that differ between the two alignments define those portions of the alignments where there exist two or more co-optimal solutions.

A simple extension of this principle to the case of multiple sequence alignment is the "Heads or Tails" (HoT) methodology [26], where the original sequence set (the Heads set) is first reversed to create a second set (the Tails set). The two sequence sets are, subsequently, aligned independently, and the two resulting alignments are compared to produce a measure of their internal consistency. While the HoT method can be applied to any MSA reconstruction method, it produces only two alignments, and its statistical power is, therefore, limited.

Here we present a more powerful extension of the HoT methodology for the case of progressive multiple sequence alignment. Progressive alignment proceeds in a series of pairwise alignments of profiles, or sub-alignments, whose order is determined by an approximate guide tree. At each of these alignment steps, the resulting sub-alignment is an arbitrary choice from among many co-optimal alternative alignments. Our extension derives a large set of alternative MSAs that explores the co-optimality envelope of the several pairwise profile alignments that can be defined for a given guide-tree.

The set of alternative alignments is then analyzed to score specific elements of the alignments by their frequency of reproduction within the set. The reproduction scores can be applied to any candidate MSA to derive a series of local reliability measures that can identify and quantify uncertainties and errors in the reconstructed MSA.

2. Methods

2.1. *Construction of the co-optimality MSA set*

We implemented the derivation of the alignment set for ClustalW [2], which uses progressive alignment. Given the ClustalW approximate guide-tree for N sequences, we define the guide-tree alignment set, $^{gt}\boldsymbol{AS}$, as follows (Fig. 1):

For each of the $(N\text{-}3)$ internal branches of the guide tree, partition the sequences into two subgroups (Fig. 1a). Construct two sub-alignments for each of the two sequence groups (Fig. 1b):

- Heads: The ClustalW alignment of the sequence subgroup.
- Tails: The ClustalW alignment of the reversed sequences, reversed to the original residue order.

Next, use the ClustalW profile alignment to align the four combinations of the sub-alignments, aligning each combination in both the head and tail directions, to yield a total of ***8*** full MSAs for each internal branch (Fig. 1c).

The process is repeated for all internal branches of the guide-tree (Fig. 1d). All in all, then, ^{gt}AS contains $8\cdot(N-3)$ alignments. These alignments differ from each other in two respects: (a) the partitioning of sequences and profiles to create the final MSA, and (b) the Heads or Tails selection of co-optimal sub-alignments and profile alignments. Any alignment in the set can be qualified as a *bona-fide* progressive alignment. Thus, the alignments in the guide-tree alignment set can be considered as equally likely alternatives that uniformly sample the co-optimality envelope.

2.2. *Local reliability measures for MSA*

Given a candidate reconstructed MSA, A, we first construct the corresponding guide-tree alignment set, ^{gt}AS, and score the elements of A by their reproduction in ^{gt}AS (Fig. 1e). For each pair of residues that are aligned as homologs in A, we define our basic reliability measure, the *residue-pair reliability measure*, $^{pair}M^{c}_{i,j}$ (where c is the column index and i,j are the sequence indices), as the proportion of alignments in ^{gt}AS that reproduce the pairing of the residue pair. The measure takes values within the interval $[0..1]$, where 1 denotes total support. Averaging of the residue-pair support gives rise to a series of reliability measures:

- The *residue reliability* is the mean of the residue-pair reliability over all pairings involving the residue:

$$^{res}M^{c}_{i} = \overline{^{pair}M^{c}_{i,*}}$$

- The *column reliability* is the mean of the residue-pair reliability over all pairs in a column:

$$^{col}M^{c} = \overline{^{pair}M^{c}_{*,*}}$$

- The *alignment reliability* is the mean of the residue-pair reliability over all residues-pairs in the alignment:

$$^{ali}M = \overline{^{pair}M^{*}_{*,*}}$$

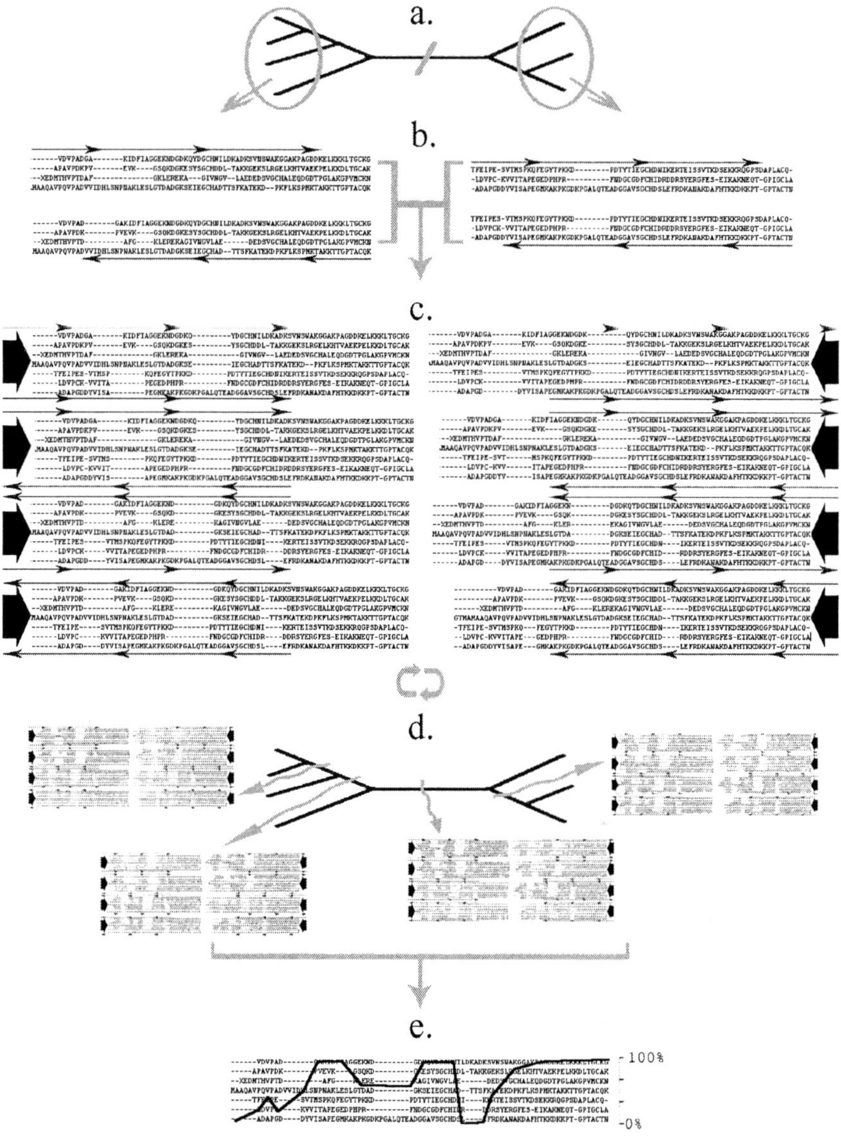

Figure 1: Construction of the guide-tree alignment set and the local reliability measures: (a) Use an internal branch of the guide tree to partition the sequences; (b) Align each subset in both heads and tails orientations, to produce 4 sub-alignments; (c) Align the four combinations of sub-alignments, in both heads and tails directions, for a total of 8 alignments; (d) Repeat a-c for each of the ***N-3*** internal branches, to produce ***8·(N-3)*** alternative alignments (***32*** for ***N***=***7***); (e) score elements of a candidate alignment by their frequency of reproduction (vertical axis) in the alignment set. (For more details, see text).

2.3. *Implementation*

Construction of the co-optimality MSA set and derivation of the local reliability measures were implemented in MATLAB scripts, available from the authors upon request.

3. Results

The local reliability measures can be used to identify and quantify errors in the reconstructed MSAs. We demonstrate their performances in a simulation study where MSAs reconstructed by ClustalW are compared to the true alignment from ROSE simulations [27]. We used 6400 datasets where the sequence evolution was simulated along a 16 taxa balanced depth-3 phylogeny, with an average branch length ranging from 0.02 to 0.30 substitutions per site, and an indel to substitution ratio of 0.015. The average sequence length was 500 nucleotides. Comparison of the true MSA to the ClustalW MSA yields rates of correct reconstruction at several resolution levels: residue-pairs, residue, column, and the entire alignment.

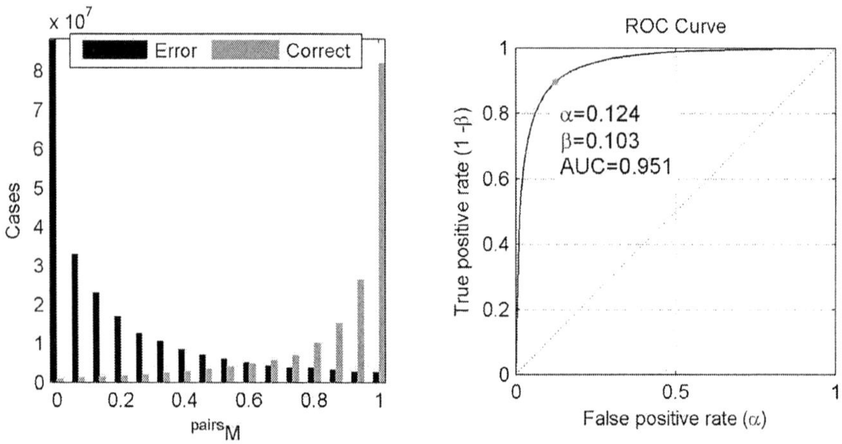

Figure 2: The residue-pairs reliability measure, ${}^{pairs}\boldsymbol{M}$, as a classifier of erroneous or correct residue-pairs in reconstructed MSAs. Histograms (left) presents the distributions of the two populations: **H0**:error (black) vs. **H1**:correct (gray). ROC curve (right) report the level of classification errors and the power of the classifier.

One use of the reliability measures is as binary classifiers of local MSA features as correct or erroneous. Figure 2 presents a receiver-operating characteristic (ROC) analysis [28] of ${}^{pairs}\boldsymbol{M}$ as a classifier of residue-pairs errors. Since the residue-pairs reconstruction rate, ${}^{pairs}\boldsymbol{R}$, is binary, the two populations, error (***H0***, black) or correct (***H1***, gray) reconstructions, are strictly defined. Our measure ${}^{pairs}\boldsymbol{M}$ is capable of separating the two populations, with a very high power (area under curve, AUC=0.95).

The most useful level of MSA scoring is the column level. Current methods employ Shannon's entropy as a measure of MSA quality, that is, column quality is judged by its residue variability. In figure 3 we compare the column reliability measure, colM to the entropy-based column quality measure reported by ClustalX, colQ [29], as classifiers of the true column errors. An ROC analysis reveals that colM separates the two populations, of erroneous and correct columns, better than colQ, with AUCs of ~0.94 and ~0.87, respectively.

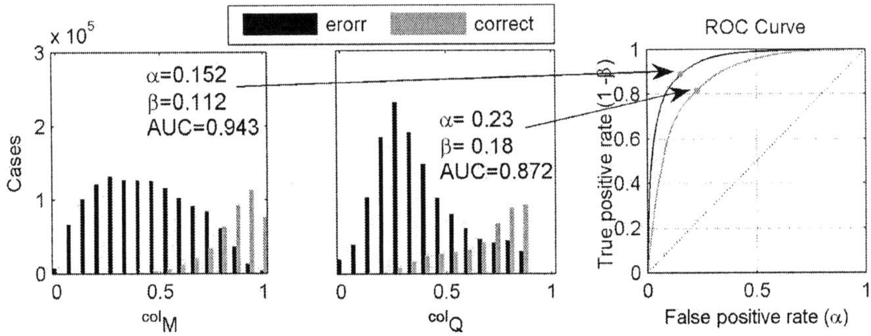

Figure 3: Comparison of two column reliability measures, ${}^{col}\boldsymbol{M}$ and ${}^{col}\boldsymbol{Q}$ as classifiers of erroneous or correct columns in reconstructed MSAs: Histograms (left) presents the different distributions of the two populations: **H0**:error (black) vs. **H1**:correct (gray). ROC curves (right) report the level of classification errors and the power of the classifier.

When interpreting the local reliability measures, ${}^{*}\boldsymbol{M}$, as estimates of the reconstruction rates, ${}^{*}\boldsymbol{R}$, we find extremely high correlations between the two types of measures, one derived from the comparison to the true MSA, ${}^{*}\boldsymbol{R}$; the other from the MSA set, ${}^{*}\boldsymbol{M}$. The correlation coefficients are $\boldsymbol{r}$ = 0.94 for the residue-base measure and $\boldsymbol{r}$ =0.87 for the column measure. Once again, the entropy-based column quality measure is inferior to our ${}^{col}\boldsymbol{M}$; the correlation between ${}^{col}\boldsymbol{Q}$ and ${}^{col}\boldsymbol{R}$, although significant, is only $\boldsymbol{r}$ = 0.66 (Fig. 4).

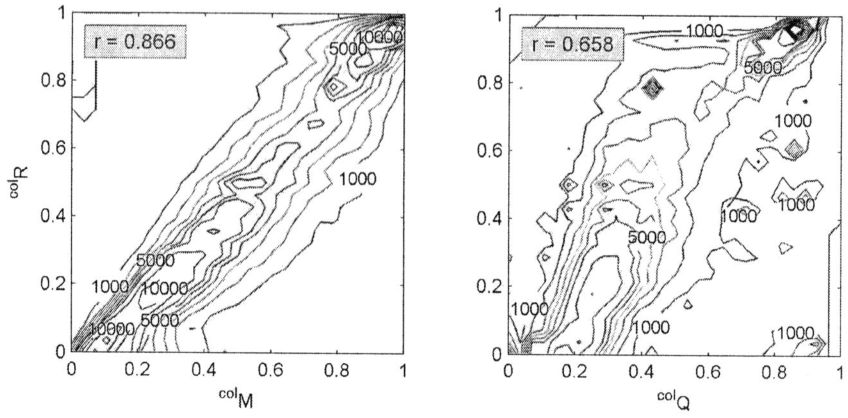

Figure 4: Comparison of two column quality measures, $^{col}\boldsymbol{M}$ and $^{col}\boldsymbol{Q}$, as estimates of the true reconstruction rates.

4. Discussion

The local reliability of reconstructed MSAs is usually viewed as related to the local divergence of the sequences. Thus, current local reliability measures are based on the column entropy or variation [e.g., 29]. While it is true that highly preserved segments of an MSA are more easily reconstructed by MSA algorithms, column entropies do not take into account the algorithmic sources of reconstruction errors. In contrast, our approach specifically addresses one common source of alignment errors, namely, the irreducible uncertainty stemming from the arbitrary choice from a set of co-optimal solutions. Hence its superiority to previous local quality measures.

The equivalence of co-optimal solutions is only one source of reconstruction errors. Two other sources of errors merit mention here: (a) the approximate nature of the guide-tree and the estimated evolutionary parameters, and (b) stochastic errors, where the true alignment is sub-optimal even when the objective function is exact [30]. It is interesting to note that although our reliability measures do not address these sources of errors directly, they do manage to correctly identify about 90% of the errors, while maintaining a low false positive rate.

The guide-tree alignment set does not exhaust the co-optimality space. In fact, it is not computationally feasible to enumerate the entire set of co-optimal alignments [23]. Even tracking every high-road low-road combination in a progressive alignment will yield a set whose size grows exponentially with the number of sequences. Our guide-tree alignment set of size $\boldsymbol{8 \cdot (N\text{-}3)}$ was designed as a practical compromise between computational feasibility and statistical power. Since the construction of the guide-tree already requires $O(N^2)$ pairwise alignment steps, the additional $O(N^2)$ steps required by our method amount to tripling the processing time.

Acknowledgments

This work was supported by NSF grant DBI-0543342.

References

1. L.J. Mullan, *BriefBioinform* **3**:303-305 (2002).
2. J.D. Thompson, D.G. Higgins, and T.J. Gibson, *Nucleic Acids Res* **22**: 4673-4680 (1994).
3. D.A. Morrison and J.T. Ellis, *Mol Biol Evol* **14**:428-441 (1997).
4. L. Florea, G. Hartzell, Z. Zhang, G.M. Rubin, and W. Miller *Genome Res.* **8**:967-974 (1998).
5. E.A. O'Brien and D.G. Higgins, *Bioinformatics* **14**:830-838 (1998).
6. R.E. Hickson, C. Simon, and S.W. Perrey, *Mol Biol Evol* **17**:530-539 (2000).
7. L. Jaroszewski L. Rychlewski, and A. Godzik, *Protein Science* **9**:1487–1496 (2000).
8. T.H. Ogden and M.S. Rosenberg, *Syst. Biol.* **55**:314–328 (2006).
9. S. Kumar and A. Filipski, *Genome Res.* **17**:127-135 (2007).
10. J.D. Thompson, F. Plewniak, and O. Poch, *Nucleic Acids Res* **27**:2682-2690 (1999).
11. A. Elofsson, *Proteins* **46**:330-339 (2002).
12. T. Lassmann and E.L. Sonnhammer, *FEBS Lett* **529**:126-130 (2002).
13. J.D. Thompson, P. Koehl, R. Ripp, and O. Poch, *Proteins* **61**:127-36 (2005).
14. Y. Chen and G.M. Crippen, *Structural bioinformatics* **22**:2087–2093 (2006).
15. P.A. Nuin, Z. Wang, and E.R. Tillier, *BMC Bioinformatics* **7**:471 (2006).
16. D. O'Callaghan, C. Cazevieille, A. Allardet-Servent, M.L. Boschiroli, G. Bourg, V. Foulongne, P. Frutos, Y. Kulakov, and M. Ramuz, *Mol. Microbiol.* **33**:1210–1220 (1999).
17. K. Kawasaki, S. Minoshima, and N. Shimizu, *J. Exp. Zool.* **288**:120-134 (2000).
18. C.M. Kullnig-Gradinger, G. Szakacs, and C.P. Kubicek, *Mycol. Res.* **106**:757-767 (2002).

19. J.L.M. Rodrigues, M.E. Silva-Stenico, J.E. Gomes, J.R.S. Lopes, and S.M. Tsai, *Applied and Environmental Microbiology* **69**:4249–4255 (2003).
20. S.B. Mohan, M. Schmid, M. Jetten AND J. Cole, *FEMS Microbiology Ecology* **49**:433–443 (2004).
21. E. Bapteste, R.L Charlebois, D. MacLeod and C. Brochier, *Genome Biology* **6**:R85 (2005).
22. M. Levisson, J. van der Oost and S.W.M. Kengen, *FEBS Journal* **274**:2832–2842 (2007).
23. D. Naor and D.L. Brutlag, *J. Comp. Biol.* **1**: 349-366 (1994).
24. D.J. States, and M.S. Boguski, *In M. Gribskov and J. Devereux, eds., Sequence Analysis Primer* pp:124–130, Oxford University Press, New York (1995).
25. T.G. Dewey, *J. Comp. Biol.* **8**: 177-190 (2001).
26. G. Landan and D. Graur, *Mol. Biol. Evol.* **24**:1380–1383 (2007).
27. J. Stoye, D. Evers, and F. Meyer, *Bioinformatics* **14**: 157-163 (1998).
28. M.H. Zweig and G. Campbell, *Clin. Chem.* **39**:561-577 (1993).
29. J.D. Thompson, T.J. Gibson, F. Plewniak, F. Jeanmougin, and D.G. Higgins, *Nucleic Acids Res.* **25**:4876-4882 (1997).
30. G. Landan, *In Zoology*, pp. 93. Tel Aviv University, Tel Aviv (2005).

THE EFFECT OF THE GUIDE TREE ON MULTIPLE SEQUENCE ALIGNMENTS AND SUBSEQUENT PHYLOGENETIC ANALYSES

S. NELESEN, K. LIU, D. ZHAO, C. R. LINDER, AND T. WARNOW *

The University of Texas at Austin
Austin, TX 78712
E-mail: {serita,kliu,wzhao,tandy}@cs.utexas.edu, rlinder@mail.utexas.edu

Many multiple sequence alignment methods (MSAs) use guide trees in conjunction with a progressive alignment technique to generate a multiple sequence alignment but use differing techniques to produce the guide tree and to perform the progressive alignment. In this paper we explore the consequences of changing the guide tree used for the alignment routine. We evaluate four leading MSA methods (ProbCons, MAFFT, Muscle, and ClustalW) as well as a new MSA method (FTA, for "Fixed Tree Alignment") which we have developed, on a wide range of simulated datasets. Although improvements in alignment accuracy *can* be obtained by providing better guide trees, in general there is little effect on the "accuracy" (measured using the SP-score) of the alignment by improving the guide tree. However, RAxML-based phylogenetic analyses of alignments based upon better guide trees tend to be *much* more accurate. This impact is particularly significant for ProbCons, one of the best MSA methods currently available, and our method, FTA. Finally, for very good guide trees, phylogenies based upon FTA alignments are more accurate than phylogenies based upon ProbCons alignments, suggesting that further improvements in phylogenetic accuracy may be obtained through algorithms of this type.

1. Introduction

Although methods are available for taking molecular sequence data and simultaneously inferring an alignment and a phylogenetic tree, the most common phylogenetic practice is a sequential, two-phase approach: first an alignment is obtained from a multiple sequence alignment (MSA) program and then a phylogeny is inferred based upon that alignment. The two-phase approach is usually preferred over simultaneous alignment and

*This work was supported by NSF under grants ITR-0331453, ITR-0121680, ITR-0114387 and EIA-0303609.

tree estimation because, to date, simultaneous methods have either been restricted to a very limited number of taxa (less than about 30) or have been shown to produce less accurate trees than the best combinations of alignment and tree inference programs, e.g., alignment with ClustalW[17] or one of the newer alignment methods such as MAFFT[5], ProbCons[1] or Muscle[2], followed by maximum likelihood methods of tree inference such as RAxML[13]. Many of the best alignment programs use dynamic programming to perform a progressive alignment, with the order of the progressive alignment determined by a guide tree. All methods produce a default guide tree and some will also accept one input by the user.

Whereas much effort has been made to assess the accuracy of phylogenetic tree reconstruction using different methods, models and parameter values, and much attention has been paid to the progressive alignment techniques, far less effort has gone into determining how different guide trees influence the quality of the alignment *per se* and the subsequent phylogeny. A limited study by Roshan et al.[10] looked at improving maximum parsimony trees by iteratively improving the guide trees used in the alignment step. However, they showed little improvement over other techniques.

We address this question more broadly in a simulation study, exploring the performance of five MSA methods (ClustalW, Muscle, ProbCons, MAFFT and FTA - a new method, which we present here) on different guide trees. We find that changes in the guide tree generally do not impact the accuracy of the estimated alignments, as measured by SP-score (Section 3.1 defines this score). *However*, some RAxML-based phylogenies, obtained using alignments estimated on more accurate guide trees, were *much* more accurate than phylogenies obtained using MSA methods on their default guide trees. Muscle and ClustalW were impacted the least by the choice of guide tree, and ProbCons and FTA were impacted the most. The improvement produced for ProbCons is particularly relevant to systematists, since it is one of the two best MSA methods currently available.

Finally, we find that using FTA as an alignment technique results in even more accurate trees than available using ProbCons when a highly accurate guide tree is input, showing the potential for even further improvements in multiple sequence alignment and phylogeny estimation.

The organization of the rest of the paper is as follows. Section 2 provides background on the multiple alignment methods we study, and includes a discussion of the design of FTA. Section 3 describes the experimental study, and the implications of these results. Finally, we discuss future work in Section 4.

2. Basics

Phylogeny and alignment estimation methods. Although there are many phylogeny estimation methods, our studies (and those of others) suggest that maximum likelihood analyses of aligned sequences produce the most accurate phylogenies. Of the various software programs for maximum likelihood analysis, RAxML and GARLI[21] are the two fastest and most accurate methods. We used RAxML for our analyses.

Of the many MSA methods, ClustalW tends to be the one most frequently used by systematists, although several new methods have been developed that have been shown to outperform ClustalW with respect to alignment accuracy. Of these, we included ProbCons, MAFFT, and Muscle. ProbCons and MAFFT are the two best performing MSA methods, and Muscle is included because it is very fast.

We also developed and tested a new MSA method, which we call FTA for "Fixed Tree Alignment." FTA is a heuristic for the "Fixed-Tree Sankoff Problem", which we now define.

The Sankoff problem. Over 30 years ago, David Sankoff proposed an approach for simultaneous estimation of trees and alignments based upon minimizing the total edit distance, which we generalize here to allow for an arbitrary edit distance function $f(\cdot,\cdot)$ as part of the input, thus defining the "Generalized Sankoff Problem"[12]:

- *Input:* A set S of sequences and a function $f(s, s')$ for the cost of an optimal alignment between s and s'.
- *Output:* A tree T, leaf-labeled by the set S, and with additional sequences labelling the internal nodes of T, so as to minimize tree-length, $\sum_{(v,w)\in E} f(s_v, s_w)$, where s_v and s_w are the sequences assigned to nodes v and w, respectively, and E is the edge set of T.

The problem thus depends upon the function $f(\cdot,\cdot)$. In this paper we follow the convention that all mismatches have unit cost, and the cost of a gap of length k is affine (i.e., equals $c_0 + c_1 * k$ for some choice of c_0 and c_1) (see[6,4]). The constants c_0 and c_1 are the "gap-open" cost and the "gap-extend" cost, respectively. The Generalized Sankoff problem is NP-hard since the special case where $c_0 = \infty$ is the maximum parsimony (MP) problem, which is NP-hard. (The problem is also called the "Generalized Tree Alignment" problem in the literature.)

In the fixed-tree version of the Sankoff problem, the tree T is given as part of the input, and the object is to assign sequences to the internal nodes of T so as to produce the minimum total edit distance. This problem is also NP-hard[19]. Exact solutions[6] which run in exponential time have been developed, but these are computationally too expensive to be used in practice. Approximation algorithms for the problem have also been developed[18,20], but their performance guarantees are not good enough for algorithms to be reliable in practice.

The FTA ("fixed tree alignment") technique. We developed a fast heuristic for the Fixed-Tree Sankoff problem[a]. We make an initial assignment of sequences to internal nodes and then attempt to improve the assignment until a local optimum is reached. To improve the assignment, we iteratively replace the sequence at an internal node by a new sequence if the new sequence reduces the total edit distance on the tree. To do this, we estimate the "median" of the sequences labelling the neighbors of v. Formally, the "median" of three sequences A, B, and C with respect to an edit distance function $f(\cdot,\cdot)$ is a sequence X such that $f(X,A)+f(X,B)+f(X,C)$ is minimized. This can be solved exactly, but the calculation takes $O(k^3)$ time[6], where k is the maximum sequence length. Since we estimate medians repeatedly ($\Omega(n)$ times for each n-leaf tree analyzed), we needed a faster estimator than these exact algorithms permit.

We designed a heuristic that is not guaranteed to produce optimal solutions for estimating the median of three sequences. The technique we picked is a simple, two-step procedure, where we compute a multiple alignment using some standard MSA technique, and then compute the majority consensus of the multiple alignment. If replacing the current sequence at the node with the consensus reduces the total treelength, then we use the new sequence; otherwise, we keep the original sequence for the node.

We tested several MSA methods (MAFFT, DCA[14], Muscle, ProbCons, and ClustalW) for use in the median estimator, and examined the performance of FTA under a wide range of model conditions and affine gap penalties. Medians based upon DCA generally gave the best performance with respect to total edit distances as well as SP-error; MAFFT-based medians were second best but less accurate. Because of the improvement in accuracy, we elected to work with DCA-based medians even though they sometimes took twice as long as MAFFT-based medians.

Selecting an affine gap penalty. We investigated the effect of affine gap penalties on alignment accuracy, using a wide range of model conditions (number of taxa, rates of indels and site substitutions, and gap length

---

[a]All modified and developed software is available upon request.

distributions). Although the best affine gap penalty (assessed by the SP-error of the alignment) varied somewhat with the model conditions, we found some gap penalties that had good performance over a wide range of model conditions. Based upon these experiments (data not shown), we chose an affine gap penalty for our analyses, with gap-open cost of 2, mismatch cost of 1, and gap-extend cost of 0.5.

3. Experimental study

Overview. We performed a simulation study to evaluate the performance of the different MSA methods we studied on each of several guide trees. We briefly describe how simulation studies can be used to evaluate two-phase techniques and give an overview of our approach.

First, a stochastic model of sequence evolution is selected (e.g., GTR, HKY, K2P, etc.[3]), and a model tree is picked or generated. A sequence of specified length is placed at the root of the tree T and evolved down the tree according to the parameters of the evolutionary process. At the end of this process, each leaf of the tree has a sequence. In addition, the true tree and the true alignment are both known and can be used later to assess the quality of alignment and phylogeny inference. The sequences are then aligned by a MSA technique and passed to the phylogeny estimation technique, thus producing an estimated alignment and an estimated tree which are scored for accuracy. If desired, the phylogeny estimation method can also be provided the true alignment, to see how it performs when alignment estimation is perfect.

In our experiment, we evolved DNA sequence datasets using the ROSE[15] software (because it produces sequences that evolve with site substitutions and also indels) under 16 different model conditions, half for 100 taxon trees and half for 25 taxon trees. For each model condition, we generated 20 different random datasets, and analyzed each using a variety of techniques. We then compared the estimated alignments and trees to the true alignments and trees, recording the SP-error and missing edge rates. The details of this experiment are described below.

3.1. *Experimental design.*

Model Trees. We generated birth-death trees of height 1.0 using the program `r8s`[11] with 100 and 25 taxa. We modified branch lengths to deviate the tree moderately from ultrametricity, using the technique used by Moret *et al.*[8] with deviation factor c set to 2.0.

Sequence Evolution. We picked a random DNA sequence of length 1000 for the root. We evolved sequences according to the K2P+Indel+Γ model of sequence evolution. For all our model trees, we set the transition/transversion ratio to 2.0, and had all sites evolve at the same rate. We varied the model conditions between experiments by varying the remaining parameters for ROSE: the mean substitution rate, the gap length distribution, and the indel rate. We set the mean substitution rate such that the edgewise average normalized Hamming distance was (approximately) between 2% and 7%. We used two single-gap-event length distributions, both geometric with finite tails. Our "short" single-gap-event length distribution had average gap length 2.00 and a standard deviation of 1.16. Our "long" single-gap-event length distribution had average gap event length 9.18 and a standard deviation of 7.19. Finally, we set insertion and deletion probabilities so as to produce different degrees of gappiness (S-gaps in the table).

The table in Figure 1 shows the parameter settings for each of the 16 model conditions, and the resultant statistics for the model conditions (MNHD=maximum normalized Hamming distance, E-ANHD=average normalized Hamming distance on the edges, S-gaps=percent of the true alignment matrix occupied by gaps, and E-gaps = average gappiness per edge); the standard error is given parenthetically.

	Model Condition Parameters				True Align Statistics			
MC	Taxa	P(gap)	P(sub)	Gap dist	MNHD	e-ANHD	S-gaps	E-gaps
1	100	0.0001	0.005	long	37.5 (.2)	1.9 (.02)	40.8 (.3)	.72 (.01)
2	100	0.0001	0.01	long	56.9 (.3)	3.2 (.04)	43.7 (.6)	.69 (.01)
3	100	0.0005	0.005	long	38.0 (.3)	2.4 (.04)	81.9 (.3)	4.1 (.07)
4	100	0.0005	0.01	long	57.4 (.4)	4.9 (.03)	83.0 (.1)	4.9 (.04)
5	100	0.0005	0.005	short	36.9 (.2)	1.9 (.04)	42.6 (.6)	.76 (.01)
6	100	0.0005	0.01	short	56.7 (.2)	4.1 (.04)	46.4 (.2)	.86 (.01)
7	100	0.0025	0.005	short	38.6 (.3)	2.2 (.04)	81.4 (.2)	4.2 (.07)
8	100	0.0025	0.01	short	56.3 (.2)	4.6 (.07)	82.4 (.2)	4.6 (.07)
9	25	0.0001	0.004	long	32.2 (.2)	2.9 (.05)	22.8 (.3)	1.2 (.02)
10	25	0.0001	0.008	long	51.3 (.2)	5.9 (.09)	25.0 (.4)	1.4 (.02)
11	25	0.0005	0.004	long	31.3 (.2)	2.2 (.03)	55.1 (.5)	4.7 (.10)
12	25	0.0005	0.008	long	50.2 (.3)	5.2 (.08)	61.3 (.5)	5.9 (.12)
13	25	0.0005	0.004	short	30.0 (.1)	2.6 (.08)	26.1 (.4)	1.4 (.04)
14	25	0.0005	0.008	short	50.0 (.1)	6.4 (.07)	28.5 (.2)	1.7 (.02)
15	25	0.0025	0.004	short	31.1 (.4)	3.2 (.06)	66.7 (.4)	7.7 (.14)
16	25	0.0025	0.008	short	50.2 (.5)	5.2 (.07)	62.4 (.4)	6.6 (.08)

Figure 1. Model condition parameters and true alignment statistics.

Methods for estimating multiple alignments and trees. We used five multiple sequence alignment programs to create alignments from raw sequences: ClustalW, Muscle, MAFFT, ProbCons and FTA. ClustalW, Muscle, MAFFT and ProbCons are publicly available, while FTA is a method we developed (see Section 2 for a description of this method). For this study, ClustalW, Muscle, MAFFT and ProbCons were each run using their default guide trees as well with guide trees that we provided. We modified ProbCons to allow it to use an input guide tree, and the authors of MAFFT provided us with a version that accepts guide trees as input. FTA does not have a default guide tree, and therefore was run using only the computed guide trees and the true tree. MAFFT has multiple alignment strategies built in, and we used each of L-INS-i, FFT-NS-i and FFT-NS-2. However, when there were difference between variants of MAFFT, FFT-NS-2 usually performed best, so we only show results using this variant. We used RAxML in its default setting.

User-input guide trees. We tested performance on four user-input guide trees. We included the true tree, and three other guide trees that we computed. The first two of the computed guide trees are UPGMA trees based upon different distance matrices. For the first UPGMA guide tree ("upgma1"), we computed a distance matrix based upon optimal pairwise alignments between all pairs of sequences, using the affine gap penalty with *gap-open* = 0, *gap-extend* = 1 and *mismatch* = 1. For the second ("upgma2"), we computed the distance matrix based upon optimal pairwise alignments between all pairs of sequences for the affine gap penalty with *gap-open* = 2, *gap-extend* = 0.5 and *mismatch* = 1. In both cases, we used custom code based on the Needleman-Wunsch algorithm with the specified gap penalty to compute the distance matrices and PAUP*[16] to compute the UPGMA trees. The third guide tree ("probtree") was obtained as follows. We used the upgma1 guide tree as input to ProbCons to estimate an alignment that was then used to estimate a tree using RAxML.

Error rates for phylogeny reconstruction methods. We used the missing edge rate, which is the percentage of the edges of the true tree that are missing in the estimated tree (also known as the false negative rate). The "true tree" is obtained by contracting the zero-event edges in the model tree; it is usually binary, but not always.

Alignment error rates. To measure alignment accuracy, we used the SP (sum-of-pairs) error rate (the complement of the SP accuracy measure), which we now define. Let $S = \{s_1, s_2, \ldots, s_n\}$, and let each s_i be a string over some alphabet Σ (e.g., $\Sigma = \{A, C, T, G\}$ for nucleotide sequences). An alignment on S inserts spaces within the sequences in S so as to create a matrix, in which each column of the matrix contains either a dash or an element of Σ. Let s_{ij} indicate the jth letter in the sequence s_i. We identify the alignment A with the set $Pairs(A)$ containing all pairs $(s_{ij}, s_{i'j'})$ for which some column in A contains s_{ij} and $s_{i'j'}$. Let A^* be the true alignment, and let $\hat{A}$ be the estimated alignment. Then the SP-error rate is $\frac{|Pairs(A^*)-Pairs(\hat{A})|}{|Pairs(A^*)|}$, expressed as a percentage; thus the SP-error is the percentage of the pairs of truly homologous nucleotides that are unpaired in the estimated alignment. However, it is possible for the SP-error rate to be 0, and yet have different alignments.

3.2. *Results.*

We first examine the guide trees with respect to their topological accuracy. As shown in Figure 2, the accuracy of guide trees differs significantly, with the ProbCons default tree generally the least accurate, and our "probtree" guide tree the most accurate; the two UPGMA guide trees have very similar accuracy levels.

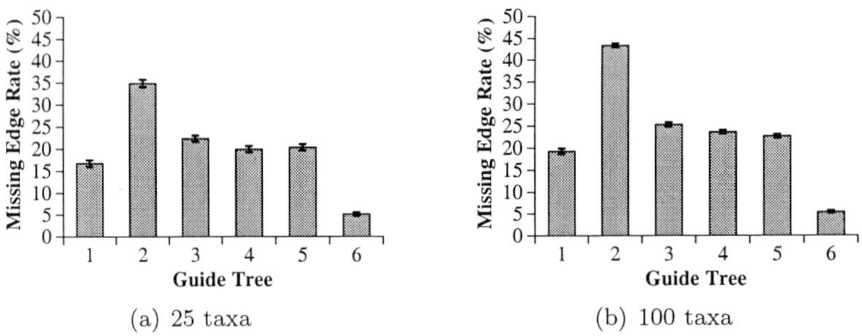

Figure 2. Guide tree topological error rates, averaged over all model conditions and replicates. (1) ClustalW default, (2) ProbCons default, (3) Muscle default, (4) upgma1, (5) upgma2, and (6) probtree.

In Figure 3 we examine the accuracy of the alignments obtained using different MSA methods on these guide trees. Surprisingly, despite the large differences in topological accuracy of the guide trees, alignment accuracy (measured using SP-error) for a particular alignment method varies relatively little between alignments estimated from different guide trees. For example, two ClustalW alignments or two Muscle alignments will have essentially the same accuracy scores, independent of the guide tree. The biggest factor impacting the SP-error of the alignment is the MSA method. Generally, ProbCons is the most accurate and ClustalW is the least.

We then examined the impact of changes in guide tree on the accuracy of the resultant RAxML-based phylogeny (see Figure 4). In all cases, for a given MSA method, phylogenetic estimations obtained when the guide

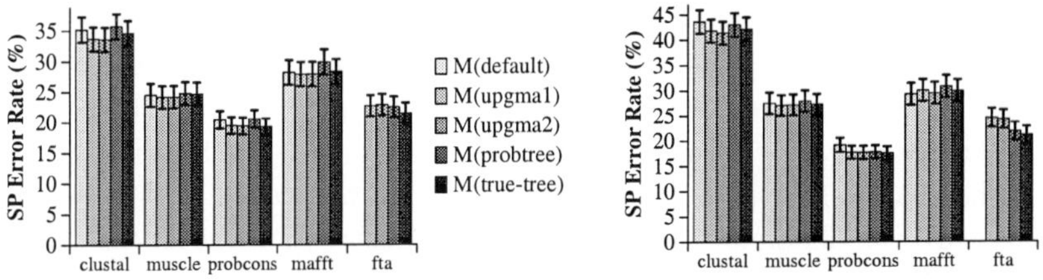

(a) 25 taxa (b) 100 taxa

Figure 3. SP-error rates of alignments. M(guide tree) indicates multiple sequence alignment generated using the indicated guide tree.

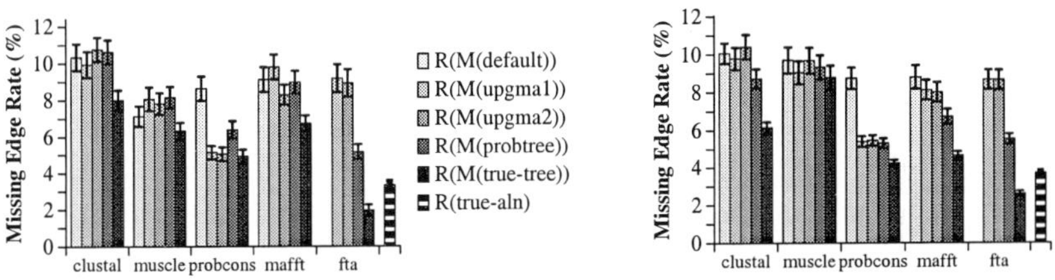

(a) 25 taxa (b) 100 taxa

Figure 4. Missing edge rate of estimated trees. R(M(guide tree) indicates RAxML run on the alignment generated by the multiple sequence alignment method using the guide tree indicated. R(true-aln) indicates the tree generated by RAxML when given the true alignment.

tree is the true tree are more accurate than for all other guide trees. However, MSA methods otherwise respond quite differently to improvements in guide trees. For example, Muscle responded very little (if at all) to improvements in the guide tree, possibly because it computes a new guide tree after the initial alignment on the input guide tree. ClustalW also responds only weakly to improvement in guide tree accuracy, often showing - for example - worse performance on the probtree guide tree compared to the other guide trees. On the other hand, ProbCons and FTA both respond positively and significantly to improvements in guide trees. This is quite interesting, since the alignments did *not* improve in terms of their SP-error rates! Furthermore, ProbCons improves quite dramatically as compared to its performance in its default setting. The performance of FTA is intriguing. It is generally worse than ProbCons on the UPGMA guide trees, but comparable to ProbCons on the probtree guide tree, and better than ProbCons on the true tree.

In fact, trees estimated using the alignment produced by FTA using the true guide tree are even better than trees estimated from the true alignment. There are several possible explanations for this phenomenon, but further study is required.

The graphs we show in Figures 3 and 4 have values that have been averaged over all model conditions and replicates (for the given number of taxa). The relative performance of the methods shown in the averages holds (with few exceptions) for each model condition. However, the magnitudes of the actual errors and amount of improvement based on a given guide tree vary. Graphs for individual model conditions are available here: http://www.cs.utexas.edu/users/serita/pubs/psb08-aux/

3.3. *Conclusions.*

Except for FTA, MSA accuracy (as measured using SP-error) is not strongly correlated with guide tree accuracy. Further, for most of these MSA methods, phylogenetic accuracy is not directly predicted by the accuracy of the guide tree (except again, in the case of FTA). Although it is common to evaluate alignments purely in terms of criteria like SP (or column score), these experiments provide clear evidence that not all errors are of equal importance, at least in terms of phylogenetic consequences. This is not completely surprising, since when Ogden and Rosenberg[9] studied the influence of tree shape on alignment and tree reconstruction accuracy they too found that alignment error did not always have a large impact on tree accuracy. Thus, although FTA alignments are often "worse" with respect to the SP-error, trees estimated from FTA alignments can be more accurate than trees estimated from other alignments with lower SP-error rates. Finally, it is important to realize that although alignments may have similar SP-error rates as compared to a true alignment, they can still be very different from each other.

The experiments show clearly that tree estimation can be improved through the use of improved guide trees, though only some alignment methods seem to be able to take advantage of these improved guide trees. It is also clear that these improvements require some additional computational effort. Is it worth it? Consider the following different methods, which we will call "Good" and "Better".

- Good: Run ProbCons in its default setting, followed by RAxML.
- Better: Run ProbCons on one of the UPGMA guide trees, followed by RAxML. (Note that this method produces the "probtree" guide

tree, if the upgmal guide tree is used.)

How much time do these methods take in our experiments?

In our experiments, run using a distributed system via Condor [7], alignment using ProbCons was the most expensive step in terms of running time. The Good technique took approximately 8 minutes on 25 taxa and slightly more than 2 hours for 100 taxa, while Better took under 9 minutes on 25 taxa and 2.5 hours for 100 taxa. In other words, for a very minimal increase in running time, substantial improvements in topological accuracy are obtainable.

4. Future Work

Our study shows clearly that improving the guide tree for MSA methods can improve estimated phylogenies, provided that appropriate multiple alignment methods are used. Furthermore, it shows that FTA can obtain better trees than the other methods tested when the guide tree is very good. Indeed, our data suggest that once the guide tree is within about 20% RF distance to the true tree, trees based upon FTA alignments will be highly accurate. Given these results, we will test an iterative approach to phylogeny and alignment estimation: begin with a good guide tree (e.g., probtree); compute FTA on the guide tree; and then compute a new guide tree for FTA by running RAxML on the resultant alignment (and then repeat the FTA/RAxML analysis). In the current experiments, RAxML and FTA were both very fast, even on the 100-taxon dataset, so the iterative approach may scale well to significantly larger numbers of taxa. Other future work will seek to develop new alignment-error metrics that better capture differences among alignments, specifically in terms of their ability to predict accuracy of subsequent phylogenetic inference.

References

1. C.B. Do, M.S.P. Mahabhashyam, M. Brudno, and S. Batzoglou. PROBCONS: Probabilistic consistency-based multiple sequence alignment. *Genome Research*, 15:330–340, 2005.
2. R. C. Edgar. MUSCLE: a multiple sequence alignment method with reduced time and space complexity. *BMC Bioinformatics*, 5(113), 2004.
3. J. Felsenstein. *Inferring Phylogenies.* Sinauer Associates, Sunderland, Massachusetts, 2004.
4. J. Fredslund, J. Hein, and T. Scharling. A large version of the small parsimony problem. In Gary Benson and Roderic Page, editors, *Algorithms in*

Bioinformatics: Third International Workshop, WABI 2003, LNCS, volume 2812, pages 417–432, Berlin, 2003. Springer-Verlag.
5. K. Katoh, K. Kuma, H. Toh, and T. Miyata. MAFFT version 5: improvement in accuracy of multiple sequence alignment. *Nucleic Acids Res.*, 33(2):511–518, 2005.
6. B. Knudsen. Optimal multiple parsimony alignment with affine gap cost using a phylogenetic tree. In G. Benson and R. Page, editors, *WABI 2003, LNBI 2812*, pages 433–446. Springer-Verlag, Berlin, 2003.
7. Michael Litzkow. Remote unix - turning idle workstations into cycle servers. In *Usenix Summer Conference*, pages 381–384, 1987.
8. B.M.E. Moret, U. Roshan, and T. Warnow. Sequence length requirements for phylogenetic methods. In *Proc. 2nd Int'l Workshop Algorithms in Bioinformatics (WABI'02)*, volume 2452 of *Lecture Notes in Computer Science*, pages 343–356. Springer-Verlag, 2002.
9. T. Heath Ogden and Michael S. Rosenberg. Multiple sequence aligment accuracy and phylogenetic inference. *Systematic Biology*, 55(2):314–328, 2006.
10. U. Roshan, D.R. Livesay, and S. Chikkagoudar. Improving progressive alignment for phylogeny reconstruction using parsimonious guide-trees. In *Proceedings of the IEEE 6th Symposium on Bioinformatics and Bioengineering.* IEEE Computer Society Press, 2006.
11. M. J. Sanderson. r8s: inferring absolute rates of molecular evolution and divergence times in the absence of a molecular clock. *Bioinformatics*, 19(2):301–302, 2003.
12. D. Sankoff. Minimal mutation trees of sequences. *SIAM J. Appl. Math.*, 28(1):35 – 42, January 1975.
13. Alexandros Stamatakis. Raxml-vi-hpc: Maximum likelihood-based phylogenetic analyses with thousands of taxa and mixed models. *Bioinformatics*, 22(21):2688–2690, 2006.
14. J. Stoye. Multiple sequence alignment with the divide-and-conquer method. *Gene*, 211:GC45–GC56, 1998.
15. J. Stoye, D. Evers, and F. Meyer. Rose: generating sequence families. *Bioinformatics*, 14(2):157–163, 1998.
16. D. Swofford. PAUP*: Phylogenetic analysis using parsimony (and other methods), version 4.0. 1996.
17. J.D. Thompson, D.G. Higgins, and T.J. Gibson. CLUSTAL W: improving the sensitivity of progressive multiple sequence alignment through sequence weighting, position-specific gap penalties and weight matrix choice. *Nucleic Acids Research*, 22:4673–4680, 1994.
18. L. Wang and D. Gusfield. Improved approximation algorithms for tree alignment. *J. Algorithms*, 25:255–273, 1997.
19. L. Wang and T. Jiang. On the complexity of multiple sequence alignment. *J. Comput. Biol.*, 1(4):337–348, 1994.
20. L. Wang, T. Jiang, and D. Gusfield. A more efficient approximation scheme for tree alignment. *SIAM J. Comput.*, 30(1):283–299, 2000.
21. D. Zwickl. GARLI download page. Website, 2006. http://www.zo.utexas.edu/faculty/antisense/Garli.html.

SENSITIVITY ANALYSIS FOR REVERSAL DISTANCE AND BREAKPOINT REUSE IN GENOME REARRANGEMENTS

AMIT U SINHA

Department of Computer Science, University of Cincinnati, Cincinnati, OH 45221, USA (sinhaam@ececs.uc.edu)

JAROSLAW MELLER

Department of Environmental Health, University of Cincinnati College of Medicine, Cincinnati, OH 45267, USA; Department of Informatics, Nicholas Copernicus University, 87-100 Torun, Poland (jmeller@cchmc.org)

Identifying syntenic regions and quantifying evolutionary relatedness between genomes by interrogating genome rearrangement events is one of the central goals of comparative genomics. However, identification of synteny blocks and the resulting assessment of genome rearrangements are dependent on the choice of conserved markers, the definition of conserved segments, and the choice of various parameters that are used to construct such segments for two genomes. In this work, we performed an extended sensitivity analysis of synteny block generation using alternative sets of markers in multiple genomes. A simple approach to synteny block aggregation is used, which depends on two principle parameters: the maximum gap (*max_gap*) between adjacent blocks to be merged, and the minimum length (*min_len*) of synteny blocks. In particular, the dependence on the choice of conserved markers and *max_gap/min_len* aggregation parameters is assessed for two important quantities that can be used to characterize evolutionary relationships between genomes, namely the reversal distance and breakpoint reuse. We observe that the number of synteny blocks depends on both parameters, while the reversal distance depends mostly on *min_len*. On the other hand, we observe that relative reversal distances between mammalian genomes, which are defined as ratios of distances between different pairs of genomes, are nearly constant for both parameters. Similarly, the breakpoint reuse rate was found to be almost constant for different data sets and a wide range of parameters. Breakpoint reuse is also strongly correlated with evolutionary distances, increasing for pairs of more divergent genomes. Finally, we demonstrate that the role of parameters may be further reduced by using a multi-way analysis that involves markers conserved in multiple genomes, which opens a way to guide the choice of a correct parameterization.

Supplementary Materials (SM) at http://cinteny.cchmc.org/doc/sensitivity.php

1. Introduction

An increasing number of newly sequenced genomes greatly enhance our ability to construct evolutionary models from their comparative analysis. One problem of central importance is the identification of blocks of genes (or other discrete markers) with evolutionary conserved order. These synteny blocks help in tracing back the evolution of genomes in terms of rearrangement events, such as inversion, translocation, fusion, fission, etc. Consequently, genome evolution and phylogenetic (phylogenomic) trees may be reconstructed from the analysis of synteny [1], [2], [3].

Nadeau and Taylor [4] argued that translocation and inversion (reversal) are the main evolutionary events that affect gene (and other markers) order. They concluded that the effect of transposition is not very significant. In fact, for the sake of computational efficiency, most of the algorithms for finding the evolutionary distance mimic translocation, fission and fusion in terms of inversions, while neglecting the effect of transpositions [5]. In particular, once two genomes are represented in terms of blocks of markers with conserved order, each genome may be transformed into a signed permutation (sign representing the strand of genes/markers). As a result, one genome may be transformed into the other by applying reversal operations, providing a model of genome rearrangements. Consequently, analyses of genome rearrangements within this model typically involve calculating the reversal distance between two genomes, which is defined as the minimum number of reversals required to sort one (signed) permutation to the other [5]. Thanks to recent algorithmic advances, the reversal distance can be computed in linear time [6].

Another quantity that we consider here is the breakpoint reuse rate (BRR), which is defined as $2d/b$, where d is the reversal distance and b is the number of breakpoints, as estimated from the observed synteny blocks. BRR can be interpreted as a simple measure of the extent to which breakpoints are used on average during rearrangement events [7]. However, this interpretation is contested by some groups [8], partly because of the divergence between alternative estimates of the numerical value of BRR, as obtained using different parameterizations of the problem, and partly because it largely disregards the mechanistic nature of rearrangement events that tend to occur within repetitive DNA fragments of certain (potentially large) length [9]. These debates clearly underscore the need for further assessment of current models of genome evolution, and methods for synteny block identification, in particular.

A set of discrete markers that represents the genome of interest in a simple model considered here, consists either of orthologous genes or conserved sequence tags (anchors). Obviously, the choice of a set of markers affects the results and attempts have been made to assess the impact of such choices [7], [10]. Another problem in identifying synteny blocks is that large potential blocks may be interrupted by local disruptions in the order of markers. However, there is no precise definition of such local disruptions. Consequently, many different methods have been devised to filter out these micro-rearrangements, using heuristics or statistical models to assess the significance of associations (co-localization) between markers [7], [11], [12], [13], [14], [15], [16], [17].

As discussed in section 2.2, many of these algorithms for constructing synteny blocks can be cast using a general framework, in which there are two principal parameters. The first parameter defines blocks to be removed from consideration if their length (either in terms of the minimum number of markers, or in terms of their physical length) is too short. The second parameter defines how adjacent blocks will be merged (effectively disregarding the markers in between that locally disrupt the order), depending on the distance (gap) between these blocks. In what follows, these parameters are referred to as the minimum length (*min_len*) of individual blocks and the maximum gap (*max_gap*) between adjacent blocks to be merged, respectively.

Since the identification of synteny blocks is a crucial step in measuring reversal distance, breakpoint reuse and other related quantities, it is important to systematically assess its sensitivity with respect to the choice of the set of markers (including the use of markers conserved in multiple genomes), *min_len* and *max_gap* parameters, and other arbitrary choices. In fact, the impact of these parameters on the analysis of evolutionary relatedness within this model has recently been highlighted in attempts to estimate breakpoint reuse rate between human and mouse genomes, leading to debates about random vs. fragile breakage model of genome evolution [11], [10].

Here, we used an efficient computational framework [18] for a comprehensive analysis of the sensitivity of the reversal distance and breakpoint reuse in multiple genomes, using both homolog and sequence tag data sets. In particular, we performed a systematic assessment of the role of critical parameters in the model. Based on our result, we suggest that using a subset of genes common to more than two related species may provide more stable results and yield improved estimates of the evolutionary relatedness. Furthermore, we find that relative measures of divergence between two pairs of genomes are less dependent on the choice of arbitrary parameters. This observation provides an additional support for the construction of robust phylogenetic (phylogenomic) trees and other analyses relying on such relative (rather than absolute) distance measures.

2. Methods

The results presented in this contribution were generated using the Cinteny server for the analysis of synteny and genome rearrangements, which is available at http://cinteny.cchmc.org/ [18]. The server allows one to use alternative data sets, including both ortholog and sequence tags (anchors)-based sets of markers in multiple genomes. It also allows the user to set parameters that affect the synteny blocks identification, as well as the computation of reversal distances and breakpoint reuse rates, enabling systematic analysis of sensitivity of the results with respect to these arbitrary choices.

2.1. *Data Sets*

While sequence tags in general provide greater coverage of the genome, the conservation of non-functional regions may not be of equal importance as gene conservation, or could simply result from spurious sequence matches, introducing noise in the model. On the other hand, the identification of orthologs is often marred by limited sensitivity of sequence searches and other annotation problems. Therefore, we used both orthologs and conserved sequence tags for a more comprehensive analysis. The orthologs from NCBI HomoloGene [19] and RoundUp Orthology Database [20] and a data set of conserved sequence tags from an earlier study by Bourque et al. [3], which will be referred to as GRIMM, were used. HomoloGene contains orthologs for human, mouse, rat, dog and chimp genomes, whereas RoundUp also contains rhesus macaque and cow. GRIMM data set has conserved markers in human, mouse and rat genomes.

2.2. *Forming Synteny Blocks*

Synteny blocks are identified as segments of the genomes in which the order of homologous markers is conserved. Typically, local rearrangement events that concern only few markers within a synteny block, and are referred to as micro-rearrangement, are being ignored. The rationale is that smaller conserved blocks do not represent a significant evolutionary signature, and might add noise to the model. This process takes the form of the aggregation of initial (entirely ordered) blocks to create larger synteny blocks, and effectively filter out these micro-rearrangements. While such an aggregation may be parameterized differently, two parameters are typically used in this context [10]:

- *max_gap*: maximum gap between blocks that are allowed to be merged;
- *min_len*: minimum length of a synteny block.

Specifically, if the gap between two adjacent synteny blocks is less than *max_gap* then they may be merged together to form a larger block. The relative order (orientation) of the two blocks has to be accounted for. This process of aggregation is continued until no more blocks may be merged. Subsequently, the blocks of length less than *min_len* are rejected.

Many algorithms for forming synteny blocks follow this paradigm, and we follow in their footsteps. For example, the GRIMM-Synteny [7] and MAUVE [15] algorithms define the parameter *min_len* as 'minimum cluster size C' and *w(cb)*, respectively. An alternative is to set a lower limit on the size of synteny blocks in terms of the number of markers within the block. This corresponds to the parameter Δ in ST-Synteny algorithm [11] and *h* in [17]. Here, *min_len* is varied to test its effect on the results. In addition, unless stated otherwise, we reject synteny blocks with less than 3 markers.

There is more ambiguity in the notion of *max_gap* [17], which prescribes how blocks should be aggregated. We define it as the threshold on the maximum gap between the two synteny blocks in *each* species that are allowed to be merged. This corresponds to the parameter 'maximum gap size G' in the GRIMM-Synteny algorithm [7] and d_c, in FISH algorithm [12], which is defined as the *sum* of gap between adjacent blocks in the two species. The parameter *MaxDist* in [14] is similar to *max_gap* as well. In some cases, *max_gap* is defined in terms of the numbers of markers, i.e., by putting a threshold on the number of out-of-order markers while merging blocks [13], [17], [16]. Some methods avoid this gap constraint by coalescing blocks after removing smaller blocks [15]. However, this behavior may still be captured by some parameterization of *max_gap*. In general, while direct comparison between different definitions may be difficult, *max_gap* has relatively small impact on measures of genome rearrangement, as we show in the results section.

2.3. *Measuring Reversal Distance*

Once the synteny blocks are identified, the relative order of blocks in two multi-chromosomal genomes is represented as numeric signed permutation. The Hannenhalli-Pevzner [5] algorithm calculates the reversal distance in linear time when used with modifications proposed by [6] and [21], which we implemented in the Cinteny server, to enable comprehensive assessment for a large range of parameters considered here. It should be noted that we do not address block or genome duplications, and we use a heuristic choice of unique markers for paralogs (see Supplementary Materials).

2.4. *Using Multiple Genomes*

Working with a set of markers conserved across multiple species instead of those conserved in individual pairs of genomes may lead to more stable results. For example, at present HomoloGene includes 16,330 orthologs for human and mouse. When using a '5-way' approach, 10,574 genes having orthologs in human, mouse, rat, dog and chimp are identified. Pairwise synteny between human and mouse can now be identified using only these 10,574 genes. The advantage of using this approach is that aggregation of synteny blocks occurs naturally, as only highly conserved segments are used. The same logic may be extended for any multi-way approach, with the hope that a subset of markers conserved and/or better annotated in multiple species may help filtering out micro-rearrangements and minimizing the effects of errors in homology prediction. A similar method was demonstrated for chromosome level comparison to yield more meaningful relationships between canine and other mammalian genomes [22].

3. Results

We used two independent ortholog data sets and a data set of conserved sequence tags, as described in the Methods section, in order to measure the variation in number of synteny blocks, reversal distance, breakpoint re-use rate, etc., by changing the parameters *max_gap* and *min_len.*

3.1. *Ortholog v/s Conserved Sequence Tags*

3.1.1. *Number of Synteny Blocks*

Figure 1 shows the variation in the Number of Synteny Blocks (NSB) due to parameters *max_gap* and *min_len* for human-mouse pair using HomoloGene data set. The parameters *max_gap* (y-axis) and *min_len* (x-axis) were increased from 0 to 1 Mb in steps of 20 Kb and NSB is plotted (z-axis). We observe that NSB decreases on increasing *max_gap* and *min_len.* When the latter is increased, more synteny blocks (of smaller size) are rejected leading to a decrease in NSB. As *max_gap* is increased, adjacent synteny blocks are aggregated and their total number decreases too, although to a smaller degree. In general, the results obtained with the RoundUp orthologs and GRIMM sequence tags were similar to those obtained with HomoloGene (SM Figure S1). However, when using sequence tags, the number of synteny blocks with small *max_gap* is large. This is because the number of sequence tags is much larger than the number of orthologs and little aggregation takes place when *max_gap* is low. As *max_gap* is increased, more aggregation takes place and there is a steep decline in the total number of synteny blocks, which becomes very close to the value observed for gene-based analysis. Similar pattern in sensitivity of NSB was observed for human-dog, human-rat, rat-mouse and other pairs of genomes (see SM Figure S2).

3.1.2. *Reversal Distance*

Once the synteny blocks are found, the disruption of the order of the blocks is measured as the Reversal Distance (RD). Figure 2 shows the variation of RD due to *min_len* for human-mouse genomes. For each value of *min_len*, the RD is calculated for different values of *max_gap* (between 60 Kb and 1 Mb) and the variation is displayed as box plots. The low heights of the boxes indicate that the variation in RD due to *max_gap* for a given value of *min_len* is limited. This is because increasing *max_gap* preferentially aggregates blocks which have a similar order in both genomes, so the reversal distance does not change much. On the other hand, there is a steep and uneven decrease in RD as *min_len* is increased but the median values start to flatten at higher values of *min_len*. Some outliers are observed for high values of *min_len* and low values of *max_gap*. Orthologs and sequence tags based data sets give similar results qualitatively. Sequence tag based analysis gives a higher RD for low values of *min_len* because the number of synteny blocks is higher. At higher values of *min_len*, the value of RD for both types of data begins to converge. The results obtained with the RoundUp orthologs were similar (see SM Figure S3).

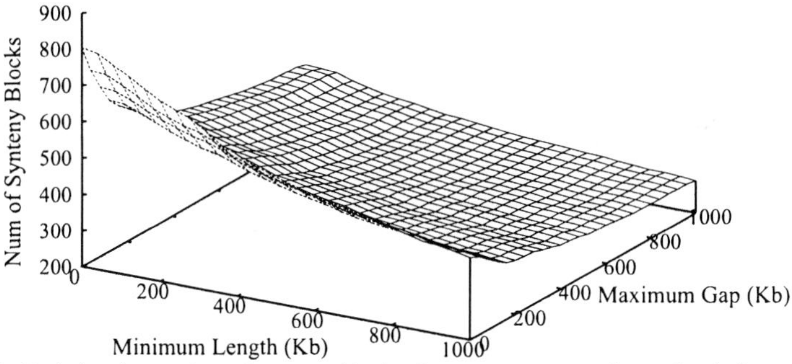

Figure 1: Variation in number of synteny blocks due to *max_gap* and *min_len* in human-mouse genomes for orthologs based analysis

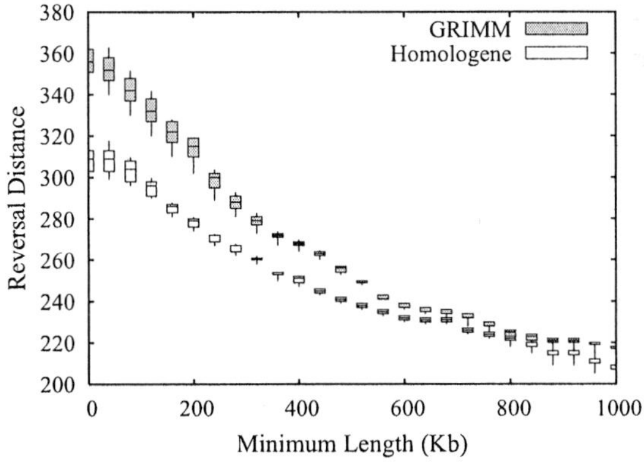

Figure 2: Variation of reversal distance due to *min_len* in human-mouse for ortholog (HomoloGene) and sequence tag (GRIMM) based analysis. The height of the boxes shows the variation in reversal distance due to *max_gap* for a given value of *min_len*.

3.2. *Breakpoint Reuse Rate*

Measurement of Breakpoint Reuse Rate (BRR) and its dependence on parameters has been debated a lot in the last few years [10], [8]. In particular, its numerical value was used as an argument in the dispute over fragile vs. random breakage model of genome evolution. We first assess the effects of the parameters on BRR for human and mouse genomes. The parameters *max_gap* and *min_len* were varied from 0 to 1 Mb in steps of 20 Kb and the BRR was calculated for different data sets. The mean and standard deviation as well as minimum and maximum values of BRR over the range of parameters are shown in Table 1. We observe that unlike RD, BRR (which is a relative quantity) shows very little variation due to the parameters or due to data sets. These results are consistent with previous findings by Peng and colleagues [10] for human-mouse genomes, for which they reported a BRR of 1.61 and 1.67 for ortholog-based and sequence-based analysis, respectively. To extend this analysis, we investigate BRR further in the next section by comparing it with other measures of evolutionary divergence.

Table 1: Breakpoint reuse rate for human-mouse genomes with varying *max_gap* and *min_len* from 0 to 1 Mb for all 3 data sets and parameterizations of the problem (see text for details)

	BRR			
Data set	**Mean**	**SD**	**Min**	**Max**
HomoloGene	1.64	0.02	1.58	1.66
RoundUp	1.63	0.02	1.59	1.67
Grimm	1.62	0.03	1.53	1.66

3.3. *Correlation of Reversal Distance and Breakpoint Reuse Rate*

One expects an increase in the number of genome rearrangement events as species evolve and diverge from their ancestral genomes. Additionally, when the number of rearrangement events is high, the chance of a breakpoint region being reused increases. Indeed, this is found to be the case for many genome pairs. Figure 3 shows BRR and RD of 5 genomes with human genome. The RD and BRR were calculated for both *min_len* and *max_gap* equal to 500 Kb. Pearson correlation coefficient was found to be 0.996 ($p < 0.001$). A correlation of 0.995 and 0.990 was found for *min_len* equal to 300 Kb and 1000 Kb, respectively, showing that the correlation stands for different values of these parameters.

There are, however, some intriguing exceptions from this general trend. For example, the human-dog and mouse-rat genomes have similar BRR (1.40 and 1.43, respectively) even though the RD is very different (150 and 71, respectively). Despite such outliers, it is evident that BRR increases, as the number of rearrangement events increases. Closely related genomes, such as human and chimp, show a BRR of 1.1, while human-mouse has a BRR of 1.64.

Similarly, mouse-rat genomes have a BRR of 1.42, while mouse-dog genomes have a BRR of 1.62. These data suggest that BRR may be used as an alternative measure of evolutionary distance, as it is largely independent of parameters.

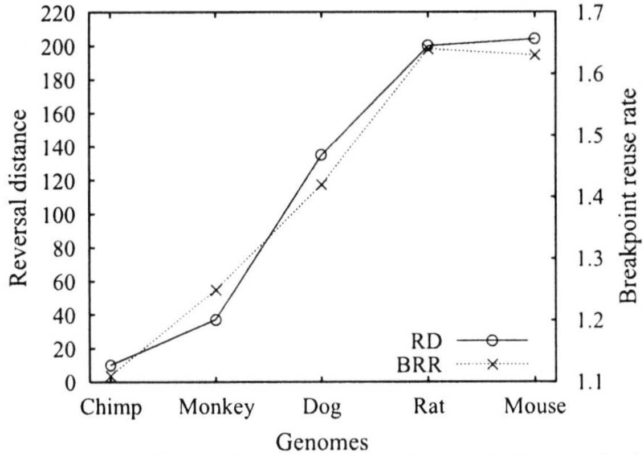

Figure 3: Correlation between breakpoint reuse rate and reversal distance for human and other genomes. The trend is independent of the parameterization of the synteny block identification.

Finally, in the context of the on-going discussion about numerical estimates of BRR and the validation of the proposed fragile breakage model [7], we would like to comment that BRR is an average quantity. In particular, it may be possible that some breakpoints are used more frequently than others, especially if they occur within large repetitive regions in the genome [9]. Since the evolutionary pathway can not be uniquely determined for a given reversal distance using Hannenhalli-Pevzner model, it is not possible to determine the actual number of breakpoints which are, in fact, reused (perhaps more than once) during the transformation of one genome to another. Consequently, the numerical value of BRR may be more informative as a relative (and weakly dependent on parameters) measure of evolutionary distances, rather than supporting (or not) one of the models of rearrangement events.

3.4. *Relative Divergence*

In light of the above conclusions regarding BRR, we investigated another relative measure of evolutionary relatedness. The absolute values of RD (reversal distance) are found to be very sensitive to the choice of *min_len*. Therefore, we define a relative divergence measure, as the ratio of RD of two different pairs of genomes. For this analysis, we measured RD and relative divergence as a function of *min_len*. The results in Table 2 show the absolute value of RD in human-mouse (H-M), rat-mouse (R-M), human-dog (H-D) and human-chimp (H-C) genomes for different choice of parameters. The table also shows the ratio of human-mouse RD with other pairs.

We observe that even though individual RD changes with the parameters, as shown earlier, the ratio of RD between pairs of genome shows negligible variation for *min_len* greater than 200 Kb. The mean relative divergence of human-mouse with respect to rat-mouse, human-dog and human-chimp genomes is almost constant at 3.29 ($\sigma = 0.05$), 1.63 ($\sigma = 0.03$) and 21.91 ($\sigma = 0.79$), respectively. This information (relative divergence) may be more useful than a simple RD as it shows very little variation due to the parameters. This perhaps bode well for attempts to use RD as a measurement of inter-genomic distances in relative terms, e.g., to construct phylogenomic trees. The ratio of NSB was also found to be constant for different choice of parameters between two pair of genomes.

Table 2: Variation of RD in different genome pairs and ratios of RD

min_len (kb)	**Reversal distances**				**Relative divergence between pairs**		
	H-M	**R-M**	**H-D**	**H-C**	**H-M/R-M**	**H-M/H-D**	**H-M/H-C**
1000	206	62	131	9	3.32	1.57	22.89
800	224	66	135	10	3.39	1.66	22.40
600	233	71	143	11	3.28	1.63	21.18
400	248	76	151	12	3.26	1.64	20.67
200	281	92	174	14	3.05	1.61	20.07
100	313	128	186	22	2.45	1.69	14.23
0	346	170	211	34	2.04	1.64	10.18

3.5. *Using Multiple Genomes*

In order to assess the behavior of more highly conserved elements, we compared the variation of reversal distance using 2-way and 5-way approaches. The former was done using the genes common to human and mouse and the latter was done using the genes common to human, mouse, dog, rat, chimp genomes. The number of orthologs for the 2-way and 5-way analysis was 16,330 and 10,574, respectively. Figure 4 shows the variation of RD due to *min_len* for human-mouse genomes. For each value of *min_len*, RD is calculated for different values of *max_gap* and the variation is displayed as a box plot.

Since fewer orthologs are used for a 5-way analysis, RD is smaller in absolute terms than in 2-way analysis. It is also evident from Figure 4 that the variation due to *max_gap* (height of the boxes) is almost negligible in the case of 5-way comparison. Furthermore, the overall variation due to *min_len* is less pronounced in 5-way comparison. This suggests that multi-way analysis reduces the role of the parameters, albeit to different degrees. We also performed an extended analysis of BRR using 5-way approach for five mammalian genomes. The results are similar to those obtained using 2-way approach, indicating again relatively low sensitivity of BRR with respect to the parameterization of the problem.

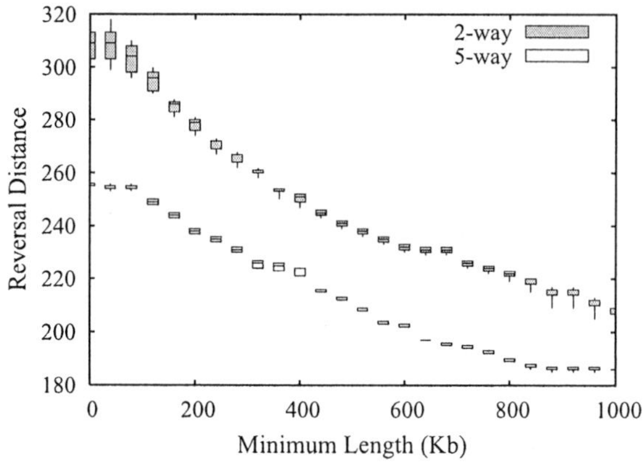

Figure 4: Variation in RD due to *min_len* in human and mouse genomes for 2-way and 5-way analysis. The height of the boxes shows the variation in reversal distance due to *max_gap* for a given value of *min_len*. The observed variation is smaller when using multiple genome approach.

4. Conclusions

Genome rearrangement analysis is often marred by the lack of a clear strategy for selecting critical parameters, choosing appropriate data sets, etc. We performed a systematic analysis of the sensitivity of genome rearrangement measures to the choice of critical parameters for several mammalian genomes. Both ortholog-based and sequence tag-based approaches are compared. Two specific parameters, i.e., the maximum allowable gap between adjacent blocks for aggregation and the minimum length of synteny blocks, are varied systematically to assess their effect. We found that the number of synteny blocks depends on both parameters, while the reversal distance depends mostly on the latter. Therefore, one needs to exercise caution while using (absolute values of) reversal distances as a measure of evolutionary relatedness. The breakpoint reuse rate was found, on the other hand, to have a negligible change due to variation in these two parameters. At the same time, it showed a strong correlation with reversal distances, indicating that high breakpoint reuse rates may simply reflect the expected higher number of inversions with increasing evolutionary divergence. This, however, opens a way to use BRR as an alternative measure of evolutionary distance, which may be more informative for inferring evolutionary relatedness, building phylogenetic trees and other applications. Another relative measure with similar properties that we consider is the relative divergence, which is defined as ratios of reversal distances between different pairs of genomes. In this context, the distance for a pair of well defined and annotated genomes, such as human and mouse, may be used to normalize all other pair-wise distances with the same parameterization. Using multiple-way comparisons decreases the dependence on parameters, when compared with two-way analysis, suggesting rational strategies to choose parameters for the identification of synteny blocks.

Acknowledgements

We would like to thank the reviewers for their insightful comments and suggestions. This work has been partially supported by NIH grant R01 AR050688.

References

1. Sankoff D, Blanchette M, In *Proc. of COCOON*, 251-63, (1997).
2. Moret BME, Wyman S, Bader DA, et al., In *Proc. of Pac Symp on Biocomputing*, 583-94, (2001).
3. Bourque G, Pevzner PA, Tesler G, *Genome Res*, 14: 507-16, (2004).
4. Nadeau JH, Taylor BA, *Proc Natl Acad Sci USA*, 81: 814-18, (1984).
5. Hannenhalli S, Pevzner PA, In *Proc. of IEEE Symp on Found. of Comp Sci*, 581-92, (1995).
6. Bader DA, Moret BME, Yan M, *J. of Comp. Bio*, 8: 483-91, (2001).
7. Pevzner PA, Tesler G, In *Proc. of RECOMB*, 247-56, (2003).
8. Sankoff D, *PLoS Comput Biol*, 2: e35, (2006).
9. Ruiz-Herrera A, Castresana J, Robinson TJ, *Genome Biol*, 7:R115, (2006).
10. Peng Q, Pevzner PA, Tesler G, *PLoS Comput Biol*, 2: e14, (2006).
11. Sankoff D, Trinh P, In *Proc. of RECOMB*, 30-35, (2004).
12. Calabrese PP, Chakravarty S, Vision TJ, *Bioinformatics*, 19 Suppl. 1: i74-i80, (2003).
13. Hampson S, McLysaght A, Gaut B, et al., *Genome Res*, 13: 999-1010, (2003).
14. Haas BJ, Delcher AL, Wortman JR, et al., *Bioinformatics*, 20: 3643-46, (2004).
15. Darling ACE, Mau B, Blattner FR, et al., *Genome Res*, 14:1394-1403, (2004).
16. Mouse Genome Sequencing Consortium, *Nature*, 420: 520-62, (2002).
17. Hoberman R, Sankoff D, Durand D, In *Proc. of RECOMB Workshop on Comparative Genomics*, 55–71, (2005).
18. Sinha AU, Meller J, *BMC Bioinformatics*, 8: 82, (2007).
19. Wheeler DL, Barrett T, Benson DA, et al., *Nuc Acids Res*, 35:D5-12, (2007).
20. Deluca TF, Wu IH, Pu J, et al., *Bioinformatics*, 22: 2044-46, (2006).
21. Tesler G, *J. of Computer and System Sciences*, 65: 587-609, (2002).
22. Andelfinger G, Hitte C, Guyon R, et al., *Genomics*, 83: 1053-62, (2004).

COMPUTATIONAL CHALLENGES IN THE STUDY OF SMALL REGULATORY RNAS

DORON BETEL

Computational and Systems Biology Center, Memorial Sloan-Kettering Cancer Center New York, NY 10065, U.S.A.

CHRISTINA LESLIE

Computational and Systems Biology Center, Memorial Sloan-Kettering Cancer Center New York, NY 10065, U.S.A.

NIKOLAUS RAJEWSKY

Max Delbrück Centrum for Molecular Medicine Berlin, Germany

1. Introduction

Small regulatory RNAs are a class of non-coding RNAs that function primarily as negative regulators of other RNA transcripts. The principal members of this class are microRNAs and siRNAs, which are involved in post-transcriptional gene silencing. These small RNAs, which in their functional form are single-stranded and ~22 nucleotides in length, guide a gene silencing complex to an mRNA by complementary base pairing, mostly at the 3' untranslated region (3'UTR)[1,2]. The association of the silencing complex to the conjugate mRNA results in silencing the gene either by translational repression or by degradation of the mRNA. The discovery of microRNAs and their regulatory mechanism has been at the center of a dogmatic shift of our view of non-coding RNAs and their biological role.

In recent years, microRNAs have emerged as a major class of regulatory genes central to a wide range of cellular activities, including stem cell maintenance, developmental timing, metabolism, host-viral interaction, apoptosis and neuronal gene expression and muscle proliferation[3]. Consequently, changes in the expression, sequence or target sites of microRNA are associated with a number of human genetic diseases[4]. Indeed, microRNAs are known to act both as tumor suppressors and oncogenes, and aberrant expression of microRNAs is associated with progression of cancer[5]. The importance of genetic regulation by microRNAs is reflected in their ubiquitous expression in almost all cell types as well as their conservation in most of the metazoan and plant species.

The molecular pathway of gene silencing by microRNAs is also the basis for RNA interference (RNAi), a powerful experimental technique that is used to selectively silence genes in living cells. This technique has gained wide use and is currently employed in a high throughput manner to investigate the effects of large scale gene repression[6].

In addition to microRNAs and siRNAs, new types of regulatory small RNAs have been identified, including rasiRNAs[7] in *Drosophila* and zebrafish, PIWI-interacting RNAs (piRNAs) in mammals[8] and 21U-RNAs in *C. elegans*[9]. Collectively, the discovery of these sequences and their regulatory role has had a profound impact on our understanding on the post-transcriptional regulation of genes, suppression of transposable elements, heterochromatin formation and programmed gene rearrangement.

2. Session papers

The accelerated pace of biochemical and functional characterization of microRNAs and other small regulatory RNAs has been facilitated by computational efforts, such as microRNA target predictions, conservation and phylogenetic analysis, microRNA gene predictions and microRNA expression profiling. The papers in this session exemplify some of the primary challenges in this field and the novel approaches used to address them.

With the advent of pyrosequencing technology investigators can now identify many of the sparse and short genomic transcripts that have previously eluded detection. Not surprisingly, pyrosequencing has become the primary method for the detection and characterization of new microRNAs[10] as well as the discovery of new regulatory RNAs such as piRNAs. One difficulty with this technology is the high rate of sequencing errors, which can be corrected to some degree by the assembly of partially overlapping fragments. The first paper in this session, by Vacic et al., addresses the problem of correcting sequencing errors in short reads that are typical in small RNA discovery where there is no fragment assembly step. They present a probabilistic framework to evaluate short reads by matching them against the genome from which the sequences are derived.

A central and still unresolved problem in the field of small regulatory RNAs is the prediction of the mRNA targets of a microRNA. Typical computational approaches search for a (near) perfect base-pairing between the 5' end of the microRNA and a complementary site in the 3' UTR of the potential target gene. Some algorithms also incorporate binding at the 3' end of the microRNA to the target or make use of conservation of target sites across species[11]. So far, these sequence-based approaches result in a large number of predictions, suggesting that more refined rules governing microRNA-mRNA interactions remain to be discovered. In the second paper in the session, Long et al. provide new results in support of their recent energy-based model for microRNA target prediction. They model the interaction between microRNA and target as a two-step hybridization reaction: (1) nucleation at an accessible target site, followed by (2) hybrid elongation to disrupt local target secondary structure and form the complete microRNA-target duplex. The authors present analysis of a set of microRNA-mRNA interactions that have been experimentally tested in mammalian systems.

Tissue-specific microRNA expression data can be also be exploited for target prediction and integrative models of microRNA gene silencing. The final paper in the session, from Huang et al., adopts such an approach in a development of their GenMiR model. Here, they integrate paired microRNA and mRNA expression data, predicted microRNA target sites, and mRNA sequence features associated with the predicted sites in a probabilistic approach for scoring candidate microRNA-mRNA target sites.

Acknowledgments

We thank all the authors who submitted papers to the session, and we gratefully acknowledge the reviewers who contributed their time and expertise to the peer review process.

References

1. D. P. Bartel, *Cell* **116** (2), 281 (2004).
2. P. D. Zamore and B. Haley, *Science (New York, N.Y* **309** (5740), 1519 (2005).
3. C. Xu, Y. Lu, Z. Pan et al., *Journal of cell science* **120** (Pt 17), 3045 (2007); M. Kapsimali, W. P. Kloosterman, E. de Bruijn et al., *Genome Biol* **8** (8), R173 (2007).
4. J. S. Mattick and I. V. Makunin, *Human molecular genetics* **15 Spec No 1**, R17 (2006).
5. G. A. Calin and C. M. Croce, *Nature reviews* **6** (11), 857 (2006).
6. Y. Pei and T. Tuschl, *Nature methods* **3** (9), 670 (2006).
7. V. V. Vagin, A. Sigova, C. Li et al., *Science (New York, N.Y* **313** (5785), 320 (2006).
8. V. N. Kim, *Genes & development* **20** (15), 1993 (2006).
9. J. G. Ruby, C. Jan, C. Player et al., *Cell* **127** (6), 1193 (2006).
10. K. Okamura, J. W. Hagen, H. Duan et al., *Cell* **130** (1), 89 (2007).
11. N. Rajewsky, *Nature genetics* **38 Suppl**, S8 (2006).

COMPARING SEQUENCE AND EXPRESSION FOR PREDICTING microRNA TARGETS USING GenMiR3

J. C. HUANG[1], B. J. FREY[1,2] AND Q. D. MORRIS[2]

[1] *Probabilistic and Statistical Inference Group, University of Toronto,*
10 King's College Rd.,
Toronto, ON, M5S 3G4, Canada
E-mail: jim,frey@psi.toronto.edu

[2] *Banting and Best Department of Medical Research, University of Toronto,*
160 College Street,
Toronto, ON, M5S 1E3, Canada
E-mail: quaid.morris@utoronto.ca

We present a new model and learning algorithm, GenMiR3, which takes into account mRNA sequence features in addition to paired mRNA and miRNA expression profiles when scoring candidate miRNA-mRNA interactions. We evaluate three candidate sequence features for predicting miRNA targets by assessing the expression support for the predictions of each feature and the consistency of Gene Ontology Biological Process annotation of their target sets. We consider as sequence features the total energy of hybridization between the microRNA and target, conservation of the target site and the context score which is a composite of five individual sequence features. We demonstrate that only the total energy of hybridization is predictive of paired miRNA and mRNA expression data and Gene Ontology enrichment but this feature adds little to the total accuracy of GenMiR3 predictions using for expression features alone.

1. Introduction

Recent research into understanding gene regulation has shed light on the significant role of microRNAs (miRNAs). These small regulatory RNAs suppress protein synthesis[1] or promote the degradation[2] of specific transcripts that contain anti-sense target sequences to which the miRNAs can hybridize with complete or partial complementarity. The catalogue of putative microRNA-target interactions predicted on the basis of genomic sequence continues to grow[3,4,5], but the most accurate computational approaches rely on the presence of a highly conserved seed in the putative target, greatly reducing their sensitivity[6]. However, even these highly selective methods appear to have low specificity[3]. Expression profiling has been proposed as a complementary method for discovering miRNA targets[7], but this can become intractable and costly when multiple miRNAs and their effects across multiple tissues must be considered.

We have recently described a probabilistic method, GenMiR++ (**Gen**erative model for **miR**NA regulation)[8,9], which incorporates miRNA and mRNA expression data with a set of candidate miRNA-target interactions to greatly improve the precision in predicting functional miRNA-target interactions. While our method was shown to be robust[8] and to improve predictive accuracy[9] according to several independent measures, it does not consider sequence-specific features of miRNA target sites beyond the presence of a highly conserved miRNA seed. Recently it has been reported that many sequence features such as secondary structure[10] or the relative positioning of sites within the target mRNA's 3'UTR[11] may play a crucial role in miRNA target recognition. We therefore set out to evaluate whether such sequence features could increase the predictive power of our model for miRNA regulation.

In this paper, we present GenMiR3, a generative model of miRNA regulation which uses sequences features to establish a prior probability of a miRNA-target interaction being functional and then uses paired expression data for miRNAs and mRNAs to compute the likelihood of a putative miRNA-target interactions. By combining these two sources of information together to compute a posterior probability of a miRNA-target relationship being functional, we score candidate miRNA-target interactions in terms of both expression support and sequence features. We evaluate several candidate sequence features by computing their predictions with the expression data and by comparing the Gene Ontology enrichment of target sets obtained using sequence and/or expression features. We then determine whether these features could be used in tandem with expression data to improve the accuracy of our miRNA target predictions.

2. The GenMiR3 model and learning algorithm

GenMiR3 makes two significant improvements over our previous model GenMiR++ [8,9]: we use sequence features to establish a prior on whether a given miRNA will bind to a target site in the 3'UTR and we use a different prior on many model parameters to give more flexibility in our posterior probability estimates. We first describe the changes to our generative model of mRNA expression and then describe how we propose to integrate sequence features

2.1. *A Bayesian model for gene and microRNA expression*

GenMiR3 is a generative model of mRNA expression levels that computes the expression support for a putative miRNA-mRNA by evaluating the degree to which the miRNA expression levels could explain the observed mRNA expression levels given of all other predicted regulators for that mRNA. Given two expression data sets profiling G mRNA transcripts and K miRNAs across T tissues, we denote by $\mathbf{x}_g = (x_{g1}x_{g2}\cdots x_{gT})^{\mathsf{T}}$ and $\mathbf{z}_k = (z_{k1}z_{k2}\cdots z_{kT})^{\mathsf{T}}$ the expression profiles over the T tissues for mRNA transcript g and miRNA k respectively. Here x_{gt} refers to the expression of the gth transcript in the tth tissue and z_{kt} refers to the expression of the kth miRNA in the same tissue.

Our model also takes as input a set of candidate miRNA-target interactions in the form of a binary matrix $\mathbf{C}$, where $c_{gk} = 1$ if transcript g is a candidate target of miRNA k and $c_{gk} = 0$ otherwise. For each (g, k) pair for which $c_{gk} = 1$, we also introduce an indicator variable s_{gk}. In our model, $s_{gk} = 1$ indicates that the candidate interaction between (g, k) is truly functional. Thus, the problem of scoring putative miRNA-target interactions can be formulated as calculating a posterior probability of $s_{gk} = 1$ given $c_{gk} = 1$.

To complete the formulation of our generative model, we introduce a set of nuisance parameters $\mathbf{\Lambda} = \{\lambda_k\}$ that each scale the regulatory effect of a given miRNA and $\mathbf{\Gamma} = \mathrm{diag}(\gamma_1, \cdots, \gamma_T)$ to account for normalization differences between the miRNA and mRNA expression levels in tissue t. We assign prior distributions $P(\mathbf{\Lambda}|\boldsymbol{\alpha})$ and $P(\mathbf{\Gamma}|\boldsymbol{\alpha})$ and we integrate over these distributions when making predictions. Having defined the above parameters and variables, we can write the probabilities of the mRNA expression profiles $\mathbf{X} = \{\mathbf{x}_g\}$ conditioned on the expression profiles of miRNAs $\mathbf{Z} = \{\mathbf{z}_k\}$, and a set of functional miRNA-target interactions, $\mathbf{S} = \{s_{gk}\}$, as

$$P(\mathbf{x}_g|\mathbf{Z}, \mathbf{S}, \mathbf{\Gamma}, \mathbf{\Lambda}, \mathbf{\Theta}) = N(\mathbf{x}_g; \boldsymbol{\mu} - \sum_k \lambda_k s_{gk} \mathbf{\Gamma} \mathbf{z}_k, \mathbf{\Sigma}) \tag{1}$$

$$P(\mathbf{\Gamma}|\boldsymbol{\alpha}) = \prod_{t=1}^{T} P(\gamma_t|m, n) = \prod_{t=1}^{T} Gamma(\gamma_t; m, n) \tag{2}$$

$$P(\mathbf{\Lambda}|\boldsymbol{\alpha}) = \prod_{k=1}^{K} P(\lambda_k|a, b) = \prod_{k=1}^{K} Gamma(\lambda_k; a, b) \tag{3}$$

where $\boldsymbol{\mu}$ is a background transcriptional rate vector and $\mathbf{\Sigma}$ is a data noise covariance matrix. Note that in the above model, we use a point-estimate of $\mathbf{\Theta} = \{\boldsymbol{\mu}, \mathbf{\Sigma}\}$. The set $\boldsymbol{\alpha} = \{a, b, m, n\}$ corresponds to fixed hyperparameters which characterize the prior distributions on the parameters $\mathbf{\Gamma}, \mathbf{\Lambda}$. In the above model, we represent the expression profile of a given mRNA transcript g as being negatively regulated by all candidate miRNAs for which $s_{gk} = 1$.

2.2. *Incorporating sequence features*

To include sequence features of the miRNA target site in the model, we introduce an N-dimensional vector $\mathbf{f}_{gk} = (f^1_{gk} f^2_{gk} \cdots f^N_{gk})$ containing a description of N sequence features associated with the miRNA-mRNA pair (g, k). We denote by $\pi_{gk} = P(s_{gk} = 1 | c_{gk} = 1, \mathbf{f}_{gk}, \mathbf{w})$ the prior probability that indicator variable $s_{gk} = 1$ given the sequence features. As a simplifying assumption, we will assume that each of the N sequence features independently contribute to π_{gk} with weight equal to $w_n, n = 1, \cdots, N$. We will also assume that the s_{gk} variables are *a priori* independent of one another. This yields

$$P(\mathbf{S}|\mathbf{C}, \mathbf{F}, \mathbf{w}) = \prod_{(g,k)} p(s_{gk}|\mathbf{C}, \mathbf{F}, \mathbf{w})$$

$$= \prod_{(g,k)|c_{gk}=0} [s_{gk} = 0] \prod_{(g,k)|c_{gk}=1} \pi_{gk}^{s_{gk}} (1 - \pi_{gk})^{1-s_{gk}} \tag{4}$$

$$\pi_{gk} = P(s_{gk} = 1 | c_{gk} = 1, \mathbf{f}_{gk}, \mathbf{w}) = \frac{1}{1 + \exp(-\mathbf{w}^{\mathsf{T}} \mathbf{f}_{gk})} \tag{5}$$

where $[H] = 1$ if H is true, otherwise $[H] = 0$.

Given the above, we can write the probabilities in our model, conditioned on the expression of miRNAs and a set of candidate miRNA targets, as

$$P(\mathbf{X}, \mathbf{S}, \mathbf{\Gamma}, \mathbf{\Lambda}|\mathbf{C}, \mathbf{F}, \mathbf{Z}, \mathbf{\Theta}, \mathbf{w}, \boldsymbol{\alpha}) = P(\mathbf{S}|\mathbf{C}, \mathbf{F}, \mathbf{w}) P(\mathbf{\Gamma}|\boldsymbol{\alpha}) P(\mathbf{\Lambda}|\boldsymbol{\alpha}) \prod_g P(\mathbf{x}_g|\mathbf{Z}, \mathbf{S}, \mathbf{\Gamma}, \mathbf{\Lambda}, \mathbf{\Theta}) \tag{6}$$

Because we have formulated our model in a Bayesian framework, we can marginalize out our nuisance parameters when calculating the likelihood of the mRNA expression data or when calculating the posterior probabilities of $s_{gk} = 1$, *e.g.*,

$$P(\mathbf{X}|\mathbf{C}, \mathbf{F}, \mathbf{Z}, \mathbf{\Theta}, \mathbf{w}, \boldsymbol{\alpha}) = \sum_{\mathbf{S}} \int_{\mathbf{\Gamma}} \int_{\mathbf{\Lambda}} P(\mathbf{X}, \mathbf{S}, \mathbf{\Gamma}, \mathbf{\Lambda}|\mathbf{C}, \mathbf{F}, \mathbf{Z}, \mathbf{\Theta}, \mathbf{w}, \boldsymbol{\alpha}) \, d\mathbf{\Lambda} \, d\mathbf{\Gamma} \tag{7}$$

Figure 1 shows the Bayesian network for our model of miRNA regulation. Under our model, each transcript g in the network is associated with a

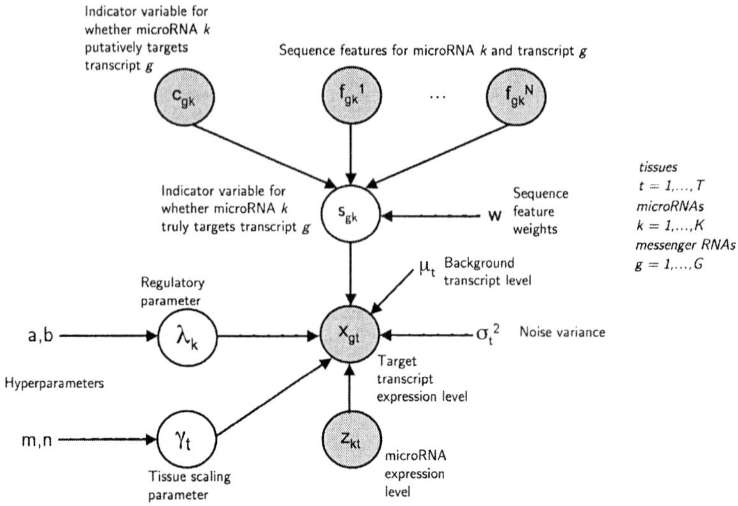

Figure 1. Bayesian network used for modelling microRNA regulation using both sequence and expression features. Nodes correspond to observed and unobserved variables as well as model parameters, with directed edges between nodes representing conditional dependencies encoded by our probability model. Each variable node and all incoming/outgoing edges associated with that node are replicated a number of times according to the number of such variables in the model. Shaded nodes correspond to observed variables and unshaded ones are unobserved. Model parameters which are estimated in a pointwise fashion are shown without nodes.

set of indicator variables $\{s_{gk'}\}, k' \in \{k|c_{gk} = 1\}$ which indicate which of its candidate miRNA regulators affect its expression level. The posterior probabilities over these variables are the predictions of the model: these posteriors are determined by combining priors over s_{gk} which are determined by examining the sequence of transcript g and miRNA k in addition to support from the expression data through our inference and learning procedure. We describe our learning method in the next section.

2.3. *Learning the model of gene and microRNA expression*

Exact Bayesian learning of our model is intractable, so we use a variational method[12,13] to derive a tractable approximation. Our learning procedure is similar to that for GenMiR++[8,9]. Here we will describe only the changes and refer the reader to our previous work[8] for the rest of the derivation. In particular, we specify the Q-distribution via a mean-field factorization with $Q(\mathbf{S},\mathbf{\Lambda},\mathbf{\Gamma}|\mathbf{C}) = Q(\mathbf{S}|\mathbf{C})Q(\mathbf{\Lambda})Q(\mathbf{\Gamma})$ such that

$$Q(\mathbf{S}|\mathbf{C}) = \prod_{g,k} Q(s_{gk}|\mathbf{C}) = \prod_{(g,k)|c_{gk}=0} [s_{gk}=0] \prod_{(g,k)|c_{gk}=1} \beta_{gk}^{s_{gk}}(1-\beta_{gk})^{1-s_{gk}}$$

$$Q(\mathbf{\Lambda}) = \prod_{k=1}^{K} Q(\lambda_k) = \prod_{k=1}^{K} Gamma(\lambda_k; a_k, b_k)$$

$$Q(\mathbf{\Gamma}) = \prod_{t=1}^{T} Q(\gamma_t) = \prod_{t=1}^{T} Gamma(\gamma_t; m_t, n_t) \tag{8}$$

where β_{gk} is the approximate posterior probability that a given miRNA-target pair (g,k) is functional given the data. Using this Q-distribution, we iteratively minimize the upper bound $L(Q)$ on the negative data likelihood with respect to the distribution over unobserved variables $Q(\mathbf{S}|\mathbf{C})$ (variational Bayes E-step), the distribution over model parameters $Q(\mathbf{\Gamma})Q(\mathbf{\Lambda})$ (variational Bayes M-step) and with respect to the regular model parameters.

2.4. *Setting the sequence-based priors using the posteriors from the gene and microRNA expression model*

The prior probability $\pi_{gk} = p(s_{gk} = 1|c_{gk} = 1, \mathbf{f}_{gk}, \mathbf{w})$ is parametrized by the weight vector $\mathbf{w}$. We estimate this weight vector by maximizing the expected log-likelihood $E_Q[\log p(\mathbf{S}|\mathbf{C},\mathbf{F},\mathbf{w})]$ of the s_{gk} variables. This then reduces to a standard logistic regression problem, with each output label set to β_{gk}, or the expected value of s_{gk} under $Q(\mathbf{S})$. We can perform the required optimization via a conjugate-gradient method, with the gradient $\nabla_\mathbf{w} E_Q[\log p(\mathbf{S}|\mathbf{C},\mathbf{F},\mathbf{w})]$ and the Hessian $\nabla\nabla_\mathbf{w} E_Q[\log p(\mathbf{S}|\mathbf{C},\mathbf{F},\mathbf{w})]$ given by

$$\nabla_\mathbf{w} E_Q[\log p(\mathbf{S}|\mathbf{C},\mathbf{F},\mathbf{w})] = \sum_{(g,k)|c_{gk}=1} \mathbf{f}_{gk}(\beta_{gk} - \pi_{gk}) \tag{9}$$

$$\nabla\nabla_\mathbf{w} E_Q[\log p(\mathbf{S}|\mathbf{C},\mathbf{F},\mathbf{w})] = \sum_{(g,k)|c_{gk}=1} \mathbf{f}_{gk}\mathbf{f}_{gk}^{\mathsf{T}}\pi_{gk}(1-\pi_{gk}) \tag{10}$$

We iteratively run the variational Bayes algorithm to estimate the approximate posterior probabilities β_{gk} and then update the weight vector $\mathbf{w}$ until convergence to a minimum of $L(Q)$. We can then assign a score to each candidate miRNA-target interaction using the log posterior odds $\log \frac{\beta_{gk}}{1-\beta_{gk}}$ so that a higher score reflects a higher posterior probability of a miRNA-target pair (g,k) being functional.

3. Results

To assess the impact of including sequence features, we downloaded human miRNA and mRNA expression data generated by [14,15] in addition to the set of TargetScanS [3] candidate human miRNA target-interactions from the UCSC Genome Browser[16] (build hg17/NCBI35) and mapped these interactions to the expression data. This yielded 6,387 candidate miRNA-target interactions between 114 human miRNAs and 890 mRNA transcripts, with patterns of expression across 88 human tissue samples. We then learned the GenMiR3 model without the sequence prior and once the algorithm converged, we selected the 100 highest and 100 lowest-scoring miRNA-target interactions and we downloaded the corresponding 3'UTR genomic sequences for each of the corresponding targeted mRNAs from the UCSC Genome Browser. The score assigned by GenMiR3 in the absence of sequence features predicts whether a given candidate miRNA-target interaction is functional based on joint patterns of expression of miRNAs and their target mRNAs across multiple tissues/cell types. We have previously shown that a similar score can distinguish functional and non-functional candidate miRNA/mRNA target pairs[9]. Here we use this "expression-only" GenMiR3 score to compare predictions made using both sequence and expression features with those made based solely on expression data. Once we evaluate the sequence features alone, we use a Gene Ontology enrichment test to evaluate the effect of combining these features with expression data using the full GenMiR3 model.

3.1. *Evaluating sequence features using cross-validation*

We evaluate three different sequence features: the *total hybridization energy*[10], a measure of the free energy of binding of the miRNA to its candidate target site that also considers any RNA secondary structure that the target site may participate in; the *context score*[11], an aggregate score combining the AU content with ± 30 bp of each miRNA target site, proximity to residues pairing to sites for coexpressed miRNAs, proximity to residues pairing to miRNA nucleotides 13-16, positioning of sites within the 3'UTR at least 15 nt from the stop codon and positioning away of sites from the center of the 3'UTR; and the *PhastCons score*, which is a measure of the conservation of the whole target site basefd on the PhastCons algorithm [17].

We calculated the total hybridization energy ΔG_{total} using a procedure related to [10]. Briefly, we set $\Delta G_{total} = \Delta G_{hybrid} - \Delta G_{disrupt}$, where ΔG_{hybrid} is the the total hybridization energy between a miRNA and its target mRNA computed by aligning the miRNA and target sequences and evaluating the total energy of hybridization using standard energy parameters. The expected disruption energy $< \Delta G_{disrupt} >$ was then obtained by using first calculating the probability that each base in the target site was paired with another base in the 3'UTR using RNAfold[18] and then using these base pair probabilities to calculate the expected hybridization energy of the target site in absence of the miRNA. If there was more than one possible site for a given miRNA in the 3'UTR, we summed ΔG_{total} over all sites.

We then downloaded the context scores from the TargetScan 4.0 website [11] and we calculated the PhastCons score by summing all of log-probabilities of conservation (obtained from the UCSC Genome Browser) over all base positions of all sites with seed matches to the mature miRNA in the target mRNA's 3'UTR. We then normalized each of these three features to be zero mean and unit variance.

We randomly split the above set of 200 high/low-scoring miRNA-target interactions under the expression-only GenMiR3 model into 1000 training and test sets of size 150/50 respectively. For each sequence feature, we trained two logistic regression models for each of the training sets: one with the feature included and a null model with the feature excluded. We evaluated the test likelihood given the learned weights and computed the average likelihood ratio between the test likelihood $L_{feature}$ for each feature and the likelihood of the null model L_{null} with no features. The median and standard deviations of the test likelihood ratios over the 1000 training/test splits are shown in Figure 2. The ΔG_{total} score is most predictive of the

Feature	$Median\left(\frac{L_{feature}}{L_{null}}\right)$	s.d.
ΔG_{total}[10]	2.3597	4.0859
PhastCons[17]	1.1262	1.9774
Context[11]	1.2129	2.0497

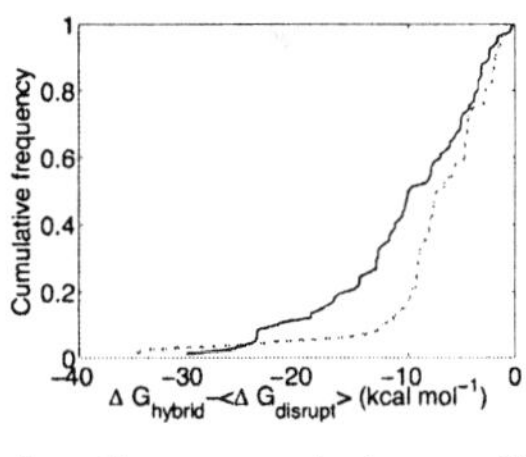

Figure 2. Sequence features and median test likelihood ratios computed over 1000 test/train splits; the total hybridization energy ΔG_{total} between a miRNA and its target mRNA transcript is shown for high GenMiR3-scoring targets (solid) and low GenMiR3-scoring targets (dashed)

three queried features, as including it in the model tends to increase the median test likelihood with respect to the null model. Neither the Phast-

Cons score nor the context score increased the median test likelihood with respect to the null model. We also found that the individual features used to compute the context score (such as AU content around the target site) did not increase the test likelihood with respect to the null model, nor was there a significant difference in these median feature values between high- and low-scoring GenMiR3 targets (data not shown). For the ΔG_{total} score however, we found that the high-scoring GenMiR3 miRNA-target interactions indeed have a lower median ΔG_{total} score than low-scoring GenMiR3 candidates ($p = 0.0138$, Wilcoxon-Mann-Whitney (WMW) test; Figure 2).

3.2. *Evaluating sequence features using functional enrichment analysis*

We have also previously shown that predicted target sets of many microRNAs are enriched for Gene Ontology Biological Process (GO-BP) categories [9]. As such, we reasoned that more accurate target predictions should show higher levels of GO-BP enrichment and we used GO-BP enrichment to assess target prediction accuracy. To calculate the different sequence features, we downloaded 3'UTR sequences for each of the mRNAs putatively targeted by a miRNA and filtered out all 3'UTR's with length greater than 5,000 bp and those without a published context score. This process yielded 410 candidate miRNA-target interactions between 89 human miRNAs and 150 mRNA transcripts. We then computed ΔG_{total} for each of these 410 candidate miRNA-target interactions and trained GenMiR3 on the expression data and ΔG_{total} as a sequence feature.

To compute GO-BP enrichment, we downloaded human GO-BP annotations from BioMart[19]. After up-propagation, we had a total of 13,003 functional annotations of which we removed annotations which were associated with less than 5 annotated Ensembl genes, leaving us with 2,021 GO-BP annotations. To establish the target sets, we selected the top 25% of candidate miRNA-target interactions for each miRNA under four scoring schemes:

(1) GenMiR3 score obtained from expression features alone
(2) GenMiR3 score obtained from both ΔG_{total} and expression features
(3) ΔG_{total} alone
(4) Context score

We computed enrichment by using Fisher's exact test to measure the statistical significance of the overlap between each GO-BP category and predicted target set of each of the 89 miRNAs in our data set (for a total of 179,869 enrichment scores). For each miRNA, we used these p-values to compute the number of significantly enriched categories (FDR < 0.05, linear step-up[20]), shown in Figure 3(a) and the maximum $-\log_{10}$ p-value across the GO-BP categories, shown in Figure 3(b). As can be seen, selecting miRNA targets on the basis of either expression alone, ΔG_{total} alone, or both, yields a higher number of enriched GO categories than selecting on the basis of the context score alone ($p = 8.2016 \times 10^{-4}, p = 2.7903 \times 10^{-5}, p = 0.0049$, respectively, Wilcoxon-Mann-Whitney). Our results also indicate, however, that adding the ΔG_{total} sequence feature to the model for expression does not significantly improve the GO enrichment GenMiR3 target sets. We will discuss possible reasons for this in the last section.

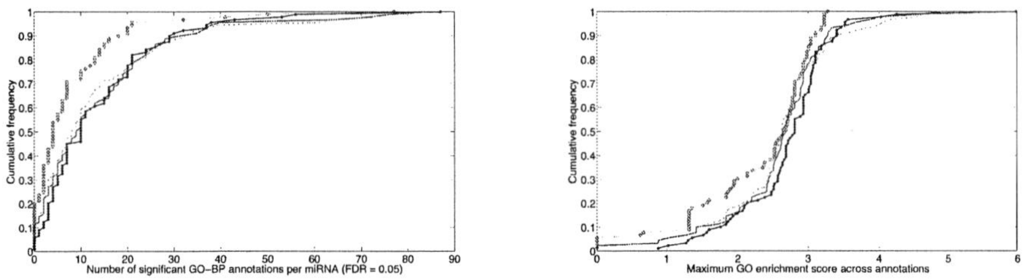

(a) (b)

Figure 3. Cumulative frequency plots of a) Number of significant GO categories per miRNA at FDR= 0.05 and b) maximum GO enrichment scores per miRNA obtained from using the GenMiR3 score obtained from expression features alone (solid), using the GenMiR3 score obtained from both ΔG_{total} and expression features (dashed), ΔG_{total} alone (star) and the context score (circle).

4. Discussion and conclusion

In this paper we have proposed the GenMiR3 probabilistic model for miRNA regulation using both sequence and expression features. We examined three sequence features: the total energy of hybridization ΔG_{total} between the microRNA and target, conservation of the target site and the context score, which itself is an aggregate score based on five sequence features. Using cross-validation, we found that the ΔG_{total} sequence feature was the best predictor of GenMiR3 score computed from expression features alone. Using a functional enrichment analysis, we found that selecting miRNA targets based on GenMiR3 score (with and without ΔG_{total}) and the ΔG_{total} score alone yielded a significantly higher number of enriched GO categories than selecting on the basis of the context score.

The relative performance of the context score[11] compared to the total hybridization score[10] was particularly surprising. Many of the features included in the context score should be predictive of whether or not the target site is likely to be single-stranded or double-stranded prior to miRNA binding, whereas the total hybridization score is a more direct indicator of this state. The results of our tests therefore suggest that single-strandedness of the miRNA target site is the most accurate sequence feature for predicting binding.

There are a number of possible explanations for the fact that adding the ΔG_{total} sequence feature to the model for expression does not improve the enrichment of GenMiR3 target sets. It is unlikely that the expression features are redundant with ΔG_{total}, as ΔG_{total} and expression-only GenMiR3 scores cease to be correlated outside of the 100 highest and lowest scoring interactions under GenMiR3 ($\rho = -0.0696, p = 0.1595$, Spearman correlation), suggesting ΔG_{total} and the expression data are making different predictions about miRNA targets. It is unclear whether ΔG_{total} or GenMiR3 are making better predictions, as we may have reached the limitation of the power of the GO analysis and require a more sensitive test. The expression signal does appear to be quite strong though because, when added to the GenMiR3 model, ΔG_{total} does not change GenMiR3 predictions: the Spearman correlation is 0.99 between the expression-only GenMiR3 posteriors and the posteriors in the GenMiR3 which also accounts for sequence data. This suggests that when expression data is limited or unavailable, the ΔG_{total} sequence prior will be a very useful addition to the GenMiR3 model, in addition to being predictive of functionality in its own right.

References

1. Ambros, V. (2004) The functions of animal microRNAs. *Nature* **431**, 350-355.
2. Bagga, S. *et al* (2005) Regulation by let-7 and lin-4 miRNAs results in target mRNA degradation. *Cell* **122**, 553-63.
3. Lewis, B.P., Burge, C.B., and Bartel, D.P. (2005) Conserved seed pairing, often flanked by adenosines, indicates that thousands of human genes are microRNA targets. *Cell* **120**, 15-20.
4. Krek, A. *et al* (2005) Combinatorial microRNA target predictions. *Nat. Gen.* **37**, 495-500.
5. Huynh, T. *et al* (2006) A pattern-based method for the identification of microRNA-target sites and their corresponding RNA/RNA complexes. *Cell* **126**, 1203-1217.
6. Sood, P. *et al* (2006) Cell-type-specific signatures of microRNAs on target mRNA expression. *Proceedings of the National Academy of Sciences (PNAS)* **103**, 2746-2751.

7. Lim, L.P. *et al* (2005) Microarray analysis shows that some microRNAs down-regulate large numbers of target mRNAs. *Nature* **433**, 769-773.
8. Huang, J.C., Morris, Q.D., and Frey, B.J. (2007) Bayesian learning of microRNA targets using sequence and expression data. *J. Comp. Bio.* **14**(5), 550-563.
9. Huang, J.C., *et al.* (2007) Using expression profiling to identify human microRNA targets. *In press.*
10. Long, D. *et al.* (2007) Potent effect of target structure on microRNA function. *Nat. Struct. Mol. Bio.* **14**, 287-294.
11. Grimson, A. it et al (2007) MicroRNA targeting specificity in mammals: determinants beyond seed pairing. *Mol. Cell* **27**, 91-105.
12. Attias, H. (1999) Inferring parameters and structure of latent variable models by variational Bayes. *Proceedings of the 15th Conference on Uncertainty in Artifical Intelligence*, 21-30.
13. Neal, R.M., and Hinton, G.E. (1998) A view of the EM algorithm that justifies incremental, sparse, and other variants, 355-368. *In* Jordan, M.I., ed., *Learning in Graphical Models*, MIT Press.
14. Lu, J. *et al.* (2005) MicroRNA expression profiles classify human cancers. *Nature* **435**, 834-8.
15. Ramaswamy, S. *et al.* (2001) Multiclass cancer diagnosis using tumor gene expression signatures. *Proceedings of the National Academy of Sciences (PNAS)* **98**, 15149-15154.
16. Karolchik, D. *et al* (2003) The UCSC Genome Browser Database. *Nucl. Acids Res.* **31**(1), 51-54.
17. Siepel, A., *et al.* (2005) Evolutionarily conserved elements in vertebrate, insect, worm, and yeast genomes. *Genome Res.* **15**, 1034-1050.
18. Hofacker, I. (2003) Vienna RNA secondary structure server. *Nucl. Acids Res.* **31**(13), 3429-3431.
19. Durinck, S. (2005) BioMart and Bioconductor: a powerful link between biological databases and microarray data analysis. *Bioinformatics* **21**(16), 3439-40.
20. Benjamini, Y., and Hochberg, Y. (1995) Controlling the false discovery rate: a practical and powerful approach to multiple testing. *J R Stat Soc Ser B Methodol* **57**, 289-300.

ANALYSIS OF MICRORNA-TARGET INTERACTIONS BY A TARGET STRUCTURE BASED HYBRIDIZATION MODEL

DANG LONG*, CHI YU CHAN*, YE DING#

Wadsworth Center, New York State Department of Health,
150 New Scotland Avenue, Albany, NY 12208

Email: dlong, yding@wadsworth.org, c.clarence@yahoo.com

MicroRNAs (miRNAs) are small non-coding RNAs that repress protein synthesis by binding to target messenger RNAs (mRNAs) in multicellular eukaryotes. The mechanism by which animal miRNAs specifically recognize their targets is not well understood. We recently developed a model for modeling the interaction between a miRNA and a target as a two-step hybridization reaction: nucleation at an accessible target site, followed by hybrid elongation to disrupt local target secondary structure and form the complete miRNA-target duplex. Nucleation potential and hybridization energy are two key energetic characteristics of the model. In this model, the role of target secondary structure on the efficacy of repression by miRNAs is considered, by employing the Sfold program to address the likelihood of a population of structures that co-exist in dynamic equilibrium for a specific mRNA molecule. This model can accurately account for the sensitivity to repression by *let-7* of both published and rationally designed mutant forms of the *Caenorhabditis elegans lin-41* 3′ UTR, and for the behavior of many other experimentally-tested miRNA-target interactions in *C. elegans* and *Drosophila melanogaster*. The model is particularly effective in accounting for certain false positive predictions obtained by other methods. In this study, we employed this model to analyze a set of miRNA-target interactions that were experimentally tested in mammalian models. These include targets for both mammalian miRNAs and viral miRNAs, and a viral target of a human miRNA. We found that our model can well account for both positive interactions and negative interactions. The model provides a unique explanation for the lack of function of a conserved seed site in the 3′ UTR of the viral target, and predicts a strong interaction that cannot be predicted by conservation-based methods. Thus, the findings from this analysis and the previous analysis suggest that target structural accessibility is generally important for miRNA function in a broad class of eukaryotic systems. The model can be combined with other algorithms to improve the specificity of predictions by these algorithms. Because the model does not involve sequence conservation, it is readily applicable to target identification for microRNAs that lack conserved sites, non-conserved human miRNAs, and poorly conserved viral mRNAs. StarMir is a new Sfold application module developed for the implementation of the structure-based model, and is available through Sfold Web server at http://sfold.wadsworth.org.

---

* Joint first authors with equal contributions

Corresponding author

1. Introduction

MicroRNAs (miRNAs) are endogenous non-coding RNAs (ncRNAs) of ~22 nt, and are among the most abundant regulatory molecules in multicellular organisms. miRNAs typically negatively regulate specific mRNA targets through essentially two mechanisms: 1) when a miRNAs is perfectly or nearly perfectly complementary to mRNA target sites, as is the case for most plant miRNAs, it causes mRNA target cleavage[1]; and 2) a miRNA with incomplete complementarity to sequences in the 3′ untranslated region (3′ UTR) of its target (as is the case for most animal miRNAs) can cause translational repression, and/or some degree of mRNA turnover[2]. miRNAs regulate diverse developmental and physiological processes in animals and plants[2-6]. Besides animals and plants, miRNAs have also been discovered in viruses[7].

The targets and functions of plant miRNAs are relatively easy to identify due to the near-perfect complementarity[1]. By contrast, the incomplete target complementarity typical of animal miRNAs implies a huge regulatory potential, but also presents a challenge for target identification. A number of algorithms have been developed for predicting animal miRNA targets. A common approach relies on a "seed" assumption, wherein the target site is assumed to form strictly Watson-Crick (WC) pairs with bases at positions 2 through 7 or 8 of the 5′ end of the miRNA. In the stricter, "conserved seed" formulation of the model, perfect conservation of the 5′ seed match in the target is required across multiple species[8,9]. One well-known exception to the seed model is interaction between *let*-7 on *lin-41*, for which G-U pair and unpaired base(s) are present in the seed regions of two binding sites with experimental support[10]. While the seed model is supported as a basis for identifying many well-conserved miRNA targets[11], two studies suggest that G-U or mismatches in the seed region can be well tolerated, and that conserved seed match does not guarantee repression[12,13]. These suggest that the seed model may represent only a subset of functional target sites, and that additional factors are involved in further defining target specificity at least for some cases with conserved seed matches. Recently, a number of features of site context have been proposed for enhancing targeting specificity[14].

For posttranscriptional gene modulation by mRNA-targeting nucleic acids, the importance of target structure and accessibility has long been established for antisense oligonucleotides and ribozymes[15,16], and evidence for this has also emerged for siRNAs[17,18]; and more recently for miRNAs[19-22]. These suggest that target accessibility can be an important parameter for target specificity.

We recently developed a model for modeling the interaction between a miRNA and a target as a two-step hybridization reaction: nucleation at an accessible target site, followed by hybrid elongation to disrupt local target secondary structure and form the complete miRNA-target duplex[19]. Nucleation potential and hybridization energy are two key energetic characteristics of the model. In this model, the role of target secondary structure on the efficacy of repression by miRNAs is taken into account, by employing the Sfold program to address the likelihood of a population of structures that co-exist in dynamic equilibrium for a specific mRNA molecule. This model can accurately account for the sensitivity to repression by *let-7* of both published and rationally designed mutant forms of the *Caenorhabditis elegans lin-41* 3′ UTR, and for the behavior of many other experimentally-tested miRNA-target interactions in *C. elegans* and *Drosophila melanogaster.* The model is particularly effective in accounting for certain false positive predictions obtained by other methods. In this study, we employed this model to analyze a set of miRNA-target interactions that were experimentally tested in mammalian models. We here report the results of the analysis and discuss implications of the findings.

2. Methods

2.1 *mRNA Secondary Structure Prediction*

The secondary structure of an mRNA molecule can influence the accessibility of that mRNA to a nucleic acid molecule that can bind to the mRNA by complementary base-pairing. Determination of mRNA secondary structure presents theoretical and experimental challenges. One major impediment to the accurate prediction of mRNA structures stems from the likelihood that a particular mRNA may not exist as a single structure, but in a population of structures in thermodynamic equilibrium[23-25]. Thus, the computational prediction of secondary structure based on free energy minimization is not well suited to the task of providing a realistic representation of mRNA structures.

An alternative to free energy minimization for charactering the ensemble of probable structures for a given RNA molecule has been developed[26]. In this approach, a statistically representative sample is drawn from the Boltzmann-weighted ensemble of RNA secondary structures for the RNA. Such samples can faithfully and reproducibly characterize structure ensembles of enormous sizes. In particular, in comparison to energy minimization, this method has been shown to make better structural predictions[27] and to better represent the likely population of mRNA structures[28], and to yield a significant correlation between predictions and data for gene inhibition by antisense oligos[29], gene knockdown by RNAi[30] and target cleavage by hammerhead ribozymes (unpublished data), and translational repression by miRNAs[19]. A sample size of 1,000 structures is sufficient to guarantee statistical reproducibility in sampling statistics and clustering features[26,28]. The structure sampling method has been implemented in the Sfold software package[31] and is used here for mRNA folding. The entire target transcript is used for folding if its length is under 7000 nts. For two targets in this study with transcript lengths over 9000 nt, we only used the UTRs (HCV and THRAP1, Table 1), so the folding could be efficiently managed.

2.2 *Two-step Hybridization Model*

We recently introduced a target-structure based hybridization model for prediction of miRNA-target interaction[19]. Here, we briefly describe this model and summarize its energetic characteristics. *In vitro* hybridization studies using antisense oligonucleotides suggested that hybridization of an oligonucleotide to a target RNA requires an accessible local target structure[32]. This requirement has been supported by various *in vivo* studies[33-35]. Such a local structure includes a site of unpaired bases for nucleation, and duplex formation progresses from the nucleation site and stops when it meets an energy barrier. In a kinetic study, it was suggested that the nucleation step is *rate-limiting*, and that it involves formation of four or five base pairs between the interacting nucleic acids[36]. Based on these and other related studies[37,38], we model the miRNA-target hybridization as a two-step process: 1) nucleation, involving four consecutive complementary nucleotides in the two RNAs (Fig. 1A), and 2) the elongation of the hybrid to form a stable intermolecular duplex (Fig. 1B).

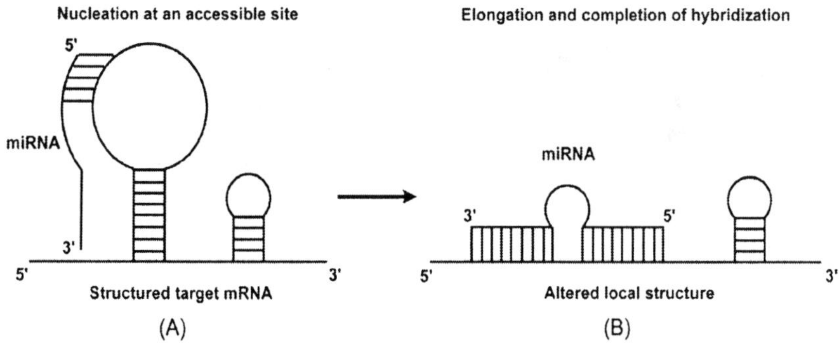

Figure 1. Two-step model of hybridization between a small (partially) complementary nucleic acid molecule and a structured mRNA: 1) nucleation at an accessible site of at least 4 or 5 unpaired bases (A); 2) elongation through "unzipping" of the nearby helix, resulting in altered local target structure (B).

The model is characterized by several energetic parameters. For a given predicted target structure, the nucleation potential, ΔG_N, is the stability of the particular single-stranded 4-bp block within the a potential mRNA binding site that would form the most stable 4-bp duplex with the miRNA (In Fig. 1, there are two 4-bp blocks for the 5-bp helix formed between the miRNA and the target). For the sample of 1000 structures predicted by Sfold for the target mRNA, the final ΔG_N is the average over the sample. The initiation energy threshold, $\Delta G_{initiation}$, is the energy cost for initiation of the interaction between two nucleic acid molecule. For two published values of $\Delta G_{initiation}$[36,39], 4.09 kcal/mol appeared to perform somewhat better in our previous study[19]. Nucleation for a potential site is considered favorable if the nucleation potential can overcome the initiation energy threshold, i.e., $\Delta G_N + \Delta G_{initiation} < 0$ kcal/mol. For a site with favorable nucleation potential, we next compute ΔG_{total}, the total energy change for the hybridization, by $\Delta G_{total} = \Delta G_{hybrid} - \Delta G_{disruption}$, where ΔG_{hybrid} is the stability of the miRNA-target hybrid as computed by the RNAhybrid program[40], and $\Delta G_{disruption}$ is the energy cost for the disruption of the local target structure (Fig. 1B), and is computed using structure sample predicted by Sfold for the target mRNA. These calculations have been incorporated into STarMir, a new application model for the Sfold package. To model the cooperative effects of multiple sites on the same 3′ UTR for either a single miRNA or multiple miRNAs, we assume energetic additivity and compute $\sum \Delta G_{total}$, where the sum is over multiple sites.

2.3 *Dataset of MicroRNA-Target Interactions*

We tried to assemble a set of high-quality and representative miRNA-target pairs in mammals. We selected reported miRNA-target interactions that were supported by at least two experimental testing using either human cells or mouse or rat models. These interactions play important roles in various biological processes. The targets also include a viral target for a cellular miRNA, and cellular targets for a viral miRNA family. The complete mRNA target sequences were typically retrieved from the Reference Sequence (RefSeq) database from the National Center for Biotechnology Information (NCBI) (http://www.ncbi.nlm.nih.gov/RefSeq). Information for these miRNA-target pairs and the references is given in Table 1. For a few reported interactions in these references, the complete transcripts were not available from the GenBank databases and thus these interactions were not included in this study.

3. Results

3.1 *Analysis of Interaction between Mammalian miRNAs and Viral Genomes*

An intriguing case worthy of particular note is the regulation of Hepatitis C virus (HCV) by miR -122[41]. In the viral RNA genome, there are a seed site in the 5′ non-coding region (NCR) and a seed site in the 3′ NCR, both are conserved among the six HCV genotypes. However, the site in the 5′ NCR was found to be essential for up-regulation of HCV replication by miR-122, whereas the site in the 3′ NCR was not. Current miRNA prediction algorithms that based on seed site conservation, e.g., TargetScan[8], PicTar[42], cannot explain the lack of function of the 3′ NCR seed site. Other algorithms that based only on the alignment and hybridization energy of miRNAs and potential binding sites, e.g., miRanda[43], RNAhybrid[40], cannot explain the difference between those two sites. We analyzed this miRNA-target pair using our interaction model that takes into account secondary structures of the target sequence. To classify an interaction as functional or nonfunctional, we previously used an empirical threshold of −10.0 kcal/mol for $\sum\Delta G_{total}$[19]. For this threshold, we predicted a functional interaction between miR-122 and the 5′ NCR, but a lack of interaction between miR-122 and the 3′ UTR, for which the $\sum\Delta G_{total}$ is merely −3.54 kcal/mol.

The energetic characteristics for potential binding sites that passed the nucleation threshold are listed below:

hsa-miR-122a:HCV 5′ NCR interaction

<u>Site 1</u>: Target site position in 5′ NCR: 21−44

```
                     G     CUC          A AU          C
Target        21      ACA     CACCAU G     CACUCC     44
                      |||     |||||| |     ||||||
miRNA         23      UGU     GUGGUA C     GUGAGG     1
                         UU         A AGU          U
```

$\Delta G_{total} = -16.70$ kcal/mol; $\Delta G_{disruption} = 6.40$ kcal/mol; $\Delta G_{hybrid} = -29.10$ kcal/mol; $\Delta G_N + \Delta G_{initiation} = -3.71$ kcal/mol.

<u>Site 2</u>: Target site position in 5′ NCR: 55−70

```
                           C         CU        A
Target        55            UACUGU   UCACGC          70
                            ||||||   ||||||
miRNA         23            GUGGUA   AGUGUG          1
                     UGUUU         AC        AGGU
```

$\Delta G_{total} = -9.981$ kcal/mol; $\Delta G_{disruption} = 7.619$ kcal/mol; $\Delta G_{hybrid} = -17.60$ kcal/mol; $\Delta G_N + \Delta G_{initiation} = -2.61$ kcal/mol.

hsa-miR-122a: HCV 3′ NCR

<u>Site 1</u>: Target site position in 3′ NCR: 9−36

```
                     C    UG AG     GGGUAA         G
Target         9      CGA   A   GUUG       ACACUCCG    36
                      |||   |   ||||       ||||||||
miRNA         23      GUU   U   UAAC       UGUGAGGU    1
                     U    UG GG     AG
```

ΔG_{total}=−3.538 kcal/mol; $\Delta G_{disruption}$=20.262 kcal/mol; ΔG_{hybrid}=−23.80 kcal/mol; $\Delta G_N + \Delta G_{initiation}$ = −3.71 kcal/mol.

The result here suggests that the lack of function for some (conserved) seed sites can be explained by poor target accessibility. In addition, for each of two single-substitution mutations (p3, p6) and a double-substitution mutation (p3-4) of the proposed seed region in the 5′ NCR[41], the HCV RNAs failed to accumulate. Our predictions for the mutants are consistent with the experimental finding, with $\sum\Delta G_{total}$ of −2.057 kcal/mol, −2.013 kcal/mol, and −1.934 kcal/mol, respectively. We note that the more energetically favorable site 1 in the 5′ NCR predicted by our model has some overlap with but is substantially different from the published binding site. This suggests an alternative binding conformation for further testing.

3.2 *Analysis of Other MicroRNA-Target Interactions*

We next analyzed 18 other validated interactions listed in Table 1. Our model accounted for 16 of the 18 (thus 17 for 19 including HCV 5′ NCR, a sensitivity of 89.5%) positive interactions. Among the two positive cases unaccounted for by our model, the interaction between miR-133a and HCN4 has a $\sum\Delta G_{total}$ of −9.5 kcal/mol, which is close to the threshold, and thus could be effective for miRNA-target hybridization. Moreover, the sum of this energy and that for the interaction between miR-1 and HCN4 is −20.304 kcal/mol, which is consistent with the combined effect by miR-133a and miR-1 on HCN4 that was reported[44]. Because miR-200c is not conserved across five vertebrate genomes, no target prediction can be made by TargetScan[8].

The regulation of HMGA2 by the *let*-7 family (all family members sharing the same seed sequence) has been reported by two studies, with *let-7a* used in one study[45], and *let-7b*, *let-7d* used in the other[46]. Data from both studies suggested functionality of multiple target sites identified by conserved seed matches. The rather large value of $\sum\Delta G_{total}$ for the interaction between HMGA2 and any of three tested *let-7* members is consistent with the understanding that a target can be efficiently regulated through multiple sites for the same miRNA.

While convincingly validated mammalian miRNA targets are limited, the functions of viral miRNAs are even less understood. Recently the regulation of several cellular targets by the KSHV-encoded miRNAs has been reported[47]. We found that our model supports the cooperativity of multiple miRNAs acting on the same target. In particular, for the well-validated target, THBS1, the $\sum\Delta G_{total}$ is rather large, a results of many binding sites on this target 3′ UTR. The results for both *let-7* and KSHV miRNAs suggest that $\sum\Delta G_{total}$ presents a promising measure for modeling the additive effects of multiple binding sites by either single or multiple mammalian or viral miRNAs.

Table 1. Target prediction based on miRNA-target interaction energy computed by $\sum\Delta G_{total}$ and local AU content of 19 positive interactions and four negative interactions (shaded)

miRNA	Target name, GenBank Accession, test system [a] and references			$\sum\Delta G_{total}$ (kcal/mol) and the prediction		Local AU content [b] and the prediction	
miR-200c	TCF8	AL831979	H[48,49]	−109.979	+	0.742	+
miR-133	RhoA	NM_016802	M[50]	−17.098	+	NA [c]	
miR-133	Cdc42	NM_009861	M[50]	−29.939	+	0.432	—
miR-133	WHSC2	NM_011914	M[50]	−18.240	+	0.461	—
miR-1	GJA1	NM_012567	R[51]	−55.805	+	0.626	+
let-7a	HMGA2	NM_003483	H[45]	−185.195	+	0.628	+
let-7b	HMGA2	NM_003483	H[46]	−300.446	+	0.628	+
let-7e	HMGA2	NM_003483	H[46]	−218.989	+	0.628	+
miR-124a	Foxa2	NM_010446	M[52]	−7.662	—	0.672	+
miR-1	HCN4	NM_021658	R[44]	−10.798	+	NA [c]	
miR-133a	HCN4	NM_021658	R[44]	−9.506	—	NA [c]	
10 KSHV miRNAs	SPP1	NM_000582	H[47]	−75.231	+	NA[d]	
10 KSHV miRNAs	SRGN	NM_002727	H[47]	−113.169	+	NA[d]	
10 KSHV miRNAs	THBS1	NM_003246	H[47]	−325.078	+	NA[d]	
miR-155	C-Maf	NM_001025 577	M[53]	−78.858	+	0.710	+
miR-208	THRAP1	NM_005121	H[5]	−45.032	+	0.895	+
miR-375	Mtpn	NM_145808	H[54]	−13.084	+	0.715	+
miR-122	CAT-1	NM_013111	R[55]	−136.82	+	0.530	—
miR-122	HCV 5′ NCR	NC_004102	H[41]	−26.681	+	0.478	—
miR-122	HCV 3′ NCR	NC_004102	H[41]	−3.538	—	0.412	—
miR-122	HCV 5′ NCR p3	NC_004102	H[41]	−2.057	—	NA [c]	
miR-122	HCV 5′ NCR p6	NC_004102	H[41]	−2.013	—	NA [c]	
miR-122	HCV 5′ NCR p3-4	NC_004102	H[41]	−1.934	—	NA [c]	

[a] H: human cells; M: mouse; R: rat. [b] as defined in Grimson et al., 2007[14]; [c] no perfect (7- or 8-mer) seed sites; [d] not calculated due to multiple miRNAs; + : predicted effective target, — : predicted ineffective target

We also calculated local AU content of seed sites of the miRNAs and targets following a scoring scheme proposed by Grimson *et al.*[14]. When there are multiple seed sites in the same 3′ UTR sequence, we report the best local AU content (Table 1). In order to correlate the local AU content to the qualitative information of miRNA activity in our dataset, we select a threshold of 0.6 for the local AU content. miRNA-target pairs having the local AU content is higher or equal 0.6 are predicted functional. This threshold is partly based on the experimental data in Grimson *et al.*[14], where the local AU content of 0.6 correlated to the average fold change of 0.89 in the mRNA level from the microarray experiment. The AU content of 0.6 is also just above the mean AU content of all possible 7-mer sites of the 3′ UTR sequences being considered here (data not shown). For this threshold, the local AU content alone can explain the positive interactions for 9 of the 13 miRNA-target pairs. For each of these 13 pairs, there is at least one seed site and only the concerned miRNA is known to be involved in regulation of the target. In comparison, we predict effective interactions for 12 of the 13 cases (Table 1). Furthermore, both of the two conserved seed sites for miRNA-122 in HCV 5′ NCR and 3′ NCR have comparable low AU content (Table 1). Therefore, the local AU content cannot explain the functional difference between the two seed sites.

4. Conclusion

In this study, we employed a recently developed target-structure based hybridization model to analyze a set of miRNA-target interactions. These interactions were experimentally tested in human cells or in animal models (mouse or rat). These include mammalian targets for both cellular miRNAs and viral RNAs, and a viral target for a cellular miRNA. Our model can well account for positive interactions, as well as negative interactions. In particular, the model can explain the difference in the interactions of miR-122 to HCV 5′ NCR and HCV3′ NCR, which could not be explained by several popular miRNA target prediction programs. In our previous analysis of repression data for worm and fly[19], we observed that the model can not only uniquely account for interactions between *let-7* and worm *lin-41* mutants that cannot be explained by other algorithms, but also explain negative experimental results for 11 of 12 targets with seed matches for *lsy-6*. These and the findings of this analysis here suggest that target structural accessibility is generally important for miRNA function in a broad class of eukaryotic systems, and that the model can be combined with other algorithms to improve the specificity of predictions by these algorithms. Our comparison of the predictions based on the interaction energies and the ones based on the local AU content suggests that the local AU content does not reflect accurately target sites' accessibility in many cases. Therefore, the interaction model considered here can more accurately account for miRNA activities. Because the model does not involve sequence conservation, it can be particularly valuable for target identification for microRNAs that lack conserved sites[56], non- or poorly-conserved human miRNAs[57] (e.g., the lack of prediction by TargetScan for miR-200c), and usually poorly conserved viral mRNAs.

Acknowledgments

The Computational Molecular Biology and Statistics Core at the Wadsworth Center is acknowledged for providing computing resources for this work. This work was supported in part by National Science Foundation grants DMS-0200970, DBI-0650991, and National Institutes of Health grant GM068726 (Y.D.).

References

1. Rhoades, M.W. et al. *Cell* **110**, 513-20 (2002).
2. Ambros, V. *Nature* **431**, 350-5 (2004).
3. Boehm, M. & Slack, F. *Science* **310**, 1954-7 (2005).
4. Dugas, D.V. & Bartel, B. *Curr Opin Plant Biol* **7**, 512-20 (2004).
5. van Rooij, E. et al. *Science* **316**, 575-9 (2007).
6. Calin, G.A. et al. *N Engl J Med* **353**, 1793-801 (2005).
7. Cullen, B.R. *Nat Genet* **38 Suppl**, S25-30 (2006).
8. Lewis, B.P., Burge, C.B. & Bartel, D.P. *Cell* **120**, 15-20 (2005).
9. Lewis, B.P., Shih, I.H., Jones-Rhoades, M.W., Bartel, D.P. & Burge, C.B. *Cell* **115**, 787-98 (2003).
10. Vella, M.C., Choi, E.Y., Lin, S.Y., Reinert, K. & Slack, F.J. *Genes Dev* **18**, 132-7 (2004).
11. Rajewsky, N. *Nat Genet* **38 Suppl**, S8-13 (2006).
12. Didiano, D. & Hobert, O. *Nat Struct Mol Biol* **13**, 849-51 (2006).
13. Miranda, K.C. et al. *Cell* **126**, 1203-17 (2006).
14. Grimson, A. et al. *Mol Cell* **27**, 91-105 (2007).
15. Vickers, T.A., Wyatt, J.R. & Freier, S.M. *Nucleic Acids Res* **28**, 1340-7 (2000).
16. Zhao, J.J. & Lemke, G. *Mol Cell Neurosci* **11**, 92-7 (1998).
17. Overhoff, M. et al. *J Mol Biol* **348**, 871-81 (2005).
18. Schubert, S., Grunweller, A., Erdmann, V.A. & Kurreck, J. *J Mol Biol* **348**, 883-93 (2005).
19. Long, D. et al. *Nat Struct Mol Biol* **14**, 287-294 (2007).
20. Zhao, Y. et al. *Cell* **129**, 303-17 (2007).
21. Zhao, Y., Samal, E. & Srivastava, D. *Nature* **436**, 214-20 (2005).
22. Robins, H., Li, Y. & Padgett, R.W. *Proc Natl Acad Sci U S A* **102**, 4006-9 (2005).
23. Christoffersen, R.E., McSwiggen, J.A. & Konings, D. *J. Mol. Structure (Theochem)* **311**, 273-284 (1994).
24. Altuvia, S., Kornitzer, D., Teff, D. & Oppenheim, A.B. *J Mol Biol* **210**, 265-80 (1989).
25. Betts, L. & Spremulli, L.L. *J Biol Chem* **269**, 26456-63 (1994).
26. Ding, Y. & Lawrence, C.E. *Nucleic Acids Res* **31**, 7280-301 (2003).
27. Ding, Y., Chan, C.Y. & Lawrence, C.E. *RNA* **11**, 1157-66 (2005).

28. Ding, Y., Chan, C.Y. & Lawrence, C.E. *J Mol Biol* **359**, 554-71 (2006).
29. Ding, Y. & Lawrence, C.E. *Nucleic Acids Res* **29**, 1034-46 (2001).
30. Shao, Y. et al. *RNA* (2007).
31. Ding, Y., Chan, C.Y. & Lawrence, C.E. *Nucleic Acids Res* **32**, W135-41 (2004).
32. Milner, N., Mir, K.U. & Southern, E.M. *Nat Biotechnol* **15**, 537-41 (1997).
33. Darnell, J.C. et al. *Genes Dev* **19**, 903-18 (2005).
34. Friebe, P., Boudet, J., Simorre, J.P. & Bartenschlager, R. *J Virol* **79**, 380-92 (2005).
35. Mikkelsen, J.G., Lund, A.H., Duch, M. & Pedersen, F.S. *J Virol* **74**, 600-10 (2000).
36. Hargittai, M.R., Gorelick, R.J., Rouzina, I. & Musier-Forsyth, K. *J Mol Biol* **337**, 951-68 (2004).
37. Paillart, J.C., Skripkin, E., Ehresmann, B., Ehresmann, C. & Marquet, R. *Proc Natl Acad Sci U S A* **93**, 5572-7 (1996).
38. Reynaldo, L.P., Vologodskii, A.V., Neri, B.P. & Lyamichev, V.I. *J Mol Biol* **297**, 511-20 (2000).
39. Xia, T. et al. *Biochemistry* **37**, 14719-35 (1998).
40. Rehmsmeier, M., Steffen, P., Hochsmann, M. & Giegerich, R. *RNA* **10**, 1507-17 (2004).
41. Jopling, C.L., Yi, M., Lancaster, A.M., Lemon, S.M. & Sarnow, P. *Science* **309**, 1577-81 (2005).
42. Krek, A. et al. *Nat Genet* **37**, 495-500 (2005).
43. Enright, A.J. et al. *Genome Biol* **5**, R1 (2003).
44. Xiao, J. et al. *J Cell Physiol* **212**, 285-92 (2007).
45. Mayr, C., Hemann, M.T. & Bartel, D.P. *Science* **315**, 1576-9 (2007).
46. Lee, Y.S. & Dutta, A. *Genes Dev* **21**, 1025-30 (2007).
47. Samols, M.A. et al. *PLoS Pathog* **3**, e65 (2007).
48. Hurteau, G.J., Carlson, J.A., Spivack, S.D. & Brock, G.J. *Cancer Res* **67**, 7972-6 (2007).
49. Hurteau, G.J., Spivack, S.D. & Brock, G.J. *Cell Cycle* **5**, 1951-6 (2006).
50. Care, A. et al. *Nat Med* **13**, 613-8 (2007).
51. Yang, B. et al. *Nat Med* **13**, 486-91 (2007).
52. Baroukh, N. et al. *J Biol Chem* **282**, 19575-88 (2007).
53. Rodriguez, A. et al. *Science* **316**, 608-11 (2007).
54. Poy, M.N. et al. *Nature* **432**, 226-30 (2004).
55. Jopling, C.L., Norman, K.L. & Sarnow, P. *Cold Spring Harb Symp Quant Biol* **71**, 369-76 (2006).
56. Farh, K.K. et al. *Science* **310**, 1817-21 (2005).
57. Bentwich, I. et al. *Nat Genet* **37**, 766-70 (2005).

A PROBABILISTIC METHOD FOR SMALL RNA FLOWGRAM MATCHING

VLADIMIR VACIC[1], HAILING JIN[2],
JIAN-KANG ZHU[3], STEFANO LONARDI[1]

[1] *Computer Science and Engineering Department*, [2] *Department of Plant Pathology*, [3] *Department of Botany and Plant Sciences, University of California, Riverside*

The 454 pyrosequencing technology is gaining popularity as an alternative to traditional Sanger sequencing. While each method has comparative advantages over the other, certain properties of the 454 method make it particularly well suited for small RNA discovery. We here describe some of the details of the 454 sequencing technique, with an emphasis on the nature of the intrinsic sequencing errors and methods for mitigating their effect. We propose a probabilistic framework for small RNA discovery, based on matching 454 flowgrams against the target genome. We formulate flowgram matching as an analog of profile matching, and adapt several profile matching techniques for the task of matching flowgrams. As a result, we are able to recover some of the hits missed by existing methods and assign probability-based scores to them.

1. Introduction

Historically, the chain termination-based Sanger sequencing[17] has been the main method to generate genomic sequence information. Alternative methods have been proposed, among which a highly parallel, high-throughput pyrophosphate-based sequencing (pyrosequencing)[16] is one of the most important. 454 Life Sciences has made pyrosequencing commercially available[11] and the resulting abundance of 454-generated sequence information has prompted a number of studies which compare 454 sequencing with the traditional Sanger method (see, e.g., [3,6,8,12,20]).

454 pyrosequencing. In the 454 technology, the highly time-consuming sequence preparation step which involves production of cloned shotgun libraries has been replaced with much faster PCR microreactor amplification. Coupled with the highly parallel nature of 454 pyrosequencing, this novel technology allows 100 times faster[8] and significantly less expensive sequencing. A detailed step by step breakdown of time required to complete the process using both methods can be found in Wicker *et al.*[20]

Recent studies by Goldberg *et al.*[8] on sequencing six marine microbial genomes and by Chen *et al.*[3] on sequencing the genome of *P. marinus* report that 454's ability to sequence throughout the regions of the genome with strong secondary structure and the lack of cloning bias represent a comparative advantage. However, the 454's shorter read lengths (100 bp on average compared to 800-1000 bp of Sanger) make it very hard if not impossible to span long repetitive genomic elements. Also, the lack of paired end reads (mate pairs) limits the assembly to contigs separated by coverage gaps. As a consequence, both studies conclude that at the present stage 454 pyrosequencing used alone is not a feasible method for *de novo* whole genome sequencing, although these two issues are being addressed in the new 454 protocol. Another problem inherent to pyrosequencing is accurate determination of the number of incorporated nucleotides in homopolymer runs, which we discuss in Section 2.

Small RNA. Since its discovery in 1998[4], gene regulation by RNA interference has received increasing attention. Several classes of non-coding RNA, typically much shorter than mRNA or ribosomal RNA, have been found to silence genes by blocking transcription, inhibiting translation or marking the mRNA intermediaries for destruction. Short interfering RNA (siRNA), micro RNA (miRNA), tiny non-coding RNA (tncRNA) and small modulatory RNA (smRNA) are examples of classes of small RNA that have been identified to date[13]. In addition to differences in genesis, evolutionary conservation, and the gene silencing mechanism they are associated with, different classes of small RNA have distinct lengths: 21-22 bp for siRNA, 19-25 bp for miRNA and 20-22 bp for tncRNA.

The process of small RNA discovery typically involves (1) sequencing RNA fragments, (2) matching the sequence against the reference genome to determine the genomic locus from which the fragment likely originated, and (3) analyzing the locus annotations in order to possibly obtain functional characterization. In this paper we focus on the second step.

Our contribution. 454 pyrosequencing appears to be particularly well-suited for small RNA discovery. The limited sequencing read length does not pose a problem given the short length of non-coding RNAs, even if we take into account lengths of adapters which are ligated on both ends on the small RNA prior to sequencing. Also, paired end reads are not required, as there is no need to assemble small RNA into larger fragments. Several projects have already used 454 to sequence non-coding RNA (see, e.g., [7,14]).

However, to the best of our knowledge, the issue of handling sequencing errors has not been addressed so far for short reads which occur in small RNA discovery. Observe that this problem could be mitigated in a scenario where an assembly step was involved – which is not the case when sequencing small RNA. In the following sections we describe the 454 sequencing model and the typical sequencing errors it produces. We propose a probabilistic matching method capable of locating some of the small RNA which would have been missed if the called sequences were matched deterministically. We adapt the enhanced suffix array[2] data structure to speed up the search process. Finally, we evaluate the proposed method on four libraries obtained by sequencing RNA fragments from stress-treated *Arabidopsis thaliana* plants and return 26.4% to 28.8% additional matches.

2. The 454 Pyrosequencing Method

In the 454 sequencing method, DNA fragments are attached to synthetic beads, one fragment per bead, and amplified using PCR to approximately 10 million copies per bead[11]. The beads are loaded into a large number of picolitre-sized reactor wells, one bead per reactor, and sequencing by synthesis is performed in parallel by cyclically flowing reagents over the DNA templates. Depending on the template sequence, each cycle can result in extending the strand complementary to the template by one or more nucleotides, or not extending it at all. Nucleotide incorporation results in the release of an associated pyrophosphate, which produces an observable light signal. The signal strength corresponds to the length of the incorporated homopolynucleotide run in the given well in that cycle. The resulting signal strengths are reported as pairs (nucleotide, signal strength), referred to as *flows*. The end result of 454 sequencing is a sequence of flows in the T, A, C, G order called a *flowgram*. Terms *positive flow* and *negative flow* denote, respectively, that at least one base has been incorporated, or that the reagent flowed in that cycle did not results in a chemical reaction, and hence that a very weak signal was observed. Every full cycle of negative flows would be called as an N, because the identity of the nucleotide could not be determined. Positive flow signal strengths for a fixed homopolynucleotide length l are reported to be normally distributed with the mean $0.98956 \cdot l + 0.0186$ and standard deviation proportional to l, while the negative flow signal strengths follow a log-normal distribution[11]. To the

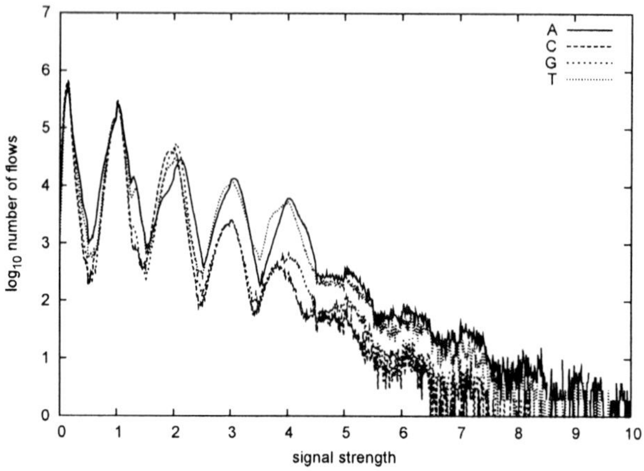

Figure 1. Distribution of signals for the *A. thaliana* pyrosequencing dataset

best of our knowledge, the other parameters of the normal and log-normal distributions have not been reported in the literature. In Section 7 we are estimating the remaining parameters from the available data.

Figure 1 shows the distribution of signal strengths for the *A. thaliana* dataset (50 million flows). Distributions of signal strengths for two additional sequencing projects performed at UC Riverside are given as Supplementary Figure 1, available on-line at http://compbio.cs.ucr.edu/flat. The overlaps between Gaussians for different polynucleotide lengths are responsible for over-calling or under-calling the lengths of incorporated nucleotide runs.

When sequencing small RNA, the 454-provided software employs a maximum likelihood strategy to call a homopolynucleotide length, with cut-off point at $l \pm 0.5$ for polynucleotide length l. This results in, for example, flows (T,2.52) and (T,3.48) both being called as TTT even though the proximity of the cut-off points indicates that the former one may have in fact come from TT and the latter one from TTTT. This could be alleviated to a degree by allowing approximate matches, where insertions or deletions would address under-calling and over-calling. However, without the knowledge of the underlying signal strengths any insertion or deletion would be arbitrary.

Also, according to the 454 procedure, a flow with signal intensity 0.49 will be treated as a negative, even though it is very close to the cut-off point for a positive flow. Consider the following example: sequence of flows (C,0.92)(G,0.34)(T,0.49)(A,0.32)(C,0.98) will be called as CNC and all information about which nucleotide was most likely to be in the middle will be lost.

These examples illustrate the intuition behind our approach: we use signal strengths to estimate probabilities of different lengths of homopolymer runs that may have induced the signal. The target genome conditions the probabilities, and the most probable explanations are returned as potential matches. The following section formally introduces the notion of *flowgram matching*.

3. Flowgram Matching

Let F be a flowgram obtained by pyrosequencing a genomic fragment originating from genome Γ, and let G be a flowspace representation of Γ derived by run-length encoding (RLE)[a] of Γ and padding the result with appropriate zero-length negative flows in a manner which simulates flowing nucleotides in the T,A,C,G order, as illustrated in Supplementary Figure 2. Let flowgram $F = \{(b_0, f_0), (b_1, f_1), \ldots, (b_{m-1}, f_{m-1})\}$ be a sequence of m flows, where b_i is the nucleotide flowed and f_i is the resulting signal strength. Let n be the length of G. Under the assumption that the occurrences of lengths of homopolynucleotide runs are independent events, the probability that a flowgram F matches a segment in G starting at position k can be expressed as

$$P(F \sim G_{k..k+m-1}) = \prod_{i=0}^{m-1} P_{b_i}(L = g_{k+i} | S = f_i) \qquad (1)$$

where L is a random variable denoting the length of the homopolynucleotide run in Γ, S is a random variable associated with the induced signal strength in the flowgram, and g_{k+i} is the length of the run at position i from the beginning of the match. For example, if the flowgram (A,0.98)(C,0.14)(G,1.86)(T,0.24)(A,3.12) is matched against AGGAAA, the run lengths for the genomic sequence are $g = \{1, 0, 2, 0, 3\}$, and the probability of matching would be $P_A(L = 1|S = 0.98) \cdot P_C(L = 0|S = 0.14) \cdot P_G(L = 2|S = 1.86) \cdot P_T(L = 0|S = 0.24) \cdot P_A(L = 3|S = 3.12)$. One of the benefits of casting the genome in flowspace is that a flowgram of length m will correspond to a segment of length m in G, whereas the corresponding segment

[a] We say that that a sequence w is a *run* of length k if $w = c^k$, where $c \in \{A, C, G, T\}$. In this case, the run-length encoding (RLE) of w is (c, k).

in Γ would have context-dependent length. Also, once the starting flows are aligned in terms of nucleotides, the remaining $m-1$ flows will be aligned as well. Using the Bayes' theorem we can rewrite equation (1) as

$$P(F \sim G_{k..k+m-1}) = \prod_{i=0}^{m-1} \frac{P_{b_i}(S = f_i | L = g_{k+i}) \cdot P_{b_i}(L = g_{k+i})}{P_{b_i}(S = f_i)} \tag{2}$$

where $P_{b_i}(L = g_{k+i})$ is the probability of observing a b_i homopolynucleotide of length g_{k+i} in Γ, and $P_{b_i}(S = f_i | L = g_{i+k})$ and $P_{b_i}(S = f_i)$ depend on the 454 sequencing model and can be estimated from the data through a combination of the called sequences and the underlying flowgrams (see Section 7).

If we assume a *null* model where homopolynucleotide runs are assigned the probabilities obtained by counting their frequencies in G, the log-odds score of the match is

$$Score(F \sim G_k) = \log \frac{P(F \sim G_{k+m-1})}{P_{null}(G_{k..k+m-1})} = \log \frac{\prod_{i=0}^{m-1} P_{b_i}(L = g_{k+i} | S = f_i)}{\prod_{i=0}^{m-1} P_{b_i}(L = g_{k+i})}$$

Rewriting the numerator using the Bayes' theorem allows us to cast flowgram matching as an analog of *profile matching* (see e.g. [2,5,19,21]), with the scoring matrix M defined as

$$M_{i,j} = \log P_{b_i}(S = f_i | L = j) - \log P_{b_i}(S = f_i)$$

The log-odds score can then be expressed as a sum of the matrix entries

$$Score(F \sim G_k) = \sum_{i=0}^{m-1} M_{i,j} \tag{3}$$

A brute-force approach for matching a flowgram F would be to align F with all m flow long segments in G and report the best alignments. This algorithm runs in $O(mn)$ time per flowgram. With typical sequence library sizes in the hundreds of thousands, flowgrams up to 100 bp and genomes in the order of billion bp, this approach is computationally not feasible.

4. Enhanced Suffix Arrays

Recently, Beckstette *et al.*[2] introduced the *enhanced suffix array* (ESA), an index structure for efficient matching of position specific scoring matrices (PSSM) against a sequence database. While providing the same functionality as suffix trees[1], enhanced suffix arrays require less memory and once precomputed they can be easily stored into a file or loaded from a file into main memory. An enhanced suffix array can be constructed in $O(n)$ time[2,9].

We employ enhanced suffix arrays to index the database of genomic sequences, with two adjustments. An ESA indexes the search space of positive flows, in the order determined by the underlying genome. To provide the view of the genomic sequences as observed by the 454 sequencer, positive flows are padded with intermediate dummy negative flows, as illustrated in Supplementary Figure 2 (available at http://compbio.cs.ucr.edu/flat). This padding does not interfere with searching for the complement of the flowgram because `CGTA`, the reverse complement of the order `TACG`, is a cyclic permutation of the original order with offset 2. Consequently the reverse complements of the dummy flows would exactly match the dummy flows inserted if the reverse complement of the RNA fragment was sequenced.

When a flowgram is being aligned along the "branches" of the suffix array, the branches are run-length encoded and negative flows are inserted where appropriate. This amounts to on-the-fly branch by branch flowspace encoding of the underlying sequence database, without sacrificing the compactness of the suffix array representation. The score of the alignment is calculated using equation (3).

The first adjustment solves the problem of intermediate negative flows. However, it can happen that the flowgram corresponding to the RNA fragment starts or ends with one or more negative flows. The second adjustment creates variants of the indexed database subsequence, where combinations of starting and ending negative flows are allowed, as illustrated in Supplementary Figure 3 (available at http://compbio.cs.ucr.edu/flat).

5. Lookahead Scoring

Flowgram matching using the index structure described in the previous section can be stopped early if the alignment does not appear to be promising. More precisely, given a threshold score t which warrants a good match of the flowgram against the sequence database, and the maximum possible score for each flow, we can discard low-scoring matches early by establishing intermediate score thresholds th_i. The final threshold for the whole flowgram, th_m, is equal to t, and the intermediate thresholds are given by $th_{i-1} = th_i - max_j(M_{i,j})$. This method, termed *lookahead scoring*, was introduced in Wu *et al.*[21], and was combined with the enhanced suffix arrays in Beckstette *et al.*[2]. The threshold score t can be estimated using statistical significance of the match (see Section 6).

Although lookahead scoring gives the same asymptotic worst case running time, in practice, it results in significant speed-ups by pruning the subtrees which start with low-scoring prefixes in the database.

6. Statistical Significance of Scores

Intuitively, a higher raw score obtained by matching a flowgram F against a segment of the sequence database should correspond to a higher likelihood that F was generated by pyrosequencing the matched genomic segment. One way to associate a probability value p with a given raw score is to compute the cumulative distribution function (cdf) over the range of scores that can be obtained by matching F against a flowspace-encoded random genomic segment. Formally, if T is a random variable denoting the score, t is the observed score, and f_T is the probability mass function, the p-value p associated with t is $P(T \geq t) = \sum_{i \geq t} f_T(i)$. The probability mass function can be computed using a dynamic programming method described in Staden *et al.*[18] and Wu *et al.*[21], using a profile matching recurrence relation adjusted for the task of flowgram matching:

$$f_T^i(t) = \sum_l f_T^{i-1}(t - M_{i,l}) \cdot P_{b_i}(L = l)$$

An improvement to this method, described in Beckstette *et al.*[2], is based on the observation that it is not necessary to compute the whole cdf, but only the part of the cdf for scores higher than or equal to the observed score t. Values of the probability mass function are computed in decreasing order of achievable scores, until threshold score t for which the sum of probabilities is greater than p is reached. Modified recurrence relation is as follows:

$$f_T^i(t) = \sum_{l \in \{l | M_{i,l} \geq max_i - d\}} f_T^{i-1}(t - d - M_{i,l}) \cdot P_{b_i}(L = l)$$

For a user specified statistical significance threshold p, this method gives a score threshold t which can be used to perform statistical *significance filtering* of the matches. The threshold score t can be used in conjunction with previously described lookahead scoring to speed up the search. In addition, a correspondence between obtained scores and p-values allows for indirect comparison between scores obtained by matching different flowgrams across different sequence databases.

The expected number of matches in a random sequence database of size n, generally known as the E-value, can be calculated as $p \cdot n$.

7. Parameter Estimation for Probability Distributions

The output of a 454 sequencer is given as a set of three files: (1) a collection of called sequences in FASTA format, (2) accompanying per-called base quality scores which are a function of the observed signal and the conditional distributions of signal strengths[11], and (3) the raw flowgram files. The 454 flowgrams start with the first observed positive flow, and signals are reported with 0.01 granularity.

We combined (1) and (3) to obtain four sets (one per nucleotide) of conditional distributions for different called lengths. Using the maximum likelihood method, we estimated means and standard deviations of the normal distributions for positive flows. Only the conditionals for $l \leq 4$ were used, as data for higher lengths becomes noisy (see Figure 1 and Supplementary Figure 1). We fit a line through the observed values for σ, and use this as an estimate for σ_l.

The signals for the negative flows are distributed according to a distribution which resembles the log-normal, but which exhibits a markedly different behavior in the tails. Most notably, as the signal intensities approach 0, the number of observed signals should also approach 0, and the observed frequencies are significantly higher. Because we have a large number of negative flow signals (no less than 3.5 million flows per library per nucleotide), we decided to use histograms for the distribution of negative flow signals on the $[0, 0.5]$ interval, and extrapolate it using an exponential function on $(0.5, \infty)$.

8. Experiments

We coded a prototype implementation of our method in C++; we called this program FLAT (for FLowgram Alignment Tool). The suffix array index was built using `mkvtree`[10].

We compared FLAT to two methods which could be used for matching small RNA against the target genome: (1) exact matching using a suffix array and (2) BLAST(version 2.2.15) with parameters optimized for finding short, near identical oligonucleotide matches (seed word size 7, E-value cutoff 1000). FLAT is matching flowgrams, whereas the other two methods are matching sequences obtained by base calling the same flowgrams which were returned by 454 Life Sciences. In all three cases, adaptors enclosing the sampled small RNA inserts were trimmed before the search. The flowgram dataset was obtained by pyrosequencing four small RNA libraries constructed from *A. thaliana* plants exposed to abiotic stress conditions: A) cold (61,685 raw flowgrams), B) drought and ABA (74,432), C) NaCl and copper (51,894), and D) heat and UV light (33,320). Reference *A. thaliana* sequences were downloaded from TAIR[15]. We matched small RNA against whole chromosome sequences as well as AGI Transcripts (cDNA, consisting of exons and UTRs) datasets, because small RNA could have been sampled before or after splicing.

All three methods were run on a 64 bit 1,594 MHz Intel Xeon processor. Searching for matches of the first library against *Arabidopsis* chromosome 1 (30.4 million bp), for example, took 6 hours 46 minutes for FLAT, 2 hours 9 minutes for BLAST and 14 minutes for exact matching using highly efficient suffix array implementation.

Results. The number of matches returned by the three methods are summarized in Figure 2. Relatively small numbers of matches compared to the sizes of the libraries is due to the high percentage (59.3-62.8%) of raw flowgrams which were shorter than 18bp once the adaptor sequences were trimmed, and hence too short to belong to a known class of small RNA.

Exact matching is the most stringent and most reliable method of the three; however, due to the number of short inserts which cannot be interpreted as small RNA candidates and due to the nature of the sequence base calling method, only a small fraction (16.0-23.9%) of the original flowgrams match the target genome.

Allowing probabilistic matching using FLAT or tolerating insertions and deletions using BLAST increases the number of matches at the expense of reliability. It is difficult to compare FLAT and BLAST directly, as they were designed with different goals in mind; furthermore, an approximate BLAST match has no grounding in the underlying flowgram signals and unlike FLAT with respect to this is completely arbitrary. However, the number of matches they return and the number of returned matches which appear also in the exactly matched dataset, given as a function of the E-value, provide an intuition about FLAT's behavior. For E-value of 10^{-3}, which in our experiments provided the best balance between the number of matches and false positives, in all four libraries, FLAT consistently returns 98.0% to 98.4% of the exact matches, while returning additional 26.4% to 28.8% matches not found exactly. At higher E-values, the relaxed matching conditions mean that less probable matches would also be included in the output.

BLAST returns nearly all exact matches at E-value 10^{-2}, at which point it returns the number of additional matches comparable to FLAT for the

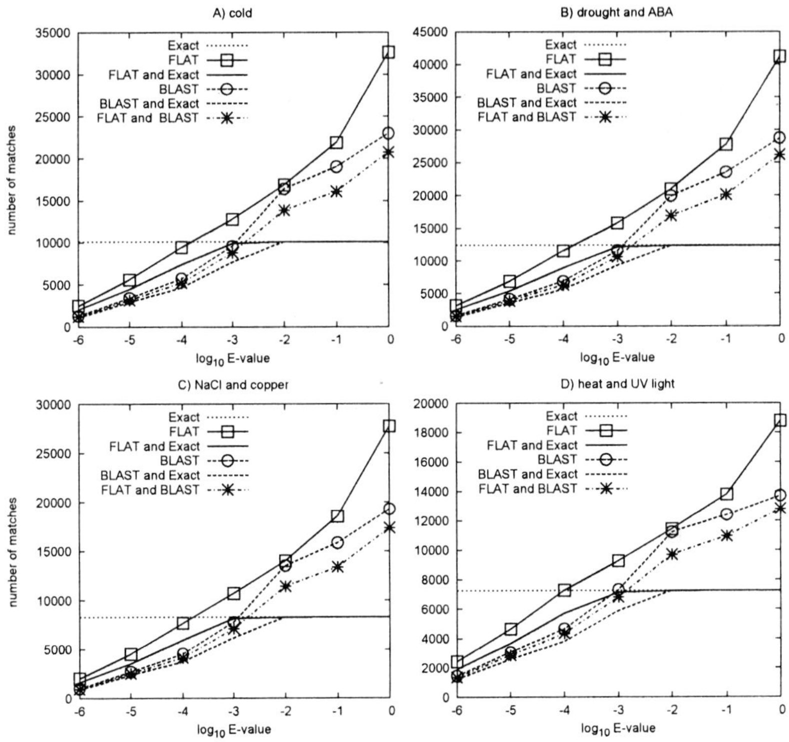

Figure 2. Comparison between the number of matches found for the four stress-induced *A. thaliana* small RNA libraries: A) cold, B) drought and AB, C) NaCl and copper, and D) heat and UV light.

same E-value. It is of interest to note that even though the number of matches is similar, not all of them are found by both methods (the dot-dashed line with star markers in Figure 2).

To illustrate some of the additional matches returned by FLAT and missed by BLAST, consider the flowgram (C,2.06)(G,1.02)(T 0.23)(A 1.53)(C 2.22)(G 0.23)(T 1.99) (A 1.13)(C 0.33)(G 0.96)(T 0.39)(A 0.19)(C 0.96)(G 0.19)(T 0.93)(A 0.10) (C 1.15)(G 0.10)(T 0.26)(A 0.90)(C 0.18)(G 1.02)(T 2.03)(A 0.22)(C 0.12) (G 2.32) for which the maximum likelihood base-called sequence is `CCGAACCTTAGCTCAGTTGG`, which does not occur in the genome. However, if we allow the first `A` flow with the intensity 1.53 to come from `A` and not `AA` we get an alternative base-called sequence `CCGACCTTAGCTCAGTTGG`, which occurs in a number of tRNA genes.

9. Discussion

In this paper, we described a procedure which makes use of the flow signal distribution model to efficiently match small RNA flowgrams against the target genome in a probabilistic framework. Depending on the user-specified statistical significance threshold, additional matches missed by exact matching of the called flowgram sequences are returned.

In principle, evaluating the biological significance as a function of the statistical significance is a challenging task. When analyzing the additional matches, most would agree that calling a flow (A,1.53) as either `A` or `AA` would make sense. However, calling a flow (A,0.20) as `A`, however less probable, is still possible under the model provided in Marguiles *et al.*[11], if less probable matches are allowed by increasing the threshold statistical significance. FLAT provides several output and filtering options which allow the user to focus on the analysis of the non-exact matches or their subset. Most promising matches, in terms of their functional analysis after the tentative genomic loci have been determined, would require additional post-processing and ultimately biological verification.

References

1. M.I. Abouelhoda et al. *Journal of Discrete Algorithms*, 2:53–86, 2004.
2. M. Beckstette et al. *BMC Bioinformatics*, 7:389, 2006.
3. F. Chen et al. In *PAG XIV Conference*, January 2006.
4. A. Fire et al. *Nature*, 391:806–11, 1998.
5. R. Fuchs. *Comput. Appl. Biosci.*, 9:587–91, 1994.
6. B. Gharizadeh et al. *Electrophoresis*, 27(15):3042–7, 2006.
7. A. Girard et al. *Nature*, 442:199–202, 2006.
8. S.M. Goldberg et al. *Proc. Natl. Acad. Sci. USA*, 203(30):11240–5, 2006.
9. J. Karkkainen and P. Sanders. In *13th ICALP*, pages 943–55, 2003.
10. S. Kurtz. http://www.vmatch.de.
11. M. Margulies et al. *Nature*, 435(7075):376–80, 2005.
12. M.J. Moore et al. *BMC Plant Biol.*, 6(17), 2006.
13. C. D. Novina and P. A. Sharp. *Nature*, 430:161–4, 2004.
14. R. Rajagopalan et al. *Genes Dev.*, 20(24):3407–25, 2006.
15. S. Rhee et al. *Nucleic Acids Research*, 31(1):224–8, 2003.
16. M. Ronaghi et al. *Anal Biochem*, 242(1):84–9, 1996.
17. F. Sanger et al. *Proc. Natl. Acad. Sci. USA*, 74:5463–7, 1977.
18. R. Staden. *Comput. Applic. Biosci.*, 5:193–211, 1989.
19. J.C. Wallace and S. Henikoff. *Comput. Applic. Biosci.*, 8:249–254, 1992.
20. T. Wicker et al. *BMC Genomics*, 7(275), 2006.
21. T. Wu et al. *Bioinformatics*, 16(3):233–44, 2000.

COMPUTATIONAL TOOLS FOR NEXT-GENERATION SEQUENCING APPLICATIONS

FRANCISCO M. DE LA VEGA

Applied Biosystems, 850 Lincoln Centre Dr.
Foster City, CA 94404, USA

GABOR T. MARTH

Department of Biology, Boston College, 140 Commonwealth Avenue
Chestnut Hill, MA 02467, USA

GRANGER SUTTON

J. Craig Venter Institute
9704 Medical Center Drive, Rockville, MD 20850, USA

Next generation, rapid, low-cost genome sequencing promises to address a broad range of genetic analysis applications including: comparative genomics, high-throughput polymorphism detection, analysis of small RNAs, identifying mutant genes in disease pathways, transcriptome profiling, methylation profiling, and chromatin remodeling. One of the ambitious goals for these technologies is to produce a complete human genome in a reasonable time frame for US$100,000, and eventually US$1,000. In order to do this, throughput must be increased dramatically. This is achieved by carrying out many parallel reactions. Despite the fact the read-length is short (down to 20-35 bp), the overall throughput is enormous, each run producing up to several hundreds of million reads and billions of base-pairs of sequence data. As the promise of these next generation sequencing (NGS) technologies becomes reality, computational methods for analyzing and managing the massive numbers of short reads produced by these platforms, are urgently needed.

The session of the Pacific Symposium on Biocomputing 2008 "Computational tools for next-generation sequencing applications" aimed to provide the first dedicated forum to discuss the particular challenges that short reads present and the tools and algorithms required for utilizing the staggering volumes of short-read data produced by the new NGS platforms. The session also aimed to establish a discussion between the academic bioinformatics community and their industry counterparts, which are engaged in the development of such platforms, through a discussion panel after the oral presentations of original contributed work. Four contributions were selected from the submissions received and accepted after peer review for inclusion in this proceedings volume and are briefly described next

Given the massive volume of data being produced by NGS platforms, data management becomes a major undertaking for those adopting this technology. New file formats with binary data representation and indexed content will be needed as text files are becoming inefficient both for routine storage and data access. The paper of Phoophakdee and Zaki presents a novel disk-based sequence indexing approach that addresses some of the problems of handling large amounts of data. Trellis+ is an indexing algorithm based on suffix-arrays that allows manipulation of sequence collections using limited amounts of main memory, facilitating NGS sequence analysis with commodity compute servers, rather than requiring specialized hardware. This algorithm can enable rapid sequence assembly and potentially other next generation sequence analysis applications.

Another challenge of analyzing NGS output is the alignment of hundreds of millions of reads coming from a single instrument run to a reference sequence in a reasonable amount of time. Traditional heuristic approaches to sequence alignment do not scale well with short-mers and dynamic programming alignment algorithms such as Smith-Waterman requires significant amount of compute time in commodity hardware, needing embarrassingly parallel approaches, or specialized accelerator chips. The contribution of Coarfa and Milosavljevic is a scalable sequence-matching algorithm based on the positional hashing method. Their current implementation, Pash 2.0, overcomes some of the limitations of positional hashing algorithms in terms of sensitivity to indels, by performing cross-diagonal collation of k-mer matches.

Beyond the (re)-sequencing of regions or whole genomes from pure DNA samples, the sheer volume of data that NGS produce should allow, in principle, tackling the more difficult task of sequencing complex or pooled samples. Sequencing of complex samples is of interest in the case of metagenomics, cancer samples, or mixtures of quickly evolving viral genomes, as well as in genetic epidemiology as a way to address the resequencing of the large number of samples that are needed. The paper of Jojic et al. addresses a significant problem in searching for sequence diversity in HIV genomes from patient samples. Since the virus is evolving rapidly in the host and combination therapy could become ineffective if certain combinations of new acquired mutations evolve, the ability to sequence and distinguish between the viral populations could have major therapeutic implications. The authors describe a method that allows recovering full viral gene sequences (haplotypes) and their frequency in the mixture down to a sensitivity of 0.01%.

Finally, the contribution of Olson et al deals with a new application that NGS enables due to the ability to generate millions of reads from a wide range of positions on the genome. In this case the authors present the tools they have developed to identify a class of small non-coding RNAs of recent relevance, the piwi-associated smallRNAs (piRNAs).

The contributions in this volume certainly address some of the "pain-points" of the utilization of NGS in diverse areas of genome research, but further work is needed. We foresee that initial infrastructural developments would be needed to address the basic analytical and data management tasks that were routine for much lower volumes of Sanger sequencing data. This should be no surprise, since a single NGS instrument can generate an amount of sequence equivalent to that of the entire GeneBank in a short period of time. As time passes and those early problems are overcome, we expect more work on application specific analysis tools to address, e.g. genome-wide gene expression, promoter, methylation and genomic rearrangement profiling. We look forward to such future developments.

TRELLIS+: AN EFFECTIVE APPROACH FOR INDEXING GENOME-SCALE SEQUENCES USING SUFFIX TREES *

BENJARATH PHOOPHAKDEE AND MOHAMMED J. ZAKI

Dept. of Computer Science, Rensselaer Polytechnic Institute, Troy, NY, 12180
E-mail: {phoopb,zaki}@cs.rpi.edu

With advances in high-throughput sequencing methods, and the corresponding exponential growth in sequence data, it has become critical to develop scalable data management techniques for sequence storage, retrieval and analysis. In this paper we present a novel disk-based suffix tree approach, called TRELLIS+, that effectively scales to massive amount of sequence data using only a limited amount of main-memory, based on a novel string buffering strategy. We show experimentally that TRELLIS+ outperforms existing suffix tree approaches; it is able to index genome-scale sequences (e.g., the entire Human genome), and it also allows rapid query processing over the disk-based index. **Availability:** TRELLIS+ source code is available online at `http://www.cs.rpi.edu/~zaki/software/trellis`

1. Introduction

Sequence data banks have been collecting and disseminating an exponentially increasing amount of sequence data. For example, the most recent release of GenBank contains over 77 Gbp (giga, i.e., 10^9, base-pairs) from over 73 million sequence entries. Anticipated advances in rapid sequencing technology, applied to metagenomics (i.e., study of genomes recovered from environmental samples) or rapid, low-cost human genome sequencing, will yield a vast amount of short sequence reads. Individual genomes can also be enormous (e.g., the *Amoeba dubia* genome is estimated to be 670 Gbp [a]). It is thus crucial to develop scalable data management techniques for storage, retrieval and analysis of complete and partial genomes.

In this paper we focus on disk-based suffix trees as the index structure for effective massive sequence data management. Suffix trees have been used to efficiently solve a variety of problems in biological sequence analysis, such as exact and approximate sequence matching, repeat finding, and sequence assembly (via all pairs suffix-prefix matching) [9], as well as anchor finding for genome alignment [4]. Suffix trees can be constructed in time and space linear in the sequence length [16], provided the tree fits entirely in the main memory. A variety of efficient in-memory suffix tree construction algorithms have been proposed [8,6]. However, these algorithms do not scale up when the input sequence is extremely large.

Several disk-based suffix tree algorithms have been proposed recently. Some of the approaches [11,12,15] completely abandon the use of suffix links

*This work was supported in part by NSF Career award IIS-0092978, and NSF grants EIA-0103708 and EMT-0432098.

[a]Database of Genome Sizes: `http://www.cbs.dtu.dk/databases/DOGS/`

and sacrifice the theoretically superior linear construction time in exchange for a quadratic time algorithm with better locality of reference. Some approaches [11,12,2] also suffer from the skewed partitions problem. They build prefix-based partitions of the suffix tree relying on a uniform distribution of prefixes, which is generally not true for sequences in nature. This results in partitions of non-uniform size, where some are very small, and others are too large to fit in memory. Methods that do not have the skew problem and that also maintain suffix links, have also been proposed [1,3]. However, these methods do not scale up to the human genome level. The only known suffix tree methods that can handle the entire human genome include TDD [15] and TRELLIS [13]. TRELLIS was shown to outperform TDD by over 3 times. However, these methods still assume that the input sequence can fit in memory, which limits their suitability for indexing massive sequence data. Other suffix trees variants [10], and other disk-based sequence indexing structures like String B-trees [7] and external suffix arrays [5,14] have also been proposed to handle large sequences. A comparison between TDD [15] and the DC3 [5] method for disk-based suffix arrays suggests that TDD is twice as fast [15].

In this paper we present a novel disk-based suffix tree indexing algorithm, called TRELLIS+, for massive sequence data. TRELLIS+ effectively handles genome-scale sequences and beyond with only a limited amount of main-memory. We show that TRELLIS+ is over twice as fast as TRELLIS, especially with restricted amount of memory. TRELLIS+ is able to index the entire human genome (approx. 3Gbp) in about 11 hours, using only 512MB of memory, and on average queries take under 0.06 seconds, over various query lengths. To the best of our knowledge these are the fastest reported time with such a limited amount of main-memory.

2. Preliminary Concepts

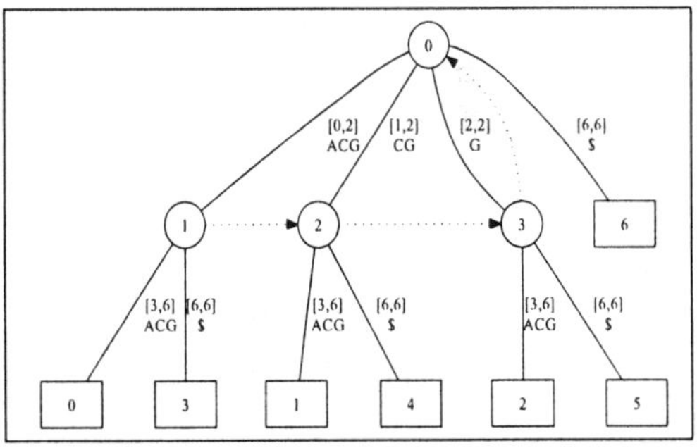

Figure 1. Suffix tree T_S for S = ACGACG$.

Let Σ denote a set of characters (or the alphabet), and let $|\Sigma|$ denote its cardinality. Let $\Sigma^\star$ be the set of all possible strings (or sequences) that can be constructed using Σ. Let $\$ \notin \Sigma$ be the *terminal* character, used to mark the end of a string. Let $S = s_0 s_1 s_2 \ldots s_{n-1}$ be the input string where $S \in \Sigma^\star$ and its length $|S| = n$. The ith suffix of S is represented as $S_i = s_i s_{i+1} s_{i+2} \cdots s_{n-1}$. For convenience, we append the terminal character to the string, and refer to it by s_n. The suffix tree of the string S, denoted as T_S, stores all the suffixes of S in a tree structure, where suffixes that share a common prefix lie on the same path from the root of the tree. A suffix tree has two kinds of nodes: internal and leaf nodes. An internal node in the suffix tree, except the root, has at least 2 children, where each edge to a child begins with a different character. Since the terminal character is unique, there are as many leaves in the suffix tree as there are suffixes, namely $n + 1$ leaves (counting \$ as the "empty" suffix). Each leaf node thus corresponds to a unique suffix S_i.

Let $\sigma(v)$ denote the substring obtained by concatenating all characters from the root to node v. Each internal node v also maintains a *suffix link* to the internal node w, where $\sigma(w)$ is the immediate suffix of $\sigma(v)$. A suffix tree example is given in Fig. 1; circles represent internal nodes, square nodes denote leaves, and dashed lines indicate suffix links. Internal nodes are labeled in depth-first order, and leaf nodes are labeled by the suffix start position. The edges are also shown in the encoded form, giving the start and end positions of the edge label.

3. The Basic Trellis+ Approach

TRELLIS+ follows the same overall approach as TRELLIS [13]. Let S denote the input sequence, which may be a single genome, or the string obtained by concatenating many sequences. TRELLIS+ follows a partitioning and merging approach to build a disk-based suffix tree. The main idea is to maintain a complete suffix tree as a collection of several prefix-based sub-trees. TRELLIS+ has three main steps: i) prefix creation, ii) partitioning, and iii) merging.

In the *prefix creation phase* TRELLIS+ creates a list of variable-length prefixes $\{P_0, P_1, \cdots, P_{m-1}\}$. Each prefix P_i is chosen so that its frequency in the input string S does not exceed a maximum frequency threshold, t_m, determined by the main-memory limit, which guarantees that the prefix-based sub-tree T_{P_i}, composed of all the suffixes beginning with P_i as a prefix, will fit in the available main-memory. The variable prefix set is computed iteratively; in each iteration prefixes up to a given length are counted (those that exceed the frequency threshold t_m in the last iteration).

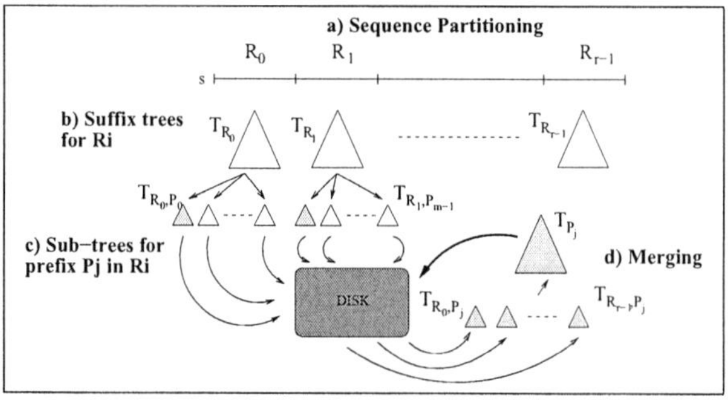

Figure 2. Overview of TRELLIS+

In the *partitioning phase*, the input string S is split into $r = \lceil \frac{n+1}{t_p} \rceil$ segments (Fig. 2, step a), where $n = |S|$ and t_p is the segment size threshold, chosen so that the resulting suffix tree T_{R_i} for each segment R_i (Fig. 2, step b) fits in main-memory. Note that T_{R_i} contains all the suffixes of S that start only in segment R_i; T_{R_i} is constructed using the in-memory Ukkonen's algorithm [16]. Each resulting suffix tree T_{R_i} from a given segment is further split into smaller subtrees T_{R_i,P_j} (Fig. 2, step c), that share a common

4.2.5. *String Buffer Summary*

As for the rest of the characters that are a buffer miss (i.e., not captured by any of the above optimizations), they are directly read from the disk. We found that the input sequence disk access pattern resulting from the buffer misses during the merge has very poor locality of reference, i.e., it is almost completely random, with the exception that short consecutive range of characters are accessed together. These short ranges represent the labels of the edges being merged. Therefore, we keep a small label buffer of size 256KB to store the characters that require a direct disk access: each disk read fetches 256KB consecutive characters at a time.

The total amount of memory required for all of the optimization constituting the string buffer can be calculated by adding the amounts of memory required for each technique: 0.5MB for the index shifting, 70MB for the guide tree, 20MB for buffering internal edge labels, $\frac{t_p}{4\times 10^6}$MB for buffering current segment, $\frac{8t_m}{10^6}$MB for leaf edge label encoding, and 0.25MB for the small label buffer. The total string buffer size is thus well under 100MB, using 512MB memory limit (using Eqs.(1) and (2) to compute t_p and t_m). Note that like TRELLIS, TRELLIS+ has $O(n)$ space and $O(n^2)$ time complexity in the worst case, due to the $O(n^2)$ worst-case merging phase time. In practice the running time is $O(n \log n)$; see [13] for a detailed complexity analysis of TRELLIS.

5. Experiments

We now present an experimental study on the performance of TRELLIS+. We compare TRELLIS+ only against TRELLIS since we showed [13] that TRELLIS outperforms other disk-based suffix methods like TDD [15], DynaCluster [3], TOPQ [1] and so on. TDD [15] was in turn shown to have much better performance than the Hunt's method [11], and even a state-of-the-art suffix array method, DC3 [5]. Note that we were not able to compare with ST-Merge [15] (an extension of TDD, designed to scale to sequences larger than memory), since its implementation is not currently available from its authors. All experiments were performed on an Apple Power Mac G5 machine with 2.7GHz processor, 512KB cache, 4GB main-memory, and 400GB disk space. The maximum amount of main-memory usage across all experiments was restricted to 512MB; this memory limit applies to all internal data structures including those for the suffix tree, memory buffers and the input string. Both TRELLIS+ and TRELLIS were compiled with the GNU g++ compiler v. 3.4.3 and were run in 32-bit mode; they produce identical suffix trees. The sequence data used in all experiments are segments of the human genome ranging from size 200Mbp to 2400Mbp, as well as the entire human genome. To study the effects of the two optimizations, we denote by TRELLIS+NB the version of TRELLIS+ that only has the large segment size optimization but no string buffer, and we denote by TRELLIS+B, the version that has both the larger segment and string buffer optimizations.

5.1. *Effect of Larger Segment Size*

Here we study the effects of the larger segment size, without the string buffer. TRELLIS+NB has larger and therefore fewer partitions than TRELLIS, since for TRELLIS the number of partitions is $O(\frac{n}{t_p})$ and the value of t_p decreases as the sequence length n increases, resulting in many partitions (as shown in Fig. 5(a)). Therefore, when indexing a very large sequence, the performance of TRELLIS suffers when t_p is small, because of a large number of partitions. In contrast, since the partitioning threshold t_p for TRELLIS+NB remains constant regardless of n, its number of partitions increases at a much slower rate, as shown in Fig. 5(b).

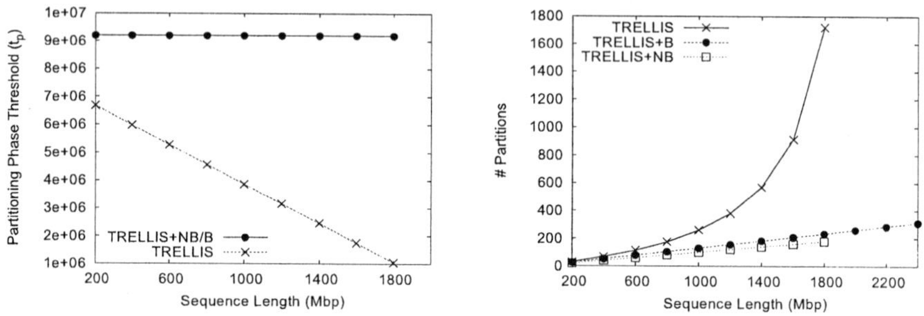

(a) Partitioning Threshold (t_p) (b) Number of Partitions

Figure 5. Effect of Larger Segment Size on Partitioning Phase

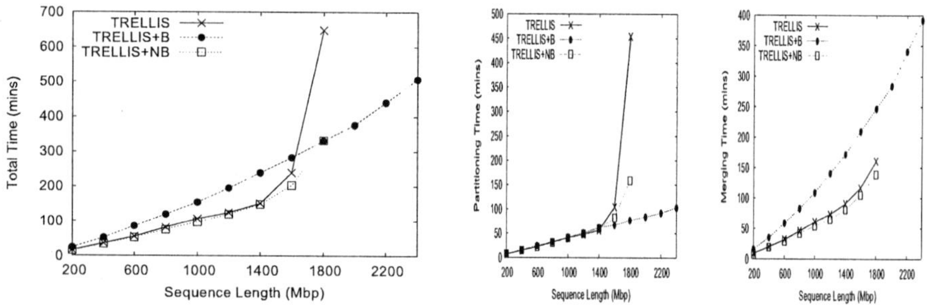

(a) Total Running Time (mins) (b) Partitioning Time (c) Merging Time

Figure 6. Running Time Comparison

The timings of TRELLIS+NB in comparison to TRELLIS are shown in Figs. 6(a), 6(b), and 6(c), which show the total time, partitioning phase time, and merging phase time for TRELLIS+NB versus TRELLIS, as we increase the sequence length from 200Mbp to 1.8Gbp. We find that TRELLIS+NB consistently outperforms TRELLIS, especially when the input sequence size is much larger than the available memory (which is only 512MB). For example, TRELLIS+NB is about twice as fast as TRELLIS for the 1.8Gbp input sequence. This is directly a consequence of the larger, fewer partitions used by TRELLIS+NB, which result in a much faster partitioning phase (see Fig. 6(b)). The impact of larger segment sizes on the merging phase is not much (see Fig. 6(c)), but TRELLIS+NB still has faster merge times, since there are fewer partitions to be merged for each prefix-based subtree T_{P_j}.

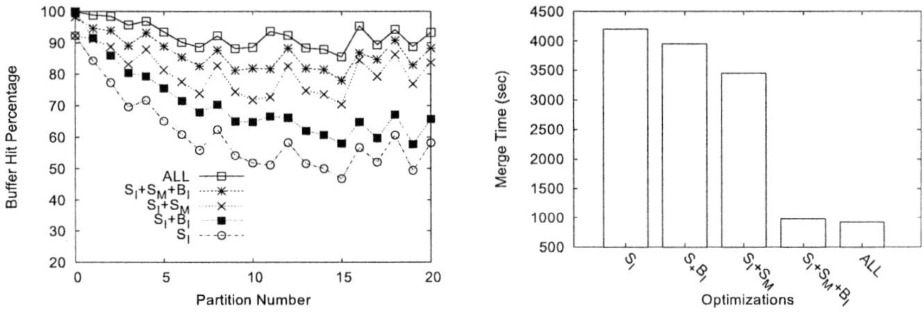

(a) Buffer Hit Rate (b) Buffer Optimizations Times

Figure 7. Effect of String Buffer Optimizations

5.2. *Effect of String Buffer*

We now investigate the effect of the string buffering strategy. First we report the difference in the buffer hit rate and merging phase time for TRELLIS+B using the different combinations of buffering optimizations. Fig. 7(a) shows the buffer hit rate for all the characters accessed during the subtree merging operations, using as input string Human Chromosome I (with length approx. 200Mbp), with the 512MB memory limit. The hit rates are shown only for the first 20 partitions, but the same trend continues for the remaining partitions. In the figure, S_I denotes the internal edge index shifting, S_M denotes index shifting during merge phase, B_I denotes buffering internal labels, and ALL denotes all the buffering optimizations. We can clearly see that internal edge index shifting alone yields a buffer hit rate of over 50%. Combination of optimizations yield higher hit rates, so that when all the optimization are combined we achieve a buffer hit rate of over 90%. Fig. 7(b) shows effect of the improved buffer hit rates on the running time of the merging phase in TRELLIS+B. All the optimizations results in a four-fold decrease in time.

Comparing the total running time, and the times for the partitioning and merging phases (shown in Figs. 6(a), 6(b), and 6(c)), we find that initially TRELLIS+NB (that does not use the string buffer) outperforms TRELLIS+B (that uses string buffer). However, as the input sequence becomes much larger, TRELLIS+NB is left with less memory to construct the tree, because it has to maintain the entire compressed input string in memory. Consequently, beyond a certain sequence length, TRELLIS+B starts to outperform TRELLIS+NB. In fact, without string buffer, we were not able to run TRELLIS+NB on an input of size larger than 1.8Gbp, whereas with the string buffer TRELLIS+B can construct the disk-based suffix tree for the entire Human genome. For a 2.4Gbp sequence, TRELLIS+B took about 8.3 hrs (500 mins, as shown in Fig. 6(a)), and *for the full Human genome (with over 3Gbp length),* TRELLIS+B *finished in about 11 hours using only 512MB memory!* [b]

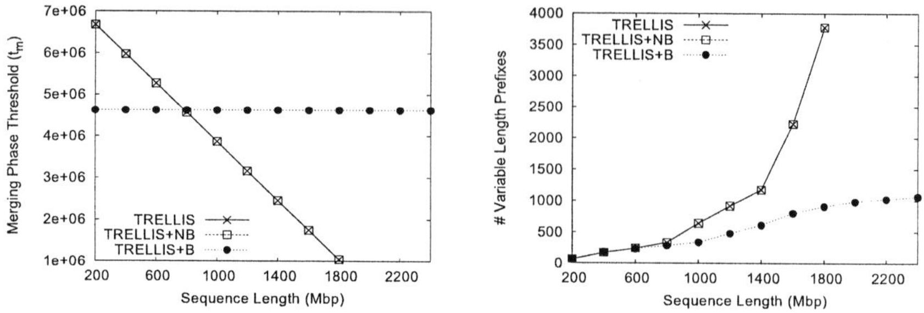

(a) Merging Threshold (t_m) (b) Number of Prefixes

Figure 8. Effect on the Merging Threshold and Number of Variable Length Prefixes

Fig. 8(a) shows the merging phase threshold t_m, and Fig. 8(b) shows the number of variable-length prefixes for TRELLIS+B and TRELLIS+NB. Since TRELLIS+NB has to retain the entire input string in memory during the merging phase, with increasing sequence length TRELLIS+NB has less amount of memory remaining, resulting in smaller t_m and many more prefixes. On the other hand, for TRELLIS+B the number of prefixes grows very slowly. Overall, as shown in Figs. 6(b) and 6(c), the suffix buffer allows TRELLIS+B to scale gracefully for sequence much larger than the available memory, whereas TRELLIS+NB could not run for an input string longer than 1.8Gbp (with 512MB memory).

5.3. *Query Times*

We now briefly discuss the query time performance on the disk-based suffix tree created by TRELLIS+ on the entire human genome (which occupies about 71GB on disk). 500 queries of different lengths ranging from 40bp to 10,000bp were generated from random starting positions in the human genome. Figure 9 shows the average query times over the 500 random queries for each query length (using 2GB memory). *The average query time for even the longest query (with length 10,000bp) took under 0.06s, showing the effectiveness of disk-based suffix tree indexing in terms of the query performance* (see [13] for more details).

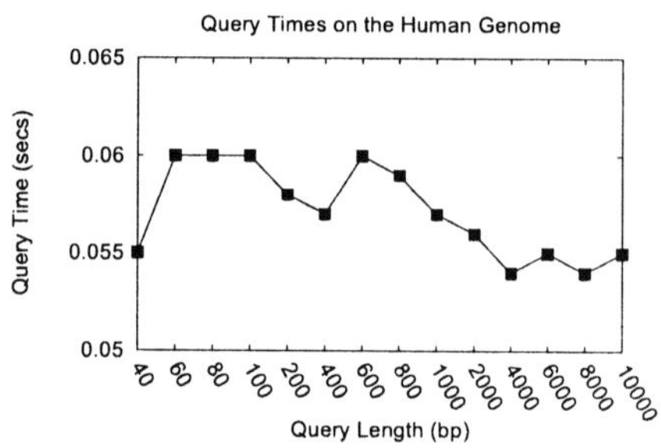

Figure 9. Average Query Times

[b] We showed earlier [13] that TRELLIS can index the entire human genome in about 4 hours with 2GB memory.

6. Conclusion

In this paper we have presented effective optimization strategies which enable TRELLIS+ to handle genome-scale sequences, using only a limited amount of main memory. TRELLIS+ is suitable for indexing entire genomes, or massive amounts of short sequence read data, such as those resulting from cheap genome sequencing and metagenomics projects. For the latter case, we simply concatenate all the short reads into a single long sequence S and index it. In addition we maintain an auxiliary index on disk that allows one to look up for each suffix position S_i, the corresponding sequence id, and offset into the short read. Using all pairs suffix-prefix matching [9], our disk based suffix tree index can enable rapid sequence assembly, and can also enable other next generation sequence analysis applications.

References

1. S.J. Bedathur and J.R. Haritsa. Engineering a fast online persistent suffix tree construction. In *20th Int'l Conference on Data Engineering*, 2004.
2. A.L. Brown. Constructing genome scale suffix trees. In *2nd Asia-Pacific Bioinformatics Conference*, 2004.
3. C.-F. Cheung, J.X. Yu, and H. Lu. Constructing suffix tree for gigabyte sequences with megabyte memory. *IEEE Transactions on Knowledge and Data Engineering*, 17(1):90–105, 2005.
4. A.L. Delcher, A. Phillippy, J. Carlton, and S.L. Salzberg. Fast algorithms for large-scale genome alignment and comparison. *Nucleic Acids Research*, 30(11):2478–2483, 2002.
5. R. Dementiev, J. Kärkkäinen, J. Mehnert, and P. Sanders. Better external memory suffix array construction. In *Workshop on Algorithm Engineering and Experiments*, 2005.
6. M. Farach-Colton, P. Ferragina, and S. Muthukrishnan. On the sorting-complexity of suffix tree construction. *J. of the ACM*, 47(6):987–1011, 2000.
7. P. Ferragina and R. Grossi. The string B-tree: a new data structure for string search in external memory and its applications. *Journal of the ACM*, 46(2):236–280, 1999.
8. R. Giegerich, S. Kurtz, and J. Stoye. Efficient implementation of lazy suffix trees. *Software Practice & Experience*, 33(11):1035–1049, 2003.
9. D. Gusfield. *Algorithms on Strings, Trees, and Sequences*. Cambridge University Press, 1997.
10. K. Heumann and H. W. Mewes. The hashed position tree (HPT): A suffix tree variant for large data sets stored on slow mass storage devices. In *3rd South American Workshop on String Processing*, 1996.
11. E. Hunt, M.P. Atkinson, and R.W. Irving. A database index to large biological sequences. In *27th Int'l Conference on Very Large Data Bases*, 2001.
12. R. Japp. The top-compressed suffix tree: A disk-resident index for large sequences. In *BNCOD Bioinformatics Workshop*, 2004.
13. B. Phoophakdee and M. J. Zaki. Genome-scale disk-based suffix tree indexing. In *ACM SIGMOD Int'l Conference on Management of Data*, 2007.
14. K. Sadakane and T. Shibuya. Indexing huge genome sequences for solving various problems. *Genome Informatics*, 12:175–183, 2001.
15. Y. Tian, S. Tata, R.A. Hankins, and J.M. Patel. Practical methods for constructing suffix trees. *VLDB Journal*, 14(3):281–299, 2005.
16. E. Ukkonen. On-line construction of suffix trees. *Algorithmica*, 14(3), 1995.

PASH 2.0: SCALEABLE SEQUENCE ANCHORING FOR NEXT-GENERATION SEQUENCING TECHNOLOGIES

CRISTIAN COARFA

[†]*Human Genome Sequencing Center, Department of Molecular and Human Genetics, Baylor College of Medicine, One Baylor Plaza Houston, Texas 77030, USA*

ALEKSANDAR MILOSAVLJEVIC

Human Genome Sequencing Center, Department of Molecular and Human Genetics, Baylor College of Medicine, One Baylor Plaza Houston, Texas 77030, USA

Many applications of next-generation sequencing technologies involve anchoring of a sequence fragment or a tag onto a corresponding position on a reference genome assembly. Positional Hashing method, implemented in the Pash 2.0 program, is specifically designed for the task of high-volume anchoring. In this article we present multi-diagonal gapped kmer collation and other improvements introduced in Pash 2.0 that further improve accuracy and speed of Positional Hashing. The goal of this article is to show that gapped kmer matching with cross-diagonal collation suffices for anchoring across close evolutionary distances and for the purpose of human resequencing. We propose a benchmark for evaluating the performance of anchoring programs that captures key parameters in specific applications, including duplicative structure of genomes of humans and other species. We demonstrate speedups of up to tenfold in large-scale anchoring experiments achieved by PASH 2.0 when compared to BLAT, another similarity search program frequently used for anchoring.

1. Introduction

Next generation sequencing technologies produce an unprecedented number of sequence fragments in the 20-300 basepair range. Many applications of next-generation sequencing require anchoring of these fragments onto a reference sequence, which involves comparison of these fragments to determine their position in the reference. Anchoring is required for the purpose of various mapping applications or for comparative sequence assembly (also referred to as comparative genome assembly and templated assembly). Anchoring is also a key step in the comparison of assembled evolutionarily related genomes. Due to the sheer number of fragments produced by next-generation sequencing technologies

†This research was partially supported by the National Human Genome Research Institute grant 5R01HG004009-02, by the National Cancer Institute grant 1R33CA114151-01A1 and the National Science Foundation grant CNS 0420984 to AM.

and the size of reference sequences, anchoring is rapidly becoming a computational bottleneck.

The de facto dominant paradigm for similarity search is that of "Seed-and-Extend" embodied in algorithms such as BLAST [1, 2], BLAT [3], SSAHA [4], PatternHunter [5, 6]. and FASTA [7, 8]. While not initially motivated by the anchoring problem, the Seed-and-Extend paradigm is employed by most current anchoring programs. We recently proposed Positional Hashing, a novel, inherently parallelizable and scaleable approach to specifically address the requirements of high-volume anchoring [9]. We first review key concepts behind Positional Hashing; then, we present the Pash 2.0 program, a new implementation which overcomes a number of deficiencies in the initial implementation of Positional Hashing. Pash 2.0 includes multidiagonal collation of gapped kmer matches to enhance accuracy in the presence of indels, and improvements that enhance speed when mapping large volumes of reads onto mammalian-sized genomes. The goal of this article is to show that gapped kmer matching with cross-diagonal collation suffices for anchoring across close evolutionary distances and for the purpose of human resequencing. To demonstrate this, we evaluate Pash by comparing its accuracy and speed against Blat, a Seed-and-Extend program that is widely used for anchoring. We determine parameters for Pash such it achieves comparable accuracy with Blat while providing several-fold speedups by avoiding basepair-level computation performed by Blat. To complement real-data experiments, we propose a simulation benchmark for evaluating performance of anchoring programs that captures key parameters in specific applications, including duplicative structure of the genomes such as that of humans. Using both real data and the simulation benchmark, we demonstrate speedups of up to tenfold without significant loss of sensitivity or accuracy in large-scale anchoring experiments when compared to BLAT.

2. Two approaches to anchoring: Seed-and-Extend vs. Positional Hashing

2.1. *The seed-and-extend paradigm*

The seed-and-extend paradigm currently dominates the field of sequence similarity search [2, 3, 4, 5, 6, 7, 10, 11]. This paradigm originally emerged to address the key problem of searching a large database using a relatively short query to detect remote homologies. A homology match to a gene of known function was used to derive a hypothesis about the function of the query sequence. The first key requirement for this application is sensitivity when

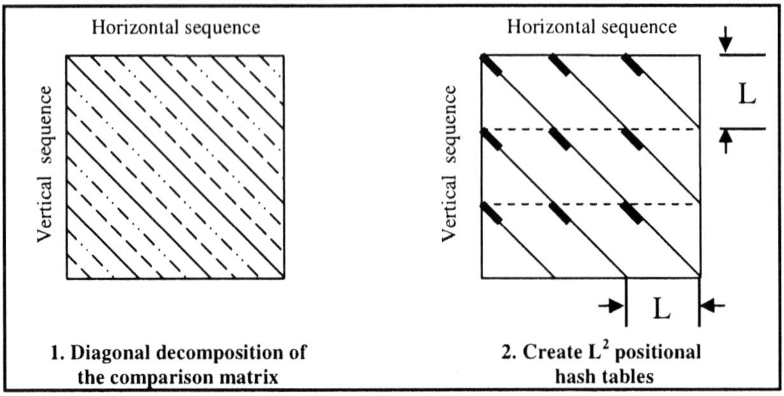

Figure 1. Positional Hashing. 1. The positional hashing scheme breaks the anchoring problem along the L diagonals of the comparison matrix; each cluster node detects and groups matches along a subset of the L diagonals. 2. Each diagonal is split into horizontal and vertical windows of size L. Short bold lines indicate positions used to calculate hash keys for positional hash table H(0,0)

comparing sequences across large evolutionary distances. The second key requirement is speed when searching a large database using a short query. The first generation seed-and-extend algorithms such as BLAST [2] and FASTA [7] employed pre-processing of the query to speed up the database search while second-generation seed-and-extend algorithms such as BLAT [3] and SSAHA [4] employed in-memory indexing of genome-sized databases for another order of magnitude of speed increase required for interactive lookup of genome loci in human genome browsers using genomic DNA sequence queries.

2.2. *Positional Hashing specifically addresses the anchoring problem*

It is important to note that the anchoring problem poses a new and unique set of requirements. First, the detection of remote homologies is less relevant for anchoring than discrimination of true orthology relations when comparing closely related genomes. Second, with the growth of the genome databases and the emergence of next-generation sequencing technologies the query itself may now contain tens of millions of fragments or several gigabases of assembled sequence. To address the requirements specific to the anchoring problem, we recently developed the Positional Hashing method [9]. The method avoids costly basepair-level matching by employing faster and more scaleable gapped kmer matching [2,5,6,9]; this is performed using distributed position-specific hash tables that are constructed from both compared sequences.

To better formulate the difference between Positional Hashing and the classical Seed-and-Extend paradigms, we first introduce a few definitions. A "seed" pattern P is defined by offsets $\{x_1,\ldots,x_w\}$. We say that a "seed" match—a gapped kmer match where k equals w-- is detected between sequences S and T in respective positions i and j if $S[i+x_1]= T[j+x_1]$, ..., and $S[i+x_w]=T[j+x_w]$. To further simplify notation, we define pattern function f_P at position i in sequence S as $f_P(S,i) = S[i+x_1]...S[i+x_w]$. Using this definition, we say that a "seed" match is detected between sequences S and T in respective positions i and j if $f_P(S,i)= f_P(T,j)$. A Seed-and-Extend method extends each seed match by local basepair alignment. The alignments that do not produce scores above a threshold of significance are discarded.

In contrast to the Seed-and-Extend paradigm, Positional Hashing groups all collinear matches—i.e., those falling along the same diagonal or, in Pash 2.0, a set of neighboring diagonals in the comparison matrix-- to produce a score. The score calculated by grouping the matches suffices for a wide range of anchoring applications, while providing significant speedup by eliminating the time-consuming local alignment at the basepair level. In further contrast to the Seed-and-Extend paradigm, Positional Hashing involves numerous position-specific hash tables, thus allowing extreme scalability through parallel computing. The positional hashing scheme breaks the anchoring problem along its natural diagonal structure, as illustrated in the Figure 1.1. Each node detects and groups matches along a subset of diagonals. More precisely, matches along diagonal d=0,1,...L-1, of the form $f_P(S,i)= f_P(T,j)$, where i=j+d (mod L) are detected and grouped in parallel on individual nodes of a computer cluster. Position-specific hash tables are defined by conceptually dividing each alignment diagonal into

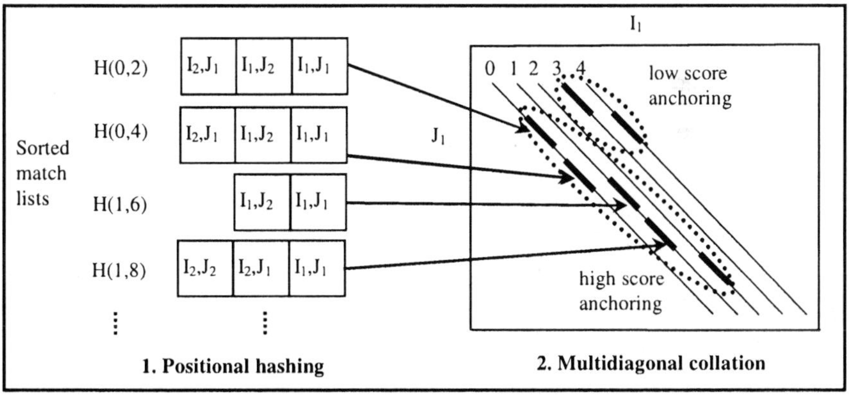

Figure 2. Positional hashing and multi-diagonal collation. 1. Lists of match positions for diagonals 0-5 induced by the appropriate hash tables are generated in the inversion step, for horizontal windows I_1 and I_2 and for vertical windows J_1 and J_2; the lists are sorted from right to left. A priority queue is used to quickly select the set of match positions within the same horizontal and vertical L-sized window, on which multidiagonal collation needs to be performed. 2. A greedy heuristic is used to determine the highest scoring anchoring across multiple diagonals; in the figure we depict matches within horizontal window I_1 and vertical window J_1, across diagonals 0-4.

non-overlapping *windows* of length L, as indicated by dashed lines in Figure 1.2. A total of L^2 positional hash tables H(d,k) are constructed for each diagonal d=0,1,...L-1 and diagonal position k=0,1,... L-1. Matches are detected by using the values of $f_P(S,i)$ and $f_P(T,j)$ as keys for storing horizontal and vertical window indices I=[i/L] and J=[j/L] into specific hash table bins. A match of the form $f_P(S,i)= f_P(T,j)$ where i=j+d (mod L) and j=k (mod L) is detected whenever I and J occur in the same bin of hash table H(d,k), as also shown in Figure 2.1. Further implementation details are described in [9].

3. Improved implementation of Positional Hashing

3.1. *Multidiagonal collation*

A key step in Pash is represented by the collation of matching kmers across diagonals. In Pash 1.0, collation was performed across a single diagonal only; an indel would split matching kmers across two or more neighboring diagonals. For Sanger reads, typically 600-800 base pairs long, Pash 1.0 could find enough information on either side of an indel to accurately anchor a read. For the shorter reads, generated by the next generation sequencing technologies, it might not be possible to find matching kmers on either side of an indel to anchor the read. The use of pyrosequencing, which causes insertion/deletion errors in the presence of homopolymer runs, further amplified this problem.

To overcome the problem, Pash 2.0 collates kmer matches across multiple diagonals. Pash detects similarities between two sequences, denoted a vertical sequence and a horizontal sequence (as indicated in Figure 1). After performing hashing and inversion for multiple diagonals, Pash generates one list of horizontal and vertical sequence positions of the matching kmers for each diagonal and positional hash table pair; these lists are sorted by the horizontal then by the vertical position of the matching kmer. Next, Pash considers simultaneously all lists of matching kmers for the set of diagonals that are being collated, and traverses them to determine all the matching positions between a horizontal and vertical window of size L (see Figure 2.1). To collate across k diagonals, Pash first selects matching positions across the same vertical and horizontal window from the kL lists of matching kmer positions. It uses a priority queue, with a two-part key: first the horizontal positions are compared, followed by the vertical position of matches, as shown in Figure 2.1. Kmers in each such set are collated, by performing banded alignment not at basepair level but at the kmer level. We used a greedy method to collate the matches across a diagonal set, and select the highest scoring match, as shown in Figure 2.2. By collating kmers across *k* diagonals, Pash is in effect anchoring across indels of size *k-1*; a user can control through command-line parameters the maximum indel size detectable by Pash. Pash 2.0 scores matches across indels using an affine indel penalty. Let m be the number of matching bases; for each indel I let $s(I)$ be the indel length. The score of an anchoring is then $2m - \sum_{I}(s(I)+1)$.

3.2. *Efficient hashing and inversion*

Pash version 1.0 was hashing both the vertical and the horizontal sequence. For comparisons against large genomes, such as mammalian genomes, hashing the whole genome during the hashing/inversion phase required significant time and memory. In Pash 2.0, only one of the sequences is hashed, namely the vertical sequence. For the horizontal sequence, instead of hashing it, Pash 2.0 traverses the horizontal kmer lists and then matches each kmer against the corresponding bin in the hash table created by hashing the vertical sequence. If a match is detected, the corresponding kmer is added to the list of matching kmers prior to proceeding to the next horizontal kmer. This improvement substantially accelerated the hashing and inversion steps.

4. Experimental Evaluation

Our experimental platform consisted of compute nodes with dual 2.2GHz AMD Opteron processors, 4GB of memory, running Linux, kernel 2.6. We used Pash 2.0, and BLAT Client/Server version 32. All experiments were run sequentially; when input was split in multiple chunks, we reported total compute time. The focus of this section is on comparing Pash 2.0 to Blat. When comparing Pash 2.0 against Pash 1.2, we determined overall speed improvements of 33%, similar accuracy for Sanger reads, and significant accuracy improvements for pyrosequencing reads. For Pash 2.0 we used the the following pattern of weight 13 and span 21: *111011011000110101011*. Code and licenses for Pash, Positional Hashing, and auxiliary scripts are available free of charge for academic use. Current access and licensing information is posted at http://www.brl.bcm.tmc.edu/.

4.1. *UD-CSD benchmark*

The choice of a program for an anchoring application depends on a number of data parameters, data volume, and computational resources available for the task. To facilitate selection of the most suitable program it would therefore be useful to test candidates on a benchmark that captures key aspects of the problem at hand. Toward this end, we developed a benchmark that includes segmental duplications, an important feature of mammalian genomes, and particularly of the genome of humans and other primates. The duplications are especially challenging because they limit the sequence uniqueness necessary for anchoring. The UD-CSD benchmark is named after five key aspects: *Unique* fraction of the genome; *Duplicated* fraction; *Coevolution* of duplicated fraction during which uniqueness is gradually developed; *Speciation*; and *Divergence* of orthologous reads. As illustrated in Figure 3, the UD-CSD benchmark is parameterized by the following four parameters: number of unique reads k; number of duplicated reads n; coevolution parameter *x;* and divergence parameter *y*; we are in fact simulating genomes as a concatenation of reads. For example, the divergence parameter y=1% may be appropriate for human-chimpanzee anchoring and y=5% anchoring of a rhesus monkey onto human. Note that in a human genome resequencing study, the divergence parameter y would be set to a very small value due to relatively small amount human polymorphism but the duplicative structure of the human genome could be captured using remaining three parameters.

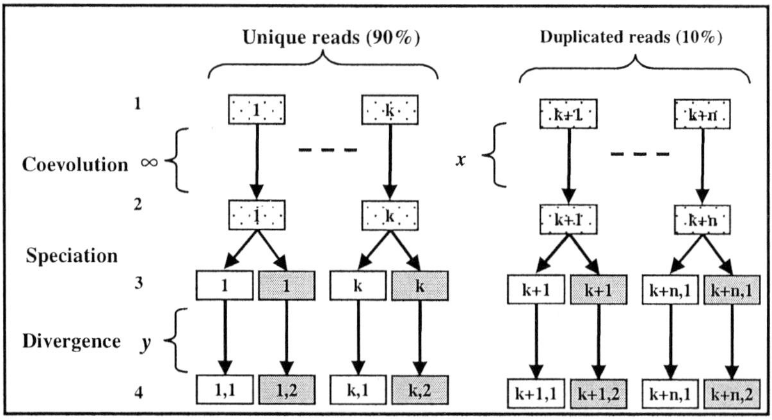

Figure 3. The *UD-CSD* (*Unique,Duplicated-Coevolution,Speciation,Divergence*) Anchoring Benchmark. 1. Randomly generate k *Unique* reads and n *Duplicated* reads. 2. *Coevolution*: each base mutates with probability x. 3. *Speciation:* Each read duplicates. 4. *Divergence*: each base mutates with probability y.

Using the UD-CSD benchmark, we evaluated the sensitivity and specificity of Pash compared to BLAT, a widely used seed-and-extend comparison algorithm. We first generated k+1 random reads of size m base pairs, then we duplicated the last read n-1 times, as illustrated in Figure 3.1, and obtained seed reads s_i, i=1,n+k. This corresponds to a genome where the k reads represent unique regions, and the n duplicated reads represent duplicated regions. Next, we evolved each read s_i, such that each base has a mutation probability of x, and each base was mutated at most once, and obtained reads r_i, i=1,n+k. Out of the mutations, 5% were indels, with half insertions and half deletions; the indel lengths were chosen using a geometric probability distribution with the parameter p=0.9, and imposing a maximum length of 10. The remaining mutations were substitutions. This process approximates a period of *coevolution* of two related species during which duplicated regions acquire uniqueness (parameterized by x) necessary for anchoring. Next, two copies of each read were generated, and one assigned to each of two simulated genomes of descendant species, as shown in Figure 3.3; this corresponds to a speciation event. Subsequently, each read evolved independently such that each base had a mutation probability of y, as illustrated in Figure 3.3; this corresponds to a period of *divergence* between the two related species. Finally, we obtained the set of reads $r_{i,1}$ and $r_{i,2}$, with i=1,n+k. We then employed Pash and BLAT to anchor the read set $\{r_{1,1},\ldots,r_{n+k,1}\}$ onto $\{r_{1,2},\ldots,r_{n+k,2}\}$, by running each program and then filtering its output such that only the top ten best matches for each read are kept. Any time a read $r_{i,1}$ is matched onto $r_{i,2}$, we consider this a true positive; we count how many true positives are found to evaluate the accuracy of the anchoring program.

One may raise objection to our considering the top ten best matches and may instead insist that only the top match counts. Our more relaxed criterion is justified by the fact that anchoring typically involves a reciprocal-best-match step. For example, a 10-reciprocal-best-match step would sieve out false matches and achieve specific anchoring as long as the correct match is among the top 10 scoring reads. Assuming random error, one may show that the expected number of false matches would remain constant (10 in our case) irrespective of the total number of reads matched. For our experiment, we chose a read length of 200 bases, and varied the total number of reads from 5,000 to 16,000,000. k and n were always chosen such that 90% of the start reads were unique, and 10% were

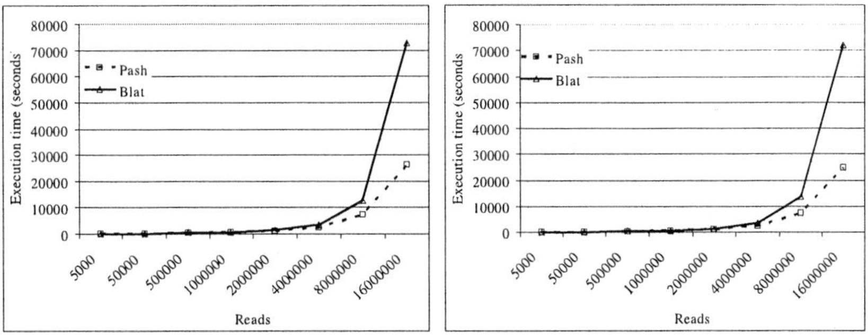

1 25% coevolution , 1% divergence　　2. 25% coevolution, 5% divergence

Figure 4.1.Anchoring times of Pash and BLAT for coevolution of 25% and divergence of 1%, which may be appropriate for comparing closely related primates such as chimpanzee and human. 2.Anchoring times of Pash and BLAT for coevolution of 25% and divergence of 5%, which may be appropriate for comparing a New World monkey such as rhesus macaque and human.

repetitive. In Figure 4.1 we present the execution times for Pash and BLAT for 25% coevolution and 1% divergence, while in Figure 4.2 we present execution times for Pash and BLAT for 25% coevolution and 5% divergence. Pash was run using a gapped pattern of weight 13 and span 21, and a kmer offset gap of 12, while for BLAT we used the default settings.

In both cases, Pash and BLAT achieve comparable sensitivity (the numbers mate pairs found are within 1% of each other). This result is significant because it indicates that time-consuming basepair-level alignments performed by BLAT are not necessary for accurate anchoring – kmer-level matching performed by Pash suffices. For up to 2 million reads, Pash and BLAT achieve comparable performance. When the number of reads increases to 4, 8, and 16 million reads, however, Pash outperforms BLAT by a factor of 1.5 to 2.7.

4.2. *Simulated Anchoring of WGS reads*

Next generation technologies enable the rapid collection of a large volume of reads, which can then be used for applications such as genome variation detection. A key step is the anchoring of such reads onto the human genome. In our experiment, we used reads obtained by randomly sampling the human genome (UCSC hg18, http://genome.ucsc.edu/downloads.html) with read sizes chosen according to the empirical distribution of read lengths observed in sequencing experiments using 454 sequencing technology. The set of reads covering the human genome at 6x sequence coverage was independently mapped back onto the reference genome using Blat and Pash. Pash anchored 73 million reads in 160 hours, using kmers of weight 13, span 21, and kmer gap offset of 12. Blat was run with default parameters; it mapped the reads from chromosomes 1 and 2 in 289 hours; this extrapolates to an overall running time of 1823 hours, for a 11.3 fold acceleration of Pash over Blat; Blat mapped only 0.3 percent more reads than Pash; this difference is caused by reads that Pash did not map because of its own default ignoring overrepresented kmers; we could improve this figure by increasing Pash's tolerance for overrepresented kmers. Next, we extracted tags of 25 base pairs from each simulated WGS read, and mapped them on the human genome using Pash and Blat. Pash anchored the tags from chromosomes 1 and 2 in 4.5 hours, while Blat anchored them in 105 hours. However, with default parameters Blat does not perform well for the 25 base pair tags, anchoring back correctly 28% of the tags for chromosome 1 and 31% for chromosome 2, compared to 77% and 85% respectively for Pash.

4.3. *Anchoring of mate pairs*

Sequenced ends of a small-insert or a long-insert clone such as a Fosmids or a Bacterial Artificial Chromosome (BAC) may be anchored onto a related reference genomic sequence. Numerous biological applications rely on this step, such as detection of cross-mammalian conservation of chromosome structure using mapping of sequenced BAC-End Sequences [13,14,15] and reconstruction of the evolution of the human genome [12]. Next-generation sequencing technologies provide a particularly economical and fast method of delineating conserved and rearranged regions using the paired-end method.

The fraction of consistently anchored paired end-sequences from a particular set depends on the accuracy of the anchoring program, making this a natural benchmark for testing anchoring programs. We obtained about 16 million Sanger reads from fosmid end sequences in the NCBI Trace Archive, for a total of 7,946,887 mate pairs, and anchored them onto the human genome with Blat and Pash 2.0. For each read we selected the top 10 matches, then looked for consistently mapped mate pairs. We counted the total number of clone ends that were anchored at a distance consistent with clone insert size (25-50 Kb) and computed their percentage of the expected number of mate pairs. Since anchoring performance also depends on the size of anchored reads, we also simulated five shorter read sizes by extracting 250bp, 100bp, 50bp, 36bp, and 25bp reads respectively from each Sanger read, generating additional sets of simulated short fosmid end sequences. We anchored each of the short read sets onto the human genome, then determined the number of clone ends consistently mapped. We summarize the results of our experiment in Table 1. We used gapped kmers of weight 13 and span 21, and kmer offsets of 12 for Sanger and 250 bp reads, of 6 for 100 bp reads, and of 4 for 50, 36, and 25 bp reads. As evident from Table 1, in all the experiments both Pash and BLAT found a comparable number of consistent mate pairs mapping, while Pash ran 4.5 to 10.2 times faster compared to BLAT. A recent option added to Blat is that of fastMap, which enables rapid mapping of queries onto highly similar targets.

Table 1. Summary of results for actual and simulated mate pair anchoring

Read Type	Pash execution time (hrs)	Percent of matepairs found	Blat execution time (hrs)	Percent of matepairs found
Sanger	102	76%	1045	76%
250 bp	45	76%	421	76%
100 bp	23	75%	154	75%
50 bp	17	68 %	92	68 %
36 bp	8	57%	85	58%
25 bp	4	56 %	154	15 %

We ran Blat with this option, but determined that it yielded very low sensitivity compared to Blat with default parameters, retrieving around 1 percent of the total number of matepairs; we argue the Blat with fastMap is not a good choice for this task. Blat with default parameters performs poorly on 25bp reads.

Pash 2.0 accelerates anchoring the most for very large input data sets. To measure this effect, we partitioned our input of 16 million reads into chunks of 0.5, 1, 2, 4, and 8 million reads each and run Pash on the whole input, computing average time per chunk. Each chunk could be run on a separate cluster node, and the parallel Pash wall time would be the maximum execution time of an input chunk. In Figure 5 we present the Pash execution time per chunk and the overall running time; our results show that while our method has a significant overhead for a small number of reads, its effectiveness improves as the number of input reads per chunk is increased. Pash 2.0 is therefore suitable for anchoring the output of high-volume, high-throughput sequencing technologies.

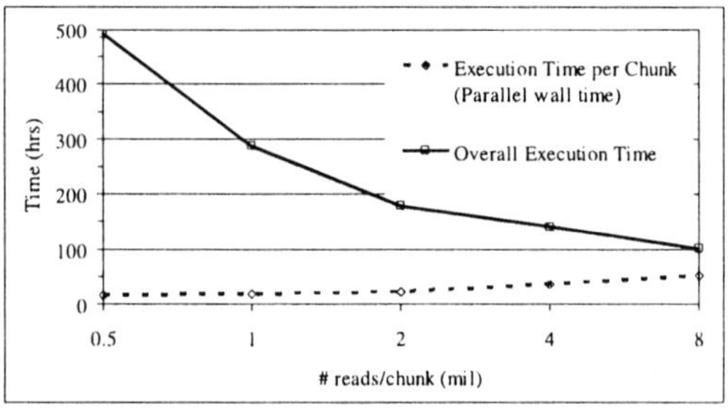

Figure 5. Anchoring time for 16 million Sanger reads onto human genome.

5. Conclusions

We demonstrate that by avoiding basepair-level comparison the Positional Hashing method accelerates sequence anchoring, a key computational step in many applications of next-generation sequencing technologies, over a large spectrum of read sizes -- from 25 to 1000 base pairs. Pash shows similar sensitivity to state-of-the-art alignment tools such as BLAT on longer reads and outperforms BLAT on very short reads, while achieving an order of magnitude speed improvement. Pash 2.0 overcomes a major limitation of previous implementations of Positional hashing, sensitivity to indels, by performing cross-diagonal collation of kmer matches. A future direction is to exploit multi-core hardware architectures by leveraging the low-level parallelism; another direction is to further optimize anchoring performance in the context of pipelines for comparative sequence assembly and other specific applications of next-generation sequencing.

Acknowledgments

We thank Andrew Jackson, Alan Harris, Yufeng Shen, and Ken Kalafus for their help.

References

1. Altschul, S.F., et al., *Basic local alignment search tool.* J Mol Biol, 1990. 215(3): p. 403-10.
2. Altschul, S.F., et al., *Gapped BLAST and PSI-BLAST: a new generation of protein database search programs.* Nucleic Acids Research, 1997. 25(17): p. 3389-402.
3. Kent, W.J., *BLAT--the BLAST-like alignment tool*, in *Genome Res*. 2002. p. 656-64.
4. Ning, Z., A.J. Cox, and J.C. Mullikin, *SSAHA: a fast search method for large DNA databases.* Genome Research, 2001. 11(10): p. 1725-9.
5. Ma, B., J. Tromp, and M. Li, *PatternHunter: faster and more sensitive homology search.* Bioinformatics, 2002. 18(3): p. 440-5.
6. Li, M., et al., *PatternHunter II: Highly Sensitive and Fast Homology Search.* Journal of Bioinformatics and Computational Biology, 2004. 2(3): p. 417-439.
7. Pearson, W.R. and D.J. Lipman, *Improved tools for biological sequence comparison.* Proc Natl Acad Sci U S A, 1988. 85(8): p. 2444-8.
8. Pearson, W.R., *Rapid and sensitive sequence comparison with FASTP and FASTA.* Methods Enzymol, 1990. 183: p. 63-98.
9. Kalafus, K.J., A.R. Jackson, and A. Milosavljevic, *Pash: Efficient Genome-Scale Sequence Anchoring by Positional Hashing.* Genome Research, 2004. 14: p. 672-678.
10. *WU-BLAST.* 2007.
11. Schwartz, S., et al., *Human-mouse alignments with BLASTZ.* Genome Res, 2003. 13(1): p. 103-7.
12. Harris, R.A., J. Rogers, and A. Milosavljevic, *Human-specific changes of genome structure detected by genomic triangulation.* Science, 2007. 316(5822): p. 235 237.
13. Fujiyama, A., et al., *Construction and analysis of a human-chimpanzee comparative clone map.* Science, 2002. 295(5552): p. 131-4.
14. Larkin, D.M., et al., *A Cattle-Human Comparative Map Built with Cattle BAC-Ends and Human Genome Sequence.* Genome Res, 2003. 13(8): p. 1966-72.
15. Poulsen, T.S. and H.E. Johnsen, *BAC end sequencing.* Methods Mol Biol, 2004. 255: p. 157-61.

POPULATION SEQUENCING USING SHORT READS: HIV AS A CASE STUDY

VLADIMIR JOJIC, TOMER HERTZ AND NEBOJSA JOJIC*

Microsoft Research, Redmond, WA 98052

**E-mail: jojic@microsoft.com*

Despite many drawbacks, traditional sequencing technologies have proven to be invaluable in modern medical research, even when the targeted genomes are highly variable. While it is often known in such cases that multiple slightly different sequences are present in the analyzed sample in concentrations that vary dramatically, the traditional techniques typically allow only the most dominant strain to be extracted from a single chromatogram. These limitations made some research directions rather difficult to pursue. For example, the analysis of HIV evolution (including the emergence of drug resistance) in a single patient is expected to benefit from a comprehensive catalog of the patient's HIV population. In this paper, we show how the new generation of sequencing technologies, based on high throughput of short reads, can be used to link site variants and reconstruct multiple full strains of the targeted gene, including those of low concentration in the sample. Our algorithm is based on a generative model of the sequencing process, and uses a tailored probabilistic inference and learning procedure to fit the model to the obtained reads.

Keywords: sequence assembly, population, HIV, epitome,rare variants, multiple strains, variant linkage

1. Introduction

Sequencing multiple different strains from a mixed sample in order to study sequence variation is often of great importance. For example, it is well known that even single mutations can sometimes lead to various diseases[1]. On the other hand, mutations in pathogen sequences such as the highly variable HIV[14] may lead to drug resistance[12,19]. At any given time, an HIV positive individual typically carries a large mixture of strains, each with a different relative frequency[21], and some over a hundred times less abundant than the dominant strains, and any one of them can become dominant if others are under greater drug pressure. The emergence of drug resistant HIV strains has lead to assembling a large list of associated single mutations[a]. However, new studies are showing that there are important linkage effects among some of these mutations[18] and that the linkage may be missed by current sequencing techniques[17].

When processing mixed samples by traditional methods[20], only a single strain can be sequenced in each sequencing attempt. Multiple DNA purifications may be costly and will usually provide accurate reconstruction only of several dominant strains. Picking the less abundant strains from the mixture is a harder problem. Recent computational approaches which infer a mixture of strains directly from the ambiguous raw chromatograms of mixed samples can deconvolve strains reliably only when their relative concentrations are higher than 20%, as the rarer variants get masked[6]. Note that unlike the problem of metagenome sequencing, where multiple species are simultaneously sequenced, the goal of multiple strain sequencing is to recover a mixture of different full sequence variants of the same species, which is complicated by the high similarity among them.

Recently, a number of alternative sequencing technologies have enabled high-throughput genome sequencing. For example, 454 sequencing[13] is based on an adaptation of the pyrosequencing procedure. Several studies have demonstrated its use for sequencing small microbial genomes, and even some larger scale genomes. One of the major advantages of pyrosequencing is that it has been shown to capture low frequency mutations. Tsibris et. al have shown that they can accurately detect low frequency mutations in the HIV env V3 loop[22]. A more recent work used pyrosequencing to detect over 50 minor variants in HIV-1 protease[2]. However, these technologies also have two important limitations. First, current sequencers can only read sequences of about 200 base pairs (and some even less). Second, sequencing errors, especially in homopolymeric regions, are high, making it potentially difficult to reconstruct multiple full sequences and estimate their frequencies. In this paper, we suggest a novel method for reconstructing the full strains from mixed samples utilizing technologies akin to 454. We formulate a statistical model of short reads and an inference algorithm which can be used to jointly reconstruct sequences from the reads and infer their frequencies. We validate our method on simulated 454 reads from the HIV sequences.

[a]see http://hivdb.stanford.edu/index.html

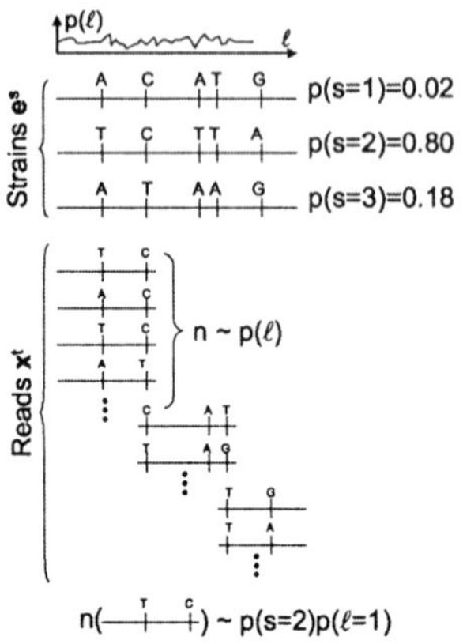

Figure 1. An illustration of population sequencing using short reads. In this toy example, three strains with five polymorphic sites are present in the sample. Short reads from various locations are taken. As the coverage depth depends on sequence content, the coverage depth will be proportional to the distribution $p(\ell)$ over the sequence location (the strains are assumed to differ little enough so that the depth of coverage of polymorphic variants of the same sequence patch are similar). The number of copies of a particular read (e.g., the TC variant shown at the bottom) depends both on the strain concentrations $p(s)$ and the depth distribution $p(\ell)$. See Section 2 for more details on notation and the full statistical model.

2. A statistical model of short sequence readouts from multiple related strains

In this section, we follow the known properties of high throughput, short read technologies, as well as the properties of populations of related sequences, e.g., a single patient's HIV population, to describe a hierarchical statistical process that leads to creation of a large number of short reads (Fig. 1). Such a generative modeling approach is natural in this case, as the process is indeed statistical and hierarchical. For example, the reads will be sampled from different strains depending on the strain concentrations in the sample, but the sampling process will include other hidden variables, such as the random insertions and deletions when the reads contain homopolymers. The statistical model will then define the optimization criterion in the form of the likelihood of the observed reads. Likelihood optimization ends up depending on two cues in the data to perform multi-strain assembly: a) different strain concentrations which lead to more frequently seen strains being responsible for more frequent reads, and b) quilting of overlapping reads to infer mutation linkage over long stretches of DNA.

We assume that the sample contains S strains $\mathbf{e}^s$ indexed by $s \in [1..S]$ with (unknown) relative concentrations $p(s)$. A single short read from the sequencer is a patch $\mathbf{x} = \{x_i\}_{i=1}^N$, with $N \approx 100$ and x_i denoting the $i-th$ nucleotide, taken from one of these strains starting from a random location ℓ. It has been shown that in 454 sequencing, a patch depth may be dependent on the patch content. We assume that different strains have highly related content in segments starting at the same location ℓ, and thus capture the expected relative concentrations of observed patches by a probability distribution $p(\ell)$, shared across the strains. This distribution will also be unknown and will be estimated from the data. Under these assumptions, a simple model of the short reads obtained by the new sequencing technologies such as 454 sequencing is described by the following sampling process:

- Sample strain s from the distribution $p(s)$
- Sample location ℓ from the distribution $p(\ell)$
- Set $x_i = e^s_{i+\ell-1}$, for $i \in [1..N]$

Here we assume that the strains $\mathbf{e}^s$ are defined as nucleotide sequences $\mathbf{e}^s = \{e^s_i\}$. However, since we will be interested in the inverse process – assembling the observed patches $\mathbf{x}^t$ into multiple strains, we make the definition of $\mathbf{e}$ softer in order to facilitate smoother inference of patch mapping in early phases of the assembly when the information necessary for this mapping is uncertain. In particular, as in our previous work concerning diversity modeling and vaccine immunogen assembly[7], we assume that each site e^s_i is a *distribution* over the letters from the alphabet (in this case the four nucleotides). Thus, we denote by $e^s_i(x_j)$ the probability of the nucleotide x_j under the distribution at coordinates (s, i) of the strain description $\mathbf{e}$. We have previously dubbed the models of this nature *epitomes* as they are a statistical model of patches contained in larger sequences. Our generative model of the patches $\mathbf{x}$ is therefore refined into:

- Sample strain s from the distribution $p(s)$
- Sample location ℓ from the distribution $p(\ell)$
- Sample x by sampling for each $i \in [1..N]$ the nucleotide x_i from the distribution $e^s_{i+\ell-1}(x)$

While the epitome distributions capture both the uncertainty about reconstructed strains and the point-wise sequencing errors, in order to model possible insertions and deletions in the patch, which are important because of the assumed strain alignment (shared ℓ), we also add another variable into the process which we call 'transformation' τ, describing the finite set of possible *minor* insertions or deletions. The insertions and deletions come from two sources: a)homopolymer issues in sequencing and b) insertions and deletions among strains. The first set of issues arise when a sequence of several nucleotides of the same kind, e.g., *AAAA* is present in the patch. In 454 sequencing, there is a chance that the number of sequenced letters in the obtained patch is not equal to the true number present in the sequence. As opposed to the indels among strains, which are usually multiples of three nucleotides to preserve translation into aminoacids, as well as consistent across the reads; the homopolymer indels are not limited in this way.

The transformation τ describes a mini alignment between the read and the epitome segment describing the appropriate strain s starting at a given location ℓ. We assume that the transformation τ will affect epitome segment just before the patch is generated by sampling from it. Thus, the statistical generative model that we assume for the rest of the paper consists of the following steps:

- Sample strain s from the distribution $p(s)$
- Sample location ℓ from the distribution $p(\ell)$
- Sample patch transformation τ from $p(\tau)$ and transform the epitome segment $\{\mathbf{e}_i^s\}_{i=\ell}^{\ell+N+\Delta}$, with Δ allowing all types of indels we want to model. This transformation provides the new set of distributions $e^s_{\tau(k)}$, where we use operator notation for τ to denote the mapping of locations.
- Sample $\mathbf{x}$ from $p(\mathbf{x}|s,\ell,\tau,\mathbf{e}) = \prod_i e^s_{\tau(i+\ell-1)}(x_i)$ by sampling for each $i \in [1..N]$ the nucleotide x_i from the distribution $e^s_{\tau(i+\ell-1)}(x)$

Each read $\mathbf{x}^t$ has a triplet of hidden variables s^t, ℓ^t, τ^t describing its unknown mapping to the catalog of probabilistic strains (epitome). In addition to hidden variables, the model has a number of parameters, including relative concentrations of the strains $p(s)$, the variable depth of coverage for different locations in the genome $p(\ell)$, and the uncertainty over the nucleotide x present at any given site i in the strain s, as captured by the distribution $e_i^s(x)$ in the epitome $\mathbf{e}$ describing the S strains. If the model is fit to the data well, the uncertainty in the epitome distributions e_i^s should contract to reflect the measurement noise (around 1%). But, if an iterative algorithm (e.g., EM) is used to jointly estimate the mapping of all reads $\mathbf{x}^t$ and the (uncertain) strains $\mathbf{e}^s$, then the uncertainty in these distribution also serves to smooth out the learning process and avoid hard decisions that are known to lead to local minima. Thus, these distributions will be uncertain early in such learning procedures and contract as the mappings become more and more consistent. In the end, each of the distributions $\mathbf{e}_i^s$ should focus most of the mass on a single letter, and the epitome $\mathbf{e}$ will simply become a catalog of the top S strains present in the sampled population. If more than S strains are present, this may be reflected by polymorphism in some of the distributions $\mathbf{e}_i^s$.

3. Strain reconstruction as probabilistic inference and learning

We now derive a simple inference algorithm consisting of the following intuitive steps:

- Initialize distributions e_i^s, strain concentrations $p(s)$ and coverage depth $p(\ell)$. More on initialization in the next section.
- Map all reads to $\mathbf{e}$ by finding the best strain s^t, location in the strain ℓ^t and the mini alignment τ that considers indels.
- Re-estimate model parameters by (appropriately) counting how many reads map to different locations ℓ and different strains s. Also count how many times each nucleotide ended up mapped to each location (s, i) in the strain reconstruction $\mathbf{e}$ and update the distributions e_i^s to reflect the relative counts.
- Iterate until convergence.

We can show that this meta algorithm corresponds to an expectation-maximization algorithm that optimizes the likelihood of obtaining the given set of reads $\mathbf{x}^t$ from the statistical generative model described in the previous section. The log likelihood of observing a given set of patches (reads) is

$$\mathcal{L} = \sum_t \log p(\mathbf{x}^t) = \sum_t \log \sum_{s^t, \ell^t, \tau^t} p(s^t)p(\ell^t)p(\tau^t)p(\mathbf{x}^t|s^t, \ell^t, \tau^t). \quad (1)$$

We note that $\mathcal{L}$ is a function of model parameters $\mathbf{e}$, $p(s)$, $p(\ell)$ and $p(\tau)$, and our goal is to maximize this likelihood wrt $\mathbf{e}$ as well as $p(s)$, as our output should be the catalog of strains, or epitome $\mathbf{e}$, present with the component concentrations $p(s)$. It is also beneficial to to maximize the log likelihood wrt other parameters, i.e. estimate the varying coverage depth for different parts of the strains as well as distribution over typical indels. Not only do these parameters may be of interest in their own right, but an appropriate fitting of these parameters increases the accuracy of the estimates of strains and their frequencies.

To express the expectation-maximization (EM)[5] algorithm for this purpose, we introduce the auxiliary distributions $q(s^t, \ell^t)$ and $q(\tau^t|s^t, \ell^t)$ that describe the posterior distribution over the hidden variables for each read $\mathbf{x}^t$, and use Jensen's inequality to bound the log likelihood:

$$\mathcal{L}(\mathbf{e}) = \sum_t \log \sum_{s^t, \ell^t, \tau^t} q(s^t, \ell^t)q(\tau^t|s^t, \ell^t)\frac{p(s^t)p(\ell^t)p(\tau^t)p(\mathbf{x}^t|s^t, \ell^t, \tau^t, \mathbf{e})}{q(s^t, \ell^t)q(\tau^t|s^t, \ell^t)},$$

$$\geq \sum_t \sum_{s^t, \ell^t, \tau^t} q(s^t, \ell^t)q(\tau^t|s^t, \ell^t) \log \frac{p(s^t)p(\ell^t)p(\tau^t)p(\mathbf{x}^t|s^t, \ell^t, \tau^t, \mathbf{e})}{q(s^t, \ell^t)q(\tau^t|s^t, \ell^t)}$$

The bound is tight when the q distribution captures the true posterior distribution $p(s^t, \ell^t, \tau^t|\mathbf{x}^t)$[16], thus the reference to q as a posterior distribution. By optimizing the bound with respect to the q distribution parameters (under the constraint that the appropriate probabilities add up to one), we can derive the E step as

$$q(\tau^t|s^t,\ell^t) \propto p(\tau^t)p(\mathbf{x}^t|s^t,\ell^t,\tau^t,\mathbf{e}) \tag{2}$$

$$q(s^t,\ell^t) \propto p(s^t)p(\ell^t)e^{\sum_{\tau^t} q(\tau^t|s^t,\ell^t)\log p(\tau^t)p(\mathbf{x}^t|s^t,\ell^t,\tau^t\mathbf{e})} \tag{3}$$

where both the computation of $q(\tau^t|s^t,\ell^t)$ and the summation over τ in the second equation are performed efficiently by dynamic programming. These operations reduce to well known HMM alignment of two sequences (in this case, one probabilistic sequence, $\{e_i^s\}_{i=\ell}^{\ell+N+\Delta}$, and one deterministic sequence, $\mathbf{x}^t$), because they estimate optimal alignment (and the distribution over alignments, and an expectation under it), in the presence of indels. In our experiments, we make an additional assumption that $q(\tau^t|s^t,\ell^t)$ puts all probability mass on one, best, alignment.

The bound simplifies the estimation of model parameters under the assumption that the q distribution is fixed. For example, the estimate of the (relative) strain concentrations and the spatially varying (relative) depth coverage is performed by

$$p(s) = \frac{\sum_t \sum_{\ell^t} q(s^t = s,\ell^t)p(\ell^t)}{\sum_t \sum_{\ell^t} \sum_{s^t} q(s^t,\ell^t)p(\ell^t)} \quad p(\ell) = \frac{\sum_t \sum_{s^t} q(s^t,\ell^t = \ell)p(s^t)}{\sum_t \sum_{\ell^t} \sum_{s^t} q(s^t,\ell^t)p(s^t)} \tag{4}$$

The estimate for the epitome probability distributions describing (with uncertainty) the strains present in the population is

$$e_i^s(x) = \frac{\sum_t \sum_{\ell^t,\tau,j|\tau(j+\ell-1)=i}[x_j = x]q(s^t = s,\ell^t)q(\tau|s^t = s,\ell^t)}{\sum_t \sum_{\ell^t,\tau,j|\tau(j+\ell-1)=i} q(s^t = s,\ell^t)q(\tau|s^t = s,\ell^t)}, \tag{5}$$

where [] denotes the indicator function. This equation simply counts how many times each nucleotide mapped to site s,i, using probabilistic counts expressed in q expectations under the possible patch alignments described by τ are again computed efficiently using dynamic programming, or, as in our experiments, they can be simplified by using the most likely alignment.

EM algorithm for our model should iterate equations (2- 5). These equations are a more precise version of the algorithm description from the beginning of the section. The iterative nature of the algorithm allows a refinement in one set of parameters to aid in refining other parameters. For example, iterating the two equations in (4) lead to estimates of strain frequency and variability in read coverage that are compatible with each other - the first equation takes into account the fact that some regions of the genome are under represented when assigning a frequency to strains based on the read counts; and the second equation discounts the effect of strain frequency on read counts in order to compute the read content dependent (approximated as genome position dependent) variability in coverage. On the other hand, the estimate of the epitome (i.e., the catalog of strains) and the strain frequency estimates are coupled through the posterior distribution q - a change in either one of these model parameters will affect the posterior distribution (2) which assigns reads to different strains, and this will in turn affect these same model parameters in the next iteration.

4. Computational cost and local minima issues

A good boost to the algorithm's performance is achieved by its hierarchical application. The epitome **e** is best initialized by an epitome **e** consisting of a smaller number learned in a previous run of the same algorithm, e.g., by repeating each of the original S strains K times and then adding small perturbations to form an initial epitome with SK strains. If the first number of strains S was insufficient, this new initial catalog of strains contains rather uncertain sites wherever the population is polymorphic, but the alignments of variables ℓ from the previous run for all patches are likely to stay the same, so that part of each distribution $q(s, \ell)$ is transferred from the previous run, and does not change much, thus making it possible to avoid search over this variable and reduce complexity. An extreme application of this recipe, that according to our experiments seems to suit HIV population sequencing, is to run the algorithm first with $S = 1$, which essentially reduces to consensus strain assembly in noisy conditions, and then increase the catalog **e** to the desired size. For a further speed up, a known consensus sequence (or a profile) can be used to initialize all strains in the epitome.

The simple inference technique described above still suffers from two setbacks. One problem is computational complexity. The number of reads can be very large, although these reads may be highly redundant at least for all practical purposes in the early iterations of the algorithm. Another, more subtle problem is the weakness of the concentration cues in inference using our model, which may cause local maxima problems. Our generative model mirrors the true data generation process closely, and thus the correct concentrations in conjunction with properly inferred strains correspond to the best likelihood. But if pure EM learning is applied, the concentration cue can be too weak to avoid local minima in **e**. Fortunately, a simple technique can be used to address both of these two issues. Reads are clustered using agglomerative clustering and the initial q distributions are estimated by mapping the cluster representations rather than all reads.

The ℓ mapping is considered reliable and fixed after that point as the described initialization makes all strains similar enough to the true solution for the purposes of ℓ mapping (but not for inferring strain index s). In the first few iterations after that, clusters are mapped to different strains, but the epitome distributions are not considered in this mapping - the assumption is made that the final set of parameters will map clusters so that all strains in the epitome are used. Each cluster mapping is iterated with updates of concentrations of $p(s)$. This results in loosely assigning read clusters with similar frequencies to the same strain. After 2-3 such iterations, epitome distributions are inferred based on the resulting q distribution, and then the full EM algorithm, over all patches, is continued. This is necessary as the agglomerative clusters may not be sufficient to infer precisely the content of all sites until individual reads are considered. It should be noted that due to the high number and overlap of reads, it is in principle possible to have a substantially lower reconstruction error than the measurement error(1%).

In our implementation, the computational cost is quadratic in the number of patches associated with particular offset in the strains, due to the agglomerative clustering step. The cost of an EM iteration is proportional to the product of the number of patches (reads) and the total length of the epitome (strain catalog).

5. Experimental validation

We assessed performance of our method on sequence data for nef and env regions of HIV. Starting with these sequences, we simulated 454 reads as 80-120 nucleotide long patches $\mathbf{x}^t$ generated by the statistical generative model described in Section 2. The generated reads, without the model parameters or results of the intermediate steps, were then analyzed using the inference technique in Section 3 to reconstruct the hidden variables, such as read-to-genome alignments ℓ^t and read-to-strain assignments s^t, and estimate the model parameters, most importantly the epitome, or strain catalog, $\mathbf{e}$, and the strain frequencies $p(s)$. These were then compared to the ground truth.

The overall error rate in 454 reads is estimated at 0.6%. For our generated reads, we set substitutions errors of 1.0%, and for homopolymers (of length at least 2 nucleotides) we set the rate of insertion at 2%, and deletion at 0.5%. The read selection probability – the probability of obtaining a read from a particular offset from a particular strain – is set to be proportional to the product of depth of coverage $p(\ell)$ at the offset ℓ and the frequency of the strain $p(s)$ (see also Fig. 1). The depth of coverage is randomly drawn from a preset range of values (and, as other parameters, it was not later provided to the inference engine, which had to reconstruct it to infer correct strain frequencies). We assume that overlap between reads is no less than 50 nucleotides.

Table 1. The fraction of nucleotides reconstructed correctly in the least frequent strain as a function of that strain's frequency and the minimum number of reads.

Min. reads \ Frequency	0.1%	0.5%	1%	2%
10	40.93%	92.59%	100%	100%
20	62.25%	95.10%	100%	100%
30	100%	100%	100%	100%

In order to assess ability of the method to reconstruct low frequency strains we first created a dataset of 10 nef strains[14]. The nef region is approximately 621 nucleotides long. We randomly picked one strain as the low frequency strain. For this lowest frequency we considered four possibilities: 0.1%, 0.5%, 1%, and 2%. For the other 9 sequences, we randomly chose frequencies between 2% and 100% and then normalized them so that the sum of frequencies is 100%, i.e., $\sum p(s) = 1$. Then, the short reads were generated from the mixture as described above. Though the depth of coverage $p(\ell)$ was randomly assigned across the region, we ensured, by scaling the total number of reads, that a minimum number of reads is guaranteed for each genome location. We experimented with three possibilities for this minimum number of reads: 10, 20, and 30. The Table 1 illustrates the impact of the minumum number of reads on our ability to reconstruct sequences with small concentrations. Even in case of the minor strain frequency of just 0.1% we were able to reconstruct all ten sequences as long as we had suitable number of reads available. Furthermore, all strain frequencies were recovered with negligible error.

We also assessed the impact of the density of viral mutations on our ability to reconstruct the full strains. We used 10 HIV env strains from MACS longitudinal study[9]. All sequences originated from the same patient and were obtained from samples collected at 10 different patient visits. The visits occurred approximately every 6 months. Whereas variable strain frequencies may help us disambiguate between frequent and infrequent strains, in case of comparable frequencies, it is the mutations which occur in the overlap between reads which enable linking of site variants and the reconstruction of full sequences. In order to assess the number and proximity of mutations in env, we analyzed sequences collected from a single patient over a number of visits spanning 8 years. These sequences contained 280 nucleotides of gp120, followed by V3 loop, followed by 330 nucleotides of gp41, total of 774 nucleotides. The entropy of these sequences at each site is shown in Figure 2. The positions with high entropy are spaced almost

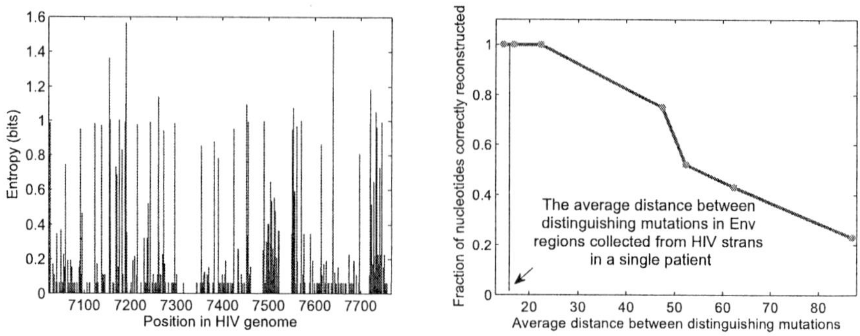

Figure 2. Left: Site entropy for an Env region, estimated over 137 sequences originated from the same patient. Note that the positions with high entropy are spaced almost uniformly throughout this region. The average distance between positions with entropy greater than 0.5 is 14.67. Right: From this dataset we selected 8 different sets of 10 sequences, each with different density of distinguishing mutable positions. We evaluated fraction of nucleotides correctly reconstructed for various densities of distinguishing mutations, represented as average distance between the distinguishing mutable positions. The vertical line traces the average distance between mutable positions in Env.

uniformly throughout this region, with separation between significantly mutable position (entropy greater than 0.5) reaching up to 57 nucleotides.

The difficulty of disambiguating strains of comparable frequency is dependent on the maximal distance between pairs of adjacent mutations. In regions where two nearest mutable positions are separated by a conserved region longer than the read length, there will be no reads spanning both of those mutable positions, and we may not be able to tell whether mutations at the two sites are occurring in the same strain or not. In these cases, we should assume that linking of mutations is correct only in parts up to and after the conserved region, but not across this region, unless the strain frequencies are sufficiently different to allow our algorithm to correctly match the separated pieces based on the frequency of site variants. Therefore, the density of the distinguishing mutable positions is a measure of difficulty of disambiguating strains of comparable frequency.

We varied the average distance between adjacent mutations in a controlled manner. More specifically, we created 8 sets of 10 Env sequence mixtures, with average distances ranging from 10-80 bases apart, and computed the percentage of correct reconstructions for each set. Figure 2 shows reconstruction accuracy as a function of mutation density, defined as an average distance between the distinguishing mutable positions.

6. Conclusion

We introduced a population sequencing method which recovers full sequences and sequence frequencies. The method leverages inherent differences in the strain frequencies, as well as the sequence differences across the strains in order to achieve perfect reconstruction under a noise model mirroring the measurement error of the 454 sequencing method. We have shown that our method can reconstruct sequences with as small a frequency as 0.01%. While our experiments have been performed on simulated (but realistic) mixes of short segments of HIV, there is no technical reason why the technique would not work for longer genomes (e.g., entire HIV sequences or longer viral sequences). For most of HIV, the density of mutable positions is so high, that the technique should work with significantly shorter reads than 200. For more information, visit www.research.microsoft.com/~jojic/popsequencing.html.

References

1. E. J. Baxter, et al. *Lancet*, 365(9464):1054–1061, Mar 2005.
2. C. Wang, et al. *Genome Res*, Jun 2007.
3. J. M. Coffin. *Science (New York, N.Y.)*, 267(5197).
4. DA. Lehman and C. Farquhar. *Rev Med Virol*, Jun 2007.
5. A. P. Dempster, N. M. Laird, et al. *Journal of the Royal Statistical Society. Series B (Methodological)*, 39(1):1–38, 1977.
6. N. Jojic. Population sequencing from chromatogram data. In *ISMB, PLOS track*. 2006.
7. N. Jojic, et al. In Y. Weiss, B. Schölkopf, et al., eds., *Advances in Neural Information Processing Systems 18*, pp. 587–594. MIT Press, Cambridge, MA, 2006.
8. D. Jones, et al. *AIDS Res Hum Retroviruses*, 21(4):319–324, Apr 2005.
9. R. A. Kaslow, et al. *Am J Epidemiol*, 126(2):310–318, Aug 1987.
10. P. Kellam and B. A. Larder. *J Virol*, 69(2):669–674, Feb 1995.
11. B. Li, et al. *J Virol*, 81(1):193–201, Jan 2007.
12. S. Lockman, et al. *N Engl J Med*, 356(2):135–147, Jan 2007.
13. M. Margulies, et al. *Nature*, 437(7057):376–380, Sep 2005.
14. C. B. Moore, et al. *Science*, 296(5572):1439–1443, May 2002.
15. S. M. Mueller, et al. *J Virol*, 81(6):2887–2898, Mar 2007.
16. R. Neal and G. Hinton. In M. I. Jordan, ed., *Learning in Graphical Models*. Kluwer, 1998.
17. S. Palmer, et al. *J Clin Microbiol*, 43(1):406–413, Jan 2005.
18. S.-Y. Rhee, et al. *PLoS Comput Biol*, 3(5):e87, May 2007.
19. T. Ridky and J. Leis. *J Biol Chem*, 270(50):29621–29623, Dec 1995.
20. F. Sanger, et al. *Biotechnology*, 24:104–108, 1992.
21. T.-K. Seo, et al. *Genetics*, 160(4):1283–1293, Apr 2002.
22. A. Tsibris, et al. In *Antivir Ther.*, vol. 11:S74 (abstract no. 66). 2006.

ANALYSIS OF LARGE-SCALE SEQUENCING OF SMALL RNAS

A. J. OLSON, J. BRENNECKE, A.A. ARAVIN, G.J. HANNON AND R. SACHIDANANDAM

Cold Spring Harbor Laboratory,
1 Bungtown Road,
Cold Spring Harbor,NY 11724, USA
E-mail: ravi.cshl.work@gmail.com

The advent of large-scale sequencing has opened up new areas of research, such as the study of Piwi-interacting small RNAs (piRNAs). piRNAs are longer than miRNAs, close to 30 nucleotides in length, involved in various functions, such as the suppression of transposons in germline [3,4,5]. Since a large number of them (many tens of thousands) are generated from a wide range of positions in the genome, large-scale sequencing is the only way to study them. The key to understanding their genesis and biological roles is efficient analysis, which is complicated by the large volumes of sequence data. Taking account of the underlying biology is also important. We describe here novel analyses techniques and tools applied to small RNAs from germ cells in D. melanogaster, that allowed us to infer mechanism and biological function.

1. Introduction

Relatively inexpensive large-scale sequencing has now become readily accessible to the masses, through the efforts of companies such as 454 and Solexa. One drawback of sequences derived from such technologies are the short read lengths, approximately 30 for Solexa and more than a 100 for 454. This does not pose a problem when the sequences can be easily identified in the genome. In fact, for small RNAs, the length is close to their size and hence such sequencing techniques are perfect for their study.

Small RNAs have been discovered to be associated with the Argonaute family of proteins. The Argonaute family is a complex one, and the nomenclature makes it even more confusing[1]. The family is further divided into two sub-classes, Argonaute and Piwi. The Argonaute sub-class is involved in the siRNA and miRNA pathways, while the Piwi sub-class is involved in piRNA processing. piRNAs tend to be longer than miRNAs and are more diverse. Their genesis is not well understood, but our analysis of piRNAs shows that they arise from clusters and are frequently repeat associated, which suggests their role in transposon silencing. Indeed, deletion of certain members of the Piwi sub-class leads to the activation of transposons.

In order to characterize the datasets (we have analyzed three such datasets [3,4,5]), as well as understand their biological role, we developed several analysis techniques and tools. Taken individually, the techniques and tools are not novel, but the combination and sequence of steps makes them a novel contribution which will be of use to others in the field. Several other large-scale sequencing projects have been published, but none involves the kind of analysis described here [2].

The aim of the experiments described here was to delineate the role of Aubergine and Ago3 in the germline in silencing transposons. We wanted to understand the transposons that are under control of this mechanism and which one of these proteins is involved in targetting the transposons and which one is involved in the maintenance of this silencing. The analysis that we describe here was arrived at by trial and error. We describe the methodology as well as our tools below.

2. Case Study:Analysis of Aubergine and Argonaute-3 associated small RNAs from *D. melanogaster*.

Small RNAs associated with Argonaute-3 and Aubergine in *D. melanogaster* were isolated by immunoprecipitation (IP), using antibodies against the proteins[4].

The analysis involves sequence processing, mapping and warehousing. In addition, it is very important to allow browsing of the data through user-friendly tools, since the patterns we look for are not readily apparent and the pattern space that has to be searched is immense. We describe our suite of web-based tools for this purpose. The exact implementation of these tools it not very important, as they are standard techniques, but the analyses allowed by the web-based tools is very relevant, since they helped us gain understanding of our datasets.

We first describe the sequence processing and then the tools. We also describe the results we obtained at each step.

2.1. *Sequence Processing*

The processing of the sequences involves clipping the sequences, mapping them to the genome, using genome annotations to identify the origin of the small RNAs and warehousing the data. We describe the steps below.

2.1.1. *Clipping*

Adaptors are ligated to the small RNA sequences for amplification and sequencing. It is essential that the parts of the adaptor sequences that are sequenced get identified and clipped. This can involve either exact matching (if the reads are short and high quality) or inexact matching, when the reads are longer and the quality might have degraded towards the end. We use a dynamic programming algorithm that scores in a position dependent manner, allowing for more relaxed matching towards the end and a stricter match towards the beginning of the sequence. This allows cleaning up sequences that do not have any inserts, and also being relatively careful about not removing sequences arbitrarily from the end. The distribution of sizes after clipping suggests the nature of the dataset (if the peak of the distribution is around 22nt, it indicates a library biased towards miRNAs while a peak closer to 30 indicates piRNAs).

2.1.2. *Warehousing the data*

The sequences from the experiment are collapsed to create a unique, non-redundant set, and the multiplicities (the number of times each fragment is sequenced in an experiment) are tracked. We use MySQL's relational database for the storage.

2.1.3. *Mapping to the genome*

Mapping of the small RNAs is an relatively easy problem, compared to mapping mRNAs, since gaps are not expected. In addition, due to the large number of sequences, the ones that do not map exactly to the genome can be ignored. We used a suffix-array based approach to find matches. This is essential in speeding up the processing of the small RNAs and highlights the importance of proper clipping.

Some small RNAs map thousands of times to the genome, while about 10% of the small RNAs map to a unique location on the genome. The unique mappers allow idenfication of the clusters which are the main sources of the small RNAs.

2.1.4. *Annotating the small RNAs*

The annotations of the underlying genome are used to annnotate the small RNAs. The annotation categories are repeats, non-coding RNAs (tRNAs, snoRNAs, miRNAs etc.) and coding mRNAs, both introns and exons. It is essential to get a good reference set of annotations or generate one from curated datasets especially for the non-coding RNAs. Upto 10 mappings of each piRNA are considered for annotation, and a majority rule is used to pick the final annotation. In addition, in case of conflicts, a hierarchy, starting with non-coding RNAs, then repeats, followed by exons and finally introns, is used to pick a unique annotation for the piRNA. The orientation of the piRNA with respect to the underlying annotation is also identified.

Result: The small RNAs in this experiment are predominantly repeat-associated. The Aubergine associated ones are mainly anti-sense to the repeats while the Argonaute-3 associated ones are sense to the repeats.

2.2. *Web-enabled tools*

We built a set of web-based tools to allow exploring the dataset. We describe the function of the tools and the conclusion reached with each tool, but not the exact implementation, since the functionality is important in understanding the nature of the piRNAs while the implementation involves fairly standard techniques. The front-end, which is the starting point of the analysis, is shown in figure 1. The front-end allows filtering the small RNAs by various criteria such as, annotation, multiplicity (number of times the sequence was sampled in the experiment), number of mappings on the genome, and location on the genome. After the filtering, the selected small RNAs can be analyzed for distribution on the genome (by using a genomic viewer that is built into the tool), for the distribution of nucleotides at various positions (by using a tool to generate weight matrices for collections of sequences), for the density distribution on the genome (by specifying a window and step size for the sliding window to a graphing tool built into the tool), and for the correlations between positions on the genome for two sets of small RNAs in a graphical format.

2.3. *Genome View*

We built a viewer to view annotations along with small RNA map positions to allow browsing the genome (Figure 2). The viewer is based on the light-weight genome viewer (lwgv [7]).

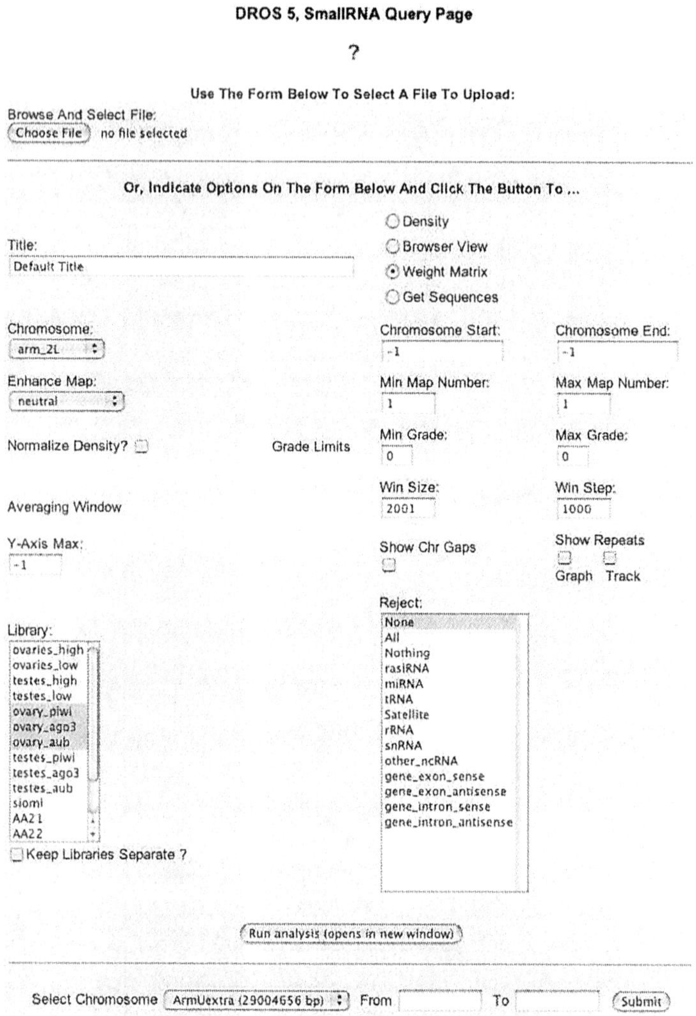

Figure 1. Front end of the tool to study small RNAs. This allows the selection of small RNAs by various filters. The filters are annotations (repeats, miRNAs etc.), number of mappings to the genome, the experimental source and by chromosome if necessary. From the selected small RNAs, (i) weight matrices (figure 3), (ii) graphical representation of the density of small RNAs in regions of the genome (figure 5), (iii) a browser view in the form of tracks (figure 2), and (iv) Position correlations between datasets (figure 4) can be generated. Each of these plays a crucial role in the analysis, as explained in the text.

Result Viewing regions of the genome with annotations of repeats and small RNAs confirms the association of the small RNAs with repeats.

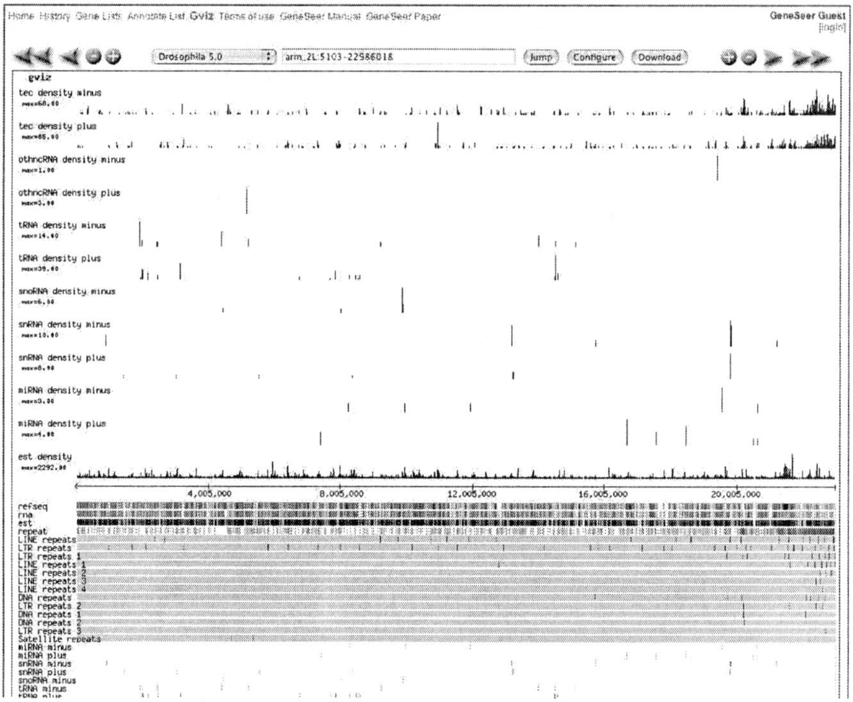

Figure 2. A view of a genomic region showing various features along with the small RNAs in our browser, based on lwgv.

2.4. *Nucleotide bias*

A standard method of characterizing collections of small sequences is to study the nucleotide bias as a function of position. Our tool generates colored images showing the frequencies of the nucleotides as rectangles, with the height of each rectangle proportional to the frequency (shown in black and white in figure 3).

Result: The Aubergine-associated small RNAs show a T-bias at position 1, which is similar to the one seen in other piRNA sets, while the Ago3-associated ones show a bias for an A at position 10. This suggests the following mechanism. Aubergine uses the small RNAs to target and cleave transposons. The cleavage occurs at position 10, which means there is an A at position 10 of the cleaved sequence from the transposon. The sequence from the transposon gets loaded into Ago3, through an unknown process, which is probably used to target the primary transcript that generates the small RNAs, setting up an amplification cycle, which explains the abundance of these small RNAs[4,6].

2.5. *Position Correlations*

The results from the nucleotide bias studies suggest that correlations between positions of small RNAs on the genome, from the two sets should reveal the connections between the two sets. The correlation between small RNAs oriented along the plus strand from set a and small RNAs oriented along the minus strand in set b at a distance Δ, $\mathrm{corr}_{ab}^{+-}(\Delta)$ is defined as

$$\mathrm{corr}_{ab}^{+-}(\Delta) = \Sigma_i \mathrm{mult}_a^{+}(x_i)\mathrm{mult}_b^{-}(x_i + \Delta) \tag{1}$$

where $\mathrm{mult}_a^{+}(x_i)$ is the multiplicity (number of times the sequence was sampled) of the sequence that maps along the plus strand of the genome to position x_i in the set a. The lengths of the small RNAs are disregarded in this analysis, only their start position is considered. The small RNAs that map a large number of times (more that 20) are excluded from this analysis. Alternatively, the multiplicity can be divided by the mapping number, so that the ones that map to multiple locations do not swamp the calculation. Either way, the result is similar to the the graph shown in figure 4.

Result: The correlation plot confirms that the small RNAs from the two set are offset from each other by 10 nucleotides and on opposite strands, further confirming the picture of the Aubergine-associated small RNAs targetting the transposons for cleavage.

2.6. *Clusters on the genome*

Density plots on the genome will show the distribution of small RNAs and highlight clusters if there are any. We use only the uniquely mapped small RNAs for this analysis. The small RNAs were binned into windows of 5Kb, which were slid over the genome in steps of 1 Kb. From the graphs of the binned distributions (figure 5) it is obvious these small RNAs arise from clusters in the genome. Computationally, the cluster boundaries are identified by the windows where the number of small RNAs is less than 5. The clusters are robust, not sensitive to the exact details of the criteria.

Result: One of the clusters on arm_X of the drosophila genome is the flamenco locus, which has been known to silence the transposons gypsy, Idefix and ZAM [4,6]. The small RNAs from this locus are responsible for the silencing, and this analysis helped understand the role of the flamenco locus in silencing these transposons.

3. Conclusions

The analyses outlined here works well in other small RNA analysis such as the study of Mili-associated small RNAs in mammals[5]. Mili is a protein belonging to the Piwi sub-class. The first steps in our analysis should be relevant to any type of large-scale sequencing project, irrespective of the source, as long as it is derived from an organism whose genome is sequenced. The correlation and cluster analyses makes sense in this context but might not be relevant in other experiments.

Further developments in the analyses will be driven by the kinds of biology that will be probed using large-scale sequencing. It is the underlying biology that will determine how the sequences will be analysed.

Acknowledgments

The authors acknowledge the help of Ted Roeder and Ankit Patel with various aspects of the front end for the web-based software and the anonymous reviewers for suggesting numerous improvements to the manuscript.

References

1. Carmell MA, Xuan Z, Zhang MQ, Hannon GJ. *Genes Dev.* 16(21):2733-42. (2002)
2. Ruby JG, Jan C, Player C, Axtell MJ, Lee W, Nusbaum C, Ge H, Bartel DP. *Cell.* 127(6):1193-207.(2006)
3. Girard A, Sachidanandam R, Hannon GJ and Carmell MA. *Nature* 442(7099):199-202. Jul 13; (2006).
4. Brennecke J, Aravin AA, Stark A, Dus M, Kellis M, Sachidanandam R and Hannon GJ. *Cell,* 128(6):1089-103 Mar 7 (2007). in Drosophila.
5. Alexei A. Aravin, Ravi Sachidanandam , Angelique Girard , Katalin Fejes-Toth, Gregory J. Hannon. *Science,*316(5825):744-7 May 4 (2007).
6. Phillip D. Zamore *Nature* **Vol 446** (7138):864-5 Apr 19,(2007)
7. J.J. Faith, A.J. Olson, T. S. Gardner and R. Sachidanandam. *BMC Bioinformatics* **8:344** Sept 18 (2007) doi:10.1186/1471-2105-8-344 Available for download from http://lwgv.sourceforge.net.

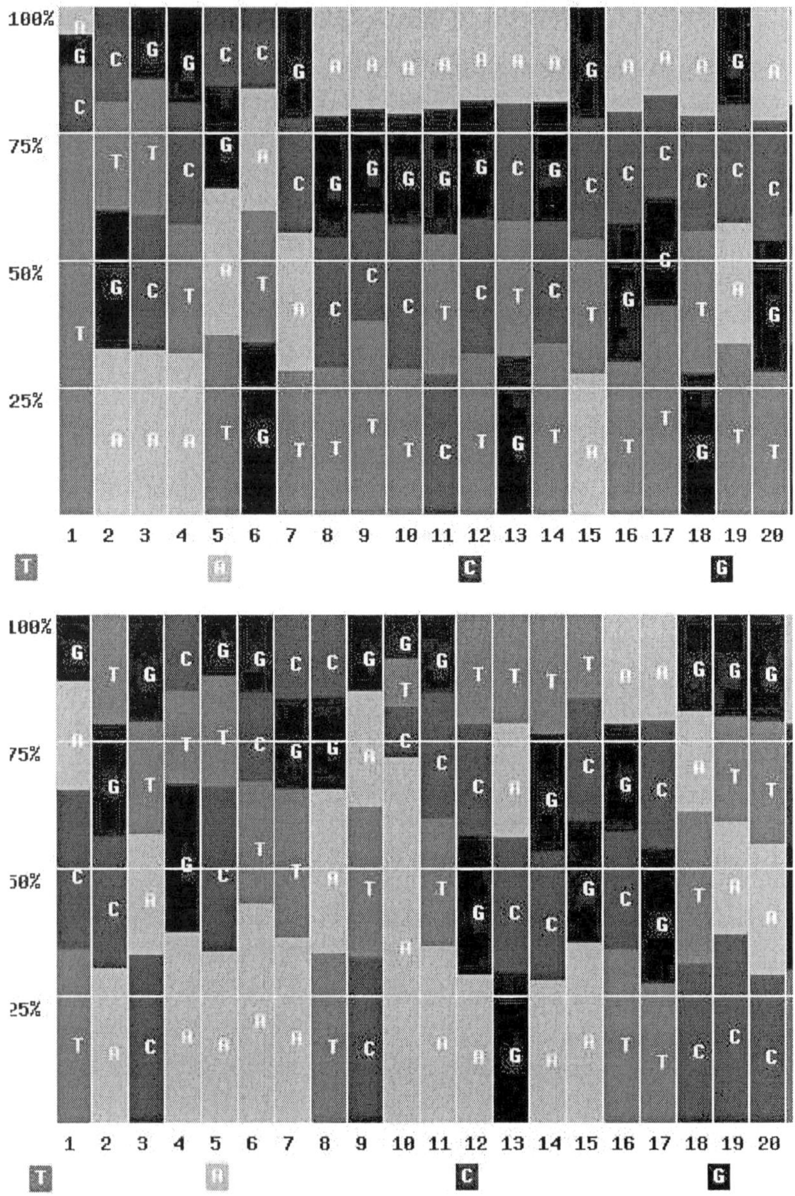

Figure 3. The top figure shows the distribution of nucleotides at various positions on the small RNAs from the Aubergine-associated set, while the bottom one depicts the distributions for the Ago3-associated set. There is a clear T bias at position 1 in the first set, while a clear A bias exists at position 10, all other positions are unremarkable. The actual figure is in color, but is shown here in grayscale.

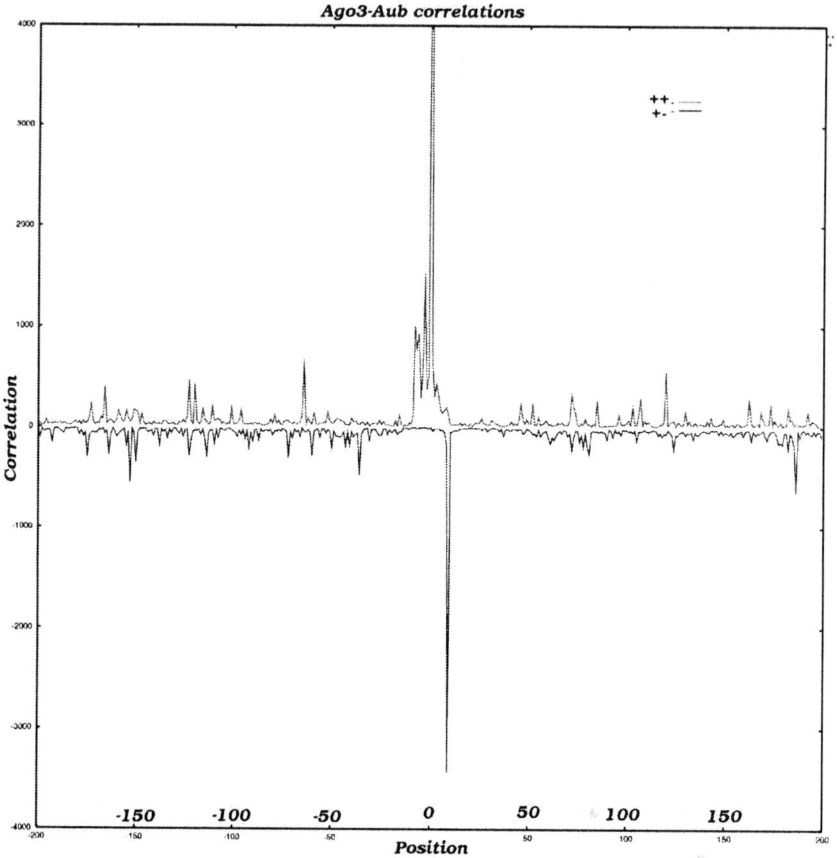

Figure 4. The correlation of the map positions of small RNAs from the Aubergine and Ago3 associated sets, calculated as discussed in Eq. (1), shows a strong peak at $\Delta = 9$ for the correlation between small RNAs from the two sets on opposite strands (+-, the plot below the x-axis), which corresponds to the 10-nt offset. The peak at zero in the ++ (the plot above the x-axis) is indicative of the origin of the small RNAs from clusters in the genome (described in figure 5).

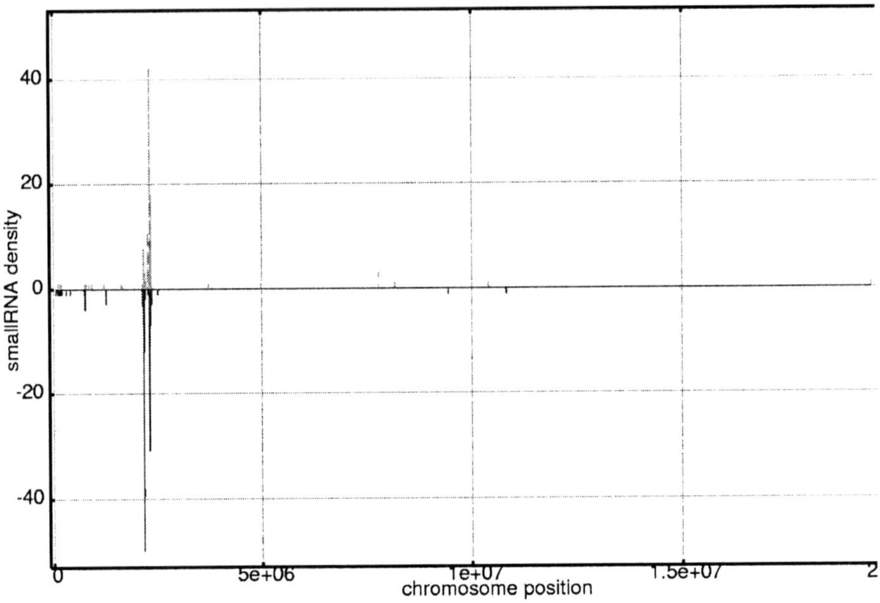

Figure 5. Density distribution of Aubergine-associated small RNAs on arm_2R. The plot above the x-axis is for the small RNAs that map to the plus strand, while the plot below the x-axis is for the small RNAs that map to the minus strand. Only small RNAs that map fewer than 5 times to the genome are considered for this plot.

KNOWLEDGE-DRIVEN ANALYSIS AND DATA INTEGRATION FOR HIGH-THROUGHPUT BIOLOGICAL DATA

M. F. OCHS*

The Sidney Kimmel Comprehensive Cancer Center, Johns Hopkins University, Baltimore, MD 21205, USA
** E-mail: mfo@jhu.edu*
http://www.cancerbiostats.onc.jhmi.edu/ochs.cfm

J. QUACKENBUSH*

Dana Farber Cancer Institute, Harvard Medical School, Boston, MA , USA
** E-mail: johnq@jimmy.harvard.edu*
http://www.hsph.harvard.edu/faculty/john-quackenbush/

R. DAVULURI*

Comprehensive Cancer Center, Ohio State University, Columbus OH 43210, USA
E-mail: Ramana.Davuluri@osumc.edu
http://www.cancergenetics.med.ohio-state.edu/2732.cfm

Keywords: Bayesian analysis, Statistical models, Statistical data analysis, Controlled vocabulary

Introduction

New high-throughput methods have led to immense data sets in biological research. However, these data sets create significant problems for data analysis. For instance, the identification of interacting genetic variants placing individuals at risk for or providing protection from the development of polygenic diseases requires identification of sets of interacting genes. The standard statistical approaches were developed for cases of many samples and few loci of interest, and they cannot achieve power in the face of the enormous growth in our knowledge of genomics. Simple calculations show that as the number of typed loci and the number of potential interactions between genes increase, it will become impossible to design a study with sufficient statistical power using present analysis methods (e.g., with 1000 loci and 4-gene interactions there are more than 40 billion potential combinations). Similar problems exist for both microarray and proteomic data, as well as for any other high-throughput data type where a comparison is made to biological samples, which tend to be limited in number.

One potentially fruitful approach to overcoming this curse of dimensionality is to guide inference on these large data sets by inclusion of prior knowledge generated over many decades by biologists and geneticists. The knowledge can be used to develop models against which experimental data can be tested, or it can be used in the design of statistical distributions for sampling techniques. One example of such treatment is the growing use of Bayesian statistical methods coupled to Markov chain Monte Carlo techniques. However, there are now many groups working on highly diverse methods to address these problems.

A second area of active research is the integration of diverse data types, which can also be considered a use of biological information to guide analysis. For instance, in the case of polygenic diseases, it would be logical to limit the loci to be analyzed to those that link in some way to differences in gene expression between cases and controls, or to genes which encode proteins in pathways of biological interest to the disease. Thus, results from analyzing one form of data, microarrays or biological pathways, serve as prior knowledge in the analysis of a second form of data, genotypes. In addition, it is often highly desirable to use data from well studied organisms, such as fruit fly or nematode, to guide inferences in higher organisms. The need to link data across data domain, such as from a single nucleotide polymorphism (SNP) to protein interaction, requires establishing ontologies or at minimum controlled vocabularies that allow automatic linking of data elements. Linking data between species requires identification of orthologs and orthologous pathways and interactions.

This session focuses on these two broad issues in integrated data analysis: the development of analysis methods that utilize multiple types of data and the establishment of methods to integrate diverse data. The papers reflect the continuum from the SGDI tool for integrating diverse data in R to analyses of the gylcan proteome and a large genotypic data set.

Papers

The first two papers in the session provide tools for data integration during analysis. SGDI, the System for Genomic Data Integration, is built on top of the widely used R/Bioconductor framework.[1] It uses the concept of assays for high-throughput data (e.g., microarrays, SNPchips) and tightly binds phenotype data to these assays (e.g., tumor stage, patient data) to track phenotypic information throughtout the analysis workflow. An important feature of the system is the inclusion of an extended version of the Sequence Ontology providing semantic integration of the data. The NARADA system leverages molecular interactions and other annotations during the analysis of networks, such as genetic regulatory networks.[2] The primary network is projected onto the annotation space, and functional networks are deduced using annotations from this primary network created by, for instance, gene expression analysis. This converts the inference of relationships between genes or proteins to inference of relationships between biological processes.

The middle three papers of the session focus on three different issues in the analysis of diverse data. The first paper addresses the problem of building classifiers from multiple data types, here microarray and proteomics data.[3] Unlike some approaches that look at mRNA species and encoded proteins, the work presented here looks to find the best discriminatory mRNA species and independently the best protein species. A subset of these are combined within a single classifier built using least squares support vector machines (LS-SVM), providing better sensitivity and specificity with fewer overall features. The second paper provides a solution to an important problem in the analysis of large genomic data sets – lists of genes carried forward during analysis rely on thresholds, leading to loss of information and questionable use of statistical tests later in the chain of analysis.[4] This work provides an integrated probabilistic framework for appropriate inference on gene sets or pathways, including gene ontology. Statistically the approach provides a full joint probability distribution from both data and annotations for estimation of biological parameters (e.g., upregulation of a pathway). The authors apply this method in a mouse model of prostate cancer. The third paper introduces a method to look at the multiscale correlation structure between different types of data, which is important since we generally do not know the length scales of interest in genomic processes.[5] In this case, correlations between histone modifications and DNase activity and between repressing and activating histone modifications are studied. The methodology relies on wavelets to calculate correlations, Kolmogorov-Smirnof statistics to test significance between different comparisons, and permutation tests to compare the results to randomized sequences.

The final two papers introduce new analysis approaches that have the potential to include significant prior information. In the first paper, a methodology for handling the large number of potential interactions between genetic variants in genome wide association studies is presented.[6] The method relies on random draws of variants at loci and the comparison of the set of variants to phenotype. A variant that is associated with phenotype should, over many random draws, obtain a higher posterior probability of association. The distribution of variants for the random draws can reflect prior knowledge of the phenotype (e.g., pathways associated with cancer) or other independent knowledge. In the final paper, a new approach for identification of biomarkers in proteomic data is presented.[7] An ongoing issue in the field is the identification of peaks that associate with a covariate of disease (e.g., age), rather than with the disease itself. The method described here first eliminates peaks from mass spectra that correlate with noninformative parameters provided by prior information, and then isolates peaks that distinguish phenotype. The technique is demonstrated by isolating a proteomic signature distinguishing hepatocellular carcinoma and chronic liver disease.

References

1. V. J. Carey, J. Gentry, R. Gentleman and S. Ramaswamy, *Pac Symp Biocomput* (2008).
2. J. Pandy, M. Koyuturk, W. Szpankowsi and A. Grama, *Pac Symp Biocomput* (2008).
3. A. Daemen, O. Gevaert, T. De Bie, A. Debucquoy, B. De Moor and K. Haustermans, *Pac Symp Biocomput* (2008).
4. M. Bhattacharjee, C. Pritchard and P. Nelson, *Pac Symp Biocomput* (2008).
5. R. E. Thurman, J. A. Stamatoyannopoulos and W. S. Noble, *Pac Symp Biocomput* (2008).
6. M. A. Province and I. B. Borecki, *Pac Symp Biocomput* (2008).
7. H. W. Ressom, R. S. Varghese, L. Goldman, C. A. Loffredo, M. Abdel-Hamid, Z. Kyselova, Y. Mechref, M. Novotny and R. Goldman, *Pac Symp Biocomput* (2008).

SGDI: SYSTEM FOR GENOMIC DATA INTEGRATION *

V. J. CAREY[†], J. GENTRY[‡], D. SARKAR[§], R. GENTLEMAN[¶], S. RAMASWAMY[‖]

Channing Laboratory, Brigham and Women's Hospital
Harvard Medical School
181 Longwood Avenue,
Boston, MA 02115, USA
E-mail: stvjc@channing.harvard.edu

This paper describes a framework for collecting, annotating, and archiving high-throughput assays from multiple experiments conducted on one or more series of samples. Specific applications include support for large-scale surveys of related transcriptional profiling studies, for investigations of the genetics of gene expression and for joint analysis of copy number variation and mRNA abundance. Our approach consists of data capture and modeling processes rooted in R/Bioconductor, sample annotation and sequence constituent ontology management based in R, secure data archiving in PostgreSQL, and browser-based workspace creation and management rooted in Zope. This effort has generated a completely transparent, extensible, and customizable interface to large archives of high-throughput assays. Sources and prototype interfaces are accessible at www.sgdi.org/software.

1. Introduction

It is becoming increasingly clear that biomarker and molecular target discovery in cancer, for example, will require the integrative analysis of multiple datasets generated in different centers, at different times, using different technology platforms. In fact, recent work suggests that integrative approaches can be highly useful for molecular target discovery [9, 11, 12], but there are still significant hurdles at the level of dataflow and data

*This work is supported in part by DFCI/HCC SPORE in Breast Cancer 2P50 CA89393-07.

†Channing Lab

‡Massachusetts General Hospital, Harvard Medical School

§Fred Hutchinson Cancer Research Center

¶Fred Hutchinson Cancer Research Center

‖Massachusetts General Hospital, Harvard Medical School

analysis workflow architecture, and deficiencies in software infrastructure, that retard progress in this research area. A very recent *Nature Reviews in Genetics* Perspectives report [8] discusses disparities between standard approaches to databasing genomic data and metadata and requirements of systems biology. Among the issues identified are deficiencies in meta-information necessary for resource discovery (by humans or by software), impoverishment of search predicate formulation options, unavailability of scalable/programmatic query resolution for queries with large payloads, non-robustness of client applications to alterations in central server data management patterns, resistance to adoption of XML markups (necessitating detailed non-generic parser development efforts), inappropriate conceptualizations (e.g., functions should be predicated of gene products, not genes, owing to splice variation) and a variety of difficulties related to communication, education, and licensing shortfalls.

To address some of these limitations, we have designed, developed, and deployed a software infrastructure for the storage and integrative analysis of biological data generated with high-thoughput tools in genomics and proteomics (www.sgdi.org/software). The proposed System for Genomic Data Integration (SGDI) is locally customizable. This is in contrast to read-only analysis-oriented repositories such as Oncomine [10], WebQTL [3], or SAGE-Genie [6], SGDI fills a critical gap in prevalent bioinformatics infrastructure, by permitting individual investigators to perform integrative analyses of unpublished data and to easily share unpublished data with colleagues, in a formally documented and auditable framework. In addition, researchers will be able to integrate their latest private data with a myriad of other publicly available data streams, thereby ensuring the greatest use of available resources. SGDI will enable integrative studies that are currently time-consuming and are difficult to standardize. It will facilitate data sharing and data reuse and will allow for data collected in one set of circumstances to be used to help test hypotheses in related areas. This system has been purpose-designed to enable sharing and analysis of private datasets that are generated either in single laboratories of through multi-investigator collaborations such as SPORE programs and program-project grants (PPGs).

While the ultimate objective of SGDI is an investigator-oriented browser-driven interface, we have adopted an approach that permits programmatic access to and manipulation of all data and metadata collected in the system. In this paper, we focus on elementary architecture and component functionalities. The first section details Bioconductor's approach to coherent container design for multiple high-throughput assays applied to fixed series of samples. The second section describes the sample annotation problem and SGDI's ontoElicitor facilities for structuring and deploying regimented vocabularies for sample characteristics. The third section describes the reporter annotation problem and SGDI's reporter query facilities. The final section provides illustrations of the integrated framework and discusses future intentions of the project.

2. Integrative data structure design in Bioconductor

Consider the problem of representing the fully preprocessed and normalized data from an experiment in genetics of gene expression, as reported in Cheung et al[4]. Let G denote the number of mRNA reporters (e.g., the number of oligonucleotide probe sets in an Affymetrix(TM) microarray), let N denote the number of samples (e.g., the number, 58, of CEPH CEU founders studied by Cheung et al.), let S denote the number of SNPs genotyped on each of the N samples, and let r denote the number of clinical, demographic, and technical variables recorded on the N samples. mRNA abundance measures are recorded in a $G \times N$ table, genotype calls (unphased) are recorded in an $S \times 2N$ table, and clinical and demographic characteristics of the N individuals are recorded in an $N \times r$ table. For the analyses reported in Cheung et al., genotyping information is condensed into SNP-specific rare allele counts, where allele rarity is reckoned relative to the source population, necessitating only an $N \times S$ table.

Some basic premises of the Bioconductor approach to dealing with high-throughput data are now described. We use the symbol `X` to name a concrete container for experimental data; the term *phenodata* is used to refer to all information gathered on samples exclusive of the assay results.

Compact representation. All the information collected in a high-throughput experiment should be available in a single object.

Tight binding of phenodata to assay data. Sample-level information should be tightly bound to assay results and should be propagated through workflows along with assay results unless intentionally excluded.

Array-like selection; closure of container type under selection. The idiom `X[G, S]` in the R programming language can be used to derive a new instance of the container type of `X` restricted to data on reporters identified in the general predicate expression `G` and to samples identified in the predicate expression `S`.

Tightly bound metadata components available. Representations allow for

Table 1. Selected methods and operators for Bioconductor containers. Most of the infrastructure for managing sample-level data is defined for the `eSet` class and is inherited to specializations.

method example	purpose	replace?
`eSet` class		
`X$n`	obtain value for all samples	yes
`X[i,j]`	restrict to selection	yes
`abstract(X)`	return main publication abstract	no
`experimentData(X)`	return MIAME schema	yes
`featureData(X)`	return reporter metadata	yes
`phenoData(X)`	return sample-level data	yes
`varMetadata(X)`	return metadata on sample attributes	yes
`ExpressionSet` class		
`exprs(X)`	return matrix of assay results	yes
`makeDataPackage(X)`	create an installable R package	no
`racExSet` class		
`snps(X)`	return matrix of rare allele counts	yes
`snpNames(X)`	return SNP identifiers	yes
`cghExSet` class		
`cloneNames(X)`	return clone identifiers	no
`cloneMeta(X)`	return clone metadata	no
`logRatios(X)`	return CGH assay results	no

storage of additional (meta)data on the experiment (following the MIAME [1] schema) and definitions of attributes defining reporters or samples.

Exemplary published experiments should be instantiated for distribution as illustrations. See the Bioconductor packages *Neve2006* (CGH+expression, discussed below) and *GGtools* (whole genome SNP+expression).

Generic workflow operations. Methods development in Bioconductor consists primarily of defining parameterized methods `f()` that interrogate and transform experimental data to support biological inference through evaluations of `f(X, ...)`. Multiassay representations should inherit type information from the constituent container types so that generic operations continue to function for the extended container type.

The main abstract class used to define high-throughput containers is called `eSet`, defined in the *Biobase* package of Bioconductor. Expression microarray assay results and allied sample and metadata are stored in instances of the `ExpressionSet` class. Table 1 sketches some of the methods/operations defined for `eSet` and some of its descendants for expression and integrative experiments.

3. Sample annotation; ontoElicitor

Careful analysis of the relationship of genomic phenomena to phenotypic or clinical condition requires detailed description of phenotypic state of the sample assayed. The data from Neve's 2006 analysis of copy number and expression variation in breast tumor cell lines [7] are a good illustration of the sort of material published in this area. Here we excerpt two records from the sample annotation:

```
> library(Neve2006); data(neveExCGH)
> pData(neveExCGH)[1:4,]
       ind cellLine geneCluster ER  PR HER2 TP53
600MPE   1   600MPE          Lu  + [-] <NA>    -
AU565    2    AU565          Lu  - [-]    + <NA>
       Source tumorType Agey Ethnicity     cultMedia
600MPE   <NA>       IDC   NA      <NA>  DMEM,10%FBS
AU565      PE        AC   43         W RPMI, 10% FBS
              cultCond commonPt reductMamm
600MPE     37c, 5% CO2        0      FALSE
AU565      37c, 5% CO2        1      FALSE
> table(neveExCGH$Source)
  AF  CWN P.Br   PE   PF   Sk
   2    1   24   19    0    1
> varMetadata(neveExCGH)["Source",]
[1] "PE = pleural effusion, P.Br = primary breast,
     Sk = skin, CWN = chest wall nodule, AF = ascites fluid"
```

This illustrates Bioconductor facilitites for accessing and interpreting sample-level data. The `pData` method extracts the R data frame of attributes on samples, the `$` operator confers direct access to variable values, and the `varMetadata` method returns a subsettable data frame with definitions of symbols used.

When different nomenclatures are used for phenotype characterization in different experiments, a problem arises for users of public microarray archives who wish to perform synthetic analyses [5]. It becomes difficult to align samples across experiments. Figure 1 illustrates the situation in a collection of 25 breast cancer microarray experiments. Sample-level data available in public archives were reviewed. The union of the sets of terms employed for sample annotation was formed, and the subset of terms related to histopathology was selected. The left margin of Figure 1 lists all the terms in this set, and the bottom margin lists the experiments. A dark square is plotted in cell (i, j) of the figure if term i is used in experiment j. It is clear that terms with similar meanings are not uniformly named, and that experimenters often do not report values of many relevant characteristics.

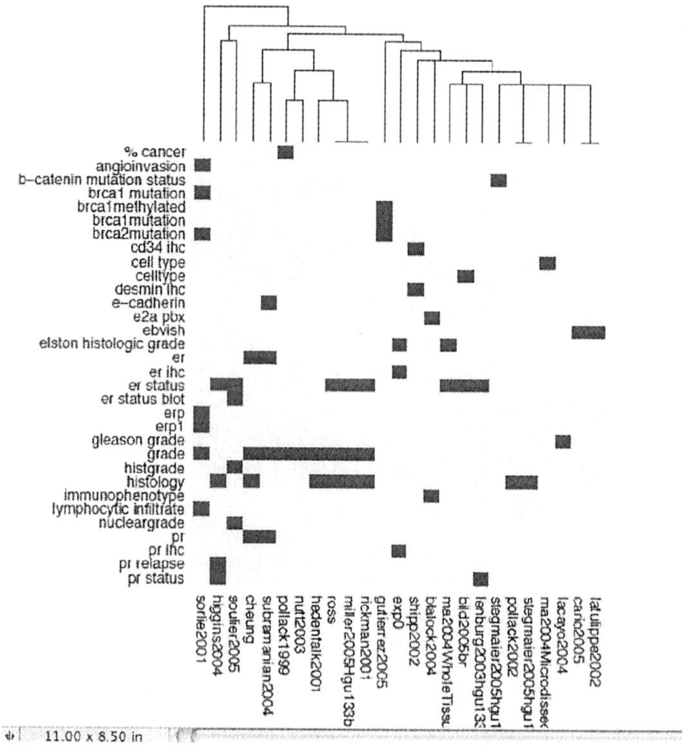

Figure 1. Rows: terms related to breast cancer histopathology. Columns: author-date tokens identifying 25 published breast cancer datasets. A dark square is plotted at location (i,j) if study i uses term j in characterizing its samples.

While Figure 1 indicates a problem with sparsity of shared annotation across independently performed experiments, it does not indicate another vulnerability: Even when experimenters do use a common term such as 'grade' in sample annotation, the values used for the term may not coincide.

SGDI has responded to this predicament with two novel tools. The first, ontoElicitor, is a simple framework for iteratively presenting and receiving feedback on a proposed structured vocabulary for sample annotation. Figure 2 illustrates a facet of the ontoElicitor for breast cancer samples.

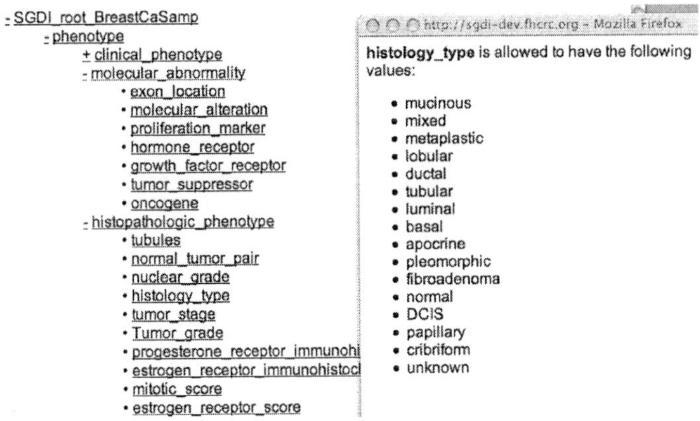

Figure 2. ontoElicitor facet for breast cancer, with expanded value set for histology type displayed.

Our *current* approach to vocabulary design and management eschews formal ontology engineering methodologies like OWL/RDF in favor of R graphs. The OWL concepts of class, property and individual are typically not familiar to experimentalists, and adaptation of OWL technology for *elicitation* and *revision* of vocabularies and valuations required in microarray archives does not seem cost-effective. We have found that practitioners are interested in working with tree-structured displays of terms, with enumerated valuations, and with valuation classes such as "numeric" or "string". Bioconductor graph structures can easily represent trees of nodes that represent terms as string literals. Because arbitrary node attributes can be attached, valuations and valuation classes can be bound directly to terms in the graph structures. These ontology graph structures, defined in the *ontoElicitor* package distributed with SGDI, can be serialized to HTML (for use in the ontoElicitor application) or CSV (for review in Excel by practitioners.) Note that we will support conversion between OWL/RDF ontology models and R ontology graphs upon adoption of a suitable RDF schematization for sample-level metadata. The *Rredland* package of Bioconductor exposes the `librdf.org` facilities for parsing, modeling, and archiving RDF.

The second tool of use in promoting adoption of uniform sample annotation is the phenoData editor application, with a demonstration instance at the SGDI portal. Given an ontoElicitor-derived ontology, the phenoData editor generates a page of fields with drop-down menus that are used to populate a sample attribute table with standardized values.

4. Reporter annotation and query facilities

Focused use of archives of high-throughput data is most convenient when genomic contexts and biological roles of reporters are easily established. In the case of SNP+expression experiments, it will be of interest to know relative locations of genotyped loci, assayed transcripts, and, e.g., locations of promoters for genes exhibiting differential expression; for CGH+expression, segmentation breakpoints need to be related to gene locations and phenotype. Substantial information on element locations is available through Bioconductor platform annotation packages and through translations of Entrez Gene and biomart-accessible annotation resources. It is frequently of interest to interrogate using higher-level concepts and gene collections. Figure 3 illustrates the interface for filtering reporters on the basis of membership in specific KEGG-catalogued pathways; GO categories and sets of HUGO symbols may be used as well. We also have recently introduced an R graph representing the KEGG orthology (a tree-structured hierarchy of KEGG pathways, package *keggorth*) and tree-based navigation of this structure will be supported.

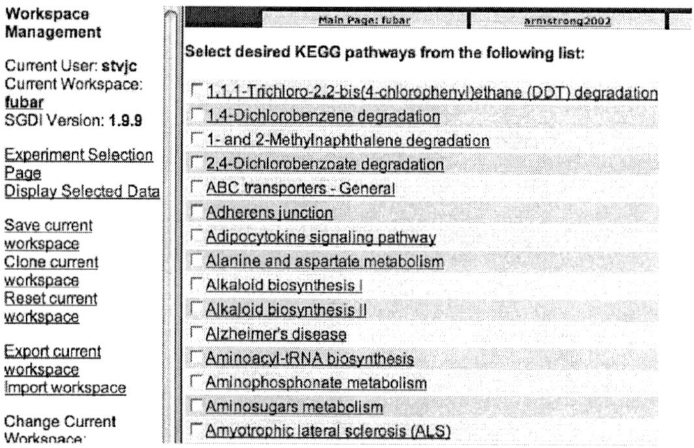

Figure 3. Selection of reporters using KEGG pathway catalog.

5. The integrated interface; use cases

The primary object that is manipulated in the SGDI framework is the workspace. This is an XML document that records all selections that have occurred. Workspaces can be exported for sharing with colleagues, can be cloned so that multiple paths with common initial segments can be explored and saved, and can be revised through rollback or continuation. In general, a user will not be concerned with the contents or structure of the *workspace document*, but will work with the system to define a data extract that will be used for downstream analysis.

Figure 4 gives a view of the workspace obtained when three experiments are in scope. armstrong2002 and blalock2004 are classical breast cancer expression array experiments; testOGTES is a test instance of expression data (obtained on the u133x3p platform) and SNP data (obtained with the Affymetrix(TM) 500K Nsp+Sty platform). Expression assay results and standard errors of estimated expression are provided in two tables; enzyme-specific tables are provided for both the genotype calls and the call confidence as measured by the *crlmm* algorithm in development by Carvalho, Irizarry and colleagues [2].

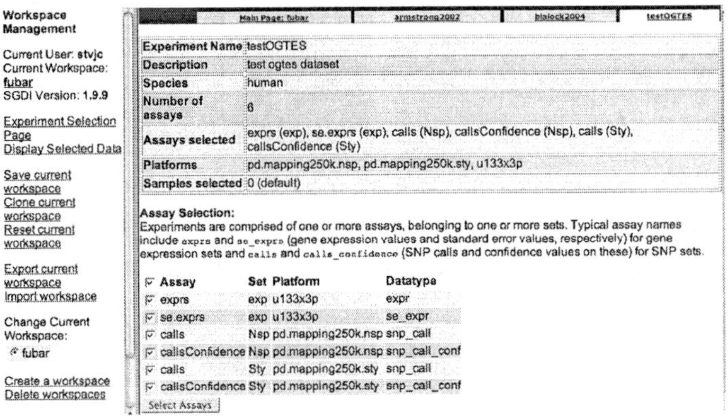

Figure 4. top level interface

Figure 5 depicts the interface to SNP selection using only physical coordinates on chromosomes. Additional facilities are available to employ annotation provided by Affymetrix detailing cytoband, harboring transcript, harboring gene, role of transcript in gene to form and condition queries. The exposition of these resources to simplify interrogation is complete for cytoband and gene relationships; more work is needed to take advantage of the detailed contextual vocabulary described in section 4 above.

Finally, a partial view of the HTML rendering of a workspace display for genotyping assays is given in Figure 6. Reporter metadata occupies the

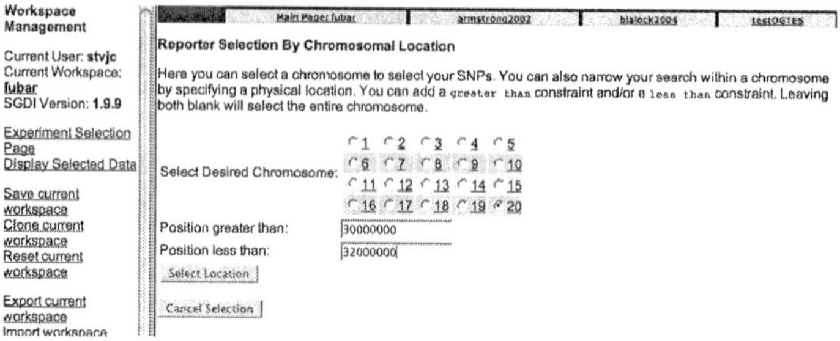

Figure 5. Selecting SNPs by location on chromosome.

first six columns, and sample characteristics occupy the first 13 rows. Some genotype calls are found at the lower right corner of the display.

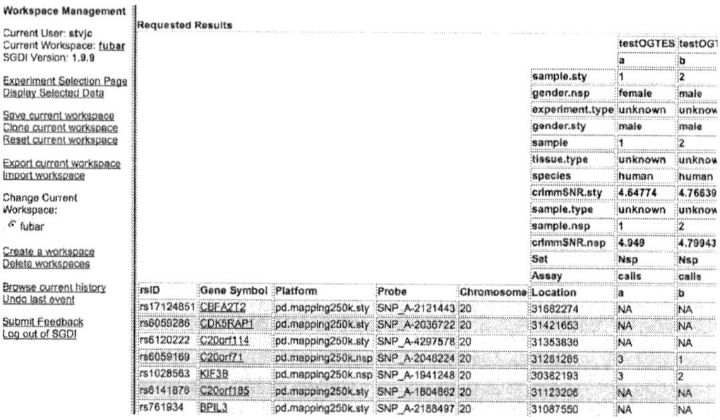

Figure 6. Reporting on selected SNPs.

6. Deployment; conclusions

One of the most significant problems tackled by SGDI is the challenge of providing fine-grained, investigator-friendly access to preprocessed and carefully annotated archives of high-throughput data. SGDI allows investigators to discover (using flexible but standardized query resolution) and extract (using a browser-based workflow) data on values of specific reporters associated with samples possessing specific phenotypic or experimental characteristics for their own local analysis. As the public instance of SGDI grows, this "read-only" facility will provide access to public datasets with high interpretability and integrability established through the use of ontoElicitor-based sample annotation.

Our open design and distribution approach helps to solve another significant problem in the management and analysis of high-throughput data. Centers and investigators are free to establish (and customize) their own instances of SGDI for use with private or pre-publication data. We have adopted a "clean room" deployment, in which all but the most basic infrastructure is wrapped in a single tarball, including specific versions of R, python, PostgreSQL, and Zope, so that intercomponent version consistencies are guaranteed. The administrator who installs the system on a reasonable unix/mac platform need only set a few Make variables, type 'make', and provide passwords when asked. The 'veil' system for securing PostgreSQL at the table access level (veil.projects.postgresql.org) is included and initialized so that group and individual access control lists can be established for any experiments. The administrator populates the system data store using code that transforms R data packages (exemplars in the ExperimentData archive at Bioconductor) into secured PostgreSQL tables. The use of R as middleware (between raw assay output files and PostgreSQL/Zope) permits extension to workflows based on other data formalisms such as MAGE-OM. The *RMAGEML* package of Bioconductor can be used to transform MAGE-ML experiment serializations into `ExpressionSet` instances, which then admit rapid incorporation into SGDI. A referree has expressed concern with R's capacity to function with very large data resources. The adoption of PostgreSQL for main data archiving and interrogation processes represents a proper matching of technology with task. When workspaces yield tables of manageable size they can be passed to R directly for numerical analysis and visualization; otherwise 'chunking' procedures can be adopted to solve many analysis problems in limited memory. At present our software has run on CentOS Linux, Suse Linux, and Mac OSX. A Windows port is believed to be feasible but has not been undertaken. Use of this software requires only a browser, but administration of the system requires familiarity with PostgreSQL, Zope, and R.

Forthcoming revisions to the software will facilitate targeting data extracts to Bioconductor using serialization of a class instance (or package, if appropriate) so that the provenance of the data extract, the associated workspace document, and the utilities to which the extract is suited are included in a self-documenting object or artifact. This will serve as a prototype for targeting other analytical systems with defined APIs.

References

1. A Brazma, P Hingamp, J Quackenbush, et al. Minimum information about a microarray experiment (miame)-toward standards for microarray data. *Nat Genet*, 29(4):365–371, Dec 2001.
2. Benilton Carvalho, Terence P. Speed, and Rafael A. Irizarry. Exploration, normalization, and genotype calls of high density oligonucleotide snp array data. *Johns Hopkins University, Dept. of Biostatistics Working Papers*, 111, 2006.
3. E J Chesler, L Lu, J Wang, R W Williams, et al. Webqtl: rapid exploratory analysis of gene expression and genetic networks for brain and behavior. *Nat Neurosci*, 7(5):485–486, May 2004.
4. V G Cheung, R S Spielman, K G Ewens, et al. Mapping determinants of human gene expression by regional and genome-wide association. *Nature*, 437(7063):1365–1369, Oct 2005.
5. R. Gentleman, M. Ruschhaupt, and W. Huber. On the synthesis of microarray experiments. *Journal de la Societe Francais de Statistique*, 146:173–194, 2005.
6. P Liang. Sage genie: a suite with panoramic view of gene expression. *Proc Natl Acad Sci U S A*, 99(18):11547–11548, Sep 2002.
7. R M Neve, K Chin, J Fridlyand, et al. A collection of breast cancer cell lines for the study of functionally distinct cancer subtypes. *Cancer Cell*, 10(6):515–527, Dec 2006.
8. S. Philippi and J. Kohler. Addressing the problems with life-science databases for traditional uses and systems biology. *Nat Rev Genet*, 7(6):482–8, 2006. 1471-0056 (Print) Journal Article Review.
9. S. Ramaswamy, K. N. Ross, E. S. Lander, and T. R. Golub. A molecular signature of metastasis in primary solid tumors. *Nat Genet*, 33(1):49–54, 2003. 1061-4036 (Print) Journal Article.
10. D R Rhodes, S Kalyana-Sundaram, V Mahavisno, et al. "oncomine 3.0: genes, pathways, and networks in a collection of 18,000 cancer gene expression profiles". *Neoplasia*, 9(2):166–180, Feb 2007.
11. E. Segal, N. Friedman, D. Koller, and A. Regev. A module map showing conditional activity of expression modules in cancer. *Nat Genet*, 36(10):1090–8, 2004. 1061-4036 (Print) Journal Article.
12. S. A. Tomlins, D. R. Rhodes, S. Perner, S. M. Dhanasekaran, R. Mehra, X. W. Sun, S. Varambally, X. Cao, J. Tchinda, R. Kuefer, C. Lee, J. E. Montie, R. B. Shah, K. J. Pienta, M. A. Rubin, and A. M. Chinnaiyan. Recurrent fusion of tmprss2 and ets transcription factor genes in prostate cancer. *Science*, 310(5748):644–8, 2005. 1095-9203 (Electronic) Journal Article.

ANNOTATING PATHWAYS IN INTERACTION NETWORKS

JAYESH PANDEY*, MEHMET KOYUTÜRK†, WOJCIECH SZPANKOWSKI AND ANANTH GRAMA

Department of Computer Science, Purdue University, West Lafayette, IN 47907, USA
**E-mail: jpandey@cs.purdue.edu*

Integrating molecular interaction data with existing knowledge of molecular function reveals mechanisms that underly cellular organization. We present NARADA, a software tool that implements a comprehensive analysis suite for functional annotation of pathways. NARADA takes as input a species-specific molecular interaction network and annotation of biomolecules in the network and provides the user with a set of pathways composed of functional attributes, which may be thought of pathway templates in the functional annotation space that recur in various contexts (different groups of specific molecules with similar functional annotation patterns) in the molecular interaction network. NARADA has its underpinnings in formal statistical measures of significance, and algorithmic bases for performance. Comprehensive evaluation on the *E. coli* transcriptional regulation and protein-protein interaction data demonstrate NARADA's ability to detect known, as well as novel pathways.

1. Introduction

Network models are commonly used to abstract biomolecular interactions. Recent research has focused on identifying common patterns in these networks, within and across species, with the expectation that such patterns reveal evolutionary design principles that underly cellular organization. Indeed, coherent topological motifs (*e.g.*, feedback and feed-forward loops) and their constituent molecules are observed to recur significantly in the protein-protein interaction and transcriptional regulatory networks of model organisms [1]. Comparative analysis of extant networks also suggests that modular subcomponents of these networks are likely to be conserved together [2,3].

These observations support the hypothesis that the organizational principles that underly interaction networks may be represented in the form of functional (sub)networks – "rules" or "templates" that recur in various contexts in the functional organization of the cell. The underlying problem of generalizing from molecular annotations, provided by libraries such as Gene Ontology [4], to subnetwork annotations is important – and forms the technical challenge addressed in this paper. Preliminary studies show that such annotations can indeed be derived;

†Present address: *Department of Electrical Engineering & Computer Science, Case Western Reserve University, Cleveland, OH, USA*

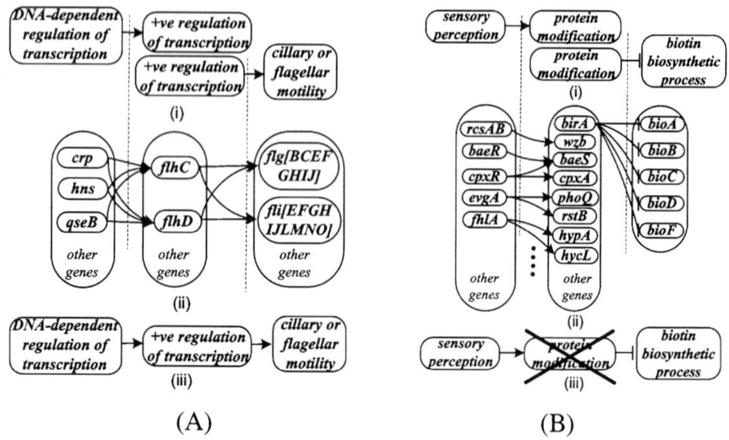

Fig. 1. (A) From interactions between functional attributes to pathways of functional attributes. (B) Significant pairwise interactions between functional attributes do not necessarily imply indirect paths.

however, they do not provide an automated methodology, or a comprehensive analytical (statistical) basis for the annotations [5–8]. Schwikowski *et. al.* [5] predict functions of proteins in *S. cerevisiae* protein-protein interaction network by hypothesizing that proteins of known function and cellular location tend to cluster together. Lee *et. al.* [6] study the *S. cerevisiae* transcriptional regulatory network with a view to understanding relationships between functional categories of genes. They observe that many transcriptional regulators within a functional category bind to transcriptional regulators that play key roles in the control of other cellular processes. For example, cell cycle activators bind to several genes that regulate metabolism, environmental response, and development. Tong *et. al.* [7] identify putative genetic interactions in yeast via synthetic genetic array (SGA) analysis and investigate the functional relevance of their results in the context of GO annotations.

These results are limited to case-specific studies that generally focus on validation or evaluation of results through simple statistical analyses – yet they provide significant insights. Generalizing these observations allows identification of standardized pathways, creation of reference databases of direct and indirect interactions between various processes, and projecting existing knowledge of model organisms to other species. What is lacking is a comprehensive set of tools that combine these sources of data (molecular annotations and interactions) to identify significantly overrepresented patterns of interaction through reliable statistical modeling with a formal computational basis.

In recent work [9], we explore the statistical and algorithmic underpinnings of this problem. In this paper, we describe a comprehensive toolkit, NARADA, for pathway annotation. NARADA can be applied to diverse abstractions (*e.g.*, gene regulatory networks, protein-protein interaction networks), and can use as reference node annotations any user-specified ontology. Users can specify functional categories of interest, query for statistically over-represented pathways in terms of these functional categories, visually manipulate and inspect these pathways, and view reflections of these pathways in "molecular" (*i.e.*, gene network) and "functional" (*i.e.*, network of functional attributes) space. NARADA evaluates the statistical significance of pathways based on a novel statistical model, which emphasizes the modularity of pathways by conditioning on the frequency of their building blocks. NARADA is implemented in Java and is available as a web applet, as well as a standalone application at http://www.cs.purdue.edu/~jpandey/narada.

2. Models

Molecular interactions are abstracted using various network models. Regulation of gene expression, for example, is commonly modeled using Boolean networks [10]. Protein-protein interactions (PPIs), on the other hand, represent various forms of physical association between proteins, including modification, transport, and complex formation [11]. NARADA is designed to handle different types of networks and different sources of data in a unifying framework. In this section, for the sake of clarity, we present the mathematical underpinnings of NARADA in the context of gene regulatory networks, and focus on identification of regulatory pathways.

The basic approach to integrating existing knowledge of gene networks and functional annotation is to (i) project nodes from the *gene space* onto the *functional attribute space*, and (ii) find significant pathways in the functional attribute space. The first step is accomplished using a reference node annotation library. A simple method for accomplishing the second step is to identify statistically abundant (significant) pairs of interacting functional attributes. For example, in Figure 1(A), the *E. coli* transcription network contains 36 activator interactions between 2 genes that take part in *positive regulation of transcription* and 18 genes that are involved in *cillary or flagellar motility*. This observation may be abstracted as a *rule* that characterizes the regulatory relationship between these two processes: *positive regulation of transcription* up-regulates *cillary or flagellar motility* in *E. coli*. Indeed, this approach is used to understand the functional organization of *S. cerevisiae* synthetic genetic array [7] and transcriptional regulatory networks [6].

Statistically significant pathway annotations cannot be directly composed from constituent pairwise annotations. This is because, each interaction between a pair of functional attributes is within a specific context (a different pair of genes) in

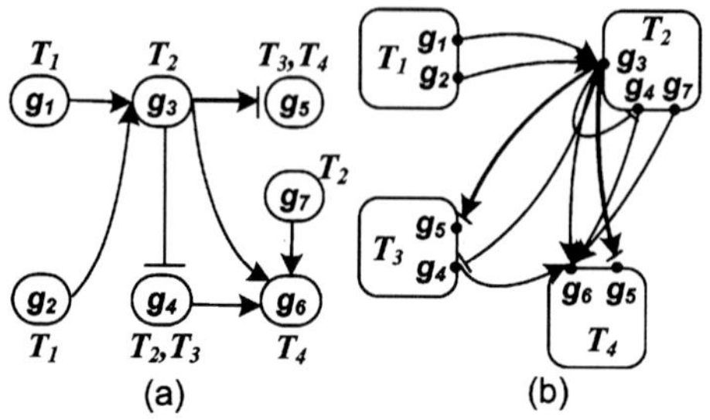

Fig. 2. (a) A sample gene regulatory network and the functional annotation of the genes in this network. Each node represents a unique gene and is tagged by the set of functional attributes attached to that gene. Activator interactions are shown by regular arrows, repressor interactions are shown by dashed arrows. (b) Functional attribute network derived from the gene regulatory network in (a). In this multigraph, nodes (functional attributes) are represented by squares and ports (genes) are represented by dark circles.

the network. This is illustrated in Figure 1. In both (A) and (B), the two regulatory interactions shown on the panel (i) are significantly frequent in the gene network. In Figure 1(A), the genes involved in *positive regulation of transcription* shown in panel (ii) are common to both interactions and the combined pathway shown in the panel (iii) is frequent. On the other hand, in Figure 1(B), the set of genes involved in *protein modification* (in panel (ii)) are different for the two interactions, so the combined pathway (in panel (iii)) does not exist in the gene network at all! However, a method that relies on assessment of only pairwise interactions would identify indirect regulation of *biotin biosynthetic process* by *sensory perception* through *protein modification* as a significant pathway, which is not a conclusion that is supported by available data.

Data Model. A gene network is modeled as a labeled directed graph with nodes representing genes and edges representing regulatory interactions. Each edge is associated with a type that specifies the mode of regulation (activation, repression or dual). Each gene in the network is associated with a set of *functional attributes*, which provide functional annotations of the gene. Without loss of generality, we use Gene Ontology (GO) [4] to annotate the genes in the network.

Given a gene network and annotation, the corresponding *functional attribute network* is defined as follows: each functional attribute T_i is represented by a *multinode*, which contains a set of *ports*, each corresponding to a gene g_j that is associated with T_i. The frequency $\phi(T_i)$ of a functional attribute is equal to the number of genes that are associated with T_i. Each *multiedge* T_iT_j corresponds to a set of edges g_kg_l in the gene network, such that g_k is associated with T_i and g_l is associated with T_j. The frequency $\phi(T_iT_j)$ of a multiedge is equal to the number of such edges in the gene network. A *multipath* of length k is a sequence of k distinct functional attributes (multinodes), $\{T_{i_1}, T_{i_2}, ..., T_{i_k}\}$. A sequence of genes $\{g_{j_1}, g_{j_2}, ..., g_{j_k}\}$ is an occurrence of multipath $\{T_{i_1}, T_{i_2}, ..., T_{i_k}\}$ if each g_{j_ℓ} is associated with the corresponding T_{i_ℓ} and there is an edge from g_{j_ℓ} to $g_{j_{\ell+1}}$ in the gene network for each ℓ. The frequency $\phi(\{T_{i_1}, T_{i_2}, ..., T_{i_k}\})$ of a multipath is the number of occurrences of that multipath in the gene network.

A sample gene network and its corresponding functional attribute network is shown in Figure 2. In Figure 2, the frequency of multipath $T_1 \rightarrow T_2 \dashv T_3$ is 4.

Statistical Model. The "interestingness" of a pathway is associated with its *modularity*, *i.e.*, the significance of the coupling of its building blocks. In statistical terms, this is achieved by conditioning the distribution of the frequency (modeled as a random variable) of a pathway on the frequency of its subpaths (modeled as fixed parameters). Note that, in this approach, statistical significance is used as an indicator of the modularity of a pathway in the functional annotation space, *i.e.*, the hypothesis that is tested here is that a pathway of functional attributes corresponds to a design template that is conserved and rediscovered through evolution [12]. Thefore, the statistical significance of a pathway should be interpreted as the likelihood that the observed pattern is biologically relevant (in Kitano's [12] terms, it may have a place in the "periodic table" of functional regulatory circuits), rather than being a measure of the pattern's biological relevance or importance.

A single interaction is the shortest pathway in a functional attribute network. We evaluate the significance of single interaction by taking into account the frequency of each functional attribute and the degree distribution of the gene network. For each functional attribute T_i, its expected in-degree β_i and out-degree δ_i are specified. Then, edges are generated by randomly selecting $n = \sum_i \beta_i = \sum_j \delta_j$ edges from $m = \sum_{T_i, T_j \in V_F} \beta_i \delta_j$ potential edges, where each of the $\beta_i \delta_j$ potential edges are between T_i and T_j. Letting $\Phi_{ij} = \Phi(T_i T_j)$ be frequency of $T_i T_j$ in the random model, we observe that Φ_{ij} is a hypergeometric random variable and obtain $p_{ij} = P(\Phi_{ij} \geq \phi_{ij}) = \sum_{\ell=\phi_{ij}}^{\min\{\beta_i\delta_j, n\}} \binom{\beta_i\delta_j}{\ell}\binom{m-\beta_i\delta_j}{n-\ell}/\binom{m}{n}$.

Now let $\Pi_{i,k}$ denote the path $\{T_{i_1}, T_{i_2}, ..., T_{i_k}\}$. For $1 < j < k$, we want to evaluate the significance of the coupling between pathways $\Pi_{1,j}$ and $\Pi_{j,k}$. Our reference model assumes that the frequency of pathways $\Pi_{1,j}$ and $\Pi_{j,k}$ is established *a-priori*. Let $\Phi_{i,k}$ and $\phi_{i,k}$ denote $\Phi(\Pi_{i,k})$ and $\phi(\Pi_{i,k})$, the random variable that represents the frequency of pathway $\Pi_{i,k}$ and its observed value, respectively. Then, the p-value of the coupling between $\Pi_{1,j}$ and $\Pi_{j,k}$ is defined as $p_{1,j,k} = P(\Phi_{1,k} \geq \phi_{1,k} | \Phi_{1,j} = \phi_{1,j}, \Phi_{j,k} = \phi_{j,k})$.We approximate this value using Chvátal's bound on hypergeometric tail [13] to obtain $p_{1,j,k} \leq \exp(\phi_{1,j}\phi_{j,k} H_{q_j}(t_{1,j,k}))$,where $t_{1,j,k} = \phi_{1,k}/\phi_{1,j}\phi_{j,k}$, $q_j = 1/\phi_j$ (ϕ_j is frequency of term T_j), and $H_q(t) = t \log \frac{q}{t} + (1-t) \log \frac{1-q}{1-t}$ denotes divergence. This estimate is Bonferroni-corrected for multiple testing, *i.e.*, it is adjusted by a factor of $\prod_{j=1}^{k} |\bigcup_{g_\ell \in T_{i_j}} \mathcal{F}(g_\ell)|$.

3. Methods and Features

NARADA is implemented in Java, and can be run as a web applet or a standalone application. It requires an installation of Java Runtime Environment version 1.4.2_14 (update 14) or later and has been tested to run of windows and linux platform. The base application ERIS is based on Cytoscape [14]. This software framework allows for development of sophisticated visualization and analysis functions through java-based plugins. The user manual and source code for NARADA are available at http://www.cs.purdue.edu/homes/jpandey/narada.

Query Interface. Currently, NARADA supports three classes of queries:

- Q_1: Given a functional attribute T, find all significant pathways that are regulated by (originate at) genes that are associated with T.
- Q_2: Given a functional attribute T, find all significant pathways that regulate (terminate at) genes that are associated with T.
- Q_3: Given a sequence of functional attributes $T_{i_1}, T_{i_2}, ..., T_{i_k}$, find all occurrences of the corresponding pathway in the gene network.

A pathway is identified as being significant if its p-value is less than a user-specified level, α. Pathways do not have repeated internal nodes, but cycles (feedback loops) are allowed, *i.e.*, the output to queries Q_1 (Q_2) may include a pathway that terminates (originates) at the queried term itself, provided each occurrence of the cycle corresponds to a cycle in the gene network.

Algorithms. For a given term, we perform an enumerative search on the functional attribute network (without explicitly constructing the network) starting from the node that corresponds to the query term. For queries of type Q_1 (Q_2), the search proceeds forwards (backwards) with respect to edge direction. Consequently, the output is a tree that is rooted at the query term.

During the course of the search process, the significance of each pathway is tested as follows: if the length of the pathway is one, *i.e.*, it is a multiedge, its significance is evaluated with respect to the baseline model. Otherwise, assume that we are trying to extend pathway $\{T_{i_1}, ..., T_{i_{k-1}}\}$ by adding multiedge $T_{i_{k-1}}T_{i_k}$, where T_{i_1} is the query term. We condition the significance of pathway $\{T_{i_1}, ..., T_{i_k}\}$ on the frequency of pathways $\{T_{i_1}, ..., T_{i_{k-1}}\}$ and $T_{i_{k-1}}T_{i_k}$. The motivation behind this is as follows: if the regulatory effect of $T_{i_{k-1}}$ on T_{i_k} is significantly coupled with pathway $\{T_{i_1}, ..., T_{i_{k-1}}\}$, *i.e.*, a significant number of its occurrences in the network are likely to be preceded by this pathway, then this may correspond to a *rule* that characterizes the regulation of T_{i_k} through a chain of regulatory interactions specified by pathway $\{T_{i_1}, ..., T_{i_k}\}$.

For queries from class Q_3, consider the sequence of functional attributes

$T_{i_1}, T_{i_2}, ..., T_{i_k}$. For such a query, NARADA finds all occurrences of the pathways $\{T_{i_1}, ..., T_{i_j}\}$, $\{T_{i_j}, ..., T_{i_k}\}$ and $\{T_{i_1}, ..., T_{i_k}\}$. To find all occurrences of a pathway, for each functional attribute, the genes that bridge the previous and next node in the sequence are identified. Then, the frequencies of these pathways are used to compute the significance as described in the previous section. By mapping all genes that are identified to the gene network, NARADA displays all occurrences of the pathway in the gene network.

Performance Enhancement and Heuristics. A major limitation of the algorithm above is that it is brute force and its time complexity is exponential in k (length of the path). The longest pathway that is of interest is a user-defined parameter in NARADA. Since pathways of biological significance are expected to be fairly short, the practical constraints posed by this exponential complexity are somewhat mitigated. However, since there exist several genes that are attached to many functional attributes and vice versa, the branching factor of the search process is quite large. For this reason, pruning heuristics that render significant pathway identification tractable for very large networks and longer pathways are still necessary. In NARADA, various heuristics that exploit *a priori* biological knowledge are implemented to accelerate the search process. We outline these heuristics below. We also note that development of efficient heuristics that integrate syntactic and semantic information remains an important open problem.

Gene Ontology hierarchy: The current release of NARADA uses Gene Ontology (GO) [4] as the default reference library for annotations. NARADA's default behavior in handling this hierarchy is to use the most specific GO term on each branch of GO hierarchy for each gene. In other words, if terms T_i and T_j are attached to gene g_ℓ and if T_j is a parent of T_i in GO hierarchy (*i.e.*, either T_i *is a* T_j or T_i is *part of* T_j), then only T_i is considered in the functional attribute network. The user is allowed to alter this behavior by selecting to annotate the genes using any specific level of the hierarchy. Each query can also be refined by moving a term in the query up or down the GO hierarchy.

Strongly significant pathways: NARADA delivers near interactive query response using a biologically motivated pruning technique. We call a pathway *strongly significant* if all of its subpaths are significant. In biological terms, a strongly significant pathway is likely to correspond to a significantly modular process, in which not only the building blocks of the pathway are significant, but are also tightly coupled. This makes it possible to extend pathway length without significant re-computation. For queries of type $\mathtt{Q}_1$ and $\mathtt{Q}_2$, the option for searching strongly significant paths is available in NARADA.

Short-circuiting common terms: The main motivation in identification of significant regulatory pathways is understanding the crosstalk between different processes, functions, and cellular components. Therefore, functions and processes that are known to play a key role in gene regulation (*e.g.*, transcription regulator activity or DNA binding) may overload the identified pathways and overwhelm other interesting patterns. However, genes that are responsible for these functions are likely to bridge regulatory interactions between different processes [6], so they cannot be ignored. For this reason, such GO terms are short-circuited, *i.e.*, if process T_i regulates T_j, which is a key process in transcription, and T_j regulates another process T_k, then the pathway $T_i \rightarrow T_j \rightarrow T_k$ is replaced with regulatory interaction $T_i \rightarrow T_k$.

Interface. A user interface with comprehensive functionality and visualization capabilities is available in NARADA. The visualization infrastructure is built using an open source library, Prefuse [15], which provides standard graph visualization functions. The current version of NARADA can handle large networks with thousands of genes and annotations. The graph views (for both gene and functional attribute space) support pan, drag, zoom, and standard layout functionalities, search by node name and node link-outs to biological databases. Screenshots from NARADA are shown in Figure 3.

The input to NARADA consists of three files:

- A molecular interaction network, in which interacting molecules and type of interaction are specified using the simple interaction file (`sif`) format [14]. Multiple networks can be loaded simultaneously, NARADA creates separate visualizations for each. These networks may belong to different organisms.
- Specification of the functional attributes and their relations (*e.g.*, Gene Ontology (GO) `obo` file). Currently, only one attribute set can be used in one session.
- Annotation file that specifies the mapping between nodes and their functional attributes. Multiple annotation files can be loaded to provide mapping for one or more networks.

Detecting pathway annotations: The interface to query significant pathways originating (or terminating) at a functional attribute allows the user to specify α parameter, the limit on pathway length, and a flag indicating whether the search is limited to strongly significant pathways. The result of a query $\mathtt{Q}_1(\mathtt{Q}_2)$ is displayed as a tree. Each path from the root to a leaf represents a significantly overrepresented pathway. The p-value of each pathway is stored at the corresponding leaf. Each pathway can also be separately viewed in a GO Path frame, which offers the user the ability to move up/down the GO hierarchy for any node in the pathway. Moreover, this interface also allows the user to view all occurrences of the pathway in the gene network. It is also possible to submit a single query to NARADA to run queries of type $\mathtt{Q}_1, \mathtt{Q}_2$ for all functional attributes in bulk. The results of such a query can be directly written to an output file. To query all occurrences of a specified pathway of attributes (query type Q_3), the user enters the sequence of GO terms, specifies the edges along with their types (*e.g.*, mode of regulation), and the output can then be explored through the GO path view.

4. Results and Discussion

We run NARADA on the *E. coli* transcriptional network and *E.coli* protein interaction network to identify core functional pathways that underly cellular regulation and signaling in *E.coli.*

We obtain the *E. coli* transcriptional network (TrN) from RegulonDB [16]. The release 5.6 of this dataset contains 1363 genes with 3159 regulatory interactions. The *E. coli* protein interaction network (PIN) is obtained from DIP [17]. The latest release (20070219) of this dataset contains 1841 proteins with 6958 interactions. We use Gene Ontology [4] as a library of functional attributes. The annotation of *E. coli* genes and proteins is obtained from the UniProt GOA Proteome 48.0 release [18]. Using the default mapping provided by GO, the gene network is mapped to functional attribute networks of the three name spaces in GO. Mapping to the biological process space provides maximum coverage in number of genes or proteins annotated. In the TrN 904 genes are mapped to one or more of 340 process terms, while for the PIN 793 proteins are mapped to one or more of 343 process terms. We discuss here results obtained by this mapping only.NARADA is equally useful for the molecular function branch of the GO, with results like *transcription factor binding − > ATP binding − > electron carrier activity*. Results relating to molecular functions and cellular components, as well as comprehensive results on pathways of biological processes for both networks, are available at http://www.cs.purdue.edu/homes/jpandey/narada/.

We use NARADA to identify all significant pathways of length 2 to 5. In order to identify these paths, we run queries Q_1 (and Q_2 for transcription network) with a critical α of 0.01 on all annotated biological processes. The number of pathways obtained using combinations of the algorithmic options described in the previous section are shown in Table 1. These results differ from previous results in [9] on account of better annotation which affects the bonferroni-correction. On a Pentium M (1.6GHz) laptop with 1.21GB RAM identification of all significantly over-represented pathways takes average of 1.2s per query for TrN, and 8s per query for PIN, for upto path length 5 and 4 respectively. For strongly significant paths, it takes less than 0.5s per query in the TrN, and less than 2s per query in the PIN for paths of length upto 5. Strongly significant pathways, *i.e.*, those obtained by extending only significant pathways, compose an important part of the highly significant pathways. This observation suggests that significantly modular pathways are also likely to be composed of significantly modular building blocks. In

Table 1. Total number of significant pathways identified by NARADA for various path lengths.

E. coli network	algorithm	2	3	4	5
transcriptional	*All significant pathways*	213	1404	3472	2251
	Strongly significant pathways	213	210	248	148
	Short-circuiting common terms	445	422	371	38
protein interaction	*All significant pathways*	208	3533	53486	-
	Strongly significant pathways	208	699	4196	36266

the TrN after short-circuiting terms related to transcription, translation, and regulation thereof, identification of all significant paths takes 0.9s per query for paths of length 5. Note that a short-circuited path of length 5 might actually correspond to a path of length upto 9 with hidden (short-circuited) nodes.

Sample results: Parts of the significant pathways that regulate phosphorylation via genes involved in transcription and DNA recombination are shown in Figure 4(a). As genes involved in transcription are abundantly present in the network, part of the pathway (*DNA recombination* → *transcription*) occurs rarely (12 times) and is not significant, but in 6 of the 12 times it occurs, the genes involved in transcription regulate phosphorylation and the complete pathway occurs 38 times ($p < 4 \times 10^{-15}$). The *fis* transcriptional regulator is responsible for regulation of *nuoA-N* operon [19], while the *fhlA* transcriptional activator regulates the *hyf* locus [20]. Indeed, it is observed that integration host factor (*ihfA,ihfB*) affects the regulation of these phosphorylation related genes (*nuoA-N, hyf,hyc*) directly and indirectly [20].

Figure 4 (b) shows a significant pathway that is composed of *translation, DNA replication* and *protein folding*, as well as the corresponding proteins and their interactions in the PPI network. This pathway recurs 20 times ($p < 3.6 \times 10^{-3}$) in the PPI network. Proteins involved in *DNA replication* are abundantly present in the network, but are connected to proteins involved in *protein folding* only 8 times. 5 out of these 8 interactions are preceded by proteins involved in translation. The dnaK chaperone system, consisting of dnaK, dnaJ and grpE are involved in remodeling and refolding of proteins, with cbpA functioning as a dnaJ-like cochaperon [21,22]. dnaA involved in DNA replication activity is protected by dnaK from reaching a self-aggregate inactive form [23]. Most of the other proteins involved in translation form part of ribosomal assembly or translational elongation factor activity [24].

A global view of E.coli functional regulatory network: A summary of all significant pathways identified on *E. coli* transcription network is shown in Figure 5. This view provides a mapping of the *E.coli* transcriptional network to the biological process space of Gene Ontology. In the figure, the top 20% significant pathways for pathway lengths 2 to 4 are shown. The edges that constitute significant pathways of length 2, 3, and 4 are shown using solid, dashed, and dotted lines, respectively. This results in a connected network of 71 functional attributes. In the figure, the font size of each GO term is proportional to its degree in this network, and thickness of an edge is proportional to its significance (or significance of the pathway it is a part of). As seen in the figure, this network is clustered into various fundamental processes. A large subnetwork consists of processes related to response to stress and stimulus, DNA repair, and negative regulation of transcription. This subnetwork is mostly composed of down-regulatory interactions. A second important group of processes that are tightly coupled in this network relates to cell motility, cytochrome assembly, flagellum and positive regulation of transcription. These processes are mostly connected via up-regulatory interactions. Observe that the regulatory interactions in these *local* subnetworks correspond to significant pathways of length 2, *i.e.*, they are direct regulatory interactions, but they also may be parts of significant indirect pathways.

The edges that are part of significant indirect pathways (those in dashed and dotted lines) form the rest of the network. These edges go through several *hub* processes (which are shown in large fonts representing their high-degree), including DNA recombination, transcription, and DNA dependent regulation of transcription. These are indeed processes that are characterized to mediate genetic regulation. The indirect pathways that go through these mediator processes connect local hubs of the clustered processes, such as response to stimulus and flagellum biogenesis, as well as other fundamental processes including various metabolic and biosynthetic processes, translation, signaling, and transport. These observations illustrate NARADA's ability to accurately capture the basic principles of genetic regulation and characterize the crosstalk between various processes through identification of indirect regulatory relationships.

For the protein interaction network a large portion of the significant pathways involve cellular protein metabolic process, cell cycle, cell division, translation, and response to antibiotic. These hub processes interact with a variety of other biological processes.

A notable problem with projecting networks to the abstract space of functional annotation is that the results are not directly testable. In other words, there is no obvious experimental method that could be used to falsify the notion that a pathway of functional attributes discovered by NARADA is biologicall relevant. These is because, by definition, the pathways identified by NARADA are abstract. Note, however, that patterns identified by NARADA can indeed be used to discover novel biological information that can be experimentally verified, and this provides an indirect method for testing the hypotheses generated by NARADA. Recent applications of frequent pathway templates in Gene Ontology space include functional annotation of individual proteins [25] and prediction of organism-specific pathways [26].

5. Conclusion

We present a comprehensive software tool, NARADA, to project molecular interaction networks to the functional attribute space. NARADA provides several interfaces to detect significantly overrepresented pathways. Based on results obtained from the *E. coli* transcription network, NARADA identifies several known, as well as novel pathways, at near-interactive query rates. Note that the current knowledge of regulatory networks is incomplete, and is limited to a few model organisms. Therefore, application of our method on currently available data does not provide a comprehensive library of regulatory network annotation. On the other hand, the partial annotation provided by our method forms a useful basis for extending our knowledge of regulatory networks beyond well-studied processes and model organisms.

References

1. S. S. Shen-Orr *et al.*, *Nature Genetics* **31**, 64 (2002).
2. M. Koyutürk *et al.*, *Journal of Computational Biology* **13**, 1299 (2006).
3. R. Sharan and T. Ideker, *Nature Biotechnology* **24**, 427 (2006).
4. M. Ashburner *et al.*, *Nat Genet* **25**, 25 (2000).
5. B. Schwikowski *et al.*, *Nat Biotechnol* **18**, 1257(Dec 2000).
6. T. I. Lee *et al.*, *Science* **298**, 799(October 2002).
7. A. H. Y. Tong *et al.*, *Science* **303**, 808(February 2004).
8. J. Gamalielsson *et al.*, A GO-based method for assessing the biological plausibility of regulatory hypotheses, in *ICCS (2)*, 2006.
9. J. Pandey *et al.*, *Bioinformatics* **23**, 377(Jul 2007).
10. S. Liang *et al.*, *Proc. Pacific Symp. Biocomp.* **3**, 18 (1998).
11. P. Uetz *et al.*, *Nature* **403**, 623 (2000).
12. H. Kitano, *Science* **295**, 1662 (2002).
13. V. Chvátal, *Discrete Mathematics* **25**, 285 (1979).
14. P. Shannon *et al.*, *Genome Res* **13**, 2498(Nov 2003).
15. J. Heer *et al.*, Prefuse: a toolkit for interactive information visualization, in *CHI '05: Proceeding of the SIGCHI*, (ACM Press, 2005).
16. H. Salgado *et al.*, *Nucleic Acids Res* **34**(January 2006).
17. L. Salwinski *et al.*, *Nucleic Acids Res* **32**, 449(Jan 2004).
18. E. Camon *et al.*, *Nucleic Acids Res* **32**, 262(Jan 2004).
19. B. Wackwitz *et al.*, *Mol Gen Genet* **262**, 876(Dec 1999).
20. S. Hopper *et al.*, *J Biol Chem* **269**, 19597(Jul 1994).
21. B. Bukau and A. L. Horwich, *Cell* **92**, 351(Feb 1998).
22. C. Chae *et al.*, *J Biol Chem* **279**, 33147(Aug 2004).
23. B. Banecki *et al.*, *Biochim Biophys Acta* **1442**, 39(Oct 1998).
24. K. Saito *et al.*, *J Mol Biol* **235**, 111(Jan 1994), Comparative Study.
25. M. Kirac and G. Ozsoyoglu (submitted).
26. A. Cakmak and G. Ozsoyoglu, *Bioinformatics* (in press).

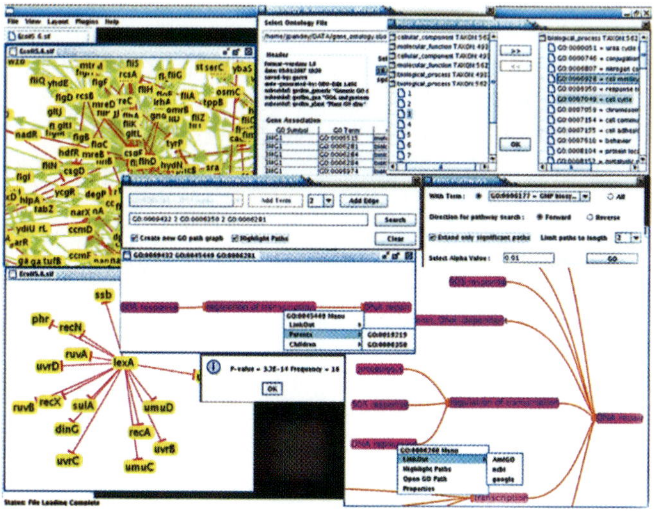

Fig. 3. Screenshots from NARADA.

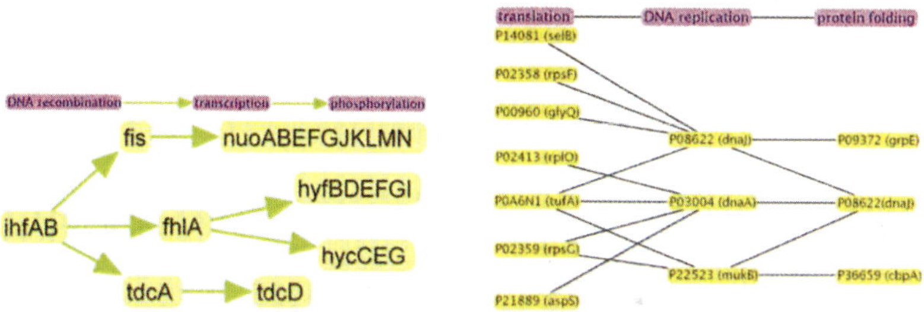

(a) (b)

Fig. 4. Sample significantly overrepresented pathways in (a) *E.coli* transcriptional network, and (b) *E.coli* protein interaction network. The pathways in functional attribute space are shown on the upper panel, their occurrences in the gene network are shown on the lower panel.

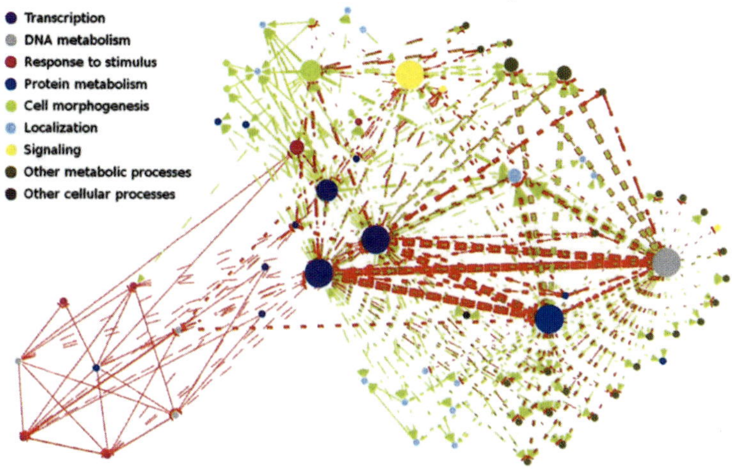

Fig. 5. A global view of *E. coli* transcriptional network mapped to cellular processes defined by GO.

INTEGRATING MICROARRAY AND PROTEOMICS DATA TO PREDICT THE RESPONSE ON CETUXIMAB IN PATIENTS WITH RECTAL CANCER

ANNELEEN DAEMEN[1,*] OLIVIER GEVAERT[1], TIJL DE BIE[2], ANNELIES DEBUCQUOY[3], JEAN-PASCAL MACHIELS[4], BART DE MOOR[1] AND KARIN HAUSTERMANS[3]

[1] *Katholieke Universiteit Leuven, Department of Electrical Engineering (ESAT), SCD-SISTA (BIOI), Kasteelpark Arenberg 10 - bus 2446, B-3001 Leuven (Heverlee), Belgium*

[2] *University of Bristol, Department of Engineering Mathematics, Queen's Building, University Walk, Bristol, BS8 1TR, UK*

[3] *Katholieke Universiteit Leuven / University Hospital Gasthuisberg Leuven, Department of Radiation Oncology and Experimental Radiation, Herestraat 49, B-3000 Leuven, Belgium*

[4] *Université Catholique de Louvain, St Luc University Hospital, Department of Medical Oncology, Ave. Hippocrate 10, B-1200 Brussels, Belgium*

To investigate the combination of cetuximab, capecitabine and radiotherapy in the preoperative treatment of patients with rectal cancer, fourty tumour samples were gathered before treatment (T_0), after one dose of cetuximab but before radiotherapy with capecitabine (T_1) and at moment of surgery (T_2). The tumour and plasma samples were subjected at all timepoints to Affymetrix microarray and Luminex proteomics analysis, respectively. At surgery, the Rectal Cancer Regression Grade (RCRG) was registered. We used a kernel-based method with Least Squares Support Vector Machines to predict RCRG based on the integration of microarray and proteomics data on T_0 and T_1. We demonstrated that combining multiple data sources improves the predictive power. The best model was based on 5 genes and 10 proteins at T_0 and T_1 and could predict the RCRG with an accuracy of 91.7%, sensitivity of 96.2% and specificity of 80%.

1. Introduction

A recent challenge for genomics is the integration of complementary views of the genome provided by various types of genome-wide data. It is likely

*To whom correspondence should be addressed: anneleen.daemen@esat.kuleuven.be

that these multiple views contain different, partly independent and complementary information. In the near future, the amount of available data will increase further (e.g. methylation, alternative splicing, metabolomics, etc). This makes data fusion an increasingly important topic in bioinformatics.

Kernel Methods and in particular Support Vector Machines (SVMs) for supervised classification are a powerful class of methods for pattern analysis, and in recent years have become a standard tool in data analysis, computational statistics, and machine learning applications.[1–2] Based on a strong theoretical framework, their rapid uptake in applications such as bioinformatics, chemoinformatics, and even computational linguistics, is due to their reliability, accuracy, computational efficiency, demonstrated in countless applications, as well as their capability to handle a very wide range of data types and to combine them (e.g. kernel methods have been used to analyze sequences, vectors, networks, phylogenetic trees, etc). Kernel methods work by mapping any kind of input items (be they sequences, numeric vectors, molecular structures, etc) into a high dimensional space. The embedding of the data into a vector space is performed by a mathematical object called a 'kernel function' that can efficiently compute the inner product between all pairs of data items in the embedding space, resulting into the so-called kernel matrix. Through these inner products, all data sets are represented by this real-valued square matrix, independent of the nature or complexity of the objects to be analyzed, which makes all types of data equally treatable and easily comparable.

Their ability to deal with complexly structured data made kernel methods ideally positioned for heterogeneous data integration. This was understood and demonstrated in 2002, when a crucial paper integrated amino-acid sequence information (and similarity statistics), expression data, protein-protein interaction data, and other types of genomic information to solve a single classification problem: the classification of transmembrane versus non transmembrane proteins.[3] Thanks to this integration of information a higher accuracy was achieved than what was possible based on any of the data sources separately. This and related approaches are now widely used in bioinformatics.[4–6]

Inspired by this idea we adapted this framework which is based on a convex optimization problem solvable with semi-definite programming (SDP). As supervised classification algorithm, we used Least Squares Support Vector Machines (LS-SVMs) instead of SVMs. LS-SVMs are easier and faster for high dimensional data because the quadratic programming problem is converted into a linear problem. Secondly, LS-SVMs are also more suitable as they contain regularization which allows tackling the problem of overfitting. We have shown that regularization seems to be very important when applying classification methods on high dimensional data.[7]

The algorithm described in this paper will be applied on data of patients with rectal cancer. To investigate the combination of cetuximab, capecitabine and radiotherapy in the preoperative treatment of patients with rectal cancer, microarray and proteomics data were gathered from fourty rectal cancer patients at three timepoints during therapy. At surgery, different outcomes were registered but here we focus on the Rectal Cancer Regression Grade (RCRG)[8], a pathological staging system based on Wheeler for irradiated rectal cancer. It includes a measurement of tumour response after preoperative therapy. In this paper, patients were divided into two groups which we would like to distinguish: the positive group (RCRG pos) contained Wheeler 1 (good responsiveness; tumour is sterilized or only microscopic foci of adenocarcinoma remain); the negative group (RCRG neg) consisted of Wheeler 2 (moderate responsiveness; marked fibrosis but with still a macroscopic tumour) and Wheeler 3 (poor responsiveness; little or no fibrosis with abundant macroscopic tumour). We refer the readers to Ref. 9 for more details about the study and the patient characteristics.

We would like to demonstrate that integrating multiple available data sources in the patient domain in an appropriate way using kernel methods increases the predictive power compared to models built only on one data set. The developed algorithm will be demonstrated on rectal cancer patient data to predict the RCRG at T_1 (= before the start of radiotherapy).

2. Data sources

Fourty patients with rectal cancer (T3-T4 and/or N+) from seven Belgian centers were enrolled in a phase I/II study investigating the combination of cetuximab, capecitabine and radiotherapy in the preoperative treatment of patients with rectal cancer.[9] Tissue and plasma samples were gathered before treatment (T_0), after one dose of cetuximab but before radiotherapy with capecitabine (T_1) and at moment of surgery (T_2). At all these three timepoints, the frozen tissues were used for Affymetrix microarray analysis while the plasma samples were used for Luminex proteomics analysis. Because we had to exclude some patients, ultimately the data set contained 36 patients.

The samples were hybridized to Affymetrix human U133 2.0 plus gene chip arrays. The resulting data was first preprocessed for each timepoint separately using RMA.[10] Secondly, the probe sets were mapped on Entrez Gene Ids by taking the median of all probe sets that matched on the same gene. Probe sets that matched on multiple genes were excluded and unknown probe sets were given an arbitrary Entrez Gene Id. This reduces the number of features from 54613 probe sets to 27650 genes. Next, one can imagine that the number of differentially expressed genes will be much lower than these 27650 genes. Therefore, a prefiltering without reference to phenotype can be used to reduce the number of genes. Taking into account the low signal-to-noise ratio of microarray data, we decided to filter out genes that show low variation across all samples. Only retaining the genes with a variance in the top 25% reduces the number of features at each timepoint to 6913 genes.

The proteomics data consist of 96 proteins, previously known to be involved in cancer, measured for all patients in a Luminex 100 instrument. Proteins that had absolute values above the detection limit in less than 20% of the samples were excluded for each timepoint separately. This results in the exclusion of six proteins at T_0, four at T_1 and six at T_2. The proteomics expression values of transforming growth factor alpha (TGFα), which had also too many values below the detection limit, were replaced by the results of ELISA tests performed at the Department of Experimental Oncology in Leuven. For the remaining proteins the missing values were replaced by half of the minimum detected for each protein over all samples, and values exceeding the upper limit were replaced by the upper limit value. Because most of the proteins had a positively skewed distribution, a log transformation (base 2) was performed.

In this paper, only the data sets at T_0 and T_1 were used because the goal of the models is to predict before start of chemoradiation the RCRG.

3. Methodology

3.1. *Kernel methods and LS-SVMs*

Kernel methods are a group of algorithms that do not depend on the nature of the data because they represent data entities through a set of pairwise comparisons called the kernel matrix. The size of this matrix is determined only by the number of data entities, whatever the nature or the complexity of these entities. For example a set of 100 patients each characterized by 6913 gene expression values is still represented by a 100×100 kernel matrix.[4] Similarly as 96 proteins characterized by their 3D structure are represented by a 100 × 100 kernel matrix. The kernel matrix can be geometrically expressed as a transformation of each data point x to a high dimensional feature space with the mapping function $\Phi(x)$. By defining a kernel function $k(x_k, x_l)$ as the inner product $\langle \Phi(x_k), \Phi(x_l) \rangle$ of two data points x_k and x_l, an explicit representation of $\Phi(x)$ in the feature space is not needed anymore. Any symmetric, positive semidefinite function is a valid kernel function, resulting in many possible kernels, e.g. linear, polynomial and diffusion kernels. They all correspond to a different transformation of the data, meaning that they extract a specific type of information from the data set. Therefore, the kernel representation can be applied to many different types of data and is not limited to vectorial or matrix form.

An example of a kernel algorithm for supervised classification is the Support Vector Machine (SVM) developed by Vapnik and others.[11] Contrary to most other classification methods and due to the way data is represented through kernels, SVMs can tackle high dimensional data (e.g. microarray data). The SVM forms a linear discriminant boundary in feature space with maximum distance between samples of the two considered classes. This corresponds to a non-linear discriminant function in the original input space. A modified version of SVM, the Least Squares Support Vector Machine (LS-SVM), was developed by Suykens *et al.*[12–13] On high dimensional data sets this modified version is much faster for classification because a linear system instead of a quadratic programming problem needs to be solved. The LS-SVM also contains regularization which tackles the problem of overfitting. In the next section we describe the use of LS-SVMs with a normalized linear kernel to predict the RCRG in rectal cancer patients based on the kernel integration of microarray and proteomics data at T_0 and T_1.

3.2. *Data fusion*

There exist three ways to learn simultaneously from multiple data sources using kernel methods: early, intermediate and late integration.[14] Figure 1 gives a global overview of these three methods in the case of two available data sets. In this paper, intermediate integration is chosen because this type of data fusion seemed to perform better than early and late integration.[14] The nature of each data set is taken into account better compared to early integration by adapting the kernel functions to each data set separately. By adding the kernel matrices before training the LS-SVM, only one predicted outcome per patient is obtained. This makes the extra decision function which was needed for late integration unnecessary.

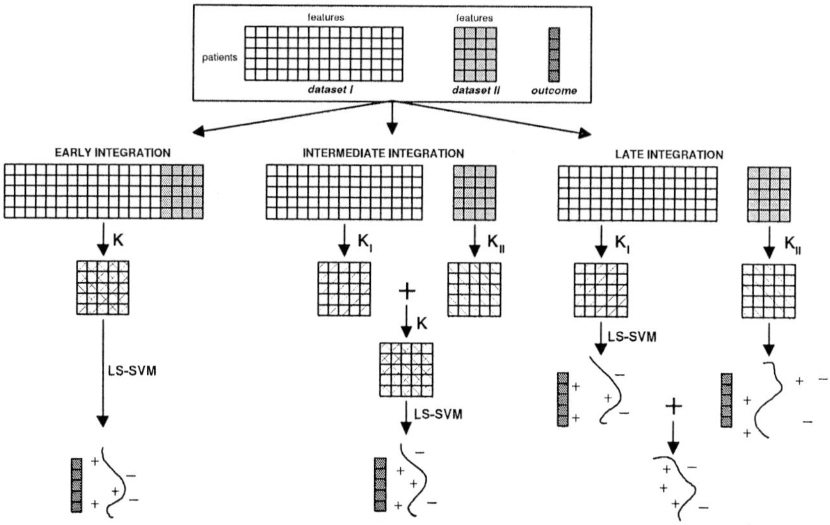

Figure 1. *Three methods to learn from multiple data sources.* In early integration, an LS-SVM is trained on the kernel matrix, computed from the concatenated data set. In intermediate integration, a kernel matrix is computed for both data sets and an LS-SVM is trained on the sum of the kernel matrices. In late integration, two LS-SVMs are trained separately for each data set. A decision function results in a single outcome for each patient.

3.3. *Model building*

In this paper, the normalized linear kernel function

$$k(x_k, x_l) = k(x_k, x_l)/\sqrt{k(x_k, x_k)k(x_l, x_l)} \tag{1}$$

with $k(x_k, x) = x_k^T x$ was used instead of the linear kernel function $k(x_k, x_l) = x_k^T x_l$. With the normalized version, the values in the kernel matrix will be bounded because the data points are projected onto the unit sphere while these elements can take very large values without normalization. Normalizing is thus required when combining multiple data sources to guarantee the same order of magnitude for the kernel matrices of the data sets.

There are four data sets that have to be combined: microarray data at T_0, at T_1 and proteomics data at T_0 and at T_1. Because each data set is represented by a kernel matrix, these data sources can be integrated in a straightforward way by adding the multiple kernel matrices according to intermediate integration explained previously. In this combination, each of the matrices is given a specific weight μ_i. The resulting kernel matrix is given in Eq. 2. Positive semidefiniteness of the linear combination of kernel matrices is guaranteed when the weights μ_i are constrained to be non-negative.

$$K = \mu_1 K_1 + \mu_2 K_2 + \mu_3 K_3 + \mu_4 K_4. \tag{2}$$

The choices of the weights are important. Previous studies have shown that the optimization of the weights only leads to a better performance when some of the available data sets are redundant or contain much noise.[3] In our case we believe that the microarray and proteomics data sets are equally reliable based on our results of LS-SVMs on each data source separately (data not shown). Therefore to avoid optimizing the weights, they were chosen equally: $\mu_1 = \mu_2 = \mu_3 = \mu_4 = 0.25$.

Due to the data set size, we chose a leave-one-out cross-validation (LOO-CV) strategy to estimate the generalization performance (see Fig. 2). Since both classes were unbalanced (26 RCRG pos and 10 RCRG neg), the minority class was resampled in each LOO iteration by randomly duplicating a sample from the minority class and adding uniform noise ([0,0.1]). This was repeated until the number of samples in the minority class was at least 70% of the majority class (chosen without optimization).

After choosing the weights fixed, three parameters are left that have to be optimized: the regularization parameter γ of the LS-SVM, the number of genes used from the microarray data sets both at T_0 and T_1 and the number of proteins used from the proteomics data sets. To accomplish this, a three-dimensional grid was defined as shown in Fig. 2 on which the parameters are optimized by maximizing a criterion on the training set. The possible values for γ on this grid range from 10^{-10} to 10^{10} on a logarithmic scale. The possible number of genes that were tested are 5, 10, 30, 50, 100, 300, 500, 1000, 3000 and all genes. The number of proteins used are 5, 10, 25, 50 and all proteins. Genes and proteins were selected by ranking these features using the Wilcoxon rank sum test. In each LOO-CV iteration, a model is built for each possible combination of parameters on the 3D-grid. Each model with the instantiated parameters is evaluated on the left out sample. This whole procedure is repeated for all samples in the set. The model with the highest accuracy is chosen. If multiple models with equal accuracy, the model with the highest sum of sensitivity and specificity is chosen.

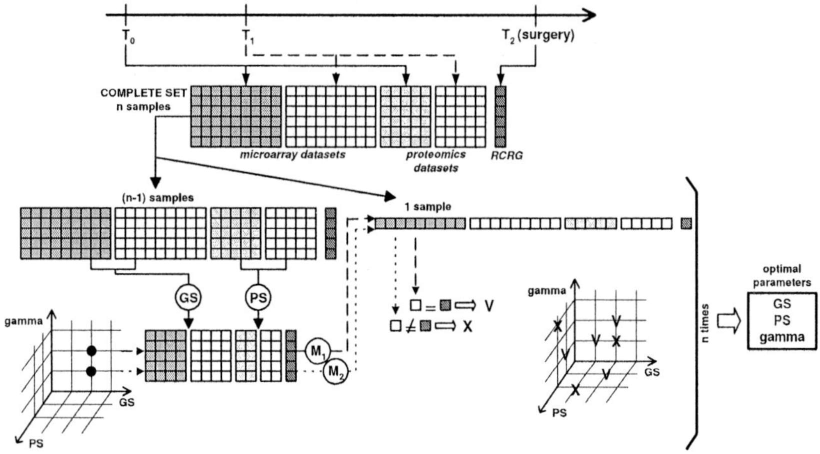

Figure 2. *Methodology for developing a classifier.* The available data contains microarray data and proteomics data both at T_0 and T_1. The regularization parameter γ and the number of genes (GS) and proteins (PS) are determined with a leave-one-out cross-validation strategy on the complete set. In each leave-one-out iteration, an LS-SVM model is trained on the most significant genes and proteins for all possible combinations of γ and the number of features. This gives a globally best parameter combination (γ,GS,PS).

4. Results

We evaluated our methodology as described in Sec. 3.3 on the rectal cancer data set to predict the Rectal Cancer Regression Grade. The model with the highest performance accuracy and an as high as possible sum of sensitivity and specificity was built on the five most significant genes and the ten most significant proteins at T_0 and T_1 according to the RCRG. From now on, we refer to this model as MPIM (Microarray and Proteomics Integration Model). To evaluate its performance, 6 other models were built on different combinations of data sources using the same model building strategy: MMT0 (Microarray Model at T_0: all microarray data at T_0), MMT1 (Microarray Model at T_1: all microarray data at T_1), MIM (Microarray Integration Model: microarray data at both timepoints), PMT0 (Proteomics Model at T_0: all proteomics data at T_0), PMT1 (Proteomics Model at T_1: all proteomics data at T_1) and PIM (Proteomics Integration Model: proteomics data at both timepoints).

Table 1 gives an overview of all these models with the number of features resulting into the best performance for each model. MPIM predicts the

RCRG correctly in 33 of the 36 patients (=91.7%). Almost all patients with RCRG positive are predicted correctly with a sensitivity of 96.2% and a positive predictive value of 0.926. Of the patients with RCRG negative, 80% are classified correctly. None of the other models performs better for one of the performance parameters shown in Table 1.

Table 1. Performance of MPIM compared to models based on different combinations of data sources.

Model	Nb genes	Nb proteins	TP	FP	FN	TN	Sens (in %)	Spec (in %)	PPV	NPV	Accuracy (in %)
MPIM	5	10	25	2	1	8	96.2	80	0.926	0.889	91.7 (33/36)
MMT0	1000	-	25	10	1	0	96.2	0	0.714	0	69.4 (25/36)
MMT1	3000	-	23	6	3	4	88.5	40	0.793	0.571	75.0 (27/36)
MIM	30	-	25	10	1	0	96.2	0	0.714	0	69.4 (25/36)
PMT0	-	all	21	4	5	6	80.8	60	0.840	0.545	75.0 (27/36)
PMT1	-	5	23	2	3	8	88.5	80	0.920	0.727	86.1 (31/36)
PIM	-	25	21	3	5	7	80.8	70	0.875	0.583	77.8 (28/36)

TP, true positive; FP, false positive; FN, false negative; TN, true negative; Sens, sensitivity; Spec, specificity; PPV, positive predictive value; NPV, negative predictive value; Accuracy, predictive accuracy.

The MPIM is built on 5 genes different for T_0 and T_1, 9 proteins different for T_0 and T_1 and 1 protein selected at both timepoints (ferritin).

Among the 5 genes at T_0 and at T_1, several were related to cancer. Bone morphogenetic protein 4 (BMP4) is involved in development, morphogenesis, cell proliferation and apoptosis. This protein, upregulated in colorectal tumours, seems to help initiate the metastasis of colorectal cancer without maintaining these metastases.[15] Integrin alpha V (ITGAV) is a receptor on cell surfaces for extracellular matrix proteins. Integrins play important roles in cell-cell and cell-matrix interactions during a.o. immune reactions, tumour growth and progression, and cell survival. ITGAV is related to many cancer types among which prostate and breast cancer for which it is important in the bone environment to the growth and pathogenesis of cancer bone metastases.[16]

Several of the proteins have known associations with rectal and colon cancer, such as ferritin, TGFα, MMP-2 and TNFα. Ferritin, the major intracellular iron storage protein, is an indicator for iron deficiency anemia. This disease is recognized as a presenting feature of right-sided colon cancer and increases in men significantly the risk of having colon cancer.[17] The transforming growth factor alpha (TGFα) is upregulated in some human cancers among which rectal cancer.[18] In colon cancer, it promotes depletion of tumour-associated macrophages and secretion of amphoterin.[19] TGFα is closely related to epidermal growth factor EGF, one of the other proteins on which MPIM is built. EGF plays an important role in the regulation of cell growth, proliferation and differentiation. The matrix metalloproteinase-2 (MMP-2), known to be implicated in rectal and colon cancer invasion and metastasis, is associated with a reduced survival of these patients when being higher expressed in the malignant epithelium and in the surrounding stroma.[20] The tumour necrosis factor TNFα has important roles in immunity and cellular remodelling and influences apoptosis and cell survival. Dysregulation and especially overproduction of TNFα have been observed to occur in colorectal cancer.[21] Some of the other proteins such as IL-4 and IL-6 are important for the immune system whose function depends for a large part on interleukins. IL-4 is involved in the proliferation of B cells and the development of T cells and mast cells. It also has an important role in allergic response. IL-6 regulates the immune response, modulates normal and cancer cell growth, differentiation and cell survival.[22] It causes increased steady-state levels of TGFα mRNA in macrophage-like cells.[23]

Several of the genes and proteins are involved in KEGG-pathways for environmental information processing (cytokine-cytokine receptor interaction, Jak-STAT signaling pathway) and for the immune system (hematopoietic cell lineage). Important functions and processes confirmed by Gene Ontology [24] are protein binding, signal transduction, multicellular organismal development, cell-cell signaling and regulation of cell proliferation.

5. Discussion

We presented a framework for the combination of multiple genome-wide data sources in disease management using a kernel-based approach (see Fig. 2). Each data set is represented by a kernel matrix based on a normalized linear kernel function. These matrices are combined according to the intermediate integration method illustrated in Fig. 1. Afterwards, an LS-SVM is trained on the combined kernel matrix. In this paper, we evaluated the resulting algorithm on our data set consisting of microarray and proteomics data of rectal cancer patients to predict the Rectal Cancer Regression Grade after a combination therapy consisting of cetuximab, capecitabine and radiotherapy. The best model (MPIM) is based on 5 genes and 10 proteins at T_0 and at T_1 and can predict the RCRG with an accuracy of 91.7%, sensitivity of 96.2% and specificity of 80%. Table 1 shows that the performance parameters of MPIM are better than or equal to the values of the other models. This demonstrates that microarray and proteomics data are partly complementary and that the performance of our algorithm in which these various views on the genome are integrated improves the prediction of response to therapy upon LS-SVMs trained on a combination of less data sources. Many of the genes and proteins on which the MPIM is built are related to rectal cancer or cancer in general.

We were inspired by the idea of Lanckriet[3] and others[4–6] to integrate multiple types of genomic information to be able to solve a single classification problem with a higher accuracy than possible based on any of the genomic information sources separately. In the framework of Lanckriet, the problem of optimal kernel combination is formulated as a convex optimization problem using SVMs and is solvable with semi-definite programming (SDP) techniques. However, LS-SVMs are easier and faster for high dimensional data because the problem is formulated as a linear problem instead of a quadratic programming problem and LS-SVMs contain regularization which tackles the problem of overfitting. Instead of applying this approach to protein function in yeast which requires the reformulation of the problem in 13 binary classification problems (equal to the number of different functional classes), we applied a modified version of this framework in the patient space where many of the prediction problems are already binary. To the author's knowledge, this is the first time that a kernel-based integration method has been applied on multiple high dimensional data sets in the patient domain for studying cancer. Our results show that using information from different levels in the central dogma improves the classification performance.

We already mentioned that kernel methods have a large scope due to their representation of the data. However, when the amount of available data will increase in the near future, the choice of the weights becomes more important, especially when applying the algorithm to problems where the reliability of the data sources differs much or is not known a priori. In this paper, we chose the weights equally. We cannot guarantee though that without optimizing the weights of the different data sources we get the most optimal model. However, this increases the computational burden significantly.

When more data sources will become available in the future, they can be easily added to this framework. Additionally, we are currently investigating ways to improve the optimization algorithm, especially for the choice of the weights. Next, we will also apply more advanced feature selection techniques. At this moment a simple statistical test is used but more advanced techniques could be applied. Finally, we will compare kernel methods with other integration frameworks (e.g. Bayesian techniques).[25]

Acknowledgments

AD is research assistant of the Fund for Scientific Research - Flanders (FWO-Vlaanderen). This work is partially supported by: **1.** Research Council KUL: GOA AMBioRICS, CoE EF/05/007 SymBioSys. **2.** Flemish Government: FWO: PhD/postdoc grants, G.0499.04 (Statistics), G.0302.07 (SVM/Kernel). **3.** Belgian Federal Science Policy Office: IUAP P6/25 (BioMaGNet, 2007-2011). **4.** EU-RTD; FP6-NoE Biopattern; FP6-IP e-Tumours, FP6-MC-EST Bioptrain.

References

1. N Cristianini and J Shawe-Taylor, *Cambridge University Press*, (2000).
2. J Shawe-Taylor and N Cristianini, *Cambridge University Press*, (2004).
3. G Lanckriet, T De Bie *et al.*, *Bioinformatics*, **20**(16), 2626 (2004).
4. B Schölkopf, K Tsuda and J-P Vert, *MIT Press*, (2004).
5. W Stafford Noble, *Nature Biotechnology*, **24**(12), 1565 (2006).
6. T De Bie, L-C Tranchevent *et al.*, *Bioinformatics*, **23**(13), i125 (2007).
7. N Pochet, F De Smet *et al.*, *Bioinformatics*, **20**(17), 3185 (2004).
8. J M D Wheeler, B F Warren *et al.*, *Dis Colon Rectum*, **45**(8), 1051 (2002).
9. J-P Machiels, C Sempoux *et al.*, *Ann Oncol*, **18**, 738 (2007).
10. R A Irizarry, B Hobbs *et al.*, *Biostatistics*, **4**, 249 (2003).
11. V Vapnik, *Wiley, New York* (1998).
12. J Suykens and J Vandewalle, *Neural Processing Letters*, **9**(3), 293 (1999).
13. J Suykens, T Van Gestel *et al.*, *World Scientific Publishing Co., Pte Ltd. Singapore* (2002).
14. P Pavlidis, J Weston *et al.*, *Proceedings of the Fifth Annual International Conference on Computational Molecular Biology*, 242 (2001).
15. H Deng, R Makizumi *et al.*, *Exp Cell Res*, **313**, 1033 (2007).
16. J A Nemeth, M L Cher *et al.*, *Clin Exp Metastasis*, **20**, 413 (2003).
17. D Raje, H Mukhtar *et al.*, *Dis Colon Rectum*, **50**, 1 (2007).
18. T Shimizu, S Tanaka *et al.*, *Oncology*, **59**, 229 (2000).
19. T Sasahira, T Sasaki and H Kuniyasu, *J Exp Clin Cancer Res*, **24**(1), 69 (2005).
20. T-D Kim, K-S Song *et al.*, *BMC Cancer*, **6**, 211 (2006).
21. K Zins, D Abraham *et al.*, *Cancer Res*, **67**(3), 1038 (2007).
22. S O Lee, J Y Chun *et al.*, *The Prostate*, **67**, 764 (2007).
23. A L Hallbeck, T M Walz and A Wasteson, *Bioscience Reports*, **21**(3), 325 (2001).
24. The Gene Ontology Consortium, *Nat Genet*, **25**, 25 (2000).
25. O Gevaert, F De Smet *et al.*, *Bioinformatics*, **22**(14), e184 (2006).

A BAYESIAN FRAMEWORK FOR DATA AND HYPOTHESES DRIVEN FUSION OF HIGH THROUGHPUT DATA: APPLICATION TO MOUSE ORGANOGENESIS

MADHUCHHANDA BHATTACHARJEE

School of Mathematics and Statistics, University of St Andrews
St Andrews, Fife, Scotland, KY16 9SS, UK

COLIN PRITCHARD & PETER NELSON

Division of Human Biology, Fred Hutchinson Cancer Research Center
Seattle, WA 98109-1024, USA

In this paper we present a framework for integrating diverse data sets under a coherent probabilistic setup. The necessity of a probabilistic modeling arises from the fact that data integration does not restrict to compiling information from data bases with data that are typically thought to be non-random. Currently wide range of experimental data is also available however rarely these data sets can be summarized in simple output data, e.g. in categorical form. Moreover it may not even be appropriate to do so. The proposed setup allows modeling not only the observed data and parameters of interest but most importantly to incorporate prior knowledge. Additionally the setup easily extends to facilitate more popular data-driven analysis.

1. Introduction

1.1. *Challenges in Data Integration*

It has been realized that in order to address biological questions more fully and to extract more knowledge from the wealth of data, researchers require tools that will allow them to integrate different datasets in a dynamic, hypothesis-driven fashion and to analyze them within a biologically meaningful framework[13].

However integration is often mistaken as making vast amount of data available to the researcher by warehousing or other methods. It is often seen that integration of large number of data sets in such a manner results in a messy incomprehensible scenario. Such output might contain vast amount of biological information however fails to generate testable hypotheses and theories that with proper validation may add to our knowledge. It is also not uncommon that a painstaking effort to integrate information sources has produced rather trivial observations on the system.

Advancement in computing abilities makes it plausible to deal with large amount of data; unfortunately it is often done in an adhoc manner. On the other hand currently more systematic approaches are also confined to amalgamating specific databases to a single experiment. There is a growing support to the idea that a more hypotheses driven choice of data sources should be made which then needs to be carefully analyzed[7].

1.2. *Challenges in Statistics Inference*

High throughput molecular biology techniques have posed new challenges for statistics, and with this we have seen some more adventurous use of statistics. Although data sets are typically large, the number of features is also large. This added to the fact that the features frequently depart from the i.i.d. set-up, makes a credible feature level inference near impossible. Moreover if this unstructured and unknown dependence is not accounted population level inference also becomes inaccurate.

An additional level of complication is introduced by the fact that the process of obtaining meaningful results involves numerous decision-making steps. Apart from a few attempts[8,12] this aspect is generally not discussed and consequently not accounted in the analysis and inference.

1.3. *Objective*

The initial findings from an analysis of a high-throughput data are often messy due to several reasons. The level of specificity of the phenotype is an important factor behind this. For most diseases there would be multiple known factors affecting the overall variability. For practical reasons it is difficult to design experiments where one would be able to account for all these factors. The hence uncontrolled factors would contribute to the observed variability. Additionally although we target to study a specific aspect of the system, as cells continue to live independent of the experiment or disease under study, changes in normal life functions also affect the results[16].

Evidently some additional information is needed to identify relevant quantities from such an inference. Most experiments are designed and carried out using prior knowledge of the conditions, diseases or treatments under study and can be used critically to control variation. We would use such experimental data, which are complete, meaningful and possibly complementary to each other. This can be thought as the hypothesis-driven part of the data fusion.

For most phenotype/disease there will be limited number of such experimental data available which are useful. The knowledge from existing databases can then be augmented to obtain a better understanding of our findings, which is the purely data driven aspect of the investigation.

We will derive an integrated modeling and inference procedure on such data, which are high dimensional data sets of varied data types from various sources. Integration of these sources utilizes existing knowledge of the phenotype of interest. We will carry out population as well as individual level inference, in presence of dependence, combining multiple decision-making steps that allows for propagation of error.

2. Data

The experimental data consists of one time-course data and three other data sets. The biological objective behind the choice of these experiments is to study developmental behavior of prostate, preferably at cell-type level, highlighting behavior of key genes like androgen regulated/responsive ones. These data sets were collected as a first part of a two-part study of Prostate cancer.

2.1. *Molecular Characteristics of Developing Mouse Prostate*

To identify genes potentially involved with prostate development, temporal expression changes were determined by measuring transcript abundance levels in cDNA libraries constructed from distinct stages of maturation. A purpose-built cDNA microarray[4,5] enriched for genes in the developing mouse prostate, which serves as a unique resource for molecular studies of prostate development were utilized.

Table 1. Summary of biological samples used for timecourse microarray experiment, where UGS: Uro-genital-sinus, DLP: dorsolateral prostate, VP: ventral prostate AP: anterior prostate.

Time point	Tissue	Developmental Process	Androgen level [1]
E15.5	♂UGS	Undifferentiated	2 days exposure
E16.5	♂UGS	Undifferentiated	3 days exposure
E17.5	♂UGS	Prostate buds	High
E18.5	♂UGS	Branching morphogenesis	High
Day 7	DLP,VP,AP	Branching morphogenesis (Peak)	Low
Day 30	DLP,VP,AP	Puberty	Very High
Day 90	DLP,VP,AP	Adult, fully differentiated	Very High

The transcriptional program of prostate development was characterized by profiling gene expression at seven time points corresponding to critical stages of prostate differentiation (Table 1). For each time point three biological samples were generated and each sample was hybridized twice for a total of 42 microarrays. The samples were hybridized with a common reference RNA consisting of embryonic age 14.5 days (E14.5) male UGS. At E14.5, UGS is undifferentiated and has not been exposed to significant levels of androgens, therefore is thought as 'start' for prostate development. Thus, the microarray ratios depict the unfolding program of prostate development in relationship to the most undifferentiated state.

2.2. *Characterizing cell-type specific expression heterogeneity*

The interpretation of temporal changes in gene expression from whole tissue is complicated by cell heterogeneity. The developing prostate can be roughly broken down into two major cell compartments: epithelium and mesenchyme/stroma. Reciprocal signaling between mesenchyme and epithelium is critical for prostate differentiation. Mesenchyme is the precursor to the adult stroma, Urogenital sinus mesenchyme (UGM) induces epithelial budding and the epithelium, in turn, stimulates mesenchymal differentiation. In subsequent tissue recombination experiments it was shown that several different types of epithelium of endodermal origin can form prostate in combination with an inducing mesenchyme. Thus, there is some inductive promiscuity in both epithelial and mesenchymal compartments.

2.3. *Androgen Response Program of the Developing Mouse Prostate*

Androgens act via the UGS mesenchyme to induce prostatic epithelial development, presumably through a paracrine mechanism. Yet no androgen-regulated genes have been identified in the UGS mesenchyme.

Using a custom cDNA microarray enriched for genes expressed in the developing mouse prostate, three in vivo strategies were adopted to identify androgen-regulated genes at the time of prostate induction. We compared (1) male UGS to female UGS, (2) female UGS dosed with testosterone *in vivo* to female dosed with placebo, and (3) wild-type male UGS to androgen receptor-deficient (*tfm*) male UGS. Each comparison is a distinct way of assessing androgen-regulated genes in the UGS. Three biological replications and two array replications were performed for each comparison at both E16.5 and E17.5 for thirty-six total microarray experiments.

2.4. *Gene-Ontology and KEGG pathways data*

The Gene Ontology consortium (GO)3 and Kyoto Encyclopedia of Genes and Genomes (KEGG)2 databases enable statistical analysis of biological processes or pathways that may be enriched or depleted in a certain experiments.

We considered 208 Biological process, 64 Cell components and 151 Molecular functions from the Gene Ontology database. Data for the present analysis was from all nodes up to level 6 that were represented by at least ten genes to ensure that functional conclusions were not drawn from very few genes. As GO terms have multiple parents, the completed trees based on these nodes consisted of 462 nodes for Biological processes, 84 nodes for Cellular Components and 185 nodes for Molecular Functions. From the KEGG database

20 specialized pathways were chosen based on their relevance to the biological problem undertaken here.

2.5. *Overall data summary*

In the overall analysis the data comes from a varied source and is schematically presented in Figure 1. The current practice of handling multiple experimental data would be to work with very crude summaries, e.g. list if differential (or otherwise interesting) genes, ignoring the fact that almost always such lists were outcome of decision-making and decisions were not taken with 100 percent confidence. Some associated measure of confidence should be utilized.

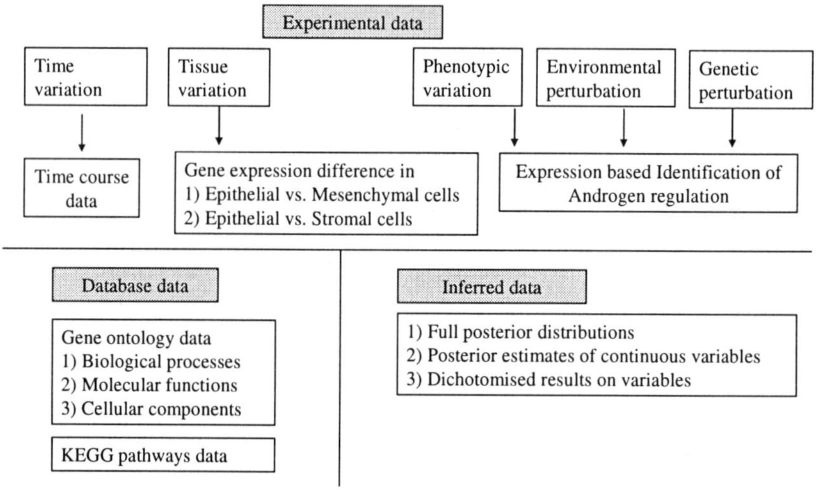

Figure 1. Schematic Presentation of all data sources used for integrated modeling

Moreover in some aspects of analysis such summarization is potentially misleading. For example, many have experienced that normalization affects subsequent biological conclusions from a microarray data analysis. One solution would be carry out a model based normalization and use the joint distribution of the (distinct) genes for further analysis[6,8]. Similarly while integrating gene characteristics from one experiment to another the joint distribution of genes with respect to the (conditioning) characteristic could be used in the subsequent experiment.

Such probabilistic summarization is denoted as inferred data and it reflects our knowledge and confidence on the data. The best possible usage of one source of data would be using the whole joint distribution, if that is too large or complex then a summary (possibly first two marginal moments for each feature) can be used. In some situations we might have to use the categorical (typically binary) summarization of a data.

3. Model

We will describe the model for the time course data. For the three additional experiments the models are similar in principle with appropriate parameterization for eliciting the desired characteristics. We will derive the model in a stepwise manner aiming to describe the parameters along with their utilities. Note that in order to exploit conjugacy we parameterized the Normal distributions using mean and precision (i.e. inverse-variance).

3.1. *Normalization of individual data set*

Normalization is carried out at block level for each array using constrained piecewise linear models (in Bayesian framework). The range of values of the log-intensities from the reference sample was divided into three windows, with breakpoints chosen to be at 5 and 7. The normalized data thus produced is highly comparable with standard loess type normalization[8], however has the advantage of being model based hence allows for propagation of error to next stage of analysis. The linear model parameters for the normalization are denoted (and described) as follows:

$\beta_{ljkm} \sim N(1, 0.1)$ and $\alpha_{ljk2} \sim N(0, 0.1)$ where $\alpha_{ljk1} = \alpha_{ljk2} + (\beta_{ljk2} - \beta_{ljk1})*5$ and $\alpha_{ljk3} = \alpha_{ljk2} + (\beta_{ljk2} - \beta_{ljk3})*7$, l:tissue/time-point, j:array, k:blocks on each array and m:1, 2 and 3 (number of windows).

Let LIR and LIE denote the Log intensities from reference and experimental samples respectively. In the rest of model description, following notation will be used, l: tissue, i: spot, j: array within l-th tissue, b(i): print tip/ block number of i-th spot, w(lij): window number for LIR_{lij} for the l-th tissue, i-th sopt on j-th array, d(i): distinct gene corresponding to i-th spot on the array. Note the arrays contain multiple spots/probes for several genes; however this was not designed in a balanced manner. The (incomplete) models for LIR and LIE experimental samples were:

$LIR_{lij} \sim N(\mu_{d(i)}, 0.1)$ and $LIE_{lij} \sim N((\alpha_{ljb(i)w(lij)} + \beta_{ljb(i)w(lij)} * LIR_{lij}),)$.

3.2. *Characterizing gene-expression behavior within an experiment:*

Assume that, a priori each gene has its own expression ratio, say θ^0_k, where $\theta^0_k \sim N(0, 0.1)$, with k = 1,..., no. of distinct genes. Let the expression ratio for the k-th gene from the l-th tissue be θ^1_{lk}, where θ^1_{lk} are assumed to be drawn from a Normal distribution with mean θ^0_k, i.e.

$\theta^1_{lk} \sim Normal(\theta^0_k, \tau^0_k)$ and let $\tau^0_k \sim Gamma(1, 1)$.

The completed model for LIE_{lij} is as follows:

$LIE_{lij} \sim Normal((\alpha_{ljb(i)w(lij)} + \beta_{ljb(i)w(lij)} (LIR_{lij} + \theta^1_{ld(i)})), \tau^1_{ld(i)})$.

The available knowledge indicated several possible profiles for cohorts of genes during mouse prostate differentiation that would be of interest. Such profiles could be explored based on posterior behavior of different functions of the parameters (θ^1_{lk} , τ^1_{lk}) and also possibly other parameters. For example for any time point upregulated genes can be identified by the posterior probability of (studentized) θ^1_{lk} exceeding a pre-specified cut-off (e.g. Normal-distribution percentile) and such probability estimates have been noted[8] to be monotonically related to the d-scores obtained using well-known SAM.

3.3. *Characterizing gene-expression behavior across experiments:*

For k-th gene let μ^1_k, μ^2_k be the expression parameter describing epithelium & mesenchyme specific behavior. Similarly v^j_k be androgen response of gene k in platform j, j=1, 2, 3. To infer whether a gene exhibits branching morphogenesis profile while being epithelium specific and androgen regulated we use posterior distribution of the variable $I(\theta^1_{1k}<\theta^1_{2k}<\theta^1_{3k}<\theta^1_{4k}<\theta^1_{5k}$ and $\theta^1_{7k}<\theta^1_{6k}$, $\mu^1_k>\varphi_{0.95}$, $v^*_k>\varphi_{0.95})$, where v^* is function of v^j's and φ is Normal percentile. If the data sets and genes exhibit varied precision then standardized parameters are used for the indicators. These may still be biased by the size (or variability) of any particular data set. Hence as a cautionary measure the information flow was allowed only from individual experimental data to the integrated analysis and not otherwise. This is easily implemented in WinBUGS framework using "cut"-function appropriately.

3.4. *Probabilistic assessment of biological processes enrichment*

For expression profile of interest the basic functional enrichment analysis is as described in (8). Apart from enrichment testing we utilized several summary assessment of functional enrichment. For example, the log-ratio of proportions of genes with certain functionality in S and Sc can be thought as similar to "expression" ratio for that functionality. These can then be visualized as heat map. These we have further analyzed using standard clustering techniques which have provided useful insight into functional pattern over time.

3.5. *Overall analysis setup*

The overall setup allows us to analyze each microarray based experiment in quite extensive manner. For each of these data sets can be analyzed individually, additionally our setup allows for modeling these data sets where desired output from one analysis along with it the uncertainty involved in decision making/inference is carried to analysis of another data.

One major achievement of this data analysis is the magnitude of the statistical problem that we succeeded in implementing using freely available software WinBUGS without having to write custom MCMC codes. This opens up numerous modeling possibilities to address complex biological questions.

To give an idea of the number of parameters being jointly monitored in this implementation, consider the following: normalization parameters 12 000+, hyperparameters 40 000+, expression parameters 220 000+, for each of 10 000+ genes 13 expression profiles from individual dataset analyses and 30 expression profiles using two datasets at a time and 20 profiles for all three data sets, 750+ GO-KEGG processes enrichment for each of these profiles. Approximately posterior distributions of 1 million variables of interest are jointly monitored under this setup. Additionally all missing data points are augmented which typically increases with size of experiments.

4. Results

The analyses of individual experimental data sets and different combination of their integration brought out many interesting results. This is due to the nature of these experiments and also due to the flexibility of the model parameterization. In the following we describe a few such outputs from the analyses.

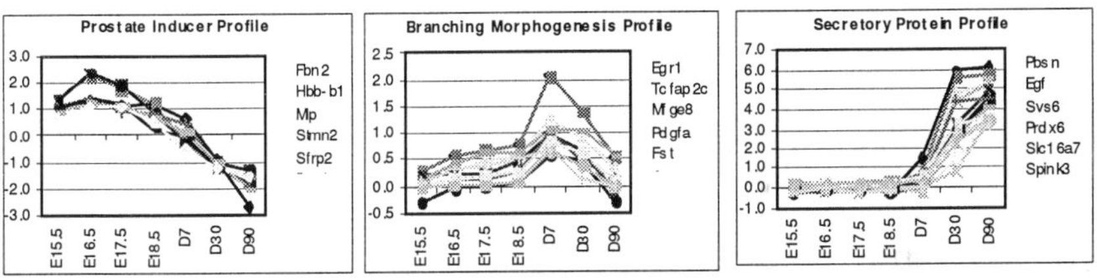

Figure 2. The critical profiles (known from pathological information) were translated into functions of expression parameters. The figures present (log) expression change of top genes with high posterior probability of having these profiles during mouse prostate differentiation.

4.1. *Analysis of time course data*

By deriving the posterior distribution of different constrained combinations of the parameters we can identify genes having specific expression profiles over time. In Figure-2 we present some of the genes estimated to have high probability of having some known profiles. Some of the genes thus identified were already known and some new ones have been verified subsequently.

In the joint analysis of the Gene Ontology information and the time course data, several functionalities clearly depict the distinctly different behavior in the two phases of life, namely embryonic and otherwise. However our model treats all time points were equally, which gives us more confidence in our findings.

By analyzing the heat map of functional enrichment on time course, we were able to identify very interesting clusters of functions whose profile correspond to the known prostate development profiles (see Figure 3).

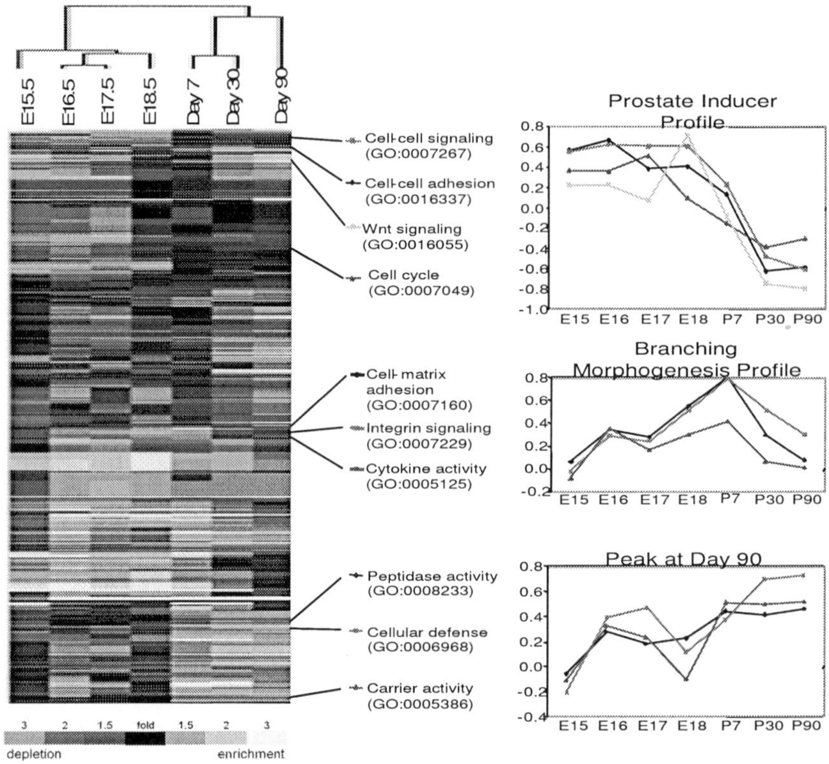

Figure 3. Examples of GO terms of interest are listed according to the time-point of peak representation and position in the heat-map, with a detailed plot of their behavior over the time-course. The data for the heatmap was generated by the methods described in Section 3.4.

4.2. *Integrated analyses of multiple experimental data sources*

The integrated analysis of the cell-type specific and the time course experiments yielded fascinating expression diversity within the developing tissues. In Table-2 we present a GO-based interpretation of these expression behaviors. The resulting findings of this integrated analysis were in high concordance with existing knowledge and our hypothesis. The distinctive nature of the two cell types is clearly visible along with their time stochastic nature. The tree structure of GO provides additional information on change in pattern over time.

By jointly analyzing all the experiments we are able to explore in each cell-type the expression profiles over time of the Androgen responsive genes. The major Androgen responsive genes showed two distinct expression profiles over time, one with highest expression observed at E16.5 and another where highest expression occurs in adult state (D90). The cell-type specific expression data indicated that the early expressing gene was in Messenchymal/Stromal cells (e.g. Sfrp2) where as the adult ones were expressed in Epithelium (e.g. Agr2 and Mmp7).

Table 2: Functional analysis of cell-type specific genes up-regulated at a certain time point. The entries represent estimated (log) change in enrichment. Upregulated functions have been highlighted in white against dark-grey-background and those downregulated in black against light grey-background ("###" indicates high negative value).

GO	Epithelium							Mesenchyme						
	E15.5	E16.5	E17.5	E18.5	D7	D30	D90	E15.5	E16.5	E17.5	E18.5	D7	D30	D90
Development	0.5	0.5	0.6	0.7	0.5	0.4	0.2	1.1	1.3	1.2	1.4	1.3	1.2	1.3
Morphogenesis	0.3	0.4	0.5	0.6	0.4	0.2	0.1	1.3	1.5	1.5	1.6	1.5	1.5	1.5
Organogenesis	0.2	0.3	0.4	0.4	0.1	-0.1	-0.3	1.4	1.6	1.6	1.7	1.7	1.6	1.7
Neurogenesis	-1.1	-0.5	0.2	-0.6	0.3	-0.1	-0.4	1.4	1.3	1.1	0.9	0.8	0.0	-1.7
axonogenesis	-1.8	-2.5	-1.8	-1.9	-2.3	-3.2	-5.3	1.7	1.5	0.9	0.6	0.3	-3.4	-3.0
morphogenesis of an epithelium	0.5	0.9	1.1	1.5	1.5	1.7	1.4	-3.9	-6.1	-7.5	-4.4	-3.3	###	-4.8
epithelial cell differentiation	-2.8	-0.1	1.3	1.7	1.7	1.4	1.4	-4.5	###	###	-5.3	-4.1	###	-4.1
epithelial to mesenchymal transition	-1.8	0.9	2.3	2.7	2.7	2.4	2.4	###	###	###	###	###	###	###
morphogenesis of embryonic epithelium	1.3	1.5	1.1	1.5	0.6	1.7	1.7	###	###	###	###	###	###	###
Growth	0.1	-0.2	0.1	0.1	0.1	0.4	0.2	1.7	1.2	1.4	1.6	1.5	1.8	1.6
cell growth	0.3	0.0	0.5	0.4	0.4	0.7	0.6	2.0	1.6	1.7	1.9	1.9	2.1	1.9
regulation of cell growth	0.5	0.2	0.7	0.6	0.6	0.9	0.8	2.1	1.6	1.7	2.0	1.9	2.1	1.8

Amongst the Androgen-responsive genes that are epithelium specific we noticed three major expression patterns (see figure 4), 1) higher expression in late embryonic state (e.g. Anxa1, Psca) 2) higher expression at infant stage (e.g. Itgb4, Sox9) and 3) high expression in adult stage (e.g. Aldh1a1, Agr2, Cldn8).

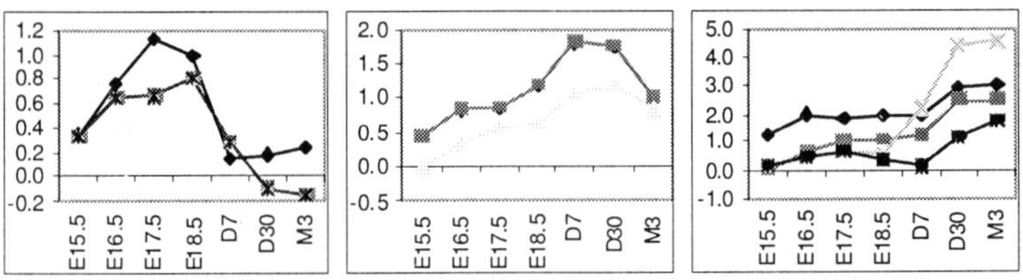

Figure 4. Developmental expression profile of Androgen-responsive genes expressed in Epithelium.

4.3. *Experimental and literature based cross-validation*

For Sfrp2 real-time RT-PCR at each of the seven time points were performed and results were highly concordant with the microarray measurements. In situ hybridization confirmed mesenchymal expression of Sfrp2. Quantitative PCR confirmed that Sfrp2 is up regulated with androgen in the UGS. Agr2 is a gene that is thought to be involved in breast cancer metastasis[14]. Subsequent analysis with prostate cancer related data has shown this gene to be significant. Mmp7 is a gene that has been previously shown to influence cancer progression[15].

Psca is a known to be androgen regulated and is associated with prostate cancer. A further experiment using whole mount in situ hybridization showed Psca was highly expressed in epithelium. Sox9 have been shown to be directly relevant for tumor suppression in Prostate[10] and in some organs domain of expression of Sox9 protein is normally known to be the distal epithelial compartment[9,11].

5. Discussion

The composite data we consider for this analysis comprises of data from several experiments, which are meaningful and complete by themselves. The hypothesis driven fusion of these enabled us to reduce the experiment size from 7x2x3 to 7+2+3 experiment for time points, cell-types and Androgen-platforms. Even if we had the resource to do the full experiment some of these may not have been biologically feasible to do in reality.

The biological hypotheses were translated in statistical framework as functions of parameters (e.g. expression) and were assessed a-posteriori. The different functional combinations of the model parameters cover a wide range of biological characteristics to be studied. This is another aspect of our modeling setup that is not easily available in the commonly used statistical tools, simply because complex hypotheses would require specialized testing procedure which may not be available readily. Complexity of individual modeling units was kept moderate to optimize computation time and parameter space of interest.

While analyzing using existing models quite often we observe that even moderate change in analysis technique for one data set & for any single step of analysis can influence overall biological conclusions. It is well-known that this happens due to not propagating uncertainties in these analysis steps to subsequent steps. The Bayesian setup proposed in (8) enables us to avoid this problem and was extended here to a much larger problem. Analytic intractability is a common consequence of such complex models. In this respect a mentionable achievement here is being able to implement this integrated model using available software, opening up varied modeling and input data-type possibilities.

Our objective was to be able to balance between quantity of data and quality of inference. Although one would be tempted to use as much data as possible we need to remember a few aspects of these data. Most experimental data come with a lot of error/noise. Using only a summary from each of these ignoring the noise potentially can (and often does) lead to non-reproducible results. This is where robust inference method is crucially needed and is provided by our method.

References

1. C. Corpechot, E.E. Baulieu and P. Robel, *Acta Endocrinol (Copenh)* 96, 127-35 (1981).
2. M. Nakao, H. Bono, S. Kawashima, T. Kamiya, K. Sato, S. Goto, and M. Kanehisa, *Genome Inform Ser Workshop Genome Inform 10*, 94-103 (1999).
3. M. Ashburner, C.A. Ball, J.A. Blake, D. Botstein, H. Butler, J.M. Cherry, A.P. Davis, K. Dolinski, S.S. Dwight, J.T. Eppig, M.A. Harris, D.P. Hill, L. Issel-Tarver, A. Kasarskis, S. Lewis, J.C. Matese, J.E. Richardson, M. Ringwald, G.M. Rubin and G. Sherlock, *Nat Genet* 25, 25-9 (2000).
4. P.S. Nelson, C.C. Pritchard, D. Abbott and N. Clegg, N. *Nucleic Acids Res* 30, 218-20 (2002).
5. D.E. Abbott, C.C. Pritchard, N.J. Clegg, C. Ferguson, R. Dumpit, R.A. Sikes and P.S. Nelson, *Genome Biol* 4, R79 (2003).
6. M. Bhattacharjee, C.C. Pritchard, M.J. Sillanpää and E. Arjas, *Proceedings of the CAMDA02 Conference*, Ed. Johnson K. and Lin, S., Kluwer-Academic Publishers (2003).
7. J.H.G.M. van Beek, Comp Funct Genom; 5: 201–204, (2004).
8. M. Bhattacharjee, C.C. Pritchard, P.S. Nelson and E. Arjas, *Bioinformatics*, 20, 2943-2953 (2004).
9. P. Blache, M. van de Wetering, I. Duluc, C. Domon, P. Berta, J.N. Freund, H. Clevers, P. Jay, *J Cell Biol*, 166 (1): 37-47, (2004).
10. R. Drivdahl, K.H. Haugk, C.C. Sprenger, P.S. Nelson, M.K. Tennant, S.R. Plymate, *Oncogene*, 23 (26): 4584-93, (2004).
11. T. Okubo, P.S. Knoepfler, R.N. Eisenman, B.L. Hogan, *Development*, 132 (6): 1363-74, (2005).
12. M. Bhattacharjee and M.J. Sillanpää, (To appear in) *Proceedings of CAMDA 2006*.
13. M. Mirona and R. Nadon, *Trends in Genetics*, 22 (2), 84-89, (2006).
14. C.L.Wilson, A.H. Sims, A. Howell, C.J. Miller, R.B. Clarke. *Endocr Relat Cancer*, 13 (2): 617-28, (2006).
15. D. Bianchi-Frias, C.C. Pritchard, B.H. Mecham, I.M. Coleman, P.S. Nelson, *Genome Biol*, 8 (6): R117, (2007).
16. T. Werner, *Mechanisms of Ageing and Development*, 128, 168–172, (2007).

GATHERING THE GOLD DUST: METHODS FOR ASSESSING THE AGGREGATE IMPACT OF SMALL EFFECT GENES IN GENOMIC SCANS*

MICHAEL A PROVINCE, PH.D.

AND

INGRID B BORECKI, PH.D.

Division of Statistical Genomics Box 8506
Center for Genome Sciences
Washington University School of Medicine
4444 Forest Park Blvd
St. Louis, MO, 63108, USA

Genomewide association scan (GWAS) data mining has found moderate-effect "gold nugget" complex trait genes. But for many traits, much of the explanatory variance may be truly polygenic, more like gold dust, whose small marginal effects are undetectable by traditional methods. Yet, their collective effects may be quite important in advancing personalized medicine. We consider a novel approach to sift out the genetic gold dust influencing quantitative (or qualitative) traits. Out of a GWAS, we randomly grab handfuls of SNPs, modeling their effects in a multiple linear (or logistic) regression. The model's significance is used to obtain an iteratively updated pseudo-Bayesian posterior probability associated with each SNP, which is repeated over many random draws until the distribution becomes stable. A stepwise procedure culls the list of SNPs to define the final set. Results from a benchmark simulation of 5 quantitative trait genes among 1,000, in 1,000 random subjects, are contrasted with marginal tests using nominal significance, Bonferroni-corrected significance, false discovery rates, as well as with serial selection methods. Random handfuls produced the best combination of sensitivity (0.95) specificity (0.99) and true positive rate (0.71) of all methods tested and better replicability in an independent subject set. From more extensive simulations, we determine which combinations of signal to noise ratios, SNP typing densities, and sample sizes are tractable with which methods to gather the gold dust.

1. Introduction

The Gold Rush of the 1840s and 50s produced a flood of prospectors in the American West. Fortified by dreams of easy discovery and driven by a desire for great fame and wealth, the only thing that separated the bold visionaries from

* This work is partially supported by NIH grants HL087700, HL088655 DK068336, and AG023746.

the reckless fools was the luck of where they staked their claims. Today, a "Genetic Gold Rush" is taking place. Scientists compete with one another to be the first to find novel complex disease genes. Using the increasingly affordable technology of Genome-Wide Association Scan (GWAS) SNP chips, new scientific prospectors are becoming inspired by the early successes such as macular degeneration[1] and obesity[2]. Like the initial 1844 discovery of gold at Sutter's Mill, these first GWASes have generated much excitement and high expectation that all such searches will be simple, straightforward, and lucrative. With talk of "low hanging fruit" there is much optimism that the dream of personalized medicine may actually be around the corner.

While the first GWASes appear to have discovered some new signals, these do not appear to explain a large portion of the variance in the target traits that our heritability estimates would indicate are in the genome. For example, the FTO gene identified by Frayerling et al., (2007)[2] is homozygous in only 16% of adults (in Caucasian populations) and increases risk for obesity by 1.67-fold (approx 3kg of weight on average). As the total heritability for BMI is roughly in the 50% range, the FTO gene is clearly only one small "gold nugget" in the entire genomic treasure. Over 40 years experience with painstaking candidate gene work in humans would indicate that there may be many genes of small effect for complex traits. For instance, the AGT gene has been studied since the 1980s as a strong candidate for hypertension, but has been estimated in large populations to explain only 0.1% of the variance[3] which may explain why some studies have detected an effect while others have not. Many genomic scanning techniques will fail to find such small effect size genes, which individually are trivial, but which in the aggregate may explain a substantial part of the variance, especially when epistatic interactions are considered[45]. Instead of concentrating only on finding the relatively few gold nuggets, perhaps we should consider ways to gather many small effect "gold dust" variants, not because any one grain is by itself important, but because in the aggregate, a pouch of gold dust can be very valuable.

2. Methods

We consider several alternative methods for "gathering the gold dust" in a genomic scan.

2.1. *Univariate Screening*

The traditional method for identifying signals in a GWAS is to screen one SNP at a time, and choose the most significant SNPs. In the univariate screening category, we considered 3 variations, choosing:

1. N = all SNPs that are nominally significant (α= 0.05)
2. B = all SNPs that are significant at the Bonferroni level (α = 0.05/M)
3. F = all SNPs that are significant using the False Discover Rate.

Intuitively, if there are no epistatic interactions between SNPs, and all signals operate additively and independently, then there is little information to be gained about the impact of any one SNP on the phenotype by considering other SNPs simultaneously. In that case, we might expect that some kind of univariate screening procedure which evaluates the marginal effects of each SNP would give the most efficient and powerful estimates of the genomic signals. However, if the actions of genetic signal variants are more complex, if there are epistatic interactions, genes which down and up regulate the action of other genes, AND/OR logic "gate-keeper" variants which are required to be present for other SNPs to have any appreciable impact, particular haplotypic combinations that increase or decrease phenotypic risk, or even environmental factors that potentiate and reveal a set of related genes in a pathway, then we may miss such signals in univariate screening, since we only examine the marginal impact of each SNP. In that case, we might prefer a method which examines SNPs in combination, using a multivariate modeling approach.

2.2. *Random Handfuls*

The random handful approach is a pseudo-Bayesian algorithm, in which we serially update information that a SNP is a signal for a phenotype, based upon the results of randomly drawn multivariate models predicting that phenotype. Let $\mathfrak{I}$ = { S_i | i =1,2, …, M} be the set of SNPs in a GWAS, of size "M." Let H* be the set of true signal SNPs for a fixed phenotype Y. SNPs in H* are either themselves causative functional variants (which would be the most extraordinary luck), or more likely, they are SNPs in LD with causative variants. Then ($\mathfrak{I}$ \ H*) is the set of noise SNPs for Y.

Let $P_0[S_i \in H^*]$ be any prior probability density function (p.d.f.) on $\mathfrak{I}$. If we wish to remain agnostic about the genetic causes of Y, then the initial priors will be flat for all SNPs, i.e. $P_1[S_i \in H^*]$ = (1/M). If we wish to incorporate prior knowledge about the genetic architecture of Y, either from some other association or linkage scan, or from a microarray experiment, or from biological knowledge of the genetic pathways, then we may suitably choose some other prior $P_1[S_i \in H^*]$.

Let H ≡ {S_j}, be a set of SNPs of a fixed size (say L), drawn at random from $\mathfrak{I}$, which we will call a "random handful". Let M(H) denote the multivariate model predicting Y from H. If Y is a continuous phenotype (e.g. lipid level), M(H) is a multivariate regression model:

$$E[Y \mid H] = \alpha^H + \Sigma_j (\beta^H{}_j S_j) \tag{1}$$

If Y is a categorical outcome (e.g. diabetes), then M(H) is a multivariate logistic regression model:

$$\text{Prob}[Y \mid H] = 1 / (1 + \exp(\alpha^H + \Sigma_j (\beta^H{}_j S_j)) \tag{2}$$

If Y is a survival outcome (e.g. age-at-onset of hypertension), then M(H) is a multivariate proportional hazards model:

$$Y(t)\, d\,\Lambda(t, H) = Y(t) \exp(\alpha^H + \Sigma_j (\beta^H{}_j S_j))\, d\,\Lambda_0(t, H) \tag{3}$$

where $\Lambda_0(t, H)$ is the baseline hazard function

In all three cases, it is trivial to include additional fixed effect covariates into these models, such as age, sex, diet, lifestyle, exposures, etc. by simply adding more linear combination terms, i.e. the extended model becomes $(\alpha^H + \Sigma_j (\beta^H{}_j S_j) + \Sigma_j \gamma_j X_j)$ for covariates X_j. Even when they are not the main focus of our research, proper modeling of important covariates can reduce unexplained variance and therefore increase our power to detect gene signals (e.g. see Province et al., 2004[6]). However, as incorporation of such covariates is simple in both the random handful and the univariate screen approaches, in order to keep the notation crisp and to maintain our focus, we will ignore this complication here and concentrate on the case of no non-genetic covariates.

Based upon the results of any model M(H) predicting Y, we can update information on the probability that each SNP S_i in H is amongst the signals for Y (i.e. that $S_i \in H^*$). If M(H) predicts Y well, then it is more likely that H contains <u>some</u> signal SNPs, so we raise the posterior probability that each S_i in H is a signal. Conversely, if M(H) predicts Y poorly, then the component SNPs in H are less likely to be signals, so that $S_i \notin H^*$ is more probable and we lower the probability that it is a signal. As we randomly pick many random handfuls of SNPs and evaluate each of those multivariate models, any given SNP S_i will be sampled many times in the context of many other background SNPs (many different Hs), so we get better and better estimates of the probability that S_i is a signal. At stage 1, we use the intial prior $P_1[S_i \in H^*]$. With each new random handful, H_k, we serially update the posterior probabilities for each SNP S_i, so that the posterior from the kth random handful becomes the prior for the next $(k+1)^{th}$ random handful, as depicted in Figure 1.

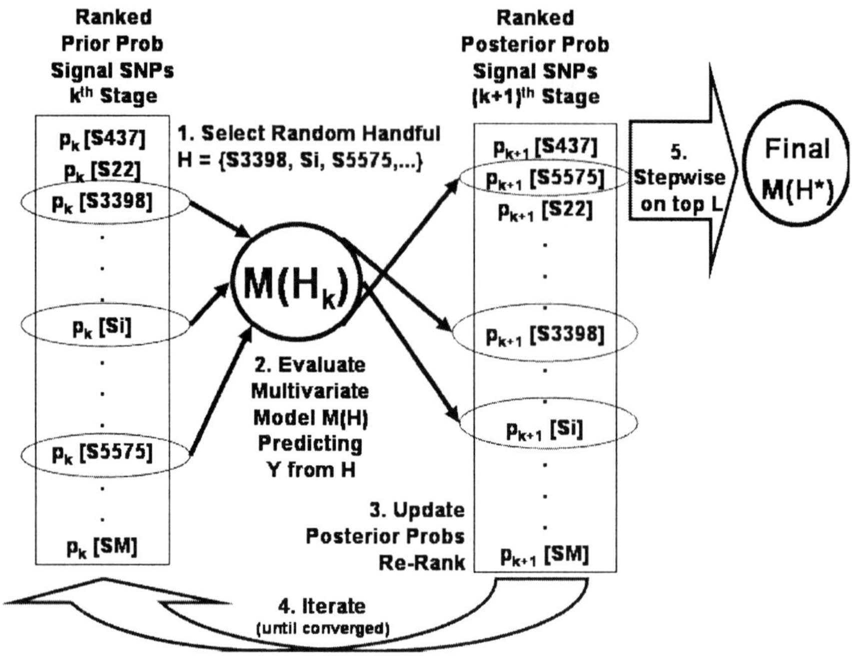

Figure 1. Random Handful Algorithm Given prior probability rankings at stage k that SNPs are signals : 1) we randomly select a handful of SNPs, Hk of size L. 2) Next, we evaluate the multivariate model M(Hk) predicting the phenotype, Y, from Hk. 3) We update the posterior probabilities at stage (k+1) to get new rankings that SNPs are signals. 4) Posterior probabilities now become the Priors for the next Hk+1 random handful. The procedure is terminated when the top L handful of SNPs is consistently and stably ranked at the top in successive updates. 5) Finally, a standard stepwise algorithm is applied to the (now stable) top L set, to select the final significant independent set of SNPs, forming the multivariate model M(H*).

Formally, at stage k, if $P_k[S_i \in H^*]$ is the current prior that S_i is a signal, then given the results of the kth multivariate model $M(H_k)$ for a random handful of SNPs containing S_i, the posterior probability that S_i is a signal SNP is:

$$P_{k+1}\left[S_i \in H^* \mid M(H_k)\right] = \frac{P\left[M(H_k) \mid S_i \in H^*\right] P_k\left[S_i \in H^*\right]}{P\left[M(H_k) \mid S_i \in H^*\right] P_k\left[S_i \in H^*\right] + P\left[M(H_k) \mid S_i \notin H^*\right] P_k\left[S_i \notin H^*\right]} \tag{4}$$

We approximate $P[M(H_k) \mid S_i \in H^*]$ by the power of the multivariate model $M(H_k)$ to detect SNP $S_i \in H^*$ since this is the probability under the alternative. In calculating this power, we assume all other SNPs in H_k are random noise, so that all of the R^2 from the multivariate model $M(H_k)$ is due to the single SNP S_i. Of course, this assumption may not be correct since H may contain many other signal SNPs as well, and we could weight the various possibilities by their corresponding probabilities (again making certain assumptions about that distribution) to obtain a "better" estimate of the desired conditional probability. However, there are 2^L possible ways for the L SNPs in H_k to be distributed as either noise or signal SNPs, which is too many to exhaustively evaluate. For the purposes of this algorithm, we are not as much interested in the exact value of the posterior probability that a SNP is a signal, as we are in approximately ranking the SNPs according to those probabilities. Again, we need not spend an inordinate amount of effort to ensure we have gathered every tiny last grain of gold, so long as we wind up with a pretty valuable pile that is mostly gold in the end.

We also use the current k^{th} stage prior distributions $P_k[S_i \in H^*]$ in selecting which SNPs to include in the corresponding random handful H_k to evaluate for the next stage. We form the cumulative distribution of the priors across all M SNPs (normalizing by the sum) and then select H_k via importance sampling according to this distribution. In this way, the SNPs with the currently highest probabilities of being a signal, are more likely to be included in the model $M(H_k)$. This strategy has two advantages. First, if the current probabilities are accurate, we are efficiently refining the posterior probabilities for the most likely signal SNPs, and thus likely to converge more quickly to the correct set of signal SNPs H^*. Second, if the higher probability for a SNP S_i is inaccurately high for some reason (e.g. by the luck of the draw, it has always accidentally been in the company of other signal SNPs in each previous H_j in which it was evaluated, even though it itself is not a signal), then preferentially sampling such a SNP in H_k gives us the opportunity to correct the probability for S_i with an additional model.

We approximate $P[M(H_k) \mid S_i \notin H^*]$ as the product of the overall p-value for the multivariate model $M(H_k)$ times the type-III SSQ p-value for the particular SNP S_i in the model (i.e. the p-value for the test of the independent contribution of S_i to Y). This is also not strictly speaking correct, as the model is not independent of one of its components. But again, we prefer a simple approximation to a more complex solution at this stage. This approximation is intuitively appealing, as we would like to include call a SNP as a signal if both the multivariate model which contains it is significant as well as if its conditional test of independent contribution is significant.

We continue iterating until the algorithm consistently ranks the top L SNPs from one stage to the next. Then we take these top L SNPS as potential signals. The final step is to do a traditional stepwise model, selecting only the significant SNPs from the top L. This insures that each SNP in the final random handful model is making an independent contribution to prediction of the phenotype.

The Random Handfuls algorithm is written as a SAS MACRO (SAS™)[7]. Within each iteration, multivariate regression is done using PROC REG and power is calculated using PROC POWER.

2.3. *Simulation*

To evaluate the performance of the methods, we conducted several Monte Carlo simulation experiments in SAS™. As large scale simulations can take excessive amounts of CPU time, our initial experiments have been relatively small, to allow us to explore a more broad space of conditions. In the first series of experiments, we simulate N=1,000 unrelated subjects on which we conduct an M=1,000 SNP scan, with 5 signal SNPs, each of which explains 2% of the variance of Y (a quantitative trait), which therefore has a total heritability of 10%. We generate SNPs without LD. The second simulation is an extension of the first, in which we add 5 pairs of epistatically interacting SNPs (none of which have any main effects). Each of these interactions has a heritability of 2%, for a total trait heritability of 20% for all main effects and interactions. For each condition, we generate 100 replications and analyze with all 4 methods.

3. Results

Results from the simulations are shown in Table 1. We tabulate the average performance across all replications, of each of the 4 methods (Nominally significant, Bonferroni significant, FDR significant, and Random Handfuls algorithm) for finding polygenic SNPs. We can classify each of the M=1,000 SNPs as real signals (including all main effect SNPs as well as each pair of SNPs involved in any interaction) vs. the noise. Thus, in each replication we can calculate agreement statistics for each screening method to capture the signals.

Not surprisingly, when there are no interactions (top half of Table 1), selecting all nominally significant SNPs at $P<0.05$ is highly sensitive (99%.), and the specificity runs at 95%, which corresponds well to the expected 5% false positives. However, the true discovery rate is only 9%, since most positives will be false. Selecting only Bonferroni or FDR significant SNPs trades much of the sensitivity (now down to 66% and 78%, respectively) for increased specificity and much improved True discovery rates (99% and 96%, respectively). There is considerably less noise in our final answer when we correct for multiple comparisons. The random handful algorithm competes well in this regard, having the high sensitivity and specificity of the Nominal criteria with a much higher True Discovery Rate. We also calculate the Kappa statistic, which is the amount of agreement beyond that expected by chance (perfect agreement yields

Table 1 Results from Monte Carlo Simulation
Comparing 4 Methods for Detecting Small Gene Polygenic Aggregate Effects
(N = 1,000 Subjects, K = 100 Replications)

M = 1,000 SNPs Total h^2 = 10%
H* = 5 Additive signal SNPs (h^2g=2% each)

Method	Sensitivity	Specificity	True Discovery Rate	Kappa	Training R^2	Test R^2
Nominal	0.99	0.95	0.09	0.16	0.26	0.01
Bonferroni	0.66	0.99	0.99	0.77	0.08	0.07
FDR	0.78	0.99	0.96	0.85	0.09	0.08
Random Handfuls	0.95	0.99	0.71	0.81	0.12	0.07

M = 1,000 SNPs Total h^2 = 60%
H* = 5 Additive SNPs (h^2g=2% each)+ 5 Epistatic Interaction SNP-pairs (h^2g=10% each)

Method	Sensitivity	Specificity	True Discovery Rate	Kappa	Training R^2	Test R^2
Nominal	0.44	0.95	0.12	0.16	0.37	0.06
Bonferroni	0.27	0.99	0.99	0.42	0.17	0.09
FDR	0.28	0.99	0.93	0.42	0.14	0.08
Random Handfuls	0.34	0.99	0.89	0.49	0.12	0.09

For each of the 4 methods (Nominal, Bonferroni, FDR and Random Handfuls) , we tabulate average agreement statistics (sensitivity, specificity, true discovery rate and Kappa) across all replications, quantifying the agreement of that method to the true classification of SNPs into signals vs. noise (signals are all main effect SNPs as well as any SNP involved in an epistatic interaction). We also show the percent of variance explained (R2) from the sum of risk variants across all SNPs chosen in the corresponding model, both in the original Training dataset (on which the SNPs were originally selected) as well as on an independent Training dataset of equal size.

Kappa=1, while chance agreement corresponds to Kappa=0). Kappa is poor for the Nominal selection method, and in the 75%-85% range for each of the Bonferroni, FDR and Random Handful methods.

We also evaluated the amount of variance explained by the selected SNPs in both the original training dataset (on which the models were developed) as well as on an independent test set of equal size. Since the true heritability in each dataset is 10%, we can see that the Nominal model overfits to noise in the training dataset, producing an R2=26%. Of course, such a model does not reproduce well in an independent test dataset, explaining only 1% of the variance. Each of the Bonferroni, FDR and Random Handful methods run much closer to the expected 10% of explained variance. Thus, the Random Handful method is as good as or sometimes exceeds the operating characteristics of the other methods, when there are no epistatic interactions.

When there are epistatic interactions without main effects (bottom half of Table 1), the pattern of results is similar. None of the methods was very successful in capturing the epistatic interacting SNPs without main effects (including the random handfuls algorithm). However, the random handfuls gave scored at least as good and sometimes better than the other algorithms across all measures of performance.

4. Discussion

Many complex traits (such as obesity, diabetes, heart disease, cancer, etc.) have heritabilities in the 30-50% range. If all of the variance in a 50% heritable trait were due to large effect genes explaining 10% of the variance or more, there would only be 5 such in the genome, and the current statistical methods for genomic scans would easily find these. However, suppose there are only two such large effect genes, explaining 20% of the variance between them, and the remaining 30% of the variance is due to 30 polygenes, each of which explains only 1% of the variance, or worse, 300 polygenes, each of which explain only 0.1% of the variance (as is the order of magnitude in our AGT – hypertension example[3]). Then the current methods which concentrate only on the marginal effects of variants will almost surely fail to find any gold dust.

It is disappointing that the Random Handfuls method was not successful in detecting interactions without main effects. We are currently incorporating a refinement of the algorithm to explicitly find such effects. There are too many possible interactions to consider all of them in a brute force way. For instance, if there are L main effects then there are

$$\binom{L}{2} = \frac{L!}{(L-2)!2!} \tag{5}$$

possible 2-way interactions. If we formally test for each of these in each random handful of size L, we can easily fit to noise. We are examining the utility of a two stage test within each random handful iteration to handle this problem. In the first stage of each iteration k, we add a single variable which is the sum of all 2-way interaction terms that can be made from the L main effects in the current random handful H_k, and test for its significance. If it is not significant, we stop at stage 1 and consider only main effects as before. If the aggregate interaction term is significant then we move to stage 2 and potentially add all pair-wise interactions in a stepwise algorithm. The conditional probabilites in the Bayesian formula for each SNP S_i are the most significant p-values of either the main effect test or of any pair-wise interaction. The idea is that the noise interaction terms will tend to cancel out, so that the aggregate interaction term provides a good 1 d.f. screen to test whether any interactions are present. If they are, then we test for which ones should be incorporated.

Our random handfuls method is similar in spirit and philosophy to Bayesian averaging and model selection methods, in which a lot of genetic work has been done recently (e.g. Viallefont et al., 2001[8]; Blangero et al., 2005[9]; among others). We do not claim that our current algorithm is the best or most optimal method for finding the "gold dust" genes. Much refinement of the technique is possible and under development. But we do believe that methods in this vein can be useful to mine the gold for complex traits, and that more investigators should consider novel ways to find the aggregate effects of small effect genes, instead of fixating on the few gold nuggets.

Acknowledgments

This work is partially supported by NIH grants HL087700, HL088655 DK068336, and AG023746.

References

---

1. Klein RJ, Zeiss C, Chew EY, Tsai JY, Sackler RS, Haynes C, Henning AK, SanGiovanni JP, Mane SM, Mayne ST, Bracken MB, Ferris FL, Ott J, Barnstable C, Hoh J. Complement factor H polymorphism in age-related macular degeneration. *Science*. 2005 Apr 15;**308(5720)**:385-9.
2. Frayling TM, Timpson NJ, Weedon MN, Zeggini E, Freathy RM, Lindgren CM, Perry JR, Elliott KS, et al. The Wellcome Trust Case Control Consortium; Hattersley AT, McCarthy MI. A Common Variant in the FTO Gene Is Associated with Body Mass Index and Predisposes to Childhood and Adult Obesity. *Science*. 2007 Apr 12; [Epub ahead of print] PMID: 17434869
3. Province MA, Boerwinkle E, Chakravarti A, Cooper R, Fornage M, Leppert M, Risch N, Ranade K. Lack of association of the angiotensinogen-6 polymorphism with blood pressure levels in the comprehensive NHLBI Family Blood Pressure Program. National Heart, Lung and Blood Institute. *J Hypertens*. 2000 Jul;**18(7)**:867-76.PMID: 10930184
4. Borecki IB, Province MA, Ludwig EH, Ellison RC, Folsom AR, Heiss G, Lalouel JM, Higgins M, Rao DC. Associations of candidate loci angiotensinogen and angiotensin-converting enzyme with severe hypertension: The NHLBI Family Heart Study. *Ann Epidemiol*. 1997 Jan;**7(1)**:13-21. PMID: 9034402
5. Ludwig EH, Borecki IB, Ellison RC, Folsom AR, Heiss G, Higgins M, Lalouel JM, Province MA, Rao DC. Associations between candidate loci angiotensin-converting enzyme and angiotensinogen with coronary heart

disease and myocardial infarction: the NHLBI Family Heart Study. *Ann Epidemiol.* 1997 Jan;**7(1)**:3-12. PMID: 9034401

6. Province MA, Rice TK, Borecki IB, Gu C, Kraja A, Rao DC. A Multivariate and multilocus variance components method, based on structural relationships to assess quantitative trait linkage via SEGPATH. *Genet Epidemiol.* 2003 Feb; **24(2)**:128-38. PMID: 12548674
7. SAS Institute, Inc. SAS/STAT© Users Guide Version 6, Fourth Edition, *SAS Institute Inc.*, Cary, NC, 1989, 846p.
8. Viallefont V, Raftery AE, Richardson S. Variable selection and Bayesian model averaging in case-control studies. *Stat Med.* 2001 Nov 15;**20(21)**:3215-30.PMID: 11746314
9. Blangero J, Goring HH, Kent JW Jr, Williams JT, Peterson CP, Almasy L, Dyer TD. Quantitative trait nucleotide analysis using Bayesian model selection. *Hum Biol.* 2005 Oct; **77(5)**:541-59. PMID: 16596940

MULTI-SCALE CORRELATIONS IN CONTINUOUS GENOMIC DATA

R. E. THURMAN, W. S. NOBLE, AND J. A. STAMATOYANNOPOULOS

Department of Genome Sciences
University of Washington
1705 NE Pacific St
Seattle, WA 98195–5065

Functional genomic quantities such as histone modifications, chromatin accessibility, and evolutionary constraint can now be measured in a nearly continuous fashion across the genome. The genome is highly heterogeneous, and the relationships between different functional annotations may be fluid. Here we present an approach for visualizing, quantifying, and determining the statistical significance of local and regional correlations between high-density continuous genomic datasets. We use wavelets to generate a multi-scale view of each component data set and calculate correlations between data types as a function of genome position over a continuous range of scales in sliding window fashion. We determine the statistical significance of correlations using a non-parametric sampling approach. We apply the wavelet correlation method to histone modification and chromatin accessibility (DNaseI sensitivity) data from the NHGRI ENCODE project. We show that DNaseI sensitivity is broadly correlated (though to differing degrees) with a number of different activating histone modifications. We examine the continuous relationship between the repressive histone modification H3K27me3 and the activating mark H3K4me2, and find these modifications to display significant duality, with both significant positively and negatively correlated genomic territories. While the former appear to recapitulate in definitive cells the so-called "bi-valent" pattern originally proposed as a signature of pluripotency, the presence of negatively correlated regions suggests that the regulatory events that underlie the observed modification patterns are complex and highly regionalized in the genome.

1. Introduction

Rapid progress in the development and application of high-density functional genomic assays has spawned a deluge of new data types. This in turn has created a significant need for computational tools to assess quantitatively the relationship between different data types as a function of genomic position, in a manner that can be related to existing genomic annotations such as genes and transcripts. Data types now available through large-scale efforts such as the NHGRI ENCODE project include various histone modifications, chromatin accessibility, DNA replication timing, bulk transcriptional/RNA output, and evolutionary conservation. Since none of these data have been available in a continuous fashion across diverse genomic regions, their interrelationships are largely unknown. For instance, how does transcriptional activity relate to replication timing? Is the relationship constant across the genome, or is it regionalized? How do histone modifications relate to chromatin accessibility and transcription, particularly in intergenic regions? How does this relationship vary over different parts of a gene, or between gene-rich and gene-poor regions? The human genome is functionally heterogenous, and the pending availability of genome-wide data sets render these questions highly relevant to our understanding of the functional architecture of the genome. The increasing scope and resolution of high throughput genomic assays encourages a multi-scale view of the genome, where some processes vary rapidly over tens or hundreds of bases, and others vary slowly over tens or hundreds of kilobases. We therefore desire to view functional genomic activity over a wide range of scales that may evince both nucleotide- and domain-level phenomena [15].

Here we present a method based on wavelet analysis for simultaneously computing and displaying correlations between different continuous genomic data types at multiple scales. Wavelets provide a mathematical framework for analyzing time series-like data at multiple scales. In the parlance of signal-processing, wavelets are a fundamental tool for *time-frequency* analysis, which in the context of genomic data means that they can describe features in data that are both scale-specific and position-specific (see Methods, below).

Briefly, our method consists first of computing the continuous wavelet transform over a range of scales for each of a pair of datasets to be compared. This gives a multi-scale representation of each dataset, as well as normalizing each pair to a common set of scales. We then correlate the wavelet-transformed results in sliding window fashion on a scale-by-scale basis. The resulting correlation patterns can be visualized in heatmap form, or in aggregate using histograms. We assess the statistical significance of these patterns using non-parametric methods including the Kolmogorov-Smirnov test and, primarily, sampling techniques.

The results and approach presented here expand on those developed in the pilot phase of the ENCODE project [8, 9], whose mission is to identify all functional elements in the human genome. A distinguishing feature of the ENCODE project is its charge to encompass a large number of diverse data types collected using high-throughput techniques (tiling DNA microarrays, high-throughput real-time PCR, and, more recently, ultra-high-throughput sequencing) that collectively expose the functional activity of the human genome in vivo.

The techniques presented here and previously [9] form a key part of the effort required to integrate these diverse functional quantities. In the present work we use improved wavelet techniques and a more rigorous statistical framework to analyze in greater detail the relationships between histone modifications and chromatin accessibility, and between different histone modification classes. The latter have attracted significant attention recently with the observation that certain modifications occurring in combination (H3k27me3 and H3k4me2/3) may have a special functional significance for cellular state.

Chromatin accessibility is a first-order indicator of chromatin structure, and it has long been measured by quantifying sequence-specific or regional DNaseI sensitivity. Epigenetic factors such as histone modifications are thought to play key roles in a number of biological processes, including initiation and propagation of transcription, and higher-order chromatin organization. Nevertheless, the interrelationships between the different histone modifications and chromatin accessibility have not been systematically studied prior to the availability of the ENCODE data.

2. Related work

Wavelets have been used in a number of bioinformatics applications to detect and analyze patterns in sequence data [11]; to de-noise microarray data [11]; and to elucidate large-scale trends in functional genomic data [15]. Wavelets have also been used to uncover sequence and gene-related correlations between prokaryotic species [1]. In these studies, correlations were measured by identifying regions (in position-scale space) of significant wavelet coefficients that were shared between datasets. By contrast, the approach presented here measures the correlation between wavelet coefficients directly, in sliding window fashion across a chromosome. In [5] the authors compare observed and randomized histograms of local correlation coefficients to relate the divergence in non-coding non-repetive DNA with the amount of repetitve DNA.

3. Results

We present two analyses to illustrate our methods. Where noted, Supplementary Information is available at `http://noble.gs.washington.edu/proj/wavecor`.

3.1. *Correlated data: DNaseI vs. H3K4me2*

The pilot phase of ENCODE [9] focused on a representative cross-section of 1% of the genome (approximately 30Mb), divided into 44 regions. One of the main conclusions drawn from the pilot analysis is that chromatin accessibility, as measured by DNaseI sensitivity, is very broadly correlated with activating histone modifications including bulk acetylation of histones H3 and H4 (H3ac, H4ac), and mono-, di-, and tri-methylation of H3 lysine 4 (H3K4me1/2/3). Figure 1 shows a heatmap depicting the correlation betweeen DNaseI and H3K4me2, which were jointly measured in the GM06990 lymphoblastoid cell line. The following steps outline the process for generating, interpreting, and assessing the statistical significance of this correlation map. The chromatin accessibility data were generated using the DNase-array method [14], and the histone modification data derived from the Sanger Institute [10] ENCODE studies.

3.1.1. *Wavelet coefficients*

The continuous wavelet transform (CWT; see Methods) coefficient for a given dataset can be computed at any position and any scale greater than the resolution of the input data. The CWT encapsulates how much the data are changing at that scale and position. For correlation analysis, we compute the CWT coefficients at a range of scales for each dataset across the regions of interest. This procedure results in a matrix of CWT coefficients for each dataset, with the x-axis representing genomic position and the y-axis representing wavelet scale. Figure 2 shows heatmap representations (or *scalograms*) of CWT matrices for DNaseI sensitivity and H3K4me2 at scales ranging from 2kb to 32kb. The wavelet family used here is an improved version of that used in [9] in its ability to capture more accurately negative correlations between the data types (see Methods).

3.1.2. *Correlation heatmaps*

The scalograms in Figure 2 show marked similarity in both position and scale; it is these similarities that we aimed to quantify. We computed the

Pearson correlation of the CWT coefficients at each scale, in a sliding window fashion across the genome. Figure 1 shows a heatmap representation of this matrix for DNaseI and H3K4me2 using a sliding window with width at any given scale equal to 2.5 times the scale (e.g., 25kb window at 10kb scale). The high percentage area of red in this figure is qualitative evidence of a high degree of positive correlation at multiple scales.

The width of the sliding window is arbitrary, and can be tailored to fit, for example, prior knowledge of the scale of effect of a biological phenomenon (e.g., the size of a nucleosome, the size of the average gene, etc.). However, wider windows may defeat the purpose of isolating local correlations, while shorter windows push correlation values towards the extremes of ± 1. This latter effect occurs also for a fixed window width as the scale increases. The scale-adaptive width used here makes correlations comparable across scales, in contrast to the fixed window size technique used previously [9].

3.1.3. *Statistical significance*

We next addressed the statistical significance of the preliminary conclusions brought forth by visual inspection of Figure 1. Specifically, how significant is the observed global positive correlation over random expectation? Is this correlation profile more or less extreme when we replace H3K4me2 by another histone modification? Do the results change if we consider all 44 ENCODE regions? To address these questions we applied non-parametric methods: the Kolmogorov-Smirnov (KS) test [6] for assessing the differences between distributions, and iterative random sampling to form empirical null distributions.

Statistical significance via KS test. Figure 3, left, shows the smoothed histogram of ENCODE-wide sliding window correlation values between DNaseI and each of the five histone modifications at the 16kb scale. The high degree of positive correlation displayed in Figure 1 is reflected in Figure 3, where the distribution for all marks is highly skewed toward +1. The distributions appear, moreover, to be ordered with respect to the degree of positive skew, with H3K4me2 most correlated, followed by H3K4me1, H3K4me3, H3ac, and H4ac in that order. Application of the one-sided KS test showed that the ordering is significant for all five marks ($p < 10^{-29}$), except for the relationship between H3K4me1 and H3K4me3, which is ambiguous. Results from [9] showed H3K4me2, H3K4me3 and H3ac most correlated with DNaseI, and that group being significantly more correlated than H3K4me1 and H4ac, but with no other significant ordering of the marks within those groups possible. Taken together, these results provide strong evidence for high though graded correlation between DNaseI sensitivity and the sampled range of histone modifications.

Statistical significance via sampling. The techniques in the previous section may be used to compare different sets of observed correlations. Next we addressed the question of comparing observed correlations to random expectation. Here, "random expectation" means relative to a null distribution formed by considering two random signals that are, in a critical sense, similar to the two observed datasets.

For the null model, rather than fitting a parametric model of the signals, which involves assumptions or simplifications of the data that may be incorrect, we pursued a non-parametric sampling approach. All available data for a given time series were concatenated into a single master series which served as a pool from which regions of fixed size were sampled (with size depending on the question being asked). Each point in the null distribution was derived by computing the correlation between regions sampled from independent positions in the two master series. This technique maintains the internal structure of the original time series while breaking any correlation between them. We obtained an empirical p-value by counting how many points in the null distribution met or exceeded the properties of the observed feature.

Figure 3, right, shows the distribution of all observed DNaseI/H3K4me2 sliding window correlation values in the 500kb region ENr132 at the 16kb scale. We found that 52% of the correlation coefficients at this scale exceed 0.7. To calculate significance, we randomly sampled blocks of size 500kb from each master series at the 16kb scale. Out of 5,000 sampled correlation profiles, only two had at least 52% of their values over 0.7, yielding a *p*-value of 0.0004 for this region. Figure 4, center, shows a plot of additional sampled correlation profiles.

There are 31 ENCODE regions of size exactly 500kb. We repeated the above analysis for each of these regions and obtained uncorrected empirical p-values ranging from 0.0000 to 0.3948 (see Supplementary Information). The variability in these results suggests region-specific differences affecting the correlations between DNaseI and histone modifications. For example, Figure 5 shows that regions of high gene density tend to have higher correlation values.

When ENCODE-wide data were considered (Figure 3, left), we observed that 56% of all 16kb DNaseI/H3K4me2 correlation values exceeded 0.7. To test the significance of findings in the wider data set, we computed a null distribution in which each point derived from sampling (for each dataset), 44 ENCODE region-sized pieces (with replacement) from the master series. Of 1,000 samples so obtained, none achieved a degree of positive correlation comparable to the observed (empirical p-value of < 0.001). This result supports the intuition that chance high correlation (pseudocorrelation) in random data is significantly harder to sustain over longer regions than shorter regions. Figure 4 shows the density of several ENCODE-wide sampled correlation profiles versus several 500kb sampled profiles, with the latter evincing far more variability. Correspondingly, Figure 4, right, shows that the size of the tails of each sampled correlation profile has a much wider distribution for the 500kb samples than for the ENCODE-wide samples.

3.2. *Uncorrelated data: H3K4me2 vs. H3K27me3*

As a further illustration of the utility of this approach, and to introduce additional methods, we next examined the relationship between two histone modifications, H3K4me2 and H3K27me3. The former is classically associated with transcriptional activation, while the latter is held to signify transcriptional or even regional chromatin repression. Due to the dual nature of these modifications we expected their profiles in lymphoblastoid cells to be largely uncorrelated, or perhaps even anticorrelated. Indeed, Figure 6, top, covering the alpha globin cluser (Chr16) shows clearly co-located peaks near position Chr16:150,000. This location contains a block of high positive correlation across multiple scales, while a number of flanking peaks for one mark or the other show no correlation or anticorrelation. Numerous examples of analogous co-located peaks occur throughout the ENCODE regions, as do examples of slightly offset peaks (see Supplementary Information for ENCODE-wide plots). The co-localization of H3K4me2/3 with H3k27me3 was first described in mouse embryonic stem cells (where it is prominent over the promoters of certain developmentally coordinated genes), and was originally thought to be a marker of pluripotency [4, 2]. However, more recent work has called this conclusion into question [3].

Figure 6 (bottom, solid line) shows the distribution of ENCODE-wide sliding window correlation values for these two marks at the 16kb scale. The plot reveals a fairly balanced distribution of positive and negative correlation values. Viewed independently, it is not clear whether this pattern is indicative of random non-correlation or if it reflects the aggregate of non-random patterns of real positive and negative correlations, including the bivalent-like domains suggested at the top of Figure 6.

To address this issue, we took two approaches. First, we performed sampling experiments to ascertain null behavior. We calculated correlation profiles from 1,000 ENCODE-wide random samples of the two sets of 16kb wavelet coefficients. Several correlation profiles are plotted as dashed lines in Figure 6. While the sampled distributions are also largely balanced between positive and negative values, the observed distribution shows extension of the tails and an offsetting central depression. Indeed, we found that no sampled distribution had the same fraction of coefficients above 0.5 and below -0.5 as the observed, for an empirical p-value of < 0.001. This provides quantitative evidence for non-random positive and negative correlations.

Next, we attempted to identify regions of local agreement between the two marks. We performed a 2-state HMM segmentation of each dataset, partitioning the ENCODE regions into "high" and "low" states based separately on wavelet smoothed versions of each mark at scales ranging from 4kb to 128kb (see [7] for methods). We then formed the intersection of the high states for both marks at each scale.

For data smoothed at the 4kb scale the intersection of the high states comprised approximately 3.4Mb ($> 10\%$) of ENCODE, which overlapped 70 annotated genes. GO analysis revealed 19 categories over-represented at p-values less than 0.01, including six transcription-related terms, terms for the regulation of cellular, physiological and biological processes, for phosphatase and enzyme activity, and for development (see Supplementary Information for full results). These categories accord with prior observations that a significant fraction of bivalent domains occur at genes encoding transcriptional regulatory factors or at the 3' ends of developmental genes [4]. They also found large bivalent domains in the Hox clusters.

To explore scale-specific effects, we repeated the HMM segmentations using 16kb, 64kb and 128kb scales. The intersection of the H3K27me3 and H3K4me2 high states at these scales covered 4.5Mb, 7.6Mb, and 8.0Mb, respectively. Almost without exception, the over-represented GO terms at each scale were a subset of the terms at the next smaller scale, and all terms were a subset of the 4kb terms. Five terms were over-represented at the 128kb scale, with transcription factor activity being the most significant ($p = 2.7 \times 10^{-7}$). See Supplementary Information for full results.

4. Discussion and conclusions

The influx of functional genomic data types collected using high-throughput methods has created a significant need for tools to integrate diverse signals into a meaningful picture of the functional structure of the human genome. The methods presented here are widely applicable to this problem. In the course of the ENCODE pilot data analyses we performed wavelet correlation and visualization analyses for scores of pairs of data types including, in addition to those discussed here, replication timing, evolutionary conservation measures, bulk RNA output, and nucleosome depletion assays (see results provided in Supplementary Information.

A key aspect to our approach is the systematic integration of locoregional analyses. These results can be used to confirm hypotheses concerning the relation between data types proposed on the basis of mechanistic relationships (e.g., the correlation between DNaseI sensitivity and activating histone modifications in gene-rich regions), and they may be applied in an exploratory mode, such as de novo identification of regions of common but generally unexpected high activity of activating (H3K4me2) and repressive (H3K27me3) modifications.

Future work will be required to elucidate the complex relationship between activating and repressive histone modifications. Indeed, it appears preliminarily that these broad labels are not sufficient to categorize behavior across all genomic terrain. Additionally, it is not clear to what extent the locoregional relationships between different modifications depend on the cell type being studied. While the particular genes or domains that evince a particular combination of marks may change between cell types, it will be interesting to determine whether the overall proportion of territory covered by that combination changes substantially. The pending availability of additional data from ENCODE as well as other large-scale chromatin profiling efforts [3, 12], including both additional cell types and additional modifications will provide an opportunity to address this sytematically. Ultimately, extension of the approach described here to encompass multiple diverse data types in addition to histone modifications and chromatin accessibility will likely hold the greatest promise for elucidating the functional landscape of the genome. On a practical level, many of the calculations required for the wavelet correlation approach are computationally expensive. As data types proliferate and expand to encompass the whole genome, a first priority will be to determine whether the distribution of correlation values observed are largely independent of data-type and dependent only on scale, which would permit realization of significant efficiencies through pre-computation of case sampled distributions.

5. Methods

Data normalization. Wavelet correlation analysis requires that both datasets be defined on a common set of equally-spaced genomic coordinates. We performed gap-filling interpolation and wavelet normalization as described previously [9].

Wavelets. In the analysis of time series-like data, wavelet analysis can be thought of as an extension of Fourier analysis. Both techniques are used to look for periodicities or strong changes in a time series at a given period, or, in the language of wavelets, *scale.* But whereas Fourier analysis is global in nature, giving a single Fourier coefficient for each period for an entire time series, wavelet analysis is local, producing a wavelet coefficient at any point in the time series that describes the strength of the change in the time series at the given scale, at that time. We use the collection of wavelet coefficients across time (genomic position, in our case) for a fixed scale to summarize the scale-specific behavior of the time series.

Other smoothing techniques (loess, sliding window averaging) could be used to also approximate scale-specific behavior. We chose wavelets because of the availability of a computational framework (R package Rwave) for simultaneously computing wavelet coefficients across multiple scales and the established role of wavelets in time-frequency analysis.

The basis for all wavelet analysis is the continuous wavelet transform (CWT) [13]. For a given equally-spaced time series $x(t)$, the CWT wavelet coefficient $W(a, s)$ for given scale a and time s is given by

$$W(a,s) \equiv \frac{1}{\sqrt{a}} \int_{-\infty}^{\infty} x(u)\psi\left(\frac{u-s}{a}\right) du,$$

where $\psi(t)$ is the wavelet function of choice, satisfying the basic properties $\int_{-\infty}^{\infty} \psi(u), du = 0$ and $\int_{-\infty}^{\infty} \psi^2(u), du = 1$. We use our own implementation of the real-valued "first derivative of Guassian," or DOG wavelet. By contrast, the analysis in [9] used the complex-valued Morlet wavelet; correlations using this wavelet required taking the absolute value of the coefficients, which masked negative correlations.

References

1. T. Allen *et al.* Long-range periodic patterns in microbial genomes indicate significant multi-scale chromosomal organization. *PLoS Computational Biology*, 2(1):13–21, January 2006.
2. V. Azuara *et al.* Chromatin signatures of pluripotent cell lines. *Nature Cell Biology*, 8:532–538, 2006.
3. A. Barski *et al.* High-resolution profiling of histone methylations in the human genome. *Cell*, 129:823–837, 2007.
4. B. E. Bernstein *et al.* A bivalent chromatin structure marks key developmental genes in embryonic stem cells. *Cell*, 125(2):315–326, 2006.
5. F. Chiaromonte *et al.* Association between divergence and interspersed repeats in mammalian noncoding genomic dna. *PNAS*, 98:14503–14508, 2001.
6. W. J. Conover. *Practical Nonparametric Statistics.* Wiley Series in Probability and Statistics. Wiley & Sons, 3rd edition, 1999.
7. N. Day *et al.* Unsupervised segmentation of continuous genomic data. *Bioinformatics*, 23:1424–1426, 2007.
8. ENCODE Consortium. The ENCODE (ENCyclopedia Of DNA Elements) project. *Science*, 306(5696):636–640, 2004.
9. ENCODE Consortium. Identification and analysis of functional elements in 1% of the human genome by the ENCODE pilot project. *Nature*, 447:799–816, 2007.
10. C. Koch *et al.* The landscape of histone modifications across 1% of the human genome in five human cell lines. *Genome Research*, 17:691–707, 2007.
11. P. Liò. Wavelets in bioinformatics and compuataional biology: state of art and perspectives. *Bioinformatics*, 19(1):2–9, 2003.
12. T. S. Mikkelsen *et al.* Genome-wide maps of chromatin state in pluripotent and lineage-commited cells. *Nature*, 448:553–560, 2007.
13. D. B. Percival and A. T. Walden. *Wavelet Methods for Time Series Analysis.* Cambridge Series in Statistical and Probabilistic Mathematics. Cambridge University Press, 2000.
14. P. J. Sabo *et al.* Genome-scale mapping of DNaseI sensitivity *in vivo* using tiling DNA microarrays. *Nature Methods*, 3:511–518, 2006.
15. R. E. Thurman *et al.* Identification of higher-order functional domains in the human ENCODE regions. *Genome Research*, 17:917–927, 2007.

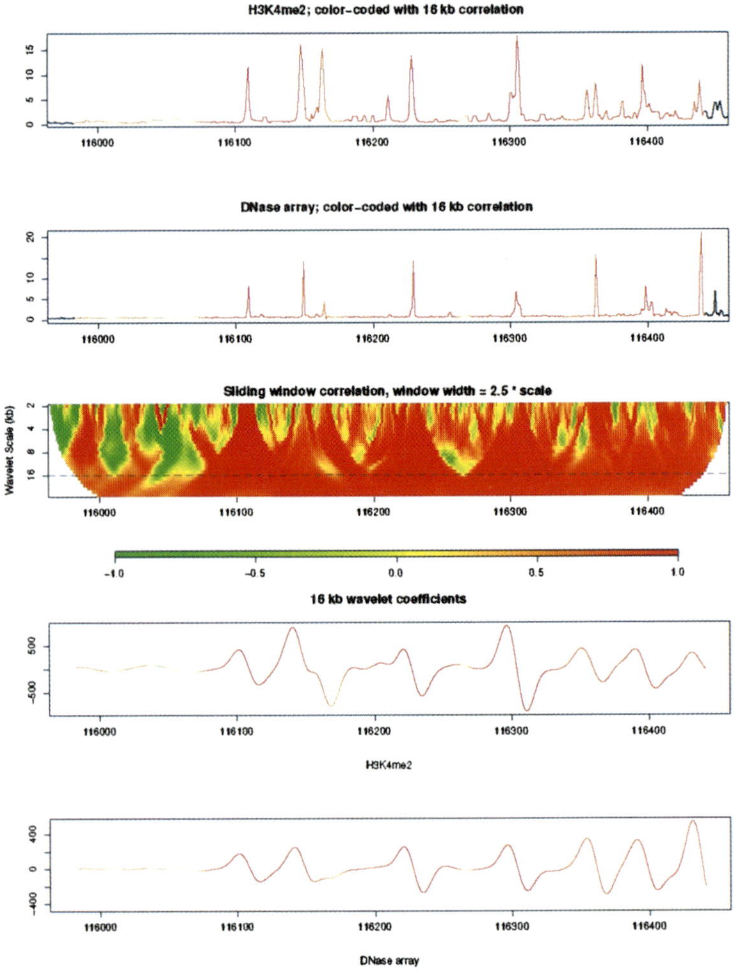

Figure 1. Correlation heatmap for H3K4me2 vs. DNaseI (GM06990) in ENCODE region ENm003. From top to bottom: raw data for H3K4me2, raw data for DNaseI, correlations heatmap, 16kb CWT coeffiecients for H3K4me2, 16kb CWT coeffiecients for DnaseI. Raw data and coefficients are colored with the correlations at the 16kb scale (dashed line).

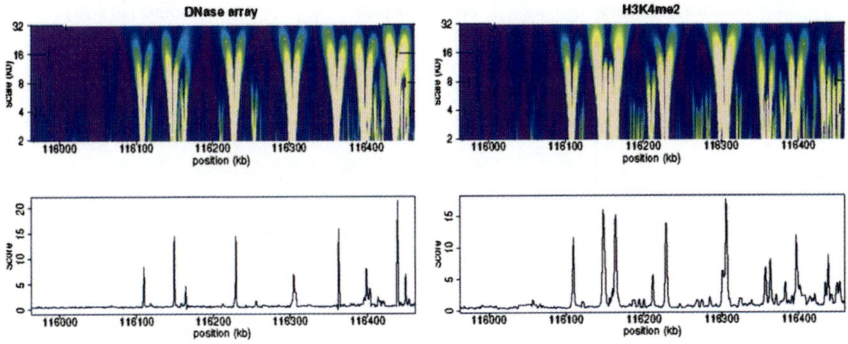

Figure 2. CWT heatmaps for DNaseI sensitivity (left) and histone modification H3K4me2 (right), in ENCODE region ENm003. The bottom plots are of the original data.

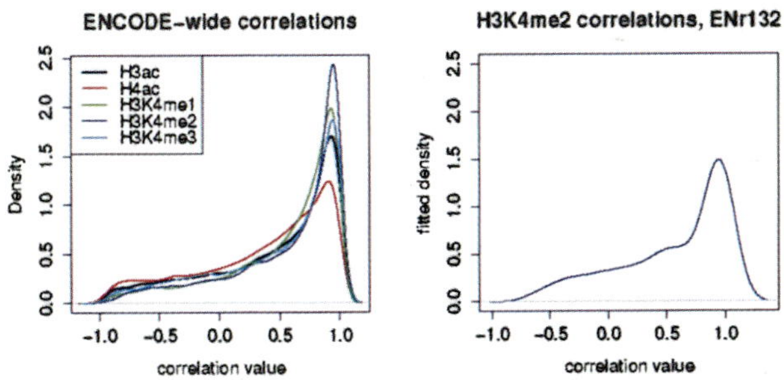

Figure 3. Distribution of correlation values at 16kb scale, DNaseI vs. Sanger activating histone marks. Left, ENCODE-wide correlations of all five marks. Right, DnaseI vs. H3K4me2 correlations in the 500kb ENCODE region ENr132.

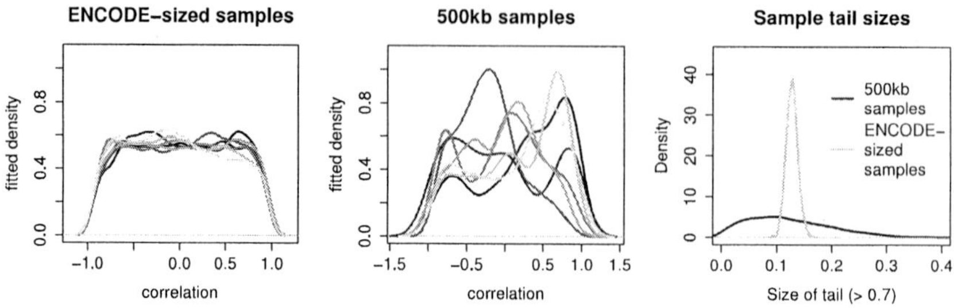

Figure 4. Distribution of several correlation profiles using ENCODE-wide samples (left) and 500kb samples (center). Right, distribution of tail sizes in sampled correlation values. For each sample we compute the fraction of the correlation values greater than 0.7. The plot summarizes those fractions in 1000 ENCODE-sized samples (red) and 5000 500kb samples (black) of simulated null correlation values.

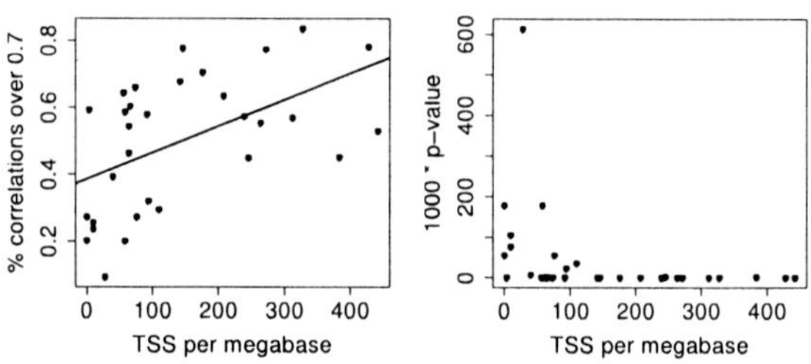

Figure 5. Gene density (transcription start sites per megabase) and DnaseI/H3K4me2 correlation in 500kb ENCODE regions. Each point corresponds to one of the thirty-one 500kb regions. For each region we computed the gene density therein, the fraction of DNaseI/H3K4me2 16kb correlation values in that region over 0.7, and the empirical p-value for that fraction. At left, gene density vs. fraction of correlation values over 0.7 (with regression line); at right, gene density vs. empirical p-values.

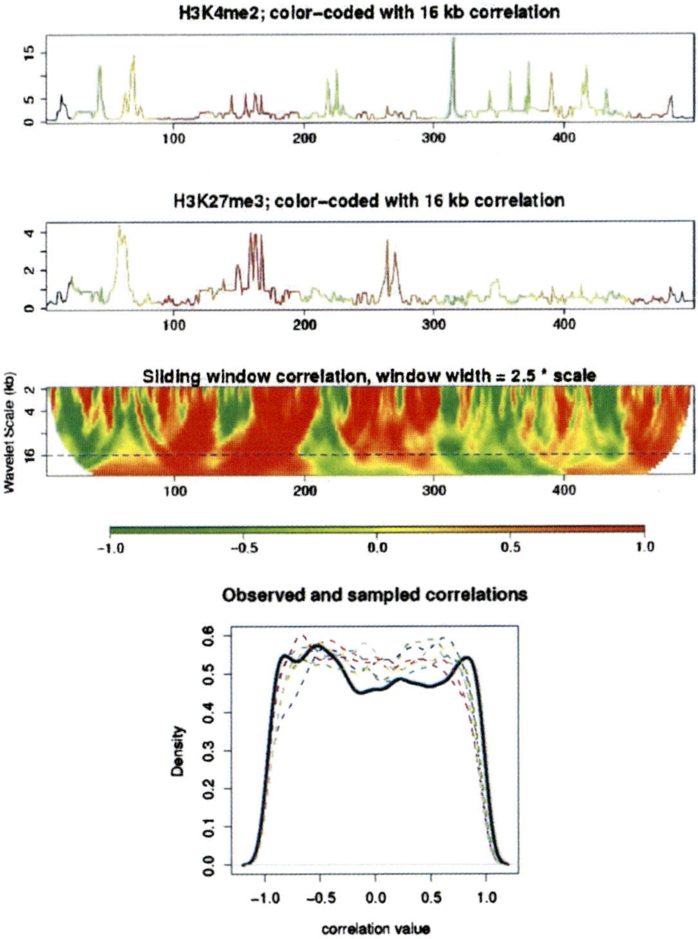

Figure 6. Correlation of H3K4me2 and H3K27me3. Top, correlation heatmap in ENCODE region ENm008. Bottom, distribution of observed (solid line) and sampled (dashed lines) ENCODE-wide correlations at the 16kb scale.

ANALYSIS OF MALDI-TOF MASS SPECTROMETRY DATA FOR DETECTION OF GLYCAN BIOMARKERS

HABTOM W. RESSOM[1†], RENCY S VARGHESE[1], LENKA GOLDMAN[1], CHRISTOPHER A LOFFREDO[1], MOHAMED ABDEL-HAMID[2], ZUZANA KYSELOVA[3], YEHIA MECHREF[3], MILOS NOVOTNY[3], RADOSLAV GOLDMAN[1]

[1]*Georgetown University, Lombardi Comprehensive Cancer Center, Washington, DC*

[2]*Minia University and Viral Hepatitis Research Laboratory, NHTMRI, Cairo, Egypt*

[3]*National Center for Glycomics and Glycoproteomics, Department of Chemistry, Bloomington, IN*

We present a computational framework for analysis of MALDI-TOF mass spectrometry data to enable quantitative comparison of glycans in serum. The proposed framework enables a systematic selection of glycan structures that have good generalization capability in distinguishing subjects from two pre-labeled groups. We applied the proposed method for a biomarker discovery study that involves 203 participants from Cairo, Egypt; 73 hepatocellular carcinoma (HCC) cases, 52 patients with chronic liver disease (CLD), and 78 healthy individuals. Glycans were enzymatically released from proteins in serum and permethylated prior to mass spectrometric quantification. A subset of the participants (35 HCC and 35 CLD cases) was used as a training set to select global and subgroup-specific peaks. The peak selection step is preceded by peak screening, where we eliminate peaks that seem to have association with covariates such as age, gender, and viral infection based on the 78 spectra from healthy individuals. To ensure that the global peaks have good generalization capability, we subjected the entire spectral preprocessing and peak selection step to a cross-validation; a randomly selected subset of the training set was used for spectral preprocessing and peak selection in multiple runs with resubstitution. In addition to global peak identification method, we describe a new approach that allows the selection of subgroup-specific glycans by searching for glycans that display differential abundance in a subgroup of patients only. The performance of the global and subgroup-specific peaks is evaluated via a blinded independent set that comprises of 38 HCC and 17 CLD cases. Further evaluation of the potential clinical utility of the selected global and subgroup-specific candidate markers is needed.

1. Introduction

Current diagnosis of hepatocellular carcinoma (HCC) relies on clinical information, liver imaging, and measurement of serum alpha-fetoprotein (AFP). The reported sensitivity (41-65%) and specificity (80-94%) of AFP is not sufficient for early diagnosis and additional markers are needed [1, 2].

Mass spectrometry (MS) provides a promising strategy for biomarker discovery. The feasibility of MS-based proteomic analysis to distinguish HCC

† Corresponding author

from cirrhosis, particularly in patients with hepatitis C virus (HCV) infection, has been studied [3-6]. Recent proteomic studies have identified potential markers of HCC including complement C3a [7], kappa and lambda immunoglobulin light chains [8], and heat-shock proteins (Hsp27, Hsp70, and GRP78) [9].

Many currently used cancer biomarkers including AFP are glycoproteins [10]. Fucosylated AFP was introduced as a marker of HCC with improved specificity [11, 12] and other glycoproteins including GP73 are currently under evaluation as markers of HCC [13, 14]. The analysis of protein glycosylation is particularly relevant to liver pathology because of the major influence of this organ on the homeostasis of blood glycoproteins [15, 16]. An alternative strategy to the analysis of glycoproteins is the analysis of protein associated glycans [17, 18]. The characterization of glycans in serum of patients with liver disease is a promising strategy for biomarker discovery [19].

Current methods allow quantitative comparison of permethylated glycan structures by matrix-assisted laser desorption/ionization time-of-flight (MALDI-TOF) MS [20], which provide a rich source of information for molecular characterization of the disease process. Although MALDI-TOF MS continuously improves in sensitivity and accuracy, it is characterized by its high dimensionality and complex patterns with substantial amount of noise. Biological variability and disease heterogeneity in human populations further complicate the MALDI-TOF MS-based biomarker discovery. While various signal processing methods have been used to reduce technical variability caused by sampling or instrument error, reducing non-disease-related biological variability remains a challenging task. For example, peaks associated to known covariates such as age, gender, smoking status, and viral infection should be eliminated; we call this preprocessing step *peak screening* [5]. In addition, robust computational methods are needed to minimize the impact of biological variability caused by unknown intrinsic biological differences.

In this paper, we present computational methods for analysis of MALDI-TOF MS to discover glycan biomarkers for the detection of HCC in patients with chronic liver disease (CLD), consisting of fibrosis and cirrhosis patients [21, 22]. The objective is to improve the diagnostic capability of a panel of "whole population" level (global) biomarkers and to investigate the extraction of subgroup-specific biomarkers that are more patient specific than the global markers. Our proposed approach involves the following two steps.

The first step searches for a panel of global peaks that distinguishes HCC from CLD at the whole population level by treating all HCC patients as one group [4, 5]. We utilize a computational method that combines ant colony optimization and support vector machine (ACO-SVM), previously described in

[5], to identify the most useful global peaks. Although these peaks may include peaks that may be attributed to subgroups of patients, neither the subgroup-specific peaks nor the subgroups are likely to be isolated due to the unknown (mostly nonlinear) interaction of the global peaks.

The second step uses a genetic algorithm (GA) to search for subgroup-specific peaks and to discover subgroups of subjects from the training set. The disease state of an unknown individual is determined by the SVM classifier built in the first step. Then, the subgroup to which the individual belongs will be determined by comparing its intensity with each of the subgroup-specific peaks defined in the second step.

The proposed hybrid method will provide the ability to capture glycans that are differentially abundant in only a subset of patients in addition to those that are differentially abundant glycans at the whole population level. This will allow us to not only identify a panel of useful global peaks that lead to good generalization, but also to offer a more patient-specific approach for the identification of glycan biomarkers.

2. Methods

2.1. *Sample collection*

HCC cases and controls were enrolled in collaboration with the National Cancer Institute of Cairo University, Egypt, from 2000 to 2002, as described previously [22]. Briefly, adults with newly diagnosed HCC aged 17 and older without a previous history of cancer were eligible for the study. Diagnosis of HCC was confirmed by pathology, cytology, imaging (CT, ultrasound), and serum AFP. Controls were recruited from the orthopedic department of Kasr El Aini Faculty of Medicine, Cairo University [22]. 17 HCC cases were classified as early (Stage I and II) and 33 HCC cases as advanced (Stage III and IV) according to the staging system [23]; for the remaining 23 HCC cases the available information was not sufficient to assign the stage. Patients with CLD were recruited from Ain Shams University Specialized Hospital and Tropical Medicine Research Institute, Cairo, Egypt during the same period. The CLD group has a biopsy confirmed 21 fibrosis and 25 cirrhosis patients; 6 individuals in the CLD group did not have sufficient clinical information. Patients negative for hepatitis B virus (HBV) infection, positive for HCV RNA, and with AFP less than 100 mg/ml were selected for the study. Blood samples were collected by a trained phlebotomist each day around 10 am and processed within a few hours according to a standard protocol. Aliquots of sera were frozen at -80 °C immediately after collection until analysis; all mass spectrometric measurements were performed on twice-thawed sera. Each patient's HBV and HCV viral infection status was assessed by enzyme immunoassay for anti-HCV, anti-HBC, and HBsAg, and by PCR for HCV RNA [22, 24].

2.2. *Sample preparation and MS data generation*

The sample preparation involved release of N-glycans from glycoproteins, extraction of N-glycans, and solid-phase permethylation as described previously [20]. The resulting permethylated glycans were spotted on a MALDI plate with DHB-matrix, MALDI plate was dried under vacuum, and mass spectra were acquired using a 4800 MALDI TOF/TOF Analyzer (Applied Biosystems Inc., Framingham, MA) equipped with a Nd:YAG 355-nm laser as described previously [17]. MALDI-spectra were recorded in positive-ion mode, since permethylation eliminates the negative charge normally associated with sialylated glycans. [25]. 203 raw spectra were exported as text files for further analysis [a]. Each spectrum consisted of approximately 121,000 m/z values with the corresponding intensities in the mass range of 1,500-5,500 Da.

2.3. *Global peak selection*

Figure 1 illustrates our approach for global peak selection, which begins by splitting the spectra into a labeled set and a blinded set. The labeled set consists of a subset of HCC cases, a subset of CLD cases, and all healthy individuals (normal). The blinded set comprises of masked HCC and CLD cases; it is used to evaluate the generalization capability of the selected peaks. Peak detection, peak screening, and peak selection are performed on the labeled set by subjecting the entire process to cross-validation. As illustrated in Figure 1, a subset of the labeled HCC and CLD spectra (~70% from each group) is randomly selected at each iteration as a training set, while the remaining HCC and CLD spectra are used as validation set. A spectrum in the training set is considered as an outlier, if its record count is more than two standard deviations away from the median record count of the spectra within the training set. Outliers are removed from the subsequent analyses. Each spectrum in the training set is binned, baseline corrected, and normalized as described previously [5]. After scaling the peak intensities to an over all maximum intensity of 100, local maximum peaks above a specified threshold are identified and peaks that fall within a pre-specified mass are coalesced into a single m/z window to account for drift in m/z location. The maximum intensity in each window is used as the variable of interest. The threshold intensity for peak detection is selected so that isotopic clusters are represented by a single peak.

---

[a] These files are available at http://microarray.georgetown.edu/web/files/psb.zip

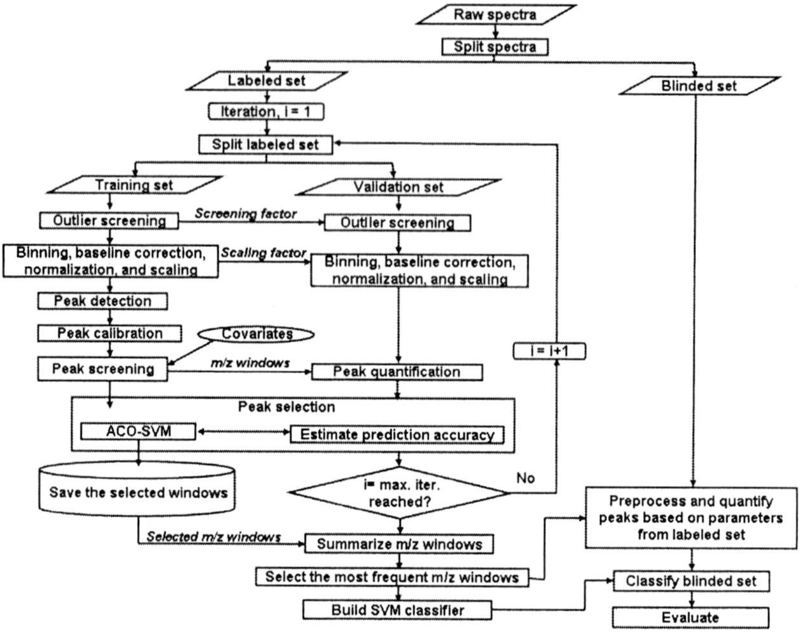

Figure 1. Methodology for global peak detection.

Logistic regression models are used to examine association of the glycans to known covariates including age, gender, smoking status, residency, HCV and HBV viral infections. This analysis is performed on the samples from healthy individuals to unambiguously isolate peaks associated to the covariates. The independent variables of a logistic regression model are the intensities of a given peak across all normal samples. The dependent variable is the status of a given covariate; all covariates in this study have binary values including age (young vs. old). The association of every peak to each covariate was determined on the basis of the corresponding statistical significance ($p<0.01$) in fitting a logistic regression model. Glycan intensities associated to the covariates are removed. From the remaining peaks, ACO-SVM selects the best peaks in terms of their ability to distinguish a subset of the HCC and CLD spectra in the validation set, which was not involved in the peak selection process. The spectra in the validation set are screened for outliers, binned, baseline corrected, normalized, and scaled on the basis of the parameters used to preprocess the spectra in the training set. The peaks in the validation set are quantified at the selected m/z windows and are presented to SVM classifier previously trained using the peaks from the training set. The performance of the SVM classifier in predicting the disease state of the subjects in the validation set is used by ACO-SVM to guide its search for the optimal peak set. The above steps are repeated multiple times by randomly splitting the labeled spectra into training and validation sets.

The peaks selected in multiple runs are summarized to determine the most frequently selected m/z windows. Note that the number of peaks detected and their m/z windows could vary at each iteration due to the change in the population set in each iteration. After obtaining all peaks selected in multiple iterations, we summarize the peaks by merging overlapping m/z windows. The optimal peak set is determined based on the frequency of occurrence of the peaks in multiple runs.

To evaluate the peak selection process further, we quantify the glycan intensities at the m/z windows of the optimal peak set in the labeled and blinded sets. Note that the blinded set is not used during the peak detection and peak selection phases, thus it serves as an independent set to evaluate the generalization capability of the selected peaks. The spectra in the blind set are outlier screened, binned, baseline corrected, normalized, and scaled on the basis of parameters used to preprocess the spectra in the labeled set. We build an SVM using the labeled set and evaluate the capability of the SVM classifier in distinguishing HCC from CLD in the blinded set in terms of sensitivity, specificity, and area under the ROC (AuROC).

2.4. *Identification of subgroup-specific peaks*

Figure 2 illustrates our proposed method to identify subgroup-specific peaks by searching for peaks that are differentially abundant in a subset of patients. The method is described here in two phases: training and operation phase.

In the training phase, for each candidate peak we search a subgroups of HCC cases in which the peak is differentially abundant. The candidate peaks are the summarized peak set from the global peak selection process. Note that this peak list includes each summarized peak regardless of its frequency of occurrence. We apply GA to search the optimal subgroup of patient for each candidate peak. A chromosome in the GA assigns a binary bit for each HCC patient in the labeled set ("1" for a patient selected in the subgroup, "0" otherwise). The algorithm starts with randomly selected binary bits. GA evolves the chromosomes with the aim of maximizing a multi-objective fitness function, which involves two parameters (1) the AuROC obtained in using the peak to separate a selected subgroup of HCC patients from patients with CLD and (2) the number of HCC patients involved in the subgroup. The goal is to search for a peak and a subgroup that not only display good separation between the HCC subgroup and patients with CLD, but also assign a reasonable number of subjects to the subgroup. During the operation phase, the label of a spectrum from the blinded set is predicted by an SVM classifier previously built using the global peaks in the labeled set. If the predicted label is HCC, its glycan intensities will be compared with the subgroup-specific peaks to determine which subgroup the individual belongs to. The subject is assigned to a previously identified subgroup, if its peaks intensity falls within the subgroup's intensity range.

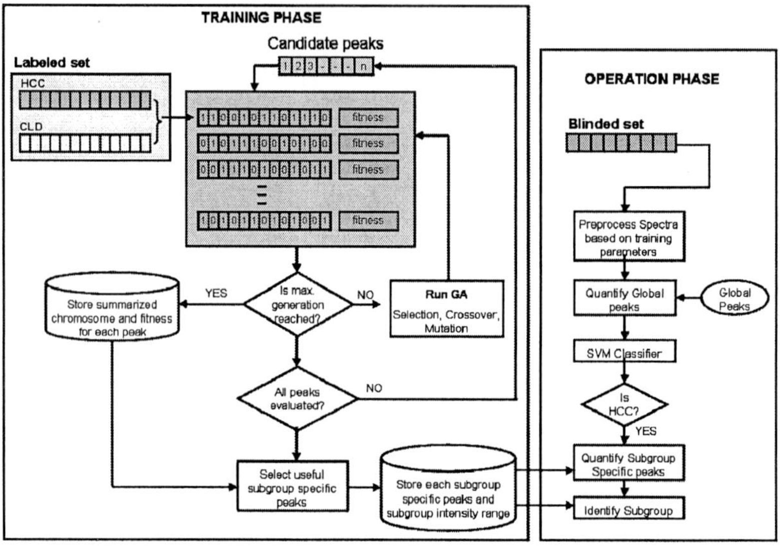

Figure 2: Methodology for subgroup-specific peak selection.

3. Results

MALDI-TOF mass spectrometric analysis of permethylated N-glycans enzymatically detached from serum proteins allowed relative quantification of about 100 oligosaccharides. We analyzed serum samples from 203 participants. Glycan analysis was performed as described previously [17, 20].

Spectral preprocessing and global peak detection were carried out following the methodology depicted in Figure 1. Briefly, we began the analysis by splitting the raw spectra into labeled set (35 HCC, 35 CLD, and 78 normal) and a blinded set (38 HCC and 17 CLD). From the labeled set, 25 HCC and 25 CLD spectra were randomly selected as a training set; the remaining 10 HCC and 10 CLD spectra were used as a validation set. Outlier screening was performed on the training set to determine if its record count of each spectrum is within two standard deviations from the median record count for the spectra within the training set. Outlier spectra were removed from the subsequent analyses. A binning algorithm reduced the dimension of each of these spectra from ~121,000 to 13,030 using a bin size of 100 ppm. The mean of the intensities within each bin was used as the bin intensity. For each binned spectrum, we estimated the baseline by obtaining the minimum value within a shifting window size of 50 bins and a step size of 50 bins. Spline approximation was applied to regress the varying baseline. The regressed baseline was smoothed using the lowess smoothing method. The resulting baseline was subtracted from the spectrum. Then, each spectrum was normalized by dividing it by its total ion current. The spectra were scaled to have a maximum intensity of 100. Local maximum peaks above a specified threshold are identified and nearby peaks within 300 ppm mass separation are coalesced into a single m/z window and the maximum intensity in each window is used as the variable of interest. We adjusted the threshold intensity and the mass separation so that isotopic clusters resolved by the high resolution reflectron acquisition were represented by one glycan peak. The isotopic cluster at 1543-1547 Da was the only cluster resolved by the procedure to three individual peaks; we grouped this cluster to one variable prior to subsequent analyses. This procedure resulted in about 100 m/z windows. After performing peak screening on the basis of the 78 normal spectra, about 20 peaks were removed. From the remaining peaks, ACO-SVM algorithm selected the three most useful peaks. The capability of these peaks to predict the labels of the spectra in the validation set was used by ACO-SVM to search for the optimal peak set. The spectra in the validation set were preprocessed in the same way as the training set. For outlier screening and scaling, the parameters used by the training set were utilized. The intensity values within the detected windows were quantified and the maximum intensities within the windows were used as input to SVM classifier built previously using the training set. The above procedure was repeated 2000 times by randomly selecting (with resubstitution) 25 HCC and 25 CLD spectra from the labeled set as a training set and using the remaining 10 HCC and 10 CLD spectra as a validation set.

The peaks selected in 2000 runs were summarized by merging overlapping windows. Figure 3 depicts a frequency plot of the summarized 66 peaks (m/z windows). As shown in the figure, two m/z windows dominated the selection, where the first and second m/z windows were selected in 76% and 35% of the runs, respectively. We quantified the peaks in the labeled set (35 HCC and 35 CLD spectra) within these two summarized windows and applied the maximum intensity values within the windows to build an SVM classifier.

To evaluate the performance of the SVM classifier, we preprocessed the spectra in the blinded set in same way as the training set and quantified the glycan intensities within the selected two summarized windows. These intensities were presented to the previously built SVM classifier, which predicted the samples with 95% sensitivity and 100% specificity; two HCC subjects in the blinded set were wrongly classified as CLD. For comparison, we repeated the entire peak selection process (Figure 1) by replacing ACO-SVM with the SVM-recursive feature elimination (SVM-RFE) method [26]. Comparing the top 10 peaks in both methods, we observed five overlaps; the peak with the highest frequency was the same in both methods. The top 10 m/z windows in both methods gave 95% sensitivity and 100% specificity in classifying the samples in the blinded set (both methods wrongly classified the same HCC subjects as CLD). However, the top two m/z windows in SVM-RFE (selected in 87% and 30% of the runs, respectively, frequency plot not shown here) distinguished the HCC cases from CLD with 92% sensitivity and 94% specificity in the blind validation set; 1 CLD patient and 3 HCC cases were misclassified.

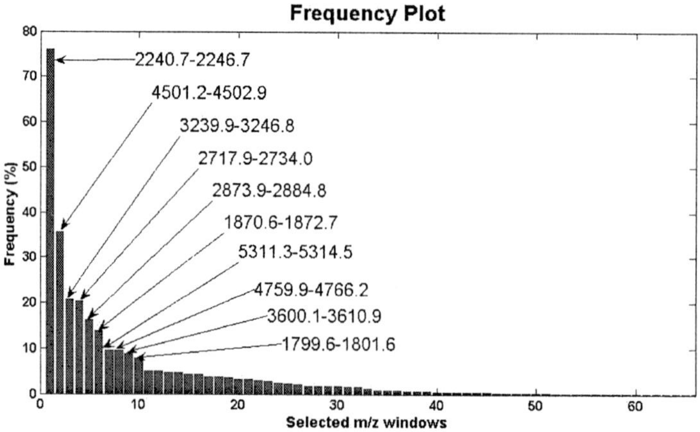

Figure 3. Frequency of occurrence of peaks selected by ACO-SVM in 2000 runs.

Glycan structures for nearly 50% of the peaks detected by the MALDI-TOF MS were determined. Out of 10 peaks selected by ACO-SVM, five have a known sugar composition. Similarly, five out of 10 peaks selected by SVM-RFE have known composition. Figure 4 depicts an overlay of the average HCC and CLD spectra. The five peaks, with known composition, identified by ACO-SVM are shown in the figure; four of these were also among the top 10 peaks selected by SVM-RFE. These five peaks yielded 87% sensitivity and 100% specificity in distinguishing HCC cases from CLD patients in the blinded set.

Finally, we used the methodology illustrated in Figure 2 to identify subgroup-specific peaks from the 66 peaks summarized from the previous 2000 ACO-SVM runs; all summarized peaks are considered as candidate peaks regardless of the frequency of occurrence in the 2000 runs. The subgroup-specific peak selection method identified four peaks that represent four subgroups (S1, S2, S3, and S4) consisting of 23, 21, 17, and 15 HCC patients, respectively. These four peaks were particularly selected, because they had better AuROC, more number of HCC patients in the subgroup they represent, and less number of overlapping subjects than other candidate peaks.

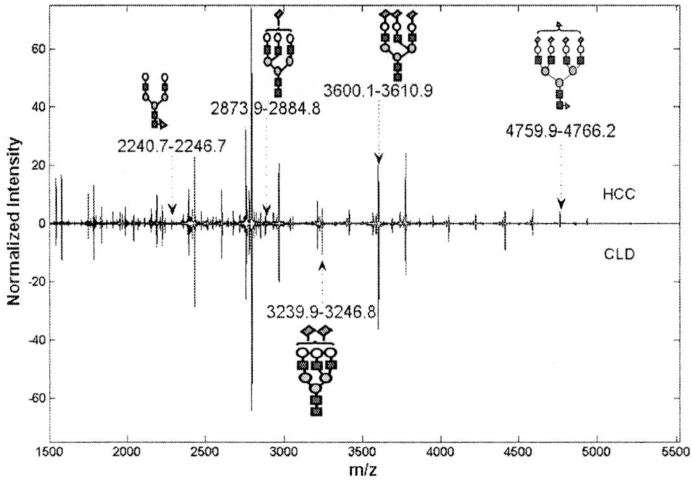

Figure 4. Mean HCC and CLD spectra and sugar composition of five peaks selected by ACO-SVM.

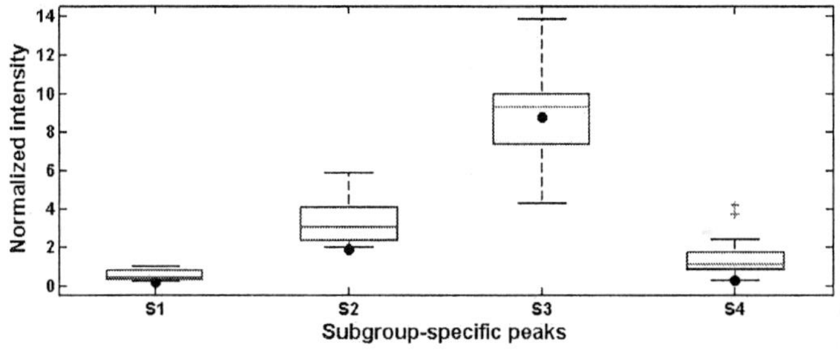

Figure 5. Box plots of peak intensities for four HCC subgroups. Dots represent glycan intensities of a blinded sample detected as HCC by a panel of global peaks. The intensity of the sample falls within the range of the peak for subgroup S3.

Figure 5 depicts box plots for the glycan intensity levels of the four subgroup-specific peaks in their respective subgroups of subjects. Note that only intensities of the subjects that belong to the subgroup the peak represents are shown by the box plots. We considered a subject from the blinded set that was correctly predicted as HCC case by the global peaks. Figure 5 shows the glycan intensities of this subject at the four subgroup-specific peaks (dots in the figure). These intensities are compared with the peak intensity distribution (box plot) of the four subgroups of HCC patients that the peaks represent. From the figure, we see that the HCC patient can be assigned to the subgroup labeled as S3.

4. Discussion

This paper introduces computational methodologies for quantitative comparison of glycans in serum and selection of biomarkers of hepatocellular carcinoma. Candidate glycan biomarkers were obtained by comparing MALDI-TOF spectra of permethylated glycan structures derived from HCC and CLD patient sera. Prior to peak selection, we removed peaks associated to covariates such as age, gender, residency, smoking, and viral infection. We showed that the algorithm has the ability to select a small set of glycan peaks that achieve high sensitivity and specificity in distinguishing HCC cases from patients with CLD in Cairo, Egypt. In addition, we proposed a method that can potentially discover subgroups of patients by searching for subgroup-specific peaks that are differentially abundant in a subset of patients only. Further analysis is needed to determine the implication of the subgroups of subjects and the subgroup-specific biomarkers. It will be interesting to see if the subgroups represent different disease stages or molecular pathways. In addition, the potential clinical utility of the selected candidate markers needs to be evaluated through independent laboratory experiments.

5. Acknowledgments

This work was supported in part by seed grants awarded to HWR and RG from NCI's Early Detection Research Network, NCI grant R03 CA119313 to HWR, and NCI grants R03 CA119288 and R01 CA115625 to RG.

References

1. J. A. Marrero: *Clin Liver Dis* 2005, **9**(2):235-251, vi.
2. S. Gupta, S. Bent, J. Kohlwes: *Ann Intern Med* 2003, **139**(1):46-50.
3. E. Orvisky, S. K. Drake, B. M. Martin, M. Abdel-Hamid, H. W. Ressom, R. S. Varghese, Y. An, D. Saha, G. L. Hortin, C. A. Loffredo *et al*: *Proteomics* 2006, **6**(9):2895-2902.
4. H. W. Ressom, R. S. Varghese, M. Abdel-Hamid, S. Abdel-Latif Eissa, D. Saha, L. Goldman, E. F. Petricoin, T. P. Conrads, T. D. Veenstra, C. A. Loffredo *et al*: *Bioinformatics* 2005, **21**(21):4039-4045.
5. H. W. Ressom, R. S. Varghese, S. K. Drake, G. L. Hortin, M. Abdel-Hamid, C. A. Loffredo, R. Goldman: *Bioinformatics* 2007, **23**(5):619-626.
6. E. E. Schwegler, L. Cazares, L. F. Steel, B. L. Adam, D. A. Johnson, O. J. Semmes, T. M. Block, J. A. Marrero, R. R. Drake: *Hepatology* 2005, **41**(3):634-642.
7. I. N. Lee, C. H. Chen, J. C. Sheu, H. S. Lee, G. T. Huang, D. S. Chen, C. Y. Yu, C. L. Wen, F. J. Lu, L. P. Chow: *Proteomics* 2006, **6**(9):2865-2873.

8. D. G. Ward, Y. Cheng, G. N'Kontchou, T. T. Thar, N. Barget, W. Wei, L. J. Billingham, A. Martin, M. Beaugrand, P. J. Johnson: *Br J Cancer* 2006, **94**(2):287-292.
9. J. M. Luk, C. T. Lam, A. F. Siu, B. Y. Lam, I. O. Ng, M. Y. Hu, C. M. Che, S. T. Fan: *Proteomics* 2006, **6**(3):1049-1057.
10. J. A. Ludwig, J. N. Weinstein: *Nat Rev Cancer* 2005, **5**(11):845-856.
11. K. Taketa, Y. Endo, C. Sekiya, K. Tanikawa, T. Koji, H. Taga, S. Satomura, S. Matsuura, T. Kawai, H. Hirai: *Cancer Res* 1993, **53**(22):5419-5423.
12. K. Shiraki, K. Takase, Y. Tameda, M. Hamada, Y. Kosaka, T. Nakano: *Hepatology* 1995, **22**(3):802-807.
13. J. A. Marrero, P. R. Romano, O. Nikolaeva, L. Steel, A. Mehta, C. J. Fimmel, M. A. Comunale, A. D'Amelio, A. S. Lok, T. M. Block: *J Hepatol* 2005, **43**(6):1007-1012.
14. M. A. Comunale, M. Lowman, R. E. Long, J. Krakover, R. Philip, S. Seeholzer, A. A. Evans, H. W. Hann, T. M. Block, A. S. Mehta: *J Proteome Res* 2006, **5**(2):308-315.
15. G. A. Turner: *Clin Chim Acta* 1992, **208**(3):149-171.
16. S. J. Lee, S. Evers, D. Roeder, A. F. Parlow, J. Risteli, L. Risteli, Y. C. Lee, T. Feizi, H. Langen, M. C. Nussenzweig: *Science* 2002, **295**(5561):1898-1901.
17. Z. Kyselova, Y. Mechref, M. M. Al Bataineh, L. E. Dobrolecki, R. J. Hickey, J. Vinson, C. J. Sweeney, M. V. Novotny: *J Proteome Res* 2007, **6**(5):1822-1832.
18. J. Zhao, W. Qiu, D. M. Simeone, D. M. Lubman: *J Proteome Res* 2007, **6**(3):1126-1138.
19. N. Callewaert, H. Van Vlierberghe, A. Van Hecke, W. Laroy, J. Delanghe, R. Contreras: *Nat Med* 2004, **10**(4):429-434.
20. P. Kang, Y. Mechref, I. Klouckova, M. V. Novotny: *Rapid Commun Mass Spectrom* 2005, **19**(23):3421-3428.
21. O. Nada, M. Abdel-Hamid, A. Ismail, L. El Shabrawy, K. F. Sidhom, N. M. El Badawy, F. A. Ghazal, M. El Daly, S. El Kafrawy, G. Esmat *et al*: *J Clin Virol* 2005, **34**(2):140-146.
22. S. Ezzat, M. Abdel-Hamid, S. A. Eissa, N. Mokhtar, N. A. Labib, L. El-Ghorory, N. N. Mikhail, A. Abdel-Hamid, T. Hifnawy, G. T. Strickland *et al*: *Int J Hyg Environ Health* 2005, **208**(5):329-339.
23. *AJCC Cancer Staging Manual, 6th Edition American College of Surgeons Philadelphia, Lippincott-Raven* 2002.
24. M. Abdel-Hamid, D. C. Edelman, W. E. Highsmith, N. T. Constantine: *J Hum Virol* 1997, **1**(1):58-65.
25. Y. Mechref, M. V. Novotny: *Anal Chem* 1998, **70**(3):455-463.
26. I. Guyon, J. Weston, S. Barnhill, V. Vapnik: *Machine learning* 2002, **46**:389-422.

MOLECULAR BIOINFORMATICS FOR DISEASE: PROTEIN INTERACTIONS AND PHENOMICS

YVES A. LUSSIER*, YOUNGHEE LEE[1]

Center for Biomedical Informatics and Section of Genetic Medicine, Dept. of Medicine and UC Cancer Research Center; The University of Chicago, IL, 60637

PREDRAG RADIVOJAC *

School of Informatics, Indiana University Bloomington, IN 47408, U.S.A.

YANAY OFRAN*

Dept. of Biochemistry & Molecular Biophysics, Columbia University New York, NY 10032, U.S.A.

MARCO PUNTA*

Biochemistry & Molecular Biophysics, New York, NY 10032, U.S.A.

ATUL BUTTE*[*]

Depts. of Pediatrics and of Medicine, Stanford University, Sanford, CA 94305, U.S.A.

MARICEL KANN *

National Center for Biotechnology Information, NIH Bethesda, MD 20894, U.S.A.

This session focuses on the emerging fields of protein interactions in diseases and phenomics: from protein-protein interactions to supracellular phenotypes. Experimental studies indicate that protein interactions play a key role in many diseases, even in some that are considered complex or multifactorial. While altered phenotypes are among the most reliable manifestations of altered gene functions, research focused on systematic analysis of phenotype relationships to study human biology is still in its infancy. In this summary, the words phenome and phenomics are used to describe the physical totality of all traits of an organism (Mahner, J Theor Biol 1997 186:55-63). The audiences targeted by this session bring together a broad audience: bioinformaticians, systems biologists, biomedical informaticians, physicians, pharmacologists, computer scientists, statisticians, members of the pharmaceutical industry and others to share their experience and scientific findings in this area.

The papers accepted for the session on Molecular Bioinformatics for Diseases comprise original research that pertain to biological scales ranging from phenomics, or

---

[1] co-author of the manuscript

[*] Session co-chairs and co-authors

relationship of whole organism's phenotypes, to proteomics. More specifically, they capitalize on novel computational methods and technological developments in bioinformatics that analyze disease or disease-associated phenotypes in a biology scale between the nanoscale where protein domains, reactions to a scale below the organism taken as a whole where disease phenotypes and phenomes are observed. .

The first two papers address human disease with phenomic datasets ranging from clinical conditions, laboratory biomarker, to the electronic medical record (EMR). **Alterovitz et al.** worked on an information theoretic framework (entropy) for discovery of novel biomarkers, integrating biofluid (e.g. blood, urine) and tissue information. This study uses 26 proteomes from 45 sources to identify candidate biofluids and biomarkers responsible for functional information transfer in the tissue domain. Among results are significant associations between tissues (e.g. cerebrospinal fluid, saliva, urine) and biomarkers (e.g. EGFR, BRCA1). A novel multipartite network of tissue-biomarker-biofluid is proposed as well as candidate biomarkers of biofluid/tissues that may have clinical applications. **Chen et al.** combined gene expression measurements and patient data from the hospital electronic medical records to examine clinical proxies for maturation and associated genes. The method was used to compare trends among different clinical laboratory test in response to an increase in age. They propose the lymphocyte count as a proxy measure for aging and infer that correlated expression of genes in the EMR are also implicated in the process of aging.

The next four papers explored associations between microarrays and an original set of approaches to the systems biology of phenotypes or literature-based statistical approaches to provide phenotypic meaning to gene expression. Since microarray data analysis often predicts many false positive genes, **Hu et al.** propose a general framework to discover the relation between two or more disease conditions in human. They applied networking pathways to association study of non-insulin dependent diabetes mellitus and obesity. Their methods involved the integration of numerous databases including microarrays with KEGG pathways. Diabetes mellitus and Obesity-associated pathways are presented. **Badea et al.** propose a clustering algorithm which is capable of simultaneously factorizing two distinct gene expression datasets. The aim of the algorithm is to uncover gene regulatory programs that are common to the two phenotypes. The methods were evaluated with gene expression profiles that are common between the more homogeneous pancreatic ductal adenocarcinoma (PDAC) and the more heterogeneous colon adenocarcinoma. The approach identified that the PDAC signature is active in a large fraction of colon adenocarcinoma. **Gevaert et al.** present an approach to integrate information from literature abstracts into probabilistic models of gene expression data in order to improve model building in gene selection. They investigated if a Bayesian network model with a text prior can be used to predict the prognosis in cancer.

Finally, these three papers consist of two proteomic studies and one DNA methylation study applied to phenomic datasets. **Sridhar et al.** developed a branch-and bound algorithm to formulate the optimal enzyme combination identification problem as an optimization problem on metabolic networks. They demonstrate that the algorithm can accurately identify the target enzymes for known successful drugs in the literature and reduce the total search time by several orders of magnitude as compared to the exhaustive search. In contrast, **Singh et al.** were interested in global alignment of multiple protein-protein interaction (PPI) networks. They developed an algorithm that maximizes the overall match across all "input" networks. It was applied to the global alignment of protein-protein interaction networks from five species: yeast, fly, worm, mouse and human. The authors propose an original way to unveil functional orthologs cover multiple (5) species. **Kim et al.** use GpG flanquing sequence composition to predict methylation susceptibility and identify susceptible methylation. sites in disease-related tissues (e.g. primary leukemia, lymphoma cells and normal blood lymphocytes).

Acknowledgements

The session co-chairs would like to thank numerous reviewers for their help in selecting the best papers among many excellent submissions and Dr. Yong Huang for his suggestions.

SYSTEM-WIDE PERIPHERAL BIOMARKER DISCOVERY USING INFORMATION THEORY*

GIL ALTEROVITZ†

Division of Health Sciences and Technology, Harvard University/Massachusetts Institute of Technology, Cambridge, MA. Children's Hospital Informatics Program, Boston, MA 02115, USA. Department of Electrical Engineering and Computer Science, Cambridge, MA 02139, USA. Harvard Partners Center for Genetics and Genomics, Harvard Medical School, Boston 02115, USA.

MICHAEL XIANG†

Division of Health Sciences and Technology, Harvard University/Massachusetts Institute of Technology, Cambridge, MA 02139, USA.

JONATHAN LIU

Department of Biology, Massachusetts Institute of Technology, Cambridge, MA 02139, USA.

AMELIA CHANG

Harvard Partners Center for Genetics and Genomics, Harvard Medical School, Boston 02115, USA.

MARCO F. RAMONI

Children's Hospital Informatics Program, Boston, MA 02115, USA. Division of Health Sciences and Technology, Harvard University/Massachusetts Institute of Technology, Cambridge, MA. Harvard Partners Center for Genetics and Genomics, Harvard Medical School, Boston 02115, USA.

The identification of reliable peripheral biomarkers for clinical diagnosis, patient prognosis, and biological functional studies would allow for access to biological information currently available only through invasive methods. Traditional approaches have so far considered aspects of tissues and biofluid markers independently. Here we introduce an information theoretic framework for biomarker discovery, integrating biofluid and tissue information. This allows us to identify tissue information in peripheral biofluids.

---

* This work was supported in part by the National Library of Medicine (NLM/NIH) under grant 5T15LM007092 and the National Human Genome Research Institute (NHGRI/NIH) under grant 1R01HG003354.

† These authors contributed equally to the work.

We treat tissue-biofluid interactions as an information channel through functional space using 26 proteomes from 45 different sources to determine quantitatively the correspondence of each biofluid for specific tissues *via* relative entropy calculation of proteomes mapped onto phenotype, function, and drug space. Next, we identify candidate biofluids and biomarkers responsible for functional information transfer ($p < 0.01$). A total of 851 unique candidate biomarkers proxies were identified. The biomarkers were found to be significant functional tissue proxies compared to random proteins ($p < 0.001$). This proxy link is found to be further enhanced by filtering the biofluid proteins to include only significant tissue-biofluid information channels and is further validated by gene expression. Furthermore, many of the candidate biomarkers are novel and have yet to be explored. In addition to characterizing proteins and their interactions with a systemic perspective, our work can be used as a roadmap to guide biomedical investigation, from suggesting biofluids for study to constraining the search for biomarkers. This work has applications in disease screening, diagnosis, and protein function studies.

1. Introduction

The rapidly increasing availability of sequenced genomes since the 1990's has made it clear that genetic analysis alone cannot fully account for organismal complexity[1]. A more complete realization focuses instead on genes' protein products. As such, the field of proteomics aims to understand protein function, structure, and interactions[1].

Proteomics has considerable clinical relevance: proteins carry out cellular functions, comprise drug targets, and often participate in or indicate disease pathogenesis. For example, a doctor may take a blood sample to perform a liver function test[2], for which certain enzymes (*e.g.*, alanine transaminase) are elevated in liver dysfunction. Recently, biomarkers for various diseases have emerged, including prostate specific antigen (PSA) for prostate cancer[3] and C-reactive protein (CRP) for heart disease[4]. Therefore, identification of clinically significant protein biomarkers of phenotype and biological function is an exciting and expanding area of research that promises to extend diagnostic capabilities.

The use of biomarkers from easily accessible biofluids (*e.g.* blood, urine) is advantageous for evaluating the state of harder-to-reach tissues and organs. Biofluids capture proteins and protein fragments released by cells in the body, either as waste or to communicate with other cells or tissues[5]. In addition, biofluids are much more readily accessible, unlike more invasive or unfeasible techniques such as tissue biopsies (e.g. brain tissue). To date, however, approaches to biomarker prediction have analyzed tissues and biofluids separately[6].

Here we propose an information theoretic framework for discovery of novel biomarkers that utilizes information from biofluid proteins that can serve as functional, phenotypic, and drug interaction proxies for the underlying tissues. In order to specify a biomarker, a researcher must identify both a biofluid (*e.g.* blood) and protein(s) in that biofluid that are relevant. Due to the presence of dozens of biofluids and many thousands of proteins, too many combinatorial possibilities exist for them to be tested individually. In this work we propose methods to identify both biofluids and specific proteins that are particularly well suited for biomarker discovery and validation.

Biofluids contain proteins from tissues and serve as effective communication/hormonal . Conceptually, the tissue acts as a transmitter of information and the biofluid (sampled by the physician) as a receiver. The informativeness of the biofluid is reliant on the fidelity of the channel. Sources of noise which decrease fidelity include addition of proteins derived from other tissues or from the biofluid itself; proteins may also be lost through the glomerular filtration process that removes proteins smaller than 45 kDa from plasma[7]. These factors can substantially bias the protein composition of a biofluid: for instance, the plasma abundances of interleukin-6 and albumin differ[8] by 10 orders of magnitude. Additionally, looking simply at protein overlap would miss information transmission that occurs through classes of proteins and protein-protein interactions. Thus, we consider not the proteins directly, but instead their projection onto functional, drug, and disease spaces, allowing the measurement of functional distance between tissues and biofluids. Closeness in these abstract spaces signifies a low level of distortion across the information channel, and hence high informativeness of the biofluid.

It turns out that information theory has already developed a robust, principled framework[9] for evaluating such a channel problem. The informativeness of a biofluid for a tissue can thus be evaluated within this framework and be used to guide disease and physiological investigations.

2. Methods

In total, 26 human proteomes[10] were obtained from 45 studies. The 16 tissue proteomes comprised brain, cartilage, cornea, heart, kidney, larynx, liver, macrophage, muscle, nose, ovary, pancreas, pituitary, platelet, skin, and stomach. The 10 biofluid proteomes comprised amniotic fluid, cerebrospinal fluid, plasma, pleural fluid, saliva, serum, sputum, synovial fluid, tear, and urine. The full human proteome was obtained from the Gene Ontology Annotation database[11]. The Gene Ontology[12], or GO (23,692 terms) was used to map proteins to functional space, employing the three hierarchies of cellular component, biological process, and molecular function. A controlled vocabulary for diseases (5,648 terms) was extracted from Online Mendelian Inheritance in

Man (OMIM)[13] to construct the disease-based ontology for mapping proteins to disease space. Similarly, the drug-based ontology (411 terms) was created using the Pharmacogenetics and Pharmacogenomics Knowledge Base (PharmGKB)[14] for mapping proteins to drug space. Of the three ontologies, GO has the largest number of terms and represents the most comprehensive distribution of information. As such, although all three ontologies were used to identify significant tissue-biofluid relationships, focus was on the GO-derived results in the identification of candidate biomarkers.

In information theory, relative entropy[15] is a measure of the distance between an unobserved distribution T (here: tissue) and an observed distribution B (here: biofluid). Lower relative entropy denotes closer correspondence between the two distributions. The relative entropy R between a tissue T and a biofluid B was determined as:

$$R(B,T) = \sum_{n=1}^{N} b(V_n) \log \frac{b(V_n)}{t(V_n)}$$

Here, $b(V_n)$ and $t(V_n)$ denote the annotation frequency of term V_n across B and T, respectively, and N is the total number of terms in the function, disease, and drug space. For example, if B = urine, with 1000 proteins, and V_n = "ion binding", then $b(V_n) = 0.01$ means that 1% (10) proteins in the urine proteome are associated with "ion binding" function. A total of 36,000 relative entropy simulations were performed between the tissues and randomly-chosen sets of proteins from the entire proteome to ascertain the significance of tissue-biofluid connections, followed by application of Bonferroni multiple test correction[16]. Thus, a biofluid B is informative of a tissue T if its relative entropy score $R(B,T)$ with that particular tissue is significantly better than the relative entropy scores of randomly-chosen protein sets with the same number of proteins as B (i.e., $p < 0.01$, after multiple test correction). The approach we proposed is diagramed in Figure 1. Connections were considered significant only if they were significantly better than random in terms of channel information faithfulness for all three spaces: function, disease, and drug.

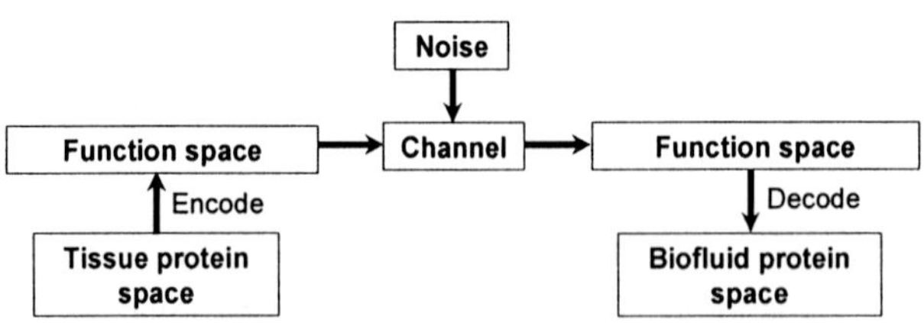

Figure 1. Information theoretic characterization of biofluid-tissue interaction

A search was done for all GO terms with representation at significantly similar frequencies between B and T. Such terms were termed "bandwidth-carrying" terms because they are primarily responsible for transfer of functional information across the tissue-biofluid channel (*i.e.*, based on relative entropy score; see above). Fisher's exact test was used to compute the probability $p(V_n, B, T \mid F)$ of selecting, from the full human proteome F, a random protein sample the same size as B sharing, with T, the same level of frequency similarity or better for V_n:

$$p(V_n, B, T \mid F) = \sum_{k=i}^{j} \frac{\binom{|F(V_n)|}{k}\binom{|F| - |F(V_n)|}{|B| - k}}{\binom{|F|}{|B|}}$$

Here, $F(V_n)$ constitutes all human proteins annotated by V_n, and set size notation is used. k ranges between $i = |B| \cdot b(V_n)$ and $j = |B| \cdot [2t(V_n) - b(V_n)]$, denoting counts of equal or higher similarity for V_n. For example, say T and B both contain 100 proteins; 40 proteins (40%) in tissue T are annotated by V_n = "ion binding" (Gene Ontology), and 38 proteins (38%) in biofluid B are annotated by V_n. In this case, $i = 38$ and $j = 42$ for an equal or more similar level of V_n frequency (between 38% and 42%; *i.e.* within 2% of 40%) with T as the specific biofluid B.

Candidate biomarkers were selected using a scoring process. For a given tissue-biofluid combination, proteins were scored by summing the Shannon information content[17] of the tissue-biofluid pair's "bandwidth-carrying" terms residing in the protein's list of GO annotations. Candidate biomarkers were chosen as biofluid proteins with $p < 0.05$ compared to the scores of randomly-chosen proteins from the full human proteome.

3. Results

3.1 Significant Tissue-Biofluid Channels

A total of 9 biofluids were found to be significantly informative for a total of 14 tissues. In all, 26 tissue-biofluid channels were significant with $p < 0.01$ after Bonferroni correction, while 10 additional tissue-biofluid channels had borderline significance of $p < 0.05$. Figure 2 displays significant channels ($p < 0.01$) between tissues (rectangles) and biofluids (circles). Two tissues, the heart and the pituitary, were not found to be significant with any biofluid tested. On the other hand, the tear biofluid was not found to be informative for any tissue.

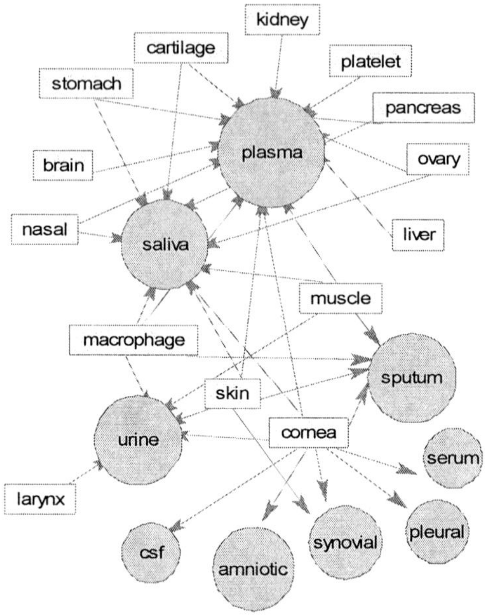

Figure 2. Significant tissue-biofluid channels

As might be expected, blood plasma was significantly informative of most tissues (exceptions were heart, larynx, and pituitary). Interestingly, saliva was the next most informative biofluid (significance across 9 tissues), followed by sputum and urine, which were informative of 4 and 5 tissues respectively. The remaining biofluids (except tear) were informative of 1-2 tissues each.

Corneal tissue shared significance with many biofluids, including cerebrospinal fluid (CSF). This connection has been noted in the literature; for instance, one study noted elevated insulin concentrations in the cornea and CSF upon intralumbar injection[18]. Moreover, topical application of insulin eye drops caused its accumulation within CSF[19]. Sclerotic diffusion could account for the detection of inflammatory response in CSF upon corneal inoculation of herpes simplex virus[20]. Interestingly, CSF was not found to be significantly informative for brain tissue, perhaps because CSF only interacts with the outer edge of the brain. Indeed, one common use of CSF is to diagnose meningitis, which is an infection of the membrane that covers the brain and not of the brain itself.

Significant connections found between the cornea and the other biofluids have also been cited in previous studies. For example, the cornea has been associated with synovial fluid; arthritis patients often display upregulation of proinflammatory cytokines in synovial fluid and corneal samples[21]. Additionally, identical bacteria can be isolated from cornea and sputum during nosocomial eye infection[22]. Literature corroboration increases confidence in our method; consequently, associations with other biofluids that have not been thoroughly explored to date can serve as useful avenues for investigation.

Another significant relationship was discovered between macrophage and sputum, which can be rationalized by macrophages' role in the removal of necrotic debris from the lungs. This tissue-biofluid link is supported by studies showing that increased inflammatory cytokine levels in sputum stimulate macrophage production of metalloproteinases[23]. Other studies have used induced sputum to determine macrophage phenotypes in airway afflictions[24]. Since macrophages are highly involved in immune diseases due to their phagocytic capacity, further elucidation of this relationship could have a variety of clinical applications.

3.2 Identification of Candidate Biomarkers

To identify actual candidate biomarkers, we discovered "bandwidth-carrying" GO terms responsible for transmitting the bulk of functional information from tissue to biofluid. Note that such "bandwidth-carrying" terms can exist between a tissue and biofluid even when the overall biofluid was not found to be informative of the overall tissue. Between the 16 tissues and the 10 biofluids, 519 "bandwidth-carrying" terms were identified with $p < 0.001$. Using these terms, 851 unique proteins were identified as candidate biomarkers for the 16 tissues in this study. Plasma was the most productive biofluid, containing an average of 269 candidate biomarkers per tissue; serum was next with an average of 112 biomarkers per tissue. Other biofluids presented varying numbers of candidate biomarkers: urine for instance had an average of 37 biomarkers per tissue, whereas tear was not found to contain any candidate biomarkers for any tissue with the sole exception of cornea. A portion of the resulting network (e.g. for ovary and biofluids) is shown in Figure 3.

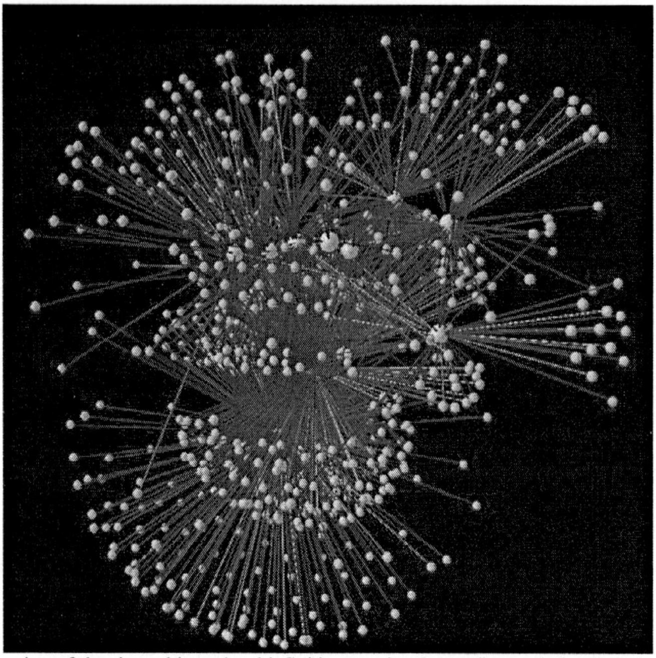

Figure 3. A portion of the tissue-biomarker-biofluid network. Ovary (large, dark sphere in center), biofluids (small spheres), and candidate protein biomarkers (large, light spheres).

Since our approach assesses function on a tissue-wide level, our candidate biomarkers are not restricted to any particular cellular or pathological process. However, their contributions to physiological state can be hypothesized from their functional annotations, and hence serve as initial candidates for screening. A quick scan of the candidate biomarkers reveals some proteins that have been discovered by traditional means. For example, in our list of potential biomarkers for measuring ovarian function, we found a number of known cancer markers. Represented in this list were epidermal growth factor receptor (EGFR), BRCA1, and Apolipoprotein E. These proteins are clinically significant markers of ovarian cancer: EGFR is a specific target for ovarian cancer therapy[25], and mutation of BRCA1 correlates with ovarian cancer risk[26]. Apolipoprotein E has been found to be upregulated in ovarian cancer[27] and also critical for cell survival and proliferation in the disease[28].

Although these ovarian cancer biomarkers have already been validated, the need for additional biomarkers is striking. Half of ovarian cancer patients initially present at Stage III or Stage IV when 5-year survival is only 20%, thus making the disease responsible for more deaths than all other gynecological cancers combined[29]. The successful identification of known ovarian cancer biomarkers confirms our approach, suggesting that the list of predicted ovary biomarkers contains promising targets for clinical investigation. The rapid, guided testing of novel biomarkers can then improve the understanding and treatment of ovarian cancer.

3.3 Establishing Biomarker Quality

The overall quality of the candidate biomarkers was assessed by measuring co-citation frequencies in PubMed (Figure 4). The co-citation frequency of each biomarker with the corresponding predicted target tissue(s) ("Predicted" bar) was compared with that of the same biomarkers but with non-corresponding (off-target) tissues, for which the biomarker was not predicted to be informative ("Non-predicted" bar). Results were then sampled for manual verification of the links within the papers as well as the underlying Medical Subject Headings (MeSH).

The median number of publications co-citing a given tissue and one of its predicted biomarkers was 24, while the median number for non-predicted biomarker/tissue combinations was 16. This difference was found to be significant by the Mann-Whitney *U*-test ($p < 5.3 \times 10^{-13}$).

Tissue specificity, and hence confidence in biomarker quality, was improved further by filtering the candidate biomarker list according to the significant tissue-biofluid channels (see section 3.1 and Figure 2). Thus, a candidate biomarker was considered only if the biofluid containing the biomarker was found to be significantly informative of the biomarker's target tissue. The filtered list, comprising 519 unique biomarkers, was about 60% of the size of the unfiltered list. However, the filtered biomarkers were even more tissue-specific: the median co-citation rate of predicted biomarkers/tissues was 28 publications, whereas the median co-citation rate of non-predicted biomarkers/tissues remained at 16 publications ($p < 5.5 \times 10^{-21}$). This increase after filtering (Figure 4) suggests that clinically relevant protein biomarkers of a tissue are likely to reside in the biofluids found to be significantly informative of that tissue, thus integrating the information channel model of tissue-biofluid interaction (*via* relative entropy) with biomarker prediction and discovery.

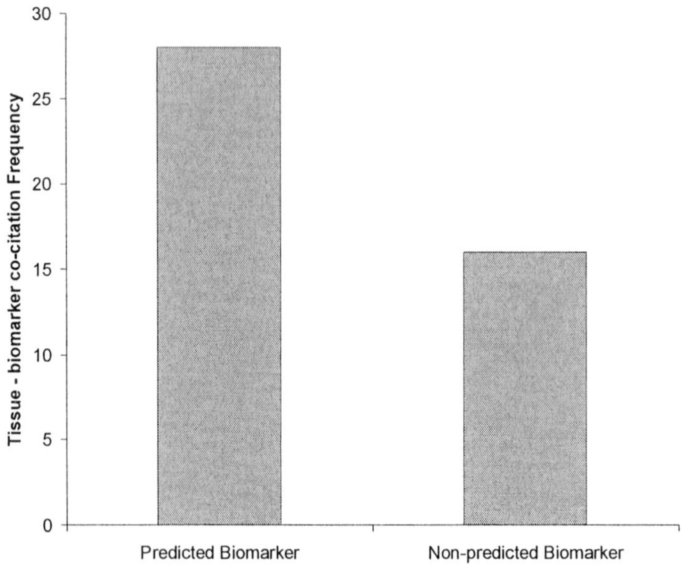

Figure 4. Tissue specificity of candidate biomarkers.

An example case study is Amyloid beta A4 protein. It was found to be modulated with ovariectomies in previous studies[30], thus helping validate the potential of evaluating ovarian function through such a biomarker. On the other hand, it was not found to been previously associated with ovarian cancer in the literature. As an indicator of function, this biomarker can be potentially informative about phenotypic state as well. To validate this, we analyzed an independently performed gene expression study of ovarian cancer[31] and found this protein was significantly upregulated ($p < 0.01$). To control the false discovery rate, we calculated the q-value[32] as 1.4×10^{-10}.

By combining protein interactions, gene expression, and PubMed with an information theoretic framework, this approach promises to allow for the discovery of novel functional and phenotypic biomarkers of internal tissue processes.

4. Discussion and Conclusion

Our framework combines biofluid and tissue information for the discovery of novel biomarkers. Unlike prior work, our approach takes advantage of functional synergy between certain biofluids and tissues with the potential for clinically significant findings not possible if tissues and biofluids were considered individually.

By conceptualizing tissue-biofluid interactions as information channels, we identified significant biofluid proxies that can be used for guided development of clinical diagnostics. We then predicted candidate biomarkers based on information transfer criteria across the tissue-biofluid channels. Significant biofluid-tissue relationships can be used to prioritize clinical validation of new biomarkers.

Some of our results have already been validated for clinical utility, increasing confidence in our findings. At the same time, many are currently novel, suggesting that multiple additional biomarkers can be experimentally confirmed regarding clinical significance. Our work provides a new approach for linking molecular bioinformatics to clinical research, with the potential to expand physiological, phenotypic, and clinical diagnostic capabilities for applications in biology and medicine.

References

1 M. Tyers and M. Mann, *Nature* **422** (6928), 193 (2003).

2 S. G. Sakka, *Current opinion in critical care* **13** (2), 207 (2007).

3 E. A. Singer, D. F. Penson, and G. S. Palapattu, *Jama* **297** (9), 949; author reply 949 (2007).

4 D. C. Crawford, C. L. Sanders, X. Qin et al., *Circulation* **114** (23), 2458 (2006).

5 L. A. Liotta, M. Ferrari, and E. Petricoin, *Nature* **425** (6961), 905 (2003).

6 Y. D. He, *Cancer Biomark* **2** (3-4), 103 (2006).

7 J. M. Jacobs, J. N. Adkins, W. J. Qian et al., *J Proteome Res* **4** (4), 1073 (2005).

8 N. L. Anderson and N. G. Anderson, *Mol Cell Proteomics* **1** (11), 845 (2002).

9 Gil Alterovitz, Michael Xiang, and Marco F. Ramoni, presented at the Proceedings of the Information Theory Applications Workshop, San Diego, CA, 2007 (unpublished).

10 S. Hu, J. A. Loo, and D. T. Wong, *Proteomics* **6** (23), 6326 (2006); M. A. Tangrea, B. S. Wallis, J. W. Gillespie et al., *Expert review of proteomics* **1** (2), 185 (2004).

11 E. Camon, M. Magrane, D. Barrell et al., *Nucleic Acids Res* **32** (Database issue), D262 (2004).

12 M. Ashburner, C. A. Ball, J. A. Blake et al., *Nature genetics* **25** (1), 25 (2000).

13 D. L. Wheeler, T. Barrett, D. A. Benson et al., *Nucleic Acids Res* **34** (Database issue), D173 (2006).

14 M. Hewett, D. E. Oliver, D. L. Rubin et al., *Nucleic Acids Res* **30** (1), 163 (2002).

[15] David J. C. MacKay, *Information theory, inference, and learning algorithms.* (Cambridge University Press, Cambridge, U.K. ; New York, 2003).

[16] Martin Bland, *An Introduction to Medical Statistics*, 3rd ed. (Oxford University Press, 2000).

[17] G. Alterovitz, M. Xiang, M. Mohan et al., *Nucleic acids research* **35** (Database issue), D322 (2007).

[18] S. B. Koevary, V. Lam, and G. Patsiopoulos, *Optometry* **75** (3), 183 (2004).

[19] S. B. Koevary, V. Lam, G. Patsiopoulos et al., *J Ocul Pharmacol Ther* **19** (4), 377 (2003).

[20] R. H. Boerman, A. C. Peters, B. R. Bloem et al., *Acta Neuropathol (Berl)* **83** (3), 300 (1992).

[21] J. Prada, B. Noelle, H. Baatz et al., *Br J Ophthalmol* **87** (5), 548 (2003).

[22] E. Hilton, A. A. Adams, A. Uliss et al., *Lancet* **1** (8337), 1318 (1983).

[23] K. F. Chung, *Curr Drug Targets* **7** (6), 675 (2006).

[24] J. Domagala-Kulawik, M. Maskey-Warzechowska, J. Hermanowicz-Salamon et al., *J Physiol Pharmacol* **57** **Suppl 4**, 75 (2006).

[25] H. Lassus, H. Sihto, A. Leminen et al., *Journal of molecular medicine (Berlin, Germany)* **84** (8), 671 (2006).

[26] M. C. King, J. H. Marks, and J. B. Mandell, *Science* **302** (5645), 643 (2003).

[27] C. D. Hough, K. R. Cho, A. B. Zonderman et al., *Cancer research* **61** (10), 3869 (2001).

[28] Y. C. Chen, G. Pohl, T. L. Wang et al., *Cancer research* **65** (1), 331 (2005).

[29] R. Gogoi, S. Srinivasan, and D. A. Fishman, *Expert review of molecular diagnostics* **6** (4), 627 (2006).

[30] S. S. Petanceska, V. Nagy, D. Frail et al., *Experimental gerontology* **35** (9-10), 1317 (2000).

[31] D. Roberts, J. Schick, S. Conway et al., *British journal of cancer* **92** (6), 1149 (2005).

[32] J. D. Storey and R. Tibshirani, *Proceedings of the National Academy of Sciences of the United States of America* **100** (16), 9440 (2003).

[33] T. I. Williams, K. L. Toups, D. A. Saggese et al., *Journal of proteome research* (2007).

NOVEL INTEGRATION OF HOSPITAL ELECTRONIC MEDICAL RECORDS AND GENE EXPRESSION MEASUREMENTS TO IDENTIFY GENETIC MARKERS OF MATURATION

DAVID P. CHEN [1], SUSAN C. WEBER [2], PHILIP S. CONSTANTINOU [2], TODD A. FERRIS [1,2], HENRY J. LOWE [2], ATUL J. BUTTE [1,3]

[1] *Stanford Medical Informatics, Department of Medicine, Stanford University School of Medicine, Stanford California 94305-5479 USA.* [2] *Information Resources and Technology, Stanford University School of Medicine, Stanford California 94305-5479 USA.* [3] *Lucile Packard Children's Hospital, Palo Alto, CA 94304 USA*

Traditionally, the elucidation of genes involved in maturation and aging has been studied in a temporal fashion by examining gene expression at different time points in an organism's life as well as by knocking out, knocking in, and mutating genes thought to be involved. Here, we propose an *in silico* method to combine clinical electronic medical record (EMR) data and gene expression measurements in the context of disease to identify genes that may be involved in the process of human maturation and aging. First we show that absolute lymphocyte count may serve as a biomarker for maturation by using statistical methods to compare trends among different clinical laboratory tests in response to an increase in age. We then propose using the rate of decay for absolute lymphocyte count across 12 diseases as a proxy for differences in aging. We correlate the differing rates with gene expression across the same diseases to find maturation/aging related genes. Among the 53 genes with strongest correlations between expression profile and change in rate of decay, we found genes previously implicated in the process of aging, including MGMT (DNA repair), TERF2 (telomere stability), POLD1 (DNA replication and repair), and POLG (mtDNA replication).

1. Introduction

The integration of bioinformatics, basic science, and statistical methods has been recognized as being essential to the progression of translational research. Advances made in the understanding of biological systems using such an integrated approach can have a direct impact at both the bench and bedside to further our understanding of human disease [1]. Techniques like gene expression microarray measurement and analysis, which has been used extensively with research involving model organisms, can be extremely informative about diseases, aging and other biological processes. There have been many innovative ways of integrating these microarrays with various data sets to identify genes and their potential function, but most of these methods have led to a reductionist approach to the study of disease, where novel subtypes and features observed in microarray analyses are used to describe singular disorders [2].

Against this reductionist trend, recent work has focused on the use of measurements across a variety of disorders to find the common elements across disease. Daniel Rhodes and colleagues searched for commonalities in cancer in 2004 [3]. After collecting 40 available sets of microarray data with over 3,700 samples of cancer, Rhodes calculated a genome-wide signature representative of neoplastic transformation. Andrea Bild and colleagues linked cellular models with disease samples to find commonly deregulated biological pathways across cancers that correspond with worsening survival [4]. Eran Segal and colleagues used microarray-based expression measurements annotated with both biological and clinical conditions to create modules which were examined across types of cancer [5]. In our previous work, we linked gene measurements, as measured by microarrays, to phenotypes and responses to environment, as represented by biomedical concepts in the Unified Medical Language System (UMLS), to create a phenome-genome network [6]. Each of these is an important example of how genome-era measurements can be used to quantitate mechanistic similarities and differences between diseases previously categorized using syndromic or anatomic descriptors.

An often-overlooked area that can contribute to translational research is clinical laboratory data. In the past, data collected during clinical care were prone to transcription errors while transferring information from paper forms to an electronic format. However, as an increasing number of institutions move towards using electronic medical records (EMR) data quality has increased due to elimination of transcription and omission errors [7]. While EMRs have created a structured environment for reporting of laboratory measurements for physicians, they rarely provide data in a manner easily accessible for translational researchers. Even when such data is available for clinical research, it is typically accessed on a disease-by-disease basis. Here, we hypothesize that these laboratory test measurements can provide an important link between gene expression measurements and the physical manifestations of patients.

One type of physical manifestation is aging. The mechanisms of aging, though still far from being determined, are thought to involve three main biological phenomena leading to cellular senescence: DNA damage, telomere shortening and mitochondrial dysfunction [8]. Research into these areas has focused almost exclusively on *in vitro* and *in vivo* models, wherein gene expression measurements for different time points in an organism's lifespan as well as the knock-out, overexpression, or mutation of genes suspected of being aging-related remain the gold standard to find such genes. However, the study of aging is difficult as there are many genetic as well as environmental influences that contribute to its progression, not to mention the fact that the mechanisms of aging in model organisms may differ from that of humans.

In this paper, we introduce a novel translational method that uses clinical laboratory measurements in conjunction with gene expression levels to elucidate genes that may be involved in the process of human maturation and aging. After using clinical laboratory measurements to find a biomarker that correlates with an increase in age, we order several diseases based on the accelerated or decelerated change in this biomarker. We then use publicly available gene expression data sets representative of these diseases to find genes changing in expression in the same profile as the rate of change in the biomarker. While we find that our set of genes correlating with the change in our aging biomarker are over-represented with known genes associated with aging, we are releasing this list in the hopes that these results will be validated through biological assays. Finally, this method of incorporating clinical laboratory data with gene expression microarray data is extensible and we believe it will be useful in deciphering and understanding many complex human diseases.

2. Methods

2.1. *Data Collection and Processing*

Quantitative clinical laboratory data, consisting of 1,104,316 measurements across 656 distinct lab tests, originally obtained at the Lucile Packard Children's Hospital, were collected in a de-identified manner from the Stanford Translational Research Integrated Database Environment (STRIDE). In total, this data represented 4,844 patients across all ages that were diagnosed with one or more of a set of 12 chronic diseases (Table 1). The use of de-identified clinical laboratory data in this manner was approved by the Institutional Review Board of the Stanford University School of Medicine.

We applied a filter to restrict laboratory measurements to only those measured between the ages of 0 and 17 years, in order to restrict our analysis to the pediatric samples making up the majority of our data. Although patients with certain diseases, like cystic fibrosis, may be seen at a children's hospital through their adult years, we felt that laboratory measurements collected after the pediatric years were not representative enough to include in our analysis. This filter resulted in 4,086 patients with a distribution of ages between 0 and 17 years, diagnosed with one or more of our set of 12 diseases.

We identified 20 microarray experiments within a 2006 snapshot of the NCBI Gene Expression Omnibus (GEO), an international repository for gene expression data, developed and maintained by the National Library of Medicine [9]. Each experiment studied one of 12 diseases using an experimental design in which normal samples were compared to disease samples. Experiments were manually examined and those lacking normal to disease comparisons as well as those not representative of the clinical diagnosis were excluded. A rank based approach was used for normalization of gene expression due to inconsistencies between microarray platforms as well as inconsistencies in submitted data. The gene expression measurements on each microarray was rank-normalized to numbers between 0 and 1, depending on the relative ranking of the expression level of a gene compared to all the other measured genes on that microarray. The mean rank expression for each gene was calculated for control and disease samples, and the difference in these mean rank-normalized expression levels was calculated and assigned to each gene. The mean rank difference for a gene between control and disease states describes relative change of expression for that gene. GEO data sets (GDS) were merged across similar series of microarray types. For example, GDS559 has the title "Inflammatory bowel disease (HG-U133A)" and GDS560 has the title "Inflammatory bowel disease (HG-U133B)". Since the A and B chips are from the same series of microarray, these two data sets were combined, and multiple measurements for a gene were averaged. In this example, the group of microarrays labeled by the submitter as "ulcerative colitis" was compared to the group of microarrays labeled as "control" and the difference in mean rank-normalized expression measurements was assigned to the disease ulcerative colitis. Finally, genes missing measurements in 2 or more of the 12 diseases were dropped. This yielded a matrix of 4,956 genes across 12 diseases.

Table 1: List of the twelve diseases, the abbreviations used in this paper, and the GEO data sets used to represent the genome-wide changes in gene expression seen in each disease.

Disease	**Abbreviation**	**GEO Data Sets**
Autoimmune polyendocrinopathy-candidiasis-ectodermal dystrophy	AP Dystrophy	167
Asthma	Asthma	13, 14, 42, 56, 58, 60
Cystic fibrosis (of pancreas)	Cystic Fibrosis	567
Juvenile spondyloarthropathy	JS	711
Crohn's disease	Crohn's	559, 560
Familial hypercholesterolemia	Fam. Hyperchol.	279
Down syndrome	Down Syndrome	681, 682
Insulin dependent diabetes mellitus	Ins. Dep. Diabetes	10
Ulcerative colitis	Ulcerative Colitis	559, 560
Duchenne muscular dystrophy	DM Dystrophy	639
HIV infection	HIV	171
Neurofibromatosis type 1	Neurofibromatosis I	604

2.2. *Finding Biomarkers for Maturation Using Analysis of Variance*

Each de-identified laboratory measurement was associated with a measurement and the age of a patient when the test was obtained, as an integer. To find a biomarker representative of maturation, we examined trends within laboratory measurements that corresponded to the age of the patient when the measurement was made. We averaged laboratory measurements across each distinct lab for individual ages within a patient's clinical history to generate a laboratory profile for that patient at that specific age. This resulted in 8,500 distinct laboratory profiles distributed between age 0 and 17 years.

We examined the variance of individual laboratory measurements within each age group binned by year (Mean Square Error, or MSE) and the variance of the laboratory measurement means between distinct age groups (Mean Square Between, or MSB) to determine whether or not maturation had an effect on the laboratory measurement. This was done for each distinct lab test separately by using one-way analysis of variance (ANOVA). In order to show that the null hypothesis was false and that a biomarker was indeed indicative of maturation, we needed to show that the MSB was significantly larger than the MSE. For each distinct laboratory we calculated F, the ratio of MSB to MSE. We also calculate a corresponding p-value along with each F as a measure of significance. The laboratory tests with the smallest p-values were taken to be our initial set of prospective biomarkers for human maturation.

2.3. *Using Diseases to Model Maturation*

Rather than looking at gene expression at different time points in an organism's life to study the effects of maturation, we examine maturation in the context of disease. We propose that different diseases exhibit different rates of maturation. Given a set of biomarkers indicative of maturation, we consider them to be proxies for aging, at least for the pediatric age group. We use the measurements of the proxy among patients within our set of 12 diseases as a surrogate for distinct disease-specific rates of maturation. For patients with multiple diagnoses, we assume that their laboratory profile at each age is associated with all previously diagnosed diseases. Multiple diagnoses were not common in these pediatric patients, as expected. We then attempt to fit the biomarker's measurements across all ages for a given disease to an exponential decay model. We arbitrarily chose two models we thought from visual inspection fit the data well, namely a linear model and exponential decay model. We found the error between an exponential decay model and the actual data is less than that of a linear model. The values of the parameters for the curve we fit represent the rate at which the biomarker changes, which we use to represent the rate of accelerating or decelerating of maturation that a disease emulates. Each disease has its own parameters, based on the curve fit to the measurements of a biomarker for patients with that disease. We take the value for these parameters and measure the correlation of these values across the 12 diseases with the changes in rank-normalized gene expression measurements (described above) across the same 12 diseases, using Spearman's rank correlation. We recorded Spearman's ρ as well as the p-value of the correlation, with the null hypothesis of no significant correlation, to inform us of the directionality of both the correlation as well as the significance. Literature search was applied to the most significant genes to determine if they were previously shown be correlated with the aging or maturation.

3. Results

3.1. *Clinical Biomarkers for Maturation*

After reducing and compiling laboratory measurement data to 8,500 patient profiles representing over 4,000 patients at different time points in their clinical history, one-way analysis of variance (ANOVA) was used to elucidate the differences between laboratory measurements at various ages. This was repeated for all lab tests individually. The result of the ANOVA returned prospective biomarkers that could be indicative of increasing age. Four of the top results were as follows: Total Bilirubin ($F = 104.54$, p-value $= 8.58 \times 10^{-279}$); Total Serum/Plasma Protein ($F = 68.66$, p-value $= 3.15 \times 10^{-193}$); Mean Corpuscular Volume ($F = 65.46$, p-value $= 1.58 \times 10^{-201}$); Absolute Lymphocyte ($F = 59.47$, p-value $= 8.57 \times 10^{-181}$). The F and p-values show a statistically significant connection between increasing age and the prospective biomarkers.

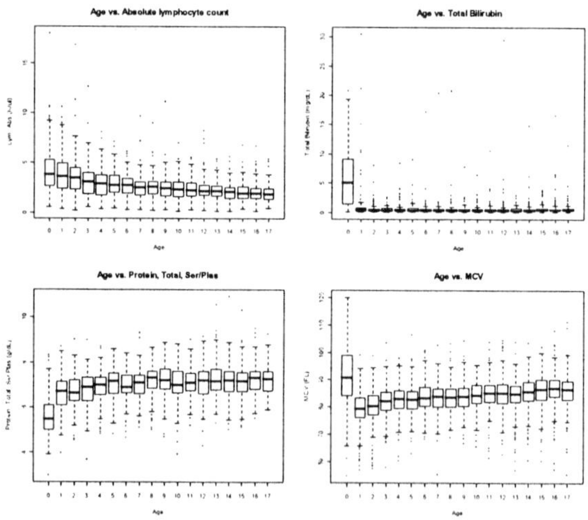

Figure 1: Boxplots showing the distribution of laboratory measurements at different ages. Top left, absolute lymphocyte count; top right, total bilirubin; bottom left, total protein; bottom right, mean corpuscular volume.

As shown in Figure 1, for three out of the four labs, the wide distribution of measurements specifically between age 0 and 1 years could unduly influence the

F and p-value. To examine this influence, an ANOVA was run again on the same data set with all measurements before age 1 excluded. The results show that Total Bilirubin (F = 1.85, p-value = 1.99 x 10^{-2}), Total Serum/Plasma Protein (F = 7.92, p-value = 1.78 x 10^{-18}), and Mean Corpuscular Volume (F=25.90, p-value = 4.57 x 10^{-85}) were influenced more than Absolute Lymphocyte (F = 44.60, p-value = 1.11 x 10^{-128}). This was also verified by applying Bonferroni correction to a pairwise t-test between all age groups. Absolute lymphocyte returned the highest number of significant pairwise comparisons. Based on these results, we selected absolute lymphocyte count as a proxy for maturation and aging.

3.2. *Finding Maturation and Aging Related Genes*

There were 4,045 distinct relations between absolute lymphocyte measurements, patient age, and disease identifier. These profiles were distributed across 12 diseases. We used nonlinear least squares to fit the absolute lymphocyte measurements for each disease across all ages to a model of exponential decay:

$$measurement = \alpha * e^{\lambda * age}$$

where α represents the magnitude and λ the rate of decay. We excluded the disease autoimmune polyendocrinopathy-candidiasis-ectodermal dystrophy, due to a paucity of measurements.

For our purpose, we ignore α and focus on λ as it represents the rate at which the absolute lymphocyte count decreases. Each disease has a distinct λ. A smaller λ, as λ is negative, represents a faster drop in the biomarker. We propose, that if our conjecture holds true and that absolute lymphocyte count is representative of maturation, a change in the biomarker rate of decline could be suggestive of a change in the rate of maturation, so that we can use the same λ to model these differences (Figure 2). We measure the correlation of the set of λ's with the change in rank-normalized expression measurements for each gene, across the same set of diseases. The correlation was done using Spearman's rank correlation. Out of 4956 genes, 53 had p-values less than 0.02 (Table 2)

Spearman's ρ represents how good the gene expression correlates with the changes in λ while also informing us of the directionality of the correlation. A positive Spearman's ρ implies that lower gene expression indicates a faster rate of maturation/aging whereas an increase in gene expression indicates a lower rate of maturation/aging. In contrast, a negative Spearman's ρ implies that a lower gene expression indicates a slower rate of maturation/aging and an increase in gene expression indicates a faster rate of maturation/aging. We investigated the significant genes returned by using literature search. Among the most biologically relevant genes were MGMT, POLD1, POLG and TERF2

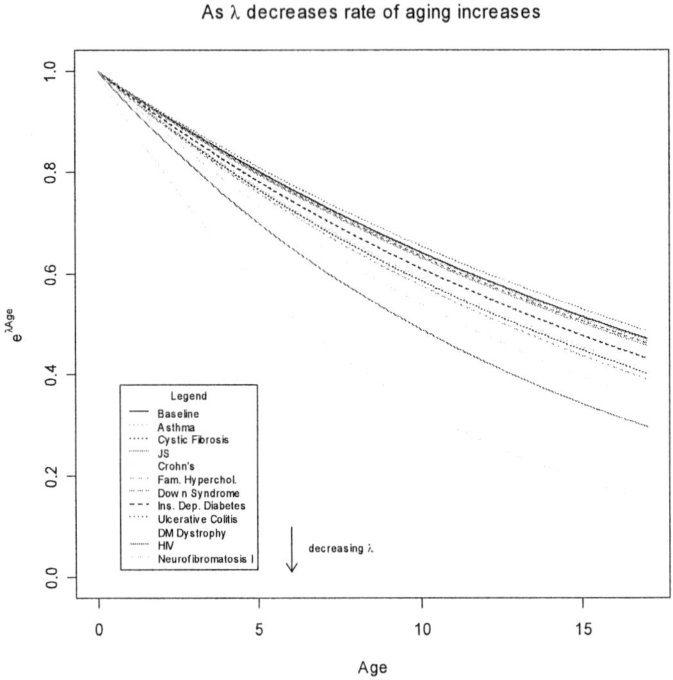

Figure 2: Comparison of different rates of decay across 11 diseases and the baseline.

Table 2: Genes with the best Spearman's Rank Correlation between λ and expression measurements (p-values < 0.02). Genes in bold are in the GenAge database, known to be involved in aging. Stars indicate genes where evidence exists for involvement in aging, yet not appearing in the GenAge database

Symbol	Gene name	p-value	Spearman's ρ
PPIC	peptidyl-prolyl isomerase c	0	-0.9182
* CYP1B1	Cytochrome P450, family 1, subfamily B, polypeptide 1	0.0025	-0.8363
TIPARP	TCDD-inducible poly(ADP-ribose) polymerase	0.0027	-0.8667
POLD1	**Polymerase (DNA directed), delta 1, catalytic subunit**	**0.0041**	**0.8091**
CES2	carboxylesterase 2	0.0044	-0.8091
* MGMT	O-6-methylguanine-DNA methyltransferase	0.0048	0.8
CENTG2	centaurin, gamma 2	0.0050	0.8303
PNN	pinin, desmosome associated protein	0.0056	0.7909
HGF	hepatocyte growth factor	0.0061	-0.7909
KLKB1	kallikrein B, plasma 1	0.0061	-0.7909
POLG	**Polymerase (DNA directed), gamma**	**0.0061**	**-0.7909**

GPD1	glycerol-3-phosphate dehydrogenase 1	0.0062	0.8182
ICT1	immature colon carcinoma transcript 1	0.0065	0.7818
DLG1	discs, large homolog 1 (Drosophila)	0.0065	0.7818
PUM2	pumilio homolog 2 (Drosophila) [	0.0065	0.7818
TNA	C-type lectin domain family 3, member b	0.0070	-0.7818
HGD	Homogentisate 1,2-dioxygenase (homogentisate oxidase)	0.0086	0.7636
RBMS2	RNA binding motif, single stranded interacting protein 2	0.0092	-0.7636
MAPK10	mitogen-activated protein kinase 10	0.0098	0.7545
PRC1	protein regulator of cytokinesis 1	0.0108	-0.8167
MYH10	myosin, heavy chain 10, non-muscle	0.0108	-0.8167
MMP15	matrix metallopeptidase 15	0.0112	0.7454
THRB	thyroid hormone receptor, beta	0.0119	-0.7454
IL10RA	interleukin 10 receptor, alpha	0.0126	0.7364
GZMM	granzyme M	0.0126	0.7364
RGS9	regulator of G-protein signalling 9	0.0126	0.7697
GUCA1A	guanylate cyclase activator 1A	0.0134	-0.7364
GFRA2	GDNF family receptor alpha 2	0.0134	-0.7364
NEUROD1	neurogenic differentiation 1	0.0134	-0.7364
UEV3	UEV and lactate/malate dehyrogenase domains	0.0135	-0.7782
LRP6	low density lipoprotein receptor-related protein 6	0.0137	-0.7697
CRYZ	crystallin, zeta	0.0138	0.8
CLCN1	chloride channel 1	0.0142	0.7273
EIF3S6	eukaryotic translation initiation factor 3, subunit 6 48kDa	0.0142	0.7273
MBNL1	Muscleblind-like (Drosophila)	0.0148	0.7576
EHD1	EH-domain containing 1	0.0148	0.7576
PTX3	pentraxin-related gene	0.0150	-0.7273
P2RY2	purinergic receptor P2Y	0.0159	0.7182
PDK1	pyruvate dehydrogenase kinase, isozyme 1	0.0159	0.7182
POU5F1	POU domain, class 5, transcription factor 1	0.0159	0.7182
SPOCK2	sparc/osteonectin, cwcv and kazal-like domains proteoglycan	0.0159	0.7182
ATP1B3	ATPase, Na+/K+ transporting, beta 3 polypeptide	0.0168	-0.7182
CYB5	Cytochrome b-5	0.0168	-0.7182
CSF1R	colony stimulating factor 1 receptor,	0.0168	-0.7182
KIAA0101	KIAA0101	0.0168	-0.7182
CYP2E1	Cytochrome P450, family 2, subfamily E, polypeptide 1	0.0168	-0.7182
EXOC7	exocyst complex component 7	0.0171	0.7454
SRP68	signal recognition particle 68kDa	0.0172	0.7833
CETN1	centrin, EF-hand protein, 1	0.0177	0.7091
NCL	Nucleolin	0.0177	0.7091
TERF2	**telomeric repeat binding factor 2**	**0.0177**	**0.7091**
CSNK1G2	casein kinase 1, gamma 2	0.0197	0.7
APOC4	Apolipoprotein C-IV	0.0197	0.7

4. Discussion

We have shown the ability to use statistical methods to infer biomarkers and predict genes implicated in maturation by integrating clinical laboratory measurements with gene expression measurements. The most significant biomarker from our analysis, absolute lymphocyte count, has not previously been shown to be a biomarker for aging. However, there is evidence that suggests a decrease in lymphocyte function, as well as a decrease in certain lymphocyte cell types, as age increases [10]. We believe that this method of using clinical laboratory measurements can be extended to find trends within complex diseases and other biological phenomena.

We acknowledge the following caveats in the way we proceeded with this research. The clinical laboratory information we used came from patients ranging in age from 0 to 17 years, which only is able to model a certain aspect of aging, namely the process of maturation. We understand that aging revolves around the complete lifespan of an organism and thus our future work will attempt to reproduce these results using a larger data set of clinical laboratory data spanning across more decades of life. We also have speculated that the rate of absolute lymphocyte change is representative of disease-specific rates of maturation, which is currently only conjecture.

There were a handful of diseases that had significantly fewer measurements than other diseases. We excluded the disease autoimmune polyendocrinopathy-candidiasis-ectodermal dystrophy but kept the others as there were enough data points to fit an exponential decay curve. Ideally we would gather more measurements from patients having these diseases. However, as these diseases tend to be rarer in comparison to conditions like asthma, they will ultimately be less represented. We also acknowledge that we binned patients with different diseases to identify biomarkers related to aging. However, as clinical data rarely consists of "normal" data we are limited to such analyses. We note, however, that the majority of absolute lymphocyte counts across patients with varying diseases all lay in the normal range. Lastly, we would hope to increase the number of diseases to more than 12 to increase the power of our correlations. We also acknowledge that the sample sizes were not large enough to enable sufficient permutation testing and q-value calculation for our Spearman's rank correlations.

The biological relevance of measuring the correlation between rate of acceleration or deceleration of maturation that a disease emulates to changes in rank-normalized gene expression measurements can be expressed via Spearman's ρ. For example, MGMT, a DNA repair gene, has been implicated in the aging of mice and trials have been underway to determine whether or not transgenic MGMT mice live longer [11]. Given the Spearman's ρ we calculated, we would predict that an increase in MGMT would slow aging and thus increase longevity. Spearman's rank correlation was used to account for the possibilities of non-linear correlations. There remain a plethora of statistical methods which can be applied to examine both linear and non-linear relationships between change in gene expression rank and rates of aging that Spearman's rank correlation may not be capturing.

Out of 53 genes that returned a p-value less than the arbitrary cut-off of 0.02, we found three that were represented in 253 aging-related genes from the curated GenAge database [12]. Using a hypergeometric distribution with 253 known genes involved in aging and the number of genes in the human genome conservatively estimated at 20,000 [13], we find that retrieving 3 aging-related genes out of 53 is statistically significant at $p = 0.023$. Although this is encouraging, a better validation strategy must be developed. The absence of the 50 remaining genes from our gold standard could be due to GenAge's lack of comprehensiveness as well as may include numerous false positives. Future work revolves around developing better validation strategies as well as increasing sample size to perform more robust analysis including false discovery rates and q-values.

We set out to use a translational approach linking basic science, clinical electronic medical records, and statistical methods to examine phenomena, maturation and aging, which have been and continue to be difficult to study. Our method returned results that were previously known to be aging-related; although that in and of itself is an accomplishment, what is more notable is the fact that we were able to integrate these disparate fields of the study of disease into a cohesive method of research that has been proselytized as being necessary for the advancement of knowledge about human health. These methods can prove to be invaluable in the future of translational research. As more clinical and hospital environments have moved towards EMRs, the amount of patient data available for translational research will only increase. This must be leveraged and used in conjunction with basic science methods in order to explain biological phenomena in humans that cannot be explained by model organisms.

Acknowledgements

The work was supported by grants from the Lucile Packard Foundation for Children's Health, National Library of Medicine (K22 LM008261 and T15 LM007033), National Institute of General Medical Sciences (R01 GM079719), Howard Hughes Medical Institute, and the Pharmaceutical Research and Manufacturers of America Foundation. Stanford Medical School provides the funding for the development of the STRIDE system. Lucile Packard Children's Hospital provides resources and ongoing operational support. We thank Alex Skrenchuk for parallel computer cluster support.

References

1. Zerhouni EA. Translational and clinical science--time for a new vision. New Engl J Med. 2005 Oct 13;353(15):1621-3.
2. Alizadeh AA, Eisen MB, Davis RE, Ma C, Lossos IS, Rosenwald A, et al. Distinct types of diffuse large B-cell lymphoma identified by gene expression profiling. Nature. 2000;403(6769):503-11.
3. Rhodes DR, Yu J, Shanker K, Deshpande N, Varambally R, Ghosh D, et al. Large-scale meta-analysis of cancer microarray data identifies common transcriptional profiles of neoplastic transformation and progression. P Natl Acad Sci USA. 2004 Jun 22;101(25):9309-14.
4. Bild AH, Yao G, Chang JT, Wang Q, Potti A, Chasse D, et al. Oncogenic pathway signatures in human cancers as a guide to targeted therapies. Nature. 2006 Jan 19;439(7074):353-7.
5. Segal E, Friedman N, Koller D, Regev A. A module map showing conditional activity of expression modules in cancer. Nat Genet. 2004 Oct;36(10):1090-8.
6. Butte AJ, Kohane IS. Creation and implications of a phenome-genome network. Nat Biotechnol. 2006 Jan;24(1):55-62.
7. Payne PR, Johnson SB, Starren JB, Tilson HH, Dowdy D. Breaking the translational barriers: the value of integrating biomedical informatics and translational research. J Invest Med. 2005 May;53(4):192-200.
8. von Zglinicki T, Burkle A, Kirkwood TB. Stress, DNA damage and ageing -- an integrative approach. Exp Gerontol. 2001 Jul;36(7):1049-62.
9. Wheeler DL, Church DM, Edgar R, Federhen S, Helmberg W, Madden TL, et al. Database resources of the National Center for Biotechnology Information: update. Nucleic Acids Res. 2004 Jan 1;32 Database issue:D35-40.
10. Linton PJ, Dorshkind K. Age-related changes in lymphocyte development and function. Nature Immunol. 2004 Feb;5(2):133-9.
11. Anisimov VN. Mutant and genetically modified mice as models for studying the relationship between aging and carcinogenesis. Mech Ageing Dev. 2001 Sep;122(12):1221-55.
12. de Magalhaes JP, Toussaint O. GenAge: a genomic and proteomic network map of human ageing. FEBS Lett. 2004 Jul 30;571(1-3):243-7.
13. Finishing the euchromatic sequence of the human genome. Nature. 2004 Oct 21;431(7011):931-45.

NETWORKING PATHWAYS UNVEILS ASSOCIATION BETWEEN OBESITY AND NON-INSULIN DEPENDENT DIABETES MELLITUS*

HAIYAN HU†

School of Informatics and Center for Computational Biology and Bioinformatics, Indiana University, 410 West 10th street, Suite 5000, Indianapolis, IN 46202, USA

XIAOMAN LI†

Division of Biostatistics, Indiana University, 410 West 10th street, Suite 5000, Indianapolis, IN 46202, USA

Genetic related health problems are often interrelated. Current practices to establish associations between diseases are expensive and rarely can reflect underlying molecular mechanisms. We propose a general framework to associate diseases by networking pathways. By applying our method on association study of non-insulin dependent diabetes mellitus (NIDDM) and obesity, we demonstrate that our method can both identify signature pathways for each disease and establish valid association of two diseases.

1. Introduction

Many diseases are interrelated. Obesity, diabetes, insulin resistance, hypertension are just a few examples. Instead of being attributed to a specific gene, these diseases are often caused by interaction among multiple genes or between genes and environment, and thus are often classified as multifactorial disease or complex disease. Great effort has been put on association studies, such as case control studies and cohort studies, to discover the potential relation between multiple disease conditions in human. Although such association studies can often produce very important information, they are either not very reliable or not efficient in terms of time and money. For example, of two large American Cancer Society cohorts, Cancer Prevention Study I (CPS-I; enrolled in 1959 and followed through 1972) and Cancer Prevention Study II (CPS-II; enrolled in 1982 and followed through 1996), one shows association of height with prostate cancer, the other does not [1]. Most importantly, from such association studies on complex diseases involving genetic factors, no matter how significant the identified associations are statistically, researchers usually cannot gain much insight of the underlying molecular mechanisms. Thus,

---

* This work is partially supported by Indiana Genomics Initiative (INGEN), by Showalter Trust award and by R01HG004359 from NHGRI.

† Corresponding authors.

efficient methods are urgently needed to identify disease associations at the molecular level.

Microarray experiments have been a very popular tool for disease study. From microarray data, gene expression signatures that can distinguish a disease phenotype from another have often been identified by implementing analytical techniques such as differential test. However, in complex diseases like cancer, it is not the individual genes but the interaction between many genes and the interaction between many genes and environment that are responsible for a certain physiological process. Therefore, dozens of suspicious genes included in an identified signature are insufficient for understanding the underlying mechanisms behind a specific disease phenotype. In order to gain deeper understanding of complex diseases from a set of differentially expressed genes, one common practice is to convert the information from gene space to structured pathway space via enrichment test of the differential expressed genes in predefined pathways [2,3]. However, unlike cancers, in which gene expression often show larger variation, for complex diseases like diabetes, obesity, and atherosclerosis, the changes in gene expression are more likely to be modest [4-6]. Yet, the genes vary subtle might be the very responsible ones for a disease phenotype [7,8]. Therefore, under such circumstances that no genes are selectable from differential tests, the traditional methods depending on identification of disease susceptibility genes have lost their power. On the other hand, analysis directly performed on pathways has been encouraging in providing deeper biological understanding compared with single-gene based methods [9-13].

Observing the success of pathway based analysis in various disease studies, we hypothesize that pathway-originated methods are also of great value in associating different disease phenotypes. In this paper, we propose a general framework to study disease association via networking pathways. As a proof of principle, we apply our methods to identify the association between obesity and Non-insulin Dependent Diabetes Mellitus (NIDDM, Type II diabetes), the two diseases that affects hundreds of millions of people worldwide with widely observed connection but with unknown association mechanisms at the molecular level. We have identified a number of pathways and gene sets with known and unknown functions that are responsible for each disease. More importantly, by networking pathways, we have also discovered a set of pathways and their interactions that are responsible for the association between obesity and NIDDM.

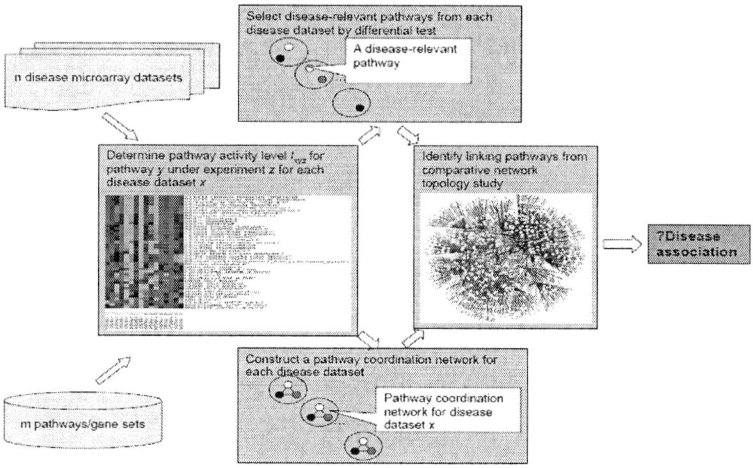

Figure 1. The pipeline of multiple disease association by networking pathways.

2. Methods

We propose a general framework to identify multiple disease association by pathway/gene set association (Figure 1). Here, a gene set is a priori defined set of genes such as a set of genes in one pathway, or a set of target genes regulated by the same transcription factor. Schematically, given *n* disease datasets and *m* predefined pathways/gene sets, we first determine its activity level under each experimental condition for each pathway/gene set. We then select differentially activated pathways/gene sets between disease and control experimental conditions in one data set, and we also construct a pathway coordination network for each disease dataset, in which each node represents a pathway/gene set and each edge connects two pathways/gene sets showing significant coordinated activities. A pathway coordination network thus converts its corresponding disease data into a relation graph depicting the interplay among various functional units (predefined pathways/gene sets in our case). By performing comparative network analysis, we finally can generate hypothesis on disease association at the molecular pathway level. The methods and techniques utilized are detailed in the following subsections.

2.1. *Microarray data sources*

We use microarray experiment data obtained from skeletal muscle. Skeletal muscle cells are the largest storage organ for glucose and considered to play the major role in glucose homoeostasis. From DGAP (Diabetes Genome Anatomy

Project), we downloaded type II diabetic human data containing 18 NIDDM samples and 17 Normal Glucose Tolerance samples generated from Human skeletal muscle samples of Swedish males for study of type II diabetes by Dr. Altshuler's Lab at MIT. From GEO (Gene Expression Omnibus) at NCBI, we also downloaded obesity skeletal muscle data (GDS268) containing 8 skeletal muscle samples from non-obese subjects and 8 skeletal muscle samples from morbidly obese subjects. All the downloaded gene expression levels were measured using Affymetrix Human U133A GeneChip platform. The same experimental platform and tissue type enables us to study obesity and NIDDM with higher signal to noise ratio.

2.2. *Compilation of pathways/gene sets for human*

We downloaded 187 pathways from KEGG [14], 263 pathways from BioCarta [15], 20 pathways related to cancer/immune signaling from NetPath website [16], 243 pathways from Rat Genome Database [17], and 1520 gene sets from mSigDB (version on Oct, 2006) [10]. Besides, we obtained another 3229 gene sets by grouping genes on AFFY-HU133A array according to their GO annotation using FatiGO [2]. Additionally, we obtained 2459 gene sets from graph clustering using MCL [18] on gene expression profiles in four microarray datasets related to NIDDM and obesity [8,19-21]. In total, 7921 pathways/gene sets were compiled for this study.

2.3. *Pathway/gene set activity level and coordination network*

We define the activity level profile of a pathway/gene set under a given set of experimental conditions using eigengenes generated from singular value decomposition (SVD) [9,22]. In detail, for each pathway/gene set containing m genes, there is one $m \times n$ matrix A consisting of the row normalized transcriptional responses of these m genes under n experiments in a microarray dataset such that the mean and standard deviation of the expression levels for each gene is 0 and 1, respectively. We then performed SVD on the matrix A to decompose A into three matrices U, S and V, i.e. $A = USV^T$. The U and V^T are commonly named as the eigenarray matrix and eigengene matrix respectively. The matrix S is a diagonal matrix with singular values of the matrix A as the diagonal elements, whose square reflect the variance of the corresponding eigengene/eigenarray. By using the top k eigengenes, with each of which accounting for no less than *(70/n)%* of the overall variability [23,24], we define activity level l of a pathway/gene set under experiment j as:

$$l_j = \sum_{i=0}^{k} V^T{}_{ij}^{2} \, . \tag{1}$$

Here i is the index of top k eigengenes we used. The intuition that a pathway's activities can be defined from eigengenes is that, linear combination of such defined pathway activity level is an optimal approximation of the transcription profile matrix A corresponding to all the genes within the pathway, as has been explained in [9]. However, unlike the pathway level analysis in [9], where the pathway activity level profile is determined from the first one eigengene (corresponding to the largest eigenvalue) only, here we utilize multiple eigengenes that can explain at least a certain percentage of variance to determine pathway activities. The advantage is evident in that the first eigengene is not always reflecting the dominant variance of the transcript levels corresponding to the genes within a pathway. Thus, given an expression value matrix, the activity level of a pathway captures the major components of the variation in the given expression matrix. After filtering out the genes not included on the Human U133A chip, 7016 out of the compiled 7921 pathways/gene sets containing at least two genes for performing SVD remained. With a pathway/gene set's activity level defined above, we define two pathways/gene sets as coordinated if the two pathways/gene sets show coordinated activity levels under a given set of conditions. Thus, for each disease microarray dataset, we can construct a pathway coordination network in which each node represents a pathway/gene set, and each edge connects two coordinated pathways/gene sets. In this paper, we measure the coordination between any two pathways/gene sets using Spearman's rank correlation. For a given pathway/gene set q, only the top 1% of the pathways/gene sets with largest correlations (larger than 0.6) with q are kept as coordinated pathways/gene sets of q.

2.4. *Differentially Activated Pathways/gene sets*

With the activity levels defined, we used SAM [25] to determine whether a pathway/gene set is activated differently between disease and control samples. SAM has been validated in a number of studies and has been shown more accurate than other differential test methods such as simple t-test [26-29]. SAM uses modified t statistic to measure the activity difference of a pathway/gene set between two types of samples as a score d. For each pathway/gene set, SAM then performs permutation test to determine the statistical significance of the d score. In our study, we chose the significant pathways/gene sets by controlling the false discovery rate (q-value) at the 0.1 level.

2.5. *Disease-relevant pathways and linking pathways*

First, if a pathway P is differentially activated between disease and control samples, then P is called an A-relevant pathway. Given two types of disease A

and B, if a pathway P is A-relevant, we define P as a linking pathway between diseases A and B if P satisfies one of the following three criteria. (1) P is directly connected to at least one of the pathways relevant to disease B; (2) P shares at least one first layer neighbor with at least one of the B-relevant pathways; (3) there is a cluster containing P shares at least one common element with a cluster containing B-relevant pathways. Any two linking pathways from different disease networks with coordinated activity profiles with each other or the common third party pathways may indicate associations between their corresponding diseases.

3. Results

3.1. *Identified obesity-relevant biological pathways/gene sets*

In total, we systematically identified 92 obesity-relevant pathway/gene sets that are differentially activated in obesity and control experiments, including 18 well defined pathways from KEGG and BioCarta database (Table 1) [14,15]. Many studies have supported the relevance of these pathways to obesity [30,31].

Table 1 - 18 well-defined pathways out of 92 pathways/gene sets that are differentially activated significantly between obesity and control experiments.

Pathway Description	Score(d)	q-value(%)
KEGG: Nicotinate and nictoinamide metabolism	1.83	0
KEGG: Glycan structures biosynthesis	1.56	6.10
mSigDB: Genes related to the insulin receptor pathway	1.59	6.10
α6-β4 Integrin Signaling Pathway	1.55	6.10
RGD: Prostaglandin and Leukotriene metabolic pathway	1.56	6.10
KEGG: Fatty acid metabolism	1.36	7.73
KEGG: Tryptophan metabolism	1.45	7.73
KEGG: Glycerophospholipid metabolism	1.50	7.73
KEGG: Arachidonic acid metabolism	1.49	7.73
KEGG: One carbon pool by folate	1.38	7.73
KEGG: MAPK signaling pathway	1.36	7.73
KEGG: mTOR signaling pathway	1.45	7.73
KEGG: Regulation of actin cytoskeleton	1.45	7.73
mSigDB: AR mouse plus testo from netaffx	1.37	7.73
mSigDB: rasPathway from Biocarta	1.43	7.73
RGD: glycerolipid metabolic pathway	1.41	7.73
Biocarta: Role of EGF Receptor Transactivation by GPCRs in Cardiac Hypertrophy	1.34	9.95
KEGG: Arginine and proline metabolism	1.34	9.95

3.2. *Identified NIDDM-relevant biological pathways/gene sets*

We identified 78 pathways/gene sets to be NIDDM-relevant, covering defined pathways in KEGG and BioCarta, expert-curated gene sets, gene sets defined by GO categories and gene sets comprised of co-expressed genes. 16 out of the 78 pathways/gene sets are well defined pathways (Table 2). Most of these pathways are related to the three components of carbohydrate catabolism: glycolysis, TCA cycle and oxidative phosphorylation, implicating the link between NIDDM and mitochondrial dysfunction [32,33].

Table 2 - 16 well-known pathways out of 78 pathways/gene sets that are significantly differentially activated between NIDDM and control experiments

Pathway Description	Score(d)	q-value(%)
mSigDB: electron transport chain	1.39	0
KEGG: Oxidative phosphorylation	1.28	0
mSigDB: Oxidative Phosphorylation	1.08	0
RGD: Oxidative Phosphorylation	1.04	0
mSigDB: Mitochondrial genes	1.04	0
KEGG: Pyruvate metabolism	0.98	0
mSigDB: Pyruvate metabolism	0.97	0
mSigDB: Role of Mitochondria in Apoptotic Signaling	0.97	0
KEGG: Citrate Cycle (TCA cycle)	0.83	2.70
RGD: Pyruvate metabolic pathway	0.82	2.70
KEGG: Propanoate metabolism pathway	0.80	4.39
RGD: glyoxylate and dicarboxylate metabolic pathway	0.80	4.39
mSigDB: Oxidative phosphorylation pathway from KEGG	0.79	4.39
mSigDB: Genes 2fold upregulated by insulin	0.76	4.39
mSigDB: krebPathway	0.75	7.19
mSigDB: Reactive oxidative species related genes	0.73	8.30

Although between NIDDM and control subjects, we have witnessed statistically significant differences at the pathway level, we have found no much difference at the individual gene level. Taking citrate cycle pathway as an example, none of the genes in this pathway is significantly differentially expressed. The genes *ACO2*, *MDH1* and *FH* are only slightly down regulated in NIDDM and with insignificant fold changes ranging from 0.8 to 0.95; the genes *SDHA* and *OGDH* only show modest increase in NIDDM (*SDHA*: fold =1.21, p-value= 0.776691; *OGDH*: fold = 1.19, p-value = 0.463948).

3.3. *Identification of association between obesity and NIDDM by networking pathways*

By comparing the defined obesity-relevant pathways and NIDDM-relevant pathways, we found that obesity-relevant pathways contains a gene set related to the insulin receptor, and coincidentally, there is a NIDDM-relevant gene set containing genes 2-fold up-regulated by insulin. Other than that, all relevant pathways in obesity and NIDDM are literally different. Besides, the genes shared by the two types of pathways are not significantly differentiated between disease and control samples and consequently provides no sufficient information to determine association between obesity and NIDDM. Thus, we proceed to associate obesity and NIDDM in the following steps. We first build a pathway coordination network for each disease. For obesity dataset, this resulted in a network containing 7016 pathway nodes and 237,226 pathway coordination edges, and for NIDDM, this generated a network with 7016 pathway nodes and 207,571 pathway coordination edges. From the two networks, we attempt to associate the two diseases by searching for linking pathways according to their three criteria defined in our methods section.

To search for linking pathways satisfying the first criteria, we examined whether there are any direct links between the two types of disease-relevant pathways. We found obesity-relevant pathways including arginine and praline metabolism pathway and fatty acid metabolism pathway, and tryptophan metabolism pathway are directly connected with NIDDM-relevant pyruvate metabolism pathway. Besides, actin cytoskeleton regulation in obesity network is linked directly to TCA cycle in NIDDM network.

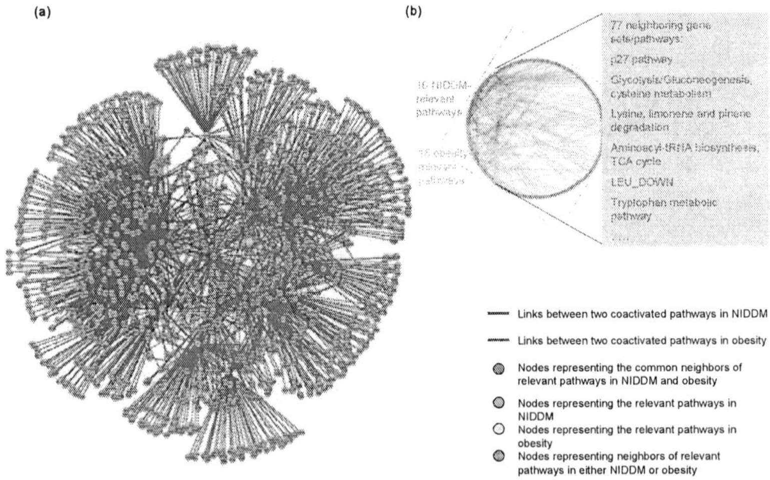

Figure 2 - Merged two pathway coordination sub-networks from obesity and NIDDM. (a) All first layer neighbors for the 18 obesity-relevant pathways and the 16 NIDDM-relevant pathways are included. (b) Only their common first layer neighbors of the 18 obesity-relevant and the 16 NIDDM-relevant pathways are included.

We next seek linking pathways satisfying the second criteria. By extracting the first-layer neighbor gene sets/pathways of the defined disease-relevant pathways from their corresponding networks (Figure 2a), we found totally 77 first-layer neighbor pathways/gene sets are in common. Figure 2b shows a sub-network from network in Figure 2a. This sub-network contains only disease-relevant pathway and their first layer neighbor pathway nodes, and 430 coordination edges linking them.

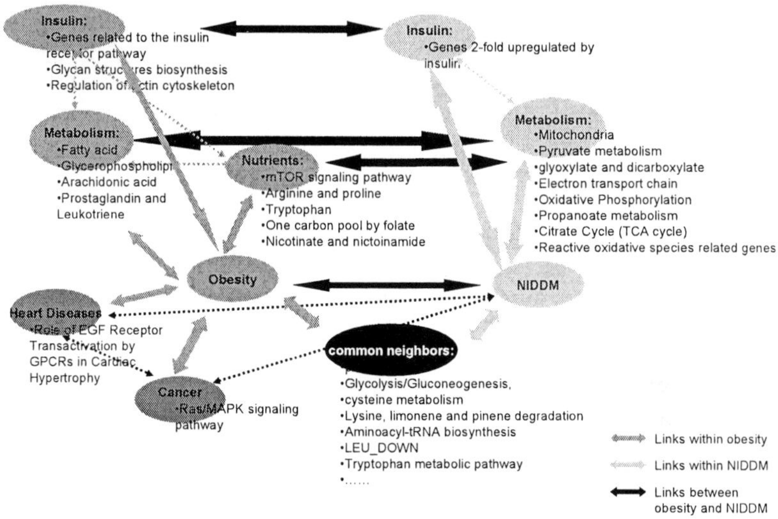

Figure 3 - Summary of identified associations between obesity and NIDDM.

Finally, we search for linking pathways satisfying the third criteria. By performing graph clustering on both networks using MCL algorithm [18], we obtained 239 clusters and 67 clusters containing more than 3 and less than 60 pathways/gene sets corresponding to obesity and NIDDM pathway coordination network respectively. We drew a comparison between the pathway clusters in obesity and NIDDM pathway coordination network. If a cluster in disease *A*'s pathway coordination network contains an *A*-relevant pathway, it is defined as an *A*-relevant cluster. It is interesting to see there are a number of obesity-relevant clusters and NIDDM-relevant clusters we identified above overlapped. For instance, a cluster in the obesity pathway network including 5 gene sets related to glycogen/glucan biosynthesis is overlapped with another cluster in the NIDDM pathway network containing 16 gene sets involving regulation of circadian rhythm, keratin sulfate metabolism. An obesity-relevant cluster comprised of 53 gene sets involving PI3K (phosphoinositide 3-kinases) and their downstream targets, and SERMs (selective estrogen receptor modulators) down regulated genes is overlapped with another NIDDM-relevant cluster containing 23 gene sets including atrial natriuretic peptide signaling pathway, lipoprotein metabolic pathway, altered lipoprotein metabolic pathway.

Taking all the findings together, we provide a summary of all these pathway associations between obesity and NIDDM (Figure 3). Many of these associations are supported by literature search [33-39].

4. Discussion and Conclusion

We have proposed a general framework for disease association by pathway analysis and networking co-activated pathways. To our knowledge, this is the first disease association method that can delineate the relationship between any two or even more disease phenotypes at the molecular pathway/pathway interaction level. In contrast to disease association by case control or cohort studies, our method is not only efficient but also can generate deeper insight about disease etiology and pathophysiology, especially for complex diseases like NIDDM and obesity where the expression differences of genes are often trivial and consequently no suspicious genes detectable by conventional methods. Besides, our strategy moves beyond single gene/pathway based study, and sets off for studying the relationship between pathways or gene sets. In order to capture the relationship between any two pathways, we first generate an activity level profile reflecting the overall response of a certain pathway under a set of experimental conditions. A unique advantage of using pathway activity levels to characterize a pathway is that pathway activity levels can be further used to establish a quantitative relation between any two pathways. After determining the coordination relationship for each pair of pathways/gene sets, a disease dataset can then be modeled as a network. The problem of associating two diseases is subsequently converted to the problem of network comparison.

By applying our approach on obesity and NIDDM, We systematically obtained important pathways that can characterize each disease phenotype and further depicted the association between obesity and NIDDM via linking pathways. The coordinated activity of disease-relevant pathway fatty acid metabolism and pyruvate metabolism in both obesity and NIDDM samples indicates that dysfunction of fatty acid metabolism is intertwined with the functioning of pyruvate metabolism, perhaps via TCA cycle. The supporting model is the early proposed cellular mechanism of glucose-fatty acid cycle in which fatty acid oxidation inhibits glucose utilization by affecting pyruvate dehydrogenase activity [34]. Our study also discovered other important associations such as insulin relevant pathway, stress related ROS genes (Reactive Oxidative Species related genes), cell growth and apoptosis and other immune related pathways.

We need to point out that the accurate interpretation of the association between diseases heavily relies on the correct definition of pathways/gene sets.

More effort on curating pathways/gene sets in order for disease association by networking pathways is essential. Besides, our present study on obesity and NIDDM is based on microarray experiments on human skeletal muscle tissue. Therefore, the conclusions we drew in this study may not reflect the pathway interaction patterns in other tissues. With more and more microarray experimental datasets become available in the near future, it will be interesting to extend our study to multiple tissue/organs such as pancreatic islets, adipose tissue, liver and kidney. It will also be interesting to compare the pathway coordination networks across different species, the results of which will make the dynamic delineation of function evolution of related pathways and pathway interactions possible.

References

1. Rodriguez C, Patel AV, Calle EE, Jacobs EJ, Chao A et al., *Cancer Epidemiol Biomarkers Prev* 10(4): 345-353.(2001)
2. Al-Shahrour F, Diaz-Uriarte R, Dopazo J, *Bioinformatics (Oxford, England)* 20(4): 578-580.(2004)
3. Segal E, Friedman N, Koller D, Regev A, *Nature genetics* 36(10): 1090-1098.(2004)
4. Yechoor VK, Patti ME, Saccone R, Kahn CR, *Proceedings of the National Academy of Sciences of the United States of America* 99(16): 10587-10592.(2002)
5. Wilson KH, Eckenrode SE, Li QZ, Ruan QG, Yang P et al., *Diabetes* 52(8): 2151-2159.(2003)
6. Garland LG, *FEMS microbiology immunology* 5(5-6): 229-237.(1992)
7. Patti ME, Butte AJ, Crunkhorn S, Cusi K, Berria R et al., *Proceedings of the National Academy of Sciences of the United States of America* 100(14): 8466-8471.(2003)
8. Mootha VK, Lindgren CM, Eriksson KF, Subramanian A, Sihag S et al., *Nature genetics* 34(3): 267-273.(2003)
9. Tomfohr J, Lu J, Kepler TB, *BMC bioinformatics* 6: 225.(2005)
10. Subramanian A, Tamayo P, Mootha VK, Mukherjee S, Ebert BL et al., *Proceedings of the National Academy of Sciences of the United States of America* 102(43): 15545-15550.(2005)
11. Pang H, Lin A, Holford M, Enerson BE, Lu B et al., *Bioinformatics (Oxford, England)* 22(16): 2028-2036.(2006)
12. Huang E, Ishida S, Pittman J, Dressman H, Bild A et al., *Nature genetics* 34(2): 226-230.(2003)
13. Shyamsundar R, Kim YH, Higgins JP, Montgomery K, Jorden M et al., *Genome biology* 6(3).(2005)
14. Kanehisa M, *Trends Genet* 13(9): 375-376.(1997)

15. BioCarta http://www.biocarta.com/genes/index.asp
16. NetPath http://www.netpath.org/.
17. Database RG http://rgd.mcw.edu/.
18. Van Dongen S, *PhD thesis, University of Utrecht*.(2000)
19. Nair S, Lee YH, Rousseau E, Cam M, Tataranni PA et al., *Diabetologia* 48(9): 1784-1788.(2005)
20. Park JJ, Berggren JR, Hulver MW, Houmard JA, Hoffman EP, *Physiological genomics* 27(2): 114-121.(2006)
21. Gunton JE, Kulkarni RN, Yim S, Okada T, Hawthorne WJ et al., *Cell* 122(3): 337-349.(2005)
22. Strang G Introduction to Linear Algebra. Cambridge: Wellesley; .(2003)
23. Raychaudhuri S, Stuart JM, Altman RB, *Pacific Symposium on Biocomputing*: 455-466.(2000)
24. Everitt BS, Dunn G, editors (1992) Applied Multivariate Data Analysis. New York, NY: Oxford University Press.
25. Tusher VG, Tibshirani R, Chu G, *Proceedings of the National Academy of Sciences of the United States of America* 98(9): 5116-5121.(2001)
26. King JY, Ferrara R, Tabibiazar R, Spin JM, Chen MM et al., *Physiological genomics* 23(1): 103-118.(2005)
27. Kittleson MM, Minhas KM, Irizarry RA, Ye SQ, Edness G et al., *Physiological genomics* 21(3): 299-307.(2005)
28. Singhal S, Kyvernitis CG, Johnson SW, Kaiser LR, Liebman MN et al., *Cancer biology & therapy* 2(4): 383-391.(2003)
29. Bullinger L, Dohner K, Bair E, Frohling S, Schlenk RF et al., *The New England journal of medicine* 350(16): 1605-1616.(2004)
30. Hausman DB, DiGirolamo M, Bartness TJ, Hausman GJ, Martin RJ, *Obes Rev* 2(4): 239-254.(2001)
31. Blaak EE, *The Proceedings of the Nutrition Society* 63(2): 323-330.(2004)
32. Petersen KF, Befroy D, Dufour S, Dziura J, Ariyan C et al., *Science* 300(5622): 1140-1142.(2003)
33. Lowell BB, Shulman GI, *Science* 307(5708): 384-387.(2005)
34. Frayn KN, *Biochemical Society transactions* 31(Pt 6): 1115-1119.(2003)
35. Patti ME, Kahn BB, *Nature medicine* 10(10): 1049-1050.(2004)
36. Ghazalpour A, Doss S, Sheth SS, Ingram-Drake LA, Schadt EE et al., *Genome biology* 6(7): R59.(2005)
37. Baum JI, O'Conner JC, Seyler JE, Anthony TG, Freund GG et al., *The American journal of physiology*(288): E86-91.(2005)
38. Kelley DE, Mintun MA, Watkins SC, Simoneau JA, Jadali F et al., *The Journal of clinical investigation* 97(12): 2705-2713.(1996)
39. Bloomgarden ZT, *Diabetes care* 23(10): 1584-1590.(2000)

EXTRACTING GENE EXPRESSION PROFILES COMMON TO COLON AND PANCREATIC ADENOCARCINOMA USING SIMULTANEOUS NONNEGATIVE MATRIX FACTORIZATION

LIVIU BADEA

AI Lab, National Institute for Research and Development in Informatics
8-10 Averescu Blvd., Bucharest, Romania, badea@ici.ro

In this paper we introduce a clustering algorithm capable of simultaneously factorizing two distinct gene expression datasets with the aim of uncovering gene regulatory programs that are common to the two phenotypes. The *siNMF* algorithm simultaneously searches for two factorizations that share the same gene expression profiles. The two key ingredients of this algorithm are the nonnegativity constraint and the offset variables, which together ensure the sparseness of the factorizations.

While cancer is a very heterogeneous disease, there is overwhelming recent evidence that the differences between cancer subtypes implicate entire pathways and biological processes involving large numbers of genes, rather than changes in single genes. We have applied our simultaneous factorization algorithm looking for gene expression profiles that are common between the more homogeneous pancreatic ductal adenocarcinoma (PDAC) and the more heterogeneous colon adenocarcinoma. The fact that the PDAC signature is active in a large fraction of colon adeocarcinoma suggests that the oncogenic mechanisms involved may be similar to those in PDAC, at least in this subset of colon samples.

There are many approaches to uncovering common mechanisms involved in different phenotypes, but most are based on comparing gene lists. The approach presented in this paper additionally takes gene expression data into account and can thus be more sensitive.

1 Introduction and motivation

Understanding cancer at the molecular level is a daunting task due to the enormous *heterogeneity* of this disease, depending not only on tissue and cell type, the progenitor cells involved, but also on the stochastic nature of genomic mutations as well as the associated local evolutionary processes.

However, not all cancers are equally heterogeneous. An ongoing microarray study of pancreatic ductal adenocarcinoma (PDAC) [6] involving 76 samples (i.e. 38 normal-tumor pairs) has revealed a surprising homogeneity of this particularly deadly type of cancer, characterized by a strong so-called "desmoplastic reaction" (fibrosis), as well as by a very high metastatic potential.

A preliminary analysis of the genes differentially expressed between tumor and control samples emphasized the essential role of the *TGF-beta pathway* in PDAC. Remarkably, the TGF-beta pathway links the two observed phenotypes: fibrosis/ extracellular matrix proliferation and the aggressive metastatic potential of PDAC, the latter being due to the fact that TGF-beta controls the so-called epithelial-mesenchymal transition (EMT).

As opposed to PDAC, sporadic colon adenocarcinoma are very heterogeneous and their best current classification based on the presence or absence of microsatellite instabilities (MSI-L, MSI-H and MSS) [1] is far from ideal from the point of view of gene expression.

To obtain a better subclassification of sporadic colon adenocarcinomas, we have applied various unsupervised clustering algorithms to a large colon cancer dataset (204 samples). Interestingly, a large colon adenocarcinoma subclass expressed a set of genes very similar to the genes differentially expressed in pancreatic ductal adenocarcinoma.

This immediately leads to the question of whether the TGF-beta related mechanism involved in PDAC is also at work in at least a subset of colon adenocarcinoma. An ad-hoc approach (like the one mentioned above) based on overlaps of gene lists is however far from satisfactory, since it entirely ignores the quantitative gene expression data available.

This paper presents a more sophisticated method of extracting the gene expression profiles common to a pair of distinct phenotypes (e.g. diseases) for which microarray studies are available. The method involves a generalization of Nonnegative Matrix Factorization (NMF) and is called "*simultaneous NMF*" (*siNMF*), since it factorizes two gene expression datasets simultaneously. More precisely, the *siNMF* algorithm searches for two factorizations (of the two gene expression datasets) *sharing the same gene expression profiles*. This allows us to discover the gene expression profiles that are common to pairs of subclasses in the two datasets.

In the special case of PDAC and sporadic colon adenocarcinoma, we found a gene expression profile highly enriched in target genes of the TGF-beta pathway that is involved in the majority of PDAC cases as well as a large subclass of colon cancers.

2 The datasets

For the present study we have used two large PDAC and sporadic colon adenocarcinoma microarray datasets, which we briefly describe below.

2.1 The pancreatic ductal adenocarcinoma dataset

The pancreatic ductal adenocarcinoma (PDAC) dataset was produced in the framework of our GENOPACT project [6]. The dataset contains microarray measurements produced with Affymetrix U133 Plus 2.0 whole genome chips for 38 pairs of PDAC and respectively control samples (76 samples in total). [1] The raw

---

[1] As far as we know, the sample size of our study is significantly larger than all published microarray studies of pancreatic ductal adenocarcinoma.

scanning data was preprocessed with the RMA normalization and summarization algorithm from the R package. (The logarithmized form of the gene expression matrix was subsequently used, since typical gene expression values are log-normally distributed.) After filtering out the probe-sets (genes) with relatively low expression as well as those with a nearly constant expression value[2], we were left with 7232 probe-sets. Finally, the Euclidean norms of the expression levels for the individual genes were normalized to 1 to disallow genes with higher absolute expression values to overshadow the other genes in the factorization.

2.2 The sporadic colon adenocarcinoma dataset

Because of the known heterogeneity of sporadic colon adenocarcinoma, a dataset much larger than the pancreatic dataset described above was needed. We combined 182 colon adenocarcinoma samples from the expO database [7] with 22 control samples from [8] to obtain a 204 sample dataset. (All of these had been measured on Affymetrix U133 Plus 2.0 chips.) After applying the same filtering step as the one used in the PDAC dataset (average expression > 100 and standard deviation > 100), we obtained a smaller set of 5617 probe-sets. The resulting gene expression matrix was also logarithmized before factorization and the Euclidean norms of the individual genes were normalized to 1.

In the following we describe the factorization algorithm in more detail before presenting its application to the two datasets.

3 Simultaneous Nonnegative Matrix Factorization with offset

SiNMF simultaneously factorizes two (non-negative) gene expression matrices $X_{sg}^{(1)}$ and $X_{sg}^{(2)}$ (the index s denotes samples, while g stands for genes) as follows:

$$X_{sg}^{(1)} \approx \sum_c A_{sc}^{(1)} \cdot S_{cg} + So_g^{(1)} \tag{1}$$

$$X_{sg}^{(2)} \approx \sum_c A_{sc}^{(2)} \cdot S_{cg} + So_g^{(2)} \tag{2}$$

with the additional nonnegativity constraints:

$$A_{sc}^{(1)} \geq 0,\ A_{sc}^{(2)} \geq 0,\ S_{cg} \geq 0,\ So_g^{(1)} \geq 0,\ So_g^{(2)} \geq 0 \tag{3}$$

where X_{sg} is the expression level of gene g in data sample s, A_{sc} the expression level of the biological process (cluster) c in sample s, S_{cg} the membership degree of gene g in c and So_g the expression offset of gene g.

Note that the gene cluster membership matrix S is common to the two factorizations, as it is influenced by both gene expression datasets $X^{(i)}$. The

---

[2] Only genes with an average expression value over 100 and with a standard deviation above 100 were retained.

nonnegativity constraints (3) express the obvious fact that expression levels, membership degrees and expression offsets cannot be negative.

More formally, the factorization (1-3) can be cast as a constrained optimization problem:

$$\min C(A^{(i)},S,So^{(i)})=\frac{1}{2}\|X^{(1)}-A^{(1)}S-e^{(1)}So^{(1)}\|_F^2+\frac{\beta}{2}\|X^{(2)}-A^{(2)}S-e^{(2)}So^{(2)}\|_F^2 \quad (4)$$

subject to the nonnegativity constraints (3) ($\|\cdot\|_F$ is the Frobenius norm of a matrix, while $e^{(i)}$ is a column of 1 of size equal to the number of samples of $X^{(i)}$).

The weight β ensures a proper balance between the two error terms and was taken in the following experiments to be $\beta=\beta_0\frac{\|X^{(1)}\|^2}{\|X^{(2)}\|^2}$ with β_0=1.

The optimization problem (4) can be solved using *multiplicative update rules*[3] in a manner similar to Lee and Seung's seminal *Nonnegative Matrix Factorization* (*NMF*) algorithm [5] (ε is a small regularization parameter):

siNMF($X^{(1)}, X^{(2)}, A_0^{(1)}, A_0^{(2)}, S_0, So_0^{(1)}, So_0^{(2)}$) → ($A^{(1)}, A^{(2)}, S$)

$A^{(1)} \leftarrow A_0^{(1)}$, $A^{(2)} \leftarrow A_0^{(2)}$, $S \leftarrow S_0$, $So^{(1)} \leftarrow So_0^{(1)}$, $So^{(2)} \leftarrow So_0^{(2)}$ (typically $A_0^{(1)}$, $A_0^{(2)}$, S_0, $So_0^{(1)}$, $So_0^{(2)}$ are initialized randomly)

loop

$$S_{cg}\leftarrow S_{cg}\frac{(A^{(1)T}\cdot X^{(1)}+\beta A^{(2)T}\cdot X^{(2)})_{cg}}{(A^{(1)T}\cdot(A^{(1)}\cdot S+e^{(1)}\cdot So^{(1)})+\beta A^{(2)T}\cdot(A^{(2)}\cdot S+e^{(2)}\cdot So^{(2)}))_{cg}+\varepsilon}$$

$$A_{sc}^{(i)}\leftarrow A_{sc}^{(i)}\frac{(X^{(i)}\cdot S^T)_{sc}}{((A^{(i)}\cdot S+e^{(i)}\cdot So^{(i)})\cdot S^T)_{sc}+\varepsilon} \qquad \text{for } i\in\{1,2\}$$

$$So_g^{(i)}\leftarrow So_g^{(i)}\frac{(e^{(i)T}\cdot X^{(i)})_g}{(e^{(i)T}\cdot(A^{(i)}\cdot S+e^{(i)}\cdot So^{(i)}))_g+\varepsilon} \qquad \text{for } i\in\{1,2\}$$

until convergence

normalize the rows of *S* to unit norm by taking advantage of the scaling invariance of the factorization: $S\leftarrow D^{-1}\cdot S$, $A^{(1)}\leftarrow A^{(1)}\cdot D$, $A^{(2)}\leftarrow A^{(2)}\cdot D$, where $D=diag\left(\sqrt{\sum_g S_{cg}^2}\right)$.

The final normalization of the rows of *S* renders the resulting clusters comparable to each other.

Note that such a factorization can be viewed as a "soft" clustering algorithm allowing for *overlapping gene clusters*, since we may have several significant S_{cg} entries on a given column *g* of *S* (so a gene *g* may "belong" to several clusters *c*). However, although overlaps are allowed, the algorithm will not produce highly

---

[3] The derivation of the above rules is very similar to the derivation of the original Lee and Seung update rules and is not reproduced here for lack of space.

overlapping clusters, due to the nonnegativity constraints and to the offset variables. This is unlike many other clustering algorithms that allow clusters to overlap, which have to resort to several parameters to keep excessive cluster overlap under control.

4 Nonnegative Matrix Factorizations with offset

Before discussing in more detail the application of *siNMF* to the adenocarcinoma datasets mentioned in the Introduction, we explain in more detail the role of the offset terms *So* in the factorizations (1-2) above.

To make things simpler, we consider a single NMF factorization with offset rather than the simultaneous one from (1-2):

$$X_{sg} \approx \sum_c A_{sc} \cdot S_{cg} + So_g \tag{5}$$

with the additional nonnegativity constraints:

$$A_{sc} \geq 0,\ S_{cg} \geq 0,\ So_g \geq 0. \tag{6}$$

The main role of the "offset" *So* is to absorb the constant expression levels of genes, thereby making the cluster samples S_{cg} "cleaner".

The associated multiplicative update rules can be easily derived using the method of Lee and Seung [5]:

$$A_{sc} \leftarrow A_{sc} \frac{\left(XS^T\right)_{sc}}{\left(\left(eS_0 + AS\right)S^T\right)_{sc} + \varepsilon}$$

$$S_{cg} \leftarrow S_{cg} \frac{\left(A^T X\right)_{cg}}{\left(A^T \left(eSo + AS\right)\right)_{cg} + \varepsilon}$$

$$So_g = So_g \frac{\left(e^T X\right)_g}{\left(e^T \left(eSo + AS\right)\right)_g + \varepsilon}$$

Figure 1 below presents a comparison between the factorizations produced by the standard NMF algorithm and its improvement NMF_{offset} on a synthetic dataset in which columns 36 to 85 are constant "genes". As can be easily seen in the Figure, these "genes" are reconstructed by the standard NMF algorithm from combinations of clusters, while NMF_{offset} uses the additional degrees of freedom *So* to produce null cluster membership degrees S_{cg} for the constant genes. Moreover, NMF_{offset} recovers with much more accuracy than standard *NMF* the original sample clusters, the standard NMF algorithm being confused by the cluster overlaps. This improvement in recovery of the original clusters is very important in our application, where we aim at a correct sub-classification of samples.

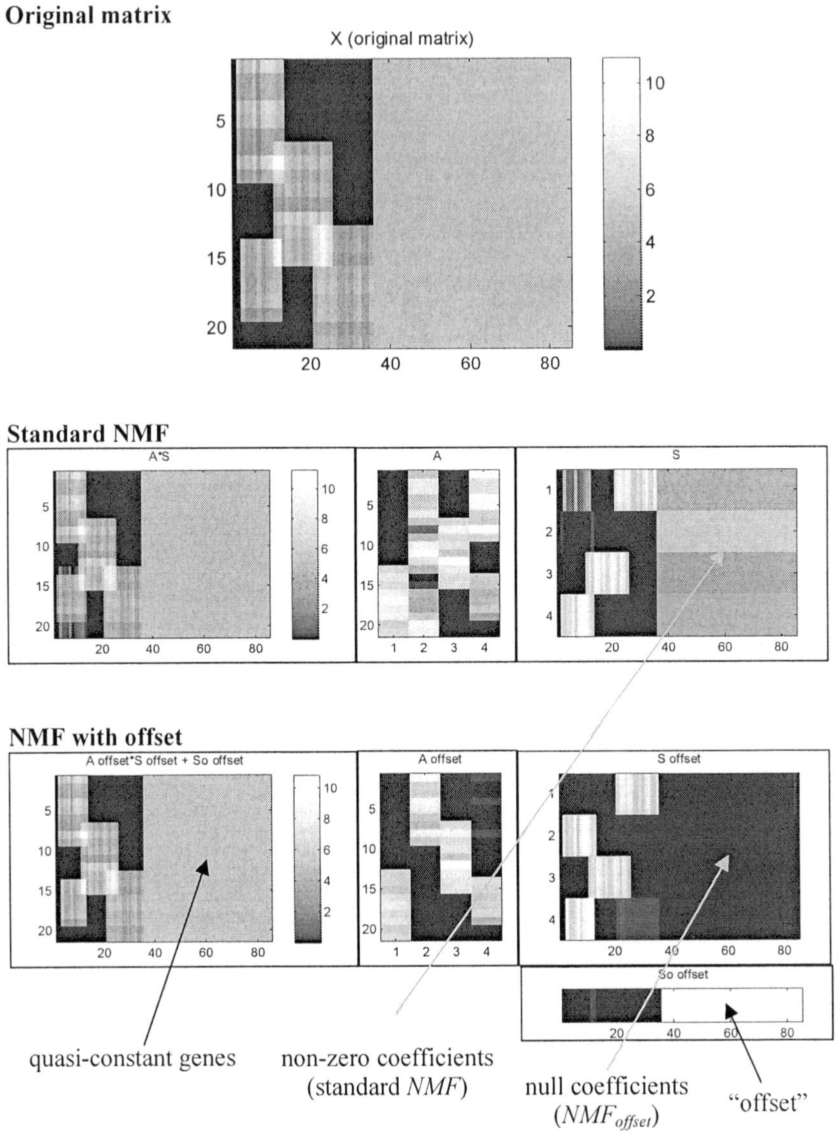

Figure 1. Comparing standard *NMF* with NMF_{offset}

5 Simultaneous factorization of the PDAC and colon adenocarcinoma datasets

In the following we describe the results obtained by applying *siNMF* to the PDAC and sporadic colon adenocarconoma datasets.

An important parameter of the factorization is its *internal dimensionality* (the number of clusters n_c). To avoid overfitting, we estimated the number of clusters n_c as the largest number of dimensions around which the change in relative error $\frac{d\varepsilon}{dn_c}$ of the factorization of the real data is still significantly larger than the change in relative error obtained for a randomized dataset [4] (similar to [9]) – see also Figure 2 below. Using this analysis we estimated the internal dimensionality of the dataset to be between 5 and 7. In the following, we used the conservative value n_c=5.

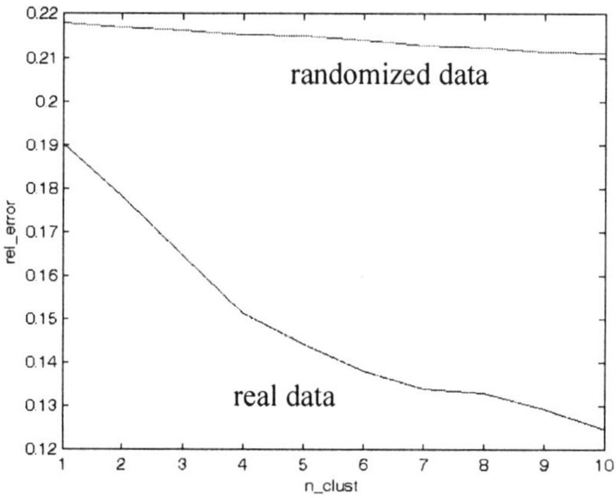

Figure 2. Determining the internal dimensionality of the datasets

We then ran the *siNMF* algorithm with n_c=5 and β_0=1 on the two datasets described previously restricted to the set of common probe-sets (4677 probe-sets).

Since the pancreatic ductal adenocarcinoma dataset is more homogeneous, we first inspected the sample cluster matrix $A^{(1)}$ to determine the cluster that best discriminates between tumor and control samples (see Figure 3 below).

---

[4] The randomized dataset was obtained by randomly permuting for each gene its expression levels in the various samples. The original distribution of the gene expression levels is thereby preserved.

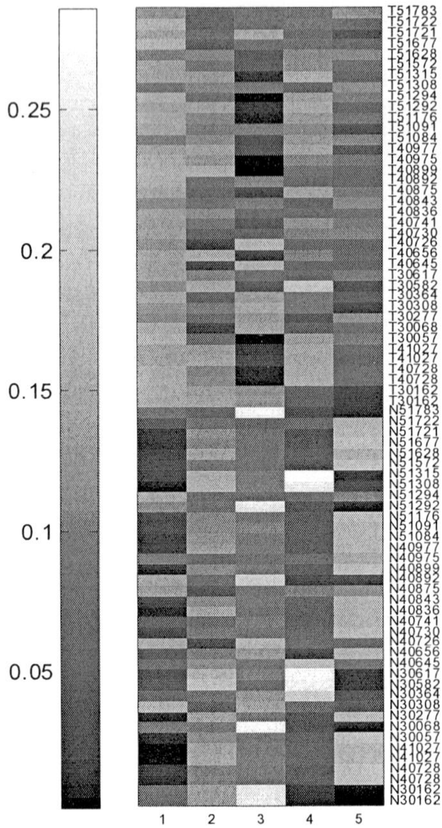

Figure 3. The normalized sample cluster matrix $A^{(1)}$ for the PDAC dataset

Note that cluster 1 recovers relatively well[5] the distinction between tumor and control samples in PDAC, although the algorithm was never provided with class information related to the samples. Similarly, cluster 5 is also significantly well correlated with the tumor-control distinction. In fact, while cluster 1 contains genes that are overexpressed in tumors, cluster 5 comprises mainly downregulated genes. The supplementary material online at www.ai.ici.ro/psb08/ contains the complete

[5] A number of 5 "control" samples (N51294, N40892, N40875, N40726 and N30308) which in our analysis are "closer" to the tumor samples than to the other control ones were later reanalyzed histologically and found to be highly fibrotic (pancreatic tumor tissue is typically very fibrotic and the respective control samples were possibly collected from a site too close to the tumoral tissue).

lists of genes for these clusters. (The threshold used for extracting gene clusters from the S matrix was $\sqrt{2}/\sqrt{n_g} = 0.0207$.)

For a more comprehensive biological interpretation of these sets of genes, we then looked for enrichment in known biological annotations using the *L2L Microarray Analysis Tool* [10]. As previously observed in the isolated analysis of PDAC, cluster 1 was enriched in TGF-beta target genes ("tgfbeta_all_up" with p-value 3.75e-29, "tgfbeta_early_up" with p-value 2.94e-25), as well as in the following Gene Ontology [11] Biological Process annotations:

GO Biological Process	p-value
response to wounding	9.44e-28
inflammatory response	2.43e-26
response to external stimulus	6.26e-26
defense response	1.74e-21
cell adhesion	8.14e-20
immune response	1.05e-18
organ development	5.13e-18
response to stress	1.10e-14
Chemotaxis	1.31e-14

The following L2L cancer gene expression modules were significantly affected: "ECM and collagens" with p-value 5.25e-82 and "Immune (humoral) and inflammatory response" with p-value 6.17e-65.

All of this is in line with the observed phenotype of PDAC, which involves an over-proliferation of the extracellular matrix (fibrosis, "desmoplastic reaction") and inflammation, supporting the view of cancer as an abnormal response to wounding.

It is impossible to present here a complete analysis of the cluster 1 genes. Some of the most significant ones are:

- INHBA (activin/inhibin betaA) – a ligand for the activin receptor (which triggers a TGF-beta-like pathway),
- POSTN (periostin), which is known to have an active role in the epithelial-mesenchymal transformation and metastasis [12] and whose over-expression promotes metastatic growth of colon cancer by augmenting cell survival via the Akt/PKB pathway [13], as well as enhancing invasion and angiogenesis,
- SULF1, which is known to regulate growth and invasion of pancreatic cancer cells by Interfering with Heparin-binding Growth Factor Signalling [14], etc.

Using the Transcriptional Regulatory Element Database TRED, we found that many of the genes in cluster 1 are controlled by the following transcription factors or a combination thereof: SP1, AP1, AP2, NF-kB, p53, ER, ETS1, SMAD family, CEBPA, etc. (Many direct and indirect TGF-beta targets are controlled by combinations of these factors.)

After having characterized the gene cluster 1 as being the main discriminator between tumor and control PDAC samples, we investigated its function in the colon adenocarcinoma dataset. Figure 4 below presents the normalized [6] sample cluster matrix $A^{(2)}$ for the colon dataset sorted with respect to the first column (cluster). The first sample cluster thus contains 91 tumor samples – half of the total number of 182 colon tumor samples! In other words, *the PDAC gene expression program that distinguishes tumors from controls is active in half of the colon adenocarcinoma we investigated and is highly expressed* [7] *in about* 12%.

On the other hand, cluster 5 is not significantly overexpressed in the normal colon samples (as it was in the PDAC) samples – this may be due either to the small number of normal colon samples, or to the differences in gene expression programs of these tissues (pancreas vs. colon).

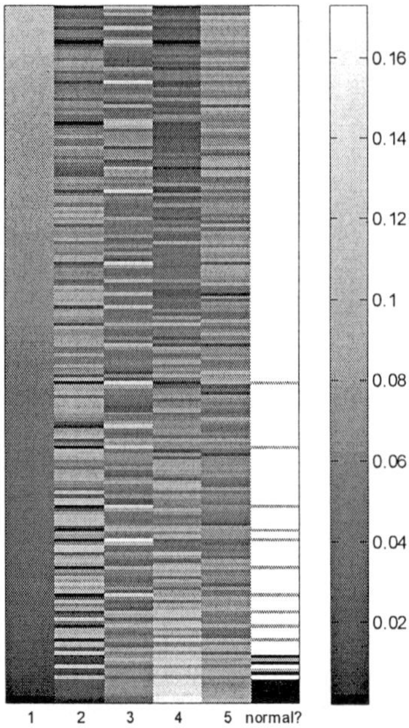

Figure 4. The normalized sample cluster matrix $A^{(2)}$ for the colon dataset (last column shows sample class: black:normal, gray:cancer susceptibility, white:tumor)

6 Conclusions and related work

Although widely used in microarray data analysis, existing clustering algorithms have serious problems, the most important one being related to the fact that biological processes are overlapping rather than isolated. In this paper we have

[6] i.e. $A_c^{(2)} / \|A_c^{(2)}\|$. We use a sample threshold $1/\sqrt{n_s} = 0.07$.

[7] over the threshold $\sqrt{2}/\sqrt{n_s} = 0.099$.

introduced a clustering algorithm capable of simultaneously clustering two distinct gene expression datasets with the aim of uncovering gene regulatory programs that are common to the two phenotypes. The two key ingredients of this algorithm are the nonnegativity constraint and the offset variables, which together ensure the sparseness of the factorizations.

Most unsupervised gene expression data analysis methods require a careful selection of genes that are "significant" for subsequent sub-class discovery. But class discovery and "significant gene" selection are tightly inter-connected and cannot be easily separated. Thus, another very important advantage of nonnegative factorization approaches with respect to other methods consists in the fact that they eliminate the need for such an explicit gene selection step prior to classification.

While cancer is a very heterogeneous disease, there is overwhelming recent evidence that the differences between cancer subtypes implicate entire pathways and biological processes involving large numbers of genes, rather than changes in single genes. This has lead us to the following strategy for discovering these processes. We have started with a relatively homogeneous cancer subtype, namely pancreatic ductal adenocarcinoma, for which we have determined the gene group that best distinguishes tumors from controls thereby verifying the homogeneity of this subtype. Then we have applied our simultaneous factorization algorithm looking for gene expression profiles that are common between the more homogeneous PDAC and the more heterogeneous colon adenocarcinoma. The fact that the PDAC signature is active in a large fraction of colon adeocarcinoma suggests that the oncogenic mechanisms involved may be similar to those in PDAC, at least in this subset of colon samples.

The *simultaneous Nonnegative Matrix Factorization* algorithm presented in this paper generalizes the simpler version introduced in [2] by estimating "offsets" for the individual genes, which produces much cleaner gene clusters. Moreover, in [2] we used siNMF to guide the factorization of gene expression data by transcription regulation data, while in this paper we are concerned with finding common mechanisms in different types of adenocarcinoma.

The *siNMF* algorithm is also related in spirit with the generalized SVD algorithm (GSVD) [3], which was applied by Alter et al. for comparing two cell cycle datasets. There, the "common part" of the decomposition is represented by the samples (rather than the genes, as in our approach).

There are many approaches to uncovering common mechanisms involved in different phenotypes, but most are based on comparing gene lists. The approach presented in this paper additionally takes gene expression data into account and can thus be more sensitive.

Of course, this work represents just a first step towards a molecular-level classification of sporadic colon adenocarcinoma, going beyond the simpler one based on microsatellite instability status [1].

Acknowledgments. This work was partially supported by the BIOINFO and GENOPACT projects. I am particularly grateful to Prof. I. Popescu for the collaboration in these projects and to the reviewers for some very useful suggestions for improving this work.

References

1. Jass JR, Biden KG, Cummings MC, Simms LA, Walsh M, Schoch E, Meltzer SJ, Wright C, Searle J, Young J and Leggett BA. Characterisation of a subtype of colorectal cancer combining features of the suppressor and mild mutator pathways. J.Clin.Pathol. 52: 455-460, 1999.
2. Badea L. Combining Gene Expression and Transcription Factor Regulation Data using Simultaneous Nonnegative Matrix Factorization. Proc. BIOCOMP'07, CSREA Press, 2007.
3. Alter O, Brown PO, Botstein D. Generalized singular value decomposition for comparative analysis of genome-scale expression data sets of two different organisms. Proc Natl Acad Sci U S A. 2003 Mar 18;100(6):3351-6.
4. Lee D.D., H.S. Seung. Learning the parts of objects by non-negative matrix factorization. Nature, vol. 401, no. 6755, pp. 788-791, 1999.
5. Lee D.D., H.S. Seung. Algorithms for non-negative matrix factorization. In Advances in Neural Information Processing 13 (Proc. NIPS*2000), MIT Press, 2001.
6. GENOPACT project (CEEX 56/2005).
7. expO. Expression Project for Oncology http://expo.intgen.org/expo/geo/goHome.do
8. Hong Y, Ho KS, Eu KW, Cheah PY. A susceptibility gene set for early onset colorectal cancer that integrates diverse signaling pathways: implication for tumorigenesis. Clin Cancer Res. 2007 Feb 15;13(4):1107-14.
9. Kim P.M., Tidor B. Subsystem identification through dimensionality reduction of large-scale gene expression data. Genome Res. 2003 Jul;13(7):1706-18.
10. L2L. L2L Microarray Analysis Tool http://depts.washington.edu/l2l/
11. Gene Ontology Consortium. Gene Ontology: tool for the unification of biology. Nature Genet. 25:25-29, 2000.
12. Yan W, Shao R. Transduction of a mesenchyme-specific gene periostin into 293T cells induces cell invasive activity through epithelial-mesenchymal transformation. J Biol Chem. 2006 Jul 14;281(28):19700-8.
13. Bao S, Ouyang G, Bai X, Huang Z, Ma C, Liu M, Shao R, Anderson RM, Rich JN, Wang XF. Periostin potently promotes metastatic growth of colon cancer by augmenting cell survival via the Akt/PKB pathway. Cancer Cell. 2004 Apr;5(4):329-39.
14. Abiatari I., J. Kleeff, J. Li, K. Felix, N.A. Giese, M.W. Büchler, H. Friess. Hsulf-1 Regulates Growth and Invasion of Pancreatic Cancer Cells by Interfering with Heparin-binding Growth Factor Signalling. J Clin Pathol. 2006 Oct;59(10):1052-8.
15. TRED. Transcriptional Regulatory Element Database. http://rulai.cshl.edu/cgi-bin/TRED/tred.cgi?process=home.
16. Brunet J.P., Tamayo P., Golub T.R., Mesirov J.P. Metagenes and molecular pattern discovery using matrix factorization. PNAS 101(12):4164-9, 2004, Mar 23.
17. Cheng Y, Church GM. Biclustering of expression data. Proc. ISMB 2000; 8:93-103.

INTEGRATION OF MICROARRAY AND TEXTUAL DATA IMPROVES THE PROGNOSIS PREDICTION OF BREAST, LUNG AND OVARIAN CANCER PATIENTS

O. GEVAERT, S. VAN VOOREN, B. DE MOOR

BioI@ESAT-SCD, Dept. Electrical Engineering
Katholieke Universiteit Leuven
Kasteelpark Arenberg 10, Leuven, B-3001, Belgium
E-mail: olivier.gevaert@esat.kuleuven.be

Microarray data are notoriously noisy such that models predicting clinically relevant outcomes often contain many false positive genes. Integration of other data sources can alleviate this problem and enhance gene selection and model building. Probabilistic models provide a natural solution to integrate information by using the prior over model space. We investigated if the use of text information from PUBMED abstracts in the structure prior of a Bayesian network could improve the prediction of the prognosis in cancer. Our results show that prediction of the outcome with the text prior was significantly better compared to not using a prior, both on a well known microarray data set and on three independent microarray data sets.

1. Introduction

Integration of data sources has become very important in bioinformatics. This is evident from the numerous publications involving multiple data sources to discover new biological knowledge[1,2,3]. This is due to the rise in publicly available databases and also the number of databases has increased significantly[4]. Still many knowledge is contained in publications in unstructured from as opposed to being deposited in public databases where they can be amenable to use in algorithms. Therefore we attempted to mine this vast resource and transform it to the gene domain such that it can be used in combination with gene expression data. Microarray data are notorious for there low signal-to-noise ratio and often suffer from a small sample size. This causes that genes are often differently expressed between clinically relevant outcomes purely by chance. Integration of prior knowledge can improve model building in general and gene selection in particular. In this paper we present an approach to integrate information from literature abstracts into probabilistic models of gene expression data. Integration of different data sources into a single framework potentially leads to more reliable models and at the same time it can reduce overfitting[2]. Probabilistic models provide a natural solution to this problem since information can be incorporated in the prior distribution over the model space. This prior is then combined with other data to form a posterior distribution over the model space which is a balance between the information incorporated in the prior and the data.

Specifically, we investigated how the use of text information as a prior of a Bayesian network can improve the prediction of prognosis in cancer when modeling expression data. Bayesian networks provide a straightforward way to integrate information in the prior distribution over the possible structures of its network. By mining abstracts we can easily represent genes as term vectors and create a gene-by-gene similarity matrix. After appropriate scaling, such a matrix can be used as a structure prior to build Bayesian networks. In this manner text information and gene expression data can be combined in a single framework. Our approach builds further on our methods for integrating prior information with Bayesian networks for other types of data[5,6] where we have shown that structure prior information improves model selection especially when few data is available.

In this study we investigated if a Bayesian network model with a text prior can be used to predict the prognosis in cancer. Bayesian networks and their combination with prior information have already been studied by others[3,7,8,9] however, to the author's knowledge, none have investigated the influence of priors in a classification setting or, more specifically, when predicting the outcome or phenotypic group of cancer patients. First, we will show how the prior performs on a well known breast cancer data set and examine the effect of the prior in more detail. Subsequently, we will validate our approach on three other data sets studying breast, lung and ovarian cancer.

2. Bayesian networks

A Bayesian network is a probabilistic model that consists of two parts: a directed acyclic graph which is called the structure of the model and local probability models[10]. The dependency structure specifies how the variables (i.e. gene expression levels) are related to each other by drawing directed edges between the variables without creating directed cycles. In our case each variable x_i models the expression of a particular gene. Such a variable or gene depends on a possibly empty set of other variables which are called the parents (i.e. their putative regulators):

$$p(x_1, ..., x_n) = \prod_{i=1}^{n} p(x_i|Pa(x_i)) \tag{1}$$

where $Pa(x_i)$ are the parents of x_i and n is the total number of variables. Usually the number of parents for each variable is small and therefore a Bayesian network is a sparse way of writing down a joint probability distribution. The second part of this model, the local probability models, specifies how the variables or gene expressions depend on their parents. We used discrete-valued Bayesian networks which means that these local probability models can be represented with Conditional Probability Tables (CPTs). Such a table specifies the probability that a variable takes a certain value given the value or state of its parents.

2.1. *Model building*

We already mentioned that a discrete valued Bayesian network consists of two parts: the structure and the local probability models. Consequently, there are two steps to be performed during model building: structure learning and learning the parameters of the CPTs. First the structure is learned using a search strategy. Since the number of possible structures increases super-exponentially with the number of variables, we used the well-known greedy search algorithm K2[11] in combination with the Bayesian Dirichlet (BD) scoring metric[11,12,13]:

$$p(S|D) \propto p(S) \prod_{i=1}^{n} \prod_{j=1}^{q_i} \left[\frac{\Gamma(N'_{ij})}{\Gamma(N'_{ij} + N_{ij})} \prod_{k=1}^{r_i} \frac{\Gamma(N'_{ijk} + N_{ijk})}{\Gamma(N'_{ijk})} \right], \tag{2}$$

with N_{ijk} the number of cases in the data set D having variable x_i in state k associated with the j-th instantiation of its parents in current structure S. Γ corresponds to the gamma distribution. Next, N_{ij} is calculated by summing over all states of a variable: $N_{ij} = \sum_{k=1}^{r_i} N_{ijk}$. In our case the state of a variable refers to the expression of the corresponding gene where each variable can have one of three states: over-expressed, under-expressed or no expression. Next, N'_{ijk} and N'_{ij} have similar meanings as N_{ijk} and N_{ij} but refer to prior knowledge for the parameters. When no knowledge is available they are estimated using[13]: $N'_{ijk} = \frac{N}{r_i q_i}$ with N the equivalent sample size, r_i the number of states of variable x_i and q_i the number of instantiations of the parents of variable x_i. K2 uses a prior ordering of the variables to restrict the number of structures that can be built. The order of the variables reflects the causal relationship between the variables, this means that regulators should come before their targets in the ordering. Because the prior ordering of the variables is not known in advance we repeat the model building process for a set of randomly drawn variable orderings and choose the model with the highest posterior BD score. The next step consists of estimating the parameters of the local probability models of each variable in the structure with the highest BD score. This amounts to filling in a CPT for every variable and every possible value of its parents using the data. This Bayesian network can then be used to predict future data.

2.2. *Structure prior*

Previously two approaches have been used to define informative prior distributions over Bayesian network structures[5]. First there are penalization methods that start from a prior structure and score structures based on the difference with the prior structure[13]. Secondly there are pairwise methods which define the prior probability of a Bayesian network structure by combining individual edge scores between variables. This method assumes that being a parent of some node is independent of any other parental relation. We have chosen the second approach to model the structure prior where the prior probability of a structure is decomposed as:

$$p(S) = \prod_{i=1}^{n} p(Pa(x_i) \rightarrow (x_i)) \tag{3}$$

The probability of a local structure (i.e. $p(Pa(x_i) \rightarrow x_i)$) is then calculated by multiplying the probability that there is an edge between the parents of x_i and, the probability there is no edge between the other variables and x_i:

$$p(Pa(x_i) \rightarrow x_i) = \prod_{p \in Pa_(x_i)} p(p \rightarrow x_i) \prod_{y \notin Pa_(x_i)} p(y \nrightarrow x_i) \tag{4}$$

where $\nrightarrow$ means no edge between y and x_i. These individual edge probabilities can be represented in a matrix. In the Text Prior Section, we will be able to derive a matrix S from the literature where the elements represent the connectedness or similarity between the genes. Rather than using these values immediately as edge probabilities, we will introduce an extra parameter, ν called the mean density, which controls the density of the networks that will be generated from the distribution. We will transform all the matrix elements in the prior with an exponent ζ such that the average of the mean number of parents per local substructure is according to the given mean density[5]. Finding the exponent, ζ that gives rise to the correct mean number of parents can be done with any single variable optimization algorithm. With this mean number of parents, we can control the complexity of the networks that will be learned.

2.3. *Inference*

After learning the model, we can use it to predict unseen data. This means that we can use a Bayesian network model to predict the value of a variable given the values of the other variables. We used the probability propagation in tree of cliques algorithm[13] to predict the state of the class variable (i.e. the prognosis in cancer). This inference algorithm was then used to evaluate the effect of using a text prior in combination with the expression data described below. To accomplish this, we used a randomization approach where we randomly distributed the data in 70% used to build a model and 30% to estimate the Area Under the ROC curve (AUC). This process was repeated 100 times to have a robust estimate of the generalization performance of the two approaches: with text prior and without text prior. Then these 100 AUCs were averaged and reported. Next a model was built using the complete data set for both methods and we investigated the possible differences between the Markov blanket variables (i.e. the set of genes which are sufficient to predict the outcome). The average AUC with and without prior are compared by calculating the p-value with a two-sided Wilcoxon rank sum test. P-values are considered statistically significant if smaller than 0.05.

3. Prior data

3.1. *Gene prior*

Since microarray data usually references thousands of genes, it is infeasible to manually construct a structure prior as described earlier. Therefore, prior construction involves methods based on an automatic elicitation of relationships between genes. In this paper, we propose the use of priors that consist of gene-by-gene similarity matrices based on biomedical literature mining. To accomplish this, genes are represented in the Vector Space Model . In the VSM model, each position of a gene vector corresponds to a term or phrase in a controlled vocabulary. In our case, we have constructed a cancer specific vocabulary which was extracted from the National Cancer Institute Thesaurus. Using a fixed vocabulary has several advantages.

Firstly, simply using all terms that occur in the corpus of literature linked to the genes involved in the microarray experiment at hand, will result in vectors of considerable size, which means genes are represented in a high dimensional space. As this 'curse of dimensionality' is detrimental to the strength of a metric, the use of only a relatively small set of concepts will improve the quality of calculated gene-to-gene distances. Further reduction of the dimensionality is accomplished by performing stemming, which will allow different terms that in essence convey a same meaning (coughing, coughs, coughed) to be treated as a single concept (cough). Secondly, the use of phrases reduces noise in the data set, as genes will only be compared to each other from a highly domain specific view. Thirdly, a structured vocabulary will enable the use of multi-word phrases as opposed to just single terms, without having to resort to co-occurrence statistics on the corpus to detect them. Fourthly, there is no need to filter out articles and stop words, as only highly specific cancer related terms are considered. The gene vectors themselves are constructed as follows. For each gene, manually curated literature references are extracted from Entrez Gene. All PUBMED abstracts linked to these genes are then indexed using the aforementioned vocabulary. As a result, all PUBMED abstracts are represented in a high dimensional vector space using IDF (Inverse Document Frequency) weights for non-zero vector positions. The resulting vectors (which represent abstracts, not genes) are normalized to bring them on the union hyper sphere in the vector space, which facilitates cosine similarity calculation. Gene vectors are then constructed by averaging the vectors of all the abstracts associated to that gene by Entrez Gene. Finally, the cosine measure is used to obtain gene-to-gene distances between 0 and 1. These gene-to-gene distances can then be represented as a symmetric matrix S which forms the structure prior for the Bayesian network modeling.

3.2. *Class variable prior*

We have already defined the way the prior is determined between the genes. Since we are developing models which predict the prognosis in cancer, the need exists for an additional variable in the model, namely the outcome class of the patients. This variable describes to which group each sample belongs, for example, good prognosis and poor prognosis. Hence, we need to define the prior relation between the class variable and the genes. To accomplish this, we used terms in the vocabulary which are related to the prediction of the prognosis of cancer, such as outcome, prognosis and metastasis. Next, we counted the number of associations each gene had with prognosis related terms and increased the gene-to-outcome similarity for every additional term the gene was associated with. Genes which had no association with either term were given a prior probability of 0.5. This information was added to the gene prior creating a structure prior for all the variables studied (i.e. genes and patient outcome). This structure prior is then, after scaling according to the mean density, used in Bayesian network learning.

4. Data

To test our approach we used publicly available microarray data on breast cancer[14] (Veer data). This data set consists of 46 patients that belonged to the poor prognosis group and 51 patients that belonged to the good prognosis group. DNA microarray analysis was used to determine the mRNA expression levels of approximately 25000 genes for each patient. Every tumour sample was hybridized against a reference pool made by pooling equal amounts of RNA from each patient. The ratio of the sample and the reference was used as a measure for the expression of the genes and they constitute the microarray data set. This data set was already background corrected, normalized and log-transformed. Preprocessing was done similarly as in[14]. This resulted in 232 genes that where correlated with the patient outcome which were used in our models.
To validate our results we used three publicly available data sets from Bild et al.[15] studying breast, lung and ovarian cancer (Bild data). These data sets contained data on 171 breast cancer patients, 147 ovarian cancer patients and 91 lung cancer patients. The three groups of tumours were analysed on different Affymetrix chips; the breast tumours were hybridized on Hu95Av2 arrays, the ovarian tumours on Hu133A arrays and the lung tumours on Human U133 2.0 plus arrays. The data were already preprocessed using RMA. For all cancer sites survival data was available and patients were split up in two groups according to the following thresholds: 53 months for breast cancer, 62 months for ovarian cancer and 36 months for lung cancer. The thresholds were chosen to make sure both classes contained approximately the same number of samples. Genes were selected similarly as in the Veer data set by selecting the top 100 genes after ranking them by their correlation with patient survival data.

4.1. *Discretization*

We have chosen discrete valued Bayesian networks therefore the microarray data has to be discretized. We specifically tried to minimize the loss of relationships between the variables by applying the algorithm of Hartemink[16]. The gene expression values were discretized in three categories or bins: baseline, over-expression and under-expression. This was done using a multivariate discretization method which minimizes the loss of mutual information between the gene expression measurements[16]. First a simple discretisation method with a large number of bins is used as a starting point (e.g. interval discretisation where the complete range of values is divided in a number of equally large bins). Then the multivariate algorithm starts and for each variable it joins the neighboring bins together which have the smallest decrease in mutual information. This is iterated until each variable has three bins. The resulting discretized data set is used as input into the Bayesian network learning algorithms.

5. Implementation

The software implementation is based on a combination of c++, java, matlab and perl. The Bayesian network algorithms were implemented in C++. Java Lucene was used for indexing Pubmed. Matlab scripts were used for discretization and to construct the structure priors. Perl was used to glue the different steps in the workflow together. A typical analysis took between 6 and 25 minutes depending on the data set size. All analysis where run on AMD dual core opteron 2.4 GHz with between 4 and 16 GB RAM memory.

6. Results and discussion

6.1. *Veer data*

First, we assessed the performance of the text prior regarding prediction of outcome on the Veer data set. We performed 100 randomizations of the data set without a prior and 100 randomizations with the text prior (as described in the Model building and testing Section in Materials and methods). We repeated the analysis for different values of the mean density to asses if this parameter had an influence on the results. Table 1 shows the mean AUC for both methods and for increasing mean density. The most important conclusion that can be drawn from Table 1 is that using the text prior significantly enhances the prediction of the outcome (P-value

$\prec 0.05$). The text prior guides model search and favors genes which have a prior record related to prognosis. This knowledge improves gene selection and most likely wards off genes which are differentially expressed by chance. Additionally, Table 1 shows that the mean density has no influence on the result in the tested range. The mean density controls the complexity of the network therefore large values should be avoided since the danger of overfitting increases. Note that the results for the mean AUC without prior are essentially the same as our previously obtained result[12].

Table 1. Results of 100 randomizations of the Veer data set with the Text prior and without prior. The mean AUCs are reported together with the p-value.

Mean Density	Text prior mean AUC	Uniform prior mean AUC	P-value
1	0.80	0.75	0.000396
2	0.80	0.75	0.000002
3	0.79	0.75	0.005770
4	0.79	0.74	0.000006

Next, we used the complete data set and we built one model with text prior and one model without the text prior, to evaluate the set of genes which are sufficient to predict the outcome (i.e. the genes in the Markov blanket of the outcome). We call the former, the TXTmodel and the latter UNImodel. Table 2 shows the gene names that appear in both models. The average text score (i.e. the probability the gene is related to patient outcome according to literature) of the genes in the TXTmodel is 0.85 compared to only 0.58 for the UNImodel. The text prior thus has its expected effect and includes genes which have a prior tendency to be associated with the prognosis of cancer. There are only 10 genes in the TXTmodel compared to 15 genes in the UNImodel which indicates that TXTmodel needs fewer genes. Moreover, the TXTmodel has many genes which have been implicated in breast cancer or cancer in general such as TP53, VEGF, MMP9[17], BIRC5, ADM[18] and CA9. Next ACADS, NEO1 and IHPK2 have a weaker link to cancer outcomes whereas MYLIP has no association. In the UNImodel, as expected, far less genes are present which have a strong link with cancer outcomes which likely increases the probability of false positives. Only WISP1, FBXO31, IGFBP5 and TP53 have a relation with breast cancer outcome. The other genes have mostly unknown function or are not related. Finally two genes appear in both set: TP53 and IHPK2.

TP53 is perhaps the best known gene to be involved in cancer. Therefore it is bound to appear in the TXTmodel and it is no surprise that it is also present in the UNImodel. IHPK2 however has a weak prior relation with prognosis in cancer therefore this gene proves that genes with a low text prior still can be selected in the TXTmodel. Additionally, genes which appear in both models can be considered more reliable.

Table 2. Genes sufficient to predict the outcome variable for the TXTmodel and the UNImodel.

TXTmodel:	MYLIP,TP53,ACADS,VEGF,ADM,NEO1,IHPK2,CA9,MMP9,BIRC5
UNImodel:	PEX12,LOC643007,WISP1,SERF1A,QSER1,ARL17P1,LGP2,IHPK2, TSPYL5,FBCO31,LAGE3,IGFBP5,AYTL2,TP53,PIB5PA

6.2. *Bild data*

Finally we validated our approach on three independent data sets on breast, ovarian and lung cancer[15] to assess if the results on the Veer data set can be confirmed. Based on the results presented in Table 1 we chose a mean density of 1 for these data sets. Again 100 randomizations of the data set with and without the text prior were performed. Table 3 shows the average AUC for the three Bild data sets and confirms that the text prior significantly improves the prediction of the prognosis on independent data sets and for other cancer sites.

Table 3. Results of 100 randomizations of the three Bild data sets with the Text prior and without prior. The mean AUCs are reported together with the p-value.

Mean Density	Text prior mean AUC	Uniform prior mean AUC	P-value
Breast	0.79	0.75	0.00020
Lung	0.69	0.63	0.00002
Ovarian	0.76	0.74	0.02540

7. Conclusions

In this paper we have shown a method to integrate information from literature abstracts with gene expression data using Bayesian network models. This prior information was integrated in the prior distribution over the possible Bayesian network structures after scaling. The results of the randomization analysis in Table 1 and 3 have shown that for both the Veer data set and the three Bild data sets the text prior improves the prediction of the prognosis of cancer patients significantly.

A possible limitation of our approach is the discretization of the data. It is inevitable that some information is lost in the process of discretization. We have chosen discrete valued Bayesian networks because the space of arbitrary continuous distributions is large. A solution could be to restrict ourselves to the use of Gaussian Bayesian networks but this class of models assumes linear interactions between the variables which, in our opinion, would restrict too much the type of relations among genes that are modeled. Moreover, by using the algorithm of Hartemink we are performing a multivariate discretization, keeping the relationships between the variables as much as possible intact.

Secondly by using text information, which is often described as highly biased, one could run the risk of focussing too much on the hot genes disregarding novel important genes. However, in our case the emphasis is not so much on biomarker discovery and more on developing models which can accurately predict the prognosis of disease. There are already many genes known to be involved in different types of cancer based on individual studies or because they are member of a cancer profile. Finding the minimal set of genes which is able to predict the prognosis of disease however, is still an open problem. Our Bayesian network framework attempts to address this issue by tackling the disadvantages of cancer microarray data sets (low signal-to-noise ratio, high dimensional, small sample size, ...) by using information from the literature as a guide.

Finally, the presented framework is complimentary to our previously published method to integrate clinical and microarray data with Bayesian networks[12]. Thus creating a Bayesian network framework which enables modeling of various data sources (i.e. clinical, microarray and text) to improve decision support of outcome (i.e phenotypic group) prediction in cancer or other genetic diseases. Moreover our definition of the structure prior makes no assumptions about the nature of prior information. Therefore other sources of information can be combined with the text prior (e.g. known protein-DNA interactions from Transfac, known pathways from KEGG or motif information). Thus, creating a white box framework that visualizes how decisions are made by a model. This is in contrast to for example a kernel framework where model parameters are not readily interpretable.

Acknowledgments

This research is supported by the Institute for the Promotion of Innovation through Science and Technology in Flanders (IWT-Vlaanderen) and GOA AMBioRICS, CoE EF/05/007 SymBioSys, FWO: G.0499.04 (Statistics), IUAP P6/25 BioMaGNet 2007-2011; FP6-NoE Biopattern; FP6-IP e-Tumours

References

1. Z. Bar-Joseph, G. K. Gerber, T. I. Lee, N. J. Rinaldi, J. Y. Yoo, F. Robert, D. B. Gordon, E. Fraenkel, T. S. Jaakkola, R. A. Young and D. K. Gifford, *Nat Biotechnol* **21**, 1337(November 2003).
2. G. R. G. Lanckriet, T. De Bie, N. Cristianini, M. I. Jordan and W. S. Noble, *Bioinformatics* **20**, 2626 (2004).
3. A. Bernard and A. J. Hartemink, *PSB* **10**, 459 (2005).
4. M. Y. Galperin, *Nucl. Acids Res.* **34**, D3 (2006).
5. P. Antal, G. Fannes, D. Timmerman, Y. Moreau and B. De Moor, *Artif Intell Med* **30**, 257 (2004).
6. O. Gevaert, F. De Smet, E. Kirk, B. Van Calster, T. Bourne, S. Van Huffel, Y. Moreau, D. Timmerman, B. De Moor and G. Condous, *Human reproduction* **21** (2006).
7. N. Friedman, M. Linial, I. Nachman and D. Pe'er, *J Comput Biol* **7**, 601 (2000).
8. N. Nariai, S. Kim, S. Imoto and S. Miyano, *PSB* **9**, 336 (2004).
9. T. Ideker, O. Ozier, B. Schwikowski and A. F. Siegel, *Bioinformatics* **18 Suppl 1** (2002).
10. J. Pearl, *Probabilistic Reasoning in Intelligent Systems: Networks of Plausible Inference* (Morgan Kaufmann Publishers, San Matteo, California, 1988).
11. G. F. Cooper and E. Herskovits, *Machine Learning* **9** (1992).
12. O. Gevaert, F. De Smet, D. Timmerman, Y. Moreau and B. De Moor, *Proceedings of the 14th Annual ISMB, Bioinformatics special issue* (2006).
13. D. Heckerman, D. Geiger and D. M. Chickering, *Machine Learning* **20**, 197 (1995).
14. L. Van 't Veer, H. Dai, M. J. Van de Vijver, U. D. He, A. M. Hart, M. Mao, H. L. Peterse, K. van der Kooy, M. J. Marton, A. T. Witteveen, G. J. Schreiber, R. M. Kerkhoven, C. Roberts, P. S. Linsley, R. Bernards and S. H. Friend, *Nature* **415**, 530 (2002).
15. A. H. Bild, G. Yao, J. T. Chang, Q. Wang, A. Potti, D. Chasse, M.-B. Joshi, D. Harpole, J. M. Lancaster, A. Berchuck, J. A. Olson, J. R. Marks, H. K. Dressman, M. West and J. R. Nevins, *Nature* (November 2005).
16. A. J. Hartemink, Principled computational methods for the validation and discovery of genetic regulatory networks, PhD thesis, MIT2001.
17. J. L. Owen, V. Iragavarapu-Charyulu and D. M. Lopez, *Breast Dis* **20**, 145 (2004).
18. M. K. Oehler, D. C. Fischer, M. Orlowska-Volk, F. Herrle, D. G. Kieback, M. C. Rees and R. Bicknell, *Br J Cancer* **89**, 1927(November 2003).

MINING METABOLIC NETWORKS FOR OPTIMAL DRUG TARGETS *

PADMAVATI SRIDHAR, BIN SONG, TAMER KAHVECI†AND SANJAY RANKA

Computer and Information Science and Engineering,
University of Florida,
Gainesville, FL, 32611
E-mail: {psridhar, bsong, tamer, ranka}@cise.ufl.edu

Recent advances in bioinformatics promote drug-design methods that aim to reduce side-effects. Efficient computational methods are required to identify the optimal enzyme-combination (i.e., drug targets) whose inhibition, will achieve the required effect of eliminating a given target set of compounds, while incurring minimal side-effects. We formulate the optimal enzyme-combination identification problem as an optimization problem on metabolic networks. We define a graph based computational damage model that encapsulates the impact of enzymes onto compounds in metabolic networks. We develop a branch-and-bound algorithm, named *OPMET*, to explore the search space dynamically. We also develop two filtering strategies to prune the search space while still guaranteeing an optimal solution. They compute an upper bound to the number of target compounds eliminated and a lower bound to the side-effect respectively. Our experiments on the human metabolic network demonstrate that the proposed algorithm can accurately identify the target enzymes for known successful drugs in the literature. Our experiments also show that OPMET can reduce the total search time by several orders of magnitude as compared to the exhaustive search.

1. Introduction

In pharmaceutics, the development of every drug mainly involves target identification, validation and lead inhibitor identification [9]. Traditional drug discovery approaches focus more on the efficacy of drugs than their toxicity (untoward side effects). Lack of predictive models that account for the complexity of the inter-relationships between the metabolic processes often leads to drug development failures. Toxicity and/or lack of efficacy can result if metabolic network components other than the intended target are affected. The current focus is on identification of biological targets (gene products, such as enzyme or protein) for drugs, which can be manipulated to produce the desired effect (of curing a disease) with minimum disruptive side-effects [23,27].

Enzymes catalyze reactions, which produce metabolites (compounds) in the metabolic networks of organisms. Enzyme malfunctions can result in the accumulation of certain compounds which may result in diseases. We term such compounds

*this work is supported in part by the national science foundation under grant itr 0325459 and dbi - 0606607. any opinions, findings, and conclusions or recommendations expressed in this material are those of the author(s) and do not necessarily reflect the views of the national science foundation.
†to whom correspondence should be addressed

as *Target Compounds* and the remaining compounds as *Non-Target compounds*. For instance, the malfunction of enzyme *phenylalanine hydroxylase* causes buildup of the amino acid, phenylalanine, resulting in phenylketonuria [26], a disease that causes mental retardation. It is, therefore, needed to locate the optimal enzyme set which can be manipulated by drugs to prevent the excess production of target compounds with minimal damage. Formally, we define *damage* of inhibiting an enzyme (or a set of enzymes) as the number of non-target compounds whose production is stopped by the inhibition of that enzyme (or set of enzymes).

Given a metabolic network and a set of target compounds, we consider the problem of identifying the set of enzymes whose inhibition eliminates the target compounds and incurs minimum damage. Evaluating all enzyme combinations is not feasible as the number of such combinations increases exponentially with the number of enzymes. Hence, more efficient computational methods are needed. In our earlier work [25], we developed a heuristic solution to this problem. Here, we propose *OPMET*, an **Opt**imal enzyme drug target identification algorithm based on **Met**abolic networks, to solve this problem optimally. This paper has two main contributions. 1) We propose a branch-and-bound algorithm, named *OPMET*, to explore the search space. Based on the damage model, OPMET dynamically updates the priorities as the search space is explored. 2) We develop two filtering approaches which are combined with the OPMET to prune the search space while still guaranteeing an optimal solution.

Our experiments on the human metabolic network demonstrates that the proposed algorithm can accurately identify the target enzymes for known successful drugs in the literature. Our experiments also show that our methods reduce the total search time by several orders of magnitude as compared to the exhaustive search. OPMET prunes 91.6 % of the search space. It generates the optimal enzyme combination within the exploration 0.005 % of the search space on average.

The rest of the paper is organized as follows. Section 2 formally defines the problem and describes our proposed cost model. Section 3 presents the proposed OPMET algorithm with filtering strategies. Section 4, discusses experimental results. Section 5 discusses the related work. Section 6 concludes the paper.

2. Problem definition

We develop a graph based representation that captures the interactions between reactions, compounds, and enzymes. Our graph representation is a variation of the boolean network model [24,16]. R, C, and E denote the set of reactions, compounds, and enzymes respectively. The vertex set consists of all the members of $R \cup C \cup E$. A vertex is labeled as reaction, compound, or enzyme based on the entity it refers to. Let V_R, V_C, and V_E denote the set of vertices from R, C, and E. A directed edge from vertex x to vertex y is then drawn if one of the following three conditions holds: (1) x represents an enzyme that catalyzes the reaction represented by y. (2) x corresponds to a substrate for the reaction represented by y. (3) x represents a reaction that produces the compound mapped to y.

Figure 1 illustrates a small hypothetical metabolic network. In this figure, C_4 is the target compound (i.e., the production of C_4 should be stopped). In order to stop the production of C_4, R_2 has to be prevented from taking place. The obvious solution is to disrupt one of its catalyzing enzymes (E_2 in this case). Another is by stopping the production of one of its reactant compounds (C_2 or C_3 in this case). If we stop the production of C_2, we need to recursively look for the enzyme which is indirectly responsible for its production (E_1 in this case). Thus, the production of the target compound can be stopped by manipulating either E_1 or E_2.

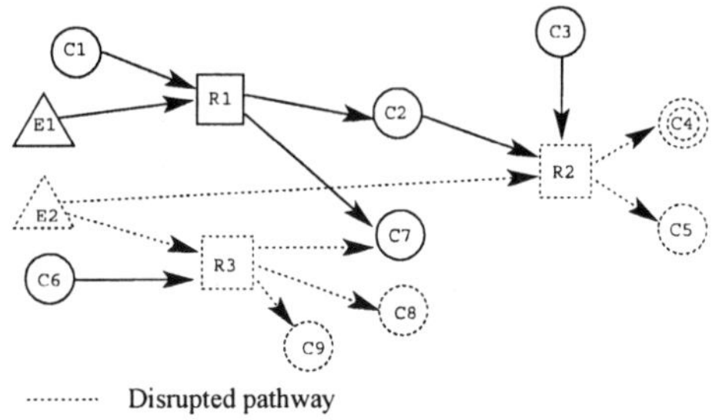

Figure 1. A graph constructed for a hypothetical metabolic network with three reactions R_1, R_2, and R_3, two enzymes E_1 and E_2, and nine compounds C_1, $\cdots$, C_9. Circles, rectangles, and triangles denote compounds, reactions, and enzymes respectively. Here, C_4 (shown by double circle) is the target compound. Dotted lines indicate the subgraph removed due to inhibition of enzyme E_2.

Figure 1 shows the disruption of E_2 and its effect on the network. Inhibiting E_2 results in the knock out of compounds C_5, C_8 and C_9 in addition to the target compound, C_4. Note that the production of C_7 is not stopped since it is produced by R_1 even after the inhibition of E_2. We define the number of non-target compounds knocked out as the *damage*, the manipulation of an enzyme set causes to the metabolic network. In this case, the damage of inhibiting E_2 is 3 (i.e., C_5, C_8 and C_9). The damage of inhibition of E_1 is 2 (i.e., C_2 and C_5). The important observation is that E_1 and E_2 both achieve the effect of disrupting the target compound, C_4. Hence, E_1 and E_2 are both potential drug targets. However, E_1 is a better drug-target than E_2 since it causes lesser damage.

Formally the optimal enzyme combination identification problem is: "*Given a set of target compounds T ($T \subset C$), find the set of enzymes X ($X \subseteq E$) with minimum damage, whose inhibition stops the production of all the compounds in T.*"

For simplicity, we assume that the input compounds to all reactions are present in the network and that there are no external inputs. Different enzymes and compounds may have varying levels of importance in the metabolic network. We consider all the enzymes and compounds to be of equal importance. This assumption can be relaxed by assigning weights to enzymes and compounds based on their role in the network. Also, we are not incorporating back-up enzyme activities [20] in this paper. This can be achieved by creating vertices for sets of enzymes in our graph representation. However, we do not discuss these extensions in this paper.

3. Proposed methods

This section proposes OPMET, a branch and bound algorithm that considerably reduces the number of possible combinations to be searched while still guaranteeing to find an optimal solution. Section 3.1 describes the basic branch and bound algorithm. Our prioritization (Section 3.2) and filtering (Section 3.3) strategies further improve this algorithm by reducing the search space.

3.1. *State space and basic search strategy of OPMET*

Let $E = \{E_i | \forall i, 1 \leq i \leq m\}$ denote the set of enzymes for a metabolic network. The search space is modeled as a tree structure. Every node of this tree corresponds to a state in the search space and it is represented by a 4-tuple $([e_{\pi_1}, e_{\pi_2}, \cdots, e_{\pi_m}], k, d, remove)$. Here, $\pi_1, \cdots, \pi_m$ is a permutation of 1, 2, $\cdots$, m. The first parameter corresponds to the state of all the enzymes (i.e., e_{π_i} corresponds to enzyme E_i). $e_{\pi_i} = 1$ if E_i is inhibited. Otherwise, $e_{\pi_i} = 0$. The parameter k indicates that the first k enzymes are considered at that search state. The decision to inhibit or not inhibit has been fixed for enzymes from 1 to $k-1$ and we now set $e_{\pi_k} = 1$ and $e_{\pi_i} = 0, \forall i, k < i \leq m$. The damage incurred due to inhibited enzymes at that state is represented by d. The final parameter, $remove$, is a boolean variable. It takes value *True* if the inhibited enzymes stop the production of all the target compounds. Otherwise, it is set to *False*. We call a node with $remove$ = True as a *true node*, and a node with $remove$ = False as a *false node*.

OPMET Algorithm: We start with the *root node* ([0, 0, $\cdots$, 0], 0, 0, False) indicating that all enzymes are present in the network. As the search space is traversed, we keep the true node with the minimum damage found so far as the *current true solution* and store the associated damage value as D, the *global cut-off threshold*. D is initialized to the number of compounds in the network. At any point, we have an *active set* of nodes A, stored in a stack structure. A contains the nodes currently being considered. Let node $N = ([e_{\pi_1}, e_{\pi_2}, \cdots, e_{\pi_m}], k, d, remove)$ be the node on top of this stack (i.e., the node to be evaluated). There are three cases:

• *Case 1:* N has damage $d > D$. In this case we prune the subtree rooted at N. We then backtrack.

• *Case 2:* N is a true node with damage $d < D$. In this case, we save N as the *current true solution* and update D with the damage value of N. We then backtrack.

• *Case 3:* N is a false node with damage $d < D$. In this case, we insert N in the active set A for backtracking purposes. We then create a new node N' by setting $e_{\pi_{k+1}} = 1$ in N (i.e., we inhibit the enzyme $E_{\pi_{k+1}}$). The resulting node is $N' = ([e_{\pi_1}, e_{\pi_2}, \cdots, e_{\pi_m}], k+1, d', remove')$. The node N' is evaluated in the next step similarly.

Backtracking involves following steps. First we pick the top node from the active nodes stack A. Let $N' = ([e'_{\pi_1}, e'_{\pi_2}, \cdots, e'_{\pi_m}], k', d', remove')$ denote this node. We then set $e_{\pi_{k'+1}} = 0$ (indicating the node we are backtracking from) and $e_{\pi_{k'+2}} = 1$ in N' (i.e., we inhibit the enzyme $e_{\pi_{k'+2}}$). The resulting node becomes the node to be evaluated in the next step. The first two cases above stop expanding the tree at the current node. The former one implies that the current node is a possible solution. The latter one implies that the current node incurs too much damage to lead to a possible solution. The third case happens when the current node does not stop production of all the target compounds, but the damage is lower than the damage of the current best solution. Such nodes may produce a possible solution with the inhibition of more enzymes. Thus, they need to be explored further to ensure that we find an optimal solution. The search terminates when there are no more nodes to explore. At this stage, the current true solution is the *optimal solution.*

3.2. *Improving the OPMET algorithm by enzyme prioritization*

In order to benefit from the pruning power of OPMET cases 1 and 2 (see Section 3.1), we need to compute the permutation $\pi_1, \cdots, \pi_m$ carefully. The earlier we place the enzymes in the optimal solution in this permutation, the better, as OPMET reaches the optimal solution earlier under such an ordering. Thus, reaching the solution with the smallest possible damage value (i.e., the *optimal solution*) increases the chances of pruning the remaining nodes of the search tree. This section develops a cost model to prioritize the enzymes dynamically.

3.2.1. *Cost model*

We develop a cost model as the basis for enzyme ordering in OPMET. This cost model takes both the observed and potential damage resulting from the inhibition of an enzyme set into the cost computation. For each enzyme $E_i \in E$, we compute a weight $W(E_i)$ as $W(E_i) = 0$ if E_i inhibited and $W(E_i) = 1$ otherwise. We assign fractional weights between 0 and 1 to the reaction and compound nodes and the edges. Intuitively, the weight of a node or edge denotes the rate at which that node or edge appears in the network. The weight of each node is calculated as follows:

• **Cost Rule 1:** Let R_j be a reaction node. Let w_i, $1 \leq i \leq k$, denote the weights of the incoming edges to R_j. We compute the weight of R_j as $W(R_j) = \min_{i=1}^{k}\{w_i\}$. This is intuitive since a reaction takes place only if all the inputs are present.

• **Cost Rule 2:** Let C_j be a compound node. Let w_i, $1 \leq i \leq k$, denote the weights of the incoming edges to C_j. We compute the weight of C_j as $W(C_j) = \frac{1}{k}\sum_{i=1}^{k}\{w_i\}$. This weight evaluates to zero only if all the reactions that produce C_j stops.

We define the weight of an edge as the weight of the node for which it is the outgoing edge. In order to compute the cost of E_i, we set the weight of E_i to zero (i.e., $W(E_i) = 0$). The weights of all the reaction and compound nodes are assigned progressively by a breadth-first search, according to the above scheme. The weights of all the nodes and edges which can be reached from E_i are recomputed to reflect the change. We define an *impact vector* for each enzyme based on the effects of its inhibition.

Definition 3.1. Given a network with n compounds, C_j, $1 \leq j \leq n$. Let $W_i(C_j)$ denote the weight of the node corresponding to C_j after the inhibition of enzyme E_i. We define the *impact vector* of E_i as $I(E_i) = [W_i(C_1), W_i(C_2), \cdots, W_i(C_n)]$. We term $W_i(C_j)$ as the *impact of* E_i *on* $C_j, \forall j$. ■

The impact vector of an enzyme approximates the amount of each compound that remains after the inhibition of that enzyme. Every entry of the impact vector is a fractional number between 0 and 1, where 0 indicates that the corresponding compound does not exist after inhibition of the corresponding enzyme. We define the cost of an enzyme as follows:

Definition 3.2. Given a network with n compounds, C_j, $1 \leq j \leq n$. Assume that the compounds $C_j, \forall j, 1 \leq j \leq k \leq n$ constitute the set of target compounds. Assume that the remaining compounds $C_j, \forall j, k+1 \leq j \leq n$ constitute the non-target compounds. Let $I(E_i) = [W_i(C_1), \cdots, W_i(C_n)]$ denote the impact vector of E_i. We define the *cost* of E_i as cost$(E_i) = I(E_i) \cdot V^T$, where $V = [v_1, \cdots, v_n]$ is the *normalization vector*: $v_i = \frac{n-k}{k}$ for $1 \leq i \leq k$, and $v_i = -(\frac{n-k}{n-k})$ for $k < i \leq n$. ■

Each target compound contributes a positive value and each non-target compound contributes a negative value to the cost of an enzyme. This is justified since the cost promotes removal of target compounds and demotes the removal of non-target compounds.

3.2.2. *Ordering of enzymes in OPMET*

Based on the impact vector of individual enzymes, we propose an incremental strategy for ordering of enzymes in OPMET. Let $R = [r_1, r_2, \cdots, r_n]$ denote the remaining fractions of compounds. Here, $r_i \in [0, 1]$ corresponds to compound C_i, $\forall i$. We initialize $r_i = 1, \forall i$ indicating that all compounds are being produced without any disruption. Let V be the normalization vector as given in Definition 3.2. Let $I(E_i)$ be the impact vector of enzyme E_i (see Definition 3.1). Assume $N = ([e_{\pi_1}, e_{\pi_2}, \cdots, e_{\pi_m}], k, d,$ *remove*) be the node currently being evaluated (i.e., the decision to inhibit or not inhibit has been fixed for $e_{\pi_1}, e_{\pi_2}, \cdots, e_{\pi_{k-1}}$). We now need to decide which enzyme has to be evaluated next. In details, for every enzyme in the remaining enzyme set (e_{π_i}, $\forall i$, $k \leq i \leq m$), we compute the new remaining fractions of compounds (R_i). This is done by a Vector Direct Product of R and the impact vector of E_i ($I(E_i)$). That is, $R_i = R \odot I(E_i)$. Vector direct product is defined as $X \odot Y = [x_1y_1, x_2y_2, \cdots, x_ny_n]$, where $X = [x_1, \cdots, x_n]$ and $Y = [y_1, \cdots, y_n]$. The resulting vector R_i is an approximation to the impact of inhibition of the enzyme E_i in addition to already inhibited enzymes. This is justified since the quantity of a compound eliminated by a combination including E_i will be at least as much as the quantity eliminated by E_i alone. A good candidate enzyme at this step is the one that ensures that lesser of the target compounds remain after its inhibition. Also, it should ensure that the non target compounds suffer the minimum possible damage. Our cost model satisfies these requirements. Then, we compute the cost of each enzyme as the dot product of R_i and V. That is, Cost$(E_i) = R_i \cdot V^T$. Suppose that the enzyme (E_j) is with the minimum cost, that is, $j = \arg\min_j \{\text{Cost}(E_j)\}$. Based on the minimum cost, we update the ratio $R = R_j$ and select E_j as the next enzyme to inhibit. Thus, this strategy chooses the next best enzyme to inhibit dynamically.

The cost of finding the best enzyme takes $O(mn)$, where m and n denote the number of enzymes and compounds in the metabolic network respectively. This is because a vector product costs $O(n)$, and $O(m)$ such products are carried out.

3.3. *Filtering strategies*

So far, we have described how OPMET traverses the state space. In this section we propose two filtering strategies to eliminate large portions of the search space quickly while still guaranteeing the optimal solution. The following theorem (proof is not given to save the space) establishes a relationship between the impact of enzymes and their damage.

Theorem 3.1. *Let $E = \{E_1, E_2, \cdots, E_r\}$ be a set of enzymes. Let C_j be a compound in the metabolic network. Let $d_i(C_j)$, $1 \leq i \leq r$, denote the impact of E_i on C_j. If the inhibition of all the enzymes in E stops the production of C_j, then $\sum_{i=1}^{r}(1 - d_i(C_j)) \geq 1$.* ■

Next, we describe our filtering strategies.

Target Filter: A combination of enzymes can not be the solution if their inhibition does not delete all the target compounds. This is the motivation behind our *Target filter*. The target filter eliminates a bulk of the search space when it is proven that there is no combination of enzymes in this space that can stop the production of all the target compounds (i.e., there is no useful drug target). This filtering strategy is based on Theorem 3.1. Formally, let node $N = ([e_{\pi_1}, e_{\pi_2}, \cdots, e_{\pi_m}], k, d, \text{False})$ be a node in the search space. Let T denote the set of target compounds. Backtrack if

$$\sum_{i=1}^{k}(1 - d_i(C))e_{\pi_i} + \sum_{i=k+1}^{n}(1 - d_i(C)) < 1, \exists C \in T.$$

In this inequality, the first term indicates the impact of enzymes, which are currently part of the solution set, on the target compounds. The second term represents the impact of the remaining enzymes on the target compounds.

Non-target Filter: This filter quickly determines if there is any solution in the subtree with a damage $d < D$, the global cut-off threshold (see Subsection 3.1). This filter utilizes Theorem 3.1 similar to the Target Filter. The idea is as follows. At a given node N, for each target compound, C, we find the minimum number of enzymes, m such that $\sum_{i=1}^{m}(1 - d_i(C))e_{\pi_i} \geq 1$. This gives us the minimum number of enzymes needed to delete C. Let m_{max} be the maximum value of m for any target compound (i.e, we will need at least m_{max} enzymes to delete the entire target compound set). Now, we sort the remaining enzymes (enzymes not considered so far) in the ascending order of their damage values. Let d_{max} be the damage of the enzyme at index m_{max}. If d_{max} in addition to the damage incurred so far is greater than D, we prune the sub tree rooted at N.

4. Experimental results

We verify the biological validity of the proposed algorithm by employing it on known existing drugs. We evaluate the performance of the OPMET algorithm using the following three criteria: *1) Number of nodes generated:* It represents the total number of enzyme combinations tested to complete the search. *2) Optimal node rank:* This indicates the number of nodes explored before the method arrives at the optimal solution.

3) Execution time: This indicates the total time taken by the method to finish the search.

We extracted the metabolic network information of *E. Coli* from KEGG [15]. The metabolic network in KEGG is divided into smaller networks according to their specific functions. We chose six of these networks for our experiments, based on the number of enzymes. We devised a labeling scheme for the networks which begins with 'N' and is followed by the number of enzymes in the network. For instance, 'N20' indicates a network with 20 enzymes. For each network, we constructed query sets of sizes one, two and four target compounds, by randomly choosing compounds from that network. Each query set contains 10 queries.

Qualitative analysis of OPMET: We first evaluate how well the proposed cost model reflects the biological process. We do this by querying well studies drugs in the literature using OPMET. KEGG contains a database of known drug molecules along with the enzymes they inhibit and their therapeutic category. We use the drugs at this database as our benchmarks. Due to space limitation, we report only four of them. The value in parenthesis that starts with letter "D", "C", or "E" (e.g., D02562) is the unique identifier assigned to the corresponding drug, compound, or enzyme respectively in KEGG.

1. Benoxaprofen (D03080). This drug inhibits arachidonate 5-lipoxygenase (E1.13.11.34) which appears in several networks including arachidonic acid metabolism network (hsa00590). In Pharmacology, 5-lipoxygenase inhibitors will decrease the biosynthesis of LTB4 (C02165), cysteinylcontaining leukotrienes LTC4 (C02166), LTD4 (C05951) and LTE4 (C05952). According to our graph model, the removal of 5-lipoxygenase eliminates three of these compounds LTB4, LTC4 and LTD4 in arachidonic acid metabolism network. Inhibition of this enzyme also eliminates five more compounds, namely 5(S)-HPETE (C05356), 5-HETE (C04805), LTA4 (C00909), 20-OH-LTB4 (C04853) and 9(S)-HPOD (C14827). These compounds can be considered as damage in our model. Running OPMET with LTB4, LTC4, LTD4 and LTE4 as the target compound finds LTA4H (E3.3.2.6) and LTC4 synthase (E4.4.1.20) as the optimal enzyme set. The inhibition of these enzymes eliminates only one non-target compound, 20-OH-LTB4 (C04853). OPMET potentially finds a better solution in this experiment than the existing drug as the same compound is eliminated by the existing drug in addition to four other compounds. Indeed, recent research supports our model since the anti-inflammatory effect of the levels of LTA4H [22] and LTC4 [29] have been observed.

2. Rasagiline (D02562). This is an antiparkinsonian drug. It inhibits amine oxidase (E.1.4.3.4) which appears in several metabolic networks. In the histidine metabolism network (hsa00340), the removal of amine oxidase eliminates the compounds Methylimidazole acetaldehyde (C05827) Methylimidazoleacetic acid (C05828) according to our graph model. Levels of pros-methylimidazoleacetic acid has correlation with severity of Parkinson's disease in patients [3,21]. This demonstrates that, our model can predict the intended target well. When OPMET is run on the same network with methylimidazoleacetic acid and the methylimidazole acetaldehyde as the target compounds it finds amine oxidase as the optimal target. This implies that Rasgiline is

Table 1. Average number of nodes generated and Optimal Node Rank of exhaustive search and OPMET with random and dynamic enzyme ordering. Optimal Node Rank is given in parentheses.

Id	Exhaustive Search	Random OPMET	Dynamic OPMET
N14	16,384	4,190 (1,220)	3,273 (3.77)
N17	131,072	58,379 (22,303)	38,973 (2.54)
N20	1,048,576	147,605 (100,845)	78,257 (11.36)

Table 2. Average number of nodes generated and Optimal Node Rank of OPMET with no-filtering (A), non-target filter (B), target filter (C), and both filters (D). Optimal Node Rank is given in parentheses. (E) shows the average execution time (millisecond) for the both filters method.

	N17	N20	N24	N28	N32
A	38973 (2.54)	78257 (11.36)	509278 (55893.25)	158989 (10834.55)	4151032 (3.61)
B	35806 (2.54)	76125 (11.36)	462980 (55103.58)	156956 (10803.00)	1512615 (3.61)
C	415 (2.54)	3496 (11.36)	55987 (56.75)	12735 (10801.90)	1049377 (3.61)
D	394 (2.54)	3428 (11.36)	55865 (6.71)	12710 (10801.90)	1044263 (3.61)
E	529.64	2619.34	46273.13	10025.50	816913.55

targeting the optimal enzyme according to our model.

For Ozagrel (D01683) and Erythromycin acistrate (D02523), running OPMET can find the same target enzyme as the actual drug. (details omitted)

Evaluation of prioritization strategies: We compare our OPMET algorithm with a random ordering of enzymes and an exhaustive search. We do not include our filtering strategies here as the goal is to focus on the enzyme ordering. We present the results only up to a network of size 20 enzymes. This is because, beyond this, the search space grows rapidly, necessitating the use of filtering strategies.

Table 1 shows the average number of nodes generated and the average optimal node rank of OPMET to that of an exhaustive search. The results show that OPMET with dynamic enzyme ordering is the best strategy for all the tested networks. It generates the least number of nodes in all the experiments. All the methods generate significantly large number of nodes for N17. This is because the number of reactions and compounds of this network is much larger than the other networks, resulting in more interactions in the network. OPMET has small Optimal Node Ranks. On an average, it arrives at the optimal solution within the generation of 0.008 % of the number of nodes possible in an exhaustive search. This is significantly better than the random ordering which arrives at the optimal solution within the generation of 11 %.

Evaluation of filtering strategies: We measure how much our filtering strategies reduce the search space. The experiments are performed using OPMET with dynamic enzyme ordering. Table 2 shows the average number of nodes generated, the average optimal node rank and the average execution time for the combined filters. The combined filters show the best pruning. On an average, the combined filters prune 91.5 % of the nodes generated in the method without filters. We also see that most of this benefit is obtained from the Target Filter (it filters 91.4 % of the nodes generated by the method without filters). The combined filter generates only 12700 nodes for N28 (0.004 % of an exhaustive search). All the methods have the same optimal node rank for networks except N24. This suggests that OPMET yielded the optimal solution as early as possible for these networks. For N24, the combined filter shows that filtering strategies can also lead to advancement in finding the optimal solution. For N24, Target filter arrives at the optimal solution 99 % earlier and the combined filters arrive at the optimal solution 99.9 % earlier than the method without filters (the additional 0.9 % improvement is obtained from the non-target filter). We observe that the target filter is more efficient than the non-target filter and the combined filter has the best performance. We observe that there is no clear correlation between the size of the target compound set and the number of nodes explored (results are not shown due to space limitation).

5. Related Work

Classical drug discovery approaches involve incorporating a large number of hypothetical targets into in-vitro or cell-based assays and performing automated high throughput screening (HTS) of vast chemical compound libraries [30,9]. Post-genomic advances in bioinformatics have fostered the development of rational drug-design methods and reduction of serious side-effects [8,5,4]. This has engendered the concept of *reverse pharmacology* [27], in which, the first step is the identification of protein targets, that may be critical intervention points in a disease process [23,1]. The reverse approach is driven by the mechanics of the disease and hence is expected to be more efficient than the classical approach [27].

Rapid identification of enzyme (or protein) targets needs a thorough understanding of the underlying metabolic network of the organism affected by a disease. The availability of fully sequenced genomes has enabled researchers to integrate the available genomic information to reconstruct and study metabolic networks [28,13]. These studies have revealed important properties of these networks [10,2,18]. The potential of an enzyme to be an effective drug target is considered to be related to its essentiality in the corresponding metabolic network [14]. Lemke et. al proposed the measure *enzyme damage* as an indicator of enzyme essentiality [17,19]. Recently, a computational approach for prioritizing potential drug targets for antimalarial drugs has been developed [31]. A choke-point analysis of *P.falciparcum* was performed to identify essential enzymes which are potential drug targets. The possibility of using enzyme inhibitors as antiparasitic drugs is being investigated through stoichiometric analysis of the metabolic networks of parasites [6,7]. These studies show the effectiveness of computational techniques in reverse pharmacological approaches.

A combination of microarray time-course data and gene-knockout data was used to study the effects of a chemical compound on a gene network [12]. An investigation of metabolite essentiality is carried out with the help of stoichiometric analysis [11]. These approaches underline the importance of studying the role of compounds (metabolites) during the pursuit of computational solutions to pharmacological problems.

6. Conclusions

In this paper, we formulated the optimal enzyme-combination identification problem as an optimization problem on metabolic networks. We proposed OPMET, a branch-and-bound algorithm to explore the search space dynamically. We also developed two filtering strategies to prune the search space while still guaranteeing an optimal solution. The filters compute an upper bound to the number of target compounds deleted and a lower bound to the side-effect respectively.

Our experiments on the human metabolic network demonstrates that the proposed model can accurately identify the target enzymes for known successful drugs in the literature. More specifically, OPMET found the same target enzyme as Rasagiline, Ozagrel, and Erythromycin acistrate when their target compounds are given as input. OPMET found a different set of enzymes than Benoxaprofen for the target compounds of Benoxaprofen. OPMET's solution in this case has a great potential to be better than Benoxaprofen since OPMET's solution damages only one non-target compound whereas Benoxaprofen damages five non-target compounds including the compound damaged by OPMET's solution. Our experiments also show that OPMET can reduce the total search time by several orders of magnitude as compared to the exhaustive search. The optimal solution is reached by OPMET within the exploration of 0.005 % of the total search space on an average, proving that our methods are effective in approximating the impact of an enzyme on a compound. OPMET with combined filters pruned 91.6 % of the search space on average.

References

1. 'Proteome Mining' can zero in on Drug Targets. Duke University medical news, Aug 2004.
2. Masanori Arita. The metabolic world of Escherichia coli is not small. *PNAS*, 101(6):1543–1547, Feb 2004.
3. P. Blandina and G. Cherici et al. Release of glutamate from striatum of freely moving rats by pros-methylimidazoleacetic acid. *Journal of Neurochemistry*, 64(2):788–793, 1995.
4. Samuel Broder and J. Craig Venter. Sequencing the Entire Genomes of Free-Living Organisms: The Foundation of Pharmacology in the New Millennium. *Annual Review of Pharmacology and Toxicology*, 40:97–132, Apr 2000.
5. Sumit K. Chanda and Jeremy S. Caldwell. Fulfilling the promise: drug discovery in the post-genomic era. *Drug Discovery Today*, 8(4):168–174, Feb 2003.
6. A. Cornish-Bowden. Why is uncompetitive inhibition so rare? : A possible explanation, with implications for the design of drugs and pesticides. *FEBS Letters*, 203(1):3–6, Jul 1986.
7. A. Cornish-Bowden and J. S. Hofmeyr. The Role of Stoichiometric Analysis in Studies of Metabolism: An Example. *Journal of Theoretical Biology*, 216:179–191, May 2002.
8. Eugene J. Davidov, Joanne M. Holland, Edward W. Marple, and Stephen Naylor. Advancing drug discovery through systems biology. *Drug Discovery Today*, 8(4):175–183, Feb 2003.
9. J Drews. Drug Discovery: A Historical Perspective. *Science*, 287(5460):1960–1964, Mar 2000.
10. Vassily Hatzimanikatis, Chunhui Li, Justin A Ionita, and Linda J Broadbelt. Metabolic networks: enzyme function and metabolite structure. *Current Opinion in Structural Biology*, 14(3):300–306, 2004.
11. Marcin Imielinski, Calin Belta, Adam Halasz, and Harvey Rubin. Investigating metabolite

essentiality through genome scale analysis of E. coli production capabilities. *Bioinformatics*, Jan 2005.

12. S. Imoto and Y. Tamada et. al. Computational Strategy for Discovering Druggable Gene Networks from Genome-Wide RNA Expression Profiles. In *PSB*, 2006.
13. H. Jeong, B. Tombor, R. Albert, Z. N. Oltvai, and A.-L. Barabasi. The large-scale organization of metabolic networks. *letters to NATURE*, 407:651–654, Oct 2000.
14. Hawoong Jeong, Zoltan N. Oltvai, and Albert-Laszlo Barabasi. Prediction of Protein Essentiality Based on Genomic Data. *ComPlexUs*, 1:19–28, 2003.
15. M Kanehisa and S Goto. KEGG: Kyoto Encyclopedia of Genes and Genomes. *Nucleic Acids Res.*, 28(1):27–30, Jan 2000.
16. S.A. Kauffman. *The Origins of Order: Self-Organization and Selection in Evolution*. Oxford University Press, 1993.
17. Ney Lemke, Fabiana Herdia, Cludia K. Barcellos, Adriana N. dos Reis, and Jos C. M. Mombach. Essentiality and damage in metabolic networks. *Bioinformatics*, 20(1):115–119, Jan 2004.
18. Hong-Wu Ma, Xue-Ming Zhao, Ying-Jin Yuan, and An-Ping Zeng. Decomposition of metabolic network into functional modules based on the global connectivity structure of reaction graph. *Bioinformatics*, 20(12):1870–1876, 2004.
19. J. C. Mombach and N. Lemke et al. Bioinformatics analysis of mycoplasma metabolism: Important enzymes, metabolic similarities, and redundancy. *Computers in Biology and Medicine*, 2005.
20. M.T.A. Ocampo and W. Chaung et. al. Targeted deletion of mNth1 reveals a novel DNA repair enzyme activity. *Mol Cell Biol.*, 22(17):6111–21, Sep 2002.
21. G. D. Prell, J. K. Khandelwal, R. S. Burns, P. Blandina, A. M. Morrishow, and J. P. Green. Levels of pros-methylimidazoleacetic acid: Correlation with severity of Parkinson's disease in CSF of patients and with the depletion of striatal dopamine and its metabolites in MPTP-treated mice. *Journal of Neural Transmission*, 3(2):1435–1463, 1991.
22. Navin L. Rao and Paul J. et al. Dunford. Anti-Inflammatory Activity of a Potent, Selective Leukotriene A4 Hydrolase Inhibitor in Comparison with the 5-Lipoxygenase Inhibitor Zileuton. *J Pharmacol Exp Ther*, 321(3):1154–1160, 2007.
23. C Smith. Hitting the target. *Nature*, 422:341–347, Mar 2003.
24. R. Somogyi and C.A. Sniegoski. Modeling the complexity of genetic networks: Understanding multi-gene and pleiotropic regulation. *Complexity*, 1:45–63, 1996.
25. Padmavati Sridhar, Tamer Kahveci, and Sanjay Ranka. An iterative algorithm for metabolic network-based drug target identification. In *PSB*, volume 12, pages 88–99, 2007.
26. R. Surtees and N. Blau. The neurochemistry of phenylketonuria. *European Journal of Pediatrics*, 159:109–13, 2000.
27. T. Takenaka. Classical vs reverse pharmacology in drug discovery. *BJU International*, 88(2):7–10, Sep 2001.
28. S. A. Teichmann and S. C. G. Rison et al. The Evolution and Structural Anatomy of the Small Molecule Metabolic Pathways in Escherichia coli. *JMB*, 311:693–708, 2001.
29. M. J. Torres-Galvan and N. Ortega et al. LTC4-synthase A-444C polymorphism: lack of association with NSAID-induced isolated periorbital angioedema in a Spanish population. *Ann Allergy Asthma Immunol.*, 87(6):506–10, 2001.
30. Gunther Wess. How to escape the bottleneck of medicinal chemistry. *Drug Discovery Today*, 7(10):533–535, May 2002.
31. I. Yeh and T. Hanekamp et al. Computational Analysis of Plasmodium falciparum Metabolism: Organizing Genomic Information to Facilitate Drug Discovery. *Genome Research*, 14:917–924, 2004.

GLOBAL ALIGNMENT OF MULTIPLE PROTEIN INTERACTION NETWORKS

ROHIT SINGH[a] JINBO XU[b] BONNIE BERGER[a*]

[a] *Computer Science and Artificial Intelligence Laboratory*
Massachusetts Institute of Technology
[b] *Toyota Technological Institute, Chicago*
E-mail: {*rsingh@mit.edu, j3xu@tti-c.org, bab@mit.edu*}

We describe an algorithm for global alignment of multiple protein-protein interaction (PPI) networks, the goal being to maximize the overall match across the input networks. The intuition behind our algorithm is that a protein in one PPI network is a good match for a protein in another network if the former's neighbors are good matches for the latter's neighbors. We encode this intuition by constructing an eigenvalue problem for every pair of input networks and then using k-partite matching to extract the final global alignment across all the species. We compute the first known global alignment of PPI networks from five species: yeast, fly, worm, mouse and human. The global alignment immediately suggests functional orthologs across these species; we believe these are the first set of functional orthologs that cover all the five species. We show that these functional orthologs compare favorably with current sequence-only orthology prediction approaches, including better prediction of orthologs for some human disease-related proteins.
Supplementary Information: `http://groups.csail.mit.edu/cb/mna`

1. Introduction

Over the past few years, the use of high-throughput experimental techniques[15,12] for discovering protein-protein interactions (PPIs) has led to a tremendous increase in the corpus of available PPI data in various species. A useful representation of this data is as a network: each node in such a network corresponds to a protein and an edge between two nodes indicates that the corresponding proteins interact. Analysis of such PPI networks has yielded some deep biological insights[9]. In this paper, we explore methods for comparing PPI networks across species. Such comparative analysis has proven to be a valuable tool. It has led, for example, to the identification of conserved functional components across various species, complementing traditional sequence-only phylogenetic analysis. It also helps in identifying errors in experimental PPI data and in transferring annotation across species.

*Corresponding author. Also in the MIT Dept. of Mathematics.

We also explore the use of such comparative analysis in improving orthology predictions across species. Identifying cross-species gene correspondences (orthologs) is a problem of fundamental biological importance— it is crucial for transferring insights and information across species.

Contributions

One of the main contributions of this paper is the first algorithm for global alignment of multiple protein interaction networks. We perform a global alignment of PPI networks from five species: yeast, fly, worm, mouse, and human. We pursue the following intuition: a node in a PPI network is a good match for a node in another network if its neighbors are good matches for the neighbors of the other node. To formalize the intuition, we construct a set of eigenvalue problems in an approach similar to Google's PageRank[18] algorithm and then use k-partite matching to compute the final alignment.

The multiple network alignment directly leads to the first comprehensive estimates of functional orthologs that incorporate both sequence and PPI data and cover all the five species mentioned previously. These estimates are more comprehensive than the two most commonly used orthology sets: Homologene[5] and Inparanoid[16]. Our list covers more genes than Homologene. Unlike Inparanoid, which considers pairs of species at a time, our method analyzes data from all input species simultaneously.

We also introduce a novel approach, *functional coherence*, for evaluating orthology predictions. Currently such predictions are evaluated by manually analyzing selected sets of orthologs. In contrast, our automated approach measures the functional similarity within each set of orthologous proteins and computes an aggregate score. Using it, we demonstrate that our algorithm makes predictions with slightly better overall quality than Homologene and Inparanoid. Also, further analysis indicates that some of the improved predictions from our method include disease-related proteins.

Related Work

PPI Network Alignment: The protein network alignment problem can be formulated either as a global or a local network alignment problem. Much of the previous work[3,11,9] in the field has focused on the problem of local network alignment (see Sec. 2). In contrast, we focus on the global alignment problem. Recently, we have proposed the first algorithm for *pairwise* global alignment of PPI networks. The multiple network alignment algorithm we present in this paper is, we believe, the first algorithm for global alignment of multiple protein networks. While this paper builds upon some of the methods presented in our previous work, there are also many significant differences between the two problems and the corresponding algorithms (see Sec. 2).

Functional Ortholog Prediction: Currently, orthology prediction is usually done by using sequence-similarity information between various genes to estimate sets of genes that have descended from a common ancestor. A key challenge here is to distinguish between orthologs and paralogs, the latter being genes that are created by duplication after the two species have diverged. We briefly describe here two commonly used orthology prediction methods: Inparanoid and Homologene (see Chen *et al.*[6] for more). Inparanoid[16] computes orthologs between pairs of species by making explicit assumptions about the relative sequence similarity scores between orthologs and paralogs. One of its drawbacks is that it is limited to pairwise orthology estimates, i.e., it cannot analyze data from multiple species simultaneously. Homologene[5] is an approach that can simultaneously compute orthologs across multiple species by using sequence-similarity scores to construct a tree of proteins and, based upon certain heuristics, grouping them into clusters of orthologous genes.

Recently, efforts have been made to integrate PPI data into the orthology prediction process, to identify sets of proteins that perform the same function. Bandyopadhyay *et al.* [2] have described the use of local network alignment results in identifying *functional orthologs* between yeast and fly. In previous work, we have described a two-way global alignment algorithm which directly suggests functional orthologs between yeast and fly; these predictions compare favorably with Bandyopadhyay *et al.*'s. This paper is the first, we believe, to present functional orthologs across *multiple* species. By integrating data from multiple species simultaneously, we should be able to improve upon predictions made from pairs of species.

2. Problem Formulation

The input to our algorithm consists of two or more protein interaction networks (one per species). Each input network can be represented as an undirected graph $G = (V, E)$ where V is the set of nodes and E is the set of edges. Furthermore, a confidence measure $w(e)$ $(0 < w(e) \leq 1)$ may be associated with each edge e in E. Additionally, the input may also consist of pairwise node similarity measures between nodes from the different networks. In this paper, these similarity measures are BLAST Bit-values, but other scores (e.g., synteny-based scores[10]) can also be used. Given these inputs, our goal is to find the best overall match (i.e, optimal global alignment) between the input networks. This will directly lead to a list of functional orthologs.

Local vs. Global Network Alignment: Network alignment problems vary in the scope of the input (two vs. multiple networks), and the kind

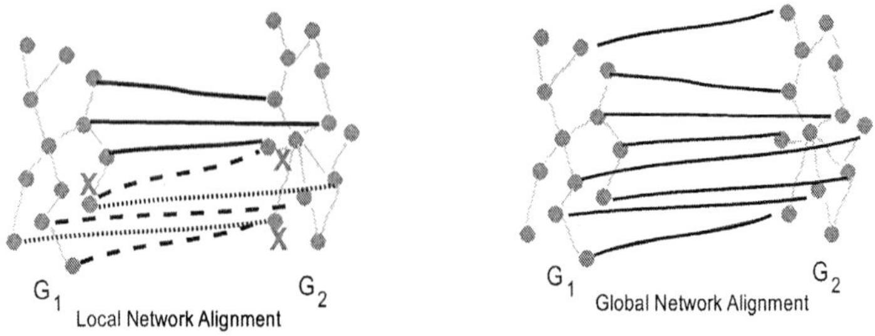

Figure 1: **Cartoon comparing global and local network alignments**: The local network alignment between G_1 and G_2 specifies three different alignments; the mappings for each are marked by a different kind of line (solid, dashed, dotted). Each alignment describes a small common subgraph. Local alignments need not be consistent in their mapping– the points marked with 'X' each have ambiguous/inconsistent mappings under different alignments. In global network alignment, the maximum common subgraph is desired. In both cases, there are 'gap' nodes for which no mappings could be predicted (here, the nodes with no incident black edges are such nodes).

of node-mapping desired. In general, the goal in all such problems is to identify one or more mappings between the nodes of the input networks and, for each mapping, the corresponding set of conserved edges. A mapping may be partial, i.e., it need not be defined for all the nodes in the networks. Each mapping implies a common subgraph between the input networks: when protein a_1 from network G_1 is mapped to proteins a_2 from G_2 and a_3 from G_3, then a_1, a_2, and a_3 refer to the same node in the common subgraph; the edges in the common subgraph correspond to the conserved edges. A key difference between our approach and many previous network alignment approaches is in the kind of mapping desired.

Much of the previous work[3,11,9] has focused on local network alignment (LNA), i.e., on finding local regions of isomorphism (i.e., same graph structure) between the input networks. Each such region implies a mapping independently of others. Many independent, high-scoring local alignments are usually possible between two input networks; in fact, the corresponding local alignments need not even be mutually consistent (i.e., a protein might be mapped differently under each alignment; see Fig. 1).

In contrast, we focus on the global network alignment (GNA) problem. The aim in GNA is to find the best overall alignment between the input networks. The mapping in a GNA should cover all the input nodes: each node in an input network is either matched to one or more nodes in other network(s) or explicitly marked as a gap node (i.e., with no match in another network). In contrast, a LNA algorithm outputs multiple, independent mappings, each corresponding to a local region of similarity. Furthermore, these partial mappings may be mutually inconsistent. The mapping corresponding to a GNA is also required to be transitive: if a_1 in G_1 is mapped to a_2 in G_2 and a_2 is mapped to nodes a_3, a_3' in G_3, then a_1 should also be mapped to a_3, a_3'. Our goal in GNA then is to find a comprehensive mapping between the nodes of the input networks such that the size of the single corresponding common subgraph is maximized. Our previous work[17] contains a more detailed comparison of the LNA and GNA problems.

A key difference between the multiple-network GNA (the focus of this paper) and pairwise GNA (the focus of our previous work[17]) is in the scope of the mapping desired. In the latter, we required that a node may be mapped to at most one node in the other network, the motivation being to find the *best* match for a node. In contrast, for the multiple networks case we allow for a node to map to multiple nodes in another network. This is necessary because gene duplication, mutation, and deletion events might make it impossible to find a valid one-to-one, transitive mapping between proteins across an arbitrary collection of species.

3. Algorithm

To describe a global alignment between input networks, we need to specify a node mapping between the input networks and the corresponding common subgraph. We focus on the computing the node mapping, since the subgraph can be easily computed once the former is known.

Our algorithm works in two stages. First, given k input networks, we create a k-partite graph $\mathcal{H}$. Each of its k parts contains nodes from one of the input networks. Edges are only allowed between nodes from different parts. The presence of an edge e_{ij} implies that node i (from G_1) can potentially be mapped to j (from G_2); the edge-weight R_{ij} indicates the strength of the potential match. In the second stage, we perform k-partite matching on $\mathcal{H}$ to group nodes into clusters. All nodes in a cluster are then mapped to each other in the corresponding GNA.

First Stage (Creating the k-partite graph): We start with the k input PPI networks and sequence similarity scores between the nodes. For every pair of input networks, we compute a score for every possible pairing between the nodes of the two networks. Let R_{ij} ($R_{ij} \geq 0$) be the score for the protein pair (i, j) where i is from network G_1 and j is from network G_2. Intuitively, R_{ij} should capture how good a match i and j are: higher R_{ij} implies a better match. In the second stage, we will use these scores to guide our algorithm towards the optimal k-partite matching of $\mathcal{H}$.

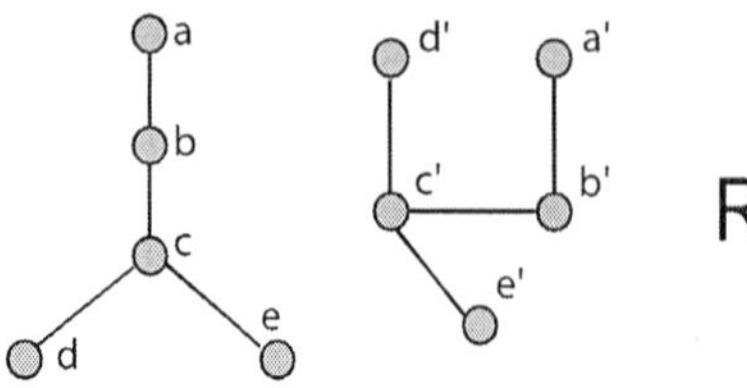

R

	a'	b'	c'	d'	e'
a	0.0312		0.0937		
b		0.1250		0.0625	0.0625
c	0.0937		0.2812		
d		0.0625		0.0312	0.0312
e		0.0625		0.0312	0.0312

$R_{aa'} = \frac{1}{4} R_{bb'}$ $R_{bb'} = \frac{1}{3} R_{ac'} + \frac{1}{3} R_{a'c} + R_{aa'} + \frac{1}{9} R_{cc'}$ $R_{dd'} = \frac{1}{9} R_{cc'}$

$R_{cc'} = \frac{1}{4} R_{bb'} + \frac{1}{2} R_{be'} + \frac{1}{2} R_{bd'} + \frac{1}{2} R_{eb'} + \frac{1}{2} R_{db'} + R_{ee'} + R_{ed'} + R_{de'} + R_{dd'}$

Figure 2: **Intuition behind the First Stage of the algorithm**: Here we show, for a pair of small, isomorphic graphs how the vector of pairwise scores (R) is computed (see Eqn. 1). Only a partial set of constraints is shown here. Here we show the vector of scores R reshaped as a table, for ease of viewing (empty cells indicate a value of zero). Observe that high values of R (e.g., $R_{cc'}$ or $R_{bb'}$) correctly indicate that the respective pairings represent good matches.

To compute R (the vector of all R_{ij}s for G_1 and G_2) we construct an eigenvalue problem. First consider the case when no sequence similarity scores are available (i.e., R_{ij} depends only on G_1 and G_2's topologies). We require that R_{ij}s satisfy the following system of constraints (for all i, j):

$$R_{ij} = \sum_{u \in N(i)} \sum_{v \in N(j)} \frac{1}{|N(u)||N(v)|} R_{uv} \quad i \in V_1,\ j \in V_2 \tag{1}$$

where $N(a)$ is the set of neighbors of node a, $|N(a)|$ is the size of this set, and V_1 and V_2 are the sets of nodes in G_1 and G_2, respectively. These constraints can be re-written as an eigenvalue equation:

$$\begin{aligned} R &= AR \\ A[i,j][u,v] &= \tfrac{1}{|N(u)||N(v)|} \end{aligned} \tag{2}$$

where A is a $|V_1||V_2| \times |V_1||V_2|$ matrix and $A[i,j][u,v]$ refers to the entry at the row (i,j) and column (u,v) (the row and column are doubly-indexed). The value of R we are interested in is the principal eigenvector of A. Typically, A is a very large matrix (about $10^8 \times 10^8$ for fly-vs.-yeast GNA). However, A is a stochastic matrix[14] and both A and R are very sparse, so R can be efficiently computed by iterative techniques, like the *power method* [14].

The intuition behind these equations is that they require that the score R_{ij} for any match (i,j) be equal to the total support provided to it by each of the $|N(i)||N(j)|$ possible matches between the neighbors of i and j. In return, each match (u,v) must distribute back its score R_{uv} equally among the $|N(u)||N(v)|$ possible matches between its neighbors (see Fig. 2 for an example). We note that these equations also capture non-local influences on the score R_{ij}: it depends on the score of neighbors of i and j and the latter, in turn, depend on the neighbors of the neighbors and so on. Also, these equations can be extended to the weighted-graph case very naturally[17].

It is straightforward to incorporate sequence similarity information, e.g. BLAST scores, into this model. Let B_{ij} denote the score between i and j; for instance, B_{ij} can be the Bit-Score of the BLAST alignment between sequences i and j. Let B be the vector of B_{ij}s. We first compute E, the normalized version of B: $E = B/|B|$. The eigenvalue equation is then modified to (this equation can also be solved by iterative techniques):

$$R = \alpha AR + (1-\alpha)E \quad \text{where} \quad 0 \leq \alpha \leq 1 \tag{3}$$

Changing α lets us control the weight of the network data (relative to sequence data) in this computation. For example, $\alpha = 0$ implies no network data will be used, while $\alpha = 1$ indicates only network data will be used.

Second Stage (K-partite matching): We construct the k-partite graph $\mathcal{H}$ as follows: for any pair of nodes i and j from different PPI networks, we add an edge e_{ij} to $\mathcal{H}$ if $R_{ij} > 0$, and set the edge-weight to R_{ij}. We now find a k-partite matching of $\mathcal{H}$ (recall that each part corresponds to nodes from one PPI network). The matching must be transitive, i.e, if i is matched to j and j is matched to l, then i must be matched to l. Furthermore, we aim to match nodes connected by high-scoring edges. More precisely, our goal is to find the maximum-weight k-partite matching of $\mathcal{H}$ where each set of matched nodes may contain upto r nodes from each of the k parts. Here, r is a user-defined parameter ($r \geq 1$). Allowing a one-to-many mapping lets us express that, for example, a particular fly protein has no corresponding yeast protein but two corresponding human proteins. In our previous work on two-way network alignment, this flexibility was not present.

The standard k-partite matching problem formulation requires that a node can match at most one node in each of the other $k-1$ parts. Our formulation thus generalizes this problem (the standard version corresponds to $r = 1$). However, the classical problem is already known to be NP-Hard[13], so our formulation is NP-Hard as well. Thus, it is unlikely that an exact solution for it can be found efficiently. Here, we present an approach that computes the matching by identifying a seed match and extending it:

- While the k-partite graph $\mathcal{H}$ has any edges left:
  (1) Select the edge (i, j) with the highest score (let i be from G_1 and j from G_2). Initialize a new match-set with i and j as its initial members.

(2) In every other species ($G_3, \ldots, G_k$), if a node l exists such that (A) R_{il} and R_{jl} are the highest scores between l and any node in G_1 and G_2, respectively and, (B) the scores $R_{ik} \geq \beta_1 R_{ij}$ and $R_{jk} \geq \beta_1 R_{ij}$, add it to the set. These set of nodes form the primary match-set; it has at most one node from each species.

(3) Add upto $r - 1$ nodes from different parts of the graph to the primary match-set. Suppose u (from G_x) is in the primary match-set. Then, a node v (from G_x) is added to the set if $R_{vw} \geq \beta_2 R_{uw}$ for each node w ($w \neq u$) in the primary set.

(4) Remove from $\mathcal{H}$ all the nodes in this match-set and their edges.

Here, the parameters r, β_1, β_2 are user-defined ($0 < \beta_2, \beta_1 < 1$); we chose their values such that the functional coherence (see Sec. 4.1) of the resulting sets of matched nodes was maximized.

Given a mapping between the nodes of the input networks, the corresponding common subgraph in the GNA can be identified relatively easily. For example, if a_1 is aligned to a_2, and b_1 is aligned to b_2, the output subgraph should contain an edge between the corresponding nodes if and only if both the input networks contain supporting edges.

4. Results

Datasets: We constructed PPI networks for five species: *S. cerevisiae*, *D. melanogaster*, *C. elegans*, *M. musculus*, and *H. sapiens*. These networks were constructed by combining data retrieved from the DIP[8], BioGRID[4] and HPRD[7] databases. The relative coverage of the PPI data varied heavily; the number of edges per species were: 36387 (human), 31899 (yeast), 25831 (fly), 4573 (worm), and 255 (mouse). Sequence data for the various proteins was retrieved from Ensembl and the BLAST Bit-values were used as the score of sequence similarity between input proteins. Even in species with relatively high PPI coverage (e.g., yeast), there were many proteins that did not occur in the PPI network. To ensure that these proteins were included in the functional ortholog lists, we added singleton (disconnected) nodes corresponding to each such protein in the respective PPI networks, thus using only sequence data.

Global Alignment of Yeast, Fly, Worm, Human and Mouse networks: When performing the alignment, we chose the following parameter settings: $\alpha = 0.6, r = 5, \beta_1 = 0.1, \beta_2 = 0.1$. These settings correspond to the node mapping with the best functional coherence (see Sec. 4.1).

We analyzed the common subgraph implied by the multiple alignment. The common subgraph has 1663 edges that are supported by edges in at least two PPI networks and 157 edges that are supported by atleast three networks. There are very few edges with support from four or more species; however, this is not surprising since the worm and mouse networks are very small. The size of the common subgraph is relatively small (only about 5% of human PPI network). One reason for the small overlap between the PPI networks, we believe, is that the current PPI data is both incomplete and noisy. As the quality and quantity of data improves, this overlap should increase further. Even with this incomplete data, we believe that the currently computed (partial) set of node-pairings is robust. In previous work[17], we have performed experiments which suggest that the eigenvalue formulation is robust to errors in PPI data, especially when sequence data is provided.

A naive approach to multiple network alignment would use current sequence-based orthology predictions to perform the mapping; however, by incorporating both sequence and network data, our algorithm performs much better. The common subgraph implied by Homologene's sequence-only mapping contains only 509 edges with support in two or more species and 40 edges with support in three or more species. Thus, the addition of network topology in computing the mappings increases the size of the common subgraph by over three-fold (from 509 to 1663). A direct comparison can not be performed against Inparanoid orthology lists because the Inparanoid's pairwise orthology lists can not be used for multiple network alignment. Instead, we evaluated the total number of conserved edges implied by Inparanoid in 10 ($= \binom{5}{2}$) pairwise network alignments. Even though this final number, 1172, likely over-counts some conserved edges, it is significantly less than the number of conserved edges implied by our algorithm.

The common subgraph in the global alignment consists of multiple components, many of which are significantly larger than those from local alignment methods. Also, unlike the latter, these subgraphs correspond to a variety of topologies: linear, complex-like, tree-shaped, etc. Some of them are also enriched in proteins involved in a specific function (see Supp. Info. for details).

4.1. Functional Coherence: Evaluating Orthology Predictions

We propose a method for scoring the quality of an ortholog list (i.e., a list which specifies sets of orthologous proteins across two or more species). The method is motivated by the lack of automated, direct measures of quality of orthology lists. Currently, the most common strategy for comparing two orthology lists is to identify pairs of proteins which are grouped differently under the two lists and perform a manual, case-by-case analysis of some pairs. Because of the manual approach, a comprehensive evaluation can be time-consuming. Recently, Chen *et al.*[6] have described a computational approach where they compare many ortholog lists to identify the list(s) with the best overall agreement with the remaining ones. However, this approach does not measure if the orthology predictions are biologically plausible.

We aim to find a direct, automated measure of ortholog quality by using functional information. The intuition behind our method is simple: given an ortholog list, we select the sets of orthologs that have many proteins with known function. For each set, we collect all the Gene Ontology[1] (GO) terms corresponding to proteins in it. We evaluate if the set is functionally coherent, i.e., if the GO terms describe similar functions. Finally, an aggregate score (across all sets) is computed. Higher scores imply higher coherence, indicating that the ortholog list groups proteins with known function together. In Supp. Info., we describe the algorithm more precisely.

In Supp. Info., we describe some experiments which demonstrate that the functional coherence scoring scheme does capture the desired biological intuition. This scoring scheme allows us to measure how similar the functions of proteins mapped to the same ortholog set are. One potential problem with this approach is that there might not be enough proteins for which GO terms are available to compute such scores. However, for both Homologene and our functional ortholog predictions, there are over 1500 sets of orthologous proteins such that functional information is available for at least 80% of the proteins in the set. We believe that this degree of coverage is sufficient to generate statistically reliable estimates of functional coherence. In Supp. Info., we also describe in greater detail these sets of orthologs: their sizes, group-wise coherence score etc.

Functional Orthologs from Multiple Network Alignment: In this paper, we present the first known set of functional orthologs (FO) across five species: yeast, fly, worm, mouse and human. The FO mapping is simply the node mapping computed by our algorithm (see Supp. Info. for the list of FOs). Of the 86932 proteins from the five species, 59539 (68.5%) of the proteins in our list were matched to atleast one protein in another species (i.e., had at least one FO). In contrast, Homologene has lower coverage, predicting at least one ortholog for only 33434 (38.5%) proteins. Also, the functional coherence of our predicted functional orthologs is comparable with that of Homologene and Inparanoid predictions. The functional coherence scores are: 0.220 (our predictions), 0.223 (Homologene) , and 0.206 (mean score across Inparanoid's pairwise ortholog sets). Homologene's slightly better score may partly be due to its use of data from many species (more than 5). Rather than relying excessively on sequence-score based heuristics, our method uses functional information (from PPI networks) to predict FOs— these scores suggest that our approach is a simpler and better way of capturing functional similarities between proteins. At the same time, our predicted FOs do not deviate drastically from sequence-only predictions: 66% of protein-pairs grouped together by Inparanoid are also grouped together by our approach.

Our predicted FOs have certain limitations. Our approach relies on PPI data to identify functionally related proteins. For many proteins, however, no PPI data is available. In such cases, the algorithm's ability to identify functionally-related sets of proteins may suffer. However, the expected increase in the availability of PPI data should help overcome this limitation.

Case-study: Functional Orthologs of two Human Disease-related Proteins: A key application of this work is in a more comprehensive prediction in of orthologs of human disease-related genes in model organisms. An accurate understanding of which genes in, say, fly are relevant in human diseases would be of significant value in directing scientific work. The human gene DHRC7 has been linked to the Smith-Lemli-Opitz syndrome. Homologene predicts only a mouse homolog for this gene. Our algorithm predicted B0250.9 (from worm), dLBR (from fly) and YNL280C (from yeast) as orthologs. Each of these proteins has been observed to perform a function similar to that of the human gene (sterol reductase). Similarly, B3GN3 is a human gene observed to be differentially expressed in colon cancer. Homologene fails to find a fly homolog of this gene; our algorithm predicts the fly gene *brn* as its homolog. This prediction is supported by the fact that both the proteins are galactosyltransferases.

Another application of the proposed algorithm is to predict a comprehensive human PPI network by combining PPI data from other species. Analysis of the connections of disease-related proteins in this large network may offer improved insights about the disease mechanisms and possible drug targets.

5. Conclusion

In this paper, we focus on the global network alignment problem and present an algorithm for computing the global alignment of multiple protein interaction networks. The algorithm is simple, yet powerful— it provides users the ability to control the relative weights of the sequence and network data in the alignment. Using the algorithm we compute the first known global alignment of PPI networks from five species: yeast, fly, worm, mouse and human. The results provide valuable insights into the conserved functional components across the various species. They also enable us to predict functional orthologs between these five species; the quality of these functional orthologs compares favorably with current sequence-only functional orthologs. Our algorithm also has some parallels with Google's PageRank algorithm, specifically in the construction of eigenvalue problem(s) (see Supp. Info.).

In future work, we intend to more deeply explore the differences and similarities between our predicted functional orthologs and currently used ortholog lists. We also intend to improve the algorithm by exploring better algorithms for k-partite matching. Finally, we plan to explore the application of this algorithm to other biological and non-biological network data.

References

1. *http://www.geneontology.org.*
2. S. Bandyopadhyay, R. Sharan, and T. Ideker. Systematic identification of functional orthologs. *Genome Research*, 16(3):428–435, 2006.
3. B. P. Kelley *et al.* PathBLAST: a tool for alignment of protein interaction networks. *Nucleic Acids Research*, 32:W83–8, 2004.
4. C. Stark *et al.* BioGRID: a general repository for interaction datasets. *Nucleic Acids Research*, 34:D535–9, 2006.
5. D. L. Wheeler *et al.* Database resources of the National Center for Biotechnology Information. *Nucleic acids research*, 35:D5–12, 2007.
6. F. Chen *et al.* Assessing performance of orthology detection strategies applied to eukaryotic genomes. *PLoS ONE*, 2:e383, 2007.
7. G. R. Mishra *et al.* Human Protein Reference Database– 2006 update. *Nucleic Acids Research*, 34:D411–4, 2006.
8. I. Xenarios *et al.* DIP, the Database of Interacting Proteins. *Nucleic Acids Research*, 30(1):303–305, 2002.
9. J. Flannick *et al.* Graemlin: general and robust alignment of multiple large interaction networks. *Genome Research*, 16(9):1169–1181, 2006.
10. M. Kellis *et al.* Methods in comparative genomics. *Journal of Computational Biology*, 11(2-3):319–355, 2004.
11. M. Koyuturk *et al.* Pairwise alignment of protein interaction networks. *Journal of Computational Biology*, 13(2):182–199, 2006.
12. P. Uetz *et al.* From ORFeomes to protein interaction maps in viruses. *Genome Research*, 14(10B):2029–2033, 2004.
13. M. R. Garey and D. S. Johnson. Computers and Intractability. *Freeman*, 1979.
14. G. H. Golub and C. Van Loan. Matrix Computations. *J.H.U. Press*, 2006.
15. T. Ito, T. Chiba, and M. Yoshida. Exploring the protein interactome. *Trends in Biotechnology*, 19(10):S23–7, 2001.
16. M. Remm, C. E. Storm, and E. L. Sonnhammer. Automatic clustering of orthologs and in-paralogs. *J of Mol Bio*, 314(5):1041–1052, 2001.
17. R. Singh, J. Xu, and B. Berger. Pairwise Alignment of Protein Interaction Networks. *Proc. of Conf on Res in Comp Mol Bio, RECOMB*, 2007.
18. L. Page *et al.* PageRank Citation System: Bringing Order to the Web. *Tech Report. Stanford Dig Lib Proj*, 1998.

PREDICTING DNA METHYLATION SUSCEPTIBILITY USING CpG FLANKING SEQUENCES

S. KIM[1,2], M. LI[3], H. PAIK[3], AND K. NEPHEW[3]

[1]*School of Informatics,* [2]*Center for Genomics and Bioinformatics,* [3]*Medical Sciences, Indiana University, Bloomington, IN 47408, USA*
E-mail: {sunkim2,menli,hyupaikknephew}@indiana.edu

H. SHI[4], R. KRAMER[5] AND D. XU[5,6]

[4]*Department of Pathology and Anatomical Sciences, Ellis Fischel Cancer Center,* [5]*Department of Computer Sciences and* [6]*Christopher S. Bond Life Sciences Center, University of Missouri, Columbia, MO 65212, USA*
E-mail: {ShiHu@health.,kramer@,xudong@}missouri.edu

T-H. HUANG[7]

[7]*Human Cancer Genetics, The Ohio State University, Columbus, OH 43210, USA; E-mail: Tim.Huang@osumc.edu*

DNA methylation is a type of chemical modification of DNA that adds a methyl group to DNA at the fifth carbon of the cytosine pyrimidine ring. In normal cells, methylation of CpG dinucleotides is extensively found across the genome. However, specific DNA regions known as the CpG islands, short CpG dinucleotide-rich stretches (500bp - 2000bp), are commonly unmethylated. During tumorigenesis, on the other hand, global de-methylation and CpG island hypermethylation are widely observed. *De novo* hypermethylation at CpG dinucleotides is typically associated with loss of expression of flanking genes, thus it is believed to be an alternative to mutation and deletion in the inactivation of tumor suppressor genes. In this paper, we report that sequences flanking CpG sites can be used for predicting DNA methylation levels. DNA methylation levels were measured by utilizing a new high throughput sequencing technology (454) to sequence bisulfite treated DNA from four types of primary leukemia and lymphoma cells and normal peripheral blood lymphocytes. After measuring methylation levels at each CpG site, we used 30 bp flanking sequences to characterize methylation susceptibility in terms of character compositions and built predictive models for DNA methylation susceptibility, achieving up to 75% prediction accuracy in 10-fold cross validation tests. Our study is first of its kind to build predictive models for methylation susceptibility by utilizing CpG site specific methylation levels.

1. Introduction

DNA methylation, the addition of a methyl group to the fifth carbon of a cytosine residue within the context of a CpG dinucleotide, is the only known epigenetic modification of DNA that can be inherited without changing the DNA sequence [1]. In normal cells, CpG methylation is extensively found across the genome and widely believed to act to silence gene expression and/or retrotransposition of parasitic repeat sequences [2]. However, significantly less CpG methylation is observed within specific regions known as the CpG islands, short CpG dinucleotide-rich stretches (500bp - 2000bp), commonly found within the promoter and first exon of active genes [2].

Patterns of epigenetic modifications that arise during tumorigenesis are quite different from normal cells [3]. These alterations include global hypomethylation of CpG dinucleotides as well as localized hypermethylation at CpG islands [4]. It is now firmly established that CpG island hypermethylation is a powerful mechanism of transcriptional repression in cancer genomes, including silencing of tumor suppressor genes [5]. However, while DNA hypermethylation of several promoter CpG islands is frequently observed in cancers, other CpG island-containing genes remain unaffected by this epigenetic modification [6]. This observation indicates that some CpG island sequences are more susceptible to aberrant methylation, while others remain resistant to alteration by DNA methylation. While the reason for this differential susceptibility to DNA methylation is unknown, recent reports suggest that DNA pattern information may play a key role in distinguishing between methylation-sensitive and -resistant CpG islands. [7,8,9,10,27,26]

A widely used experimental method to measure DNA methylation is Methylation Specific PCR (MSP), a bisulfite conversion based PCR technique [11]. The target DNA is first modified with sodium bisulfite which converts un-methylated cytosine to uracil while methylated cytosine remains 5-methyl cytosine. The technique is accurate, however limited to detecting only highly specific regions of individual genes. Several genome-wide methylation detection methods have been developed in the past few years, such as differential methylation hybridization (DMH) [12] and methylation DNA immunoprecipitation (MeDIP) [13]. Both methods are microarray based experiments. DMH uses methylation specific restriction enzymes and MeDIP uses 5-methyl cytosine antibody to distinguish methylated and unmethylated DNA. These methods can be used for genome-wide methylation detection, allowing researchers to quickly profile methylation pattern alteration. However, technical issues, such as hybridization efficiency for microarrays and antibody efficiency, affect their accuracy. Usually, MSP or bisulfite sequencing is performed to further validate the methylation levels of the genes of interest.

In the past, due to the limitation of sequencing technology, bisulfite sequencing has been performed only for determining specific regions of individual genes. With the development of high throughput sequencing technology, we are now able to perform methylation profiling on genome-wide scale. In particular, the 454 sequencing technology (`454.com`) combines emulsion PCR and pyrosequencing technique and it can determine up to 100Mbp in a single biological experiment with an average read length of 250 bp. The approach produces sequences of a very high quality with an accuracy over 99% [30] and resolution of single 5-methylcytosine, thus highly reliable for profiling methylation patterns. In this paper, we utilized the methylation data measured using the new high throughput sequencing technique to build predictive models for methylation susceptibility.

2. Related work and Motivation

There has recently been a significant research development in predicting DNA methylation susceptibility based on DNA patterns. Feltus et al [24] showed that a classification function based on the frequency of seven sequence patterns was able to discriminate methylation-prone from methylation-resistant CpG island sequences with over 80% accuracy in an experiment designed using over-expressed DNMT1. Feng et al [27] developed a support vector machine classifier for predicting methylation status of CpG islands and showed relationship between nucleotide sequence contents and transcription factor binding sites. Bock et al [26] showed that CpG island methylation in human lymphocytes was highly correlated with DNA sequence, repeats, and predicted DNA structure.

All previous studies used rather coarse-grained methylation information in long DNA regions, rather than CpG site specific information. With such coarse grained information, patterns as short as 3bp were used to build predictive models. Use of such short patterns utilizes frequencies of patterns, not specific patterns, for the predictive models. For example, a pattern of length 4bp, CCGC, is over-represented in unmethylated CpG Islands with a p-value of 5.18×10^{10} in Bock et al [26]. This means that CCGC occurs both in methylation susceptible and resistant sequences, but their occurrence frequencies in susceptible and resistant sequences are significantly different. On the other hand, Handa and Jeltsch's analysis [20] reported that flanking sequences of up to +/-four base-pairs surrounding the central CG site that are characteristic of high (5'-CTTGCGCAAG-3') and low (5'-TGTTCGGTGG-3') levels of methylation in human genomic DNA.

In this paper, we investigated on whether specific DNA sequences, not just their frequencies, can be used for predicting methylation susceptibility. We used CpG flanking sequences with CpG site specific methylation information measured by sequencing bisulfite treated DNA from four types of primary leukemia and lymphoma cells and normal peripheral blood lymphocytes with a new high-throughput sequencing technology (454). This study is first of its kind that uses CpG site specific methylation information to build predictive models.

3. Research Goal

Once we measured methylation level at each CpG site (explained in "Method" Section), we investigated two research questions:

(1) Is there any significant difference in DNA character composition of CpG flanking sequences of methylation susceptible sites and methylation resistant sites?
(2) Is it possible to use the CpG flanking sequence composition to build predictive models for methylation susceptibility?

The first research question is directly motivated by using the high throughput sequencing technique. Indeed, the answer to the first question is positive as shown in Section 6.2. In addition to the significant difference in DNA character composition of CpG flanking sequences, we also observe that CpG sites in the same region of a gene are quite different in terms of methylation level as shown in Figure 5. If methylation levels of multiple CpG sites in the same genomic region are different, we may be able to distinguish methylation susceptible CpG sites from resistant sites using sequence-specific features, especially CpG flanking sequences. Thus the second question of modeling methylation susceptibility using machine learning techniques was investigated in this paper.

4. Data

A massively parallel sequencing (454-sequencing) experiment was designed on 25 gene-related CpG islands in four different tumor types, such as acute lymphoblastic leukemia (ALL), chronic lymphocytic leukemia (CLL), follicular lymphoma (FL), and mantle cell lymphoma (MCL), and normal peripheral blood lymphocytes (PBL) [31]. These 25 genes, listed below, were previously reported to be highly methylated in leukemia and lymphoma.

CYP27B1 (Chr12: 56446851-56447152), *PON3* (Chr7: 94863816-4864092), *KCNK2* (Chr1: 213322021-213322250), *PCDHGA12* (Chr5: 140790654-140790834), *DDX51* (Chr12: 131193754-131194006), *PTPN6* (Chr12: 6930334-6930791), *DAPK* (Chr9: 89302286-89302749), *CDKN2B* (Chr9: 21998564-21998941), *TP53* (Chr17: 7531401-7531767), *CDKN1C* (Chr11: 2863438-2863596), *TRIM36* (Chr5: 114543737-114544155), *ZNF677* (Chr19: 58449649-58450000), *LRP1B* (Chr2: 142604668-142604910), *LHX4* (Chr1: 178469412-178469574), *NKX2-3* (Chr10: 101282587-101282884), *ALDH1L1* (Chr3: 127381582-127381801), *EFNA5* (Chr5: 107035339-107035619), *CCND1* (Chr11: 69165258-69165515), *DLC-1* (Chr8: 13034845-13035136), *TGFB2* (Chr1: 216586512-216586822), *ZNF566* (Chr19: 41671827-41672234), *ADAM12* (Chr10: 128066859-128067044), *MYOD1* (Chr11: 17697405-17697613), *MME* (Chr3: 156280330-156280527), and *MGMT* (Chr10: 131155100-131155259).

Prior to sequencing, bisulfite treatment was performed. Bisulfite treatment converts all unmethylated cytosines to uracil while methylated cytosines remain unaltered after the treatment. Thus, by aligning the sequences of the bisulfite treated DNA and comparing altered/unaltered cytosines, we can measure DNA methylation level. 454 pyro-sequencing on the bisulfite treated DNA is the most accurate method that can be used to measure DNA methylation. As a result, a total of 294,631 sequences was generated with an average read length of 131 bp (range 35-300bp).

From the sequences of bisulfite treated cells, we collected sequences of 30 bases centered around each CpG site and grouped the sequences into two classes: *methylation susceptible site sequences (MS)* and *methylation resistant site sequences (RS)*. See "Method" Section for detail. There were 41 CpG sites, thus 41 sequences in MS. We randomly selected 41 sequences (41 CpG sites) out of 84 sequences (81 CpG sites) in RS and used the sequence sets to build predictive models.

5. Method

5.1. *Estimating methylation level of a CpG site*

The methylation level of a CpG site was estimated by counting the number of `C`'s and `T`'s in two columns that are predominantly either `CG` or `TG`. For example, such columns are highlighted in the alignment in Figure 1.

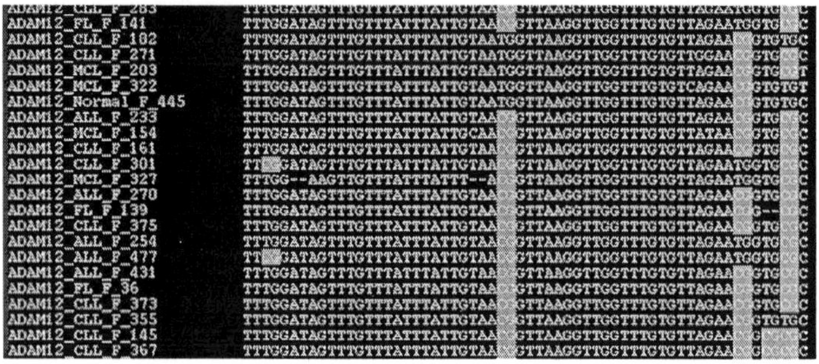

Figure 1. An alignment of bisulfite treated sequences and identification of methylated sites.

Intuitively, counting characters in an alignment of sequences will give us a good estimation of methylation level of a CpG site. There are two problems with this simple approach. First, aligning 2,000 to 3,000 sequences for each sequenced region of 25 genes is very time consuming. Second, even though we use a high performance machine to align the sequences, it is only an estimation of methylation level of a CpG site. In particular, there are sequencing errors which makes the estimation of methylation level of a CpG site more complicated. We used a sequence sampling technique to estimate methylation level as follows:

(1) Sample 20% of sequences in each DNA region. This results in a set of 400 to 500 sequences, which is large enough to estimate methylation level and also can be aligned using ClustalW [22] in a reasonable amount of time.
(2) Then look for two columns where predominantly either CG (methylated) or TG (unmethylated). Estimation of methylation level of a CpG site is computed by counting the number of CG's.

We repeated the sampling task 25 times, so there were 25 estimated methylation level per each CpG site. Then we defined a CpG site as a *methylation susceptible site* when the estimated methylation level $X \geq$

$T_{susceptible}$ with a p-value less than 0.01, assuming X following a Gaussian distribution. We define a CpG site as a *methylation resistant site* when the estimated methylation level $X \leq T_{resistant}$ with a p-value less than 0.01, assuming X following a Gaussian distribution. The two threshold values are set to $T_{susceptible} = 0.5$ and $T_{resistant} = 0.01$.

5.2. *Preparing input data to prediction algorithms*

We collected sequences of 30 bases around each CpG site and grouped the sequences into two classes: methylation susceptible site sequences (MS) and methylation resistant site sequences (RS). An alignment of sequences in MS has 30 columns, each of which becomes an attribute. Each sequence in MS is designated as "UP" class label. We used the WEKA package [23], thus each attribute has four possible values:

```
@attribute C1 {A,T,G,C}
```

and each sequence of 30 bases in MS is represented as

```
T,T,T,A,T,T,T,A,T,T,G,T,A,A,C,G,G,T,T,A,A,G,G,T,T,G,G,T,T,T,UP
```

Sequences in RS were represented in the same way as those in MS, except the class label being designated as "DOWN."

For classification tests, we used four machine learning algorithms in WEKA: SMO that implements John C. Platt's sequential minimal optimization algorithm [21] for training a support vector classifier using polynomial or RBF kernels; IBk-type classifier [17] that is a simple distance measure to find the training instance closest to the given test instance, and predicts the same class as this training instance; Multilayer Perceptron that uses backpropagation to classify instances (all nodes are sigmoid); and a Naive Bayes classifier [18].

5.3. *Character composition analysis*

To analyze character composition of CpG flanking sequences, we compared sequences in MS and RS using Weblogo[19] (`http://weblogo.berkeley.edu`) and the two sample logo [14] (`http://www.twosamplelogo.org/`). Weblogo uses only one sequence input file and compare its character frequencies to a random model, thus we used MS and RS separately and generated two Weblogos. The two sample logo uses two input sequence sets simultaneously, thus the name "two sample", and effectively highlighted character composition difference in MS and RS.

6. Result

6.1. *CpG methylation is not uniform*

Our data analysis of methylation level showed that methylation level was not uniform even in the upstream region of a single gene (see Figure 5 for *CYP27B* gene). Thus we conjectured that there should be biological mechanisms, possibly sequence specific ones, for DNA methylation, which inspired us to investigate into building predictive models for methylation susceptibility.

6.2. *Analysis of character composition*

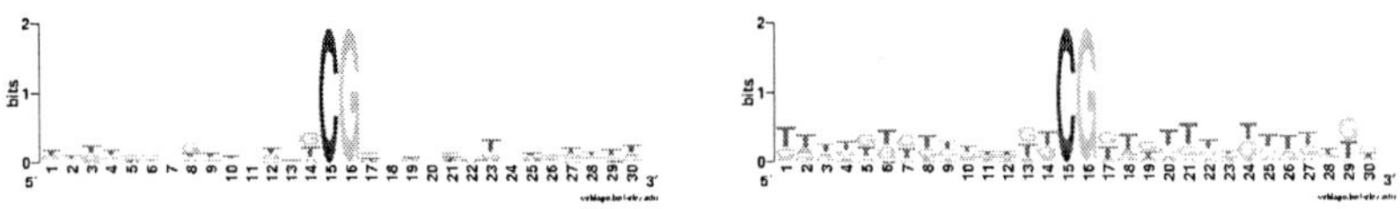

(A) methylation susceptible (B) resistant

Figure 2. Relative entropy of CG flanking sequences using WebLogo. Logos of methylation susceptible sequences and Logos of methylation resistant sequences.

We computed logos (relative entropy with respect to a random model) of CG flanking sequences using WebLogo. As shown in Figure 2, flanking sequences of susceptible CpG sites (MS) are more like random characters while flanking sequences of methylation resistant CpG sites (RS) consistently lack cytosine (C) with adenine (A), guanine (G), and thymine (T) over-represented.

To further investigate how different character composition of sequences in MS and RS, we used the two sample logo technique [14] (`http://www.twosamplelogo.org/`).

The two sample logo in Figure 3 highlights character composition difference clearly: characters in the upper panel for methylation susceptible sequences and characters in the lower panel for methylation resistant sequences. The over-represented characters in the two sample logo analysis agree well with the methylation susceptibility experiments using DNMT1 in Handa and Jeltsch's analysis [20], which reported flanking sequences of up to +/-four base-pairs surrounding the central CG site that were characteristic of high (5'-CTTGCGCAAG-3') and low (5'-TGTTCGGTGG-3') levels of methylation in human genomic DNA. Five positions of the two sample logo (11, 12, 14, 17, and 18) agreed with Handa and Jeltsch's analysis. Only two

(13 and 20 in the two sample logo) out of eight flanking positions disagree with Handa and Jeltsch analysis [20]. However, in Handa's analysis, G is prominent in both methylation susceptible and resistant sequences. In our analysis, T is prominent in methylation susceptible sequences, which may be worth further investigation. In summary, among eight CpG flanking characters, only one position (13 in the two sample logo) disagrees with a published methylation susceptibility analysis.

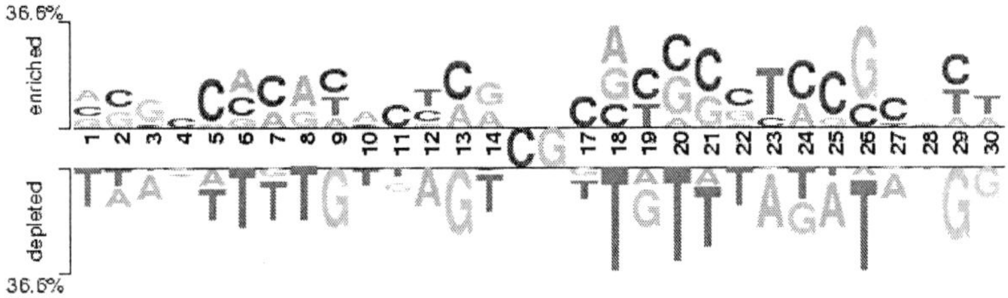

Figure 3. Two sample logo plot showing the DNA character composition difference.

6.3. *Predicitive model for methylation susceptibility*

Since this is to determine either of two classes, UP and DOWN, we used four classification methods in WEKA: SMO, Multilayer Perceptron (MP), Naïve Bayes (NB), Instance Based Classifier (IBk). The prediction accuracy was measured with the standard 10 fold cross validation. As shown in Figure 4, all four algorithms achieved over 70% accuracy with 30bp flanking sequences. We expect that the prediction accuracy drops as the length of the flanking sequences decreases. To measure the effect of flanking sequence length, we used flanking sequences of 30 bp to 2bp with a 2bp decrease in length (one from the left end and the other from the right end). The prediction accuracy of SMO, NB, and MP did not drop much until the flanking sequence length was reduced to 10bp, which agrees with the experimental result in Handa and Jeltsch's analysis.

6.4. *Is this cancer specific?*

Given that methylation levels are measured for four types of primary lymphoma and leukemia cells and normal peripheral blood lymphocytes, it is natural to ask whether there is difference in methylation level between cancer types and normal cells. Our quick, initial analysis was not able to find

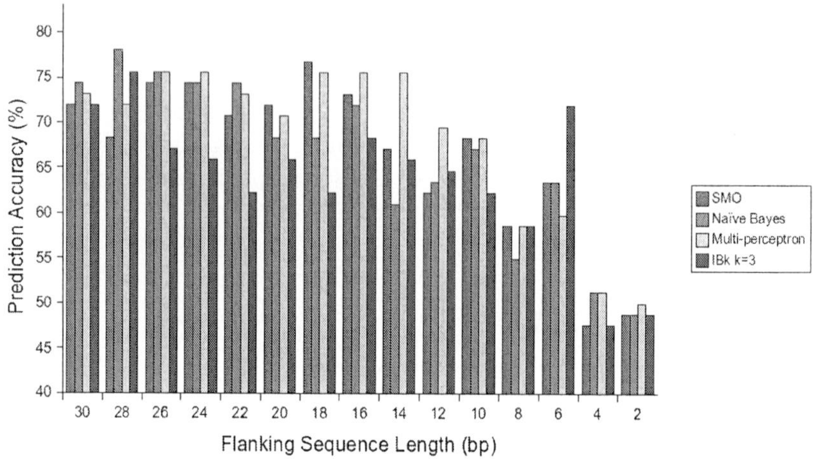

Figure 4. Prediction accuracy of four classification models using 10 fold cross validation. X-axis is the length of CpG flanking sequences.

distinctive methylation patterns between four leukemia cells and one normal cells, as shown in Figure 5 for *CYP27B* gene. In general, we observed that methylation levels in leukemia cell lines were higher than in the normal cell lines, as expected. However, this was not true at all CpG sites. For example, some CpG sites in the upstream region of the *CYP27B* gene showed higher methylation in the normal cell line compared to the cancer cell lines. We plan to investigate this question further with new data sets.

7. Discussion

In this paper, we utilized CpG site specific methylation information to characterize CpG site methylation susceptibility. First, we showed that there was significant difference in DNA character composition between methylation susceptible and resistant sequences. In particular, comparison of methylation susceptible and resistant sequences using the two sample logo technique showed that over-represented characters in methylation susceptible sequences are in agreement with the analysis by Handa and Jeltsch showing CpG flanking sequence specificity for methylation susceptibility. Secondly, we used the CpG flanking sequences to build predictive models for methylation susceptibility and achieved over 75% prediction accuracy in 10 fold cross validation tests. This study is first of its kind that uses CpG site specific methylation information to build predictive models.

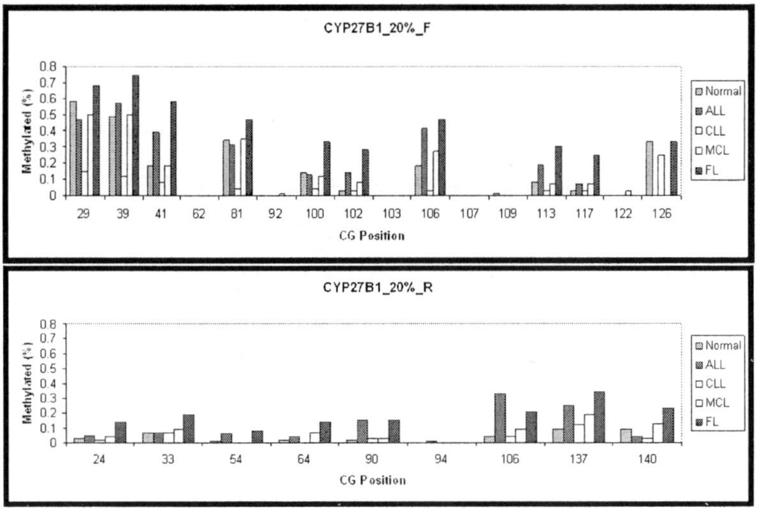

Figure 5. Methylation level of each CpG sites in the upstream region of *CYP27B* gene. Two fragments, Forward (upper panel) and Reverse (lower panel), were sequenced after bisulfite treatment.

Further study includes characterization of leukemia specific methylation pattern signatures and related sequence and machine learning analysis.

Acknowledgement

This work is supported by the National Cancer Institute grants U54 CA11300 and R01 CA85289.

References

1. Jaenisch, R. and Bird, A. *Nat Genet*, 33 Suppl: 245-254, 2003.
2. Feinberg AP, Tycko B *Nature Reviews Cancer* 2004 4: 143-153
3. Herman JG, Baylin SB *New England Journal of Medicine* 2003 349:2042-54
4. Toyota M, Issa JP *Semin Oncol* 2005 32:521-30
5. Nephew KP, Huang TH *Cancer Letters* 2003 190:125-33
6. Jones P A *Semin Hematol*, 2005 42: S3-8, 2005.
7. Keshet I, Schlesinger Y, Farkash S, Rand E, Hecht M, et al. *Nat Genet* 2006 38: 149-153.
8. Goh L, Murphy SK, Muhkerjee S, Furey TS *Bioinformatics* 2007 23: 281-288.
9. Fang F, Fan S, Zhang X, Zhang MQ *Bioinformatics* 2006 22: 2204-2209.
10. Bock C, Walter J, Paulsen M, Lengauer T *PLoS Comput Biol* 2007 3: e110.
11. Herman JG, Graff JR, Myohanen S, Nelkin BD, Baylin S *Proc Natl Acad Sci U S A* 1996 Sep 3;93(18):9821-6

12. Yan PS, Chen CM, Shi H, Rahmatpanah F, Wei SH, Caldwell CW, Huang TH. *Cancer Res* 2001; 61:8375-80.
13. Weber M, Davies JJ, Wittig D, Oakeley EJ, Haase M, Lam WL, Schubeler D. *Nat Genet* 2005; 37:853-62.
14. Vacic V., Iakoucheva L.M., and Radivojac P. *Bioinformatics*, 22(12): 1536-1537. 2006
15. Taylor K.H., Kramer R.S., Davis J.W., Xu D., Caldwell1 C.W., and Shi H., *Cancer Research*, in press. 2007
16. Zheng Z. and Webb G., *Machine Learning*, 41(1): 53-84. 2000
17. Aha D. and Kibler D., *Machine Learning*, 6:37-66. 1991
18. John G.H. and Langley P. *Proceedings of the Eleventh Conference on Uncertainty in Artificial Intelligence.* pp. 338-345. 1995
19. Crooks G.E., Hon G., Chandonia J.M., Brenner S.E. *Genome Res.*, 14:1188-1190, 2004
20. Handa V. and Jeltsch A. *J Mol Biol.* 348(5):1103-12. 2005
21. Platt J. *Advances in Kernel Methods - Support Vector Learning.* 1998
22. Chenna R., Sugawara H., Koike T., Lopez R., Gibson T.J., Higgins D.G., Thompson J.D. *Nucleic Acids Res.* 31(13):3497-500. 2003
23. Witten I. H. and Frank E. *Data Mining: Practical Machine Learning Tools and Techniques* Morgan Kaufmann, San Francisco, 2 edition, 2005.
24. Feltus F.A., Lee E.K., Costello J.F., Plass C. and Vertino P.M., *Proc Natl Acad Sci U S A* 100(21):12253-8. 2003.
25. Feltus F.A., Lee E.K., Costello J.F., Plass C., Vertino P.M., *Genomics* 87(5):572-9. 2006.
26. Bock C., Paulsen M., Tierling S., Mikeska T., Lengauer T., Walter J., *PLoS Genet.* **2(3)** e26, 2006.
27. Fang F., Fan S., Zhang X., Zhang M.Q., *Bioinformatics.* 22(18):2204-9. 2006.
28. Das R., Dimitrova N., Xuan Z., Rollins R.A., Haghighi F., Edwards J.R., Ju J., Bestor T.H., Zhang M.Q., *Proc Natl Acad Sci U S A.* 103(28):10713-6. 2006.
29. Goh L., Murphy S.K., Muhkerjee S., Furey T.S., *Bioinformatics.* 23(3):281-8, 2006.
30. Margulies, M. Eghold, M. et al. *Nature* 2005 Sep 15; 437(7057):326-7
31. Taylor KH, Kramer RS, Davis JW, Guo J, Duff DJ, Xu D, Caldwell CW, Shi H. *Cancer Res* 2007 67: 8511-8518

MULTISCALE MODELING AND SIMULATION SESSION: FROM MOLECULES TO CELLS TO ORGANISMS

JUNG-CHI LIAO
JEFF REINBOLT

Simbios, NIH Center for Biomedical Computing
Stanford University
Palo Alto, CA, 94305

ROY KERCKHOFFS
ANUSHKA MICHAILOVA

Cardiac Mechanics Research Group
University of California San Diego
La Jolla, CA 92093

PETER ARZBERGER

National Biomedical Computation Resource
University of California San Diego
La Jolla, CA 92093

1. Session Background and Motivation

Today's biology and biomedical research is increasing inundated with data, based in part on improved instrumentation and techniques that automate the data capture and storage. Genomics and proteomics efforts are producing data at an increasing rate, the data are more of a descriptive nature and do not provide information on functional and structural integration and interactions of the parts. Improved microscopy and imaging, along with other types of instrumentation, are also producing data on components of larger living system. But to better understand human physiology and to allow for predictive capabilities of disease prevention and treatment it is crucial to develop multiscale models and simulation systems that can operate at and across various scales, across the length scale from nanometers for molecules to meters for human bodies, as well as across time scale from nano-seconds for molecular interactions to minutes, hours and years for human life. For example, new experimental techniques are able to observe the cellular dynamics and the localization of intracellular proteins at the same time. A multiscale modeling approach incorporating molecular and cellular mechanisms is needed to explain these data.

A number of workshops and panel recommendations have recently recognized and addressed the importance and the difficulties in interpreting experimental results that are cross-scale in space, time and state. The multiscale modeling consortium, along with the participation of the Interagency Modeling and Analysis Group (IMAG) of a number of federal agencies that include the National Institutes of Health (NIH), National Science Foundation (NSF), National Aviation and Space Agency (NASA), Department of Energy, (DOE), Department of Defense (DoD) and the United States Department of Agriculture (USDA), aims to promote the development and exchange of tools, models, data and standards for the MultiScale Modeling (MSM) community. The NSF blue ribbon panel recognizes that Simulation Based Engineering Science (SBES) [1] applied to the multiscale study of biological systems and clinical medicine, or simulation based medicine, may bring us closer to the realization of P4 medicine (predictive, preventative, personalized, and participatory). The June 2005 President's Information Technology Advisory Committee (PITAC) report on computational science [2] and the 2005 National Research Council report [3], both specifically recommended increased and sustained support for infrastructure development to meet the computational challenges ahead.

2. Session Summary

This session includes an invited talk, six reviewed oral presentation, three additional accepted papers, and an associated tutorial. We structured the session to include work at various levels (molecular, cellular, tissue, organ, and whole body) and across these levels. The presentations reflect a variety of approaches, from molecular dynamics, numerical analysis, mesh generation, Markov Chain modeling, and use of ontologies. The variety of approaches will continue developing and will bring together new data types and models. This will result in a better understanding of the relationship between scales of biology, and ultimately will enhance biological understanding. The area of multiscale modeling is rich indeed, and will provide challenges for years to come.

The invited talk by **Kamm *et al.*** discusses the role that "mechanical signaling" plays in regulating biological function in health and disease. His approach considers how forces are transmitted through the various load-bearing structures within the cell and how these forces act to create conformational change in critical signaling proteins or protein complexes. He considers a hierarchical structure of transmission of stress across the membrane, and uses a variety of simulation methods (molecular dynamics, to Brownian dynamics to finite element methods) to predict the distribution of stress and their consequences. Experimental data on specific systems (models that focus on forces acting through the focal adhesion complex and transmitted throughout the actin cytoskeleton) are used as input or to validate the computational models.

Two papers focus on either predicting protein structure or identifying functional sites in proteins. **Glazer *et al.*** describe an approach to help identify function of the increasing number of solved protein structure with unknown function, produced by the structural genomics initiative. In particular they combine molecular dynamics simulation (to produce a variety of snapshots of the protein) along with their previously described machine-learning algorithm FEATURE can provide an improvement in the recognition of functional sites in proteins. Treating the molecules as dynamic entities improves the ability of structure-based function prediction methods to annotate possible function sites.

Li and Goddard a novel method to predict transmembrane proteins structures in a heuristic fashion. They focus on the G protein – coupled receptors (GPCRs) system, which is important since it mediates the sense of vision, smell, taste and pain, and they are involved in cell recognition and communication processes. The authors developed a first principles methods for predicting structures and functions of GPCR and apply these methods to two receptor systems. Their approach is to model the entire GPCR with molecular dynamics and then use this model of the receptor to design a target ligand.

Four papers present models of cardiac cell function and structure, or organ behavior. **Deremigio *et al.*** presents a Markov Chain model for coupled intracellular Ca^{2+} channel modeling. This specific approach uses a Kronecker structure representation for the Ca^{2+} release site models. The authors show that the Kronecker structured representation can take advantage of a number of off-the-shelf solvers. The paper provides benchmarks using numerical iterative solution techniques and shows that convergence can be much faster than the traditional methods of Monte Carlo simulation.

Rice *et al.* explore widely used approach to computational modeling of the cardiac muscle contraction (global feedback on Ca^{2+} binding affinity). The results suggest that this approach produces hysteresis in the steady-state force-Ca^{2+} responses when sufficient positive feedback is employed to replicate the steep Ca^{2+} sensitivity found in real muscle. The resulting hysteresis is quite pronounced and disagrees with experimental characterizations in cardiac muscle that generally show little if any hysteresis. This result will impact future modeling in this area.

Lumens *et al.* present a lumped model (CircAdapt model) of the closed-loop cardiovascular system that includes ventricular interaction. Direct ventricular interaction via the interventricular septum plays an important role in ventricular hemodynamics and mechanics. The authors describe the model and the inclusion of septum geometry. They then compare their model output with experimental data. They are able to realistically represent the left ventricle right ventricle interaction through the septum. This allows the authors to claim the broader applicability of physiological application of the CircAdapt model.

Sachse *et al.* describe an approach to develop anatomical models of the cardiac cell, with input the confocal imaging of living ventricular myocytes with sub-micrometer resolution. The method includes generation of dense triangular surface meshes representing the sub-cellular structures (transverse tubular systems). The modeling approach can be applied to computational studies of the cell and sub-cellular physical behavior and physiology and is more broadly applicable to cardiac function associated with changes in anatomy or protein distribution.

Three other papers were accepted to the session as part of the final publication of the meeting. The paper by **Agarwal and Roychowdhury** introduce concepts from electronic circuit design, namely automated nonlinear phase macromodel extraction techniques, to model circadian rhythm in both mammalian and Drosophila systems. It has been shown that such "PPV" (Perturbation Projection Vector) phase macromodels are able to accurately capture the gamut of phase/frequency related dynamics of oscillators. These techniques provide fast/accurate simulations of oscillator systems, predicting synchronization and resetting in circadian rhythms via injection locking cued by light inputs. In addition, PPV waveforms provide direct insight into the effect of light on phases of the oscillating rhythms.

The paper by **Gennari *et al.*** describes an approach to merge computational models of sub-systems into larger integrated models. The approach is illustrated by combining three independently-coded models of overlapping parts of the cardiovascular regulatory system. They demonstrate approach to annotating these models with ontologies that enables the merging of the three models into a multi-scale model that can answer questions beyond the scope of any single model.

Finally, a paper by **Rader and Harrell** present a method of classifying protein structures based on a simple model of the protein's dynamics, using correlation analysis of mode shapes as the guideline of protein classification. The approach is based on a Gaussian Network Model (GNM). The method is based on a coarse-grained model, where each protein residue is represented by a single point (at the residue's alpha-carbon), and residue centers within 7 angstroms are connect by a harmonic potential. The proteins are classified based on similarity of the low frequency eigenvectors of the model's harmonic interaction matrix.

Acknowledgments

The session organizers would like to thank the set of reviewers for their efforts in evaluating the many papers that were submitted. We also acknowledge the assistance of many individuals who helped support the notion of holding a special session on Multiscale Modeling. The previous special session in this area was held in 1999, with a session on Computer Modeling in Physiology: From Cell to Tissue. We felt it was time to re-engage the participants of this symposium with a broader set of modern biological and biomedical challenges. We hope that it will not be another 8 years before the richness of this area is seen again at the Pacific Symposium on Biocomputing.

Finally, the authors wish to acknowledge the support from their individual NIH centers on computational issues in multiscale modeling and simulation, the National Biomedical Computation Resource (NBCR, RR 08605) led from the University of California San Diego and National NIH Center for Physics-Based Simulation of Biological Structures (Simbios) at Stanford.

References

1. Oden JT, Belytschko T, Fish J, Hughes TJ, Johnson C, et al. (2006) Revolutionizing Engineering Science through Simulation
2. Computational Science: Ensuring America's Competitiveness. In: Committee PsITA, editor. National Coordination Office for Networking and Information Technology Research and Development.
3. Catalyzing Inquiry at the Interface of Computing and Biology. In: Wooley JC, Lin HS, editors. National Research Council, National Academy Press 2005.

COMBINING MOLECULAR DYNAMICS AND MACHINE LEARNING TO IMPROVE PROTEIN FUNCTION RECOGNITION

DARIYA S. GLAZER

Department of Genetics, Stanford University, 318 Campus Drive Clark Center S240 Stanford, CA 94305, USA

RANDALL J. RADMER

SIMBIOS National Center, 318 Campus Drive Clark Center S231 Stanford, CA 94305, USA

RUSS B. ALTMAN

Departments of Bioengineering and Genetics, Stanford University 318 Campus Drive Clark Center S170, Stanford, CA 94305, USA

As structural genomics efforts succeed in solving protein structures with novel folds, the number of proteins with known structures but unknown functions increases. Although experimental assays can determine the functions of some of these molecules, they can be expensive and time consuming. Computational approaches can assist in identifying potential functions of these molecules. Possible functions can be predicted based on sequence similarity, genomic context, expression patterns, structure similarity, and combinations of these. We investigated whether simulations of protein dynamics can expose functional sites that are not apparent to the structure-based function prediction methods in static crystal structures. Focusing on Ca^{2+} binding, we coupled a machine learning tool that recognizes functional sites, FEATURE, with Molecular Dynamics (MD) simulations. Treating molecules as dynamic entities can improve the ability of structure-based function prediction methods to annotate possible functional sites.

1. Introduction:

The problem of function prediction is central to bioinformatics. Recently, the number of approaches dealing with solving this problem has increased dramatically. Some methods use interaction data collected from genomic and microarray experiments [1]. Sequence based approaches use sequence conservation through analysis of related sequences [2]. Some methods recognize sequence motifs, compiled into databases such as PROSITE [3] and PRINTS [4], and analyze their patterns of co-occurrence in the related sequences. There are also aggregate approaches that apply several methods at once to examine a given structure [5, 6].

Structural genomics efforts specifically attempt to solve the structures of proteins with novel folds [7, 8]. As such, there is a pressing need for reliable structure-based function prediction methods. These methods rely on conserved geometric context of sites of interest. They range from global, identifying possible substrate binding pockets, to local, concentrating on the particular atoms coordinating ligand binding [9-12]. Because three-dimensional (3D) environment is more conserved than sequence, these methods may have more success when sequence similarity is too low to detect 1D motifs or even overall similarity.

Methods that identify calcium (Ca^{2+}) binding sites in protein structures have had variable success. These methods employed artificial neural networks [13], graph theory and geometric similarity [14], bond-valence calculations [15, 16] and distribution of distances between C_α atoms of residues [17]. We have previously described FEATURE [18], a machine learning algorithm which employs Bayesian scoring scheme, and its ability to identify Ca^{2+} binding sites [19]. FEATURE uses models built by examination of local physico-chemical environments to predict whether a site of interest has a potential for a particular function of interest. The chief advantages of FEATURE are its generality, extending to many types of sites [18], and its focus on 3D environments which allows it to recognize divergent binding sites, without depending on sequence or structure similarity.

Until now, FEATURE has only been applied to static structures. However, protein molecules are dynamic and examining their behavior over time may improve the performance of structure-based function prediction methods. Molecular Dynamics (MD) uses physical principles to simulate the motion of protein molecules, and has been applied for many purposes, including structure refinement, drug docking, protein engineering, and protein folding [8,9]. We propose a novel application of MD simulations: generating structural diversity in order to improve our ability to detect functional sites. For a single protein, there can now be many structural examples that can reveal its functions. In order to test this idea, we asked whether MD simulations coupled with FEATURE analysis could reproduce and further improve performance of FEATURE alone. We present our preliminary results on a single protein, parvalbumin β.

2. Methods:

2.1 ***Structures:*** From the Protein Data Bank we obtained two structures for parvalbumin β (10) (PDB IDs 1B9A and 1B8C). Since we were only interested in the monomeric protein structures and associated ions we used only the coordinates of the structure and Ca^{2+} atoms for 1B9A (HOLO) and the coordinates of the first monomer and associated Mg^{2+} atom for 1B8C (APO).

2.2 *Molecular Dynamics Simulations:* Using software suit GROMACS version 3.3.1 [20], we set up two simulation systems – one for each structure. In case of 1B9A, the structure was solvated in 4,411 simple point charge (SPC) [21] water molecules. 1B8C structure was solvated in 4,479 SPC water molecules. The solvent buffer zone comprised 1nm. Addition of one Na^+ atom and three Na^+ atoms neutralized the 1B9A and 1B8C systems, respectively.

Each simulation started with a 200-step energy minimization run using the steepest descent algorithm and was followed by a 10 picosecond (ps) simulation with harmonic position restraints applied to all protein atoms to allow relaxation of the solvent molecules and added ions. The use of LINCS [22] algorithm and GROMOS96 [23] force field (GROMOS96 43a1) allowed for a 2 femtosecond (fs) integration time step. The systems were coupled to external temperature baths [24] of 300K with a coupling constant tau_T = 0.1ps separately for the protein and the solvent with added ions. Electrostatic and Van der Waals interactions and neighborlist cut-offs were set at 1nm. Finally, each of the systems underwent free dynamic simulations for 1 nanosecond (ns), at constant temperature, as above, and constant pressure, kept at 1 bar by weak coupling to pressure baths [24] with tau_P = 0.5ps. We obtained snapshots of the simulations every 2.5ps, 401 total for each simulation, to examine the generated structures further with FEATURE.

RMS fluctuations per residue were calculated by averaging the atomic RMS fluctuations of each residue, as determined by GROMACS.

2.3 *FEATURE Scanning:* Using FEATURE [18] version 1.9 and Ca^{2+} binding site model [19] we analyzed original PDB structures and those generated by the simulations in two ways: using a 6 x 6 x 6 $Å^3$ grid with point density of 1Å (local grid) placed directly over the Ca^{2+} binding sites, and a grid that encompassed the whole protein with point density of 1Å (global grid). The local grid was centered on an equivalent atom in each of the structures obtained from the simulations, identified as the closest atom to the Ca^{2+} of interest in the HOLO PDB structure file. For both 1B9A and 1B8C structures this atom was defined as ASP_{90} $OD1_{675}$ for EF loop binding site and PHE_{57} O_{407} for CD loop binding site. At each grid point, FEATURE generated a score, representing the likelihood of there being a potential Ca^{2+} binding site centered at that point. The threshold of the Ca^{2+} binding model is 50, therefore, any point scoring 50 or above is considered a plausible center for a Ca^{2+} binding site. From the local grid scanning, only the coordinates of the highest scoring point were kept for further visualization and analysis. Coordinates of all the points scoring above model threshold in the global grid scanning were kept for further visualization.

For negative controls, local grids were centered on two points located far away from the true Ca^{2+} binding sites in the HOLO structure with the following criteria: the atom density of the environment near the points and the distance of the points from the surface were comparable to the local grid centers of the true Ca^{2+} binding sites. These points were LEU_{15} N_{93} and ASP_{24} C_{295}. Coordinates for the top scoring points were kept, just as in the local grid scanning of Ca^{2+} binding sites.

2.4 ***Viewing of the Structures:*** Visual Molecular Dynamics (VMD) [25] allowed visual analysis of simulation trajectories and illustrations of the structures and results of FEATURE scanning. Molecular images were generated using Tahyon Parallel/Multiprocess Ray Tracer System (Fig. 1,2,4) [26, 27].

3. Results:

3.1 ***Molecular Dynamics Simulations:*** On Dual Core AMD Opteron™ Processor 880, 2.4 GHz, MD simulations of the HOLO system took 8.698 hours and of the APO system took 8.768 hours. Analysis of potential and kinetic energies of the systems over the course of the simulations revealed that both systems were stable throughout. Backbone root mean square deviation (RMSD) to the original crystal structure continued to increase slowly over the course of the simulation for 1B9A, reaching a maximum value of 4.5Å at the end of 1ns. In the case of 1B8C, backbone RMSD to the original crystal structure seemed to stabilize around 2.5Å towards the end of 1ns. Average RMSD between the consecutive structures generated by the simulations every 2.5ps was ~0.5Å for both systems, demonstrating that the systems were evolving slowly over time.

3.2 ***FEATURE Scanning:*** On Dual Core AMD Opteron™ Processor 880, 2.4 GHz, local grid analysis took 7.117 minutes for HOLO and 7.167 minutes for APO, while global grid analysis took 3.225 hours for HOLO and 3.942 hours for APO. With the local grid scanning approach, points located in the vicinity of the Ca^{2+} binding sites present in the original HOLO and original APO structures, respectively, scored above the model threshold of 50. HOLO EF loop and APO EF loop binding sites showed persistent Ca^{2+} binding conformations over the course of MD simulation, obtaining scores of 50 or greater at several different time points. The local grid scan detected the Ca^{2+} binding signal in the HOLO CD loop binding site in the beginning of the simulation, but over time the signal was lost. At numerous time points, some conformations achieved higher scores than the original starting structures. Negative controls had scores clustered around zero, never exceeding fifteen, while showing similar extent of structural diversity and sampling (Fig. 2).

FEATURE correctly identified all of the Ca^{2+} binding sites expected in the HOLO and APO structures in the global grid scanning analysis approach. Figure

4 shows all the points which scored over the model threshold of 50 in the structures at all of the examined time points of the simulations. The structures from all the time points were aligned globally to minimize the total RMSD among them and only the starting structures of the HOLO and APO 1ns simulations are shown.

4. Discussion:

Parvalbumin β is a small, 10-12kDa member of EF-hand Ca^{2+} binding proteins. Structurally, it consists of 108 residues, which form six α-helices and intermittent loops. There are two Ca^{2+} binding sites in wild type parvalbumin β: one located in the loop between helices C and D (CD loop binding site) and one located in the loop between helices E and F (EF loop binding site). Several wild type and mutant structures exist for parvalbumin β in the Protein Data Bank (PDB). We used two structures taken from PDB entries 1B9A and 1B8C [28]. These two structures are not of wild type parvalbumin, but share two mutations: F102W and D51A. The first mutation allows experimental monitoring of metal-ion presence by following the magnitude of tryptophan fluorescence upon metal binding. This mutation did not affect metal binding properties of the molecule. The second mutation was introduced to study the cooperative effects of the CD loop on the EF loop binding site. Interestingly, this mutation did not change the binding properties of the EF loop site, but severely decreased Ca^{2+} affinity for the CD loop site. Nevertheless, Ca^{2+} binding was observed in both of these sites in the original 1B9A structure [28].

There are several differences between the 1B9A and 1B8C structures. Firstly, 1B9A crystallized with Ca^{2+} present in the binding sites (HOLO), while 1B8C crystallized without Ca^{2+} in the binding sites (APO). Secondly, the 1B8C original structure contains a third mutation, E101D, introduced to study the role of the glutamic acid at the last coordinating position of the Ca^{2+} binding loop. This mutation stabilizes EF loop, since the side chain of aspartate is shorter and less mobile than the side chain of glutamate. This mutation had similar effects on the two Ca^{2+} binding sites in the molecule, as measured experimentally. The Ca^{2+} affinity of EF loop site was reduced 10-fold. The affinity of the CD loop site was also reduced, such that in combination with the D51A mutation, Ca^{2+} binding was no longer observed at this site, even when crystallization media did contain Ca^{2+} [28]. Therefore, while HOLO structure retains the two Ca^{2+} binding sites present in the wild type parvalbumin, only one Ca^{2+} binding site remains in the APO structure.

MD simulations provided conformations alternative to the original HOLO and APO structures, allowing to generate conclusions on the basis of a set of snapshot structures for HOLO and APO systems, rather than on a single static structure. The structures generated by the 1ns simulation seemed to be sufficiently diverse for the purposes of this study. The backbone RMS deviations over 1ns were no more than 4.5 Å, and were sufficient to find conformations that achieved a varied range of scores. Figure 1 presents the structural diversity sampled by the MD simulations, as given by structures created by the simulations at the 401 analyzed time points.

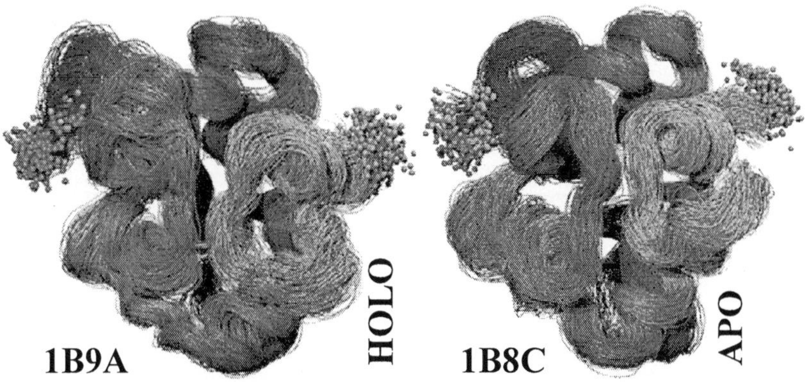

Figure 1: Visited structural conformations over the course of the MD simulations for the HOLO and APO systems. Grey balls represent the first and the last atoms in each of the chains. Greater flexibility of the HOLO structure is noticeable.

FEATURE examines the local environment of functional sites, compares it to a set of non-sites, and builds a model to identify the same functional sites in other structures. Using a Ca^{2+} binding site model [19], FEATURE correctly identified all of the Ca^{2+} binding sites present and expected in the original static HOLO and APO structures. Although the original training set for this Ca^{2+} binding site model contained other parvalbumin structures, these formed a small proportion of the total number of structures used to create the general model. Therefore, this model does not specifically bias results towards Ca^{2+} binding sites of parvalbumin. As such, this HOLO-APO pair formed a good test case for determining whether simulations of molecular dynamics can be applied to structure based function prediction.

The coupling of FEATURE scanning to structures generated by the MD simulations is promising. First, using local grid scanning, we observed local structural conformations in the vicinity of Ca^{2+} binding sites that were both high scoring and low scoring for Ca^{2+} binding (Fig. 2). Such dynamic FEATURE score profiles may be useful in assessing the presence, stability and persistence of Ca^{2+} binding sites. For the HOLO structures, local grid scanning revealed that EF loop binding site was persistent and stable (Fig. 2a). The scores repeatedly

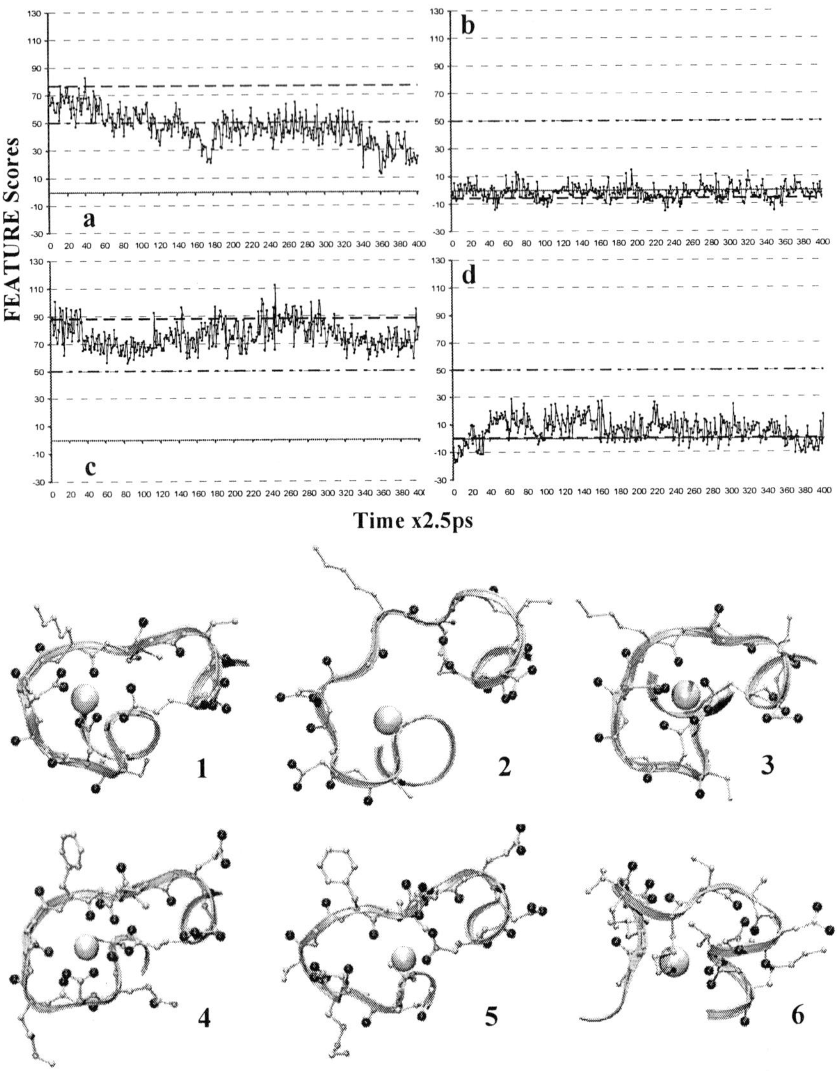

Figure 2: FEATURE scores profiles of local grid analysis and associated structures. 1ns Simulation: ——— . Original FEATURE Score: — • • — • . Model Threshold: — — —. **a**) 1B9A EF loop binding site. **b**) 1B9A $ASP_{24}C_{295}$, negative control. **c**) 1B8C EF loop binding site, positive control. **d**) 1B8C CD loop binding site, which was abolished by mutations. In the structures, location of the binding site is made obvious by the close association of oxygens, black balls, whether terminal ones from ASPs and GLUs or main-chains ones, near the highest-scoring points, white balls. **1**) 1B9A EF loop binding site in the original static structure, which scores ~77. **2**) 1B9A EF loop binding site at 445ps, which scores ~20. **3**) 1B8C EF loop binding site at 615ps, which scores ~112. **4**) 1B9A CD loop binding site at 184ps, which scores ~52. **5**) 1B8C CD loop binding site, abolished by the mutations, at 187.5ps, which scores ~13.5. **6**) 1B9A negative control centered on $ASP_{24}C_{295}$ at 805ps, which scores ~13.5.

surpassed the model threshold of 50. CD loop showed good Ca^{2+} binding site signal in the beginning of the simulation, but over time that signal was lost, due to structural rearrangements within the loop. The APO structures yielded similar results. The EF loop binding site exhibited stability and persistence even more than in the HOLO case (Fig. 2c). In essence, this result could be considered a positive control: an environment created and shown experimentally to be more stable obtained higher FEATURE scores. On the other hand, the CD loop binding site did not at any point in the simulation attain favorable Ca^{2+} binding conformations (Fig. 2d). This site, in essence, could be considered as a negative control, since based on experimental evidence, there was no Ca^{2+} binding site to be found in the CD loop of the APO structure.

In order to understand how the E101D mutation present in the APO molecule stimulates a difference in structural behaviors of the HOLO and APO systems, we examined RMS fluctuations per residue over the course of the two simulations (Fig. 3). The two grey rectangles on the *Residue Number* axis mark where the two Ca^{2+} binding sites are in the sequence. The small lightning bolt symbol points to the location of the concerned mutation. It is interesting to note that the change from glutamine to aspartate at position 101 dampens

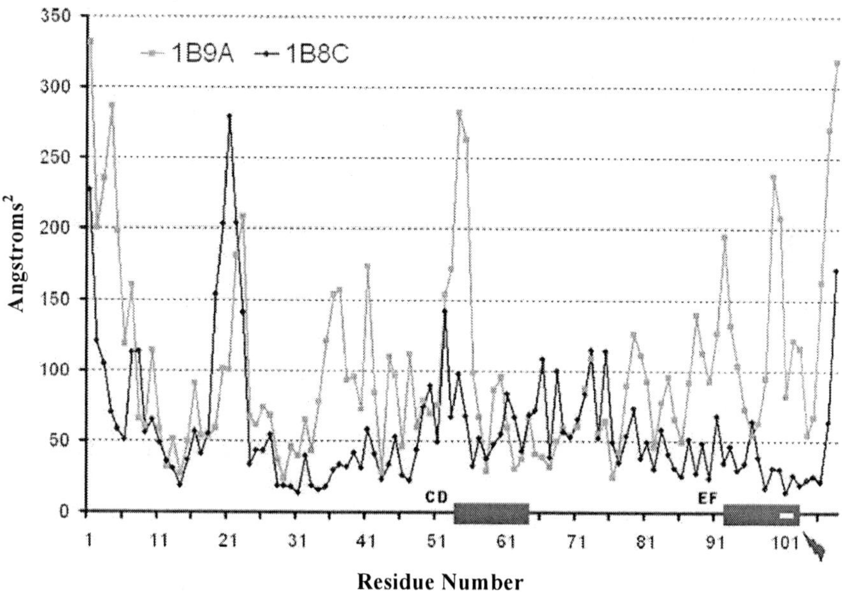

Figure 3: RMS fluctuations per residue over the course of the simulations for both HOLO and APO systems. Grey boxes along the *Residue Number* axis depict location of CD loop and EF loop Ca^{2+} biding sites. White rectangle in the EF loop grey box as well as the small lightning bolt depict the position of the mutation that abolishes CD loop binding site in the APO structure.

dramatically not only the flexibility of the EF loop surrounding this residue, as expected, but also the flexibility of the CD loop and its immediate upstream surroundings. Such a reduction in the possible motion space may be the reason why CD loop binding site is abolished in the APO molecule – it may require greater structural fluctuations, such as the ones observed in the HOLO case, to form properly.

In order to explore further the potential of MD simulations coupled with FEATURE analysis to discern true positive Ca^{2+} binding sites, we performed local grid scanning analysis centered on two other points in the HOLO structure. These points were chosen to contain in their environment the same number of atoms as FEATURE sees at the centers of the local grids in the true Ca^{2+} binding sites in HOLO and APO structures. Additionally, these points resided as close to the surface of the protein, as the centers of the grids chosen for the true Ca^{2+} binding sites, and far away from the true Ca^{2+} binding sites, to avoid aberrant influences. FEATURE score profiles of these negative sites showed that the frequency and magnitude of the structural changes distinguished by FEATURE were comparable to the magnitude observed in the true Ca^{2+} binding sites. However, the scores at these sites were much lower (Fig. 2b). The scores observed for LEU_{15} N_{93} site were even lower than the ones observed for ASP_{24} C_{295}. As such, the FEATURE scores profiles of the negative controls underscored the lack of Ca^{2+} binding site in the CD loop of 1B8C (Fig. 2d).

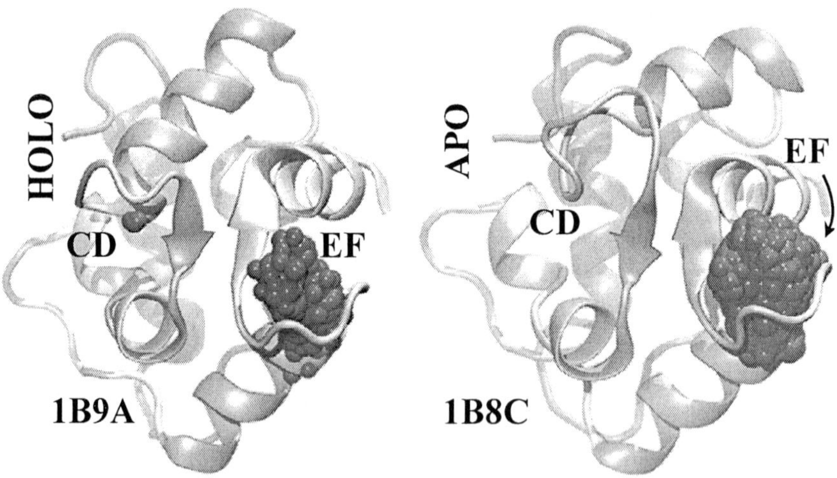

Figure 4: Structures of 1B9A, HOLO, and 1B8C, APO, analyzed by global grid scanning. All of the structures generated by the simulations were aligned for each molecule, and global backbone RMSD was minimized. Shown are the initial structures used in the 1ns simulation in a backbone cartoon representation, as well all of the points that scored over 50 during the simulation as grey balls.

Global grid analysis confirmed further the results observed with the local grid scanning (Fig. 4). Over the course of the simulations, FEATURE recognized only the points in the vicinity of the true sites as favorable centers for Ca^{2+} binding. Thus, EF loop and CD loop of 1B9A and EF loop of 1B8C were correctly identified as Ca^{2+} binding sites. Since only the points at the true Ca^{2+} binding sites scored over the model threshold, the rest of the points encompassing the whole protein could be thought of as correctly predicted negative controls.

The profiles of FEATURE scores obtained by local grid analysis revealed that some of the conformations generated during the simulations achieved scores that were higher than in the original crystal structures. These structures offer intriguing alternative local conformations that the protein may adopt in order to facilitate Ca^{2+} binding. These conformations may be worth studying to investigate shared features that contribute to the details of Ca^{2+} binding. In addition, the ability of MD to create alternative conformations, in which the score at the Ca^{2+} binding site is significantly different from the score of the crystal structure, indicates that it should be able to discover Ca^{2+} binding sites in simulations of proteins whose Ca^{2+} sites happen not to be sufficiently well-formed in their crystalline state. In fact, even if FEATURE were not to recognize Ca^{2+} binding sites in the original structures, correct assignment of Ca^{2+} binding sites would have been possible for parvalbumin β based on the HOLO and APO conformations and associated FEATURE score distributions generated by the simulations. Furthermore, using the global grid approach, it was possible to identify *de novo* potential Ca^{2+} binding sites, when a structure of the molecule without bound Ca^{2+} was used. Lacking knowledge of the HOLO structure, APO structure would have been correctly annotated to have one Ca^{2+} binding site based on our results.

It is interesting to note that the FEATURE scores change significantly even across relatively small time steps, thus indicating that it is very sensitive to the detailed position of atoms during the simulation. This is a good sign that FEATURE will be sensitive to small conformational changes that might affect the ability to bind calcium. Based on this, we are optimistic that MD will be able to sample conformational diversity sufficiently to improve performance of FEATURE when it misses false negative sites in crystal structures that are within 2 – 4Å of local atom RMSD from conformations that show the missed function. This estimate is based on the amount of conformational change seen in our simulations and the associated variation in the FEATURE scores. It is also reassuring that the true negative control sites never achieved scores close to the range sampled by the true positives.

An intriguing question emerges from our results: is there a correlation between the frequency of high scores and binding affinity of the site? A study involving a larger set of diverse Ca^{2+} binding proteins is necessary to pursue the answer to this. In addition, we only simulated protein dynamics for 1ns. It is likely that with longer simulations, other conformations would be visited that would broaden the distribution of scores to reach both higher and lower than the scores of the conformations sampled in 1ns. We are keen to explore the conformational diversity at different lengths of MD simulations and to asses the value of longer simulations for function recognition purposes.

The results of this study are limited to a single protein and a single function, and thus they can not be generalized to all proteins and functions. Experiments are underway to explore further the potential of MD to improve FEATURE predictions of Ca^{2+} binding sites using more HOLO – APO pairs. These include examples of sites that FEATURE can not identify by itself as Ca^{2+} binding. Furthermore, we plan to examine functions other than Ca^{2+} binding, as well as couple together different methods to sample conformational space and to identify functional sites in 3D structures, that are less expensive computationally and thus can be applied to larger datasets..

Coupling MD simulations with FEATURE showed potential in being able to allow better annotation of novel structures with unknown function and of structures where functions have been already assigned.

5. Acknowledgements:

This work was supported by Stanford Genome Training Grant (NIH 5 T32 HG00044) and FEATURE Grant (NIH LM-05652) and the Simbios National Center for Biomedical Computing (http://simbios.stanford.edu/, NIH GM072970). FEATURE is available online at http://feature.stanford.edu/ and http://simtk.org.

6. References:

1. Walker, M.G., Volkmuth, W., and Spinzak, E., *Genome Research*, 1999. **9**: p. 1198.
2. Sjölander, K., *Bioinformatics*, 2004. **20**: p. 170.
3. Hulo, N., Sigrist, C., Saux, V.L., Langendijk-Genevaux, P., Bordoli, L., Gattiker, A., Castro, E.D., Bucher, P., and Bairoch, A., *Nucleic Acids Research*, 2004. **32**: p. D134.
4. Attwood, T., Bradley, P., Flower, D., Gaulton, A., Maudling, N., Mitchell, A., Moulton, G., Nordle, A., Paine, K., Taylor, P., Uddin, A., and Zygouri, C., *Nucleid Acids Research*, 2003. **31**: p. 400.
5. Laskowski, R.A., Watson, J.D., and Thornton, J.M., *Nucleic Acids Research*, 2005. **33**(Web Server issues): p. W89.

6. Pal, D. and Eisenberg, D., *Structure*, 2005. **13**: p. 1.
7. Chandonia, J.M. and Brenner, S.E., *Science*, 2006. **311**(5795): p. 347.
8. Terwilliger, T.C., *Nature Structural and Molecular Biology*, 2004. **11**(4): p. 296.
9. Holm, L. and Sander, C., *Journal of Molecular Biology*, 1993. **233**: p. 123.
10. Fetrow, J.S., Godzik, A., and Skolnick, J., *Journal of Molecular Biology*, 1998. **282**: p. 703.
11. Laskowski, R.A., Watson, J.D., and Thornton, J.M., *Journal of Molecular Biology*, 2005. **351**: p. 614.
12. Krissinel, E. and Henrick, K., *Acta Crystallographica*, 2004. **D**(60): p. 2256.
13. Sodhi, J.S., Bryson, K., McGuffin, L.J., Ward, J.J., Wernisch, L., and Jones, D.T., *Journal of Molecular Biology*, 2004. **342**(307-320): p. 307.
14. Deng, H., Chen, G., Yang, W., and Yang, J.J., *Proteins: Structure, Function, and Bioinformatics*, 2006. **64**(34-42): p. 34.
15. Müller, P., Köpke, S., and Sheldrick, G.M., *Acta Crystallographica* 2003. **D59**: p. 32.
16. Nayal, M. and Cera, E.D., *Proc. Natl. Acad. Sci. USA*, 1994. **91**: p. 817.
17. Asaoka, T., Ando, T., Meguro, T., and Yamato, I., *Chem-Bio Informatics Journal*, 2003. **3**(3): p. 96.
18. Wei, L. and Altman, R.B., *Recognizing Protein Binding Sites Using Statistical Descritions of Their 3d Environments*, in *PSB: Pacific Symposium of Biocomputing*. 1998: Maui, HI. p. 497.
19. Wei, L. and Altman, R.B., *Journal of Bioinformatics and Computational Biology*, 2003. **1**(1): p. 119.
20. Lindahl, E., Hess, B., and van der Spoel, D., *Journal of Molecular Modeling*, 2001. **7**: p. 306.
21. Berendsen, H.J.C., Postma, J.P.M., van Gunsteren, W.F., and Hermans, J., in *Intermolecular Forces*, Pullman, B., Editor. 1981, D. Reidel Publishing Company: Dordrecht, The Netherlands. p. 331.
22. Hess, B., Bekker, H., Berendsen, H.J.C., and Fraaije, J.G.E.M., *Journal of Computational Chemistry*, 1997. **18**: p. 1463
23. van Gunsteren, W.F., Billeter, S.R., Eising, A.A., Hünenberger, P.H., Krüger, P., Mark, A.E., Scott, W.R.P., and Tironi, I.G., *Biomolecular Simulation: The Gromos96 Manual and User Guide*. 1996, Zürich, Switzerland: Hochschulverlag AG an der ETH Zürich.
24. Berendsen, H.J.C., Postma, J.P.M., van Gunsteren, W.F., Nola, A.D., and Haak, J.R., *Journal of Chemical Physics*, 1984. **81**(8): p. 3684.
25. Humphrey, W., Dalke, A., and Schulten, K., *Journal of Molecular Graphics*, 1996. **14**(1): p. 33.
26. Stone, J., *An Efficient Library for Parallel Ray Tracing and Animation*, in *Computer Science Department*. 1998, University of Missouri: Rolla.
27. Frishman, D. and Argos, P., *Proteins: Structure, Function and Genetics*, 1995. **23**: p. 566.
28. Cates, M.S., Barry, M.B., Ho, E.L., Li, Q., Potter, J.D., and George N. Phillips, J., *Structure*, 1999. **7**: p. 1269.

PREDICTION OF STRUCTURE OF G-PROTEIN COUPLED RECEPTORS AND OF BOUND LIGANDS, WITH APPLICATIONS FOR DRUG DESIGN

YOUYONG LI, WILLIAM A. GODDARD III[†]

Materials and Process Simulation Center (MC 139-74),
California Institute of Technology, Pasadena CA 91125

G protein-coupled receptors (GPCRs) mediate the senses of vision, smell, taste, and pain. They are also involved in cell recognition and communication processes, making them a prominent superfamily for drug targets. Unfortunately, the atomic-level structure is available from experiment only for bovine rhodopsin. We report here improvements in methods (MembStruk and HierDock) for predicting the structures of GPCRs, including bound ligands, with applications to prostanoid and Urotensin GPCRs. We find that the predicted binding sites are consistent with available mutation and SAR data, suggesting that the predicted structures are sufficiently accurate for drug design applications.

1. Introduction

G protein-coupled receptors (GPCRs) mediate senses such as odor, taste, vision, and pain in mammals.[1] In addition, important cell recognition and communication processes often involve GPCRs. The GPCR superfamily is diverse with over 850 genes in the human genome. The diversity of the GPCRs is matched by the variety of ligands that activate them, including odorants, taste ligands, light, metals, biogenic amines, fatty acids, amino acids, peptides, proteins, nucleotides, lipids, Krebs-cycle intermediates, and steroids.

Indeed, many diseases involve malfunction of GPCRs,[2] making them important targets for drug development. Greater than 30 % of all marketed therapeutics act on those receptors.[3] Prostanoids (prostaglandins (PG) and thromboxanes (TX), both metabolites of arachidonic acid)[4] play important physiological roles in the cardiovascular and immune systems and in pain sensation in peripheral systems. They exert a variety of actions in the body through binding to specific cell surface prostanoid receptors and mediate many processes that are inhibited by non-steroidal anti-inflammatory drugs (NSAIDs). Urotensin II (U-II) is the most potent vasoconstrictor known and plays an important role in cardiovascular regulation upon the Urotensin II receptor (UT2R). U-II is also a neuropeptide and may play a role in tumor development.

The development of subtype specific agonists and antagonists has been hampered by the lack of 3D structures for GPCR receptors. Here we report the

[†] Work partially supported by NIH (R21- MH073910-01-A1) and Sanofi-Aventis Corporation.

application of MembStruk[5-14] and HierDock to study 3D structure and function of prostanoid receptors and Urotensin receptors. We obtain the binding sites consistent with available mutation and SAR data, validating the predicted structures.

2. Methods

We predicted the three-dimensional structures of prostanoid receptors and Urotensin receptors using MembStruk4.1 computational method summarized here (see Figure 1).

***2.1 Prediction of the TM regions and hydrophobic centers*:** The TM regions were predicted using the TM2ndS method described elsewhere[9]. Here we selected 43 sequences from the GPCR database having sequence identities to the target sequences ranging above 20%. Then multiple sequence alignments among these sequences were performed using clustalW and inputted into the TM2ndS procedure [9] to predict the TM domains. This procedure predicts the centroid for each helix which is used to position the 7 helices with respect to a common xy plane (the midpoint of the lipid). This *centroid* for each helix is chosen as the residue that partitions equally the area under the hydrophobicity curve.

2.2 *Prediction of the 3D structure*: Based on the predicted TM regions and the TM centroids, the MembStruk method was used to build and optimize the 3-D structure. The steps of MembStruk and the predicted structure are described below.

1. *Helix packing*: First, canonical α-helices were built for each TM domain. These α-helix structures were then bundled together as follows. The predicted helix centroid is placed on the xy plane using x,y coordinates based on the low resolution (7.5 Å) electron density map of frog rhodopsin. The orientation of each helix about its z-axis (the χ angle) is chosen so that its helical face with the maximum hydrophobic moment points outwards to contact the lipid. In this analysis, we calculate the hydrophobic moment over the full helix but including only the half of the residues that would face outward. Then each helix is tilted about the point at which the central axis intersects the xy plane to match the tilt angles (θ,φ) from frog rhodopsin.
2. *Helix bending*: Next, molecular dynamics (MD) simulations were performed (200 picoseconds) for each individual helix, allowing the helix to attain its equilibrium structure (in some cases it bends or kinks). We found that 200ps is enough to equilibrate individual helix. Then we chose the structure with the lowest potential energy for each helix and assembled it back into the bundle so that the average axis coincides with the original axis. The side chains were then optimized using SCWRL[15,16] and the total energy minimized (conjugate gradients).
3. *RotMin*. This initial packed structure was minimized and then the individual packing interactions were optimized as follows. Each helix was independently rotated (χ) by +5° and -5°, the side chains repositioned using SCWRL, and then all atoms of the bundle optimized. If either new angle was lower, it was selected.
4. *Lipid Insertion*: At this point, we inserted the 7-helix bundle into a lipid framework ending up with 48 lipids molecules arranged as a bilayer. These lipid molecules were optimized using rigid body dynamics.
5. *RotScan*. Starting from the final RotMin structure, we performed a full 360-degree rotational scan (χ) on each of the helices in 5° increments. For each angle, the side chains were re-assigned with SCWRL and full bundle re-minimized. Multiple minima based on energy and interhelical hydrogen bonds were chosen for each helix. Combination of

multiple minima for each helix leads to an ensemble of conformations which were then sorted by the number of interhelical hydrogen bonds and then by total energy.

2.3 Prediction of the Extra-Cellular (EC) and Intra-Cellular (IC) loop structure

We took the best structure from the previous step and added the three EC and IC loops. We expect the loops of GPCRs to be quite flexible and strongly affected by the solvent, which is treated only implicitly in MembStruk. Thus to provide initial loop structures for our MD studies, we used the alignment of our predicted structure with bovine rhodopsin and then homology threaded the loops to the crystal structure of bovine rhodopsin (1L9H.pdb). Then we carried out minimization and dynamics on the loops with fixed helix bundle atoms.

In the crystal structure of bovine rhodopsin the ECII loop (connecting TM4 and TM5) is closed over the 7-TM barrel, contributing to the binding of 11-cis-retinal. This ECII loop has a disulfide bond to TM3 (C105-C183), which is highly conserved among the rhodopsin super family of GPCRs. Thus, we included this disulfide bond in our loop structures. It is generally believed that the disulfide bond plays critical role in the folding of 7 helices and in the closing of the ECII loop over the 7-TM barrel.[17] Since the rhodopsin in the crystal study is in the inactive form, it is possible that substantial changes occur in ECII and in other loops upon activation.

2.4 Molecular Dynamics Simulation

The MembStruk procedure uses an implicit description of the lipid and water, which is the only practical way to cover many 1000s of packing of the TM domains and the ~100,000 positions for the ligands in the binding site. However after predicting the best packing of the TM domains and of the ligand interacting with these domains we inserted this ligand-bound bundle into an infinite lipid bilayer (POPC: 1-palmytoil-2-oleoyl-sn-glycero-3 -Phosphatidylcholine) immersed in a periodic box of water. In this process, we eliminated lipid molecules within 5 Å of the protein. Then we inserted this in a box of water molecules and eliminated waters within 5 Å of the lipid and protein. Then keeping the protein fixed we allowed the lipid and water to relax using minimization.

Starting with this periodic structure including 100 lipid molecules and 6617 water molecules per periodic cell (with ~ 33000 atoms per cell), we carried out molecular dynamics (MD) simulations (using NAMD) of the predicted structure with and without ligand for 1ns in explicit lipid bilayer and water.[13,18] These calculations were carried out for both the ligand bound and apo protein systems.

2.5 HierDock Method: ***Scan the entire receptor for binding sites***

To predict the binding site and conformation of the ligand within the receptor, we used the HierDock method. The first step is to scan all void regions in the entire receptor structure to locate putative binding regions for the ligand. The void region in the entire receptor structure was partitioned into 27 regions and the HierDock method was used to dock the ligand in each box. Here we examined the best binding sites that have at least 80% buried surface area. Subsequently we docked the ligand in this putative binding region using the HierDock2.0 method.

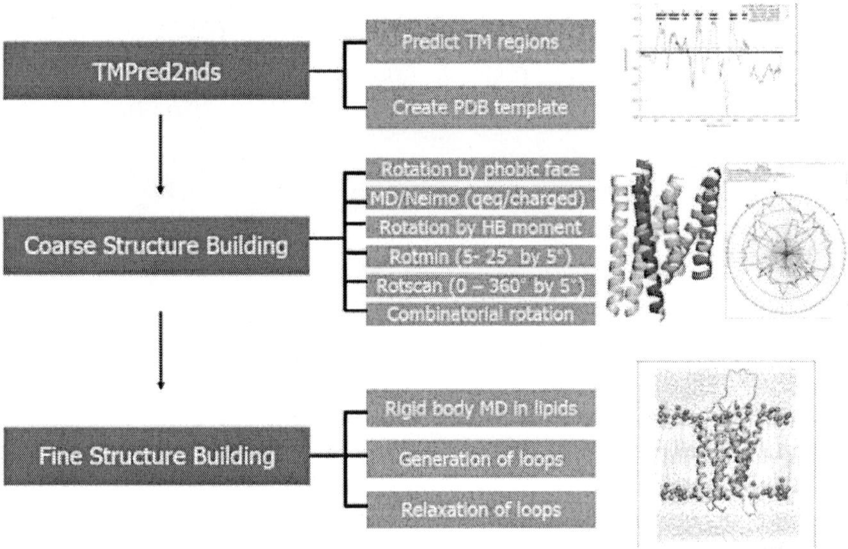

Figure 1. The MembStruk procedures consists of 1) TM prediction, 2) Coarse structure building, and 3) fine structure building in explicit membrane and water

3.0 Results and discussion

3.1Predicted human DP structure and binding modes with PGD2 and antagonists.

The DP receptor of prostanoid receptors lacks some of the well-conserved motifs present in class A GPCRs. For example, the DRY motif on TM3 is ECW, the well-conserved Trp on TM4 becomes Leu, the WXP motif on TM6 becomes SXP, and the NPXXY motif on TM7 is a DPWXF in the DP receptor. Thus, we can expect that the DP receptor might have a different set of stabilizing interhelical hydrogen bonds from rhodopsin. The predicted 3D structure of human apo-DP receptor is shown in Figure 2.

We find an interhelical hydrogen bond between N34(1) and D72(2). N34(1) and D72(2) are conserved in the rhodopsin family A including DP, but the conserved Asn of the NPXXY motif in TM7 is a DPWXF motif in the DP receptor. S316(7), which is not a conserved residue, makes a hydrogen bond with the N34(1) and D72(2). D319(7) makes a hydrogen bond with S119(3). D72(2) also forms a strong salt bridge with the K76(2) on the same helix. K76(2) is a conservative replacement in other prostaglandin receptors except for thromboxane receptors. We also find a hydrogen bond between R310(7) and Y87(2), where R310(7) is conserved across all prostaglandin receptors while Y87 is present only in DP receptors.

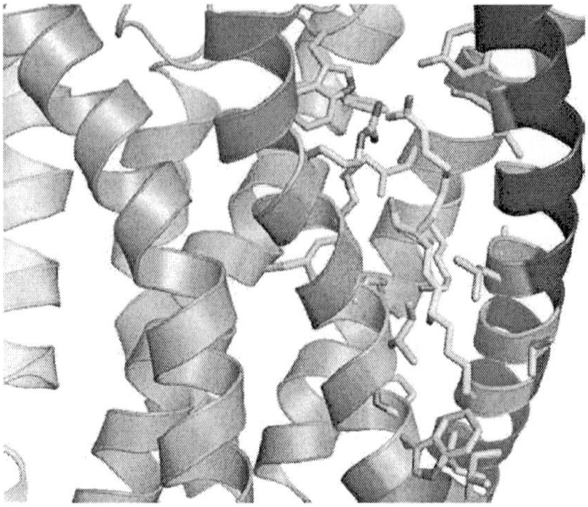

Figure 2. Predicted human DP receptor and binding mode with PGD2.

The predicted binding site of PGD2 is shown in Figure 2. PGD2 is located between the TM 1, 2, 3, 7 helices and is covered by the ECII loop.

We find favorable hydrophobic interactions of the α chain with L26(1) and F27(1). The α chain of PGD2 points up toward the EC region with the ω chain pointing down between TM1 and TM7. The critical elements of bonding are the carboxylic acid interacts with R310(7); the carbonyl on the cyclopentane ring of PGD2 has a hydrogen bond with K76(2); the hydroxyl on the ω chain interacts with S316(7) and K76(2); 9-OH forms a hydrogen bond with S313(7); a hydrophobic pocket surrounds the α chain with M22(1), G23(1), Y87(2), W182(ECII), L309(7), R310(7), L312(7), and S313(7) within 6 Å; a hydrophobic pocket surrounds the ω chain with L26(1), G30(1), I317(7), P320(7), and W321(7) within 6 Å.

The prediction that the carboxylic acid group of PGD2 interacts with R310(7) is confirmed strongly by various experiments. The carboxylic acid group and the hydroxyl group on the ω−chain are present in all the prostanoid compounds. R310(7) is 100% conserved among prostanoid receptor family and K76(2) is not. Other hydrophobic residues P320(7), W321(7) interacting with ω chain are also 100% conserved in prostanoid receptor family. Structure activity relationship studies of PGE2 show that the carboxylic acid group, both the ω chain itself and the hydroxyl group in the ω chain are critical for agonist activity.[19] The hydroxyl and carbonyl groups on the cyclopentane ring are *not* present in all prostanoid compounds and these groups offer receptor selectivity to the ligand as discussed next.

DP receptor binds to PGD2 and shows at least 2 orders of magnitude lower affinity to other prostanoid compounds. However, the IP receptor binds to PGE1 and PGI analogs (iloprost), but it does not bind PGE2. Assuming that these other prostanoid compounds bind to the *h*DP receptor in similar binding mode as PGD2, we can explain the how the DP receptor prefers PGD2 to other prostanoid compounds like PGF2α, PGE2, and PGI2 as shown in Figure 3.

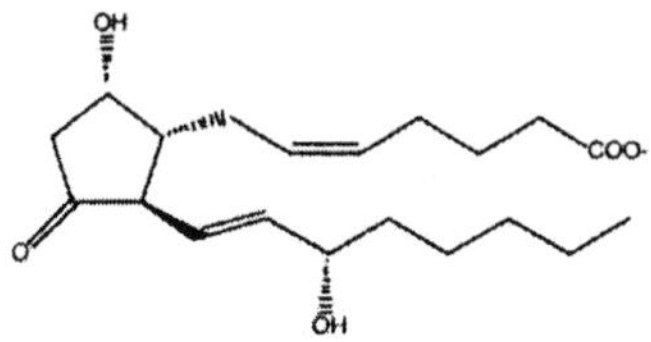

(a) PGD2

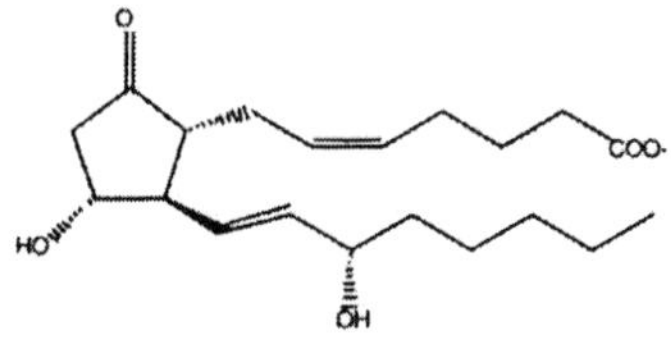

(b) PGE2

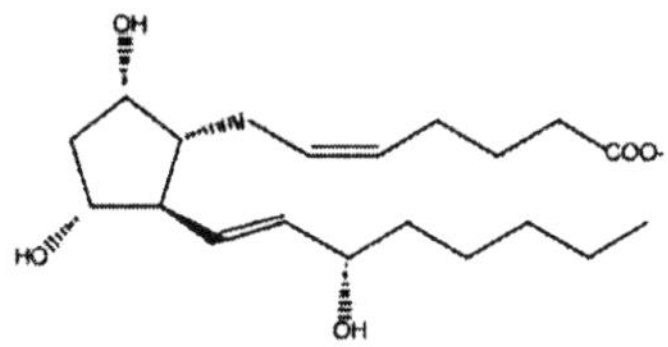

(c) PGF2a

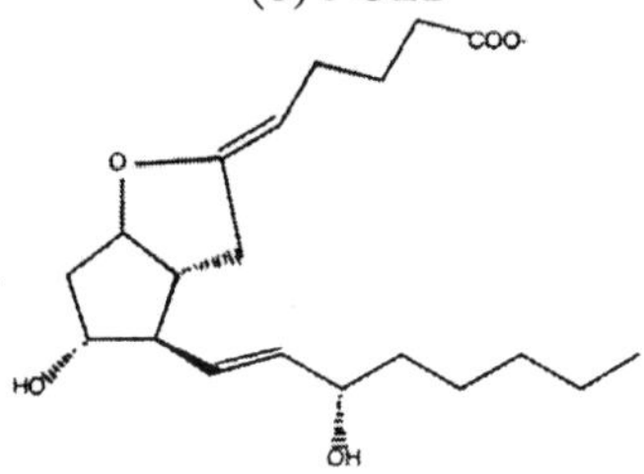

(d) PGI2

Figure 3. Prostanoid compounds

Based on our predicted structure of human DP, we collaborated with Sanofi-Aventis to predict the binding of several families of antagonists and used these results to optimize these families. One of the compounds is currently under pre-clinic trials.

Here we illustrate the results for the cyclopentanoindole (CPI) family shown in Figure 4, which Merck has recently filed[20,21] as an Investigation Lipid-Modifier (Cardaptive^{TM}). CPI is predicted to be located among the TM1237 region, which is similar our predicted structure for the endogenous ligand (PGD2), The binding mode of CPI in Figure 4b correlates very well with published SAR data, validating the accuracy of the predicted structures. However, we find that the Molecular Dynamics (MD) motion for CPI bound to hDP are quite different to that found for PGD2 bound hDP. This MD studies illustrate the functional differences between the effect of binging agonists and antagonists.

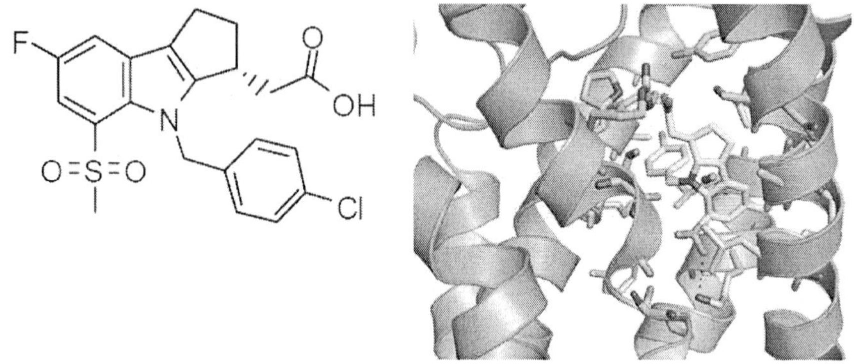

(a) (b)

Figure 4. Cyclopentanoindole (CPI) compound and predicted binding mode

3.2Predicted human UT2R structure and binding mode with peptide CFWKYC

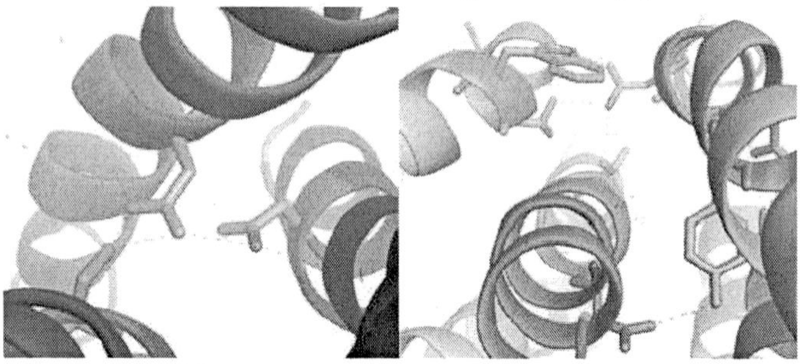

(a) (b)

Figure 5. MembStruk Predicted structure of human UT2R. It has conserved interhelical hydrogen bond networks among (a) TM127; and (b) TM234.

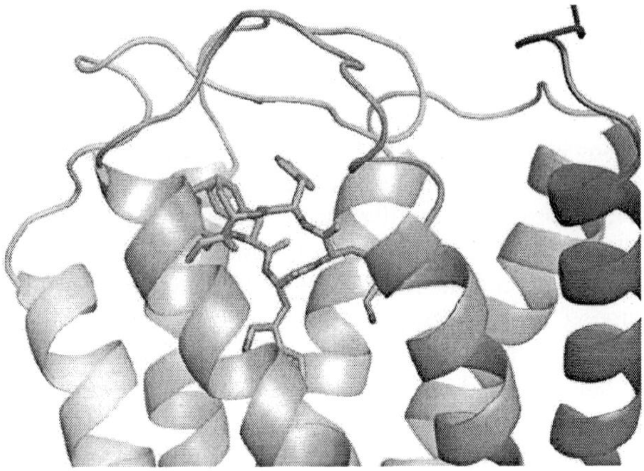

Figure 6. Predicted structure and binding mode human UT2R with endogenous agonist, CFWKYC

Urotensin II (UII), an urophysical peptide, is a cyclic dodecapeptide (AGTADCFWKYCV). The composition of UII ranges from 11 amino acids in humans to 14 amino acids in mice, always with the conserved cysteine-linked macrocycle CFWKYC, which is essential for the biological activity.[22] Indeed, CFWKYC itself has been shown to bind hU-II.[23]

First, we predicted the 3D structure of human Urotensin II GPRC (UT2R) using MembStruk. As shown in Figure 5, the predicted structure has strong interhelical interactions among TM127 (Figure 5a) and TM234 (Figure 5b) as in other GPCRs. Those important interhelical interactions are found automatically at the RotScan step of MembStruk.

Then we docked CFWKYC into predicted structure of hU-II using HierDock. The docked result is shown in Figure 6, where CFWKYC is located in TM3456 region and covered by ECII loop.

In the predicted binding mode, the Lys in the middle of CFWKYC forms a salt bridge with D116(3). This is consistent with SAR data,[24] where K=>A causes > 1000 fold decrease in activity .

In the predicted binding mode, the Tyr of CFWKYC interacts strongly with C109(3), F113(3), L184, C185, L186(EcII), which form a hydrophobic pocket. This explains that Y=>F retains activity.[24]

In the predicted binding mode, the Trp of CFWKYC is located among F202(6), F257(7), F117(3), F113(3), an aromatic pocket. This is consistent with SAR data,[24] where W=>A or 2Nal, causes a dramatic loss in potency.

In the predicted binding mode, Phe of CFWKYC interacts directly with V170(4), which is consistent with photo-labeling experiment results.[25]

In summary, the binding conformation predicted with HierDock explains SAR data and photo-labeling experiment results. This validates both the MembStruk and HierDock protocols, suggesting that the predicted structures are adequate for predicting the binding sites for new molecules, making these methods useful for drug discovery and optimization.

Summary

We report the structure and function of human DP receptor and human UT2R receptor from using the MembStruk and HierDock methods. Both receptors play important role in the cardiovascular and immune systems of the human body.

The predicted structure and function for these two receptors agrees with currently available experimental data. The predicted binding position of PGD2 is located among TM127 region. It has important interactions with R310(7), S316(7), K76(2), and S313(7). These results suggest that site-directed mutagenesis studies on these residues would be useful to test the predicted structure and function of GPCR.

The predicted binding mode of PGD2 provides the structural basis for understanding of the selectivity of prostanoid receptors. Indeed the predicted structure of human DP, was used in a collaboration with Sanofi-Aventis to predict the binding of new antagonist families and to optimize them. One of these compounds is currently on pre-clinic trials.

We also obtained high quality structure of human UT2R structure and binding structure with core peptide segment CFWKYC. The peptide is predicted to be located in TM3456 region covered by ECII and agrees with available SAR data.

(1)Dong, X.; Han, S. K.; Zylka, M. J.; Simon, M. I.; Anderson, D. J. *Cell* **2001**, *106*, 619.

(2)Wilson, S.; Bergsma, D. *Pharmaceutical News* **2000**, *7*, 3.

(3)Schlyer, S.; Horuk, R. *Drug Discovery Today* **2006**, *11*, 481.

(4)Narumiya, S.; Sugimoto, Y.; Ushikubi, F. *Physiological Reviews* **1999**, *79*, 1193.

(5)Hall, S. E.; Floriano, W. B.; Vaidehi, N.; Goddard III, W. A. *Chem. Senses* **2004**, *29*, 595.

(6)Floriano, W. B.; Vaidehi, N.; Goddard, W. A., III. *Chemical Senses* **2004**, *29*, 269.

(7)Kalani, M. Y.; Vaidehi, N.; Hall, S. E.; Trabanino, R.; Freddolino, P.; Kalani, M. A.; Floriano, W. B.; Kam, V.; Goddard, W. A., III. *Proc. Natl. Acad. Sci.* **2004**, *101*, 3815.

(8)Freddolino, P.; Kalani, M. Y.; Vaidehi, N.; Floriano, W.; Hall, S. E.; Trabanino, R.; Kam, V. W. T.; Goddard, W. A. *Proc. Natl. Acad. Sci.* **2004**, *101*, 2736.

(9)Trabanino, R.; Hall, S. E.; Vaidehi, N.; Floriano, W.; Goddard, W. A. *Biophys. J.* **2004**, *86*, 1904.

(10)Hummel, P.; Vaidehi, N.; Floriano, W. B.; Hall, S. E.; Goddard, W. A. *Protein Science* **2005**, *14*, 703.

(11)Peng, J.; Vaidehi, N.; Hall, S.; Goddard III, W. A. *ChemMedChem* **2006**, *2006*, in press.
(12)Vaidehi, N.; Schlyer, S.; Trabanino, R.; Kochanny, M.; Abrol, R.; Koovakat, S.; Dunning, L.; Liang, M.; Sharma, S.; Fox, J. M.; Floriano, W. B.; Mendonça, F. L. d.; Pease, J. E.; III, W. A. G.; Horuk, R. *J Biol Chem* **2006**, *281*, 27613.
(13)Spijker, P.; Vaidehi, N.; Freddolino, P.; Hilbers, P.; Goddard, W. *Proc. Natl. Acad. Sci.* **2006**, *103*, 4882.
(14)Heo, J.; Han, S. K.; Vaidehi, N.; Wendel, J.; Kekenes-Huskey, P.; Goddard, W. A. *Chembiochem* **2007**, *8*.
(15)Canutescu, A.; Shelenkov, A.; Dunbrack, R. *Protein Science* **2003**, *12*, 2001.
(16)Altschul, S.; al., e. *Journal of Molecular Biology* **1990**, *215*, 403.
(17)Palczewski, K.; Kumasaka, T.; Hori, T.; Behnke, C.; Motoshima, H.; Fox, B.; Trong, I.; Teller, D.; Okada, T.; Stenkamp, R.; Yamamoto, M.; Miyano, M. *science* **2000**, *289*, 739.
(18)Philips, J.; Braun, R.; Wang, W.; Gumbart, J.; Tajkhorshid, E.; Villa, E.; Chipot, C.; Skeel, R.; Kale, L.; Shulten, K. *Journal of computational chemistry* **2005**, *26*, 1781.
(19)Ungrin, M. K.; Carriere, M.-C.; Denis, D.; Lamontagne, S.; Sawyer, N.; Stocco, R.; Tremblay, N.; Metters, K. M.; Abramovitz, M. *Molecular Pharmacology* **2001**, *59*, 1446.
(20)Sturino, C. F.; Lachance, N.; Boyd, M.; al., e. *Bioorg. Med. Chem. Lett.* **2006**, *16*, 3043.
(21)Sturino, C. F.; O'Neill, G.; Lachance, N.; al., e. *J. Med. Chem.* **2007**, *50*, 794.
(22)Davenport, A. P.; Maguire, J. J. *Trends Pharmacol. Sci.* **2000**, *21*, 80.
(23)Foister, S.; Taylor, L. L.; Feng, J.-J.; Chen, W.-L.; Lin, A.; Cheng, F.-C.; Smith, A. B.; Hirschmann, R. *Organic Letters* **2006**, *8*, 1799.
(24)Kinney, W. A.; Almond, H. R.; Qi, J.; Smith, C. E.; Santulli, R. J.; de Garavilla, L.; Andrade-Gordon, P.; Cho, D. S.; Everson, A. M.; Feinstein, M. A.; Leung, P. A.; Maryanoff, B. E. *Angew. Chem. Int. Ed.* **2002**, *41*, 2940.
(25)Boucard, A. A.; Sauve, S. S.; Guillemette, G.; Escher, E.; Leduc, R. *BioChem. J.* **2003**, *370*, 829.

MARKOV CHAIN MODELS OF COUPLED INTRACELLULAR CALCIUM CHANNELS: KRONECKER STRUCTURED REPRESENTATIONS AND BENCHMARK STATIONARY DISTRIBUTION CALCULATIONS

HILARY DEREMIGIO[†], PETER KEMPER[‡],
M. DREW LAMAR[†], and GREGORY D. SMITH[†]

[†] *Department of Applied Science,* [‡] *Department of Computer Science,*
The College of William and Mary, Williamsburg, VA 23187

Mathematical models of calcium release sites derived from Markov chain models of intracellular calcium channels exhibit collective gating reminiscent of the experimentally observed phenomenon of stochastic calcium excitability (i.e., calcium puffs and sparks). We present a Kronecker structured representation for calcium release site models and perform benchmark stationary distribution calculations using numerical iterative solution techniques that leverage this structure. In this context we find multi-level methods and certain preconditioned projection methods superior to simple Gauss-Seidel type iterations. Response measures such as the number of channels in a particular state converge more quickly using these numerical iterative methods than occupation measures calculated via Monte Carlo simulation.

1. Introduction

The stochastic gating of voltage- and ligand-gated ion channels in biological membranes that is observed by single channel recording techniques is often modeled using discrete-state continuous-time Markov chains (CTMCs).[1,2] While these single channel models can be relatively simple (e.g., two physicochemically distinct states) or complex (hundreds of states), most include only two conductance levels (closed and open). For example, a transition state diagram for a three-state calcium (Ca^{2+})-regulated channel activated by sequential binding of two Ca^{2+}ions is given by

$$C_1 \underset{k_a^-}{\overset{k_a^+ c}{\rightleftharpoons}} C_2 \underset{k_b^-}{\overset{k_b^+ c}{\rightleftharpoons}} O_1 \qquad (1)$$

where $k_i^+ c$ and k_i^- with $i \in \{a, b\}$ are transition rates with units of reciprocal time, k_i^+ is an association rate constant with units of conc^{-1} time^{-1}, and c is the $[Ca^{2+}]$ near the channel. If this local $[Ca^{2+}]$ is specified, the transition-state diagram of the channel (1) defines a CTMC that takes on values in the state-space (C_1, C_2, O_1). The experimentally observable conductance of this stochastically gating channel is the aggregated process of transitions between the closed and open classes of states: $\mathcal{C} = \{C_1, C_2\}$ and $\mathcal{O} = \{O_1\}$.

The scientific literature developing stochastic models for the behavior of ion channels is largely focused on single channels or populations of independent channels. One notable exception is the work of Ball and colleagues analyzing interacting aggregated CTMC models of membrane patches containing several ion channels.[3,4] A second example, the subject of this paper, are simulations of clusters of intracellular Ca^{2+}-regulated Ca^{2+} channels—inositol 1,4,5-trisphosphate receptors (IP_3Rs) and ryanodine receptors (RyRs)—located on the surface of the endoplasmic reticulum or sarcoplasmic reticulum membrane—that give rise to localized intracellular $[Ca^{2+}]$ elevations known as Ca^{2+} puffs and sparks.[5–9]

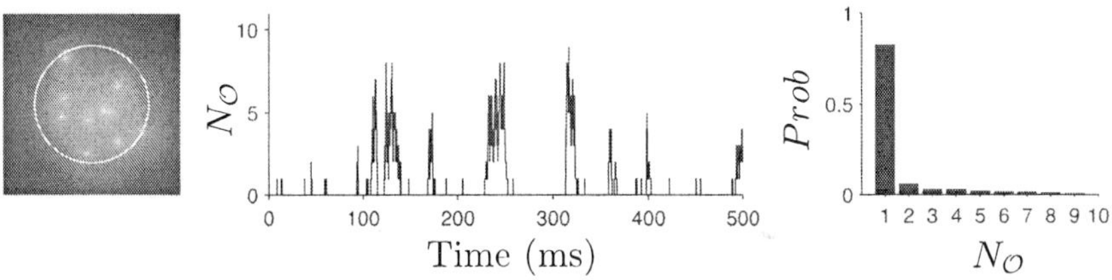

Fig. 1. Left: Local $[Ca^{2+}]$ near 3×3 μm ER membrane with Ca^{2+} channels modeled as 0.05 pA point sources with positions randomly chosen from a uniform distribution on a disc of radius 2 μm. Buffered Ca^{2+} diffusion is modeled as in Ref. 10. Middle: Stochastic Ca^{2+} excitability reminiscent of Ca^{2+} puffs/sparks. Right: Probability distribution of the number of open channels leading to a puff/spark *Score* of 0.39.

When Markov chain models of Ca^{2+}-regulated Ca^{2+} channels such as (1) are coupled via a mathematical representation of buffered diffusion of intracellular Ca^{2+}, simulated Ca^{2+} release sites may exhibit the phenomenon of "stochastic Ca^{2+} excitability" where the IP_3Rs or RyRs open and close in a concerted fashion[10,11] (see Fig. 1 for representative simulation) . Such models are stochastic automata networks (SANs) that involve a large number of functional transitions, that is, the transition probabilities of one automata (i.e., an individual channel) may depend on the local $[Ca^{2+}]$ and thus the state of the other channels. The experimentally observable quantity is either the local $[Ca^{2+}]$ or the number of channels in the open class of states, $N_{\mathcal{O}}(t)$ (see Fig. 1, middle).

The relationship between single channel kinetics of Ca^{2+}-regulated channels and the emergent phenomenon of Ca^{2+} puffs and sparks is not well understood. However, if each release site configuration is known, several informative response measures can be determined from the steady-state probability distribution. For example, the so-called puff/spark *Score*[10] given by $\text{Var}[f_{\mathcal{O}}]/\text{E}[f_{\mathcal{O}}]$ is the index of dispersion of the steady-state fraction of open channels, $f_{\mathcal{O}} = N_{\mathcal{O}}/N$ (see Fig. 1, right). This response measure takes values between 0 and 1, and a puff/spark *Score* of greater than approximately 0.3 indicates the presence of Ca^{2+} excitability. However, Ca^{2+} release sites are composed of 5–250 channels and this leads to a state-space explosion that makes numerical calculation of the stationary distribution of model Ca^{2+} release sites difficult.

2. Formulation of Model

In this paper we consider two single channel models: the three-state Ca^{2+}-activated channel described above (1) and a six-state model that includes both fast Ca^{2+} activation and slow Ca^{2+} inactivation, processes that are important aspects of the dynamics of both IP_3Rs and RyRs. The six-state model assumes two identical channel subunits that both require Ca^{2+} binding to enter a permissive state and include a second Ca^{2+}-mediated transition into a long-lived non-permissive state (for transition state diagram and parameters see Ref. 12).

In both the three- and six-state models, Ca^{2+}-mediated transitions out of open states can be accelerated due to the increase in local $[Ca^{2+}]$ when a Ca^{2+}-regulated Ca^{2+} channel is open.[13,14] Assuming the formation and collapse of Ca^{2+} microdomains are fast compared to channel gating, we can denote the background and domain $[Ca^{2+}]$ experienced by the channel when closed and open as c_∞ and c_d, respectively. With this assumption the three- and six-state single channel models take the form $Q = K_- + (c_\infty I + c_d I_{\mathcal{O}}) K_+$ where K_- and K_+ are $M \times M$ matrices that collect the unimolecular (k_i^-) and bimolecular (k_i^+) transition rates, $I_{\mathcal{O}} = \text{diag}\{\boldsymbol{e}_{\mathcal{O}}\}$, and $\boldsymbol{e}_{\mathcal{O}}$ is a $M \times 1$ vector indicating open states of the single channel model.[10] In our model formulation, the interaction between channels is mediated through the buffered diffusion of intracellular Ca^{2+} (see Ref. 10 for a complete description). For the purposes of this paper we do not assume any particular cell type with known release site ultrastructure (e.g., cardiac myocytes with channels arranged in a dyad) and instead consider that the N channels at the Ca^{2+} release site have positions chosen from a two-dimensional uniform distribution on a disc of radius 0.1–2.0 μm (i.e., constant surface density; see Fig. 1, left). When in the open state, each channel contributes to the landscape of $[Ca^{2+}]$ throughout the Ca^{2+} release site and influences the local $[Ca^{2+}]$ experienced by other channels. For simplicity we assume that the formation and collapse of individual peaks within the Ca^{2+} microdomain occur quickly compared to channel gating. We also assume the presence of a single high concentration Ca^{2+} buffer and the validity of superposing local $[Ca^{2+}]$ increases due to each of the N channels.[15,16] Thus, channel interactions can be summarized by an $N \times N$ 'coupling matrix' $C = (c_{ij})$ that gives the increase over c_∞ experienced by channel j when channel i is open.

2.1. *Instantaneous Coupling of Two* Ca^{2+}*-Regulated* Ca^{2+} *Channels*

In the case of two identical Ca^{2+}-regulated Ca^{2+} channels the interaction matrix takes the form

$$C = \begin{pmatrix} c_d & c_{12} \\ c_{21} & c_d \end{pmatrix}$$

and the expanded generator matrix is given by $Q^{(2)} = Q_-^{(2)} + Q_+^{(2)}$ where

$$Q_-^{(2)} = K_- \otimes I + I \otimes K_- \tag{2}$$

collects the unimolecular transition rates and $\otimes$ denotes the Kronecker product (see Ch. 9 in Ref. 17). The transition rates involving Ca^{2+} take the form

$$Q_+^{(2)} = D_1^{(2)} \left(K_+ \otimes I \right) + D_2^{(2)} \left(I \otimes K_+ \right), \tag{3}$$

where the two terms represent Ca^{2+}-mediated transitions of each channel. The diagonal matrices $D_1^{(2)}$ and $D_2^{(2)}$ give the $[Ca^{2+}]$ experienced by channel 1 and 2, respectively, in every configuration of the release site, that is,

$$\begin{aligned} D_1^{(2)} &= \text{diag}\left\{ c_\infty \left(\boldsymbol{e} \otimes \boldsymbol{e} \right) + c_d \left(\boldsymbol{e}_\mathcal{O} \otimes \boldsymbol{e} \right) + c_{21} \left(\boldsymbol{e} \otimes \boldsymbol{e}_\mathcal{O} \right) \right\} \\ &= c_\infty \left(I \otimes I \right) + c_d \left(I_\mathcal{O} \otimes I \right) + c_{21} \left(I \otimes I_\mathcal{O} \right) \end{aligned}$$

and similarly for $D_2^{(2)}$. Using the Kronecker identities such as $(I \otimes I_\mathcal{O})(I \otimes K_+) = I \otimes I_\mathcal{O} K_+$, Eq. 3 can be rearranged as

$$\begin{aligned} Q_+^{(2)} = c_\infty K_+^{(2)} &+ c_d \left(I_\mathcal{O} K_+ \otimes I \right) + c_{12} \left(I_\mathcal{O} \otimes K_+ \right) \\ &+ c_{21} \left(K_+ \otimes I_\mathcal{O} \right) + c_d \left(I \otimes I_\mathcal{O} K_+ \right) \end{aligned} \tag{4}$$

where $K_+^{(2)} = K_+ \otimes I + I \otimes K_+$. Combining Eqs. 2 and 4 and simplifying, $Q^{(2)}$ can be written compactly as

$$Q^{(2)} = A_d \otimes I + I_O \otimes A_{12} + A_{21} \otimes I_O + I \otimes A_d \tag{5}$$

where $A_d = K_- + c_\infty K_+ + c_d I_O K_+$, and $A_{ij} = c_{ij} K_+$.

2.2. *Instantaneous Coupling of N* Ca^{2+}*-Regulated* Ca^{2+} *Channels*

In the case of N channels coupled at the Ca^{2+} release site, the expanded generator matrix—i.e., the SAN descriptor—is given by

$$Q^{(N)} = Q_-^{(N)} + Q_+^{(N)} \tag{6}$$

$$Q_-^{(N)} = \bigoplus_{n=1}^{N} K_- = \sum_{n=1}^{N} I^{(n-1)} \otimes K_- \otimes I^{(N-n)} \tag{7}$$

$$Q_+^{(N)} = c_\infty K_+^{(N)} + \sum_{i,j=1}^{N} c_{ij} \left(Y_{ij}^1 Z_{ij}^1 \otimes \cdots \otimes Y_{ij}^N Z_{ij}^N \right) \tag{8}$$

$$Y_{ij}^n = \begin{cases} I_O & \text{for } i = n \\ I & \text{otherwise} \end{cases} \qquad Z_{ij}^n = \begin{cases} K_+ & \text{for } j = n \\ I & \text{otherwise} \end{cases} \tag{9}$$

where $I^{(n)}$ is an identity matrix of size M^n and $K_+^{(N)} = \oplus_{n=1}^{N} K_+$. Combining Eqs. 7 and 8 and simplifying, $Q^{(N)}$ can be written as

$$Q^{(N)} = \sum_{i,j=1}^{N} X_{ij}^1 \otimes \cdots \otimes X_{ij}^N \tag{10}$$

$$X_{ii}^n = \begin{cases} A_d & \text{for } i = n \\ I & \text{otherwise.} \end{cases} \qquad \text{and} \qquad X_{ij}^n = \begin{cases} I_O & \text{for } i = n \\ A_{ij} & \text{for } j = n \\ I & \text{otherwise} \end{cases} \quad \text{for } i \neq j$$

where $A_d = K_- + c_\infty K_+ + c_d I_O K_+$, and $A_{ij} = c_{ij} K_+$. Note that all states of the expanded Markov chain $Q^{(N)}$ are reachable, the matrices I, I_O, A_d, A_{ij}, and X_{ij}^n are all $M \times M$, and $2N^2 - N$ of the N^3 matrices denoted by X_{ij}^n are *not* identity matrices.

3. Stationary Distribution Calculations

The limiting probability distribution of a finite irreducible CTMC is the unique stationary distribution $\boldsymbol{\pi}^{(N)}$ satisfying global balance,[17] that is,

$$\boldsymbol{\pi}^{(N)} Q^{(N)} = \mathbf{0} \quad \text{subject to} \quad \boldsymbol{\pi}^{(N)} \mathbf{e}^{(N)} = 1 \tag{11}$$

where $Q^{(N)}$ is the Ca^{2+} release site SAN descriptor (Eq. 10) and $\boldsymbol{e}^{(N)}$ is an $M^N \times 1$ column vector of ones. Although Monte Carlo simulation techniques such as Gillespie's Method[18] can be implemented to estimate response measures such as the puff/spark *Score*, this is an inefficient approach when the convergence of the occupation measures to the limiting probability distribution is slow. This problem is compounded by the state-space explosion that occurs when the number of channels (N) or number of states per channel (M) is large (i.e., physiologically realistic). Both space requirements and quality of results can be addressed using the Kronecker representation (Eq. 10) and various iterative numerical methods to solve for $\boldsymbol{\pi}^{(N)}$.

Many methods are available to solve Eq. 11 with different ranges of applicability (see Ref. 17 for review). For larger models, a variety of iterative methods are applicable including the method of Jacobi, and Gauss-Seidel, along with variants that use relaxation, e.g., Gauss-Seidel with relaxation (`SOR`). Such methods require space for iteration vectors and $Q^{(N)}$ but usually converge quickly. More sophisticated projection methods—e.g., the generalized minimum residual method (`GMRES`) and the method of Arnoldi (`ARNOLDI`)—have better convergence properties but require more space. While the best method for a particular Markov chain is unclear in general, several options are available for exploration including the iterative methods described above that can be also enhanced with preconditioning, aggregation-disaggregation (`AD`), or Kronecker-specific multi-level (`ML`) methods that are inspired by multigrid and `AD` techniques. Unfortunately, we cannot acknowledge all relevant work on iterative methods due to limited space.[19,20]

A number of software tools are available that implement methods for Kronecker representations, and we selected the APNN toolbox[21] and its numerical solution package Nsolve for its rich variety of numerical techniques for the steady state analysis of Markov chains. Nsolve provides more than 70 different methods and comes with an ASCII file format for a SAN descriptor easily interfaced with our MATLAB modeling environment. Nsolve mainly supports hierarchical Markovian models that include a trivial hierarchy with a single macrostate such as Eq. 10 as a special case (see Refs. 21–24).

4. Results

In order to investigate which numerical techniques work best for the Kronecker representation of our Ca^{2+} release site models, we wrote a script for the matrix computing tool MATLAB that takes a specific Ca^{2+} release site model—defined by $K_+, K_-, \boldsymbol{e}_{\mathcal{O}}$, c_∞, and C—and produces the input files needed to interface with Nsolve. Using 10 three-state channels (1) we performed a preliminary study to determine which of the 70-plus numerical methods implemented in Nsolve were compatible with Eq. 10.

4.1. *Benchmark Stationary Distribution Calculations*

Table 1 lists those solvers that converged in less than 20 minutes CPU time with a maximum residual less than 10^{-12} for one configuration of 10 three-state channels. For each method we report the maximum and sum of the residuals, the CPU and wall clock times (in seconds), and the total number of iterations performed. We find that traditional relaxation methods (e.g., `JOR`, `RSOR`) work well for this problem with $3^{10} = 59,049$ states, but the addition of `AD` steps is not particularly helpful. `AD` steps do however greatly improve the performance of the `GMRES` solver and to a smaller extent the `DQGMRES` and `ARNOLDI` methods. The separable preconditioner (`PRE`) of Buchholz[23] and the `BSOR` preconditioner are very effective and help to reduce solution times to less than 50 seconds for several projection methods. Among `ML` solvers, a `JOR` smoother gives the best results and dynamic (`DYN`) or cyclic (`CYC`) ordering is better than a fixed (`FIX`) order where `V`, `W`, or `F` indicate the type of cycle used.[19,20]

4.2. *Problem Size and Method Performance*

In Sec. 4.1 we benchmarked the efficiency of several different algorithms that can be used to solve for the stationary distribution of Ca^{2+} release site models. To determine if this result depends strongly on problem size, we chose representatives of four classes of solvers (`JOR`, `PRE_ARNOLDI`, `BSOR_BICGSTAB`, and `ML_JOR_F_DYN`) that worked well for release sites composed of 10 three-state channels (see Table 1). Using these four methods, Fig. 2 shows the wall clock time required for convergence of $\boldsymbol{\pi}^{(N)}$ as a function of the number of channels (N) for both the three- and six-state models (circles and squares, respectively). Because the N channels in each Ca^{2+} release site simulation have randomly chosen positions that may influence the time to convergence, Fig. 2 shows both the mean and standard deviation (error bars) of the wall clock time for five different release site configurations. Note that for each value of N in Fig. 2, the radius of each Ca^{2+} release site was chosen so that stochastic Ca^{2+} excitability was observed. Due to irregular release site ultrastructure, these calculations can not be simplified using spatial symmetries.

Figure 2 shows that the time until convergence is shorter when the Ca^{2+}

Table 1. Benchmark calculations for 10 three-state channels computed using Linux PCs with dual core 3.8GHz EM64T Xeon processors and 8GB RAM solving Eq. 10.

Solver	Max Res	Sum Res	CPU	Wall	Iters
JOR	9.49E-13	5.16E-12	279	279	1840
SOR	9.49E-13	5.16E-12	435	436	1840
RSOR	8.76E-13	2.40E-12	1190	1197	990
JOR_AD	9.44E-13	5.13E-12	415	415	1550
SOR_AD	9.44E-13	5.13E-12	413	414	1550
DQGMRES	9.87E-13	6.78E-10	490	492	2940
ARNOLDI	2.42E-13	4.04E-11	214	215	1440
BICGSTAB	8.66E-13	4.89E-11	146	148	602
GMRES_AD	6.43E-13	3.61E-11	88	89	900
DQGMRES_AD	1.03E-12	1.84E-10	184	184	2008
ARNOLDI_AD	7.23E-13	7.60E-11	109	109	1280
PRE_POWER	9.37E-13	5.27E-12	246	247	1670
PRE_GMRES	8.62E-15	3.73E-12	45	46	180
PRE_ARNOLDI	8.62E-15	1.82E-12	26	27	160
PRE_BICGSTAB	4.44E-16	2.49E-14	28	28	188
BSOR_BICGSTAB	8.22E-15	5.29E-13	19	19	52
BSOR_GMRES	3.05E-13	7.73E-12	20	20	49
BSOR_TFQMR	1.83E-13	1.39E-12	17	17	48
PRE_GMRES_AD	1.29E-13	1.52E-11	36	36	140
PRE_ARNOLDI_AD	4.32E-13	7.18E-12	27	28	140
ML_JOR_V_FIX	9.69E-13	3.54E-11	105	105	372
ML_JOR_W_FIX	9.12E-13	1.14E-10	156	157	326
ML_JOR_F_FIX	9.93E-13	1.01E-10	146	146	330
ML_JOR_V_CYC	8.35E-13	6.36E-12	42	43	168
ML_JOR_W_CYC	4.36E-13	5.41E-11	26	26	38
ML_JOR_F_CYC	6.76E-13	1.39E-11	18	19	56
ML_JOR_V_DYN	8.07E-13	6.09E-12	58	59	152
ML_JOR_W_DYN	2.81E-13	5.15E-11	14	15	38
ML_JOR_F_DYN	5.87E-13	1.68E-10	15	15	46

release site is composed of three-state as opposed to six-state channels regardless of the numerical method used (compare circles to squares). Consistent with Table 1 we find that for large values of N the ML_JOR_F_DYN (black) method requires the shortest amount of time, followed by BSOR_BICGSTAB (dark gray), PRE_ARNOLDI (light gray), and finally JOR (white). Though there are important differences in the speed of the four solvers, the wall clock time until convergence is approximately proportional to the number of states (M^N), that is, the slope of each line in Fig. 2 is nearly $M = 3$ or 6 depending on the single channel model used.

We also experienced substantial differences in the amount of memory needed to run those solvers. While simple methods like JOR and SOR allocate space mainly for a few iteration vectors, Krylov subspace methods like

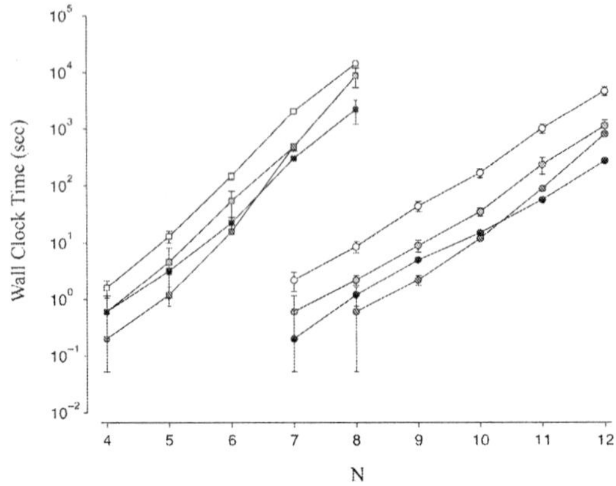

Fig. 2. *Circles* and error bars show the mean ± SD of wall clock time for five release site configurations of the three-state model (1) using: `JOR` (*white*), `PRE_ARNOLDI` (*light gray*), `BSOR_BICGSTAB` (*dark gray*), and `ML_JOR_F_DYN` (*black*). Three-state model parameters: $k_a^+ = 1.5\ \mu\text{M}^{-1}\ \text{ms}^{-1}$, $k_a^- = 50\ \text{ms}^{-1}$, $k_b^+ = 150\ \mu\text{M}^{-1}\ \text{ms}^{-1}$, $k_b^- = 1.5\ \text{ms}^{-1}$. *Squares* and error bars give results for the six-state model (parameters as in Ref. 12). Calculations performed using 2.66 GHz Dual-Core Intel Xeon processors and 2 GB RAM.

`GMRES`, `DQGMRES` and `ARNOLDI` use more vectors (20 in the default Nsolve configuration), and this can be prohibitive for large models. For projection methods that operate on a fixed and small set of vectors like `TFQMR` and `BICGSTAB`, we observe that the space for auxiliary data structures and vectors is on the order of 7–10 iteration vectors for these models. In general we find that the iterative numerical methods that incorporate pre-conditioning techniques are quite fast compared to more traditional relaxation techniques such as `JOR`. However, the power of pre-conditioning is only evident when problem size is less than some threshold that depends upon memory limitations. On the other hand, `ML` methods are constructed to take advantage of the Kronecker representation and to have very modest memory requirements. This is consistent with our experiments that indicate `ML` methods have the greatest potential to scale well with problem size, whether that be an increase in the number of channels (N) or the number of states per channel (M).

4.3. *Comparison of Iterative Methods and Monte Carlo Simulation*

Although there may be problem size limitations, we expected that the stationary distribution of our Ca^{2+} release site models could be found more quickly using iterative methods than Monte Carlo simulation. This is confirmed in the convergence results of Fig. 3 using a release site composed of

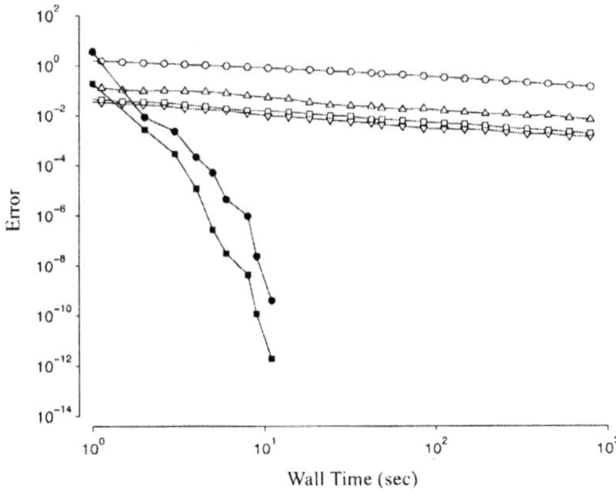

Fig. 3. Convergence of response measures for a release site composed of 10 three-state channels using ML_JOR_F_DYN and Monte Carlo (*filled* and *open* symbols, respectively). *Circles* and *squares* give 1- and ∞-norms of the residual errors, *upper pointing triangles* give the relative error in the puff/spark *Score* for Monte Carlo (mean of 50 simulations shown) compared with the *Score* given by ML_JOR_F_DYN upon convergence. Similarly, the *lower pointing triangles* give the relative error in the probability that all N channels are closed. Parameters as in Fig. 1.

10 three-state channels for both ML_JOR_F_DYN (filled symbols) and Monte Carlo simulation (open symbols). We run a Monte Carlo simulation to estimate the stationary distribution and that estimate depends on the length of the simulation measured in seconds of wall clock time (our implementation averaged 1,260 transitions per second). The simulation starts with all N channels in state C_1—chosen because it is the most likely state at the background $[\text{Ca}^{2+}]$ (c_∞). Figure 3 shows the maximum and sum of 1- and ∞-norms of the residuals averaged over 50 simulations. As expected, the residuals associated with the Monte Carlo simulations converge much slower than those obtained with ML_JOR_F_DYN. Interestingly, Fig. 3 shows that even coarse response measures can be more quickly obtained using numerical iterative methods than Monte Carlo simulation. We find that the relative errors of the puff/spark *Score* (upwards pointing triangles) and the probability that all N channels were closed (downwards pointing triangles) obtained via Monte Carlo simulation did not converge significantly faster than the maximum residual error (open squares).

5. Conclusions

We have presented a Kronecker structured representation for Ca^{2+} release sites composed of Ca^{2+}-regulated Ca^{2+} channels under the assumption that these channels interact instantaneously via the buffered diffusion of intracellular Ca^{2+} (Sec. 2). Because informative response measures such as the puff/spark *Score* can be determined if the steady-state probability of each release site configuration is known, we have identified numerical interative solution techniques that perform well in this biophysical context.

The benchmark stationary distribution calculations presented here indicate significant performance differences among iterative solution methods. Multi-level methods provide excellent convergence with modest additional memory requirements for the Kronecker representation of our Ca^{2+} release site models. When the available main memory permits, `BSOR`-preconditioned projection methods such as `TFQMR` and `BICGSTAB` are also effective, as is the method of Arnoldi combined with a simple preconditioner. In case of tight memory constraints, Jacobi and Gauss-Seidel iterations are also possible (but slower). When numerical iterative methods apply, they outperform our implementation of Monte Carlo simulation for estimates of response measures such as the puff/spark *Score* and the probability of a number of channels being in a particular state.

Single channel models of IP_3Rs and RyRs can be significantly more complicated than the three- and six-state models that are the focus of this manuscript. For example, the well-known DeYoung-Keizer IP_3R model includes four eight-state subunits per channel for a total of 330 distinguishable states.[25] Because biophysically realistic Ca^{2+} release site simulations can involve tens or even hundreds of intracellular channels, we expect that the development of approximate methods for our SAN descriptor (Eq. 10) will be an important aspect of future work. Of course, some puff and spark statistics—such as puff/spark duration and inter-event interval distributions—cannot be determined from the Ca^{2+} release site stationary distribution. Consequently, it will be important to determine if transient analysis can also be accelerated by leveraging the Kronecker structure of Ca^{2+} release sites composed of instantanteously coupled Ca^{2+}-regulated Ca^{2+} channels. Furthermore, although the SAN conceptual framework and its associated analysis techniques presented in this manuscript have focused solely on the emergent dynamics of Ca^{2+} release sites, it is also important to note that these techniques should be generally applicable to our understanding of signaling complexes of other kinds.[26,27]

Acknowledgments

The authors thank Buchholz and Dayar for sharing their implementation of Nsolve. This material is based upon work supported by the National Science Foundation under Grants No. 0133132 and 0443843.

References

1. D. Colquhoun and A. Hawkes, A Q-matrix cookbook: how to write only one program to calculate the sigle-channel and macroscopic predictions for any kinetic mechanism, in *Single-Channel Recording*, eds. B. Sakmann and E. Neher (Plenum Press, New York, 1995) pp. 589–633.
2. G. Smith, Modeling the stochastic gating of ion channels, in *Computational Cell Biology*, eds. C. Fall, E. Marland, J. Wagner and J. Tyson (Springer-Verlag, 2002) pp. 291–325.
3. F. Ball, R. Milne, I. Tame and G. Yeo, *Advances in App Prob* **29**, 56 (1997).
4. F. Ball and G. Yeo, *Methodology and Computing in App Prob* **2**, 93 (1999).
5. H. Cheng, W. Lederer and M. Cannell, *Science* **262**, 740 (1993).
6. H. Cheng, M. Lederer, W. Lederer and M. Cannell, *Am J Physiol* **270**, C148 (1996).
7. Y. Yao, J. Choi and I. Parker, *J Physiol* **482**, 533 (1995).
8. I. Parker, J. Choi and Y. Yao, *Cell Calcium* **20**, 105 (1996).
9. M. Berridge, *J Physiol (London)* **499**, 291 (1997).
10. V. Nguyen, R. Mathias and G. Smith, *Bull. Math. Biol.* **67**, 393 (2005).
11. S. Swillens, G. Dupont, L. Combettes and P. Champeil, *Proc Natl Acad Sci USA* **96**, 13750(Nov 1999).
12. H. DeRemigio, P. Kemper, M. LaMar and G. Smith, *Technical Report* **WM-CS-2007-06** (2007).
13. G. Smith, An extended DeYoung-Keizer-like IP_3 receptor model that accounts for domain Ca^{2+}-mediated inactivation, in *Recent Research Developments in Biophysical Chemistry, Vol. II*, eds. C. Condat and A. Baruzzi (Research Signpost, 2002).
14. I. Bezprozvanny, *Cell Calcium* **16**, 151 (1994).
15. M. Naraghi and E. Neher, *J Neurosci* **17**, p. 6961(6973 1997).
16. G. Smith, L. Dai, R. Muira and A. Sherman, *SIAM J Appl Math* **61**, 1816 (2001).
17. W. Stewart, *Introduction to the Numerical Solution of Markov Chains* (Princeton University Press, Princeton, 1994).
18. D. Gillespie, *J Comp Phys* **22**, 403 (1976).
19. P. Buchholz and T. Dayar, *Computing* **73**, 349 (2004).
20. P. Buchholz and T. Dayar, *SIAM Matrix Analysis and App (to appear)* (2007).
21. P. Buchholz and P. Kemper, A toolbox for the analysis of discrete event dynamic systems., in *CAV. LNCS 1633*, 1999.
22. P. Buchholz and T. Dayar, *SIAM J. Sci. Comput.* **26**, 1289 (2005).
23. P. Buchholz, Projection methods for the analysis of stochastic automata networks, in *Numerical Solution of Markov Chains*, eds. B. Plateau, W. Stewart and M. Silva (Prensas Univerversitarias de Zaragoza, 1999) pp. 149–168.
24. P. Buchholz, *Prob in the Eng and Informational Sci* **11**, 229 (1997).
25. G. De Young and J. Keizer, *Proc Natl Acad Sci USA* **89**, 9895 (1992).
26. J. Schlessinger, *Cell* **103**, 211 (2000).
27. H. Husi, M. A. Ward, J. S. Choudhary, W. P. Blackstock and S. G. Grant, *Nat Neurosci* **3**, 661 (2000).

SPATIALLY-COMPRESSED CARDIAC MYOFILAMENT MODELS GENERATE HYSTERESIS THAT IS NOT FOUND IN REAL MUSCLE

JOHN JEREMY RICE
YUHAI TU

IBM T.J. Watson Research Center
Yorktown Heights, NY 10598, USA

CORRADO POGGESI‡

Dipartimento di Scienze Fisiologiche,
Viale Morgagni 63, I-50134 Firenze, Italy

PIETER P. DE TOMBE†

Department of Physiology and Biophysics,
University of Illinois Chicago, Chicago, IL 60612, USA

In the field of cardiac modeling, calcium- (Ca-) based activation is often described by sets of ordinary differential equations that do not explicitly represent spatial interactions of regulatory proteins or crossbridge attachment. These spatially compressed models are most often mean-field representations as opposed to methods that explicitly compute the surrounding field (or equivalently, the surrounding environment) of individual regulatory units and crossbridges. Instead, a mean value is used to represent the whole population. Almost universally, the mean-field approach assumes developed force produces positive feedback to globally increase the mean binding affinity of the regulatory proteins. We show that this approach produces hysteresis in the steady-state Force-Ca responses when developed force increases Ca-affinity troponin to the degree that is observed in real muscle. Specifically, multiple stable solutions exist as a function of Ca level that could be alternatively reached depending on stimulus history. The resulting hysteresis is quite pronounced and disagrees with experimental characterizations in cardiac muscle that generally show little if any hysteresis. Moreover, we provide data showing that hysteresis does not occur in carefully controlled myofibril preparations. Hence, we suggest that the most widely used methods to produce multiscale models of cardiac force generation show bistability and hysteresis effects that are not seen in real muscle responses

‡Work partially supported by MUR (PRIN2006) and Università di Firenze (ex-60%).
† Work partially supported by NIH grants HL-62426 (project 4), HL-75494 and HL73828.

1. Introduction

As described in a previous review [1], there are still difficulties in developing predictive myofilament models given that the underlying muscle biophysics has yet to be fully resolved. Another difficulty lies in trying to compress the spatial aspects of myofilaments at the molecular level into a tractable system of equations. Partial differential equations or Monte Carlo approaches are typically required for explicit consideration of the spatial interactions, whereas spatially-compressed sets of ordinary differential equations (ODEs) are required for computational efficiency to allow large-scale multicellular models.

The spatially compressed models can be termed mean-field as opposed to methods that explicitly compute the surrounding field (environment) of individual regulatory units and/or crossbridges; instead, a mean value is used to represent the whole population. The most widely used approach is that force/activation level produces positive feedback to globally increase the mean binding affinity of the regulatory unit (troponin/tropomyosin). The mean-field approach is used in almost all ODE-based modeling efforts from diverse research groups. Recent examples are refinements of earlier models (e.g., [2-4]). We construct a generic version of this approach and show that hysteresis and bistability can result from this construction.

2. Method

Most myofilament models contain a strong positive feedback of muscle activation to increase Ca binding to regulatory units. This feedback plays a dual role in both simulating experimental observed increases in Ca affinity and providing a mechanism to produce steep Ca sensitivity and high apparent cooperativity (often in conjunction with other mechanisms). A typical mean-field approach to modeling cardiac myofilaments is shown in Fig. 1A. Here the state names are coded with 0 for no Ca bound or 1 for Ca bound in the first character. The second character is W for weakly-bound (non-force generating) or S for strongly-bound (force generating) crossbridges. Activation occurs as increasing [Ca] will cause transition from the rest state (0W) to a Ca-bound state that is still weakly bound (1W). Transitions between weakly- and strongly-bound states are controlled by constants f and g that represent apparent weak-to-strong binding transition as typically defined in two-state crossbridge schemes. Note that the right-hand side has only the crossbridge detachment step that illustrates an implicit assumption which is that crossbridges do not strongly bind and generate force when no Ca is bound to the associated regulatory proteins.

Very similar approaches have been employed and explained in depth elsewhere (e.g., [5, 6]). For the remainder of the paper, we will refer to the approach as global feedback on Ca-binding affinity (GFCA).

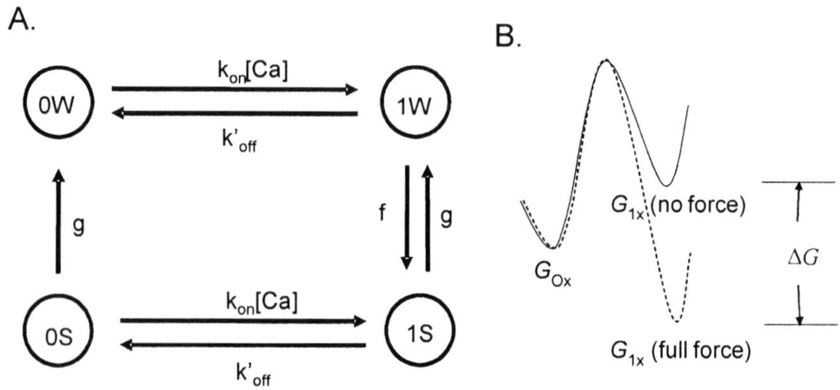

Figure 1: Generic model of force feedback on Ca binding. A. State diagram with transition rates. B. Schematic of assumed energy diagram for Ca binding where the free energy of the Ca-bound state is assumed to decrease as the model transitions from no force to full force.

For the model shown, the normalized developed force can be computed as fraction of strong bound crossbridges as shown below

$$FractSBXB = \frac{0S + 1S}{f/(f+g)} \tag{1}$$

where 0S and 1S refer to the fractional occupancy of the respective states and the denominator is the theoretical fraction of strongly–bound crossbridge for the limiting case of high [Ca] conditions (hence, only states 1W and 1S are populated).

The Ca binding is described by the left to right transitions. Ca binding is assumed to be more complicated than a simple buffer in that the dissociation constant is a function of the developed force. Specifically we assume the following formulation

$$K_d' = \frac{k_{off}'}{k_{on}} = \frac{k_{off}}{k_{on}} e^{-\frac{\Delta G}{RT} FractSBXB} \tag{2}$$

where ΔG is the change in free energy of the Ca-bound state as the system transitions from no force to fully developed force (see Fig. 1B). The other constants are the Universal Gas Constant (R) and the absolute temperature (T).

The forward transition k_{on} is assumed fixed because Ca binding is generally assumed to be diffusion limited. In contrast, the backward transition k'_{off} is assumed to be a function of developed force. As *FractSBXB* transitions from a minimum value of 0 to a maximum value of 1, k'_{off} will decrease from the default value equal k_{off} to the minimum value of k_{off} exp(-ΔG/RT). For the simulation shown the default values of the parameters are k_{on} = 50 μM^{-1} s^{-1}, k_{off} = 500 s^{-1}, f = 40 s^{-1} and g = 10 s^{-1}. Similar values are used in previous studies and are justified elsewhere [5, 6].

3. Results

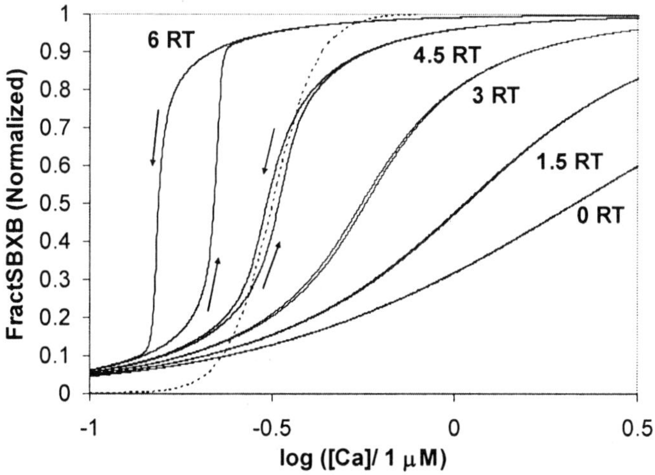

Figure 2: Pseudo-steady-state responses for the generic GFCA model for varying levels of ΔG as labeled. Increasing ΔG produces both an increase in apparent cooperativity and also leads to hysteresis. The dashed trace shows a true Hill function with N_H = 7 similar to what is measured experimentally for real muscle [7, 8].

3.1. *Pseudo-steady-state solution*

The system shown in Fig. 1 can be solved for the approximate steady-state response by slowly increasing the [Ca] (-3 to 3 in units of log([Ca]/1 μM)) over 160 s so that the model is in approximate steady-state conditions. The [Ca] is lowered from the maximum values over the next 160 s to check for hysteresis. As shown in Fig. 2, the steady-state Force-Ca (F-Ca) relation increase in steepness and apparent cooperativity with ΔG. When the ΔG = 4.5RT, the middle part of the curve has a steepness that approximates real cardiac muscle that has a Hill coefficient (N_H) of approximately 7 for sarcomere lengths (SLs) in the range of 1.85-2.15 μm [7]. Note that the shape of the model F-Ca relation deviates from that of a true Hill function ($N_H = 7$) as shown by the dashed line. Specifically, the model response shows relatively little apparent cooperativity in the low [Ca] and high [Ca] regimes with the most steepness near the mid-force regions. In fact, increasing the level of GFCA by setting $\Delta G = 6$ RT will increase the steepness in the mid-force regions but does little to increase the apparent cooperativity outside this regime. Such behavior has been described before as generic behavior of the GFCA models, and thus may hamper their appropriateness model for simulating real muscle responses [1]. However, the focus of the paper here is on the hysteresis that can occur when the GFCA is strong. Note that little or no hysteresis is seen for $\Delta G = 3.0$ RT, 1.5 RT or 0 RT. However these lower values do not generate steep enough F-Ca relations to replicate real muscle response as seen in the literature (e.g., [7-9]).

3.2. *True steady-state solution*

The hysteresis behavior shown in Fig. 2 could potentially be an artifact of not reaching steady state in the traditional sense of $t \rightarrow \infty$. Similar effects can often be seen in models when [Ca] is changed too quickly. We analyzed the true steady-state response using AUTO as part of the XPPAUT software package[§]. Briefly, AUTO implements continuation methods that compute a family of fixed points of a non-linear system as one or more parameters are varied. Commonly, continuation methods start at an initial fixed point and then use the system Jacobian to extend the solution as parameters are varied. Iteration produces a continuous family of fixed points, and Jacobian singularities signal bifurcations.

Figure 3B shows that true steady-state hysteresis does occur for values of ΔG in the range of 5.791 to 6.048 RT. The data correspond to one [Ca] level, but the general behavior is found for [Ca] values near the Ca_{50} with similar values of ΔG. The effects of the two stable solutions are illustrated in Fig. 3B where the model is started at either high or low force levels. The upper trace for $\Delta G = 6$ RT is producing more force for lower [Ca] compared to the lower trace which is started at the lower force level. Moreover, the lower traces $\Delta G = 6$ RT show extreme parameter sensitivity as the steady-state solution may change branches on the bifurcation diagram in Fig. 3A.

For lesser values of ΔG, while true steady-state hysteresis is not found, one can still observe that the model takes several seconds to reach steady-state. As

---

[§] http://www.math.pitt.edu/~bard/xpp/xpp.html

shown in Fig. 3B, the $\Delta G = 4.5$ RT take relatively long to settle to a steady-state value near 50% force. The long time to reach steady state is not intuitive given that all model rate constants are $\geq 10\ s^{-1}$, suggesting a time constant of relaxation on the order of 100 ms. Moreover, the delay also shows why hysteresis effects can appear in the $\Delta G = 4.5$ RT trace in Fig. 2. Note that hysteresis appears in pseudo-steady-state response in Fig. 2 but not true steady-state in Fig. 3. Hence, for biological systems that have finite lifetimes (especially for *in vitro* preparations where data collection is limited), a long-time to settle to steady state may produce hysteresis-like effects even if not in the traditional sense of $t \rightarrow \infty$.

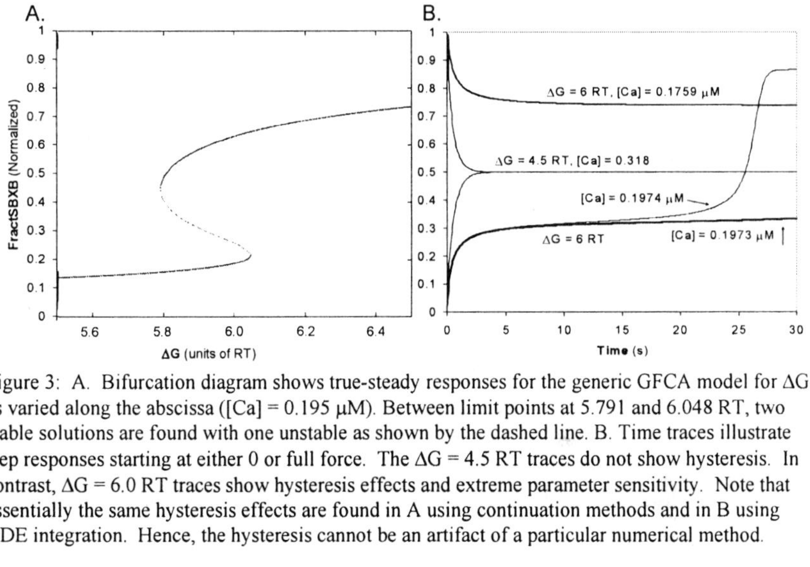

Figure 3: A. Bifurcation diagram shows true-steady responses for the generic GFCA model for ΔG as varied along the abscissa ([Ca] = 0.195 μM). Between limit points at 5.791 and 6.048 RT, two stable solutions are found with one unstable as shown by the dashed line. B. Time traces illustrate step responses starting at either 0 or full force. The ΔG = 4.5 RT traces do not show hysteresis. In contrast, ΔG = 6.0 RT traces show hysteresis effects and extreme parameter sensitivity. Note that essentially the same hysteresis effects are found in A using continuation methods and in B using ODE integration. Hence, the hysteresis cannot be an artifact of a particular numerical method.

3.3. *Comparison to other models*

Several published models show similar behavior to the generic model developed above. As an example, Fig. 4 shows a simulated F-Ca relation for the model proposed in [4] for SL = 1.8, 2.0 and 2.2 μm. In this model, the actual change in Ca-binding affinity is roughly 20 fold. In addition, the change in Ca-affinity is assumed to increase using a Hill-like function ($N_H = 3.5$) of the concentration of strongly-bound crossbridges. Note that a 20-fold change in affinity corresponds to $\Delta G = 3$ RT which does not produce substantial hysteresis in the generic model (see Fig. 2). However, the additional nonlinearity in the Hill function generates the higher level of hysteresis seen in Fig. 4. While we have not reprinted data here, the model in [4] shows true steady-state hysteresis when stepped to different levels of [Ca] (compare SL = 1.7 μm trace in Fig. 6 in [4] with the data in Fig. 3B; the SL = 2.2 μm trace is operating higher on the F-Ca relation where hysteresis is not seen).

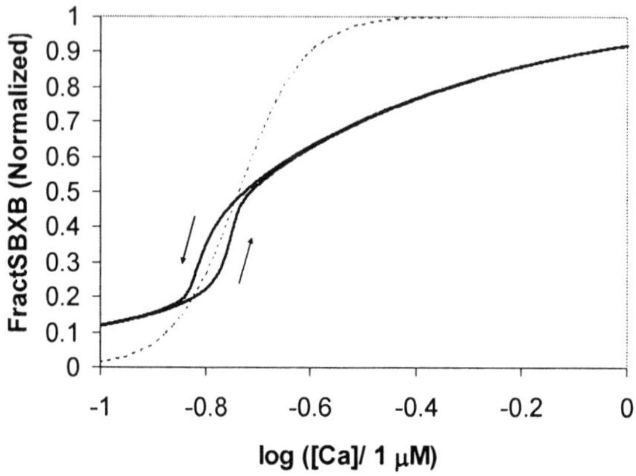

Figure 4: The F-Ca relation of the Yaniv et al. model [4] for SL = 1.8, 2.0 and 2.2 μm (traces essentially overlap) show clear hysteresis as seen in the generic model in Fig. 1. The protocol is the same as in Fig. 2. The dashed trace shows a true Hill function with $N_H = 7$ similar to what is measured experimentally for real muscle [7, 8].

4. Discussion

4.1. *Implications of modeling results*

The modeling formalism, shown in Fig. 1, is developed here as the most generic formulation of GFCA. As shown in the previous sections, the approach generalizes to published models and also represents the mean-field approximation of the spatially-explicit approaches [10]. However, GFCA produces steady-state F-Ca relations that deviate substantially from true Hill functions in ways that real muscles do not, i.e., are too steep at mid-level [Ca] and not cooperative enough at low and high [Ca] regimes (see Figs. 2 and 4).

If GFCA is the only cooperative mechanism in a model, then the assumed change in Ca-binding affinity is much larger than experimental estimates. Specifically, experimental estimates suggest a maximal affinity change of 15-20 fold [1, 5, 6]. In contrast, the results in Fig. 2 suggest a change of approximately 90 fold is required to replicate the degree of cooperativity seen in real F-Ca relations. This finding casts doubt on the ability to produce models with realistic Ca sensitivity when GFCA is the only cooperative mechanism, as is the case for some published models.

While GFCA may be insufficient to generate steep enough F-Ca relations, one might assume that other cooperative mechanisms could be added to improve the steepness of F-Ca relations. This approach has been tried in many published modeling efforts (e.g., [3, 5, 11]). However, adding even a small amount of GFCA can produce the undesirable effects of increasing apparent cooperativity at mid-level [Ca] but not at low and high [Ca] regimes [1, 5]. As a specific example, compare F-Ca results for models with GFCA (M1 and M2) with the model without (M5) in Fig. 5 of Schneider et al. [11]. Only models without GFCA seem to be able to produce F-Ca relations that resemble a Hill function as seen in real muscle. While a complete analysis of all published models is not possible here, we suspect that adding GFCA with other cooperative mechanisms can also produce marked bistability in the F-Ca relation (e.g., model in Fig. 4 has additional Hill-like cooperative effects in Ca binding). As the next section discusses, high levels of bistability do not generally agree with experiment results.

The modeling results here are for steady-state [Ca] and fixed muscle lengths. In a real contracting ventricle, both [Ca] and muscle length will be varying with time so that hysteresis effects may be masked. However, the dynamic responses of muscle are strongly affected by the steady-state Ca sensitivity, and GFCA has been proposed to produce activation and relaxation kinetics that are slower in models than in real muscle [1, 5]. Figure 3B explicitly shows this slowing for a step response in Ca level. We envision pathological conditions (e.g., congestive heart failure) for which a prolonged Ca transient and/or increased diastolic transient could unmask the hysteresis.

4.2. *Experimental evidence of hysteresis*

Experimental evidence for hysteresis in the activation of the myofilament was first reported in single muscle fibers of the barnacle by Ridgway *et al.* [12]. The fibers were either micro injected with aquorin to measure intracellular calcium and electrically stimulated or chemically permeabilized (skinned) by treatment with detergent. In both cases, these investigators found larger force at equivalent levels of activator Ca when the muscle had first experienced a higher level of contractile activation. Brief periods of full relaxation, on the other hand, were sufficient to eliminate this "memory" or hysteresis effect. A follow up study by Brandt et al., however, failed to confirm these results in skinned vertebrate skeletal muscle fibers [13]. Another phenomenon that may, or may not, be related to hysteresis is stretch activation in skeletal muscle, first described by Edman et al. [14]. Here, a tetanized single skeletal muscle is stretched, relatively slowly, for a brief period and then returned to the original muscle length. The stretch resulted in a change in tetanic force precisely as predicted by the active force-length relation. However, sustained elevated tetanic force is found only following the brief stretch-release maneuver on the descending limb of the force-SL relationship, and hence, is unlikely to occur in cardiac tissue that does not operate on the descending limb (see [1]).

The most comprehensive, and to our knowledge only, study on myofilament activation hysteresis has been reported by Harrison *et al.* [15]. In that study, skinned rat myocardium was sequentially immersed into solutions containing varying amounts of activator Ca. Similar to the Ridgway *et al.* study, prior exposure to a high [Ca] led to an apparent left shift of the F-Ca relationship consistent with an increase in overall myofilament Ca sensitivity. Interestingly, this phenomenon was most pronounced at short SLs and virtually disappeared at SL>2.1 μm (i.e., lengths for which actin double overlap is no longer present). Moreover, osmotic compression of the myofilament lattice by application of dextran, a high molecular weight compound that cannot enter the space between contractile filaments [16, 17], eliminated hysteresis. Based on the SL dependence of hysteresis and its elimination by osmotic compression, these authors speculated that prior activation at the higher Ca levels induced a persistent reduction in inter-filament spacing to increase Ca sensitivity. Although not the specific focus of our studies, we have nevertheless not found evidence for hysteresis in inter-filament spacing as measured by x-ray diffraction in either intact or skinned isolated skeletal or cardiac muscle [16-19].

Studies on both intact [20-22] and skinned [7, 23] myocardium do not find evidence for hysteresis in F-Ca relationships, albeit hysteresis was not the primary focus of these studies. Likewise, intact cardiac trabeculae with pharmacologically slowed Ca transients show prolonged relaxations that occur along a single F-Ca relation that is independent of the preceding developed force (see Figs. 5-6 in [9] and Fig. 6 in [24]). Finally, hysteresis of the type referred to above as "stretch activation" is expected to lead to a significant phase shift in sinusoidal perturbation analysis experiment at frequencies close to DC. Although there is some indication in skeletal muscle for such a phenomenon [25, 26], this has not been observed in isolated cardiac muscle [27-30].

We propose that the controversial hysteresis finding above may result from inadequate control of the ionic environment surrounding the myofilaments.

Specifically, diffusion delays in activation-relaxation dynamics are a significant limitation associated with the study of large isolated fibers (such as the barnacle single fiber) or multi-cellular isolated cardiac muscle. Hence, rapid changes in [Ca] in the bathing solution surrounding these muscles do not translate in equal changes in activator Ca as sensed by troponin. For this reason, the single myofibril rapid solution change technique has been widely adopted to study skeletal and cardiac muscle activation-relaxation dynamics [29, 31-36]. This technique employs single myofibrils or small bundles of myofibrils (~1-5 μm average diameters) that are mounted between two glass micro-pipettes; the ionic environment can be altered within ~5 ms by rapid solution switching. The short diffusion pathway coupled with continuous superfusion produce essentially no ambiguity in the ionic environment surrounding the myofilaments.

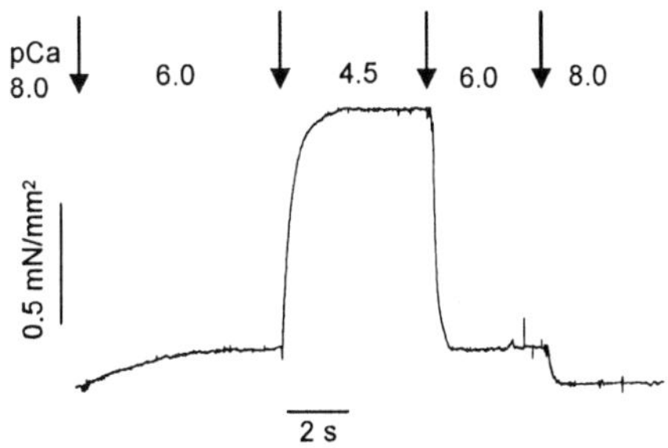

Figure 5: Activation-relaxation cycles recorded in human atrium cardiac muscle (15°C; initial SL = ~2.2 μm). Activator [Ca] is altered rapidly (within 5 ms) by rapid solution switching techniques. The actual [Ca] applied is as indicated in the figure in pCa units (pCa = -log([Ca]/1 M). Similar to previous studies in skeletal muscle [33, 34, 36, 37], there is no apparent hysteresis in myofilament steady-state force level. Unpublished results from the laboratory of C. Poggesi.

As seen in Fig. 5, hysteresis in myofilament steady-state activation level cannot be readily detected and hence is small if extant. These data can be qualitatively compared to Fig. 4B that shows pronounced history dependence for $\Delta G = 6$ RT (see also Fig. 6 in [4]). Also, there is no variation in the kinetic parameters of myofilament activation. Thus, the rate of force development is a direct function of the [Ca], being faster at higher [Ca], regardless of the activation history that precedes the switch to a particular [Ca]. Furthermore, the rate of force relaxation is relatively slow and not affected by the level of Ca activation from which relaxation is initiated [33, 34, 36, 37]. Overall, these experiments suggest that there is little, if any, hysteresis in myofilament Ca activation.

5. Conclusion

The paper has shown that bistability and hysteresis in the F-Ca response is an inherent behavior in models with high levels of GFCA. We have also shown that such behaviors can result with lesser amounts of GFCA when other cooperative mechanisms are represented. In contrast, experimental data suggests little or no hysteresis in real muscle responses. Hence, one should consider these effects when using spatially compressed ODE-based models that include GFCA. Moreover, the ODE-based models are often developed to combine single cells into multiscale tissue-level models. If bistability and hysteresis exist in the single cells, one could envision situations in which the stability of larger scale models could be adversely affected because individual cells can reach multiple stable steady-state forces depending on small changes in the stimulus and environment histories of each cell.

References

1. J.J. Rice & P.P. de Tombe, Prog Biophys Mol Biol. 85, No. 2-3, 179-95 (2004).
2. L.B. Katsnelson & V.S. Markhasin, J Mol Cell Cardiol. 28, No. 3, 475-86. (1996).
3. S.A. Niederer, P.J. Hunter & N.P. Smith, Biophys J. 90, No. 5, 1697-722 (2006).
4. Y. Yaniv, R. Sivan & A. Landesberg, Am J Physiol Heart Circ Physiol. 288, No. 1, H389-99. (2005).
5. J.J. Rice, R.L. Winslow & W.C. Hunter, Am J Physiol. 276, No. 5 Pt 2, H1734-54 (1999).
6. A. Landesberg & S. Sideman, Am J Physiol. 266, No. 3 Pt 2, H1260-71 (1994).
7. D.P. Dobesh, J.P. Konhilas & P.P. de Tombe, Am J Physiol Heart Circ Physiol. 282, No. 3, H1055-62 (2002).
8. J.C. Kentish & A. Wrzosek, J Physiol. 506, No. Pt 2, 431-44. (1998).
9. L.E. Dobrunz, P.H. Backx & D.T. Yue, Biophys J. 69, No. 1, 189-201 (1995).
10. J.S. Shiner & R.J. Solaro, Biophys J. 46, No. 4, 541-3. (1984).
11. N.S. Schneider, T. Shimayoshi, A. Amano & T. Matsuda, J Mol Cell Cardiol. 41, No. 3, 522-36 (2006).
12. E.B. Ridgway, A.M. Gordon & D.A. Martyn, Science. 219, No. 4588, 1075-7. (1983).
13. P.W. Brandt, B. Gluck, M. Mini & C. Cerri, J Muscle Res Cell Motil. 6, No. 2, 197-205. (1985).

14. K.A. Edman, G. Elzinga & M.I. Noble, J Gen Physiol. 80, No. 5, 769-84. (1982).
15. S.M. Harrison, C. Lamont & D.J. Miller, J Physiol. 401, No., 115-43. (1988).
16. J.P. Konhilas, T.C. Irving & P.P. de Tombe, Circ Res. 90, No. 1, 59-65. (2002).
17. G.P. Farman, J.S. Walker, P.P. de Tombe & T.C. Irving, Am J Physiol Heart Circ Physiol. 291, No. 4, H1847-55 (2006).
18. G.P. Farman, E.J. Allen, D. Gore, T.C. Irving & P.P. de Tombe, Biophys J. 92, No. 9, L73-5 (2007).
19. T.C. Irving, J. Konhilas, D. Perry, R. Fischetti & P.P. de Tombe, Am J Physiol Heart Circ Physiol. 279, No. 5, H2568-73. (2000).
20. H.E. ter Keurs, W.H. Rijnsburger, R. van Heuningen & M.J. Nagelsmit, Circ Res. 46, No. 5, 703-14. (1980).
21. P.P. de Tombe & H.E. ter Keurs, J Physiol. 454, No., 619-42 (1992).
22. P.P. de Tombe & H.E. ter Keurs, Circ Res. 66, No. 5, 1239-54. (1990).
23. J.C. Kentish, H.E. ter Keurs, L. Ricciardi, J.J. Bucx & M.I. Noble, Circ Res. 58, No. 6, 755-68 (1986).
24. P.H. Backx, W.D. Gao, M.D. Azan-Backx & E. Marban, J Gen Physiol. 105, No. 1, 1-19 (1995).
25. M. Kawai & P.W. Brandt, J.Muscle.Res.Cell Motil. 1, No., 279-303 (1980).
26. M. Kawai & Y. Zhao, Biophysical Journal. 65, No., 638-51 (1993).
27. T. Wannenburg, G.H. Heijne, J.H. Geerdink, H.W. Van Den Dool, P.M. Janssen & P.P. De Tombe, Am J Physiol Heart Circ Physiol. 279, No. 2, H779-90 (2000).
28. K.B. Campbell, M.V. Razumova, R.D. Kirkpatrick & B.K. Slinker, Biophys J. 81, No. 4, 2278-96 (2001).
29. M. Chandra, M.L. Tschirgi, S.J. Ford, B.K. Slinker & K.B. Campbell, Am J Physiol Regul Integr Comp Physiol. [Epub ahead of print], No. (2007).
30. M. Kawai, Y. Saeki & Y. Zhao, Circ Res. 73, No. 1, 35-50. (1993).
31. R. Stehle, M. Kruger & G. Pfitzer, Biophys J. 83, No. 4, 2152-61. (2002).
32. P.P. de Tombe, A. Belus, N. Piroddi, B. Scellini, J.S. Walker, A.F. Martin, C. Tesi & C. Poggesi, Am J Physiol Regul Integr Comp Physiol. 292, No. 3, R1129-36 (2007).
33. C. Tesi, F. Colomo, S. Nencini, N. Piroddi & C. Poggesi, Biophys J. 78, No. 6, 3081-92 (2000).
34. C. Tesi, N. Piroddi, F. Colomo & C. Poggesi, Biophys J. 83, No. 4, 2142-51 (2002).
35. K.B. Campbell, M.V. Razumova, R.D. Kirkpatrick & B.K. Slinker, Ann Biomed Eng. 29, No. 5, 384-405 (2001).
36. C. Poggesi, C. Tesi & R. Stehle, Pflugers Arch. 449, No. 6, 505-17 (2005).
37. C. Tesi, F. Colomo, S. Nencini, N. Piroddi & C. Poggesi, J Physiol. 516, No. Pt 3, 847-53. (1999).

MODELING VENTRICULAR INTERACTION: A MULTISCALE APPROACH FROM SARCOMERE MECHANICS TO CARDIOVASCULAR SYSTEM HEMODYNAMICS

JOOST LUMENS[1,2], TAMMO DELHAAS[1], BORUT KIRN[2], THEO ARTS[2]

Departments of [1]Physiology and [2]Biophysics, Maastricht University, Universiteitssingel 50, Maastricht, P.O. Box 616, The Netherlands

E-mail: j.lumens@fys.unimaas.nl; t.delhaas@fys.unimaas.nl; b.kirn@bf.unimaas.nl; t.arts@bf.unimaas.nl

Direct ventricular interaction via the interventricular septum plays an important role in ventricular hemodynamics and mechanics. A large amount of experimental data demonstrates that left and right ventricular pump mechanics influence each other and that septal geometry and motion depend on transmural pressure. We present a lumped model of ventricular mechanics consisting of three wall segments that are coupled on the basis of balance laws stating mechanical equilibrium at the intersection of the three walls. The input consists of left and right ventricular volumes and an estimate of septal wall geometry. Wall segment geometry is expressed as area and curvature and is related to sarcomere extension. With constitutive equations of the sarcomere, myofiber stress is calculated. The force exerted by each wall segment on the intersection, as a result of wall tension, is derived from myofiber stress. Finally, septal geometry and ventricular pressures are solved by achieving balance of forces. We implemented this ventricular module in a lumped model of the closed-loop cardiovascular system (CircAdapt model). The resulting multiscale model enables dynamic simulation of myofiber mechanics, ventricular cavity mechanics, and cardiovascular system hemodynamics. The model was tested by performing simulations with synchronous and asynchronous mechanical activation of the wall segments. The simulated results of ventricular mechanics and hemodynamics were compared with experimental data obtained before and after acute induction of left bundle branch block (LBBB) in dogs. The changes in simulated ventricular mechanics and septal motion as a result of the introduction of mechanical asynchrony were very similar to those measured in the animal experiments. In conclusion, the module presented describes ventricular mechanics including direct ventricular interaction realistically and thereby extends the physiological application range of the CircAdapt model.

1. Introduction

The left (LV) and right ventricle (RV) of the heart are pumping blood in the systemic and pulmonary circulation, respectively. Although both ventricular cavities are completely separated, there is a strong mechanical interaction between the ventricles, because they share the same septal wall, separating the cavities. A vast amount of evidence demonstrates that septal shape and motion depend on transseptal pressure [1, 2]. Also, a change in pressure or volume load of one ventricle influences pumping characteristics of the other ventricle [3-5].

Various mathematical models have been designed to describe the consequences of mechanical left-right coupling by the septum for ventricular geometry and hemodynamics [6-11]. Commonly, interaction is assumed to be global and linear, using coupling coefficients for pressures, volumes or compliances. An exception was found in the model by Beyar *et al.* [6], which was based on the balance of forces between free walls and septum. The latter model was primarily designed for diastolic interaction and was not suited to implement the dynamic mechanics of myocardial contraction.

The CircAdapt model [12] has been developed to simulate cardiovascular dynamics and hemodynamics of the closed-loop circulation. The model is configured as a network, composed of four types of modules, i.e., cardiac chamber, blood vessel, valve and flow resistance. The number of required independent input parameters was reduced tremendously by incorporating adaptation of geometry, e.g., size of ventricular cavities and thickness of walls, to mechanical load so that stresses and strains in the walls were normalized to physiological standard levels. Ventricular interaction was modeled as an outer myocardial wall, encapsulating both ventricles, and an inner wall around the LV cavity accommodating the pressure difference between LV and RV. This description is reasonable, as long as LV pressure largely exceeds RV pressure. However, for high RV pressures, the description is not accurate anymore.

Because of the need to describe pathologic circumstances with high RV pressure, a new model of left to right ventricular interaction was designed. This model should be symmetric in design, allowing RV pressure to exceed LV pressure. Furthermore, the new model should satisfy the following requirements to fit in the CircAdapt framework. 1) For given LV and RV volumes as input, LV and RV pressures should be calculated as a result. 2) The model should incorporate dynamic myofiber mechanics, responsible for pump action. 3) The model should satisfy conservation of energy, i.e., the total amount of contractile work, as generated by the myofibers, should equal the total amount of hydraulic pump work, as delivered by the ventricles.

In the present study, a model setup was found, satisfying abovementioned requirements. The LV and RV cavities are formed between an LV free wall segment and a septal wall segment and between the septal wall segment and an RV free wall segment, respectively. The area of each wall segment depends on myofiber length in that wall. Pressures are generated by wall tension in the curved wall segments. Equilibria of mechanical forces are used to restrict degrees of freedom for geometry.

The model was tested by manipulating timing of mechanical activation of the various wall segments. Consequences of left bundle branch block (LBBB) have been simulated for septal motion and timing of LV and RV pressure development. Model results were compared with experimental results reported earlier [2, 13-17].

2. Methods

2.1. *Model design*

In the model, LV and RV cavities are enclosed by an LV (L) and an RV (R) free wall segment, respectively. The cavities are separated by a shared septal wall segment (S) (Fig. 1). The wall segments are modeled as thick-walled spherical segments. The segments are assumed to be mechanically coupled at midwall. Midwall surface is defined to divide the wall in two spherical shells of equal wall volume. Midwall geometry of a wall segment depends on two variables, i.e., the bulge height of the spherical segment (x_s), and the radius of the midwall boundary circle (y) (Fig. 1). Midwall curvature, area, and volume of a wall segment can be expressed as a function of these two variables. Since all three wall segments share the same circle of intersection, four variables are needed to describe complete ventricular geometry, i.e., x_R, x_S, x_L, and y.

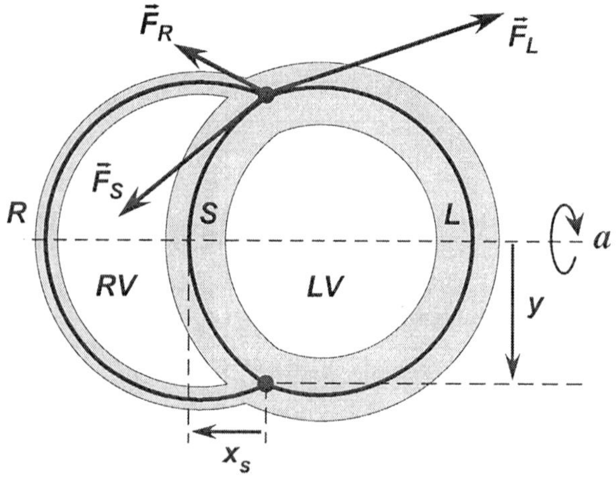

Figure 1: A cross-section of the model of ventricular mechanics. Three thick-walled spherical segments (shaded), i.e., the LV free wall segment (L), the RV free wall segment (R), and the septal wall segment (S) are coupled mechanically. The resulting ventricular composition is rotationally symmetric around axis a and has a midwall intersection circle crossing this image plane perpendicularly through the thick points. Midwall geometry of the septal wall segment is expressed by bulge height (x_s) and the radius (y) of the midwall intersection circle. In this intersection each wall segment exerts a force (F) caused by wall tension.

Moreover, in Fig. 4, experiment and simulation appear in agreement on the increase of asymmetry of the RV pressure curve with LBBB.

Figure 5 shows LV SFWD as derived from typical M-Mode echocardiograms acquired in a dog before (Control) and after induction of LBBB [14]. During LBBB, the experimental LV SFWD curve shows the same typical motion pattern of the septum early in systole as seen in the LBBB simulation.

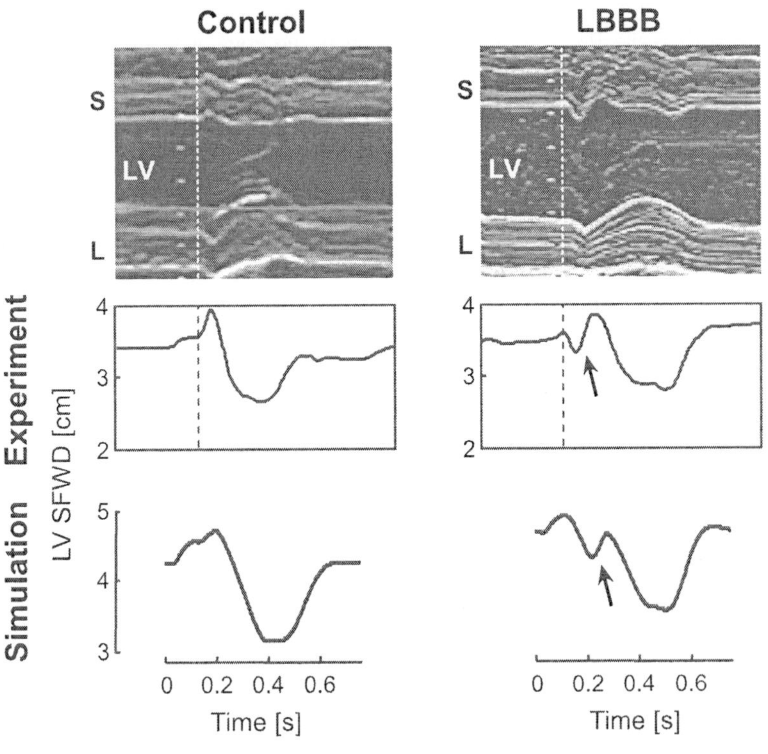

Figure 5: Left ventricular septal-to-free wall distance (LV SFWD) as derived from M-Mode echocardiograms of the left ventricle (LV) in the dog. Adapted from Liu *et al.* (2002) [14]. The septal wall and LV free wall are indicated by S and L, respectively. The left panel was acquired with synchronously contracting ventricles (Control) and the right image after induction of left bundle branch block (LBBB). Start of the QRS complex is indicated by vertical dashed lines. The arrows indicate the early systolic leftward motion of the septum, followed by the paradoxical rightward motion. The simulated curves of LV SFWD, as shown in the bottom panels, appear similar.

Figure 6 shows simulated LV and RV pressure-volume loops and myofiber stress-strain loops of all three wall segments. Stroke volumes do not change because cardiac output and heart rate were fixed in both simulations. In the LBBB simulation, the LV pressure-volume loop is shifted rightward, indicating ventricular dilatation that is generally considered representative for loss of cardiac contractile function. The areas of the stress-strain loops indicate contractile work of the myofibers per unit of tissue volume in the different wall segments. In Control circumstances, myocardial stroke work per unit of tissue volume is similar in all three segments, i.e., 5.5, 4.7, and 4.6 kPa for LV free wall, septal wall, and RV free wall, respectively. With LBBB, the early activated RV free wall generates clearly less work per unit of tissue volume (4.2 kPa) than the late activated LV free wall (7.8 kPa). Although the septum is later activated than the RV free wall, the septal tissue generates far less work (0.9 kPa).

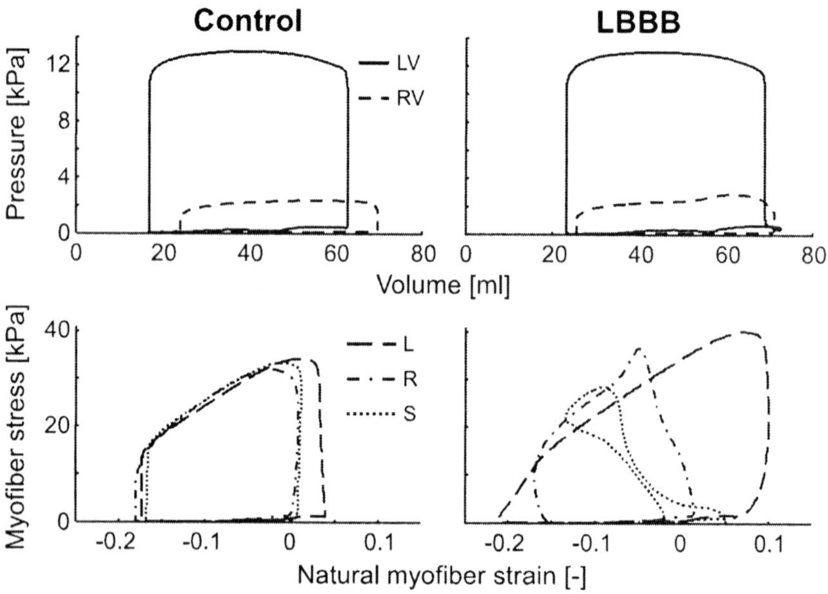

Figure 6: Simulated pressure-volume loops of the left ventricular (LV) and right ventricular (RV) cavities (top panels) and myofiber stress-strain loops of the left ventricular free wall (L), septal wall (S), and the right ventricular free wall (R) (bottom panels). The left panels show results of the Control simulation and the right panels that of the LBBB simulation.

4. Discussion

A lumped module was designed, describing ventricular mechanics with direct ventricular interaction. The ventricular cavities were considered to be formed between three wall segments, being the LV free wall, septum and RV free wall. Mechanical interaction between the walls caused mutual dependency of LV and RV pump function. The three-segment ventricular module was incorporated in the closed-loop CircAdapt model of the complete circulation. Size and weight of the constituting wall segments were determined by adaptation of the myocardial tissue to imposed mechanical load. A comparison with experimental data [14, 16, 17] demonstrated that simulation results of ventricular mechanics and hemodynamics at baseline and LBBB conditions were surprisingly realistic.

In the model, the atrioventricular valves could close only when the following two conditions were satisfied: 1) ventricular pressure exceeded atrial pressure and 2) the distal ventricular wall segments were mechanically activated. The latter condition mimicked papillary muscle function preventing valvular prolapse when ventricular pressure exceeded pressure in the proximal atrium. In the LBBB simulation, mitral backflow occurred because LV pressure rose above left atrial pressure before mechanical activation of the LV free wall. As soon as the LV free wall was activated, the mitral valve closed. Patient studies have shown that LBBB patients often have mitral regurgitation possibly as a result of late activation of papillary muscles [20].

Figure 6 showed remarkable changes in the amount of myofiber work done by early and late activated wall segments of the LV. The same qualitative changes in LV regional myofiber work density have been observed in animal experiments in which regional LV pump work was derived from strain analysis of short-axis MR tagging images and simultaneous invasive pressure measurements [17]. In chronic LBBB, these regional differences in work density may be responsible for asymmetric remodeling of the LV wall [19].

A crucial step in the calculation procedure was the estimation of sarcomere extension. The one-fiber model by Arts *et al.* [21], related sarcomere extension to the ratio of cavity volume to wall volume. This model has previously been shown to be applicable to an anisotropic thick-walled structure like a myocardial wall when assuming rotational symmetry and homogeneity of mechanical load in the wall. In our new model, the relation between midwall area and sarcomere extension was derived by applying the one-fiber model to a closed spherical cavity. The resulting relation was then extended to a partial segment of the sphere by considering a fraction of the wall, having the same curvature, wall tension, and transmural pressure difference. The one-fiber model has been shown to be rather insensitive to wall geometry [21]. We expected the present relation between midwall area, curvature, and transmural pressure also to be quite insensitive to actual geometry. However, this fact has not been proven.

The simulation results demonstrated that ventricular interaction through the septum is one very important mechanism for the hemodynamic changes associated with abnormal mechanical activation of the ventricular wall segments. However, another important potential mechanism might be changes in contractility due to asynchronous contraction within each wall segment. Due to its lumped character, this model did not allow description of regional interactions within each wall segment but was limited to the description of its average sarcomere mechanics.

Experimental data show a decrease of cardiac output by approximately 30% after induction of LBBB [16, 19]. In our simulations, however, cardiac output was the same in the Control and LBBB simulations. In the model, cardiac output affects the forces in the intersection of the three wall segments proportionally, provided LV and RV stroke volumes are the same. Thus, a change of cardiac output as observed in the experiments will only affect the amplitude of septal wall motion (Fig. 5) but not its characteristic course in time.

The mechanical coupling of the three spherical wall segments resulted in a circle of intersection with two degrees of freedom, namely, radial and axial displacement. This ventricular composition resulted in simple equations relating wall segment geometry to sarcomere behavior. Implementation of this ventricular module in the CircAdapt model resulted in a closed-loop system model that related fiber mechanics within the cardiac and vascular walls to hemodynamics realistically. Calculation time was limited to 6 seconds per cardiac cycle on a regular personal computer. Furthermore, the model behaved symmetrically around zero septal curvature, so that inversion of transseptal pressure and septal bulging could be handled. In conclusion, the resulting ventricular module satisfied all requirements mentioned in the introduction.

5. Conclusion

In the lumped CircAdapt model of the complete circulation, a new module was incorporated, representing the heart with realistic left to right ventricular interaction. The ventricular part of the heart was designed as a composition of the LV free wall, the septum, and the RV free wall, encapsulating the LV and RV cavities. In a test simulation, ventricular hemodynamics and septal motion during normal synchronous activation was compared with these variables during left bundle branch block. Simulated time courses of ventricular pressures and septal motion were in close agreement with experimental findings. The newly developed three-segment module, describing ventricular mechanics with direct ventricular interaction, is a promising tool in realistic simulation of right heart function and septal motion under normal as well as pathologic circumstances, using the framework of the CircAdapt model.

Acknowledgments

This research was financially supported by Actelion Pharmaceuticals Nederland B.V. (Woerden, The Netherlands).

References

1. I. Kingma, J. V. Tyberg, and E. R. Smith, *Circulation* **68**, 1304 (1983).
2. W. C. Little, R. C. Reeves, J. Arciniegas, R. E. Katholi, and E. W. Rogers, *Circulation* **65**, 1486 (1982).
3. A. E. Baker, R. Dani, E. R. Smith, J. V. Tyberg, and I. Belenkie, *Am J Physiol* **275**, H476 (1998).
4. C. O. Olsen, G. S. Tyson, G. W. Maier, J. A. Spratt, J. W. Davis, and J. S. Rankin, *Circ Res* **52**, 85 (1983).
5. B. K. Slinker and S. A. Glantz, *Am J Physiol* **251**, H1062 (1986).
6. R. Beyar, S. J. Dong, E. R. Smith, I. Belenkie, and J. V. Tyberg, *Am J Physiol* **265**, H2044 (1993).
7. D. C. Chung, S. C. Niranjan, J. W. Clark, Jr., A. Bidani, W. E. Johnston, J. B. Zwischenberger, and D. L. Traber, *Am J Physiol* **272**, H2942 (1997).
8. J. B. Olansen, J. W. Clark, D. Khoury, F. Ghorbel, and A. Bidani, *Comput Biomed Res* **33**, 260 (2000).
9. W. P. Santamore and D. Burkhoff, *Am J Physiol* **260**, H146 (1991).
10. B. W. Smith, J. G. Chase, G. M. Shaw, and R. I. Nokes, *Physiol Meas* **27**, 165 (2006).
11. Y. Sun, M. Beshara, R. J. Lucariello, and S. A. Chiaramida, *Am J Physiol* **272**, H1499 (1997).
12. T. Arts, T. Delhaas, P. Bovendeerd, X. Verbeek, and F. W. Prinzen, *Am J Physiol Heart Circ Physiol* **288**, H1943 (2005).
13. A. S. Abbasi, L. M. Eber, R. N. MacAlpin, and A. A. Kattus, *Circulation* **49**, 423 (1974).
14. L. Liu, B. Tockman, S. Girouard, J. Pastore, G. Walcott, B. KenKnight, and J. Spinelli, *Am J Physiol Heart Circ Physiol* **282**, H2238 (2002).
15. I. G. McDonald, *Circulation* **48**, 272 (1973).
16. X. A. Verbeek, K. Vernooy, M. Peschar, T. Van Der Nagel, A. Van Hunnik, and F. W. Prinzen, *Am J Physiol Heart Circ Physiol* **283**, H1370 (2002).
17. F. W. Prinzen, W. C. Hunter, B. T. Wyman, and E. R. McVeigh, *J Am Coll Cardiol* **33**, 1735 (1999).
18. X. A. Verbeek, A. Auricchio, Y. Yu, J. Ding, T. Pochet, K. Vernooy, A. Kramer, J. Spinelli, and F. W. Prinzen, *Am J Physiol Heart Circ Physiol* **290**, H968 (2006).
19. K. Vernooy, X. A. Verbeek, M. Peschar, H. J. Crijns, T. Arts, R. N. Cornelussen, and F. W. Prinzen, *Eur Heart J* **26**, 91 (2005).
20. A. Soyama, T. Kono, T. Mishima, H. Morita, T. Ito, M. Suwa, and Y. Kitaura, *J Card Fail* **11**, 631 (2005).
21. T. Arts, P. H. Bovendeerd, F. W. Prinzen, and R. S. Reneman, *Biophys J* **59**, 93 (1991).

SUB-MICROMETER ANATOMICAL MODELS OF THE SARCOLEMMA OF CARDIAC MYOCYTES BASED ON CONFOCAL IMAGING

FRANK B. SACHSE[1,2], ELEONORA SAVIO-GALIMBERTI[1], JOSHUA I. GOLDHABER[4], AND JOHN H. B. BRIDGE[1,2,3] *

[1] *Nora Eccles Harrison Cardiovascular Research and Training Institute,* [2] *Bioengineering Department, and* [3] *Division of Cardiology, University of Utah, Salt Lake City, UT 84112, USA*

[4] *David Geffen School of Medicine, University of California, Los Angeles, CA 90095, USA*

We describe an approach to develop anatomical models of cardiac cells. The approach is based on confocal imaging of living ventricular myocytes with sub-micrometer resolution, digital image processing of three-dimensional stacks with high data volume, and generation of dense triangular surface meshes representing the sarcolemma including the transverse tubular system. The image processing includes methods for deconvolution, filtering and segmentation. We introduce and visualize models of the sarcolemma of whole ventricular myocytes and single transversal tubules. These models can be applied for computational studies of cell and sub-cellular physical behavior and physiology, in particular cell signaling. Furthermore, the approach is applicable for studying effects of cardiac development, aging and diseases, which are associated with changes of cell anatomy and protein distributions.

1. Introduction

Computational simulations of physical behavior and physiology of biological tissues have given valuable scientific insights, which are applied in drug research, development of medical instrumentation and clinical medicine to improve diagnosis and therapy of patients. In the cardiac field, for example, computational simulations have been carried out to understand effects of drugs and mutations of ion channels on cellular electrophysiology,

*Work supported by the Richard A. and Nora Eccles Harrison endowment, awards from the Nora Eccles Treadwell Foundation, and the National Institutes of Health research grants no. HL62690 and no. HL70828.

Figure 1. Pipeline for generating anatomical models of cardiac myocytes.

metabolism and mechanics. Furthermore, the simulations helped to improve pacemaker and defibrillator efficacy, and to understand and prevent arrhythmogenesis.

Frequently, detailed anatomical models are applied in these simulations[15]. These models describe geometry of tissues and their microscopic properties such as fiber orientation and lamination. Commonly, these anatomical models were created by digital image processing of computer tomographic and magnetic resonance imaging. Eventually, the computational models are generated by extending the anatomical models with descriptions of physical and physiological properties.

In this work, we will address first steps in the generation of realistic detailed anatomical models of heart cells (Fig. 1). Our focus is on describing the geometry of the sarcolemma of ventricular myocytes with sub-micrometer resolution. The sarcolemma represents a semi-permeable barrier delimiting the extracellular from the intracellular space. The sarcolemma is built up primarily by a phospholipid bilayer with a thickness of $3 - 5\ nm$. The bilayer contains peripheral proteins attached to the surface of the sarcolemma and transmembrane proteins spanning over the sarcolemma. The proteins are responsible e.g. for signaling and cell-adhesion. Important transmembrane proteins are ion channels, exchangers, and ion pumps as well as gap junctions and receptors. Control of intracellular ion concentrations and cellular signaling in myocytes is mostly governed by these proteins in the sarcolemma.

In mammalian ventricular myocytes, the sarcolemma invaginates into the cytosol forming the so-called transverse tubular system (t-system)[5,2]. The t-system is composed of transversal tubules (t-tubules), which enter the myocyte primarily adjacent to Z disks[3]. The t-system occupies a large area of the sarcolemma. The ratio of t-system to sarcolemma area is species specific[1]. For instance, 42% and 33% of the sarcolemma comprise the t-system in rabbit and rat ventricular myocytes, respectively[11]. The t-system supports fast propagation of electrical excitation into the cell interior. Various proteins are associated with the t-system[10,14]. Morphological changes of the t-system have been associated with cardiac development, hypertrophy and heart failure[7,23].

Our modeling of the sarcolemma and t-system started by obtaining three-dimensional images of isolated cardiac myocytes and cell segments with scanning confocal microscopy. Usually, this technique is applied with fluorescent indicator dyes or antibodies tagged to a suitable fluophore, which permits specific labeling of compartments and proteins. For our modeling, we used a fluophore conjugated to membrane-impermeable dextran (excitation wave length: 488 nm, emission wave length: 524 nm, Invitrogen, Carlsbad, CA) to label the extracellular space.

Major processing steps in our modeling were image deconvolution and segmentation. We deconvolved the three-dimensional image datasets with the Richardson-Lucy algorithm using point spread functions (PSFs), which characterize the optical properties of our two confocal microscopic imaging systems. PSFs were extracted from images of fluorescent beads, which were suspended in agar to avoid Brownian-type motion. After deconvolution, the extra- and intracellular space were segmented in the images with methods of digital image processing. Furthermore, the t-system was decomposed into its components.

We identified the border between the extra- and intracellular segment with the sarcolemma and represented it by triangle meshes. Similarly, single t-tubules of various shapes and topologies were described with triangle meshes. This representation of the sarcolemma and t-tubules with triangle meshes permits application of standard tools for generation of computational models, such as volumetric mesh generators and automated annotation of mesh elements with protein density data. The resulting anatomical models provide a basis for computational studies of various physiological and pathophysiological processes at cellular level.

2. Methods

2.1. *Preparation and Imaging of Cardiomyocytes*

Our approach for preparation and imaging of alive cardiac cells was previously described in more detail[16,17]. In short, ventricular myocytes were isolated from adult rabbit hearts by retrograde Langendorff perfusion with a recirculating enzyme solution. After isolation, myocytes were stored at room temperature in a modified Tyrodes solution. Imaging of whole cells or segments of them was performed 4-8 h after isolation. Cells were superfused with membrane impermeant dextran conjugated to fluorescein and then transferred to a coverslip. Either a BioRad MRC-1024 laser-scanning confocal microscope (BioRad, Hercules, CA, USA) with a 63x oil

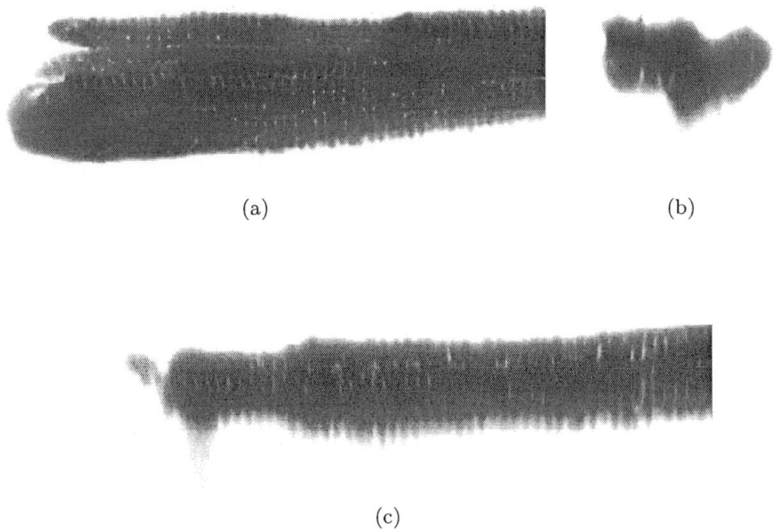

Figure 2. Exemplary image of ventricular myocyte segment. The high intensity of the extracellular space results from staining with a fluophore conjugated to membrane impermeable dextran. Dots and lines of high intensity Inside of the myocyte label the t-system. The dataset describes a hexahedral region with a size of 102 μm x 34 μm x 26 μm by a lattice of 768 x 256 x 193 cubic voxels. Intensity distributions are shown in the central (a) XY, (b) XZ and (c) YZ plane.

immersion objective lens (NA: 1.4, Nikon, Tokyo, Japan) or a Zeiss LSM 5 confocal microscope (Carl Zeiss, Jena, Germany) together with a 60x oil immersion objective lens (NA: 1.4) was used for imaging. It resulted in three-dimensional image stacks consisting of cubic voxels with a volume of $(133\ \mathrm{nm})^3$ and $(100\ \mathrm{nm})^3$, respectively (Fig. 2). The dimension of the stacks varied with size of the region of interest. The data volume of the stacks ranged from 20 to 250 million voxels.

2.2. *Image Processing*

The image processing was carried out in three dimensions and consisted of the following tasks:

- Correction of depth-dependent attenuation
- Image deconvolution
- Segmentation of intra- and extracellular space
- Decomposition of the t-system
- Surface extraction

- Visualization

Our approach for correction of depth-dependent intensity attenuation was a-posteriori using information from each individual image stack: Average intensities were slice-wise calculated in regions filled only with dye. A 3rd order polynomial P was fitted to the averages by least squares. For each slice z a scaling factor s was determined by:

$$s(z) = \frac{Max_{i \in [0,\ldots,N-1]} P(i)}{P(z)} \tag{1}$$

with the average background intensity P and the number of slice N. The scaling factor s was used for correction of each slice.

We applied the iterative Richardson-Lucy algorithm to reconstruct the source image f from the response g of the confocal imaging system[13,4]:

$$g_{n+1} = g_n \left(\frac{g_0}{g_n \star h} \otimes h \right) \tag{2}$$

with the PSF h, cross-correlation operator $\otimes$, convolution operator $\star$, and $g_o \equiv g$. We determined the PSF h by imaging fluorescent beads with a diameter of 100 nm in agar. 10 images of single beads were extracted in $\approx$ 10 nm distance to the coverslip, aligned and averaged yielding the PSF h. Specific care was given to detection and suppression of ringing artefacts, which are a common problem associated with this deconvolution method. We applied edge tapering methods to avoid intensity jumps at image borders. Furthermore, we cropped images manually to remove regions related to the coverslip and in excessive distance to the myocyte.

We segmented the extracellular space with morphological operators and the region-growing technique in the median filtered deconvolved image data[6,15]. Subsequently, the extracellular segment was applied as a mask to extract a segment containing the myocyte together with the t-system. Single t-tubules were segmented with the region-growing technique in the latter segment and with seed points determined by thresholding in a high-pass filtered image.

2.3. *Surface Mesh Generation and Visualization*

A modified marching-cube algorithm was applied to reconstruct the sarcolemma by creating surface meshes with sub-voxel resolution[9]. The algorithm generated meshes of triangular elements approximating iso-intensity surfaces in the three-dimensional image stacks. Modifications of the original algorithm assured closeness of the generated surfaces and permitted sub-voxel resolution by adjusting positions of mesh nodes based on edge-wise interpolation of intensities[8]. Meshes were visualized with software based on OpenInventor and can be exported in the VRML format[22].

We used the triangular meshes together with node-wise calculated surface normals for three-dimensional visualization of the sarcolemma. The normals were determined from gradients in averaged images stacks.

3. Results

We applied the foregoing methods to create and visualize anatomical models of 6 cells and 3064 t-tubules. The cells were from the left ventricle of rabbits and selected from an image library of more than 250 cells.

An exemplary model created from a living ventricular myocyte is shown in Fig. 3. The image dataset includes 1000 x 376 x 252 cubic voxels and describes a volume of 100 μm x 37.6 μm x 252 μm. The segmentation assigned 21 % of the voxels to the myocyte and the remainder to the extracellular space. The shape of the myocyte appears to be horizontally flattened and has sharp edges particularly at its endings. The sarcolemma exhibits a partly regular pattern of indentations, which refer to mouths of t-tubules.

An enlargement of an area at the cell bottom shows two rows of three mouths of t-tubules (Fig. 4a). Distances between the mouths are $\approx 1.5\mu m$ and $\approx$ 3.1 μm in row and column direction. Application of the marching cube algorithm led to a surface represented by a triangular mesh (Fig. 4b).

A single t-tubule is visualized in Fig. 5. The t-tubule has a length of $\approx$ 2.6 μm and is of simple topology without branching and lateral connections, so-called anastomoses. Constrictions of the t-tubule diameter are visible close to the mouth and slightly above the middle. The triangular mesh representing the sarcolemma is shown in Figs. 5b and d.

In our set of 3064 t-tubule models extracted from 6 cells, lengths varied between 1 and 7 μm, with mean values of 2.8 μm. The occurrence of constrictions was correlated with t-tubule length. The t-tubule diameter was in average $\approx$400 nm.

(a)

(b)

(c)

(d)

Figure 3. Three-dimensional visualization of single myocyte from different perspective. The myocyte is shown from (a) above, (b) below, (c) lateral and (d) lateral-below.

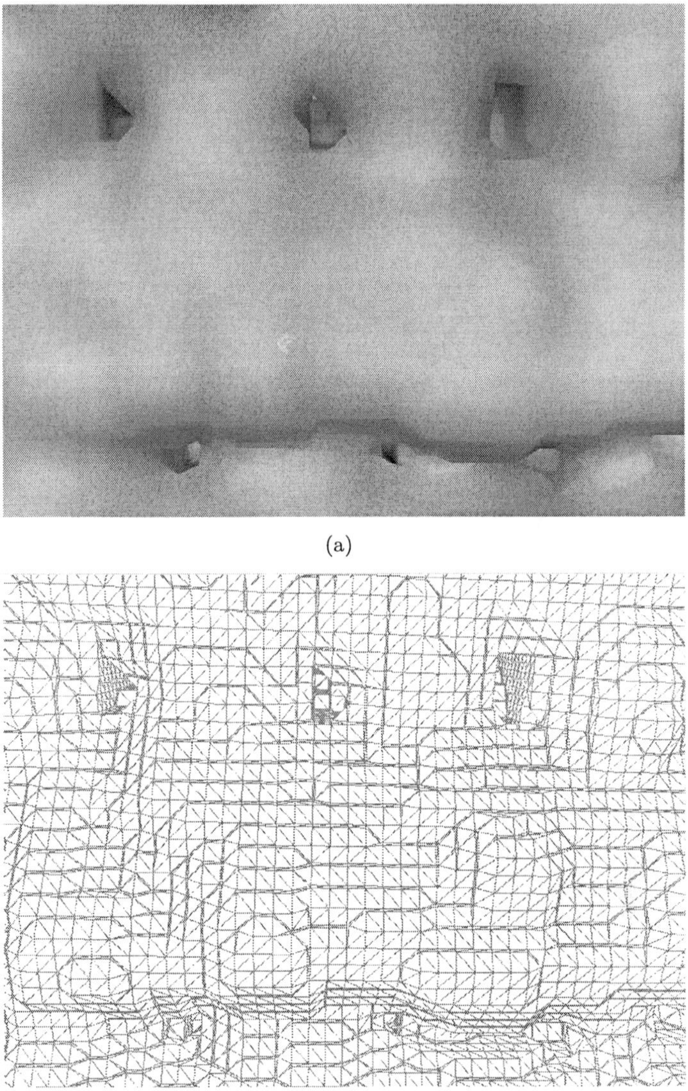

(a)

(b)

Figure 4. Visualization of sarcolemma segment with mouthes of t-tubules. The surface was generated with the marching cube algorithm and is shown with (a) filled triangles and (b) edges only.

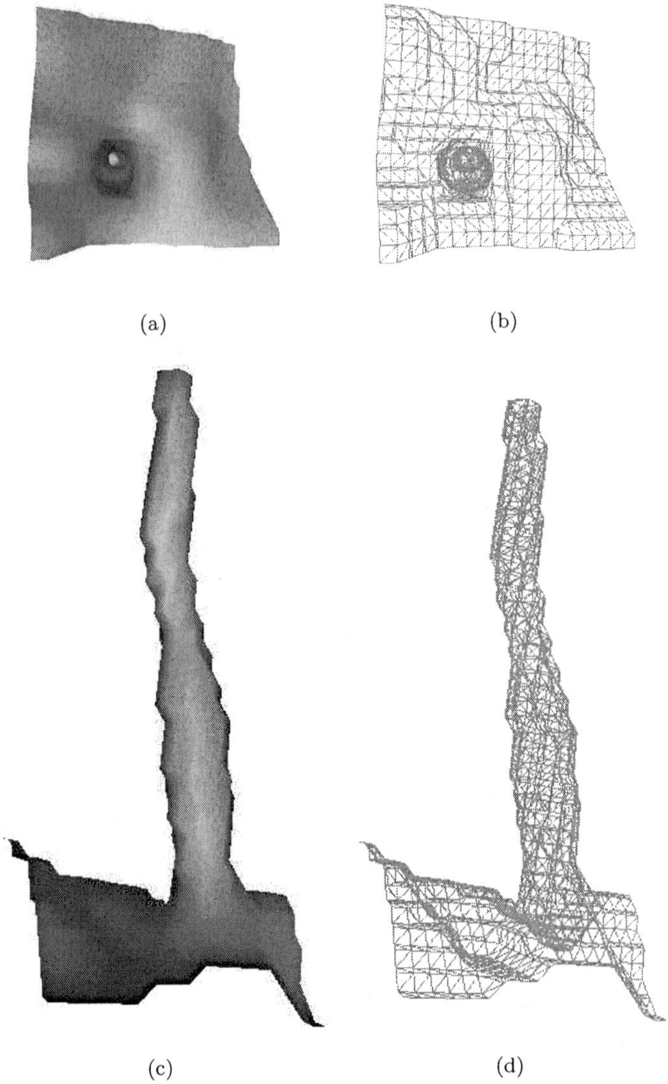

Figure 5. Visualization of single t-tubule (a,b) through mouth into cavity and (c,d) from lateral. The surface is shown with (a,c) filled triangles and (c,d) edges only.

4. Discussion and Conclusions

We presented an approach to generate anatomical models of cardiac cells. The models describe with sub-micrometer resolution the sarcolemma including the t-system by processing of confocal images. Our approach complements analytical methods of cell surface modeling such similar as those introduced by Stinstra et al[20] and provides realistic geometrical data for their approach.

Our focus on modeling the sarcolemma is motivated by its central role as a border between the intra- and extracellular environment as well as for cell signaling. The sarcolemma comprises various proteins for cellular signaling such as controlling inward and outward flows of ions. Annotation of our anatomical models with published information of sarcolemmal protein density distributions is straightforward and will allow us to generate novel computational models of cellular physiology.

Our methodology is related to work of Soeller and Cannell[19], who used confocal microscopy and methods for digital image processing to characterize the topology of the transverse tubular system (t-system) in rat ventricular cardiac myocytes. In this work, we focused on generation of anatomical models, which are applicable in computational studies. The t-tubule diameter in our study on rabbit ventricular cells was in average $\approx$400 nm and thus mostly above the resolution of the confocal imaging system. The t-tubule diameter was much larger in rabbit than in rat, which corresponds to the reported differences of t-system surface area between the two species[11]. The large diameter allowed us to apply the surface meshing method not only for generation of models of the outer sarcolemma but also for modeling of the t-system.

Of particular interest for us is extending the models with information on distributions of ion channels, exchanger and pumps, which would permit to study electrophysiological processes at nanometer level. Resulting from recent advantages in confocal imaging technology, this information can be gained by using combinations of multiple fluorescent labels. In currently ongoing work, we are exploring dual labeling methods to relate proteins involved in excitation-contraction coupling to regions of the sarcolemma and t-system. Here, one label is associated with a specific type of ion channel and imaged together with another for labeling the extracellular space.

An application of our models can be found in studying ion diffusion in the t-system. In previous simulation studies of Shepherd and McDonough[18] and Swift et al[21], t-tubule geometry was simplified and diffusion approximated in one dimension. The presented models would allow us to gain insights into the significance of morphology and topology of the t-system for ion diffusion, particularly the role of constrictions in t-tubules, anastomoses and rete-like structures. We suggest that our models can be applied in computational studies of ion diffusion in the t-system by volume meshing of the t-tubule cavity and numerical solvers for partial differential equations describing diffusion[12].

Our approach can also be applied for modeling cells during development and aging as well as affected by cardiac diseases. Morphological changes of the t-system of myocytes have been described for diseased human ventricles[23] and in addition to changes of protein densities for tachycardia induced heart failure[7]. Effects of these changes are difficult to assess at cellular and tissue level with traditional experimental and analytical approaches. Computational studies based on realistic models of cell anatomy might give insights in these effects and thus complement the traditional approaches.

References

1. D. M. Bers. *Excitation-Contraction Coupling and Cardiac Contractile Force.* Kluwer Academic Publishers, Dordrecht, Netherlands, 1991.
2. B. A. Block, T. Imagawa, K. P. Campbell, and C. Franzini-Armstrong. Structural evidence for direct interaction between the molecular components of the transverse tubule/sarcoplasmic reticulum junction in skeletal muscle. *J. Cell Biol.*, 107(6):2587–2600, 1988.
3. F. Brette and C. Orchard. T-tubule function in mammalian cardiac myocyte. *Circ. Res.*, 92:1182–1192, 2003.
4. J. B. de Monvel, S. Le Calvez, and M. Ulfendahl. Image restoration for confocal microscopy: Improving the limits of deconvolution, with application to the visualization of the mammalian hearing organ. *Biophys J.*, 80:2455–2470, 2001.
5. D. W. Fawcett and N. S. McNutt. The ultrastructure of cat myocardium. I. ventricular papillary muscle. *Cell Biol.*, 42:1–45, 1969.
6. R. C. Gonzalez and R. E. Woods. *Digital Image Processing.* Addison-Wesley, Reading, Massachusetts; Menlo Park, California, 1992.
7. J. He, M. W. Conklin, J. D. Foell, M. R. Wolff, R. A. Haworth, R. Coronado, and T. J. Kamp. Reduction in density of transverse tubules and l-type ca(2+) channels in canine tachycardia-induced heart failure. *Cardiovasc Res*, 49(2):298–307, 2001.
8. W. Heiden, T. Goetze, and J. Brickmann. 'Marching-Cube'-Algorithmen zur schnellen Generierung von Isoflächen auf der Basis dreidimensionaler Daten-

felder. In M. Frühauf and Martina Göbel, editors, *Visualisierung von Volumendaten*, pages 112–117. Springer, Berlin, Heidelberg, New York, 1991.

9. W. E. Lorensen and H. E. Cline. Marching cubes: A high resolution 3D surface construction algorithm. *Computer Graphics*, 21(4):163–169, 1987.
10. P. J. Mohler, J. Q. Davis, and V. Bennett. Ankyrin-B coordinates the Na/K ATPase, Na/Ca exchanger, and $Insp_3$ receptor in a cardiac t-tubule/SR microdomain. *PLoS Biology*, 3(12):2158–2167, 2005.
11. E. Page and M. Surdyk-Droske. Distribution, surface density, and membrane area of diadic junctional contacts between plasma membrane and terminal cisterns in mammalian ventricle. *Circ. Res.*, 45(2):260–267, 1979.
12. W. H. Press, S. A. Teukolsky, W. T. Vetterling, and B. P. Flannery. *Numerical Recipes in C.* Cambridge University Press, Cambridge, New York, Melbourne, 2 edition, 1992.
13. W. H. Richardson. Bayesian-based iterative method of image restoration. *J Opt Soc Am*, 62:55–59, 1972.
14. V. G. Robu, E. S. Pfeiffer, S. L. Robia, R. C. Balijepalli, Y. Pi, T. J. Kamp, and J. W. Walker. Localization of functional endothelin receptor signaling complexes in cardiac transverse tubules. *J Biol Chem*, 278(48):48154–48161, 2003.
15. F. B. Sachse. *Computational Cardiology: Modeling of Anatomy, Electrophysiology, and Mechanics*, volume 2966 of *Lecture Notes in Computer Science.* Springer, Heidelberg, 2004.
16. E. Savio, J. Frank, M. Inoue, J. I. Goldhaber, M. B. Cannell, J. H. B. Bridge, and F. B. Sachse. High-resolution three-dimensional confocal microscopy reveals novel structures in rabbit ventricular myocyte t-tubules. In *Biophys. J (Annual Meeting Abstracts)*, 2007.
17. E. Savio, J. I. Goldhaber, J. H. B. Bridge, and F. B. Sachse. A framework for analyzing confocal images of transversal tubules in cardiomyocytes. In F. B. Sachse and G. Seemann, editors, *Lecture Notes in Computer Science*, volume 4466, pages 110–119. Springer, 2007.
18. N. Shepherd and H. B. McDonough. Ionic diffusion in transverse tubules of cardiac ventricular myocytes. *Am J Physiol Heart Circ Physiol*, 275:852–860, 1998.
19. C. Soeller and M. B. Cannell. Examination of the transverse tubular system in living cardiac rat myocytes by 2-photon microscopy and digital image-processing techniques. *Circ Res*, 84:266–275, 1999.
20. J. G. Stinstra, B. Hopenfeld, and R. S. MacLeod. On the passive cardiac conductivity. *Ann Biomed Eng*, 33(12):1743–51, 2005.
21. F. Swift, T. A. Stromme, B. Amundsen, O. Sejersted, and I. Sjaastad. Slow diffusion of K^+ in the T tubules of rat cardiomyocytes. *J Appl Physiol*, 101:1170–1176, 2006.
22. J. Wernecke. *The Inventor Mentor: Programming Object-Oriented 3D Graphics with Open Inventor.* Addison-Wesley Professional, 1 edition, 1994.
23. C. Wong, C. Soeller, L. Burton, and M. B. Cannell. Changes in transverse-tubular system architecture in myocytes from diseased human ventricles. In *Biophys. J (Annual Meeting Abstracts)*, number 82, page a588, 2002.

EFFICIENT MULTISCALE SIMULATION OF CIRCADIAN RHYTHMS USING AUTOMATED PHASE MACROMODELLING TECHNIQUES

SHATAM AGARWAL
Indian Institute of Technology, Kanpur
shatam@iitk.ac.in

JAIJEET ROYCHOWDHURY
University of Minnesota, Twin Cities
jr@umn.edu

Circadian rhythm mechanisms involve multi-scale interactions between endogenous DNA-transcription oscillators. We present the application of efficient, numerically well-conditioned algorithms for abstracting (potentially large) systems of differential equation models of circadian oscillators into compact, accurate phase-only macromodels. We apply and validate our auto-extracted phase macromodelling technique on mammalian and *Drosophila* circadian systems, obtaining speedups of $9-13\times$ over conventional time-course simulation, with insignificant loss of accuracy, for single oscillators being synchronized by day/night light variations. Further, we apply the macromodels to simulate a system of 400 coupled circadian oscillators, achieving speedups of $240\times$ and accurately reproducing synchronization and locking phenomena amongst the oscillators. We also present the use of parameterized phase macromodels for these circadian systems, and elucidate insights into circadian timing effects directly provided by our auto-extracted macromodels.

1. Introduction

Circadian rhythms are amongst the most fundamental of physiological processes. They are found in virtually all organisms, ranging from unicellular (*e.g.*, amœbæ, bacteria) to complex multicellular higher organisms (*e.g.*, human beings). These daily rhythms, of period about 24 hours, are associated with periodic changes in hormones controlling sleep/wakefulness, body temperature, blood pressure, heart rate and other physiological variables. Importantly, circadian rhythms are *endogenous* or *autonomous*; however, they are typically influenced by external cues, such as light. Progress in quantitative biology has established that such rhythms stem fundamentally from the molecular level,[1,2] involving complex chains of biochemical reactions featuring a number of key proteins/hormones (such as melatonin and melanopsin), whose levels rise and fall during the course of the day. These biochemical reactions, which take place both within individual cells and at an extracellular level, function as *biological oscillators or body clocks*.[3]

Quantitative understanding, simulation and control of circadian rhythms is of great practical importance. Applications include devising medical remedies for rhythm disorders (*e.g.*, insomnia, fatigue, jet lag, *etc.*), synthetic biology (where a goal is to "program" artificial rhythms that are biologically viable), artificially extending periods of wakefulness/alertness (*e.g.*, for military purposes), and so on. Improved understanding of circadian rhythm mechanisms has led to increased awareness of how pervasively they affect virtually every aspect of the life of an organism. Hence, their simulation/analysis is an important endeavour in the biological domain.[1,2]

Although individual oscillators constitute the fundamental core of circadian rhythm mechanisms, the rich circadian functionality of multicellular organisms results from the *interactions of many oscillators* over multiple temporal and spatial scales. Observations of periodicity in behavior, metabolism, body temperature, *etc.*, indicate that coupling/coherence mechanisms play a key role. Hierarchical organization of the circadian system, from the fundamental DNA transcription/translation level to endocrine system levels, involves complex oscillator interactions. The complex connectivity and high dimensionality of such coupled oscillator networks, which lead to unique effects such as *synchronization* and *injection locking/pulling*,[4,5] make them difficult to understand at the intuitive or analytical level, thus engendering the need for efficient and powerful simulation and analysis tools with multiscale capabilities.

Several oscillatory mathematical models are available for circadian rhythms[1,2] that capture the dynamics of the relevant molecular biochemical reactions (see Section 2 for details). These models are in the form of systems of differential-algebraic equations (DAEs) or ordinary differential equations (ODEs). The prevalent technique today for their simulation is to run initial value simulations. While such "time-course integration" of ODEs/DAEs has the advantage of generality, it suffers from serious disadvantages for oscillators, which are inherently *marginally stable.*[6] For initial-value simulations, marginally stable systems tend to require orders of magnitude more computation for a specified accuracy, particularly *phase/timing accuracy*, than stable systems; even for individual oscillators, very small timesteps (*e.g.*, many hundreds per oscillation cycle) are typically needed, leading to high computational cost. The situation worsens for coupled oscillator systems, which typically feature multiple time scales; *e.g.*, *envelopes*[7,8] typically feature much longer time scales than individual oscillation cycles.

In electronic circuit design, *automated nonlinear phase macromodel extraction* techniques[5,6,9] have proven effective in solving sucsuch oscillatory problems. Given any oscillator as a system of DAEs or ODEs (however complicated), efficient and well-conditioned numerical techniques extract a *scalar nonlinear differential equation*, the phase macromodel. This macromodel captures the dynamic response of the oscillator's phase or timing characteristics to external influences. It has been shown that such "PPV" (Perturbation Projection Vector) phase macromodels are able to accurately capture the gamut of phase/frequency-related dynamics of oscillators; most importantly locking, synchronization and phase noise (timing jitter) effects.[6,10] Using the PPV macromodel instead of the original DAEs/ODEs confers important advantages: large simulation speedups due to system size reduction, the ability to use larger timesteps than for the original system, abstraction to the phase or timing level, precise insight about timing influences without the need for simulation, *etc.*. These advantages are especially pronounced for systems of many coupled oscillators spanning different temporal and spatial scales.[11]

In this work, we present the first application of PPV-based automated nonlinear time-shifted macromodelling methods to biological systems, focussing on circadian rhythms. We use PPV phase macromodels[6,12] to model circadian oscillators and show that they are considerably more efficient than standard "time course" simulations. PPV models alleviate the lack of accuracy and general applicability of a widely used prior phase model (Kuramoto's model, see below), while retaining its advantage of relative simplicity and computational efficiency. PPVs provide direct insights into the effects of external stimuli, such as slowing down/speeding up of circadian rhythms; for example, it is easy to determine when and how to apply a light pulse for greatest de-synchronization. Using PPV macromodels, we are able to efficiently produce plots of circadian lock range vs amplitude of external stimuli; this is valuable for guiding experiments, explaining observations, and designing new ("synthetic") DNA/protein based biological clock networks. Furthermore, we also present the application of *parameterized* PPV macromodels,[13] which directly incorporate *effects of parameter changes* into our phase-only models of circadian rhythms. Being able to directly and quantitatively predict the impact of parameter changes on phase, frequency and timing behaviour is of significant biological value.

Indeed, the PPV constitutes a rigorous, Floquet-theoretic generalization of Winfree's seminal concept of timing maps and phase sensitivity functions,[14] used within "phase-only models" of oscillators popular in computational biology. Kuramoto[15] applied the theory of asymptotics to find *sinusoidal* expressions for Winfree's phase sensitivity functions; these are widely used for phase-only models, since they are capable of capturing synchronization effects. However, the sinusoidal simplifications inherent in Kuramoto's model lead to significant inaccuracies for non-sinusoidal oscillators (including circadian ones). These inaccuracies compound in large networks and often lead to qualitatively incorrect conclusions about, *e.g.*, collective synchronization.[11,16] From a utilitarian point of view, the usefulness of PPV macromodels over Kuramoto is twofold: firstly, PPVs represent the correct (often highly non-sinusoidal) phase sensitivity functions of the oscillator; secondly, the PPV macromodel is generated via an *algorithmic computational procedure* from the oscillator's DAE/ODE description, typically taking seconds or less to reduce systems 1000s of equations in size.

We apply PPV macromodels to two different circadian rhythm models: one for mammals[1,17] and one for *Drosophila melanogaster* (the fruit fly, shown in Fig. 1).[2] We show that the PPV macromodels are significantly faster to simulate than the original equation systems even for single oscillators (9x speedup for the mammalian clock, and 13x speedup for the *Drosophila* clock). Modelling light as an external input impinging upon circadian oscillators, we confirm injection locking using PPV macromodels and obtain plots of lock range vs amplitude of the light signal. We comment on the biological significance of the shapes of important components of the PPV. We then use PPV macromodels to rapidly explore synchronization behaviour in a network of 400 coupled oscillators, obtaining speedups of about 240× over standard time-course simulation. Finally, using parameterized PPV macromodels, we predict the effects of varying a number of model parameters on oscillation frequency and lock range.

Fig. 1: Male *Drosophila* (fruit fly).

The remainder of the paper is organized as follows. In Section 2.1, we provide background on circadian rhythms and their mechanisms, followed by a review of mathematical models for circadian rhythms in Section 2.2. Oscillator and phase macromodels are then introduced in Section 3; a brief review of PPV macromodels, injection locking analysis and parameterized PPVs is provided. Finally, in Section 4, results and speedups are presented.

2. Background and Previous Work

2.1. *Circadian rhythms*

Circadian rhythms are generated by "clock genes", which encode genetic instructions that produce certain proteins whose levels oscillate during the course of the day. These oscillating biochemical signals control various functions, such as sleep/waking cycles – in other words, they constitute our "internal biological clock", which adapts to the daily cycle of day and night. However, the natural period of this internal clock is not exactly 24 hours; it is typically longer if the organism is kept isolated and away from external cues,[18] most importantly light (these cues are called *Zeitgebers*). Therefore, the internal clock needs to be "reset" every day, in order to keep the organism's bodily rhythms synchronized with the external world's day/night cycle.

Higher organisms are often composed of billions of cells. The nucleus of each cell contains the genetic material DNA, a long chain-like linear molecule built up of many links. RNA, also a nucleic acid polymer, serves as a DNA template for the *translation* of genes into proteins. The process of formation of an RNA molecule from a particular DNA is called *transcription*. Unlike DNA, RNA is capable of leaving the nucleus and moving into the cytoplasm. There, with the help of enzymes, specific RNA strings get converted to specific proteins responsible for different bodily functions. Some of these proteins return to the nucleus, forming complexes by binding to other proteins, some of which inhibit the expression of their own genes, giving rise to oscillatory patterns in protein concentrations and hence, *circadian rhythms*.[1,2,18]

In mammals, core clock genes include *Per*, *Cry*, *Bmal1* and *Clock* genes. Their proteins act by inhibiting or stimulating transcriptions of other core clock genes. The proteins of the *Bmal1* and *Clock* genes, namely BMAL1 and CLOCK, form a complex CLOCK-BMAL1 inside the nucleus. This complex activates the transcription of the *Per* and the *Cry* genes. In the cytoplasm, *Per* and *Cry* RNA translate to their respective proteins, PER and CRY. Some of these proteins dimerize to form the complex PER-CRY which returns to the nucleus where, by binding to the CLOCK-BMAL1 complex, it prevents the further transcription of *Per* and *Cry* genes. Thus a negative feedback loop is created; PER and CRY proteins blocking the transcription of their own genes. The above mechanism for the mammalian clock is illustrated in Fig. 2.

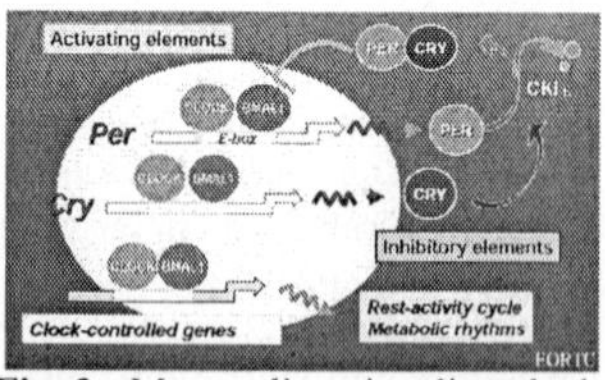

Fig. 2: Mammalian circadian clock mechanism (Francis Levi, EORTC Chronotherapy Group).

2.2. *Models of Drosophila and mammal circadian rhythms*

Computational models are available for circadian rhythms in *Drosophila*, *Neurospora*, cyanobacteria and mammalian systems.[1,2] These models, useful for computing concentrations of core clock genes, take into account the processes of transcription, translation and phosphorylation.

The mammalian circadian clock model we use[1] consists of 16 variables (hence 16 differential equations) and 52 parameters. It incorporates the effect of negative autoregulation of *Per*/*Cry* gene expression by their own proteins. The *Drosophila* circadian model we use,[2] consisting of only 5 variables (5 equations) and 18 parameters, is small enough to be reproduced here:

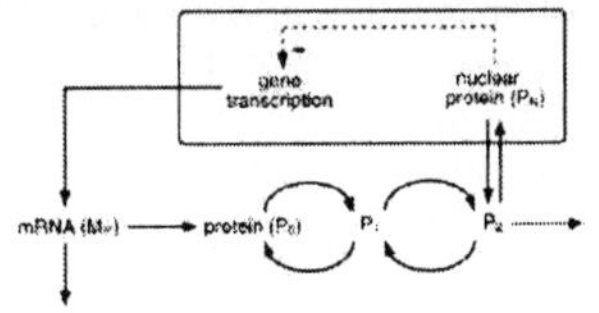

Fig. 3: *Drosophila* clock mechanism[2].

$$\begin{aligned}
\frac{dM_p}{dt} &= v_s \frac{K_I^n}{K_I^n + P_N^n} - v_m \frac{M_p}{K_m + M_p}, \\
\frac{dP_0}{dt} &= k_s M_p - v_1 \frac{P_0}{K_1 + P_0} + v_2 \frac{P_1}{K_2 + P_1}, \\
\frac{dP_1}{dt} &= v_1 \frac{P_0}{K_1 + P_0} - v_2 \frac{P_1}{K_2 + P_1} - v_3 \frac{P_1}{K_3 + P_1} + v_4 \frac{P_2}{K_4 + P_2}, \\
\frac{dP_2}{dt} &= v_3 \frac{P_1}{K_3 + P_1} - v_4 \frac{P_2}{K_4 + P_2} - v_d \frac{P_2}{K_d + P_2} - k_1 P_2 + k_2 P_N, \\
\frac{dP_N}{dt} &= k_1 P_2 - k_2 P_N.
\end{aligned} \tag{1}$$

The variables in Eq. 1 are the same as those shown in Fig. 3; parameters that lead to endogenous oscillations are taken from Gonze/Goldbetter.[2] Note that Eq. 1 is in the canonical nonlinear ODE/DAE system form

$$\frac{d}{dt}\vec{q}(\vec{x}(t)) + \vec{f}(\vec{x}(t)) + \vec{b}(t) = 0, \tag{2}$$

with $\vec{q}(\vec{x}) \equiv \vec{x}$ and $\vec{b}(t) \equiv 0$. $\vec{b}(t)$ represents the influence of external inputs, such as light, which affects the transcription rate of the *Per* gene in both mammals and *Drosophila*. The effect of light is modelled by including a parameter in the rate equations of the *Per* gene (v_s in Eq. 1), which we recast in the form of Eq. 2 with $\vec{b}(t) \neq 0$ (details in Section 4).

3. Oscillators and PPV phase macromodels

The quantitative study and design of oscillators has a rich history in engineering, particularly in electronics: oscillators are fundamental components in virtually all electronic systems. For example, they are widely used in communication systems for frequency translation of information signals; phase locked loops (PLLs) for clock generation and frequency synthesis, *etc.*. As noted earlier, phase macromodelling techniques are widely used to improve simulation efficiency and accuracy[5,19] in electronics. In particular, the Perturbation Projection Vector (PPV) phase macromodel[6,9] is well established on account of its rigorous Floquet-theoretic underpinnings, broad applicability, effective numerical extraction procedures, large simulation speedups and extensive validation. We have already noted its advantages in Section 1; here, we summarize mathematical details of the model. For expositional convenience, we assume an ODE form for an oscillator under external perturbation:

$$\frac{d}{dt}\vec{x}(t) + \vec{f}(\vec{x}(t)) = \vec{b}(t). \tag{3}$$

$\vec{b}(t)$ is the vector of perturbations applied to the free running oscillator; $\vec{x}(t)$ and $\vec{f}(\vec{x}(t))$ have their usual meanings, as in Eq. 2. The solution of this perturbed oscillator can be shown[6] to be in the form

$$\vec{x}_p(t) = \vec{x}_s(t + \alpha(t)) + \vec{y}(t + \alpha(t)), \tag{4}$$

where $\vec{x}_s(t)$ is the periodic, oscillatory solution of the unperturbed oscillator and $\alpha(t)$ is a *phase deviation* caused by the external perturbation $\vec{b}(t)$. $\vec{y}(t + \alpha(t))$ is an amplitude variation; it is typically very small in circadian oscillators[2] and is therefore of secondary importance compared to the phase deviation $\alpha(t)$. Using a nonlinear extension of Floquet theory, Demir et al[6] proved that $\alpha(t)$ is governed by the scalar, nonlinear, time-shifted differential equation

$$\dot{\alpha}(t) = v_1^T(t+\alpha(t)) \cdot b(t), \tag{5}$$

where $v_1^T(t)$ is a periodic vector known as the *perturbation projection vector* or PPV. Importantly, they also showed that the PPV can be calculated efficiently via simple post-processing steps following time- or frequency-domain steady-state computation.[6,9] Each component of the PPV waveform represents the oscillator's "nonlinear phase sensitivity" to perturbations of that component. The PPV needs to be extracted only once from Eq. 1 (even if parameters change, see the description of parameterized PPVs below); once extracted, Eq. 5 is used for simulations.

3.1. *Using the PPV macromodel for systems of coupled oscillators*

By employing $\vec{b}(t)$ in Eq. 5 to capture coupling, PPV macromodels can be composed to represent systems of many coupled oscillators with different characteristics. For purposes of illustration, we outline the procedure for N identical oscillators coupling via only one component of $b(t)$. This results in the following set of governing equations for the coupled system:

$$\dot{\alpha}_i(t) = v^T(t+\alpha_i(t)) \cdot \gamma_i(t), \qquad i \in 1, \cdots, N, \tag{6}$$

where $\alpha_i(t)$ is the phase shift of oscillator i, $v(t)$ is the phase sensitivity of the node on which coupling occurs and $\gamma_i(t)$ is the perturbation resulting on oscillator i due to coupling from other oscillators. If the coupling $\gamma_i(t)$ and phase sensitivity $v(t)$ are purely sinusoidal, it is easy to show that Eq. 6 is equivalent to Kuramoto's model.[15] In general, however, Eq. 6 is far more accurate since it considers all harmonics of the PPVs. We use the coupling function model given in To et al[20] as $\gamma_i(t)$ in Eq. 6, and solve for the phase dynamics of a 20×20 network of coupled oscillators.

3.2. *Injection locking analysis*

When an external signal of frequency f is injected into an oscillator with a central frequency f_o close to f, the oscillator can lock to the injected signal both in phase and frequency. This phenomenon is known as injection locking and can be very easily captured by the PPV macromodel of the oscillator.[4] It has been shown[5] that when injection locked, an oscillator's phase shift $\alpha(t)$ varies linearly with time as

$$\alpha(t) = \frac{\Delta\omega}{\omega_o}t + \frac{\theta(t)}{\omega_o}, \tag{7}$$

where ω_o is the natural frequency of the unperturbed oscillator and $\Delta\omega$ the difference between the frequencies of the injected signal and the unperturbed oscillator. $\theta(t)$ represents a *bounded, periodic* phase difference function, the exact form of which can be determined via time-course or steady-state simulation[5,8] of Eq. 5. The presence of injection locking can therefore be detected by comparing the time-average of $\dot{\alpha}(t)$ with $\frac{\Delta\omega}{\omega_o}$.

3.3. *Parameterized PPV macromodels*

Circadian rhythm models typically involve large numbers of model parameters. For example, there are 18 parameters in the *Drosophila* clock model,[2] while the mammalian clock model[1] has 52 parameters. The values of these parameters are chosen so that the model's predictions best fit experimental observations. Leloup/Goldbetter[17] have noted that circadian rhythm properties (particularly frequency) are sensitive to variations in several parameters. The conventional approach to assessing the effect of parameter variations involves brute-force time-course simulation of circadian models, a process that is not only expensive but can also generate numerical inaccuracies in phase.[5]

We, instead, use an extended form of Eq. 5 that directly incorporates parameter variations – we call this the *parameterized PPV macromodel.*[13] The key advantage of the parameterized PPV macromodel is that it does not involve re-extracting the PPV when parameters change - this leads to huge speedups when, *e.g.*, many coupled oscillators with different parameters are involved. The parameterized PPV equation is given by

$$\dot{\alpha}(t) = v_1^T(t+\alpha(t)) \cdot (b(t) - S_p(t+\alpha(t))\Delta p), \tag{8}$$

where Δp is a vector containing parameter variation terms and $S_p(t)$ is a periodic, time-varying matrix function given by

$$S_p(t) = \frac{\partial f}{\partial p}|_{x_s(t),p^*}. \tag{9}$$

In Eq. 9, $x_s(t)$ denotes the natural periodic solution of the unperturbed oscillator; p^* represents the vector containing nominal (basal) parameter values. This extra term captures phase deviations due to parameter variations, without having to re-extract the PPV when the parameters change. It also enables the study of the effects of multiple parameters varying at the same time.

4. Simulation of mammalian and *Drosophila melanogaster* circadian rhythms using PPV macromodels

In this section, we present results obtained by applying PPV macromodelling, described in Section 3, to mammalian and *Drosophila* circadian rhythm models.[1,2] We first extract PPV macromodels for both circadian systems at nominal parameters[9] and then simulate for phase deviation with external perturbation to demonstrate injection locking. We model the external perturbations as changes in external light intensity by first assigning a constant value to the light sensitive parameter v_s (signifying darkness) and then applying an external light signal of intensity

$$L(t) = A + Asin(\omega t) W/m^2, \tag{10}$$

where $\omega = 2\pi f$, f being the frequency of the light/dark cycles, *i.e.*, corresponding to 1 cycle in 24 hours. Often, light is modelled as a step function for simulations in biological systems (*i.e.*, constant values for light and dark conditions respectively). However, to correspond more closely with continuously changing light intensities in reality and to illustrate the generality of the PPV model, we apply sinusoidal intensity waveform around an average value.[21] (Note that any other shape, including step function shapes, can be handled equally easily). We assume the experimental setup used by Usui/Okazaki,[21] where the illuminance of light is varied from 20 lux to 0.01 lux (*i.e.*, variation in light intensity from 0.15 W/m^2 to 0.00009 W/m^2), giving A $\sim$ 0.05W/m^2 in Eq. 10. Moreover, Eq. 10 multiplied by a constant gives the term $b(t)$ of Eq. 2, where the constant signifies the change in *Per* gene concentration for 1 W/m^2 of light intensity. In this paper, we assume the constant to be equal to 1 $nM/(W/m^2)$. The constant can be modelled accurately in future experiments.

We also extract parameterized PPV macromodels to study the effect of parameter variations in two cases – with and without external light variations. In the absence of external light variations, phase deviations from the parameterized PPV macromodel are useful for predicting changes in free-running frequency. When external light perturbations are included in parameter-varying PPV simulations, lock range information is also generated.

Finally, we put the above single-oscillator PPV macromodels together to model a locally-coupled 20×20 network of oscillators – a simple representation of a spatially multiscale, coupled circadian system. We use this model to demonstrate synchronization behavior, obtaining speedups of about 240× over traditional time-course simulation.

4.1. *Time-course simulations using full ODE models*

For reference and validation, we first perform time-course simulations of the two ODE circadian rhythm models directly, to obtain concentration waveforms for all clock proteins and mRNAs in the model. The waveforms thus obtained are shown in Fig. 4(a) and Fig. 4(b). We observe an anti-phase relationship between the concentrations of the *Per*/*Cry* and *Bmal1* mRNAs, as expected from theory.[1] The period of the oscillating waveforms is equal to 23.8 hrs for the mammalian clock and 22.4 hrs for the *Drosophila* clock.

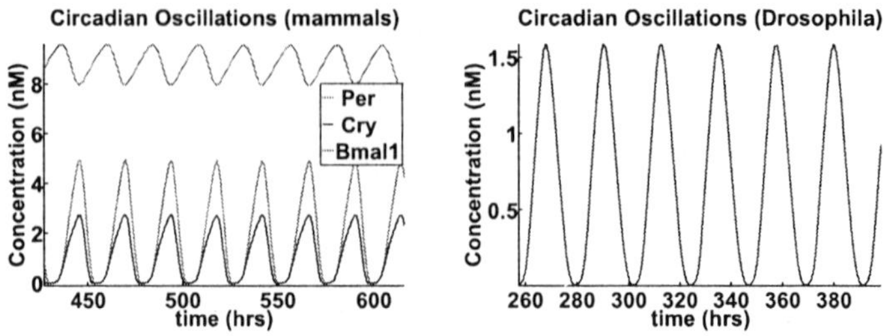

(a) ***Per, Cry* and *Bmal1* Gene Concentrations (Mammals)** (b) ***Per* Gene Concentration (*Drosophila*)**

Fig. 4. (a) Plot of core clock gene concentrations (*Per*, *Cry* and *Bmal1*) in mammals vs. time. The concentrations are oscillatory and there is an antiphase relationship between the *Per*/*Cry* and *Bmal1* concentrations. (b) Plot of the *Per* gene concentration in *Drosophila* vs. time

4.2. *Circadian PPV macromodels*

In this section, we extract the PPV macromodel of the circadian oscillator for both models. Fig. 5(a) and Fig. 5(b) show the PPV waveforms of *Per* gene concentrations. This waveform gives the phase sensitivity of the concentration at each time instant and can be directly used to find the new concentration waveform under the effect of an external perturbation. It is equivalent to the phase response curve described by Winfree,[14] with the only exception that PPV waveforms do not involve sinusoidal simplifications,[15] implying greater accuracy, as already noted previously. By inspecting the phase sensitivity at each time instant, it becomes possible to determine the time at which light should be applied to shift the oscillator's time-keeping forward or backward. At zero crossings of the PPV phase sensitivity function, for example, a light pulse will have no effect on the phase/frequency characteristics of the oscillator.

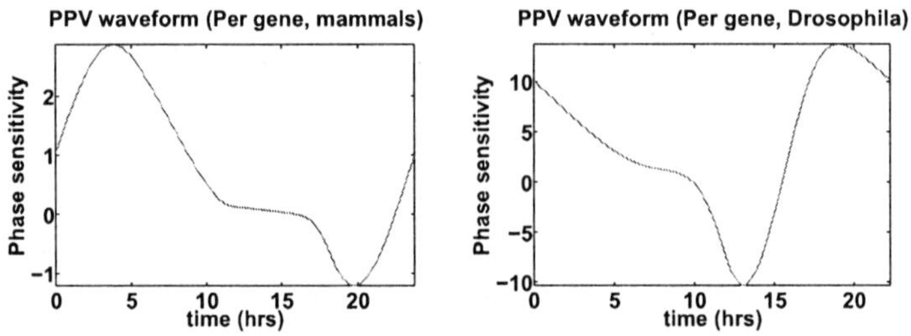

(a) **PPV waveform for *Per* mRNA (gene) concentration (Mammals)** (b) **PPV waveform for *Per* mRNA (gene) Concentration (*Drosophila*)**

Fig. 5. **(a) and (b) Plot of PPV phase sensitivities vs. time for the *Per* gene concentration.**

Speedups: (a) For the mammalian clock model: full time course simulation using the Backward Euler (BE) integration method requires 18 seconds of computer time, while finding the free-running steady state via harmonic balance analysis takes about 6 seconds. This is followed by the PPV extraction algorithm, which takes around 1.5 seconds; the total time required for PPV extraction is about 7.5 seconds, representing a speedup of some 2.5 $\times$. (b) For the *Drosophila* model: full time course simulation takes about 13 secs; harmonic balance analysis takes $\sim$ 4 seconds, PPV extraction $\sim$ 0.5 seconds; resulting in a speedup of $\sim 3\times$.

To gauge the accuracy of the PPV macromodels, we plot concentration waveforms for the *Per* gene, obtained from time course and PPV simulations, in Fig. 6. Fig. 6(a) shows waveforms for an locked oscillator (distinct frequencies), while Fig. 6(b) shows waveforms for a locked oscillator.

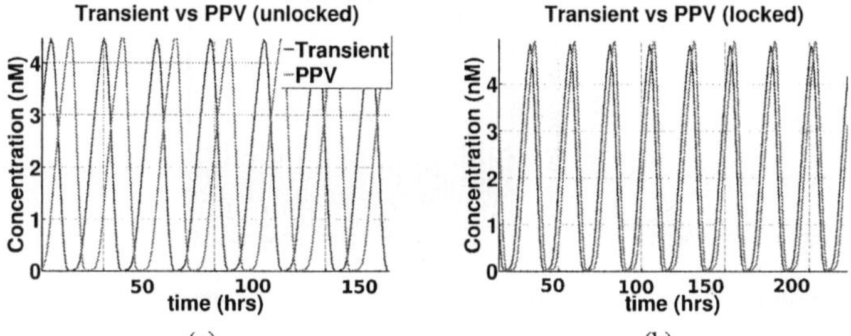

(a) (b)

Fig. 6. **Plots of *Per* gene (mRNA) concentration obtained from transient and PPV macromodel simulations. (a) Unlocked case. (b) Locked case.**

4.3. *Simulation of injection locking*

In order to study the effects of external perturbations on circadian rhythms, we calculate phase deviations due to external perturbations (Eq. 10) by solving Eq. 7. If the period of the externally applied signal is close enough to the oscillator's free-running frequency, *entrainment* or *injection locking* occurs.

4.3.1. *Mammalian clock model*

The free-running frequency of the mammalian circadian clock is $f_o = 4.19 \times 10^{-2} hr^{-1}$. We apply an external signal with $\frac{\Delta f}{f_o} = -0.00623$. Fig. 7(a) depicts injection locking with a light input of 0.009 + 0.009sin(ωt) W/m^2, where the locking starts around 690 hours ($\sim$ 30 cycles). Fig. 7(b) shows the same curve for light input of 0.05 + 0.05sin(ωt) W/m^2; locking time reduces to about 260 hrs ($\sim$ 11 cycles). From these results, we can infer that with smaller light intensities, the resetting phenomenon takes a longer time.

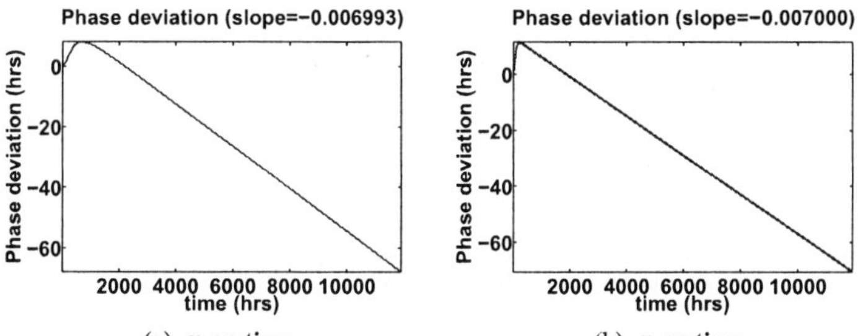

(a) α_t vs time (b) α_t vs time

Fig. 7. (a) Plot of phase deviation $\alpha(t)$ vs. time for the mammalian clock model, with light input $0.009 + 0.009\sin(\omega t) W/m^2$. The slope of -0.0069 indicates injection locking, with lock reached in about 690 hours. (b) Plot of $\alpha(t)$ vs. time for the mammalian clock model, with light input $0.05 + 0.05\sin(\omega t) W/m^2$. The slope is -0.007; lock is reached in about 260 hours.

4.3.2. *Drosophila clock model*

The free-running frequency of the *Drosophila* circadian oscillator was $f_o = 4.48x10^{-2} hr^{-1}$; the frequency of the injected light signal was $f = 4.16x10^{-2} hr^{-1}$, leading to $\frac{\Delta f}{f_o} = -0.071$. The intensity of the applied light is given by Eq. 10, with $A = 0.05 W/m^2$. Fig. 8(a) shows the phase deviation vs. time; its slope is 0.071, equal to the relative frequency difference, thus confirming that the oscillator is locked to the injection frequency.

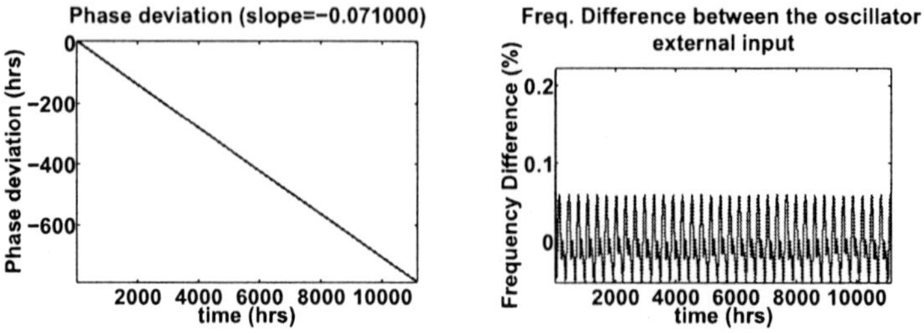

(a) α_t vs time (b) **Frequency Deviation vs time**

Fig. 8. **(a) Plot of phase deviation $\alpha(t)$ vs. time for the *Drosophila* clock model. The slope is** -0.071 **(injection locked) (b) Plot of frequency deviation vs. time.**

4.4. *Lock range vs. injection amplitude*

We also calculate the lock range (frequency range over which the oscillator remains locked to the external signal) for the mammalian clock and plot it as a function of injection amplitude. We find that the lock range increases roughly linearly with injection amplitude, as can be seen in Fig. 9. However, at higher amplitudes, the linearity between lock range and injection amplitude collapses. By calculating the lock range for a given light amplitude, we can infer whether the system would lose its rhythmicity or not on exposure to that particular light. Conversely, one can calculate the light amplitude required to synchronize the free running oscillators.

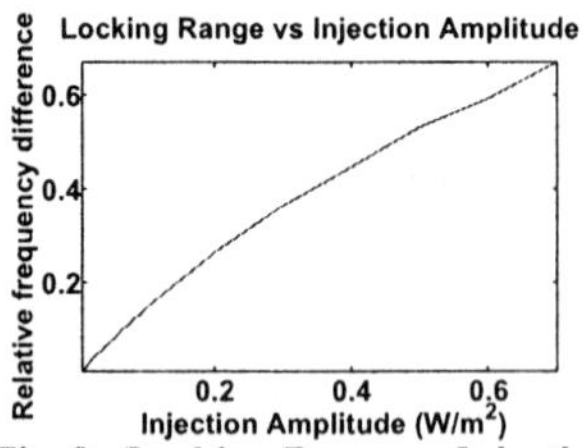

Fig. 9: **Locking Range vs Injection Amplitude (Mammalian Clock**

Speedups: (a) Mammalian clock model: Time course simulations take $\sim$ 18 seconds. PPV macromodel simulations take $\sim$ 2 seconds after PPV extraction, resulting in a speedup of $\sim 9\times$. (b) *Drosophila* clock model: Time course simulations take $\sim$ 13 seconds; PPV macromodel simulations take $\sim$ 1 second; resulting in a speedup of $\sim 13\times$.

4.5. *Parameter variation simulations*

To study the effect of parameter variations on circadian rhythms, we first simulate the phase deviations given by Eq. 8 with $b(t) = 0$. As an example, we vary all parameter values by 10% of their nominal values; *i.e.*, $\Delta p = 0.1p$. The slope of the phase deviation curve gives the relative change in frequency due to the change in parameters. As is evident from Eq. 8, we can study the effects of all possible combinations of parameter variations.

For the mammalian clock model, the relative frequency change is found to equal 0.186; *i.e.*, the new frequency is equal to 20.1 hours (Fig. 10(a)). For the *Drosophila* clock model, the relative frequency change equals -0.114, *i.e.*, the new frequency equals 25.2 hrs (Fig. 10(a)). It is evident that even small changes in parameter values can affect rhythm frequency significantly.

Next, we combine parameter variations with external perturbations to the oscillator. Using the slope of the phase deviation curve, we can calculate the range over which parameters can be varied while keeping the oscillator oscillator still locked to the injection signal frequency for the same light input. As an example, we vary all parameters simultaneously and find the respective variation range for each model. In case of the *Drosophila* model, parameters can be varied from $-5 to 10\%$ without loss of lock; while for the mammalian model, the range is smaller, $-5 to 1\%$ variation. The light input in both the cases is given by Eq. 10 with $A = 0.05 W/m^2$.

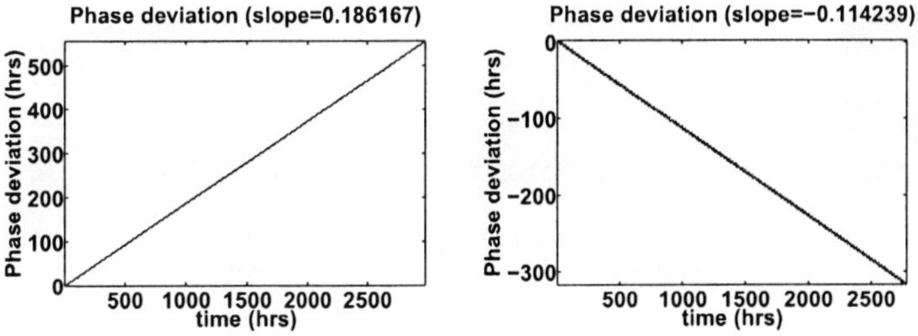

(a) α_t vs time

(b) α_t vs time

Fig. 10. **(a) Plot of the phase deviation $\alpha(t)$ vs. time for the mammalian clock model with 10% variation in the parameter values. The slope of the curve = 0.186 implying that the new frequency of oscillations equals to 20.1 hrs. (b) Plot between the phase deviation $\alpha(t)$ and time for the *Drosophila* Clock Model with 10% variation in the parameter values. The slope = -0.114 and the new frequency of oscillations is hence equal to 25.2 hrs.**

4.6. *Synchronization of coupled oscillators*

In this section, we extend the single oscillator analysis to a system of many coupled oscillators (*i.e.*, a system of several interacting biological cells, each behaving as an individual oscillator and oscillating with a period of ~ 24 hrs). We consider a system of 400 mammalian clock oscillators arranged in a 20x20 grid, as shown in Fig. 11.[1] The oscillators are identical in all respects except for their free-running frequencies, which are selected randomly from a uniform distribution. Each oscillator is modelled by a system of 16 ODEs (as used before for single oscillator analyses). In order to introduce coupling between the oscillators, we use a recently proposed coupling model given in To et al,[20] wherein neurotransmitters act as synchronizing agents between the cells. Then, using a PPV macromodel for each oscillator augmented by coupling equations, we simulate the entire oscillator system.

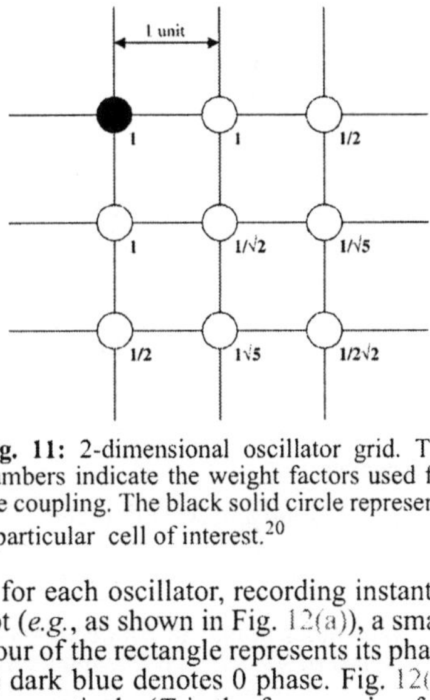

Fig. 11: 2-dimensional oscillator grid. The numbers indicate the weight factors used for the coupling. The black solid circle represents a particular cell of interest.[20]

We use Eq. 6 to calculate the phase deviations for each oscillator, recording instantaneous phases at regular intervals. In every phase plot (*e.g.*, as shown in Fig. 12(a)), a small rectangle represents an individual oscillator; the colour of the rectangle represents its phase visually; *e.g.*, dark red denotes a phase of π, while dark blue denotes 0 phase. Fig. 12(a) and Fig. 12(b) show phase plots at $t = 1T$ and $5T$ respectively (T is the free running frequency of an oscillator) in the absence of coupling. The absence of coupling can easily be surmised, from the random nature of the plots (absence of any pattern, *i.e.*, unsynchronized phases). For the coupled case, Fig. 12(d) shows the phases at $0.5T$, when all the oscillators start synchronizing to the same phase (and frequency). Fig. 12(e) and Fig. 12(f) are the phase plots at later stages, confirming synchronization amongst the coupled oscillators. We have also varied the random center frequency distributions of the oscillators, and found that with the same coupling strength, the oscillators cease to lock to each other for deviations greater than 0.5 of the free-running circadian period

Speedups: (a) Time course simulations require ~ 12 hours for full simulations, including the time required for the formation of the coupling matrix. (b) PPV simulations require ~ 158 seconds for complete simulations. Hence, we obtain a speedup of $\sim 240\times$. If the system size is larger and the oscillator model is more complex, the speedups will be greater.

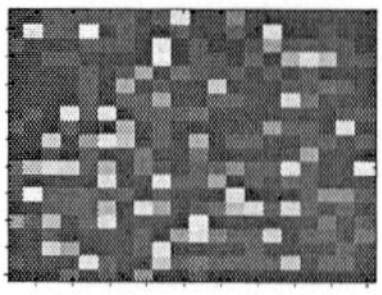

(a) **Phase Plot at t = 1T (No Coupling)**

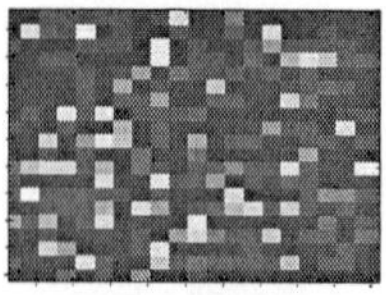

(b) **Phase Plot at t = 5T (No Coupling)**

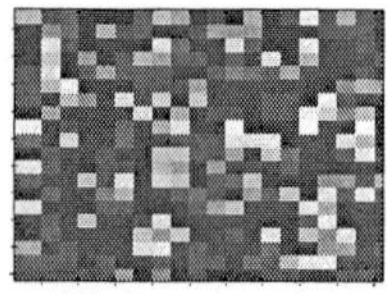

(c) **Phase Plot at t = 0 (Coupled)**

(d) **Phase Plot at t = 0.5T (Coupled)**

(e) **Phase Plot at t = 0.75T (Coupled)**

(f) **Phase Plot at t = 1.25T (Coupled)**

Fig. 12. **(a) and (b) Phase plots in case of no intercellular coupling between individual oscillators. (c) - (f) Phase plots showing the synchronization of coupled oscillators (all oscillators at the same phase)**

5. Conclusion

We have applied PPV phase macromodelling techniques to mammalian and *Drosophila* circadian rhythms, for the first time. These techniques provide fast/accurate simulations of oscillator systems, predicting synchronization and resetting in circadian rhythms via injection locking cued by light inputs. In addition, PPV waveforms provide direct insight into the effect of light on phases of the oscillating rhythms. We have accurately predicted synchronization in a coupled multi-scale system of 400 circadian oscillators using PPV macromodels. Finally, the efficacy of parameterized PPV macromodels for circadian problems has also been demonstrated.

References

1. J.C. Leloup and A. Goldbetter, "Toward a detailed computational model for the mammalian circadian clock" in *Proceedings of the National Academy of Sciences of the United States of America* , 7051(June 2003).
2. J. D. Gonze and A. Goldbetter, "Robustness of circadian rhythms with respect to molecular noise" in *Proceedings of the National Academy of Sciences of the United States of America* , 673(January 2002).
3. Y. Touitou, *Biological Clocks: Mechanisms and Applications (Proceedings of the International Congress on Chronobiology)* (Elsevier, Paris, France, 1997).
4. R. Adler, "A study of locking phenomenon in oscillators" in *Proceedings of the I.R.E. and Waves and Electrons* **34**, 351 (1946).
5. X. Lai and J. Roychowdhury, "Capturing Oscillator Injection Locking via Nonlinear Phase-Domain Macromodels" in *IEEE Trans. MTT* **52**, 2251(September 2004).
6. A. Demir, A. Mehrotra and J. Roychowdhury, "Phase noise in oscillators: a unifying theory and numerical methods for characterization" in *IEEE Trans. Ckts. Syst. – I: Fund. Th. Appl.* **47**, 655(May 2000).
7. T. Mei and J. Roychowdhury, "A Robust Envelope Following Method Applicalbe to both Non-autonomous and Oscillatory Circuits" in(July 2006).
8. T. Mei and J. Roychowdhury, "An Efficient and Robust Technique for Tracking Amplitude and Frequency Envelopes in Oscillators" in(November 2005).
9. A. Demir and J. Roychowdhury, "A Reliable and Efficient Procedure for Oscilla-

tor PPV Computation, with Phase Noise Macromodelling Applications" in *IEEE Trans. Ckts. Syst. – I: Fund. Th. Appl.* , 188(February 2003).

10. X. Lai and J. Roychowdhury, "Fast, accurate prediction of PLL jitter induced by power grid noise" in(May 2004).
11. X. Lai and J. Roychowdhury, "Fast Simulations to Large Networks of Nanotechnological and Biochemical Oscillators for Investigating Self-Organization Phenomenon" in *Proc. IEEE ASP-DAC* (2006).
12. A. Demir and J. Roychowdhury, "A reliable and efficient procedure for oscillator PPV computation, with phase noise macromodelling applications." in *IEEE Transactions on Computer-Aided Design* **22**, 188(February 2003).
13. X. Z. Wang and J. Roychowdhury, "PV-PPV: Parameter Variability Aware, Automatically Extracted, Nonlinear Time-Shifted Oscillator Macromodels" in(June 2007).
14. A. Winfree, "Biological Rhythms and the Behavior of Populations of Coupled Oscillators" in *Theoretical Biology* **16**, 15 (1967).
15. Y. Kuramoto, "Chemical osillations, waves and turbulence" in *Springer* (1984).
16. S.H. Strogatz, "From Kuramoto to Crawford: exploring the onset of synchronization in populations of coupled oscillators" in *Elsevier* **143**, 1 (2000).
17. J.C. Leloup and A. Goldbetter, "Modeling the mammalian circadian clock: Sensitivity analysis and multiplicity of oscillatory mechanisms" in *Theoretical Biology* , 541(April 2004).
18. Circadian rhythms laboratory homepage http://www.circadian.org/main.html.
19. X. Lai and J. Roychowdhury, "Macromodelling Oscillators Using Krylov-Subspace Methods" in *Proc. IEEE ASP-DAC* (January 2006).
20. T.L. To, M.A. Henson, E.D. Herzog and F.J. Doyle III, "A Molecular Model for Intercellular Synchronization in the Mammalian Circadian Clock" in *Biophysical Journal* **92**, 3792 (2007).
21. Y. S. Usui and T. Okazaki, "Range of entrainment of rat circadian rhythms to sinusoidal light-intensity cycles" in *AJP-Regulatory, Integrative and Comparitive Physiology* **278**, R1148(May 2000).

INTEGRATION OF MULTI-SCALE BIOSIMULATION MODELS VIA LIGHT-WEIGHT SEMANTICS

JOHN H. GENNARI[1], MAXWELL L. NEAL[1], BRIAN E. CARLSON[2], & DANIEL L. COOK[3]

[1]Biomedical & Health Informatics, [2]Bioengineering, [3]Physiology & Biophysics, University of Washington, Seattle, WA, 98195, USA

Currently, biosimulation researchers use a variety of computational environments and languages to model biological processes. Ideally, researchers should be able to semi-automatically merge models to more effectively build larger, multi-scale models. However, current modeling methods do not capture the underlying semantics of these models sufficiently to support this type of model construction. In this paper, we both propose a general approach to solve this problem, and we provide a specific example that demonstrates the benefits of our methodology. In particular, we describe three biosimulation models: (1) a cardio-vascular fluid dynamics model, (2) a model of heart rate regulation via baroreceptor control, and (3) a sub-cellular-level model of the arteriolar smooth muscle. Within a light-weight ontological framework, we leverage *reference ontologies* to match concepts across models. The light-weight ontology then helps us combine our three models into a merged model that can answer questions beyond the scope of any single model.

1. Semantics for biosimulation modeling

Biomedical simulation modeling is an essential tool for understanding and exploring the mechanics and dynamics of complex biological processes. To this end, researchers have developed a wide variety of simulation models that are written in a variety of languages (SBML, CellML, etc.) and are designed for a variety of computational environments (JSim, MatLab, Gepasi, Jarnac, etc.). Unfortunately, these models are not currently interoperable, nor are they annotated in a sufficiently consistent manner to support intelligent searching or integration of available models.

In the extreme case, a biosimulation model contains no explicit information about what it represents— it is only a system of mathematical equations encoded in a computational language. The biological system that is the subject of the model is implicit in the code; the code is an abstraction of that system into mathematical variables and equations that must be interpreted by a researcher. If one researcher wishes to understand or use a model created by another, he or she must (usually) communicate directly with those that created the model. For complex, multi-scale models, this problem is a bottleneck to further progress—if models could be archived, re-used, and connected together computationally, we would avoid a great deal of work spent "re-creating the wheel", by leveraging more directly the work of others.

Recognizing this problem, there are on-going efforts to build repositories of annotated biosimulation models [1-4]. However, these annotations are predominantly human-interpretable and depend on local semantics. For example, repositories of JSim models [4] and CellML models [1] rely on in-line code annotations to explain mathematical equations—annotations that are not machine-interpretable. The BioModels repository [3] of SBML-encoded models uses XML-based annotations, but, we argue, these still lack the strong semantics required for computer-aided integration. (This library is also restricted to the scales of cellular and biomolecular problems). Given that the goal of multi-scale modeling is the flexible reuse and integration of models to solve large-scale modeling problems, we argue that a much stronger, machine-interpretable semantic framework needs to be applied to these biosimulation models.

In this paper, we propose a flexible solution that will allow biosimulation models to be re-used and re-combined in a plug-n-play manner. The thrust of our approach is to build light-weight ontological models of biological systems for annotating model variables in terms of the physical properties and the anatomical entities to which they refer, and for explicitly representing how these property variables depend upon each other. More concretely, we demonstrate how our ontologies can represent the semantics of three models, and then use this information to help merge these into a larger, multi-scale biosimulation model.

We begin by describing the three source models that make up a driving use-case for our research, and then show how each model is semantically mapped to our light-weight ApplModel Ontology framework (section 2). We can then analyze and visualize the semantics of the models using available software tools (Prompt [5], see section 3). Such tools help us merge the models, and we show that our merged model can answer multi-scale questions that are not answerable by single component models (section 4).

1.1 Motivating use-case: Arteriolar calcium uptake & heart rate

Our driving biological problem is to create a multi-scale cardiovascular model from three independently-coded models that contain overlapping parts of the cardiovascular regulatory system. Figure 1 provides both a view of our three 'source' models (top half) and our 'target'—a merged, multi-scale model (bottom half). Our use-case goal is to employ the merged model to answer a multi-scale, systems-level question such as "How do heart rate and blood pressure depend on calcium uptake into arteriolar smooth muscle cells?"—a question that cannot be answered by the individual source models.

The three source models at the top of figure 1 are each a lumped-parameter

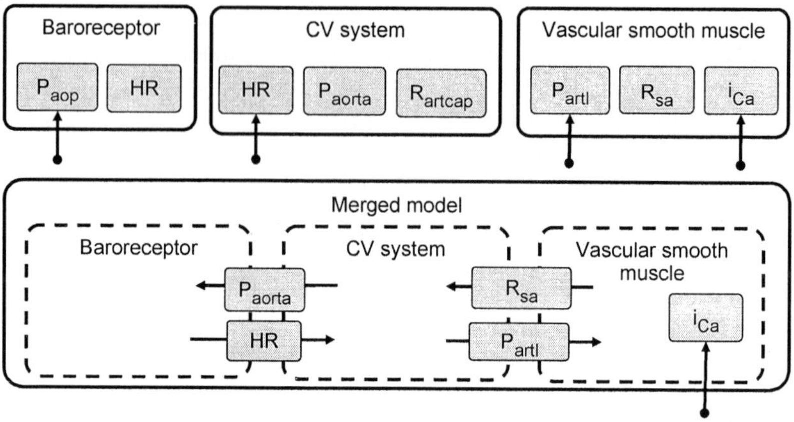

Figure 1. A simple overview of our use-case and computational goals. We are building an infrastructure for querying, interpreting and merging biosimulation model such as the three models at the top of the figure, into larger, multi-scale models, such as shown on the bottom.

model independently encoded in the JSim simulation environment[6].[a] A cardiovascular model (CV) was coded by the second author and is a condensed version of a previously published model [7]. Using a constant heart rate input (HR) and other parameters, the CV model computes time-varying blood pressures and flows in a 4-chambered heart and in the pulmonary and systemic vessels. Our baroreceptor model (BARO) was originally coded by Daniel Beard and is based on Lu and Clark [8] and Spickler et al. [9]. The BARO model takes aortic blood pressure as input and computes a time-varying heart rate as a feedback signal to control blood pressure. A vascular smooth muscle model (VSM) was coded by the third and fourth authors to model the effect of Ca++ ion uptake into arteriolar smooth muscles cells and its consequent effect on arteriolar flow resistance. In section 4, we provide details about how we created the merged model, as well as descriptions of the parameters and variables listed in figure 1.

As one measure of the challenges inherent in merging these models, our combined source models include over 190 named variables and parameters whose biophysical meanings are buried in code annotations (where available) that are specific to each model. To merge these models appropriately, we need to consider three sorts of challenges. First, we must discover identical biophysical entities. For example, heart rate is only coincidentally encoded as HR in both the CV and BARO models and in fact, represents the same biophysical entity. Sec-

[a] Full source code for these three models are available at http://trac.biostr.washington.edu/trac/wiki/JSimModels

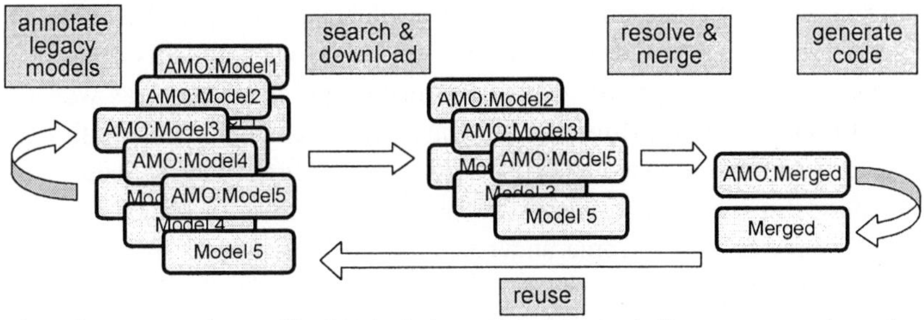

Figure 2. An approach to making biosimulation models "plug-n-play": annotate, search, resolve, merge, encode, and ultimately reuse.

ond, we must discover and resolve variables that are related, but not identical. For example, Rsa represents the arteriolar fluid resistance in VSM but the arterioles are only part of the systemic arterial vasculature whose fluid resistance is represented as Rartcap (arteries, arterioles and capillaries) in CV. Third, we must discover and resolve variable dependencies. HR in the CV model is an input or controlled variable whereas in BARO it is an output or computed variable that depends ultimately on aortic blood pressure (Paop). Thus, the HR variables from CV and BARO should be merged into a single variable, so that the heart rate calculated by BARO becomes an input to the CV model.

1.2 A solution: Light-weight ontological annotation

The above challenges all revolve around defining the biophysical semantics of the variables and parameters within models. As we describe in the next section, our solution begins by annotating biosimulation models with light-weight semantics, as provided by our Application Model Ontology (AMO, see also section 2.2). The AMO is small, and we envision tool support to make annotation as easy as possible for simulation modelers. More broadly, figure 2 shows how this annotation step is part of a more general architecture for reusable biosimulation models. Once models are annotated with AMO, model libraries can be more intelligently searched for relevant models. As we show in section 3, once selected, AMO annotations can help with the tasks of resolving differences between models to create merged models. Next, from the merged models, we plan to generate code in a variety of simulation languages using code-generation methods with which we have experience [10]. Ultimately, as with software reuse, merged models can be returned to the library for reuse by others.

2. Semantic annotation via ontologies

Computer-interpretable semantics are best captured by formal ontologies. In recent years, a wide variety of ontologies for biology have become available.

Prominent among these are the ontologies available at the Open Biological Ontologies (OBO), and its OBO foundry project (at www.obofoundry.org). These ontologies cover a variety of levels of formality and abstraction, as well as a variety of domain topics. However, although ontologies of physical entities such as genes, species, and anatomy have been well-developed, the domain of biosimulation also requires *properties* of anatomical entities (such as volume or fluid pressure) as well as some understanding of the *processes* by which these properties change over time.

In general, we posit that although formal, abstract, "heavy" ontologies are essential for unambiguous, machine interpretable annotation, end-users need a light-weight methodology for semantic annotation. Thus, we advocate using two sorts of ontologies: (1) *reference* ontologies, that allow us to ground our work in the formal semantics of structural biology and physics, and (2) *application model* ontologies that are tailored for the specific semantics of particular biosimulation models.

2.1 Reference ontologies: FMA and OPB

For our example, we use two reference ontologies: the Foundational Model of Anatomy (FMA, [11]), a mature reference ontology of human anatomy, and the Ontology of Physics in Biology (OPB), an ontology of classical physics designed for the physics of biological systems. The FMA is a nearly complete structural description of a canonical human body. Its taxonomy of *Anatomical entities* is organized according to kind (e.g., *Organ* system, *Organ*, *Cell*, *Cell part*) with parthood relations so that, for example, the *Cardiovascular System* has parts such as *Heart*, *Aorta*, *Artery*, and *Arteriole*. Parts are also related by other structural relations so that, for example, the *Aorta* is *connected_to* the *Heart* and the *Blood in aorta* is *contained_in* the *Aorta*.

The Ontology of Physics for Biology (OPB) is a scale-free, multi-domain ontology of classical physics based on systems dynamics theory [12-15]. It thus distinguishes among four *Physical property* superclasses for lumped-parameter systems: *Force, Flow, Displacement,* and *Momentum*. As shown in figure 3A, each of these *Physical property* classes has subclasses in seven "energy domains": *Fluid mechanics, Solid mechanics, Electricity, Chemical kinetics, Particle diffusion, Heat transfer,* and *Magnetism*. The OPB also encodes *Physical dependency* relations that include *Theorems of physics* (e.g., *Conservation of energy*) and *Constitutive property dependencies* (shown in figure 3B) such as the *Fluid capacitive dependency* relation that governs, say, how ventricular volume depends on ventricular blood pressure.

By combining the knowledge in the FMA and the OPB one can unambigu-

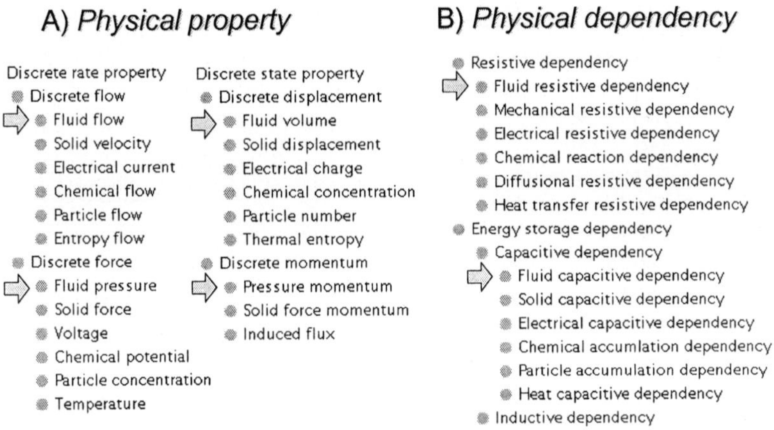

Figure 3. Main classes of the Ontology of Physics for Biology (OPB). The classes highlighted with arrows indicate the fluid mechanics aspects for both physical properties and dependencies.

ously annotate model variables by associating an FMA:Anatomical entity with an OPB:Physical property, creating duples such as [FMA:Blood of aorta :: OPB:Fluid pressure]. Thus, for modeling multi-scale biological systems, the FMA and OPB offer a wealth of machine-accessible anatomical and biophysical knowledge that can be leveraged for annotating biosimulation code.

2.2 *The application model ontology*

Our goal in developing the Application Model Ontology (AMO) is to provide an ontological framework for creating reusable, lightweight ontological annotations of biosimulation models—ApplModels (for application models). The fundamental idea of the AMO is to allow researchers to build models that use only very small subsets of very large and complex reference ontologies. A biosimulation researcher does not care or want to know about all of the anatomic entities in the FMA nor about theorems across all seven of the energy domains in the OPB. Thus, ApplModels exploit, but do not depend on, external reference ontologies, and yet can be "lightweight" and customized to represent idiosyncratic biophysical entities and relations. AMO classes are formally defined according to the principles espoused by the OBO foundry, and are created and edited within the Protégé environment [16].

Figure 4a shows a screenshot of a portion of the AMO base classes in Protégé and some examples of how these classes are filled in by the BARO biosimulation model we described earlier. The higher-level classes such as *physical entity* or *physical property* are basic AMO classes, while the leaf nodes

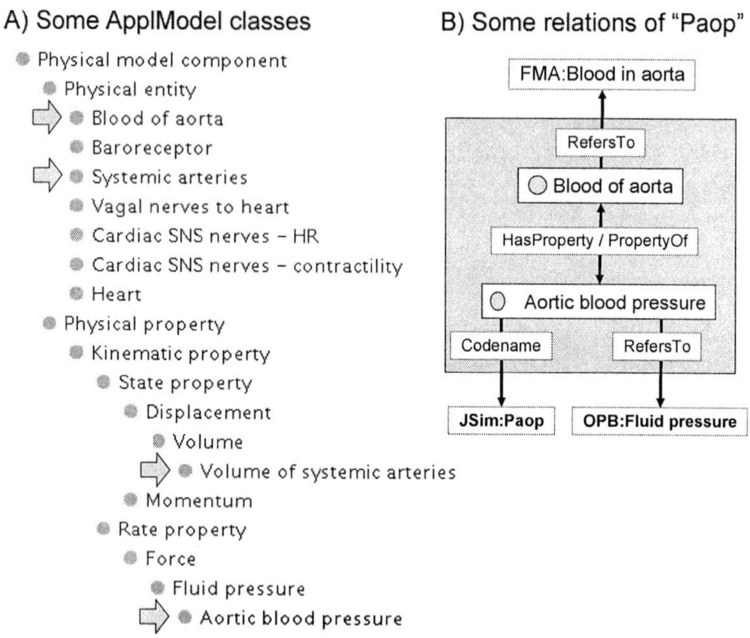

Figure 4. The Application Model Ontology, as filled in for the BARO biosimulation model.

show how AMO was filled in for the BARO model. Figure 4b shows some detail of the annotation for the BARO variable "Paop", including links to reference ontologies. To capture the semantics for this variable, we first created ApplModel classes that refer to the corresponding reference ontology concepts (FMA: *Blood in aorta* and OPB:*Fluid pressure*) and then the specific class that represents one particular fluid pressure, namely "Paop". Figure 4a shows the entire set of "physical entities" for the BARO model; not shown are the constitutive relationships and dependencies that are represented by equations in the model code.

Wherever possible, users should refer directly to reference ontology classes—such ontologies make model integration possible, by enforcing a common semantics to particular terms. However, users can also create special-purpose (or idiosyncratic) subclasses for particular biosimulation models. For example, the CV model variable "Rartcap" is the resistance in a single entity that lumps systemic arteries and capillaries together; such an entity does not exist in the FMA reference ontology. However, with the ApplModel annotations, we can easily create a special subclass of *Physical thing* such as *Systemic-Arteries-Capillaries* that uses *AMO:HasPart* relations to the *Systemic arteries* and *Systemic capillaries* classes that are available in the FMA.

It is the ability to integrate idiosyncratic semantics with reference ontologies and other external knowledge resources that makes light-weight ApplModels a powerful approach to biosimulation model annotation. Once annotated with such semantics, we can both better understand these models, and use existing ontology analysis tools to help with the model integration task.

3. Comparing and merging models

When merging biosimulation models there are usually significant semantic differences that must be resolved. While some of these differences may be obvious and easy to find, automated analysis tools can greatly help researchers find and resolve such differences. One major advantage of annotating biosimulation models with ontologies, is that there are pre-existing tools to help with these sorts of tasks. In our case, we have employed Protégé's Prompt plug-in tool for ontology comparison and merging tasks [5].

Prompt is designed for interactive, semi-automatic model merging. Given two ontologies, Prompt analyzes the classes and relationships in the two models, and then suggests a set of mappings that connect concepts between the two models. The user can inspect these candidate matches, and confirm all or some of these matches. Prompt then uses this information to suggest additional matches, and this interactive cycle repeats.

For our use-case, when we gave Prompt the ApplModel ontologies for the BARO and CV models, it was able to recognize that, for example, "systemic arteries" is a shared concept—regardless of how it was coded in the source models—because it was annotated with the common FMA reference ontology term. Furthermore, the Prompt visualization tools reveal that there are significantly different relationships around "systemic-arteries" across the two models. Figure 5 shows the "neighborhood view" of nearby semantic relationships as presented by Prompt when it proposes the match for "systemic-arteries". Figure 5a shows that the BARO model links resistance as a direct property of systemic arteries, whereas the CV model (figure 5b) uses the set "systemic arteries and capillaries", which has a resistance. As figure 1 shows, there is a similar discrepancy about resistance between the CV model (Rartcap) and the VSM model, which only considers the systemic arterioles (Rsa). To appropriately merge models, researchers must resolve these sorts of semantic discrepancies.

Even when the underlying anatomy is consistent, models may use the properties of those entities in different manners. In our case, heart rate (HR in figure 1) is defined consistently across the BARO and CV models, but in the BARO model it is an output, whereas the CV model uses HR as one of its inputs. The difference can be readily visualized in Prompt, because the ApplModel ontology

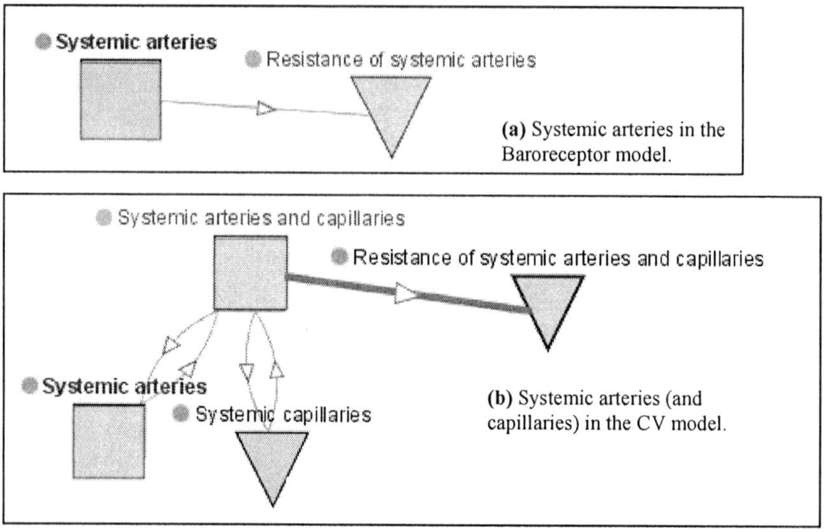

Figure 5. The two uses of the concept "Systemic arteries" in the BARO and CV models. These views were produced by Prompt when suggesting mappings between the models. (Arcs show relations, such as "part-of" between entities. Squares are "fully expanded" entities, whereas triangles are entities that can be further expanded.)

includes the relationships "dependsDirectlyOn" and "affectsDirectly". (Such relationships are shown as colored arcs in the Prompt visualization.)

As we show in figure 2, our overall expectation is that Prompt can be used as a step in the overall process of building multi-scale biosimulation models. However, even if researchers do not immediately expect to combine models directly, a Prompt comparison of closely related models can used to reveal model semantics and physiological relationships that are otherwise implicit in the mathematical code. By visualizing graphically the set of relationships among anatomic entities and their physical properties, biosimulation researchers can better understand how two models are and are not the same. In addition, Prompt may actually help de-bug biosimulation models by making it visually apparent when relationships are missing, problematic, or incorrect.

4. Status and Results

As a summary of our results so far, we have (a) annotated three related source models with the Application Model Ontology, (b) used the Prompt tool to analyze these models, helping us visualize and understand the differences across models, and (c) hand-coded a merged model into JSim that implements the decisions made during the merge step. Thus, our product is an integrated, executable model that can indeed answer our original driving question: "How do heart rate and blood pressure depend on calcium uptake into arteriolar smooth muscle

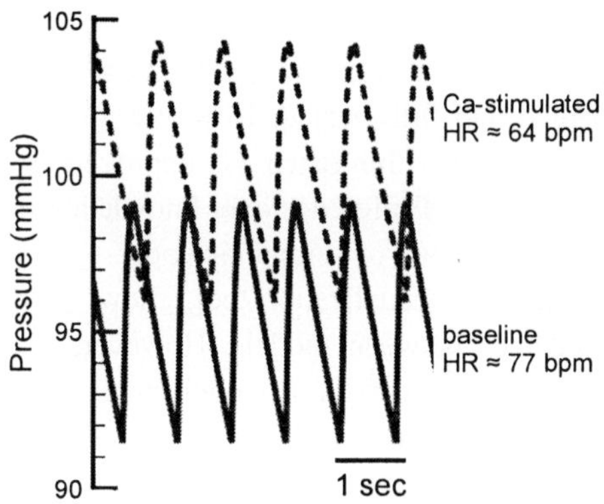

Figure 6. The result of increased Ca++ uptake, as an output from our merged JSim model.

cells?" Figure 6 shows an annotated output from the JSim execution of our model, showing the expected increase in blood pressure and decrease in heart rate when Ca uptake is increased.

We began integration of the JSim multi-scale model by first merging the CV and BARO models. To do this, we merged two shared concepts: heart rate and aortic blood pressure (see figure 1: HR, Paop, and Paorta).

- We changed the BARO term Paop from an independent input to a time-dependent variable output and set it equal to Paorta, the aortic blood pressure variable from the CV model.
- We removed the independent HR input from the CV model so that cardiac activation would depend on the BARO model's variable HR.
- We added a new discrete HR variable (HRdiscrete) that only updates at the end of the cardiac cycle to prevent intra-beat fluctuations in heart rate. (To do this, we needed to add some procedural code to the merged model.)

Next, we merged the result with the VSM model by combining representations of resistance:

- Given the high proportion of vascular resistance in the arterioles, we assumed that the time-varying arteriolar resistance (Rsa) computed in the VSM model to be the same as the resistance of arteries and capillaries from the CV model (Rartcap, a constant). Therefore, we changed Rartcap to a time-dependent variable equal to Rsa.
- To couple arteriolar resistance with the dynamics of the CV model, we changed the arteriolar blood pressure input (Partl) to a time-dependent variable equal to the average pressure between the CV model's arterial/capillary and venous compartments.

The resulting multi-scale model includes 65 algebraic and 25 ordinary differential equations. Although more detailed models of the cardiovascular system and smooth muscle cell dynamics exist, our system produces physiologically normal steady state averages for circulatory and smooth muscle cell dynamics and allows investigations into the influence of subcellular activity on tissue-level dynamics (as in figure 6).

5. Discussion and future work

There remains much work to do before our broad ideas of model integration (as shown in figure 2) can be implemented and fully tested. However, given our work to date, three aspects of our vision seem within reach: (1) improved use of Prompt for model merging, (2) improved use of inference over knowledge from reference ontologies, and (3) automatic generation of simulation code.

To date, we have only used Prompt as a visualization tool, allowing us to see discrepancies and understand linkages between models. However, as we have described, Prompt is designed to actually carry out model merging in an interactive manner. Furthermore, Prompt is designed with a plug-in architecture, which means it can be easily custom-tailored to meet our needs. Therefore, we will be able to use Prompt to carry out most of the model merging, although some parts of the work will remain manual (e.g., the addition of procedural code described earlier around the "HRdiscrete" variable).

As reference ontologies, the FMA and OPB both contain a wealth of knowledge that could be used to more intelligently guide model merging. For example, Prompt cannot currently notice that the diameter of an arteriole, a variable in the VSM model is related to arterial blood volume. However, the FMA knows that the arterioles are part of the systemic arterial tree, and the OBP knows that the diameter (along with the length) can determine the volume of an arteriole. We should therefore be able to use this sort of reference ontology knowledge to improve Prompt so that it can suggest mappings between variables such as arteriolar diameter and arterial/capillary blood volume.

We designed our ontologies and semantic markup methods to be independent of any particular biosimulation modeling language. We have so far worked exclusively with JSim models, but we believe that our ideas apply equally well to SBML and other simulation languages. We do not yet have a system for automatic code generation from our ApplModels, but we do have prior experience generating JSim code[10], and thus, we aim to build a code-generator for at least two targets: JSim and SBML. Such a tool would allow us to explore code-level semantic differences that might affect merging SBML models with JSim models. We hope that the semantic annotations provided by our ApplModel ontologies will help clarify these differences, but this intuition must be verified.

Our results represent a novel application of ontology-based semantics to help understand the deep biophysical meanings of terms used in biosimulation models. We have then used these semantics to facilitate merging models into larger, multi-scale biosimulations across very different physiological domains.

Acknowledgments

We thank Natasha Noy for helping with (and for improving) the Prompt tools. Thanks to Jim Brinkley and Onard Mejino for help refining our ideas of reference ontologies and the AMO. This work was partially funded by the NIH: for BEC, #T32EB001650-03, for MLN: #T15 LM007442-06, and for JHG & DLC: #R01HL087706-01.

References

1. CellML. *Model Repository -- CellML*. http://www.cellml.org/models/
2. MATLAB. *MATLAB Central File Exchange*. http://www.mathworks.com/matlabcentral/fileexchange/loadCategory.do
3. Le Novere N, Bornstein B, Broicher A, Courtot M, Donizelli M, Dharuri H, et al. BioModels Database: a free, centralized database of curated, published, quantitative kinetic models of biochemical and cellular systems. *Nucleic Acids Res* 2006;34(Database issue):D689-91.
4. NSR-Physiome. *NSR Physiome Model Wiki*. http://www.physiome.org/model/
5. Noy NF, Musen, MA. The PROMPT suite: Interactive tools for ontology merging & mapping. *International Journal of Human-Computer Studies* 2003;59(6):983-1024.
6. JSim. *The JSim Home Page at NSR*. http://physiome.org/jsim/index.html
7. Kerckhoffs RC, Neal ML, Gu Q, Bassingthwaighte JB, Omens JH, McCulloch AD. Coupling of a 3D finite element model of cardiac ventricular mechanics to lumped systems models of the systemic and pulmonic circulation. *Ann Biomed Eng* 2007;35(1):1-18.
8. Lu K, Clark JW, Jr., Ghorbel FH, Ware DL, Bidani A. A human cardiopulmonary system model applied to the analysis of the Valsalva maneuver. *Am J Physiol Heart Circ Physiol* 2001;281(6):H2661-79.
9. Spickler JW, Kezdi P, Geller E. Transfer characteristics of the carotid sinus pressure control system. In: Kezdi P, editor. Baroreceptors and Hypertension. Dayton, OH: Pergamon. p. 31-40.
10. Cook DL, Gennari JH, Wiley JC. Chalkboard: Ontology-Based Pathway Modeling And Qualitative Inference Of Disease Mechanisms. *Pac Symp Biocomput* 2007;12:16-27.
11. Rosse C, Mejino JL, Jr. A reference ontology for biomedical informatics: the Foundational Model of Anatomy. *J Biomed Inform* 2003;36(6):478-500.
12. Riggs DS. *Control theory and physiological feedback mechanisms*; 1970.
13. Borst P, Akkermans J, Pos A, Top J. The PhysSys ontology for physical systems. In: Bredeweg B, editor. Working Papers of the Ninth International Workshop on Qualitative Reasoning QR'95; 1995: University of Amsterdam; p. 11-21.
14. Karnopp D, Margolis DL, Rosenberg RC. *System dynamics: a unified approach*. 2nd ed. New York: Wiley; 1990.
15. Mikulecky DC. Network thermodynamics: a candidate for a common language for theoretical and experimental biology. *Am J Physiol* 1983;245(1):R1-9.
16. Gennari JH, Musen MA, Ferguson RW, Grosso WE, Crubezy M, Eriksson H, et al. The evolution of Protege: an environment for knowledge-based systems development. *Int. J. Human–Computer Studies* 2003;58:89-123.

COMPARISONS OF PROTEIN FAMILY DYNAMICS

A. J. RADER* and JOSHUA T. HARRELL

Department of Physics, Indiana University Purdue University Indianapolis,
402 N. Blackford St., LD156D
Indianapolis, IN 46219, USA
**E-mail: ajrader@iupui.edu*
www.physics.iupui.edu/~ajrader/

Similarities between different protein structures have led to the identification of protein families based upon some measure of structural similarity. Using these similarities one can classify proteins into structural families and higher-order groupings from which inferred function can be transferred. When taken for a large number of proteins, these schemes point to evolutionary relationships between organisms. We propose a novel classification scheme based upon the structurally-inspired dynamics of each protein. This classification scheme has the advantages of being quantitative, automatically assigned, and able to also make distinctions within protein families. Results are presented for five protein families illustrating the correct identification of previously un-classified structures and sources of intrafamily distinctions.

Keywords: GNM; protein dynamics; conformations; families.

1. Introduction

The comparison of proteins from different organisms relies heavily upon the paradigm that sequence encodes for protein structure which in turn determines protein function. Often protein function is not a easily definable quantity[1] making some associations unreliable. More directly, proteins can be grouped into families based upon shared structural characteristics since structural changes are generally more conservative than sequence changes. Two widely used structurally-based classification systems are SCOP (Structural Classification of Proteins)[2] and CATH (Class, Architecture, Topology, and Homologous superfamily).[3] Both classification systems require some manual intervention and depend upon the additional step of defining the domains within a protein structure. The assignment of such domains is not a unique process and adds another layer of complication to such classifications.

We contend that structurally-inspired information, specifically protein *dynamics*, are important for making the correct *functional* assignment of proteins.[4] Information regarding dynamics is absent in both of the two structural classification systems mentioned above. In this paper we present an automatic assignment criteria for grouping protein families based upon their entire structure rather than the added step of domain identification. Thus although the analysis presented here is similar to the SCOP and CATH classifications it differs by considering the dynamics of complete protein structures.

The Gaussian Network Model (GNM)[5,6] provies an efficient calculation of protein dynamics by representing the protein structure by an elastic network of residues. This creates a coarse-grained representation of the structure and its dynamics. As a result, comparison of low frequency (global) modes of motion from GNM to proteins with a similar Rossmann-fold displayed a striking similarity.[7] A related study applied to the globin family observed a similar trend that similar protein structures exhibited similar dynamics.[8] Preliminary analysis at the superfamily level found that regions with high mobility also demonstrated high levels of evolutionary fluctuations.[9] In this work we quantify the degree of similarity in dynamics with the aim of exploring how these dynamics play a role in defining protein function. We generalize these comparisons to families of proteins based upon the SCOP classification schemes, allowing a new automatic classification of each protein in terms of their GNM-defined dynamic similarities.

2. Methods

2.1. *Protein Family Selection*

The Protein Data Bank (PDB)[10] was used in conjunction with the iGNM database[11] to select the families of proteins used in this study. The iGNM database is an online resource of pre-computed GNM analysis for all structures deposited in the PDB. The low frequency eigenvectors, termed slow modes, were used in the analysis presented here because these slow modes have previously been associated with global motions and likely (large-scale) functional motions.[12]

The SCOP (v 1.71) classification was used as the inital basis for familial groupings. Five families were chosen from the SCOP database site such that the proteins in these families represented different functions, architectures and number of residues. A family was considered if it had more than 25 member structures with the same number of residues. Since the number of structurally resolved residues does not always match the protein sequence length, the list of SCOP family member PDB structures was checked against the iGNM database to determine the number of nodes (residues) present in each structure. Only structures with the same number of residues were selected for use in this study. Requiring the proteins to have the same length allowed a direct comparison of them against each other using the dot product of their modes of motion (see below). Once this number of residues was determined, an additional set of structures was obtained by retrieving all structures present in the iGNM with this number of residues. Thus each protein family studied had a set of structures already deemed part of the (SCOP) family and a second set of (non-family) structures each having the same length as those in the family. The analysis was carried out for the five SCOP families listed in Table 1.

Table 1. List of protein families tested.

Abbreviation: Name	Residue count	Family	Non-family
FABP: Fatty acid binding protein-like	131	30	42
Glob: Globins	153	84	58
CytC: monodomain cytochrome c	108	28	70
DHFR: Dihydrofolate reductases	159	27	46
PoBP: Phosphate binding protein-like	517	30	6

2.2. *GNM*

The GNM[5,6] treats the structure as an elastic network model where amino acid residues within a cutoff distance, r_c are connected by springs with a uniform force constant. In this model, the C^α atom positions of each residue serve as the nodes. Denoting R_{ij} as the distance between residues i and j, a Kirchhoff or connectivity matrix, $\mathbf{\Gamma}$, is constructed such that off-diagonal elements are -1 when $R_{ij} \leq r_c$ and 0 when $R_{ij} > r_c$; while the diagonal elements are the sum of off-diagonal elements. The normal modes characterizing the motion of this network are found by eigenvalue decomposition of the Kirchhoff matrix according to Eq. (1)

$$\mathbf{\Gamma} = \mathbf{U\Lambda U}^T \tag{1}$$

where $\mathbf{U}$ is a matrix composed of eigenvectors, $\mathbf{u}_i$ $(1 \leq i \leq N)$, and Λ is the diagonal matrix of the eigenvalues λ_i. Despite being a purely topological model, GNM and related models have been widely used to characterize functionally relevant motions in terms of a few low frequency (small λ) modes.[4]

For each of the structures in the protein families, the 20 slowest mode (lowest frequency) eigenvectors were downloaded from the iGNM database. The correlation between residue fluctuations ($\Delta\mathbf{R}_i$) due to a specific mode is calculated according to Eq. (2).

$$[\Delta\mathbf{R}_i \cdot \Delta\mathbf{R}_j]_k = \frac{3k_bT}{\gamma}\lambda_k^{-1}[\mathbf{u}_k]_i[\mathbf{u}_k]_j \tag{2}$$

Here $[\mathbf{u}_k]_i$ is the ith element of the eigenvector $\mathbf{u}_k$, λ_k is the eigenvalue, T is the absolute temperature and k_b is the Boltzmann constant.

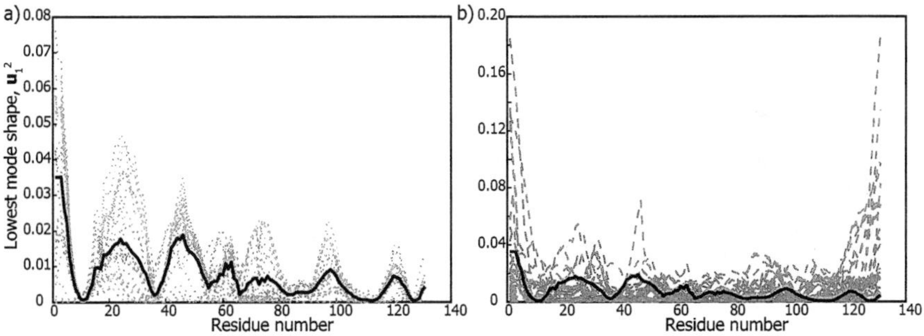

Fig. 1. Relationships between mode shapes. a) The lowest mode shape $[\mathbf{u}_1]^2$ is plotted for each family structure with gray dashed lines. The thick black line highlights the average mode shape. b) The lowest mode shape for each non-family structure is plotted in gray dashed lines and contrasted with the average family mode shape in black.

Figure 1 illustrates the lowest mode shape $[\mathbf{u}_1]^2$ plotted against residue number. By inverting Eq. (2) one can see that this is proportional to the lowest mode self-fluctuation or mobility. In Fig. 1a) the results are plotted for each of the *family* protein structures as gray dashed lines. One can see that each structure has a slightly different degree of mobility in this plot. Clearly there are some regions of qualitative agreement such as the coincidence of minima, highlighted by the average curve in black. It has previously been demonstrated that these minima serve as *hinge* sites that correlate with binding and/or catalytic sites.[12] Complimenting this previous insight, we observe that the largest variations occur not in these hinge sites but in the mobile regions between such hinges. As shown here, the general mode shape illustrates the ability of GNM to cluster groups of structures by their dynamics. Additionally the variations in the degree of modal mobility points to the ability of GNM to differentiate between similar structures. Figure 1b) shows the same average slow mode mobility in black compared results for each of the non-family structures (dashed lines). Unlike the case of structures from a family, there is no observable trend for mobility among non-family structures. Beyond this qualitative observation, we desire to develop a more quantitative comparison for a large number of structures.

2.3. *Quantitative Mode Comparisons*

Letting $\mathbf{x}_\alpha$ and $\mathbf{y}_\beta$ represent the αth and β^{th} eigenvectors of proteins x and y respectively, one can define the dot product between eigenvectors of different proteins according to Eq. (3).

$$P^{xy}_{\alpha\beta} = \mathbf{x}_\alpha \cdot \mathbf{y}_\beta = \sum_{i=1}^{N} x_{\alpha i} y_{\beta i} \tag{3}$$

Using the fact that these are eigenvectors, we ignore elements of Eq. (3) corresponding to the same protein. Thus if k represents the number of eigenvectors being considered we define a $(k \times k)$ matrix for each pair of proteins in the dataset given by Eq. (4).

$$_k\mathbf{P}(x,y) = \begin{bmatrix} P^{xy}_{11} & P^{xy}_{12} & \cdots & P^{xy}_{1k} \\ P^{xy}_{21} & P^{xy}_{22} & \cdots & P^{xy}_{2k} \\ \vdots & \vdots & \ddots & \vdots \\ P^{xy}_{k1} & P^{xy}_{k2} & \cdots & P^{xy}_{kk} \end{bmatrix} \tag{4}$$

Thus when we compare L proteins, we can define a large $(kL \times kL)$ matrix comprised of smaller $(k \times k)$ matrices, which compare the individual k lowest modes of each protein against the k slowest modes of the other proteins in the family set. The amount of correlation data contained here makes it hard to recognize the correlations between proteins. In order to succintly compare the data, two correlation metrics are introduced as functions of the number of eigenvectors being compared, denoted by k. The first correlation, $M_k(x,y)$, defines the maximum dot product between the slowest k modes for each pair of structures (Eq. (5)) and the second (Eq. (6)) calculates the sum of the maximum values for each column, j, in the smaller $(k \times k)$ matrix.

$$M_k(x,y) = \max\left(|_k\mathbf{P}(x,y)|\right) \tag{5}$$

$$S_k(x,y) = \sum_{j=1}^{k} M_j(x,y) \tag{6}$$

Two different measures of correlation were introduced because it is not clear how correlations between different modes in different structures should be computed *a priori*. The M_k correlation measure is concerned with identifying *any* two highly correlated eigenvectors between the two proteins without regard for the specific order of these low frequency modes. This acccounts for the possibility of mode *mixing* which occurs when the lowest eigenvector from one protein is highly correlated with an eigenvector from another protein that is not the lowest frequency mode. In contrast, the second correlation measure, S_k, focuses on identifying how well the *entire* subspace spanned by the low frequency eigenvectors of one protein matches the subspace spanned by the low frequency eigenvectors of another protein.

By averaging these measures over the set of family structures we can determine the average amount of correlation a structure has with respect to a protein family. For a family with l_f proteins, the family averaged, M_k value for the xth protein, $\langle M_k(x)\rangle_f$, is defined by Eq. (7) along with a family averaged standard deviation, $\langle \sigma_k^M \rangle_f$.

$$\langle M_k(x)\rangle_f = \frac{1}{l_f}\sum_{y=1}^{l_f} M_k(x,y) \tag{7}$$

Similarly, the family averaged, S_k value for the xth protein, $\langle S_k(x)\rangle_f$, is defined in Eq. (8) along with an average standard deviation, $\langle \sigma_k^M \rangle_f$.

$$\langle S_k(x)\rangle_f = \frac{1}{l_f}\sum_{y=1}^{l_f} S_k(x,y) \tag{8}$$

3. Results

3.1. *Classification of Protein Families by Dynamics*

In order to determine a suitable number of modes to consider we performed calculations for all values of $k \leq 20$. Taking the average of the family averaged metrics in Eq. (7) and Eq. (8), defines an overall correlation value for each protein family.

$$\langle\langle M_k\rangle_f\rangle = \frac{1}{l_f}\sum_{x=1}^{l_f} \langle M_k(x)\rangle_f \tag{9}$$

Figure 2 plots these family-averaged values ($\langle\langle M_k\rangle_f\rangle$ and $\langle\langle S_k\rangle_f\rangle$) as a function of the number of modes, k. The M_k averages when $k = 1$ are relatively low (0.67) in the case of FABP. This is due to the fact that using only one eigenvector from each protein prevents considering the correlation

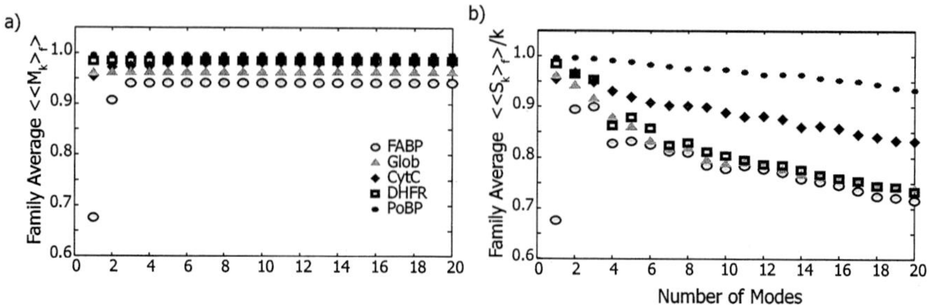

Fig. 2. The family averaged metrics calculated for different numbers of modes. a) $\langle\langle M_k\rangle_f\rangle$ b) $\langle\langle S_k\rangle_f\rangle/k$

between potentially mixed modes. However, as more modes are included, this situation is quickly remedied. By the time $k = 5$, the value of $\langle\langle M_k\rangle\rangle$ has reached an asymptote and so $k = 5$ was chosen for the results presented here. Due to the high average family correlation expressed by M_5, this measure suggests a means to distinguish family from non-family structures.

The trend for $\langle\langle S_k\rangle_f\rangle/k$ in Fig. 2b) is different, generally reflecting the a greater disparity for larger values of k. However, as illustrated by FABP, more than one mode is required to account for potential mode mixing. The overall lower correlation of S_k when compared to M_k makes S_k more appropriate for monitoring distinctions *within* a family. To keep the analysis consistent with M_k and allow for intrafamily distinctions, we set $k = 5$ for the S_k results presented here.

We plot the $M_k(x, y)$ correlation values between each pair of protein structures using a scheme that runs from no (zero) correlation in blue to maximal (one) correlation in red. Figure 3 shows this plotted for FABP, Glob and CytC in panels a) through c) respectively with each row and column corresponding to a specific protein structure. The proteins identified by SCOP as family members are plotted first in each case. Similarly, we plot the $S_k(x, y)$ values with a color scheme ranging from no correlation in blue to maximal correlation in red. Figure 3d) through f) plots this correlation for the family members of these three families. Results for the other two protein families (DHFR and PoBP) are similar and thus not shown explicitly. As mentioned above, these results are shown for $k = 5$ although the calculations were repeated for all values of $k \leq 20$.

To understand the significance of these plots, consider CytC (Fig. 3c) as an example. The first 28 rows and columns correspond to the 28 CytC family proteins and are shown in shades of red to signify their high corre-

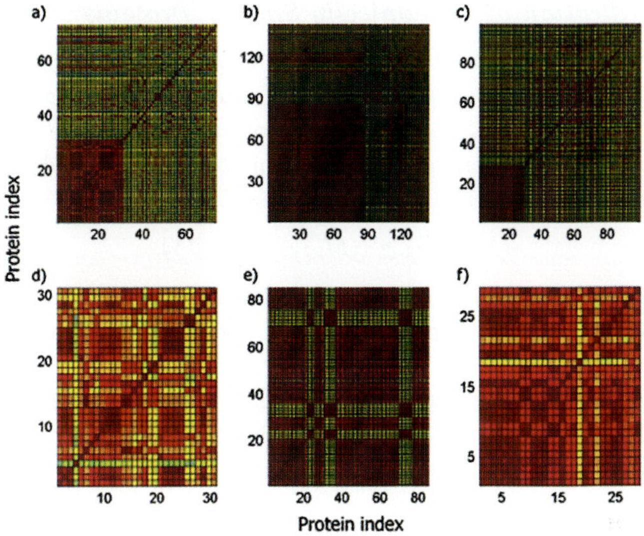

Fig. 3. Correlation measure ($k = 5$) plots for three protein families. a) M_5 for FABP b) M_5 for Glob c) M_5 for CytC. The known family members are listed first followed by other proteins outside the family. In each case there are clear distinctions between family and non-family structures. High correlation corresponds to red and low correlation is in blue. For the S_5 plots, only the subset corresponding to family members is plotted to show the intra-family distinctions. d) S_5 for FABP e) S_5 for Glob f) S_5 for CytC.

lation. By contrast, the last 70, non-family structures (rows and columns) are in shades of greens and yellows indicating low correlation (0.3 to 0.6). This distinction in color clearly shows a difference between the modes of proteins within the SCOP family compared to the modes of proteins not in the family. This trend is observed for each of the five protein families studied. Such a trend suggests M_k correlations of protein dynamics can be used as a classifying technique.

The clear distinction between M_k values for proteins in the family versus non-family proteins makes idenfication of potential *candidates* for inclusion in the family relatively simple. There are some proteins in Fig. 3a) and b) not classified as being part of the SCOP family that share the same color pattern as those within the family. This indicates a high degree of correlation between the eigenvectors from these structures and those in the family implying that they should be considered as part of the family.

3.2. *Identification of Dynamically Similar Proteins*

The numerical values behind the similar color patterns are used to identify these candidate structures with respect to the family average $\langle M_k(x)\rangle_f$ values defined in Eq. (7). Specifically, we consider a structure as a family *candidate* if its family averaged value is within three standard deviations of the mean family averaged value as defined in Eq. (10).

$$\langle M_k(x)\rangle_f \geq \langle\langle M_k\rangle_f\rangle - 3\langle\sigma_k^M\rangle_f \tag{10}$$

Thus instead of looking at the correlation color patterns illustrated in Fig. 3, we can plot the family averaged, $\langle M_k(x)\rangle_f$, values against protein index as in Fig. 4a). Indicatting the $3\sigma_5^M$ limits by gray dashed lines quickly identifies three potential candidates for the FABP family according to the criteria in Eq. (10). These three candidate structures have protein indices 57, 66 and 68 corresponding to PDBids 1t8v, 1yiv and 2a0a.

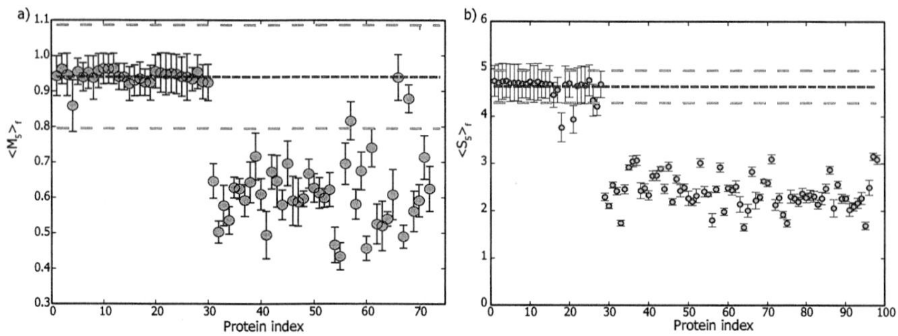

Fig. 4. The family averaged correlation measures for different proteins in a family illustrate the candidate and outlier criteria. (a) $\langle M_5(x)\rangle_f$ values for each FABP family and non-family structure. The mean family value, $\langle\langle M_5\rangle_f\rangle$, is shown by a black dashed line and the candidate $3\sigma_5^M$ range is indicated by gray dashed lines. (b) $\langle S_5(x)\rangle_f$ values for each CytC family and non-family structure. The mean family value, $\langle\langle S_5\rangle_f\rangle$, is shown by a black dashed line and the outlier σ_5^S range is indicated by gray dashed lines.

Table 2 summarizes the mean family averaged values as well as the number of candidates for each family. In the case of FABP, none of these three structures are part of the SCOP family. 1t8v is a fairly recent PDB structure which is annotated as a fatty acid binding protein, but due to its recent deposition it has not been included in the SCOP database. The other two structures, 1yiv and 2a0a, are annotated as a myelin protein and dust mite allergen respectively. Visual inspection of these structures confirms that they contain the dominant β-barrel structure of FABP structures and

Table 2. Family averaged correlation metrics and standard deviations

Family	$\langle\langle M_5\rangle_f\rangle$	$\langle\sigma_5^M\rangle_f$	Candidates	$\langle\langle S_5\rangle_f\rangle$	$\langle\sigma_5^S\rangle_f$	Outliers
FABP	0.9411	0.0484	3	4.1579	0.4818	2
Glob	0.9617	0.0377	10	4.2958	0.5221	17
CytC	0.9804	0.0267	0	4.5958	0.3373	3
DHFR	0.9852	0.0104	0	4.3925	0.4174	0
PoBP	0.9967	0.0083	0	4.4045	0.0710	0

suggests a new potential functional mechanism for these structures, namely as a fatty acid binding protein.

Analysis of the Glob family indicated ten candiate structures all of which are recent myoglobin structures which are not included in the SCOP database. Correct identification of these structures as part of the Glob family by comparison of their dynamic modes serves to confirm the applicability of this method to distinguish protein families. The other three families: CytC, DHFR and PoBP do not have any candidate proteins.

3.3. *Intrafamily Distinctions*

After demonstrating the ability of this method to distinguish family from non-family structures, the ability to distinguish variations among structures within a family was also investigated. As can be seen in Fig. 3d) through f), correlations among structures within a family are not uniform but have variations. Regions with lower correlation (greens, yellows and oranges in these panels) correspond to structures that are potential family *outliers*. Similar to the definition of candidates in Eq. (10), we define such outliers as being more than one standard deviation ($\langle\sigma_k^S\rangle_f$) away from the mean family averaged S_k value as in Eq. (11).

$$\langle S_k(x)\rangle_f \leq \langle\langle S_k\rangle_f\rangle - \langle\sigma_k^S\rangle_f \tag{11}$$

Again the actual values of $\langle\langle S_5^S\rangle_f\rangle$ and $\langle\sigma_5^S\rangle_f$ used for each of the families are listed in Table 2. Using this outlier criteria we are able to pick out structures that may be structurally and/or functionally distinct from within a family in an automated fashion.

Beginning with CytC as an example case, one can see two green-yellow-orange bands in Fig. 3c) representing protein indices 18 and 21 that are less correlated with other CytC structures in general. Figure 4b) plots the family averaged correlation measures, $\langle S_5(x)\rangle_f$, for each protein along with the outlier criteria from Table 2. In this plot one can see that all non-family members fall below the $\langle\sigma_5^S\rangle_f$ level shown with a dashed gray line because there were no candidates this family. More importantly, there are three outliers corresponding to protein indices 18,21 and 27. These indices refer to PDBids 1fhb, 1nmi and 1yic. Since these *outliers* have the same overall family structure, the differences identified by this analysis of the dynamics are due to some other factor(s). Using similar analysis, FABP had two outliers: PDBids 1ael and 1tou (indices 4 and 25 in Fig. 3a) and Glob had 17 outliers identified (protein indices 21–24, 30–35 and 69–75 in Fig. 3b).

Examining the structures that were deemed to be outliers as a whole we are able to determine a few reasonable explanations for these differences. The outliers for FABP and CytC were structures that were determined by NMR rather than X-ray crystallography. Although these were not the only NMR structures in the datasets representing these families, it suggests that structures that are not forced to conform to the Ramachandran phi-psi plot may adopt a "looser" structure with measurable differences in dynamics. Further supporting this claim is the fact that one of the CytC outliers, 1nmi, is an averaged NMR structure which would not necessarily reflect the true dyanamics of the CytC family. The 17 outliers in the Glob family correspond to *all* of the structures of one type of globin: leghemoglobin from a specific species: yellow lupin. In this case, these structures serve to form a sub-family within Glob that can be seen visually by the green-yellow bands in Fig. 3b).DHFR and PoBP had no outliers according to the criteria of Eq. (11). However examination of the most varied structures support the claims of sub-family organization by speciation and differences due to ligand-binding state (data not shown).

4. Conclusions

We have demonstrated an automatic family classification scheme for protein structures based upon their computed dynamics. Comparisions using the low frequency eigenvectors of structures accurately assigns these structures to a unique protein family. Using this precomputed data, Eq. (10) provides a measure for assigning newly determined structures as candidates to a particular protein family.

In addition, this method provides a quantitative measure of the differences within protein families. These differences can be investigated in terms of outliers or most dynamically different as indicated in the text. Examination of the outliers indicates that differences within the familes can be attributed to some combination of differences in ligand-binding state, method of structural determination and sequence. These factors are an initial list of possible explanations, and more research on a larger set of structures needs to be done in order to obtain a more complete understanding of what these variations in dynamics correspond to universally.

Finally we introduce a new direction for protein classification schemes that is both automatic and relies upon the dynamics of protein structures. In the version presented here, the comparisons were only made for structures that had the same number of residues. Admittedly, this restricted the number of families this analysis is applicable to. However, it is clear that dynamics can be used to make meaningful distinctions both between and within protein families. In the future we anticipate generalizing this method to allow comparisons between structures having different numbers of residues. It is expected that such an extension will allow for more family comparisons and thus more general results regarding the relationships between protein structures, dynamics and functions to surface. This method can also provide an automatic classification scheme to aid in the identification of functions for so called "hypothetical proteins" produced by structural genomics initiatives.

Acknowledgments

The authors thank the IUPUI Department of Physics for funding this work.

References

1. J. C. Whisstock and A. M. Lesk, *Quart. Rev. Biophys.* **36**, 307, (2003).
2. L. Lo Conte, et.al., *Nuc. Acid Res.* **28**, 257, (2000).
3. C. A. Orengo, et.al., *Structure* **5**, 1093, (1997).
4. I. Bahar and A.J. Rader, *Curr. Opin. Str. Biol.* **15**, 1 (2005).
5. T. Haliloglu, I. Bahar and B. Erman, *Phys. Rev. Lett.* **79**, 3090, (1997).
6. I. Bahar, A. R. Atilgan, and B. Erman, *Fold. Des.* **2**, 173, (1997).
7. O. Keskin, R. L. Jernigan and I. Bahar, *Biophys. J.* **78**, 2093, (2000).
8. S. Maguid, et.al., *Biophys. J.* **89**, 3, (2005).
9. A. Leo-Macias, et.al., *Biophys. J.* **8**, 1291, (2006).
10. H. M. Berman, et.al.,*Nuc. Acids Res.* **28**, 235 (2000).
11. L. W. Yang, et.al., *Bioinformatics* **20**, 2978, (2005).
12. C. Chennubhotla, et.al., *Phys. Biol.* **2**, S173 (2005).

PROTEIN-NUCLEIC ACID INTERACTIONS: INTEGRATING STRUCTURE, SEQUENCE, AND FUNCTION

MARTHA L. BULYK
Brigham & Women's Hospital and Harvard Medical School
Boston, MA 02115

ALEXANDER J. HARTEMINK
Duke University
Durham, NC 27708

ERNEST FRAENKEL
Massachusetts Institute of Technology
Cambridge, MA 02139

YAEL MANDEL-GUTFREUND
Technion-Israel Institute of Technology
Haifa, Israel 32000

Over the last several years, various groups have been developing methods to incorporate information about the three-dimensional structure of proteins, DNA, and RNA into algorithms for analyzing high-throughput genomic and proteomic data. In particular, these methods have been shown to significantly improve predictions of a wide range of functional properties, including the regulatory targets, of nucleic acid binding proteins. These approaches are likely to become increasingly important in analyzing the many different types of data that can now be collected on a genome-wide and proteome-wide scale, including DNA sequence from various genomes, gene expression data, protein-protein and protein-ligand interactions, and protein-DNA and protein-RNA binding data.

This emerging paradigm builds on recent technological advances in data collection and computational developments in diverse areas including DNA binding site motif discovery, modeling of transcriptional regulatory networks, multiple sequence alignments, structural genomics, and structural and evolutionary studies of proteins and nucleic acids. While each of these specific aspects of protein-nucleic acid interactions has been studied previously, these different aspects have just recently begun to be considered together. This PSB session focuses on methods that bridge structure, sequence, and function to infer previously undiscovered associations between these different aspects of protein-nucleic acid interactions.

Methods that employ structure and sequence as they relate to function have several key advantages. First, structural data alone often do not permit the inference of biological function. Second, experimental genomic datasets often contain errors or noise due to imperfections in the applied technology. Third, functional studies typically do not connect function to structure. Indeed, only a small body of work addresses how to take advantage of these currently separate areas of research on protein-nucleic acid interactions. We anticipate that combining these different types of data will allow us to identify essential biological associations, and ultimately to model and predict these interactions.

This year there are six papers in this session. In the first paper, McCord and Bulyk reveal that numerous families of transcription factors from bacteria, yeast, fly, and mouse that contain the same type of DNA binding domain have similar functions and/or regulate genes of similar function. The observed correlations between transcription factor structural classes and the regulatory roles of the transcription factors themselves suggest that structural information could be useful for predicting the functions of transcription factors and their regulatory targets.

In their paper, Gordân and Hartemink report that experimentally determined transcription factors' DNA binding sites in yeast are significantly biased toward regions of higher predicted DNA duplex stability. By incorporating information about helix destabilization energy—which can be calculated directly from DNA sequences—as a Bayesian prior, they are able to markedly improve the accuracy of transcription factor DNA binding site motif discovery.

Lusk and Eisen introduce an evolutionarily based approach for choosing an appropriate position weight matrix cutoff when identifying transcription factor binding site motif matches in genomic DNA. They find that yeast transcription factors appear to fall into different categories of cutoff stringency, suggesting that different transcription factors may have been under pressure to maintain binding sites of varying stringency.

Pan and coauthors introduce a parametric mixture model for estimating the targets of a transcription factor genome-wide by combining evidence from assays of transcription factors' DNA binding (such as from ChIP-chip experiments), assays of target co-expression, and presence of transcription factors' DNA binding site motifs in target promoters. By combining this evidence in a joint mixture model, they present a method that is at once both simple and effective.

Two of the papers in this session examine methods for predicting the protein residues that make contact with DNA and RNA. Kauffman and Karypis use mutual information to systematically analyze the relationship between various sequence and structural properties of amino acids and their role in binding DNA in a set of almost 250 protein-DNA complexes. Lee and colleagues combine threading and machine learning methods to identify residues that contact RNA and DNA in the catalytic subunit of human and yeast telomerase.

Progress in these areas may further improve the ability to predict the functions, targets, and regulatory mechanisms of DNA- and RNA-binding proteins. In addition, numerous other challenges remain in this nascent research area aside from those addressed in the accepted papers for this session. Future work will need to address questions such as:

- Do certain types of domains of DNA/RNA binding proteins confer particular biophysical properties, either in terms of kinetics or ligand specificity?
- How is RNA structure involved in interaction with proteins, and what are the regulatory or other functional consequences of those interactions?
- How are affinities of protein-DNA interactions tied to function?
- What are the relative contributions of biophysical constraints and evolutionary history in shaping the functional roles of proteins sharing a common domain structure?
- Can a fully predictive (energetic) model of protein-nucleic acid interactions be developed?

As more types of data become widely available, integrative approaches will become increasingly important in computational approaches for understanding regulation.

Acknowledgments

We are grateful to those who submitted manuscripts for consideration for inclusion in this session, and we thank the numerous reviewers for their valuable expertise and time throughout the peer review process.

FUNCTIONAL TRENDS IN STRUCTURAL CLASSES OF THE DNA BINDING DOMAINS OF REGULATORY TRANSCRIPTION FACTORS

RACHEL PATTON MCCORD[1,4] AND MARTHA L. BULYK[*,1,2,3,4]

[1]Division of Genetics, Department of Medicine, [2]Department of Pathology, Brigham & Women's Hospital and Harvard Medical School, Boston, MA 02115
[3]Harvard/MIT Division of Health Sciences & Technology (HST), Harvard Medical School, Boston, MA 02115
[4]Harvard University Graduate Biophysics Program, Cambridge, MA 02138
Email: rpmccord@fas.harvard.edu, mlbulyk@receptor.med.harvard.edu

The DNA-binding domain (DBD) structure of a regulatory transcription factor (TF) is important in determining its DNA sequence specificity, but it is unclear whether a relationship exists between DBD structure and general TF biological function or regulatory mechanism. We observed moderate enrichment of functional annotation terms among TFs of the same structural class in *Escherichia coli*, *Saccharomyces cerevisiae*, *Drosophila melanogaster*, or *Mus musculus*, suggesting some preference for TFs of similar structures in the regulation of similar processes. In yeast, we also found trends among TF structural classes in phenomena including gene expression coherence, DNA binding site motif similarity, the general or specific nature of TFs' regulatory roles, and the position of a TF in a gene regulatory network. These results suggest that the biophysical constraints of different TF structural classes play a role in their gene regulatory mechanisms.

1. Introduction

The concepts that structure leads to function and that form follows function are common principles throughout biology[1]. In the study of gene regulation, TFs can be classified based on the structures of their DBDs, domains that mediate their interaction with specific DNA sequences[2,3]. These structural class designations have been used to infer the sequence specificity of a TF, predict binding sites and potential target genes, and infer biological function based on these target genes[4-7]. Since TF sequence specificities have been used to infer TF functional properties, it follows that members of a given TF structural class might have similar biological roles, and that the structure of a DBD could be used directly to predict the functions of uncharacterized TFs. Indeed, previous studies have identified instances of enrichment of a particular TF structural class in the regulation of a certain biological process. For example, homeodomains are enriched within genes involved in *C. elegans* neuronal function[8]. However, a

[*] Corresponding author

large-scale analysis to determine the extent of functional enrichment within different TF structural classes has not been described previously.

TFs of the same class might also share other gene regulatory properties, such as their position in gene regulatory networks, the similarity or divergence and information content of their DNA binding site motifs, or co-expression across diverse conditions. Analysis of such regulatory features will elucidate ways in which the biophysical properties of a DBD structure might inform its modes of regulation.

Here, we investigate enrichment for common biological function among members of different TF structural classes in *E. coli, S. cerevisiae, D. melanogaster,* and *M. musculus*. We find several examples of modest functional enrichment among TFs of the same structural class in bacteria, yeast, fly, or mouse. Target genes of yeast TFs within some structural classes are also observed to share similar functions. In a few cases, the biological functions enriched for a particular structural class appear to be conserved across species. Using numerous genome- and proteome-wide datasets available in *S. cerevisiae*, we relate this observed functional enrichment to other regulatory mechanisms. Our results suggest that different modes of gene regulation are used by different TF structural classes. The functional relationships found here identify cases in which DBD structure could be used to predict TF biological function, suggest different ways in which structural classes partition functional roles, and inform future studies of the link between TF structure and function and the evolution of TF regulatory roles.

2. Methods

2.1. Data Sets Used in This Study

TFs and DBD Structural Classes

The TFs and structural class assignments for *E. coli* were obtained from GenProtEC[9], last updated on Dec 7, 2004. The structural classes of 421 known and predicted *S. cerevisiae* TFs[10] were assigned based on annotation in Pfam[11] and DBD[12] databases. For subsequent analyses, we considered only the subset of TFs from this initial list that belonged to known DBD structural classes with 4 or more members. *D. melanogaster* TFs and structural classifications were downloaded from FlyBase on July 11, 2006[13]. Mouse TF information and DBD assignments were derived from a set of known TFs listed in Gray *et al.*[14]. All TFs and structural class assignments are listed in Supplementary Table 1[†].

Functional Annotations

Each *E. coli* protein was assigned MultiFun classifications according to the GenProtEC database, last updated on February 1, 2007[9]. Specific annotations were divided into corresponding broader categories (i.e., a protein annotated "1.3.5: Fermentation" would also be given the annotations "1: Metabolism" and "1.3: Energy metabolism (carbon)"). Multiple sources of gene annotations, including the Gene Ontology (GO)[15] and MIPS database[16], last updated in June 2005, were used to annotate yeast target genes. We used GO annotations for yeast, fly, and mouse TFs that were last updated on September 12, 2007[17]. To avoid circularity and annotation bias, we eliminated all GO annotations that were inferred from structure or from a non-traceable author statement (GO Evidence Codes ISS and NAS, respectively)[15].

Genome-wide Yeast Datasets

Yeast TF binding site motif sequences, target gene information, and motif information content values (IC; a measure of the specific vs. degenerate nature of the DNA sequences recognized by a TF) for 82 TFs were derived from a re-analysis[18] by MacIsaac *et al.* of the single most comprehensive set of yeast ChIP-chip data[19]. We considered TF binding sites identified at $p<0.005$ binding threshold in ChIP-chip that were also conserved in at least 2 other yeast species. We considered only those structural classes with at least 3 TFs with greater than 5 target genes in our target gene analyses. Yeast gene regulatory interaction data were derived from networks compiled by Yu *et al.*[20]. The 1,327 publicly available gene expression microarray datasets were compiled by McCord *et al.*[21]

2.2. Statistical Approaches

Functional Enrichment Evaluation

To evaluate functional enrichment among groups of TFs or their target genes in bacteria, yeast, fly and mouse, we calculated p-values using the hypergeometric distribution:

Eqn. (1): $$P = 1 - \sum_{i=0}^{k-1} \frac{\binom{C}{i}\binom{G-C}{n-i}}{\binom{G}{n}}$$

[†] All supplementary files, figures, and scripts (implemented in Perl and Matlab) are available on our lab website at http://the_brain.bwh.harvard.edu/TFstr/

where G is the number of genes in the entire genome or in a defined background gene set, C is the number of genes in this background set with a particular functional attribute, and n is the size of the query set of TFs or target genes, of which k are known to possess the functional attribute.

We evaluated functional enrichment within DBD structural classes in mouse, fly, and yeast with respect to all TFs using the FuncAssociate algorithm[17], which estimates an adjusted p-value (p_{adj}) by comparing the enrichment in the query gene set to the frequency of this degree of enrichment among 1,000 randomly generated gene sets. We report results at $p_{adj}<0.05$. Our implementation of the hypergeometric distribution for *E. coli* allowed us to search for functional enrichment at many levels of the MultiFun[22] annotation hierarchy. Our threshold for functional enrichment in *E. coli* was an uncorrected $p<0.05$, but we also report p-values from a stringent Bonferroni correction.

To evaluate TF target gene functional enrichment in yeast, we employed the Funspec algorithm[23] to calculate p-values of target gene functional enrichment for each TF (p_{TF}) in a class, and then calculated the geometric mean of the p-values for each annotation term over all TFs in a structural class (p_{avg}). We controlled for a potential inflationary effect on this functional enrichment, resulting from the existence of paralogous TFs due to the ancient yeast genome duplication, by calculating filtered p_{avg} values that excluded paralogous gene pairs. Specifically, for classes containing paralogous pairs, we calculated all possible filtered p_{avg} values resulting from averaging p-values over all but one TF of a structural class by leaving out, one at a time, members of literature-defined paralogous TFs[24]. We report results for which the least significant of these filtered average p-values (max filtered p_{avg}) was less than 0.05.

Coherence Scores

Co-expression of a set of TFs or target genes in yeast was measured by *expression coherence (EC)*[25]. Briefly, we calculated the Pearson correlation coefficient between the expression profiles of every pair of yeast genes over 1,327 expression conditions[21]. Then, the EC was calculated as the fraction of correlation coefficients between foreground genes (TFs or target genes in a DBD class) that were in the top 5th percentile of correlations among background genes (all TFs or all genes). In the case of TF target genes, we considered only the pairwise correlations between targets of different TFs within the structural class to ensure that high expression coherence was not attributable solely to regulation of targets of a single TF. A p-value was estimated by calculating the EC scores of 10,000 randomly generated sets of genes identical in size to the foreground set and then calculating the fraction of random sets with an EC greater than that of the foreground set of interest.

The similarity of DNA binding site motifs recognized by TFs in a structural class was measured by a metric we developed termed *"motif coherence"*, which we modeled after the expression coherence metric described above. The pairwise correlation coefficients between all motifs were calculated by the CompareACE algorithm[26], and then the motif coherence was calculated as the fraction of motif correlations within a structural class in the top 5th percentile of all motif correlations. A p-value for this coherence was estimated as for expression coherence, but here we considered 10 million random sets in order to allow estimation of p-values as low as 1.0×10^{-7} and thus to provide finer distinctions in the degree of motif coherence among structural classes with highly similar DNA binding domains.

Bottlenecks and Hubs

We classified yeast TFs as *"hubs"* if they were in the top 20% of the regulatory network degree distribution and as *"bottlenecks"* if they were in the top 20% of the betweenness distribution, as in Yu *et al.*[20]. The hypergeometric distribution (Eqn. 1) was used to assign a p-value to hub/bottleneck enrichment within a structural class by comparing the fraction of hubs/bottlenecks within a structural class to the fraction of hubs/bottlenecks over all TFs.

3. Results and Discussion

3.1 Functional Enrichment by TF Structural Class

We first searched for functional enrichment within a structural class by examining gene annotation terms assigned to the TFs themselves. Modest functional enrichment was seen for some structural classes in all 4 organisms, (see Table 1 for highlights of enriched annotations and Supplementary Table 2 for full results) though some classes in each organism showed enrichment for no biological functions, or only those common to most transcriptional regulatory proteins (*e.g.* "transcription, DNA dependent"). In *E. coli*, most classes showed some degree of functional enrichment; winged-helix TFs are enriched for roles in amino acid biosynthesis, while proteins with lambda repressor DBDs are enriched for carbohydrate metabolism functions. In fly, 40% of classes showed no specific enrichment, but classes like the HLH TFs and homeodomains are enriched for roles in the development of various systems. The minimal enrichment observed for 40% of mouse TF classes may be due to a lack of comprehensive GO annotation for most mammalian genes. However, as in fly,

Table 1. Highlighted examples of enriched functional annotation terms for DBD structural classes. k = number of TFs in structural class with the indicated annotation term; C = number of genes in background set (all TFs) with the indicated annotation term; p = p-value of functional annotation enrichment calculated using the hypergeometric distribution. p_{adj} = adjusted p-value calculated as described in Methods.

Annotation	*k*	*C*	*p*	p_{adj}
E. coli **(N_{TFs} = 225)**				
Winged Helix (FKH-like) (n=111):				
Repressor	69	101	2.2E-07	2.8E-05
Building block biosynthesis	10	12	0.015	1.9
Amino acid biosynthesis	8	9	0.016	2.0
Lambda repressor like (n=24):				
Carbohydrate catabolism	8	24	0.0012	0.16
Metabolism	15	79	0.0036	0.46
Yeast (N_{TFs} = 346)				
GATA (n=10):				
Regulation of nitrogen metabolism	4	4	3.6E-07	<0.001
Homeodomains (n=7):				
Mating-type specific transcriptional control	3	4	2.0E-05	0.003
G1/S-specific transcription in cell cycle	3	9	0.00041	0.05
Forkhead (n = 4)				
Positive regulation of progression through cell cycle	2	2	1.0E-04	0.018
G2/M-specific transcription in cell cycle	2	3	3.0E-04	0.031
Fly (N_{TFs} = 573)				
Homeodomains (n=102):				
System development	73	186	2.4E-16	<0.001
Pattern specification process	32	69	4.6E-08	<0.001
Cell fate specification	13	25	0.00022	0.022
Helix-Loop-Helix (HLH) (n=54):				
Sensory organ development	17	58	3.7E-07	0.004
Mouse (N_{TFs}=1,160)				
Homeodomains (n=221):				
Organ development	108	303	9.3E-16	<0.001
Central nervous system development	38	74	1.3E-10	<0.001
Endocrine system development	16	29	1.4E-05	0.004
Cell migration/motility	20	46	0.00011	0.018
Forkhead (FKH) (n=40):				
Organ development	25	303	1.2E-06	<0.001
Cell proliferation	11	72	1.5E-05	0.002
E2F (n=8):				
Regulation of progression through cell cycle	8	48	5.4E-12	<0.001
G1/S transition of mitotic cell cycle	3	4	9.0E-07	<0.001

some structural classes in mouse, such as homeodomains and forkhead TFs, are enriched for roles in organism development, and, as expected, the E2F TFs showed enrichment for roles in cell cycle control[27]. In *S. cerevisiae*, some structural classes (HLH, HSF, and others) showed no functional enrichment. Other classes are enriched for regulation of specific biological pathways, including GATA factors for regulation of nitrogen utilization, forkhead TFs in

Table 2: Highlighted examples of enriched functional annotation terms among target genes (N_{tar}) of yeast TFs. The values of p_{avg} and *max filtered* p_{avg} were calculated as defined in Methods. All genes in the *S. cerevisiae* genome (N_{bg}) were used as the background gene set in the *p*-value calculations.

Annotation	p_{avg}	*max filtered* p_{avg}
Yeast Target Genes (N_{bg} = 6,267)		
GATA (N_{TF} = 6; N_{tar} = 177):		
Nitrogen and sulfur metabolism	0.0004	5.1E-05
Myb (N_{TF} = 5; N_{tar} = 525):		
Cell growth and/or maintenance	0.0009	0.00090
Protein metabolism	0.0019	0.0019
Ribosome biogenesis	0.0037	0.0037
APSES (N_{TF} = 4; N_{tar} = 530):		
Cell wall	0.0003	0.0013
Mitotic cell cycle and cell cycle control	0.0043	0.0016
Cell fate	0.0003	0.00050
Cys_2His_2-type Zinc Finger (Zf-C2H2) (N_{TF} = 13; N_{tar} = 824):		
Stress response	0.0083	0.059

cell cycle progression, and homeodomain factors in mating type determination and the cell cycle.

The availability of ChIP-chip data for many yeast TFs allowed us to extend our analysis to the annotations of target genes of yeast TFs (see Table 2 for highlights and Supplementary Table 3 for full results). We observed that the GATA TFs and their target genes are both enriched for the same biological functions: nitrogen and sulfur metabolism. Consideration of target genes also provided additional functional information for several classes, including cell cycle and cell fate target gene enrichment for the APSES TFs, stress response for the C2H2 zinc finger (Zf-C2H2) TFs, and cell growth and protein biosynthesis for the Myb factors. We found that most of the enriched annotations were robust to paralog removal, so functional enrichment is not solely attributable to paralogous TFs resulting from the ancient yeast whole genome duplication[24].

We observed a few instances of functional enrichments that were consistent across organisms. In particular, homeodomain TFs in yeast are enriched for roles in the mating type determination, and the homeodomain TFs in fly and in mouse are enriched for roles in similar cell fate specification and development. Additionally, some basic transcription-related processes are shared across species: HMG factors are enriched for roles in chromatin architecture in both yeast and mouse. However, conservation of functional enrichment for members of a TF structural class is small, suggesting that, in most cases, functional specialization of structural classes arose according to different selective pressures in each of these organisms' evolutionary histories.

3.2 TF and Target Gene Expression Coherence (EC)

Observable functional enrichment within TF structural classes in several organisms suggests that other regulatory features of TFs might relate to this functional enrichment and vary across DBD structures. Since co-expression is often used to infer functional relationships between genes, we hypothesized that structural classes exhibiting functional annotation enrichment might also be co-expressed or exhibit co-expression of their target genes. Thus, we evaluated the EC of TFs or target genes within each structural class in yeast over 1,327 expression conditions (Figure 1). We found a range of EC across TF structural classes, suggesting further distinctions in the regulatory roles of different structural classes. As predicted, many classes with functional enrichment (Zf-C2H2, GATA, Myb, and Forkhead) do show strong EC, particularly among target genes. However, other TFs with enriched functional annotations (APSES, homeodomains) do not exhibit significant EC.

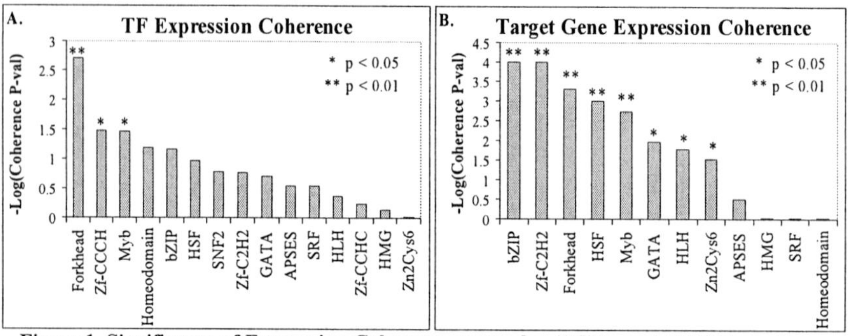

Figure 1. Significance of Expression Coherence scores for A) TFs, and B) TF target genes across structural classes in yeast.

3.3 Regulatory Bottlenecks

Functional enrichment without EC within a structural class may indicate that members of this structural class regulate different phases of the same biological process. Alternatively, lack of EC among targets of the same structural class may arise from regulatory network complexity. We searched for significant trends in network topology among members

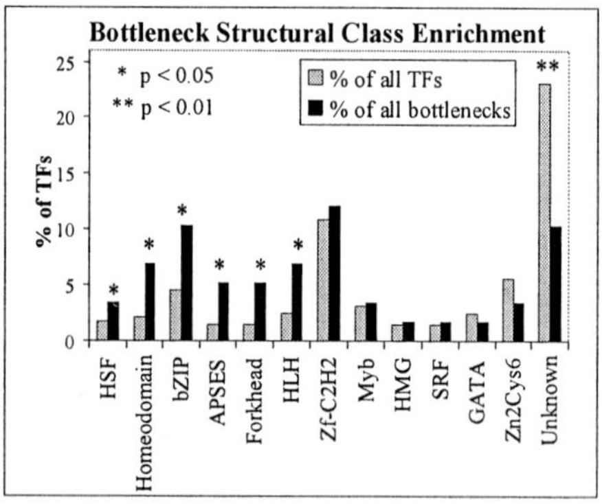

Figure 2. Bottleneck TFs within structural classes. Classes are ordered left to right from most enriched for bottlenecks to most depleted.

of a structural class within experimentally derived regulatory networks. Recent work has shown that "bottleneck" status (a measure of "betweenness", *i.e.*, how often regulatory pathways pass through a particular protein in a network graph) is a meaningful measure of the role of a TF in a regulatory network[20]. We found that certain TF structural classes are significantly enriched ($p<0.05$) for bottlenecks (Figure 2). Interestingly, APSES and homeodomain TFs, two classes that showed functional enrichment but insignificant EC, are among those enriched for bottlenecks. Since bottleneck proteins often connect multiple biological modules[20], TFs in these classes may regulate genes within different specific pathways expressed at different times, but which all contribute to similar biological functions. Such a regulatory mode could explain the functional enrichment without significant EC observed for these TFs.

3.4 *Motif Coherence (MC)*

We hypothesized that TFs within structural classes that show functional enrichment should exhibit similarity in their DNA binding site motifs[28]. We observe variation in the degree of MC from one TF structural class to another. Structural classes with strong functional enrichment, even some that do not show significant EC, tend to have highly significant within-class MC (Figure 3). However, some classes with functional enrichment (Myb, forkhead, homeodomain) do not have significant MC, suggesting that motif similarity is not the only factor contributing to similarity in function.

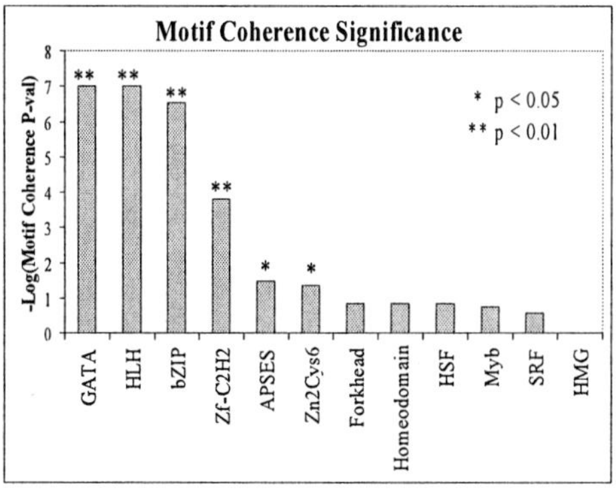

Figure 3. Motif coherence by TF structural class.

3.5 *General vs. Specific Regulation*

The binding mechanism of a particular DBD structure might be well-suited for a certain type of regulation, and thus, certain

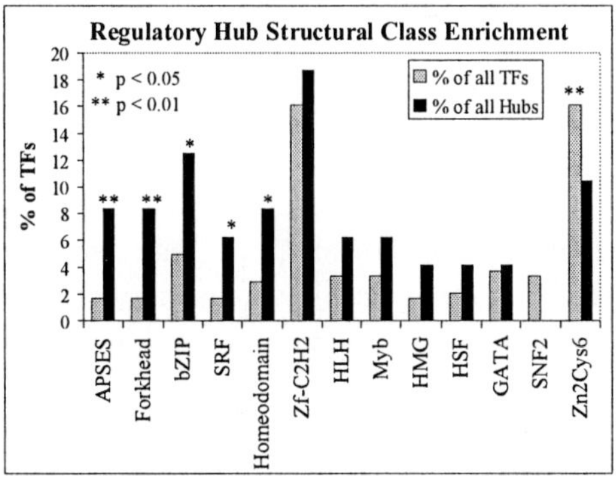

Figure 4. Regulatory hub enrichment within structural classes. Classes are ordered left to right from most enriched for hubs to most depleted.

biological processes. For example, structures that bind more degenerate sequences and/or have many potential binding sites in the genome might be utilized for general, housekeeping functions while structures that recognize highly specific binding sites might be used for processes requiring carefully restricted regulation. We examined trends in the information content (IC; a measure of motif specificity vs. degeneracy) and number of target genes recognized by TFs of each structural class[18]. We observed only modest variation in average motif IC between structural classes, but note that such variation tends to be anti-correlated with the average number of genes identified as bound in ChIP-chip experiments by TFs of the same class, as expected (Supplementary Figure 1). A clearer distinction between classes exists in the enrichment for regulatory hubs (proteins with the most connections in the regulatory network) within each structural class (Figure 4). Structural classes containing well-known "global" TFs (*i.e.*, those regulating many genes for broadly important functions) like the bZIP protein Gcn4 are significantly enriched for regulatory hubs, while those containing known "local" TFs (*i.e.*, those regulating a few genes for a specific function) like the Zn_2Cys_6 TF Gal4 are significantly depleted for such hubs. Thus, the global vs. local nature of these TFs appears to be a general feature of their structural class. Interestingly, structural classes with many regulatory hubs tend to be enriched for cell fate and cell cycle functions while those with fewer regulatory hubs tend to be involved in regulating the metabolism of specific nutrients such as nitrogen and carbohydrates.

4. Conclusions and Future Directions

We have found evidence for biological function enrichment among TFs in various structural classes in a wide range of organisms. We observed differences across structural classes in terms of regulatory features that may relate to this functional enrichment, including expression coherence, motif similarity, and regulatory network position. In addition to suggesting explanations for the observed functional enrichments, such regulatory feature differences indicate that different structural classes may have fundamentally different modes of gene regulation. Specifically, the data presented here suggest that different TF structural classes achieve regulatory specificity and avoid crosstalk in different ways. The combination of low motif coherence, low expression coherence, and lack of functional enrichment within some structural classes suggests that diversity in DNA recognition motifs allows different TFs of the same DBD class to participate in different biological functions and regulate distinct sets of target genes. In other structural classes, similar recognition motifs, high expression coherence, and functional enrichment suggest that harmful crosstalk is avoided as TFs within a class act redundantly or supplementarily in the regulation of similar processes, as has been previously hypothesized in studies of the function of TFs with similar motifs[28]. Functional enrichment and high motif coherence paired with low expression coherence and an enrichment for regulatory bottlenecks suggests that, in yet other classes, TF function is partitioned into different modules so that all TFs in a class. Thus, though they bind similar motifs and participate in similar biological processes, they perform unique roles in the cell with precise functional specificity determined by their regulatory partners in the overall network.

These results offer a set of interesting correlations and potential distinctions in regulatory mechanism by structural class, but do not provide a mechanistic explanation for the existence of these correlations nor elucidate the causality or order of events that led to functional enrichment within certain TF structural classes. We can, however, note that certain structural classes, like the C2H2 zinc finger TFs, have retained their paralogs after yeast whole genome duplication at a much higher than average rate (Supplementary Figure 2). Interestingly, C2H2 zinc finger TFs have undergone expansion and neofunctionalization within diverse lineages[29,30]. Thus, we can hypothesize that the structural properties and corresponding regulatory mechanisms of certain structural classes made them more suited for neofunctionalization and expansion over evolutionary time.

The regulatory trends for different DBD structural classes could be used to improve gene function prediction. DBD structure is already used indirectly to predict TF function when biological roles are inferred from target genes that were in turn identified using binding sites predicted by structural homology[4,6]. The results presented here indicate that for certain TF structural classes, such as homeodomains in mouse, fly, and yeast, TF function prediction based on DBD structure is likely to be informative. For other TF classes, such as Myb domains in both fly and mouse, however, functional inferences from structure must be interpreted with caution. Likewise, our observed correlations of certain DBD structural classes with various regulatory properties suggest that such regulatory properties could also be included in predictions of TFs' regulatory roles. The resulting predictions of gene function could then be tested by directed experimentation. Beyond experimental testing to validate the predicted functions for novel or poorly characterized TFs, any TFs whose regulatory properties fall outside the general trends presented here could be investigated further to determine whether existing data and annotations have missed certain regulatory aspects of TF function that are expected for members of its structural class.

The trends we observed here may have been affected by incomplete or biased annotations. In the future, as more precise data on the DNA binding specificities of TFs from each structural class and the biological processes they regulate become available[31], more concrete relationships between these features might be revealed. Analysis of other regulatory features, such as co-regulation within and between classes, other domains associated with a structural class, and the variability of TF and target gene expression could also further elucidate the role of DBD structure in TF function and regulatory mechanism.

5. Acknowledgments

The authors thank Gabriel Berriz for advice regarding FuncAssociate. This work was supported in part by NIH/NHGRI grant # R01 HG002966 (M.L.B.). R.P.M. was supported by a National Science Foundation Graduate Research Fellowship.

References

1. Gaskell, W.H. *J Physiol* **7,** 1-80.9 (1886).
2. Narlikar, L. and Hartemink, A.J. *Bioinformatics* **22,** 157-63 (2006).
3. Luscombe, N.M., Austin, S.E., Berman, H.M., *et al. Genome Biol* **1,** REVIEWS001 (2000).
4. Tan, K., McCue, L.A., and Stormo, G.D. *Genome Res* **15,** 312-20 (2005).
5. Siggers, T.W. and Honig, B. *Nucleic Acids Res* **35,** 1085-97 (2007).
6. Kaplan, T., Friedman, N., and Margalit, H. *PLoS Comput Biol* **1,** e1 (2005).
7. Narlikar, L., Gordan, R., Ohler, U., *et al. Bioinformatics* **22,** e384-92 (2006).
8. Vermeirssen, V., Barrasa, M.I., Hidalgo, C.A., *et al. Genome Res* **17,** 1061-71 (2007).
9. Serres, M.H., Goswami, S., and Riley, M. *Nucleic Acids Res* **32,** D300-2 (2004).
10. Hu, Y., Rolfs, A., Bhullar, B., *et al. Genome Res* **17,** 536-43 (2007).
11. Bateman, A., Coin, L., Durbin, R., *et al. Nucleic Acids Res* **32,** D138-41 (2004).
12. Kummerfeld, S.K. and Teichmann, S.A. *Nucleic Acids Res* **34,** D74-81 (2006).
13. Grumbling, G. and Strelets, V. *Nucleic Acids Res* **34,** D484-8 (2006).
14. Gray, P.A., Fu, H., Luo, P., *et al. Science* **306,** 2255-7 (2004).
15. Harris, M.A., Clark, J., Ireland, A., *et al. Nucleic Acids Res* **32,** D258-61 (2004).
16. Mewes, H.W., Frishman, D., Guldener, U., *et al. Nucleic Acids Res* **30,** 31-4 (2002).
17. Berriz, G.F., King, O.D., Bryant, B., *et al. Bioinformatics* **19,** 2502-4 (2003).
18. MacIsaac, K.D., Wang, T., Gordon, D.B., *et al. BMC Bioinformatics* **7,** 113 (2006).
19. Harbison, C.T., Gordon, D.B., Lee, T.I., *et al. Nature* **431,** 99-104 (2004).
20. Yu, H., Kim, P.M., Sprecher, E., *et al. PLoS Comput Biol* **3,** e59 (2007).
21. McCord, R.P., Berger, M.F., Philippakis, A.A., *et al. Mol Syst Biol* **3,** 100 (2007).
22. Serres, M.H. and Riley, M. *Microb Comp Genomics* **5,** 205-22 (2000).
23. Robinson, M.D., Grigull, J., Mohammad, N., *et al. BMC Bioinformatics* **3,** 35 (2002).
24. Kellis, M., Birren, B.W., and Lander, E.S. *Nature* **428,** 617-24 (2004).
25. Pilpel, Y., Sudarsanam, P., and Church, G.M. *Nat Genet* **29,** 153-9 (2001).
26. Roth, F.P., Hughes, J.D., Estep, P.W., *et al. Nat Biotechnol* **16,** 939-45 (1998).
27. Kusek, J.C., Greene, R.M., Nugent, P., *et al. Int J Dev Biol* **44,** 267-77 (2000).
28. Itzkovitz, S., Tlusty, T., and Alon, U. *BMC Genomics* **7,** 239 (2006).
29. Huntley, S., Baggott, D.M., Hamilton, A.T., *et al. Genome Res* **16,** 669-77 (2006).
30. Chung, H.R., Lohr, U., and Jackle, H. *Mol Biol Evol* **24,** 1934-43 (2007).
31. Bulyk, M.L. *Curr Opin Biotechnol* **17,** 422-30 (2006).

USING DNA DUPLEX STABILITY INFORMATION FOR TRANSCRIPTION FACTOR BINDING SITE DISCOVERY

RALUCA GORDÂN, ALEXANDER J. HARTEMINK

Duke University, Dept. of Computer Science,
Box 90129, Durham, NC 27708, USA
E-mail: {raluca,amink}@cs.duke.edu

Transcription factor (TF) binding site discovery is an important step in understanding transcriptional regulation. Many computational tools have already been developed, but their success in detecting TF motifs is still limited. We believe one of the main reasons for the low accuracy of current methods is that they do not take into account the structural aspects of TF-DNA interaction. We have previously shown that knowledge about the structural class of the TF and information about nucleosome occupancy can be used to improve motif discovery. Here, we demonstrate the benefits of using information about the DNA double-helical stability for motif discovery. We notice that, in general, the energy needed to destabilize the DNA double helix is higher at TF binding sites than at random DNA sites. We use this information to derive informative positional priors that we incorporate into a motif finding algorithm. When applied to yeast ChIP-chip data, the new informative priors improve the performance of the motif finder significantly when compared to priors that do not use the energetic stability information.

1. Introduction

An important step in deciphering eukaryotic transcriptional regulatory control is the discovery of TF binding sites. Although the amount of TF binding data and the number of *de novo* motif discovery tools have been increasing over the last few years, the problem of finding and characterizing TF binding sites is far from being solved. Most DNA motif discovery tools focus on finding overrepresented motifs in sets of sequences believed to be bound by certain TFs. Recent tools also use cross-species conservation information, and thus look for overrepresented and conserved motifs. However, these tools do not take into account structural aspects of the physical interaction between DNA molecules and TFs.

We have shown previously that using structural information, such as the structural class of the TF[1] or nucleosome occupancy information[2] can significantly improve the accuracy of motif finders. In this paper, we explore another aspect of the TF-DNA interaction: the stability of the DNA double helix. During transcription, the two DNA strands must be separated so that the RNA polymerase can slide along the DNA molecule and synthesize a nascent protein. Since proximal promoter regions, containing the TATA box and binding sites for *general* TFs, are located immediately upstream of the transcribed gene where transcription is initiated, one would expect these regions to have a low DNA duplex stability. It is not clear, however, whether a low or high DNA duplex stability at *specific* TF binding sites would be more beneficial for transcription initiation.

Some regulatory proteins bind DNA in a single-strand specific manner (*e.g.* the FBP protein in human[3]). However, the crystal structure of many TF-DNA complexes reveals interactions between TFs and both strands of DNA. This suggests that destabilization of the double helix could actually prevent the TFs from binding to their specific sites on the DNA.

Taking this into account, we hypothesis that TF binding sites occur preferentially in regions with high DNA duplex stability. To test this hypothesis, we consider a set of high-confidence TF binding sites in yeast and compare the duplex stability of these binding sites against the stability of randomly selected sites from the same genomic regions. As a measure of stability we use the *helix destabilization profiles* of Bi and Benham.[5] These profiles contain, for each position in a DNA molecule, the incremental free energy needed to separate the base-paired nucleotides at that position.

We will show that the distribution of the average energy needed to separate the base pairs in TF binding sites is significantly different than the distribution of the average energy needed to destabilize random sites, so we use these distributions to derive informative positional priors that we incorporate into our framework for DNA motif discovery, PRIORITY.[1] Intuitively, the first prior simply guides the search towards DNA sites that have a high energy of destabilization, while the second prior gives more weight to motifs with a higher energy of destabilization in the set of bound sequences than in the genome overall. We show that both energy-based priors significantly improve the performance of motif finding.

2. Data and methods

2.1. *TF binding data*

We use the *Saccharomyces cerevisiae* chromatin immunoprecipitation (ChIP-chip) data published by Harbison *et al.*,[7] who profiled 203 TFs in several environmental conditions. For each TF profiled under each condition, we define its bound sequence-set to be those intergenic sequences (probes) reported to be bound with p-value ≤ 0.001. Of the 307 resulting sequence-sets, we use only the 156 sets that contain at least 10 sequences each, and correspond to 80 TFs with known binding sites (as summarized by Harbison *et al.*,[7] or as reported earlier[11,12]). Each sequence-set is identified as *TF_condition* (*e.g.* Mbp1_YPD).

2.2. *DNA duplex stability data*

The B-form structure of the DNA double helix is not invariant. At specific sites, local DNA strand separation must occur for certain processes to take place (*e.g.* initiation of transcription or replication). The problems of characterizing the duplex stability of DNA molecules and finding the locations most susceptible to strand separation have been studied intensively by Benham and collaborators.[4,5,6]

Although eukaryotic chromosomes are linear, it is easier to understand the process of duplex destabilization in the context of circular DNA. These molecules have a constant *linking number*, defined as the number of times either strand links through the closed circle formed by the other strand.[5] All conformational rearrangements that do not break the strands must preserve this constant. The case of linear DNA molecules is similar because they are partitioned into topological domains consisting of closed loops within a chromosome, and these loops have fixed linking numbers in the relaxed state.[5]

Due to transient strand breakage and re-ligation, the actual linking number of a DNA molecule can deviate from the linking number in the relaxed state, a phenomenon known as *DNA superhelicity*. In general, DNA superhelicity is negative *in vivo* (*i.e.* the actual linking number is smaller than the linking number in the relaxed state) and therefore imposes untwisting torsional stresses on the DNA that can destabilize the double helix at specific sites, a phenomenon called *SIDD (stress-induced duplex destabilization)*.[5]

Bi and Benham[5] developed an approximate method for analyzing local destabilization in superhelically stressed DNA molecules. The method uses statistical mechanics and nearest neighbor energetics of local denaturation to find all states with free energy below a certain threshold, among the 2^N possible states for a DNA molecule of size N. Each state can be viewed as a binary array of size N, with each position indicating the state of the base pair at that position (denatured or not). Next, the authors use the ensemble of low energy states to derive a measure of destabilization called the *(helix) destabilization profile*. For each position j in a DNA molecule $\mathcal{X}$, the destabilization profile $G(\mathcal{X}, j)$ represents the incremental free energy needed to separate the base pair at that position.

We use Bi and Benham's online tool WebSIDD[6] to compute the destabilization profiles for all 6140 DNA probes in the yeast TF binding data. Accurate estimation of the energy profile requires that it be computed within a larger genomic context, because the stacking interactions of neighboring base pairs may have non-local influence on the energy profile. For this reason, when computing the profile for each probe, we include 1000 base pairs upstream and downstream.

2.3. *Average destabilization energy at TF binding sites vs. random sites*

To compute the average energy of destabilization at TF binding sites we use the 4312 high-confidence sites reported by MacIsaac *et al.*[8] The width of these binding sites varies from 5 to 13 nucleotides. Since in our study we primarily search for motifs of size 8—whose length can be refined later using criteria such as information content—we restrict our attention to the 2740 binding sites of size 7 to 9 nucleotides. For every resulting binding site $\mathcal{B}$ we compute the energy of destabilization $G(\mathcal{B})$ as the average of the destabilization profiles $G(\mathcal{B}, j)$ for all positions j in the site.

We build a histogram of the energies of the 2740 binding sites, normalize the values to get a valid probability distribution, and then use a moving average to obtain a smooth distribution of energy values, plotted as a CDF in Figure 1. For every energy value e, this distribution represents the probability of a DNA site $\mathcal{S}$ having that energy, given that $\mathcal{S}$ is a true TF binding site, *i.e.* $P(G(\mathcal{S}) = e \mid \mathcal{S} \in \text{TFBS})$, where TFBS is the set of all binding sites.

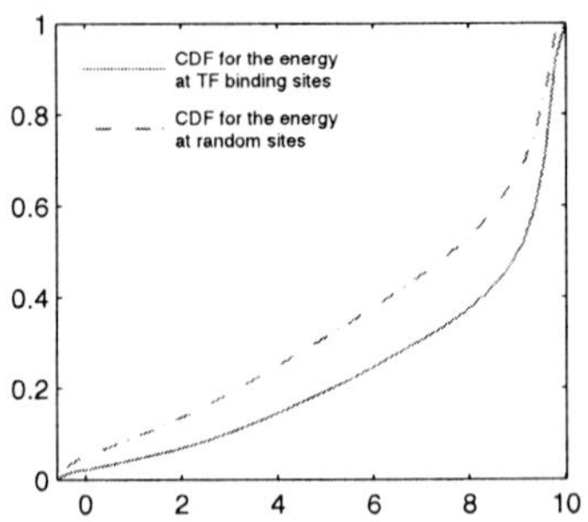

Figure 1. The cumulative distribution functions (CDFs) for the average energy of destabilization at TF binding sites (solid) versus random DNA sites (dashed). A two-sample Kolmogorov-Smirnov test indicates these two distributions to be different at a p-value of 2×10^{-68}.

Next, for each high-confidence binding site $\mathcal{B}$ of size 7 to 9 nucleotides we randomly select 20 DNA sites of the same size, from the same intergenic sequence as $\mathcal{B}$. We compute the energy of destabilization for each of the 54,800 random sites, and use these values to build the distribution of energies for random DNA sites, plotted as a CDF in Figure 1. For every energy value e, this distribution gives us the probability of a DNA site $\mathcal{S}$ having that energy, *i.e.* $P(G(\mathcal{S}) = e)$.

We can now use Bayes rule to compute the probability that a DNA site $\mathcal{S}$ is a TF binding site, given its energy:

$$P(\mathcal{S} \in \text{TFBS} \mid G(\mathcal{S})) = \frac{P(G(\mathcal{S}) \mid \mathcal{S} \in \text{TFBS}) \times P(\mathcal{S} \in \text{TFBS})}{P(G(\mathcal{S}))} \quad (1)$$

The only unknown term on the right side of Eq. (1) is the prior probability of $\mathcal{S}$ being a TF binding site. We estimate this term using the frequency of random DNA sites that have a significant overlap with any of the known TF binding sites, as reported by MacIsaac *et al.*[8]

Given that the distributions of the average energy of destabilization are significantly different for true TF binding sites compared to random sites, we can leverage this information to improve TF binding site discovery. More precisely, we use $P(\mathcal{S} \in \text{TFBS} \mid G(\mathcal{S}))$, as defined in Eq. (1), to derive informative positional priors that we incorporate into PRIORITY,[1] our generative framework for identifying motifs in sets of DNA sequences.

2.4. *The PRIORITY framework*

Let $\boldsymbol{X} = \{\boldsymbol{X}_1, \ldots, \boldsymbol{X}_n\}$ be a set of n DNA sequences reported to be bound by the same TF. For simplicity, we assume that each DNA sequence contains at most one binding site of the TF, and we use a vector $\boldsymbol{Z}$ to denote the starting location of the binding site in each sequence: $Z_i = j$ if there is a binding site starting at location j in $\boldsymbol{X}_i$. Since the TF binding data may have been affected by experimental errors, we also allow for the DNA sequences to contain no binding sites, and in this case we adopt the convention that $Z_i = 0$.

We model the TF binding sites as position-specific scoring matrices (PSSMs) of length W parameterized by ϕ, and we assume that the rest of the sequence follows some background model parameterized by ϕ_0. We fixed the length W of the binding sites to be 8, and the background model ϕ_0 to be a third order Markov model trained on all intergenic regions in yeast.

The goal of our motif finding algorithm is to find the ϕ and $\boldsymbol{Z}$ that maximize the joint posterior distribution of all the unknowns given the data. Assuming independent priors $P(\phi)$ and $P(\boldsymbol{Z})$ over ϕ and $\boldsymbol{Z}$ respectively, our objective is:

$$\arg\max_{\phi,\boldsymbol{Z}} P(\phi, \boldsymbol{Z} \mid \boldsymbol{X}, \phi_0) = \arg\max_{\phi,\boldsymbol{Z}} P(\boldsymbol{X} \mid \phi, \boldsymbol{Z}, \phi_0) \times P(\phi) \times P(\boldsymbol{Z}) \quad (2)$$

We use Gibbs sampling to sample repeatedly from the posterior over ϕ and $\boldsymbol{Z}$ with the hope that we are going to visit those values of ϕ and $\boldsymbol{Z}$ that maximize the posterior probability. Gibbs sampling is a Markov chain Monte Carlo (MCMC) method that approximates sampling from a joint posterior distribution by sampling iteratively from individual conditional distributions.[9] For faster convergence, we apply collapsed Gibbs sampling[10] and integrate out ϕ to sample only the Z_i:

$$\begin{aligned} P(Z_i \mid \boldsymbol{Z}_{[-i]}, \boldsymbol{X}, \phi_0) &= P(\boldsymbol{X} \mid Z_i, \boldsymbol{Z}_{[-i]}, \phi_0) \times P(Z_i)/P(\boldsymbol{X} \mid \boldsymbol{Z}_{[-i]}, \phi_0) \\ &\propto P(\boldsymbol{X}_i \mid \boldsymbol{Z}, \phi_0) \times P(Z_i) \end{aligned} \quad (3)$$

Most motif discovery algorithms based on Gibbs sampling strategies implicitly assume a uniform prior over the possible starting locations Z_i of a binding site in each sequence $\boldsymbol{X}_i$, and thus sample only according to the likelihood term. Our algorithm has a great advantage over other motif finders: it allows the incorporation of informative positional priors.

2.5. *Building an energy-based positional prior*

Given a DNA sequence $\boldsymbol{X}_i$ and the energy profile $G(\boldsymbol{X}_i, j)$ we derive an informative positional prior in two steps. First, for each W-mer $X_{i,j}^W$ that starts at position j in sequence $\boldsymbol{X}_i$ we compute an energy-based score that reflects the prior probability of the W-mer being a TF binding site:

$$S_{\mathcal{E}}(\boldsymbol{X}_i, j) = P\left(X_{i,j}^W \in \text{TFBS} \mid \langle G(\boldsymbol{X}_i, j)\rangle_W\right) \tag{4}$$

where $\langle G(\boldsymbol{X}_i, j)\rangle_W$ is the average energy of destabilization for the W-mer that starts at position j in sequence $\boldsymbol{X}_i$:

$$\langle G(\boldsymbol{X}_i, j)\rangle_W = \frac{1}{W}\sum_{k=0}^{W-1} G(\boldsymbol{X}_i, j+k) \tag{5}$$

The score $S_{\mathcal{E}}$ can then be calculated from the distributions of the average energy of destabilization, as described in Eq. (1).

The second step in the derivation of the positional prior is to build a valid probability distribution $P(Z_i = j)$ using the energy-based score $S_{\mathcal{E}}$. Note that the values $S_{\mathcal{E}}(\boldsymbol{X}_i, j)$ themselves do not define a probability distribution over j, as they may not sum to 1. In addition, according to our model, we allow for the sequence $\boldsymbol{X}_i$ to contain no binding sites. In this case, none of the positions in $\boldsymbol{X}_i$ can be the starting locations of binding sites, so we must have:

$$P(Z_i = 0) \propto \prod_{u=1}^{l_i - W+1} (1 - S_{\mathcal{E}}(\boldsymbol{X}_i, u)) \tag{6}$$

where l_i is the length of sequence $\boldsymbol{X}_i$. On the other hand, if $\boldsymbol{X}_i$ has one binding site at position j, not only must a binding site start at location j but also no such binding site should start at any of the other locations in $\boldsymbol{X}_i$. Formally, we write:

$$P(Z_i = j) \propto S_{\mathcal{E}}(\boldsymbol{X}_i, j) \prod_{\substack{u=1 \\ u\neq j}}^{l_i - W+1} (1 - S_{\mathcal{E}}(\boldsymbol{X}_i, u)) \quad \text{for} \quad 1 \leq j \leq l_i - W + 1 \tag{7}$$

We then normalize $P(Z_i)$ using the same proportionality constant in Eqs. (6) and (7), so that under the assumptions of our model we have: $\sum_{j=0}^{l_i - W+1} P(Z_i = j) = 1$, for $1 \leq i \leq n$. Finally, we incorporate this energy-based positional prior into our search algorithm PRIORITY, and we refer to the resulting algorithm as PRIORITY-$\mathcal{E}$.

To visualize how the positional prior $\mathcal{E}$ can improve TF binding site discovery, we show in Figure 2 the score $S_{\mathcal{E}}$ from which the prior $\mathcal{E}$ is computed, over four DNA probes from the sequence-set corresponding to TF Mbp1 profiled in YPD. We notice that most of the Mbp1 sites, depicted as black boxes on the DNA

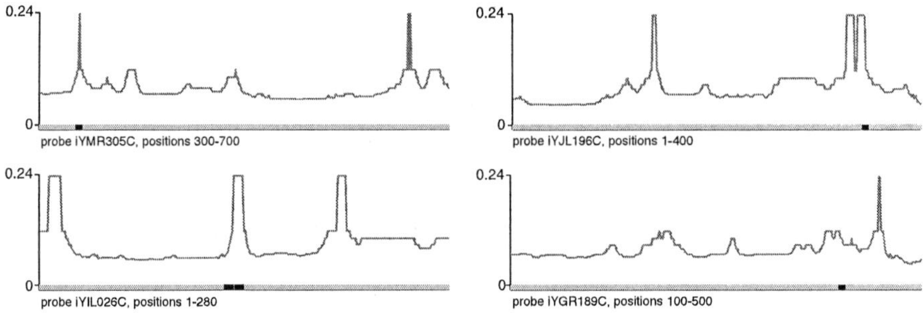

Figure 2. The energy-based score $S_{\mathcal{E}}$ used to compute the $\mathcal{E}$ prior. The x-axes represent DNA probes from the sequence-set Mbp1_YPD. The black boxes on the DNA sequences represent matches to the Mbp1 motif, ACGCGT.

sequences in Figure 2, correspond to peaks of the energy score $S_{\mathcal{E}}$, so they also correspond to peaks of the prior $P(Z_i = j)$. Thus, when prior $\mathcal{E}$ is used for sampling the starting locations of putative binding sites (see Eq. (3)), the locations of the true Mbp1 sites already have a high weight, even before the likelihood information is taken into account.

2.6. *Building a discriminative energy-based positional prior*

In Figure 2 we notice that matches to the Mbp1 motif correspond to peaks of the energy-based score. However, $S_{\mathcal{E}}$ has a number of other peaks that do not correspond to Mbp1 sites. This is not surprising since we cannot expect all the high-energy sites in these DNA sequences to be binding sites of the profiled TF, Mbp1. The other peaks may correspond to binding sites of other TFs, or to other DNA elements that have a high energy of destabilization. To address this issue we build a second informative prior, $\mathcal{DE}$, which uses the energy profiles in a discriminative manner. To do this we need, in addition to the set $\boldsymbol{X}$ of bound sequences, another set $\boldsymbol{Y}$ that contains sequences believed *not* to be bound by the TF in question. Both sets of sequences can be obtained from large-scale experimental methods like ChIP-chip.

The prior $\mathcal{DE}$ is derived similarly to the derivation of the simple energy prior $\mathcal{E}$, but using a new score that takes into account the energy of putative binding sites in both the positive (bound) and the negative (unbound) sequences. For a W-mer $X_{i,j}^W$ starting at position j in sequence $\boldsymbol{X}_i$, the discriminative energy score is defined as the ratio between the sum of the simple energy score for the occurrences of $X_{i,j}^W$ in the positive set, and the sum of the energy score for the occurrences of

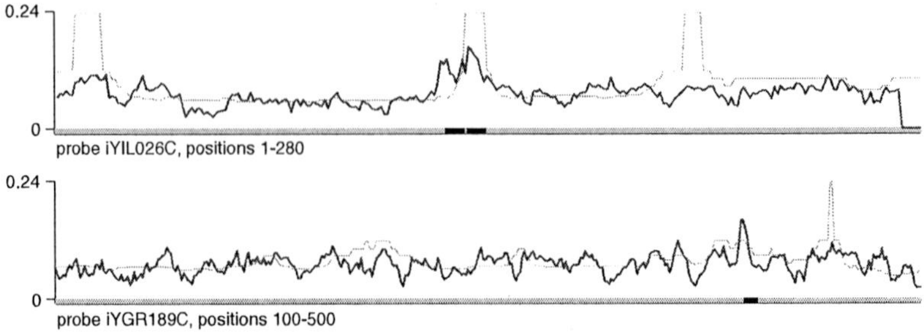

Figure 3. The discriminative energy score $S_{\mathcal{DE}}$ used to compute the $\mathcal{DE}$ prior. The x-axes represent DNA probes from the sequence-set Mbp1_YPD. The lighter curves represent the simple energy score $S_{\mathcal{E}}$ over the same DNA sequences. The black boxes on the DNA sequences represent matches to the Mbp1 motif, ACGCGT.

the same W-mer in both the positive and negative sets:

$$S_{\mathcal{DE}}(\boldsymbol{X}_i, j) = \frac{\sum_{(k,l):X_{kl}^W = X_{ij}^W} S_{\mathcal{E}}(\boldsymbol{X}_k, l)}{\sum_{(k,l):X_{kl}^W = X_{ij}^W} S_{\mathcal{E}}(\boldsymbol{X}_k, l) + \sum_{(k,l):Y_{kl}^W = X_{ij}^W} S_{\mathcal{E}}(\boldsymbol{Y}_k, l)} \tag{8}$$

Using the discriminative score $S_{\mathcal{DE}}$ instead of the simple score $S_{\mathcal{E}}$ we build a valid probability distribution $P(Z_i = j)$, as described in Section 2.5. We call the new prior $\mathcal{DE}$, and we refer to our algorithm with this informative prior PRIORITY-$\mathcal{DE}$.

To illustrate the advantages of the new discriminative prior over the simple energy prior, we show in Figure 3 the score $S_{\mathcal{DE}}$ over the last two DNA sequences in Figure 2 (see the Supplementary Material for plots of $S_{\mathcal{DE}}$ over all four DNA sequences). We notice that in both sequences the highest $S_{\mathcal{DE}}$ peaks correspond to Mbp1 sites. In the first sequence, the simple score $S_{\mathcal{E}}$ has two peaks that do not correspond to Mbp1 sites: the peak on the left corresponds to a Mot3 motif, and the peak on the right to a Swi5 motif. The score $S_{\mathcal{DE}}$ does not contain these two peaks because of its specificity for the profiled TF, which in this case is Mbp1. In the second sequence, the highest peak of $S_{\mathcal{E}}$ is misleading: it corresponds to an imperfect match to the Swi4/Swi6 motif. $S_{\mathcal{DE}}$, however, does not have a peak at this position. Instead, it indicates the correct location of the Mbp1 binding site.

The energy-based priors $\mathcal{E}$ and $\mathcal{DE}$ are derived from distributions of the average energy of destabilization for both known TF binding sites and random DNA sites. When using these priors to find the binding motif of a certain TF, one might worry that occurrences of this motif may have been included in the training data (*i.e.* the set of known binding sites) and therefore the algorithms may be successful simply because they are being tested on some of the data that was used for training. One way to overcome this issue is to remove all the binding sites of the TF

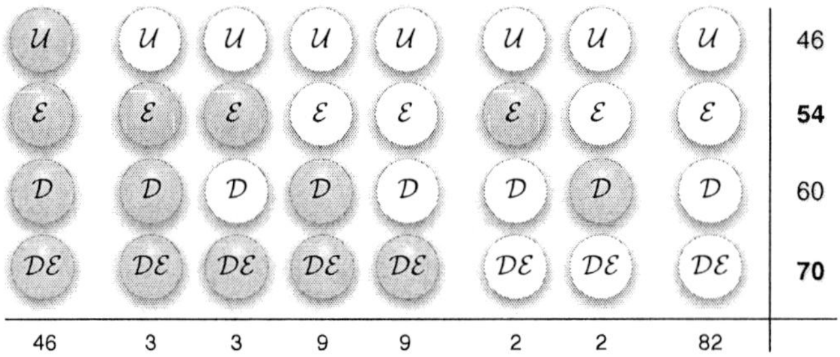

Figure 4. Summary of the results obtained by PRIORITY with priors $\mathcal{U}$, $\mathcal{E}$, $\mathcal{D}$, and $\mathcal{DE}$. Each column represents a possible combination of successes (filled balls) and failures (empty balls) for the four priors. Out of the 16 possible combinations, we only depict those that occur in at least one of the 156 sequence-sets. The number of sequence-sets falling into each category is indicated below the respective column. The last column contains the total number of successes for each algorithm.

in question from the set of known binding sites, compute the two energy distributions, derive the priors, and then apply the algorithms for that TF. We did exactly this and noticed that the two energy distributions were virtually unchanged. This makes sense since the set of binding sites is very large (2740 sites), so leaving out the sites of a particular TF does not influence the distribution of average energy significantly.

3. Results

To assess the performance of PRIORITY-$\mathcal{E}$ and PRIORITY-$\mathcal{DE}$ we use the 156 sequence-sets compiled from the ChIP-chip data of Harbison *et al.*[7] (see Section 2.1). For each sequence-set we run the algorithms 10 times from different random starting points for 10,000 sampling iterations and report the top-scoring motif among the 10 runs. We consider an algorithm to be successful for a sequence-set only if the top-scoring motif is at a distance less than 0.25 from the literature consensus. For details about the distance function, see Narlikar *et al.*[2]

We first compare the performance of the energy-based positional priors with that of a uniform prior $\mathcal{U}$ and a simple discriminative prior $\mathcal{D}$. These two priors are similar to $\mathcal{E}$ and $\mathcal{DE}$, respectively, except that they do not use information about the destabilization energy. We build the uniform prior using a flat score $S_{\mathcal{U}} = 0.5$. The simple discriminative prior $\mathcal{D}$ is calculated similarly to $\mathcal{DE}$, but using the uniform score $S_{\mathcal{U}}$ instead of the energy score $S_{\mathcal{E}}$ in Eq. (8). We incorporate the priors into our framework PRIORITY and refer to the new algorithms as PRIORITY-$\mathcal{U}$ and PRIORITY-$\mathcal{D}$. The results of the four algorithms on the 156 sequence-sets are summarized in Figure 4 and presented in detail in the Supplementary Material.

3.1. *Energy-based priors perform better than uniform prior*

An accurate quantification of the extent to which the energy-based priors improve motif discovery can be obtained by comparing PRIORITY-$\mathcal{E}$ and PRIORITY-$\mathcal{DE}$ with PRIORITY-$\mathcal{U}$.

We notice that PRIORITY-$\mathcal{E}$ is able to find 54 correct motifs, an improvement of 17% over the uniform prior. PRIORITY-$\mathcal{DE}$ performs even better: it finds the correct motif in 70 sequence-sets, 52% more than the uniform prior. Furthermore, we notice that in all the sequence-sets where PRIORITY-$\mathcal{U}$ succeeds, the energy-based priors also succeed, so they are never detrimental to motif discovery. We also mention that in the sequence-set Mbp1_YPD, from which the DNA sequences depicted in Figures 2 and 3 were extracted, PRIORITY-$\mathcal{U}$ is unable to find the correct Mbp1 motif, while both PRIORITY-$\mathcal{E}$ and PRIORITY-$\mathcal{DE}$ succeed.

The improvement of PRIORITY-$\mathcal{DE}$ over PRIORITY-$\mathcal{U}$ is remarkable: 70 correctly found motifs versus 46. We note, however, that this improvement is not due solely to the energy information, but also to the discriminative information. Out of the 24 motifs found by PRIORITY-$\mathcal{DE}$ and not found by PRIORITY-$\mathcal{U}$, 9 motifs are only detected when using the discriminative priors, so it is probably the discriminative information that causes the improvement in these cases. In 9 other sequence-sets, though, the $\mathcal{DE}$ prior is the only one to find the correct motif. This suggests that neither $\mathcal{E}$ nor $\mathcal{D}$ alone contains enough information to identify the true motif, though the combination $\mathcal{DE}$ is successful.

Figure 4 also reveals that there are four cases in which either $\mathcal{E}$ or $\mathcal{D}$ succeeds in finding the correct motif, but $\mathcal{DE}$ fails. We next discuss these cases in more detail. The two sequence-sets where PRIORITY-$\mathcal{E}$ is the only one that finds the correct motif are Met32_SM and Sip4_YPD. In both cases we notice that the occurrences of the true motif in the bound set have a high energy of destabilization, which explains the success of PRIORITY-$\mathcal{E}$, but the two motifs also have a high energy of destabilization overall in the genome, which explains why PRIORITY-$\mathcal{DE}$ fails. We also notice that the sequence-sets Met32_SM and Sip4_YPD contain very few occurrences of the Met32 and Sip4 motifs, respectively. We believe it is possible that some high-energy occurrences of these motifs in the unbound sets are in fact binding sites of the profiled TFs, but were not bound in the particular environmental conditions of the ChIP-chip experiments.

In two sequence-sets, the $\mathcal{D}$ prior succeeds while both energy-based priors fail: Skn7_H2O2Lo and Msn2_H2O2Hi. In the case of Skn7_H2O2Lo, both $\mathcal{E}$ and $\mathcal{DE}$ fail because they get stuck in local optima. If we score the motif found by $\mathcal{D}$ according to the posteriors obtained using $\mathcal{E}$ and $\mathcal{DE}$, we get significantly higher scores than the ones reported by PRIORITY-$\mathcal{E}$ and PRIORITY-$\mathcal{DE}$, respectively, for their top motifs (which do not match the literature consensus). In the case of Msn2_H2O2Hi, the fact that PRIORITY-$\mathcal{DE}$ does not find the correct motif is due to the motif size, which by default is 8. If we set it to 6—the true size of the Msn2 motif—PRIORITY-$\mathcal{DE}$ succeeds. For the same sequence-set Msn2_H2O2Hi, the failure of PRIORITY-$\mathcal{E}$ seems to be the result of the algorithm getting stuck in a local optimum.

3.2. *Comparison with popular motif finders*

Finally, we present a comparison between the results of our algorithm with energy-based positional priors and the results of six popular motif finders, as reported by Harbison *et al.*[7]: AlignACE,[13] MEME,[14] MDscan,[15] and three methods that use evolutionary conservation information (MEME_c,[7] a method of Kellis *et al.*,[16] and Converge[7]). We emphasize, however, that the goal of this paper is not to introduce a new motif discovery tool, but to show that structural information typically disregarded by motif finders can significantly improve their performance.

Out of the 156 sequence-sets, AlignACE is successful in 16, MEME in 35, MDscan in 54, MEME_c in 49, the method of Kellis *et al.* in 50, and Converge in 56, so our algorithm PRIORITY-$\mathcal{DE}$ outperforms all six methods, with a total of 70 correctly identified motifs. Furthermore, even the simpler PRIORITY-$\mathcal{E}$ outperforms five of the six methods.

4. Discussion

In this paper we demonstrate the benefits of using information about the DNA double-helical stability to detect TF binding sites. Using the energy profiles of Bi and Benham[5] as a measure of stability, we notice that in general more incremental free energy is needed to separate the DNA strands at TF binding sites compared to random sites across the genome. This is not surprising since TF binding sites are usually GC-rich. We stress, however, that the energy profiles we used in our analysis were computed using a complex method that takes into account not only individual base pairs, but also the neighboring effects of other base pairs in the same DNA region. Although there is some correlation between the energy profiles and the GC content of the DNA sequences, using an informative positional prior similar to $\mathcal{E}$ but derived from GC content instead of destabilization profiles did not show any improvement over the uniform prior.

One limitation of using helix destabilization energy is that the only eukaryotic organism whose profile has been made available is yeast. The online tool WebSIDD[5] could in principle be used to compute energy profiles for other eukaryotic genomes, but it is limited to sequences a few kilobases long and a downloadable version of the software is not currently available.

The improvement obtained using the energy-based priors demonstrates, once again, the importance of incorporating structural information into motif discovery algorithms; whenever structural information can be translated into a prior over sequence positions, it can be straightforwardly incorporated into our PRIORITY framework for DNA motif discovery. We have shown that useful positional priors can be derived from knowledge of TF structural class,[1] from nucleosome occupancy information,[2] and now from profiles of helix destabilization energy. The usefulness of each of these sources of information leads naturally to the question of the degree of redundancy among them; for instance, the positioning of nucleosomes may be correlated with DNA duplex stability. However, we observe that only some priors are successful on certain sequence-sets. As one example, although both the discriminative nucleosome prior $\mathcal{DN}^2$ and the discriminative energy prior $\mathcal{DE}$ succeed on 70 sequence-sets, in 10 of these sets, only one of the two succeeds, suggesting that combining the informative priors in a principled way—which is not a trivial task—has the potential to further improve motif discovery using informative positional priors.

Supplementary material is available at *www.cs.duke.edu/∼amink/*.

Acknowledgments

This project began during a course taught by Bruce Donald, whom R.G. wishes to thank for his early advice. A.J.H. gratefully acknowledges funding for this work from an NSF CAREER award, an Alfred P. Sloan Fellowship, and awards from NIEHS and NIGMS.

References

1. L.Narlikar, R.Gordân, U.Ohler, A.Hartemink, *Bioinformatics* **22**, e384 (2006).
2. L.Narlikar, R.Gordân, A.Hartemink, *PLoS Comp. Bio.*, in press (2007).
3. R.Duncan *et al.*, *Genes Dev.* **8**, 465 (1994).
4. C.J.Benham, PSB 2001, 103 (2001).
5. C.Bi, C.J.Benham, CSB2003 (2003).
6. C.Bi, C.J.Benham, *Bioinformatics* **20**, 1477 (2004).
7. C.T.Harbison *et al.*, *Nature* **431**, 99–104 (2004).
8. K.D.MacIsaac *et al.*, *BMC Bioinformatics* **7**, 113 (2006).
9. A.Gelfand, A.Smith, *J. Amer. Statistical Assoc.* **85**, 398–409 (1990).
10. J.Liu, *J. Amer. Statistical Assoc.* **89**, 958–966 (1994).
11. R.A. Dorrington, T.G. Cooper, *Nucleic Acids Res.* **21**, 3777–3784 (1993).
12. Y. Jia *et al.*, *Mol. Cell. Biol.* **17**, 1110–1117 (1993).
13. F.Roth *et al.*, *Nature Biotech.* **16**, 939–945 (1998).
14. T.Bailey, C.Elkan, *ISMB '94*, AAAI Press, Menlo Park, pp. 28–36 (1994).
15. X.Liu, D.Brutlag, J.Liu, *Nature Biotech.* **20**, 835–839 (2002).
16. M.Kellis *et al.*, *Nature* **432**, 241–254 (2003).

A PARAMETRIC JOINT MODEL OF DNA-PROTEIN BINDING, GENE EXPRESSION AND DNA SEQUENCE DATA TO DETECT TARGET GENES OF A TRANSCRIPTION FACTOR

WEI PAN[1], PENG WEI[1], ARKADY KHODURSKY[2]

[1]*Division of Biostatistics, School of Public Health,* [2] *Department of Biochemistry, Molecular Biology and Biophysics, University of Minnesota*

This paper concerns with predicting the regulatory targets of a transcription factor (TF). We propose and study a joint model that combines the use of DNA-protein binding, gene expression and DNA sequence data simultaneously; a parametric mixture model is used to realize unsupervised learning, which however can be extended to semi-supervised learning too. We applied the methods to an E coli dataset to identify the target genes of LexA, which, along with applications to simulated data, demonstrated potential gains of jointly modeling multiple types of data over using only one type of data.

1. Introduction

This paper concerns with identifying the transcriptionally regulated target genes of a transcription factor (TF). The task is commonly approached based on one of the three data types: DNA-protein binding data (also called ChIP-chip data or genome-wide location analysis) surveying genome-wide DNA-TF interactions [11,12],microarray gene expression data comparing expression changes before and after perturbing the function of, e.g. by knocking-out, a TF-coding gene [5], and DNA sequence data which are aligned and scanned to find specific binding sites or motifs of a TF [1,13], Because of relatively high noise levels with high-throughput data, using only one data source may result in high false positives or false negatives. As a compensation, it is now widely recognized that an integrative analysis of multiple types of data should be more efficient in identifying the target genes of a TF [2,4,19,26]. With the ever-increasing availability of various types of high-throughput data, a main challenge is how to integrate them effectively. In the literature, there are several classes of the approaches. First, one can use one type of data to validate results from analyses of other types of data [24]. Second, one first conducts a separate analysis on each type of data and then combine their results [27]. Third, regression analyses of one type of data (e.g. gene expression) on another type (e.g. DNA sequence) [3,4,23]. Fourth, one uses one type of data to generate priors or hypotheses for analyzing other types of data; e.g. Liu et al [14] used binding data to generate candidate binding regions, then used DNA sequence data to locate binding sites; Xie et al [29] used expression data to generate a prior list of potential binding targets, which was then utilized to analyze binding data. Finally, a joint model of multiple types of data can be employed to use all the data simultaneously to draw inference or make predictions, which is presumably more efficient than many other alternatives; our method belongs to this class, which also includes the following ones for detecting the targets of a TF. Wang et al [26] proposed a parametric mixture model for both DNA sequence data and binding (or expression) data; our method is similar to theirs except that we used three data sources and a different format of DNA sequence data. Pan et al [18] proposed a nonparametric mixture model; it requires duplicated arrays, not applicable to the E coli expression data to be analyzed here. Xie [28] proposed a fully parametric Bayesian approach using binding, expression and DNA data; because of analytically intractable posterior probability calculations, computationally intensive simulation methods (MCMC) were used to draw inference. Our work here shows that a simple parametric mixture model similar to that of Wang et al [26] works well, even when some parametric modeling assumptions are moderately violated, while accommodating more than two sources of data; furthermore, we extend the method from unsupervised learning to semi-supervised learning.

This paper is organized as follows. We first introduce our joint model as a parametric mixture model, then we outline an EM algorithm to estimate the parameters in the model and thus obtain posterior probabilities to draw inference. We present an application of the methods to an E coli dataset to identify the targets of LexA, comparing the results with the known and putative targets listed in regulonDB (v5.5) [21] and in Wade et al [25]. We also show results of simulation studies to demonstrate statistical efficiency gains from joint modeling over using only one data source. We end with a short discussion on some possible future work.

2. Methods

2.1. *A Joint Model*

Our goal is to identify which genes in a genome are the targets of a given TF. To be concrete, we consider three data sources corresponding to DNA-protein binding, gene expression and DNA sequence data, as to be used for an E coli example. We assume that the three data sources can be summarized as (X_i, Y_i, Z_i) for each gene i, $i = 1, ...G$: X_i is a summary or test statistic measuring the relative abundance of the TF binding to gene i, or the statistical significance of rejecting a null hypothesis that gene i is not bound by the TF; Y_i is a test statistic for differential expression of gene i when the TF-coding gene's function is perturbed; Z_i is a score measuring the degree to which one of its subsequences matches a known motif for the TF. Depending on whether gene i is a target or not, we have $T_i = 1$ or $T_i = 0$ respectively. To realize unsupervised learning, it is natural to assume that (X_i, Y_i, Z_i) comes from a mixture distribution: $f(x, y, z) = \pi f_1(x, y, z) + (1 - \pi) f_0(x, y, z)$, each component corresponding to the subpopulation of the genes with $T_i = 1$ or $T_i = 0$ respectively, and π is the prior proportion of the target genes. Further, we assume that conditional on T_i, the three data sources are independent; that is

$$\begin{aligned} f(x, y, z) = \ & \pi f_{11}(x; \theta_{11}) f_{12}(y; \theta_{12}) f_{13}(z; \theta_{13}) + \\ & (1 - \pi) f_{01}(x; \theta_{01}) f_{02}(y; \theta_{02}) f_{03}(z; \theta_{03}), \end{aligned}$$

where θ_{jk}'s are the (unknown) parameters for distributions f_{jk}. To infer whether gene i is a target, we use the posterior probability

$$Pr(T_i = 1 | X_i, Y_i, Z_i) = \frac{\pi f_{11}(X_i; \theta_{11}) f_{12}(Y_i; \theta_{12}) f_{13}(Z_i; \theta_{13})}{f(X_i, Y_i, Z_i)}.$$

Here we use $f_{jk} = \phi(.; \mu_{jk}, \sigma_{jk})$, a normal probability density function with mean μ_{jk} and variance σ^2_{jk}.

2.2. *Estimation via EM*

An EM algorithm [6] can be derived to estimate the unknown θ_{jk}'s. Given T_i, the complete data log-likelihood is

$$\log L_c = \sum_{i=1}^{G} T_i \log \pi f_1(X_i, Y_i, Z_i) + (1 - T_i) \log(1 - \pi) f_0(X_i, Y_i, Z_i).$$

The E-step is to calculate the conditional expectation

$$Q = E(\log L_c|Data) = \tau_i[\log \pi + \log f_1(X_i, Y_i, Z_i)] + \\ (1-\tau_i)[\log(1-\pi) + \log f_0(X_i, Y_i, Z_i)],$$

where $\tau_i = Pr(T_i = 1|X_i, Y_i, Z_i)$. The M-step maximizes the above Q with respect to the unknown parameters. We use the generic notation $\theta^{(m)}$ to denote the updated estimate of θ in iteration m; it is easy to verify that, at iteration $m+1$,

$$\tau_i^{(m+1)} = \frac{\pi^{(m)} f_1^{(m)}(X_i, Y_i, Z_i)}{\pi^{(m)} f_1^{(m)}(X_i, Y_i, Z_i) + (1-\pi^{(m)}) f_0^{(m)}(X_i, Y_i, Z_i)}$$

where

$$f_j^{(m)}(X_i, Y_i, Z_i) = \phi(X_i; \mu_{j1}^{(m)}, \sigma_{j1}^{(m)})\phi(Y_i; \mu_{j2}^{(m)}, \sigma_{j2}^{(m)})\phi(Z_i; \mu_{j3}^{(m)}, \sigma_{j3}^{(m)}),$$

for $j = 1, 2$, and

$$\mu_{11}^{(m+1)} = \sum_{i=1}^{G} \tau_i^{(m)} X_i / \sum_{i=1}^{G} \tau_i^{(m)}, \quad \sigma_{11}^{2,(m+1)} = \sum_{i=1}^{G} \tau_i^{(m)} (X_i - \mu_{11}^{(m)})^2 / \sum_{i=1}^{G} \tau_i^{(m)},$$

$$\mu_{10}^{(m+1)} = \sum_{i=1}^{G} (1-\tau_i^{(m)}) X_i / \sum_{i=1}^{G} (1-\tau_i^{(m)}),$$

$$\sigma_{10}^{2,(m+1)} = \sum_{i=1}^{G} (1-\tau_i^{(m)})(X_i - \mu_{10}^{(m)})^2 / \sum_{i=1}^{G} (1-\tau_i^{(m)}),$$

and $\pi^{(m+1)} = \sum_{i=1}^{G} \tau_i^{(m+1)}/G$, where the updates for other μ_{jk} and σ_{jk}'s are similar and omitted. The above iterations are continued until convergence. Because the EM may converge to a local maximum point, multiple starting values are needed, and the one with the maximum log-likelihood is chosen. The resulting estimates are maximum likelihood estimates (MLEs); the final τ_i are used to rank the genes for their likelihoods of being a target.

2.3. *Other Models*

The above joint model is for three data sources; it is straightforward to have a model for more or less than three data sources, and its corresponding EM updates for parameter estimation. For example, if we use only one source of data, say X_i's, we can have a corresponding mixture model

$$f(x) = \pi f_1(x; \theta_1) + (1-\pi) f_0(x; \theta_0),$$

and the posterior probability $Pr(T_i = 1|X_i) = \pi f_1(X_i; \theta_1)/f(X_i)$. The EM updates are

$$\tau_i^{(m+1)} = \frac{\pi^{(m)}\phi(X_i; \mu_{11}^{(m)}, \sigma_{11}^{(m)})}{\pi^{(m)}\phi(X_i; \mu_{11}^{(m)}, \sigma_{11}^{(m)}) + (1-\pi^{(m)})\phi(X_i; \mu_{01}^{(m)}, \sigma_{01}^{(m)})},$$

and the updates for $\mu_{11}, \mu_{01}, \sigma_{11}, \sigma_{01}$ and π are exactly the same as before. Again at the convergence, we use the posterior probabilities τ_i to rank the genes.

2.4. *Extensions to Semi-supervised Learning*

The approaches taken so far are unsupervised learning, assuming that no known targets for the TF, which is not usually true. Supervised learning approaches have been proposed [32], which however may not work well if there are only few known targets for the TF, e.g., for LexA. We can extend our proposal to semi-supervised learning, combining the strengths of unsupervised and supervised learning, which is an advantage of the mixture model [15]. Suppose that the first G_1 genes are known targets while the remaining ones may or may not be. The models are the same as before. The parameter estimation procedures are also similar except that $\tau_i = 1$ for $i = 1, ..., G_1$.

Although in general semi-supervised learning improves over unsupervised or supervised learning, for our example, because there were only few known targets of LexA, the results of semi-supervised learning were similar to that of unsupervised learning. We will skip the discussion of semi-supervised learning. Nevertheless, we expect that this semi-supervised learning will be useful for other TFs and other types of data.

3. Results

3.1. *E coli data*

We extracted the DNA-protein binding data [25] and gene expression data [5] from the authors' supplied web sites respectively, and DNA sequence data from the NCBI and Affymetrix web sites.

The binding data contained two LexA samples (called LexA1 and LexA2 respectively) and two control samples (one Gal4 and one MelR (no Ab, no antibody) samples) hybridized on four Affymetrix Antisense Genome Arrays respectively. We downloaded the raw intensity data (i.e. CEL files) from the authors' supplied web page. Largely following Wade et al [25], we processed the data in below steps. First, we used the Bioconductor R package `affy` to pre-process the data, including background correction with MAS 5 algorithm, and quantile normalization. Second, we calculated four $\log_2$ intensity ratios (LIRs), corresponding to the four combinations of any two arrays, for each probe: $\text{LexA}_1/\text{Gal4}$, $\text{LexA}_1/\text{no Ab}$, $\text{LexA}_2/\text{Gal4}$, $\text{LexA}_2/\text{no Ab}$; a large LIR indicated a locus containing enriched LexA. Third, we mapped each probe to a genome position based on the Affymetrix Ecoli_ASv2 annotation file. Fourth, for each of the four array combinations, we smoothed the LIRs over all probes with a sliding window of 1250 bp. Fifth, for each gene in each array combination, we identified its LIR peak among the probes belonging to the gene's coding and intergenic regions (if any) separately. Finally, each gene i's binding score or signal X_i was taken to be the average of its four LIR peaks from its coding region, or if there were probes from its intergenic region, X_i was the larger one of i) the average of its four LIR peaks from its coding region and ii) that from its intergenic region. The final step differed from that in Wade et al: they had an extra step to identify a candidate LexA-bound region/block containing ≥ 20 consecutive probes with all LIRs ≥ 0.17; they calculated the average of the four peaks only for the genes with such blocks, which were taken as candidate binding targets of LexA; they identified about 50 such binding targets. Because for our purpose, we would like to obtain a binding score for every gene, obviously we could not follow their route. This procedural difference contributed to some differences in X_i's between theirs and ours.

The expression data were drawn from four cDNA microarrays profiling gene expression levels for the wild type before and 20-minute after UV treatment, and for the lexA mutant before and 20-minute after UV treatment, respectively; a common control sample was used for each array. Two-channel intensities on each array were normalized using the loess local smoother to eliminate dye bias, as implemented in the R package `sma` [30] Suppose that normalized log-ratios of the two-channel intensities for gene i on the four arrays were M_{1i},...,M_{4i} respectively, then we used the summary statistics for gene expression data as $Y_i = (M_{1i} - M_{2i}) - (M_{3i} - M_{4i})$. Because LexA is known to be a repressor of some "SOS response" genes, it is expected that the transcriptional targets of LexA should have larger values of Y_i's (i.e. expression changes).

To extract DNA sequence data, on July 21, 2006, we downloaded ten known binding sites of LexA from regulonDB (v4.0), involving nine genes each with a binding site except two binding sites for gene lexA [20]. We input either these ten binding sites or five of them (in the order of #2, #4,...,#10 as ranked by MEME) into MEME [1] to find a top motif (Table 1). We then used scanACE [19] to scan the whole genome with a very low threshold such that at least one subsequence matching the motif could be obtained for most genes; we assigned the maximum of all the matching scores for gene i as Z_i, the summary statistic for the sequence data. Depending on whether the ten or five known binding sites were used to obtain the top motif, the resulting sequence data were denoted as Seq 1 (S1) or Seq 2 (S2).

After combining the three data sources and deleting genes with any missing values, we obtained $G = 3779$ genes in the combined data.

3.2. *Analysis*

Table 1. Ranks given by various methods for known (marked by *) and putative targets annotated in regulonDB (and in our data). Seq1 and Seq2 were the sequence scores obtained from the top motif using the nine and four known targets (marked by * and **) respectively.

Gene	Bind	Expr	B+E	Seq1	Seq2	B+E+S1	B+E+S2
polB	156	114	135	153	1593	127	146
phrB	1346	1826	2083	530*	81**	1516	452
uvrB	48	172	92	31*	6**	78	46
dinG	96	448	213	138	143	169	171
ftsK	75	3757	223	127	303	173	199
sulA	11	12	1	17*	728	1	1
umuD	31	29	1	19	8	1	1
umuC	192	12	1	3454	3652	34	37
ydjM	30	111	53	70	74	49	44
ruvB	2780	313	509	1471	2966	645	708
ruvA	127	147	141	10	38	94	108
uvrC	3015	3104	3646	3008	796	3377	2692
uvrY	3538	3473	3679	3008	796	3384	2685
recN	7	5	1	33	36	1	1
oraA	82	50	54	1220	871	61	59
recA	12	15	1	23*	4**	1	1
rpsU	464	1214	766	1097*	304	896	572
dnaG	2906	3621	3451	782	177	2620	954
rpoD	2906	3749	3455	782	177	2621	953
t150	2121	175	262	50	76	176	178
uvrD	263	245	274	4*	50	106	160
lexA	15	61	1	7*	1**	1	1
dinF	2549	217	323	7	1	118	77
uvrA	41	169	77	14*	114	58	72
ssb	41	143	74	14*	114	54	68

We considered using binding data alone, expression data alone, sequence data alone (either using the motif found from the 10 or 5 known binding sites), both binding and expression data, and all the three data sources. For each type of data, a two-component mixture model was fitted, and the posterior probabilities were used to rank the genes, as discussed earlier. To motivate the mixture model for each data source, we fitted a two-component normal mixture model to each data source separately; it appeared that the mixture model fitted each data source well (not shown). The parameter estimates $(\hat{\pi}_j, \hat{\mu}_{j1}, \hat{\mu}_{j0}, \hat{\sigma}^2_{j1}, \hat{\sigma}^2_{j0})$ with $j = 1, ..., 4$ for the four data sources were $(0.063, 0.11, 0.89, 0.07, 0.64)$, $(0.174, 0.02, 0.20, 0.16, 2.81)$, $(0.278, 12.8, 15.3, 2.5, 7.3)$ and $(0.196, 17.1, 19.7, 2.4, 7.3)$. As a comparison, the joint model with the binding, expression and Seq 1 data resulted in estimates $\hat{\pi} = 0.122$, and $(\hat{\mu}_{j1}, \hat{\mu}_{j0}, \hat{\sigma}^2_{j1}, \hat{\sigma}^2_{j0})$ with $j = 1, ..., 3$ as $(0.11, 0.51, 0.07, 0.51)$, $(0.01, 0.29, 0.21, 3.53)$, and $(13.3, 14.9, 4.0, 11.2)$ for the three types of the data respectively.

Table 1 gave the results for all known/putative targets listed in regulonDB (v5.5) [21], downloaded in November 1, 2006, and in our combined data. In general, combining multiple types of data increased the chance of detecting the true targets as compared to using only binding data alone; for example, the ranks based on binding and expression data, or based on the three data sources, were higher, in some cases much higher, than those based on using binding data alone. This was due to the fact that our joint analysis combined the evidence from all the three data sources. For example, each of umuD, recA and lexA was ranked relatively high (but not highest) based on each of the three data sources alone, and combining any two or three sources of data led to a highest ranking (i.e. tied at the 1st with posterior probability equal to 1); umuC was ranked only at the 192nd based on the binding data alone, with the incorporation of the expression data its rank improved to a tied 1st.

We also obtained the results (not shown) for the putative targets with a common motif (Class II) and without any common motif (Class III) identified based on only the binding data by Wade et al. For the genes in class III, because no common motifs were found in the DNA sequences of the genes, it was not surprising that a separate or combined use of sequence data gave the genes lower ranks than those based on the binding data alone. More surprisingly, for most genes, a combined analysis using both the binding and expression data also gave lower ranks than that based on the binding data alone, due to low level expression changes.

3.3. *Simulation*

To further evaluate and compare the methods with various sources of data, we did a simulation study; the simulated data were generated from the fitted models for the real data to mimic real situations. Four simulation set-ups were considered. 1) Case I: we assumed that the joint model fitted to the three data sources (with Seq 1) was correct, and simulated data from the fitted joint model; this represented an ideal scenario for the joint analysis. 2) Case II: we assumed that the binding data came from its component from the fitted joint model as in 1), but each of the other two data sources came from a two-component normal mixture model as fitted to each data source separately (Figure 1); because there were a higher proportion of the genes in the first component for the expression data and sequence data, the joint model did not hold: in particular, the second component f_{02} and f_{03} were not a single normal distribution, but a mixture of two normals. This was a scenario for which a two-component mixture model for binding data alone was correct but a joint model was not. 3) Case III was similar to Case I except that some between-gene correlations were introduced for the binding data (which might arise when the probe intensities were smoothed as in the real data). Specifically, the genes were randomly divided into blocks with size about 10, then we added some noises drawn from $N(0, \sigma_{10})$ to binding data (as generated in Case I) such that the genes within each block had correlated X_i's. Hence all the methods had an incorrect independence assumption. 4) Case IV was a combination of Cases II and III: some between-gene correlations as in Case III were introduced to the binding data while other aspects were the same as in Case II. For each case, 100 independent datasets were generated; the realized false discovery rates (FDRs) were averaged over the 100 replicates for each method in each case.

Figure 1 summarized the results for using binding data alone, using both binding and expression data, and using all three types of data. It was clear that, compared to using only binding data, using more than one data source largely reduced the FDR; that is, at any given number of estimated positives (i.e. claimed targets), the joint model could identify a much larger number of true targets (and hence fewer false negatives). Although using three data sources improved over using two data sources, because of limited information available from sequence data (as measured by the small difference between the two component distributions for the sequence data), the improvement was not dramatic.

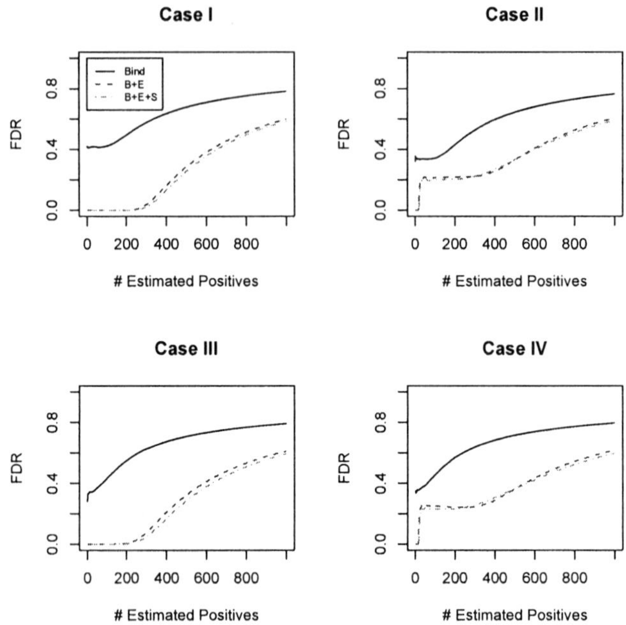

Figure 1. *Comparison of the FDRs from the three methods for simulated data.*

4. Discussion

We have demonstrated possible efficiency gains with a parametric mixture model to jointly combine multiple types of data for target detection. A key feature of our joint model is its simplicity, however, this does not exclude some possible modifications or extensions. First, rather than using a single normal distribution f_{jk} for each component for each data source, a more flexible choice is to use a mixture distribution for each f_{jk}; for the E coli data here, we considered this idea for binding data but it did not lead to much improvement, perhaps due to the goodness-of-fit of a single normal distribution to each component. We emphasize that, with some appropriate transformation, such as the Z-transformation [7], the normality of some component distributions is expected; furthermore, McLachlan et al [16] demonstrated that a two-component normal mixture model worked quite well for several typical expression datasets. Second, rather than using a sequence score for each gene, we may supply the sequence best matching the motif, and use a multinomial model for each component of the sequence data f_{3k}, as done in Wang et al [33]. This could possibly help refine the motif model. We may also consider using multiple motifs for lexA and their multiple matching sequences for each gene. Third, an advantage of the mixture model is to use the estimated posterior probabilities to estimate FDR and false non-discovery rate (FNR) [16], However, such a use critically depends on the correctness of the assumed mixture model [18]. Because here we aim to use a simple parametric model which may or may not hold exactly, we did not pursue the task of estimating FDR or FNR, which however is important in practice. To relax the possibly too strong parametric assumption, we may consider the use of a more flexible mixture model approach as outlined above, alleviating the issue of FDR/FNR estimates' dependence on correct modeling. These are all interesting topics for future research.

Acknowledgments

This research was supported by NIH grants HL65462 (WP and PW) and GM066098 (AK), and a UM AHC Faculty development grant (WP, PW and AK).

References

1. Bailey T.L. and Elkan C. (1995). *Machine Learning*, **21**, 51-80.
2. Bar-Joseph, Z., gerber, G.K., Lee, T.I., Rinaldi, N.J., Yoo, J.Y., Robert, F., Gordon, D.B., Fraenkel, E., Jaakkola, T.S., Young, R.A. and Gifford, D.K. (2003). *Nature Biotechnology*, **21**, 1337-1342.
3. Bussemaker, H.J., Li, H. and Siggia, E.D. (2001). *Nat. Genet.* **27**, 167-171.
4. Conlon, E.M., Liu, X.S., Lieb, J.D. and Liu, J.S. (2003). *Proc. Natl. Acad. Sci. USA* **100**, 3339-3344.
5. Courcelle J, Khodursky A, Peter B, Brown PO, Hanawalt PC (2001). *Genetics*, **158**, 41-64.
6. Dempster, A.P., Laird, N.M. and Rubin, D.B. (1977). *J. R. Statist. Soc. B,* **39**, 1-38.
7. Efron, B. (2004). *J. Amer. Statist. Assoc.,* **99**, 96-104.
8. Efron, B., Tibshirani, R., Storey, J.D. and Tusher, V.G. (2001). *J. Amer. Statist. Assoc.,* **96**, 1151-1160.
9. Holmes, I. and Bruno, W.J. (2000). *Proc. Int. Conf. Intell. Syst. Mol. Biol.* **8**, 202-210.
10. Ihaka, R. and Gentleman, R. (1996). *Journal of Computational and Graphical Statistics* **5**, 299-314.
11. Lee, M.-L., Bulyk, M., Whitmore, G., and Church, G. (2002). *Biometrics*, **58**, 981-988.
12. Lee, T., Rinaldi, N., Robert, F., Odom, D., Bar-Joseph, Z., Gerber, G., Hannett, N., Harbison, C., Thompson, C., Simon, I., et al. (2002b). *Science*, **298**, 799-804.

13. Liu JS, Neuwald AF, and Lawrence CE (1999). *J. Amer. Statist. Assoc.* **94**, 1-15.
14. Liu, X.S., Brutlag, D.L. and Liu, J.S. (2002). *Nat. Biotechnol.* **20**, 835-839.
15. McLachlan, G.J. and Peel, D. (2002) Finite mixture model. New York. John Wiley & Sons, Inc.
16. McLachlan, G.J., Bean, R.W., Jones, L.B.-T. (2006). *Bioinformatics* **22**, 1608-1615.
17. Newton, M.A., Noueiry, A., Sarkar, D. and Ahlquist, P. (2004). *Biostatistics*, **5**, 155-176.
18. Pan W, Jeong KS, Xie Y, Khodursky A. (2006). To appear *Statistica Sinica.*
19. Roth, F.P., Hughes, J.D., Estep, P.W., and Church, G.M. (1998). *Nat. Biotech.*, **16**, 939-945.
20. Salgado H, Gama-Castro S, Martinez-Antonio A, Diaz-Peredo E, Sanchez-Solano F, Peralta-Gil M, Garcia-Alonso D, Jimenez-Jacinto V, Santos-Zavaleta A, Bonavides-Martinez C, Collado-Vides J. (2004). *Nucleic Acids Res.*, **32**, D303-306.
21. Salgado H, Gama-Castro S, Peralta-Gil M, Diaz-Peredo E, Sanchez-Solano F, Santos-Zavaleta A, Martinez-Flores I, Jimenez-Jacinto V, Bonavides-Martinez C, Segura-Salazar J, Martinez-Antonio A, Collado-Vides J. (2006). *Nucleic Acids Res.*, **34**, D394-D397.
22. Storey, J.D. and Tibshirani, R. (2003). *Proc. Natl Acad. Sci. USA*, **100**, 9440-9445.
23. Sun N, Carroll RJ, Zhao H. (2006). *Proc Natl Acad Sci USA*, **103**, 7988-7993.
24. von Mering, C., Krause, R., Snel, B. et al. (2002). *Nature* **417**, 399-403.
25. Wade JT, Reppas NB, Church GM, Struhl K. (2005). *Genes and Development*, **19**, 2619-2630.
26. Wang, W., Cherry, J.M., Nochomovitz, Y., Jolly, E., Botstein, D. and Li, H. (2005). *Proc. Nat. Acad. Sci. USA,* **102**, 1998-2003.
27. Xiao G and Pan W. (2005). *Journal of Bioinformatics and Computational Biology* **3**, 1371-1389.
28. Xie, Y. (2006) Statistical analysis for microarray data: false discovery rate estimation, statistical testing and integrated analysis. PhD dissertation, University of Minnesota, Minneapolis, MN, USA.
29. Xie, Y., Pan, W., Jeong, K.S., Khodursky, A. (2007). *Statistics in Medicine* **26**, 2258-2275.
30. Yang, Y.H., Dudoit, S., Luu, P., and Speed, T. (2002). *Nucleic Acids Research* **30**, e15.
31. Zhao, H., Wu, B. and Sun, N. (2003). In Goldstein, D.R. (ed.), *Science and Statistics: A Festschrift for Terry Speed*, IMS Lecture Notes-Monograph Series, Vol. 40, p. 259-274.
32. Hong, P., Liu, X.S., Zhou, Q., Lu, X., Liu, J.S. and Wong, W.H. (2005). *Bioinformatics* **21**, 2636-2643.
33. Wang, W., Cherry, J.M., Botstein, D. and Li, H. (2002). *Proc. Nat. Acad. Sci. USA,* **99**, 16893-16898.

AN ANALYSIS OF INFORMATION CONTENT PRESENT IN PROTEIN-DNA INTERACTIONS

CHRIS KAUFFMAN AND GEORGE KARYPIS*

Department of Computer Science, University of Minnesota
117 Pleasant St. SE
Minneapolis, MN 55455, USA
E-mail: {kauffman,karypis}@cs.umn.edu

Understanding the role proteins play in regulating DNA replication is essential to forming a complete picture of how the genome manifests itself. In this work, we examine the feasibility of predicting the residues of a protein essential to binding by analyzing protein-DNA interactions from an information theoretic perspective. Through the lens of mutual information, we explore which properties of protein sequence and structure are most useful in determining binding residues with a particular focus on sequence features. We find that the quantity of information carried in most features is small with respect to DNA-contacting residues, the bulk being provided by sequence features along with a select few structural features. Supplemental information for this article is available at `http://www.cs.umn.edu/~kauffman/supplements/psb2008`

1. Introduction

Complex behaviors of the genome are now beginning to be understood in terms of feedback network models in which regulatory elements promote or inhibit transcription of genes and are themselves affected by the transcription of other elements. Key to this system are interactions between DNA, the main storage unit for genetic information, and proteins, which are both products and managers of transcription. To that end, a plethora of computational methods have been presented to predict which proteins will bind to DNA[1,15], what parts of a protein will bind to DNA[2,11,17,18], and which segments of DNA a protein will favor for binding. These methods have yet to reach a performance plateau and researchers continue to apply machine learning and statistical techniques in an attempt reach the

*Work supported by the NIH Training for Future Biotechnology Development grant, NIH T32GM008347

highest accuracy and sensitivity supported by available information.

We endeavor in this study to provide some insight into the inherent difficulty of predicting protein-DNA interactions. From a thermodynamic perspective, the interactions have been found to be quite sensitive. Binding is marginally favored when considering the whole complex[7]. This leaves very little in the way of individual contributions for each residue requiring methods that predict binding residues to make shrewd use of any available features to achieve accuracy. Predicting binding residues would benefit genome studies as mutating them to less favorable analogues gives a mechanism to affect a protein's role in the system. In particular, prediction of binding residues from sequence alone is desirable as it would open the door to a wide variety of experiments involving transcription regulatory elements which have not been co-crystallized with DNA and for which CHIP-Chip experiments[10] are not feasible.

In this paper we focus on sequence and structure features of single protein residues and how they may describe a residue's contributions to the DNA-binding event. We lay out an information theoretic framework in which to conduct the study, illustrate the features of interest, and report the most likely candidates for use in prediction methods.

2. Methods and Materials

2.1. *Mutual Information (MI)*

The main tool we employ for analysis is mutual information (MI)[5,14]. The MI between two random variables is a measure of how easily the value of one may be predicted given the other's value. That is, mutual information measures how much information two variables carry about one another. In the discrete case, it is defined for random variables X and Y as

$$I(X;Y) = \sum_{x,y} p(x,y) \log_2 \frac{p(x,y)}{p(x)p(y)} \tag{1}$$

where x and y are the discrete values or classes which random variables X and Y can take on and $p(x,y)$ is the probability of x and y occurring together. Due to the base-two logarithm, mutual information in this paper is reported in bits.

2.2. *Features*

In our setting, each residue of a protein has associated with it features that are represented by random variables. The first feature considered is always whether the residue is DNA-contacting or not, a binary feature, while the second feature is varied. The MI between the DNA-contacting feature and other features gives us an idea of how informative these other features will be for predicting binding residues.

The features we consider are described in Table 1 and include sequence and structure properties. Only a few of them have a natural discrete definition (such as the 20 amino acids). Solvent accessible surface area (SASA) and information per position (IPP), both single continuous values, were discretized by choosing boundaries to divide the values into bins. These boundaries were chosen by a grid search so that the resulting class definitions maximized mutual information with the DNA contacting classes. Residues were assigned as either DNA contacting or non-contacting based on distance cutoffs which were varied by 0.25 angstroms. The SASA and IPP class boundaries were varied in increments of 0.01 and boundaries that achieved high MI across several DNA-contacting cutoffs were further considered. The values selected for these boundaries are shown in the rightmost column of Table 1.

In order to discretize the remaining vector-valued features we employed clustering techniques. The toolkit CLUTO[9], version 2.1.2, was used with default options to create various numbers of clusters. Each cluster is then one of the discrete values this feature takes on when calculating mutual information. Some experimentation was done using similarity measures other than the default cosine measure, but none yielded a significant change.

A sensible prediction method will employ a variety of features to decide whether a residue contacts DNA. To partially address this, we explore joint features, combinations of two single features, whose values represent every possible combination of the values of the single features. The size of the joint feature is the product of the sizes of the two single features, e.g., amino acids may take on 20 values, secondary structure 3 values, and their joint feature may take on 60 values.

As it is central to the whole study, the definition of DNA binding and non-binding residues is treated with special attention. Distances are calculated between each atom of a residue in a protein and each atom in the DNA structures of each data file. The minimum distance of these is taken as the residue-DNA distance. When computing mutual information, the cutoff distance is varied in increments of 0.2 Å which defines the DNA contacting and non-contacting residues. This allows us to plot a curve for each feature showing characteristics of the signal separating contacting and non-contacting residues. If any combination of feature values does not occur, mutual information becomes undefined. This frequently happens at low

Table 1. Residue Features Considered for Mutual Information with DNA-contacting classes.

Feature	Description	Discrete Values
Amino Acid	Amino acid type of the residue	20 values
Positive, Negative, Neutral Amino Acids	The 20 amino acids divided into 3 classes for their charge. Divisions taken from Cline et al.[4]	Pos: Arg, Lys His Neg: Asp, Glu Neu: All others
Profiles	Combination of the position specific scoring matrix (PSSM) and position specific frequency matrix (PSFM) generated from 3-iterations of PSI-BLAST[3] against the NCBI NR sequence database.	5, 10, and 20 clusters
Concatenated Profiles	A sliding window of size 5 around each residue was used to concatenate the full profiles of adjacent residues. End residues without enough sequence neighbors were assigned 0 in each column of the profile for a missing residue.	5, 10, and 20 clusters
PSSMs	Only the PSSM from the PSI-BLAST profile.	5, 10, 20 clusters
Concatenated PSSMs	Only the PSSMs of residues within a sliding window of size 5 concatenated together.	5, 10, and 20 clusters
Information Per Position (IPP)	The second to last column in PSI-BLAST profiles, gives an account of the sequence diversity in a column of the profile. Low values indicate a strong preference for certain amino acids in that column.	2-value: 0.0-0.62, >0.62 3-value: 0.0-0.48, 0.48-1.0, >1.0 4-value: 0.0-0.48, 0.48-0.81, 0.81-1.27, >1.27
Solvent Accessible Surface Area (SASA)	Surface area of a residue accessible to solvent (water) molecules, normalized based on the maximum SASA of a residue in Gly-X-Gly. Calculated using DSSP[8] and normalized using the values of Miller et al.[13].	2-value: 0.0-0.09, >0.09 3-value: 0.0-0.09, 0.09-0.20, >0.20 4-value: 0.0-0.01, 0.01-0.07, 0.07-0.20, >0.20
Structural Neighbors	Sum of amino acid types within a 14 Å sphere and with sequence distance ≥ 3; distance is between alpha carbons.	5, 10, and 20 clusters
Structural Neighbor PSSMs	Sum of the PSSMs of structural neighbors.	5, 10, and 20 clusters
Secondary Structure	The secondary structure assigned to a residue, by DSSP and mapped into 3 values for helix, strand, and coil	3 values, DSSP letters H,G,I are helix, E is strand, and all others are coil
Physical Quantities	Features of Wang and Brown[17] which are pK_a, a measure of the acidity of side-chains (7 for neutral side-chains), hydropathy according to the scale of Kyte and Doolittle[12], and molecular mass. A sliding window of size 11 around each residue was used to create features which were then used in clustering.	5, 10, and 20 clusters

and high distance cutoff values, especially for features which take on many values. In the plots shown subsequently, undefined MI is set artificially to 0.

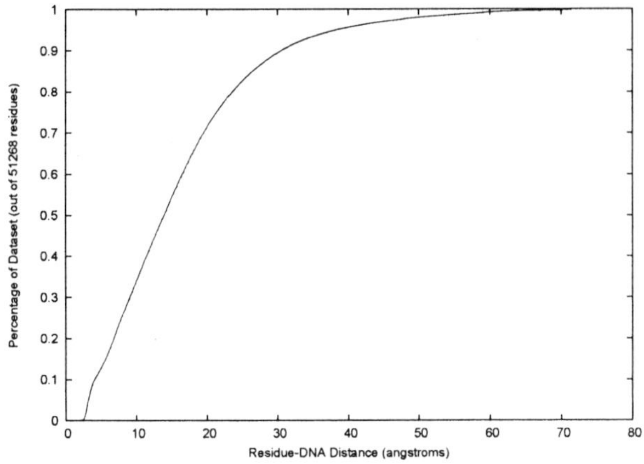

Figure 1. Percentage of Contacting Residues vs. Distance Cutoff.

2.3. *Data Sets*

The data that we employ is derived from that used by Tjong and Zhou[15] with further culling. Beginning with their 264 PDB files, we separated each into protein chains according to the PDB chain identifier. Within protein-DNA co-crystal PDB files, there may exist several chains with identical sequence. This type of duplication may cause an unfair bias in calculating mutual information so the chains were submitted to the PISCES server[16] to be culled to less than 30% sequence identity. The remaining data set comprises 246 chains from 218 different PDB files and includes 51268 residues. Figure 2.3 illustrates the percentage of residues classified as DNA-contacting according to a sliding distance cutoff. The full list of PDB chains used and their associated data is available in the online supplement.

2.4. *Corrections for Small Sample Size*

Calculations of mutual information must be done with care as they may yield an artificially high estimate particularly with small sample sizes. Two approaches taken in the literature to overcome this have been to use bootstrap sampling[6] and to calculate the excess mutual information over a random shuffling of the data[4]. We employ the latter method on single features by leaving the DNA-contacting classes fixed and randomly permuting the values of the second feature. This shuffling preserves the background probabilities of each value of the feature. Calculating mutual information with these shuffled values gives an idea of what MI we can expect to get at random for the background probabilities and number of values for the feature. We compute the average MI over 200 permutations of each feature. Subtracting this quantity led to only a slight drop in MI, about 1% for single features in the worst case. Based on this, we report raw MIs for the rest of the paper.

Joint features pose a problem as they are likely to be more inflated due to the large number of values they take on. We find this difficult to correct as random permutation of class values often leads to zero probability of some combinations and an undefined MI. We report raw values for joint classes here and will attempt to estimate the bias in future works through sampling methods.

3. Results

3.1. *Single Features*

None of the features we explore yield a large magnitude of mutual information with the DNA-binding feature. The most informative features are on the order of hundredths of bits for both single and joint features. This is the same order of magnitude at which previous works have shown contact potentials[4] and aspects of sequence-structure correlations[6] to reside.

For features discretized via clustering, an increased number of clusters leads to an increase in mutual information. In order to give a basis of comparison to the largest natural set of values, amino acids with 20 discrete values, we consider 5, 10, and 20 clusters per feature.

Table 2 summarizes the calculated values for single features while Figure 3.1 illustrates how mutual information for some of the features alters as the distance cutoff defining DNA-contacting residues is altered. The single features yielding the most information on contact vs. non-contact residues are entirely sequence based. Amino acid sequence alone yields a maximum of 0.029 bits at a distance cutoff of 3.37 Å. This is modestly exceeded by PSSMs with 20 clusters (0.032 bits at 4.97 Å cutoff) and profiles (0.032 bits at 4.97 Å cutoff) and is succeeded in information by 10 clusters of profiles (0.027 bits at 4.77 Å cutoff). Using a sliding window of PSSMs or profiles did not improve mutual information: 20 clusters generated using a sliding window of 5 full profiles gives a maximum of 0.020 bits at 5.77 Å while using only the PSSM in clustering yields 0.016 bits at 5.17 Å. Dividing the 20 amino acids into three classes for positive, negative, and neutral residues significantly reduces the information content to a maximum 0.016 bits at 3.57 Å.

Table 2. Mutual Information of Single Features. The mutual information is with the DNA-contacting/non-contacting class (binary) and the distance cutoff is at the maximum MI achieved by the feature. The table is sorted by MI. The column N_{val} is the number of discrete values the feature may take.

Feature	N_{val}	MI	Dist. Cutoff
PSSMs	20	3.1933e-02	4.97
Profiles	20	3.1856e-02	4.97
Amino Acids	20	2.9465e-02	3.37
Profiles	10	2.6765e-02	4.77
Struct. neighbor PSSMs	20	2.6379e-02	10.17
PSSMs	10	2.4402e-02	4.97
Struct neighbors	20	2.2810e-02	8.57
Concat. profiles	20	2.0252e-02	5.77
Struct. neighbor PSSMs	10	1.9237e-02	9.57
PSSMs	5	1.8971e-02	4.97
Struct neighbors	10	1.8597e-02	7.17
Concat. PSSMs	20	1.6257e-02	5.17
Pos/Neg/Neut Amino Acids	3	1.5879e-02	3.57
Solv. Acc. Surf. Area	20	1.5125e-02	3.97
Concat. PSSMs	10	1.4767e-02	4.97
Struct. neighbors	5	1.4166e-02	6.97
Solv. Acc. Surf. Area	4	1.4060e-02	3.77
Concat. profiles	10	1.3289e-02	5.17
Solv. Acc. Surf. Area	2	1.2471e-02	3.97
Info per position	4	1.1519e-02	9.57
Profiles	5	1.1500e-02	3.57
Concat. PSSMs	5	1.1399e-02	4.97
Struct. neighbor PSSMs	5	1.1114e-02	9.57
Info per position	3	1.0934e-02	9.57
Concat. profiles	5	1.0788e-02	5.17
Info per position	2	9.4190e-03	13.97
pK_a/hydropathy/mass	20	3.0624e-03	5.17
pK_a/hydropathy/mass	10	2.7191e-03	5.17
Secondary structure	3	2.4700e-03	5.77
pK_a/hydropathy/mass	5	2.1319e-03	7.17

The lowest information content for single features came from secondary structure assignment (max of 0.002 bits at 5.77 Å) and clusters formed from the combination of pK_a, hydropathy, and molecular mass in sliding window of 11 residues (20 clusters, max of 0.003 bits at 5.17 Å).

3.2. *Joint Features*

The large number of combinations prevents a full discussion of joint features. For brevity, we mention a few interesting cases and include the full numerical results in the online supplement. These cases are summarized in Table 3 and Figure 3. Unsurprisingly, combinations of the most informa-

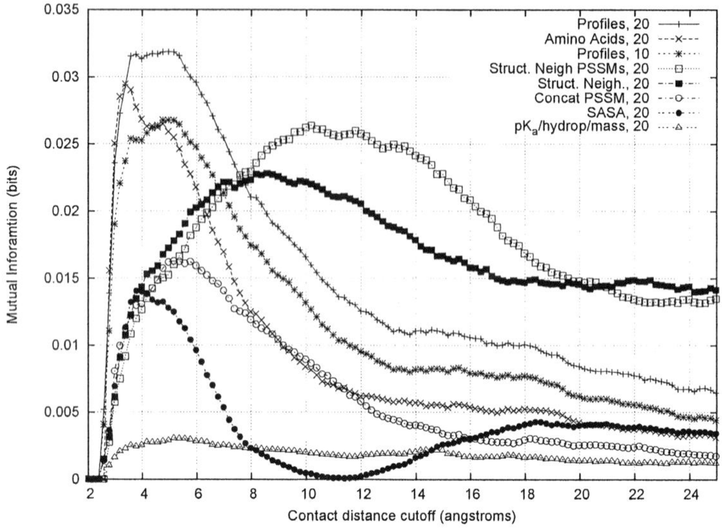

Figure 2. Single Features: Distance Cutoff for DNA-contacting residues versus Mutual Information. The cutoff distance which defines DNA-contacting versus non-contacting residues is varied by small increments to show the character of some single features and their mutual information with the DNA-contacting classes.

Table 3. Selected Mutual Information of Joint Features. The mutual information is with the DNA–contacting/not-contacting class (binary) and the distance cutoff is at the maximum MI achieved by the joint features. N_{val1} and N_{val2} are the number of discrete values features 1 and 2 may take on respectively while N_{tot} is their product, the number of discrete values the joint feature may take.

Feature 1	N_{val1}	Feature 2	N_{val2}	N_{tot}	MI	Dist. Cutoff
PSSMs	20	Struct neighbors	20	400	5.2781e-02	5.77
PSSMs	20	Struct neighbors	10	200	4.7563e-02	5.37
Profiles	20	Struct neighbors	10	200	4.6912e-02	6.57
Profiles	10	Struct neighbors	20	200	4.5558e-02	5.97
PSSMs	20	Info. per pos.	4	80	4.4948e-02	4.97
Profiles	20	SASA	3	60	4.4649e-02	4.17
Amino Acids	20	SASA	4	80	4.0379e-02	3.77
PSSMs	20	SASA	4	80	4.3894e-02	3.97
Amino Acids	20	Info. per position	4	80	4.2580e-02	3.57
Profiles	10	Info. per position	4	40	4.2397e-02	5.37
Profiles	20	Struct. neigh. PSSMs	5	100	3.9513e-02	5.37
Profiles	20	Second. Struct.	3	60	3.6432e-02	4.97
Concat. PSSMs	10	Struct. neigh. PSSMs	20	200	3.3650e-02	6.97
Amino Acids	20	Second. struct.	3	60	3.2341e-02	3.57
Struct neighbors	20	pK$_a$/hydropathy/mass	5	100	2.3224e-02	13.77

Figure 3. Joint Features: Distance Cutoff for DNA-contacting residues versus Mutual Information. The cutoff distance which defines DNA-contacting versus non-contacting residues is varied by small increments to show the character of some joint features and their mutual information with the DNA-contacting classes.

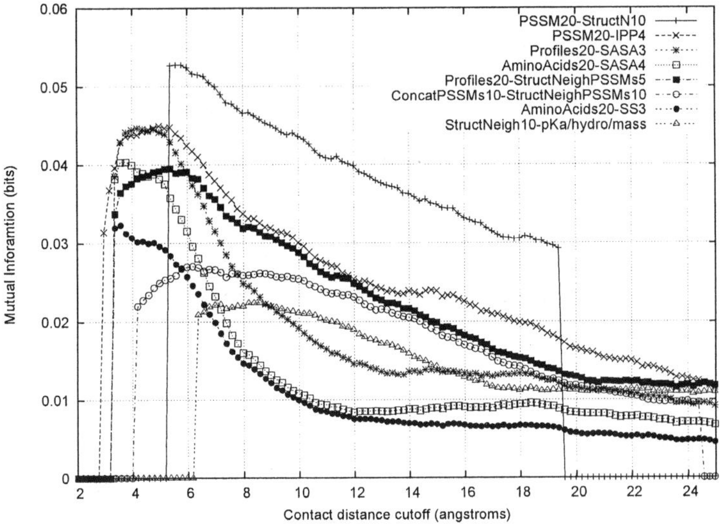

tive single features lead to the highest MIs, the best pairs being PSSMs or profiles with structural neighbors (first rows of Table 3). The next major combination that proved fruitful was between PSSMs, profiles, or sequence with SASA. Combining information per position with sequence or profiles provides the next highest mutual information followed by combinations of profiles or sequence with the PSSMs of structural neighbors. The lower quality single features result mostly in low joint MI, profiles with secondary structure being one exception.

4. Discussion

Most significant among the results are the contributions of sequence based features. Utilizing PSSMs, full profiles, or even simply sequence yields the most information about the differences between residues with high propensities for contacting DNA. It is well known that the negatively charged phosphate backbone of DNA prefers proximity to residues which have a positive charge such as arginine and lysine rather than neutral or positive alternatives. However, limiting the division of amino acids to simply positive, negative, and neutral types severely diminishes MI, giving only 0.016

bits versus 0.029 bits for all amino acid. Counter to intuition, the use of a sliding window with concatenated profiles does not increase MI over the single profile column. The reasons for this are unclear and are worth investigating further. Information per position, when combined with a PSSM, provides a surprisingly informative joint feature. The two together likely amplify the conservation signal present in many DNA contacting residues. With the majority of the information present coming from sequence sources, we can begin to understand why the performance of sequence-based methods such as Ahmad and Sarai[2] have produced prediction results that are nearly as good as those incorporating structure features.

The poor mutual information given by structural features such as SASA and secondary structure class may seem surprising as it is expected that most DNA-contacting residues at least have a high SASA and probably prefer a helix (a common binding motif is helix-turn-helix). However, considering that there are many surface residues with high SASA which do not contact DNA and that helices are a very common secondary structure element, these features are quite noisy. Combining profile information with SASA improves MI significantly, underscoring their reinforcement of one another.

Structural features which do carry information appear to come in the form of the local environment, i.e., descriptions of other residues proximal in space. This is evidenced by the relatively high MI of the structural neighbor feature. Information of this sort is used in a number of DNA-protein prediction methods[1,11,15] and seems to improve performance though not spectacularly. From the standpoint of sequence only predictions, these properties would need to be predicted in order to be used for DNA-contact predictions. Based on the fact they carry a moderate amount of information, there may be some hope that using predicted values would yield improvement.

The physical features of pK_a, hydropathy, and molecular mass did not yield much information and were uniformly lowest both on their own and in combinations. Wang and Brown report quite promising results using support vector machines with only these features[17] indicating that the clustering method used to discretize the feature may not be appropriate. We will explore alternatives in the future to verify that a signal is indeed present in these features as they are some of the easiest to utilize in the protein-DNA interaction prediction.

The literature pertaining to binding residue prediction has defined the binding class using cutoffs in the range of 3.5-5.0 Å. The ideal cutoff distances for both single and joint features seem to support this definition with preference towards the higher end.

5. Conclusion

Armed with the knowledge that signals pertaining to DNA proximity are weak but present, we can understand why prediction methods have enjoyed only marginal success thus far. Incorporating additional features that have not, as of yet, been explored may be the only way to boost performance. From the structure standpoint, this likely involves more complicated geometric information about residues or the consideration of multiple residues interacting with DNA simultaneously. This direction precludes DNA-binding protein with no available structure information. Including features of the DNA being contacted might be the only route as yet unexplored for sequence-only features. Training prediction methods with the knowledge that residues with specific characteristics favor a specific DNA sequence may lead to visible improvements. Approaching the problem from this side will also allow us to incorporate knowledge generated by DNA-binding motif studies.

As for an immediate extension of the present work, we plan to expand the study to account for several shortcomings. Previously mentioned is the issue of properly estimating bias in mutual information for the case of joint features with many values. Sampling techniques and additional compute time are likely to provide the remedy. Also, we have not yet incorporated truly non-contacting residues, only those that are in a DNA-binding protein but far from the interaction site. Adding proteins known not to bind to DNA, especially if they bind to something else such as a small molecule or another protein, will solve this problem and give a better assessment of those characteristics separating DNA-contacting residues from general interaction sites. Finally, the techniques applied here need not be limited to DNA but can also be applied to RNA interactions with proteins.

References

1. Shandar Ahmad, M. Michael Gromiha, and Akinori Sarai. Analysis and prediction of dna-binding proteins and their binding residues based on composition, sequence and structural information. *Bioinformatics*, 20(4):477–486, Mar 2004.
2. Shandar Ahmad and Akinori Sarai. Pssm-based prediction of dna binding sites in proteins. *BMC Bioinformatics*, 6:33, 2005.
3. SF Altschul, TL Madden, AA Schaffer, J Zhang, Z Zhang, W Miller, and DJ Lipman. Gapped blast and psi-blast: a new generation of protein database search programs. *Nucl. Acids Res.*, 25(17):3389–3402, 1997.

4. Melissa S Cline, Kevin Karplus, Richard H Lathrop, Temple F Smith, Robert G Rogers, and David Haussler. Information-theoretic dissection of pairwise contact potentials. *Proteins*, 49(1):7–14, Oct 2002.
5. Thomas M. Cover and Joy A. Thomas. *Elements of Information Theory*. Wiley, 2006.
6. Gavin E. Crooks, Jason Wolfe, and Steven E. Brenner. Measurements of protein sequence-structure correlations. *Proteins: Structure, Function, and Bioinformatics*, 57:804–810, 2004.
7. B. Jayaram, K. McConnell, S. B. Dixit, A. Das, and D. L. Beveridge. Free-energy component analysis of 40 protein-dna complexes: a consensus view on the thermodynamics of binding at the molecular level. *J Comput Chem*, 23(1):1–14, Jan 2002.
8. Wolfgang Kabsch and Chris Sander. Dictionary of protein secondary structure: pattern recognition of hydrogen-bonded and geometrical features. *Biopolymers*, 22:2577–637, 1983.
9. George Karypis. Cluto: A clustering toolkit. Online at `http://www.cs.umn.edu/~karypis/cluto`, 2007.
10. Tae Hoon Kim and Bing Ren. Genome-wide analysis of protein-dna interactions. *Annu Rev Genomics Hum Genet*, 7:81–102, 2006.
11. Igor B. Kuznetsov, Zhenkun Gou, Run Li, and Seungwoo Hwang. Using evolutionary and structural information to predict dna-binding sites on dna-binding proteins. *Proteins*, 64:19–27, 2006.
12. Jack Kyte and Russell F. Doolittle. A simple method for displaying the hydropathic character of a protein. *Journal of Molecular Biology*, 157:105–132, May 1982.
13. Susan Miller, Joel Janin, Arthur M. Lesk, and Cyrus Chothia. Interior and surface of monomeric proteins. *Journal of Molecular Biology*, 196:641–656, Aug 1987.
14. C. E. Shannon. A mathematical theory of communication. *Bell System Technical Journal*, 27:pp. 379–423 and 623–656, 1948.
15. Harianto Tjong and Huan-Xiang Zhou. Displar: an accurate method for predicting dna-binding sites on protein surfaces. *Nucl. Acids Res.*, 35(5):1465–1477, 2007.
16. Guoli Wang and Jr Dunbrack, Roland L. Pisces: recent improvements to a pdb sequence culling server. *Nucl. Acids Res.*, 33:W94–98, 2005.
17. Liangjiang Wang and Susan J Brown. Bindn: a web-based tool for efficient prediction of dna and rna binding sites in amino acid sequences. *Nucleic Acids Res*, 34(Web Server issue):W243–W248, Jul 2006.
18. Changhui Yan, Michael Terribilini, Feihong Wu, Robert L Jernigan, Drena Dobbs, and Vasant Honavar. Predicting dna-binding sites of proteins from amino acid sequence. *BMC Bioinformatics*, 7:262, 2006.

USE OF AN EVOLUTIONARY MODEL TO PROVIDE EVIDENCE FOR A WIDE HETEROGENEITY OF REQUIRED AFFINITIES BETWEEN TRANSCRIPTION FACTORS AND THEIR BINDING SITES IN YEAST

RICHARD W. LUSK

Department of Molecular and Cell Biology, University of California, Berkeley
Berkeley, California 94720, USA
E-mail: lusk@berkeley.edu
www.berkeley.edu

MICHAEL B. EISEN

Genomics Division, Lawrence Berkeley National Laboratory, Department of Molecular and Cell Biology, University of California, Berkeley
Berkeley, California 94720, USA
E-mail: mbeisen@lbl.gov
www.lbl.gov

Keywords: binding sites, evolution, PWM, ChIP-chip, affinity

1. Abstract

The identification of transcription factor binding sites commonly relies on the interpretation of scores generated by a position weight matrix. These scores are presumed to reflect on the affinity of the transcription factor for the bound sequence. In almost all applications, a cutoff score is chosen to distinguish between functional and non-functional binding sites. This cutoff is generally based on statistical rather than biological criteria. Furthermore, given the variety of transcription factors, it is unlikely that the use of a common statistical threshold for all transcription factors is appropriate. In order to incorporate biological information into the choice of cutoff score, we developed a simple evolutionary model that assumes that transcription factor binding sites evolve to maintain an affinity greater than some factor-specific threshold. We then compared patterns of substitution in binding sites predicted by this model at different thresholds to patterns of substitution observed at sites bound *in vivo* by transcription factors in *S. cerevisiae.* Assuming that the cutoff value that gives the best fit between the observed and predicted values will optimally distinguish functional and non-functional sites, we discovered substantial heterogeneity for appropriate cutoff values among factors. While commonly used thresholds seem appropriate for many factors, some factors appear to function at cutoffs satisfied commonly in the genome. This evidence was corroborated by local patterns of rate variation for examples of stringent and lenient p-value cutoffs. Our analysis further highlights the necessity of taking a factor-specific approach to binding site identification.

2. Introduction

A gene's expression is governed largely by the differential recruitment of the basal transcription machinery by bound transcription factors.[1,2] In this way, transcription factor binding sites are fundamental components of the regulatory code, and this code's decipherment is partially a problem of recognizing their location and affinity.[3] These are usually determined using position weight matrices, although a number of more recently developed methods are beginning to become adopted.[4] We use position weight matrices here due to their ease of use with evolutionary analysis and their established theoretical ties with biochemistry. A position weight matrix generates a score comprising the log odds of a given subsequence being drawn from a binding site distribution of nucleotide frequencies vs. an analogous background distribution.[5] The score's p-value is used to determine the location of binding sites: subsequence scores above a predetermined cutoff designate that subsequence to be a binding site, and subsequence scores below the cutoff designate the subsequence to be ignored.

The interpretation of regulatory regions is thus dependent on the choice of the p-value cutoff. However, this choice is not straightforward, although it is commonly made to conform to established but biologically arbitrary statistical standards, e.g. $p < .001$. In addition to assuming that this particular p-value is appropriate, the user here also assumes that a single p-value is appropriate for all transcription factors. Being that score shares an approximately monotonic relationship with affinity,[6,7] this implies that the nature of the interaction between different transcription factors and their binding sites is the same. This may not be the case. For example, some transcription factors may require a stronger binding site to compensate for weaker interactions with other transcription machinery, and so a lenient cutoff would be inappropriate. Conversely, the choice of a stringent cutoff could eliminate viable sites of factors that commonly rely on cooperative interactions with other proteins to be recruited to the DNA. A single common standard of significance is a compromise that may not be reasonable.

Ideally, biological information should inform the choice of a p-value and its consequent ramifications in the determination of function. Several recent approaches have well used expression[8] and ChIP-chip[9] data towards understanding binding specificity. Here we take advantage of selective pressure as a third source of information. Tracking selective pressure has the advantage of directly interpreting sequence in terms of its value to the organism in its environment; to a degree, function can be inferred by observing the impact of selection. To this end, we propose a simple selective model of binding site evolution. Selection prevents the fixation of low affinity sites that may not affect expression to a satisfactory level and does not maintain unnecessary high affinity sites. We train the model on the ChIP-chip data available in yeast, and we find evidence for a wide heterogeneity in required binding site affinity between factors. Supporting recent work by Tanay,[10] many factors appear to require only weak affinity for function, and we find some evidence that these may rely on cooperative binding to achieve specificity.

3. Results and Discussion

3.1. *Definition and training of the affinity-threshold model*

In order to use selection as a means to investigate function, a model must be defined to describe how selection acts on functional and non-functional binding site sequence. Our model was created to be the simplest possible for our purposes. We assume that binding sites evolve independently from other sites in their promoter, but that all sites that bind the same factor evolve equivalently. We interpret a binding site's function in a binary manner: our model supposes that there exists a satisfactory level of expression and that binding site polymorphisms that are able to drive this expression level or greater have equal fitness, while binding site polymorphisms that cannot are deleterious. By assuming that this deleterious effect is large enough to preclude fixation in *S. cerevisiae*, our model imposes an effective threshold on permitted affinity: it does not allow a substitution to occur if it drops the position weight matrix score beneath a given boundary. Analogous reasoning lets us treat repressors identically. By imposing a threshold on permitted affinity and by relying on the assumption that position weight matrix score shares a monotonic relationship with affinity,[6] we impose a threshold weight matrix score.

Our purpose in training the model is to find where that threshold lies for each factor, which we accomplish using simulation. For any given threshold and matrix, we simulate the relative rates of substitution that would be expected, and then we compare these rates to empirically determined rates to choose the most appropriate threshold. The simulation is run as follows: we start with the matrix's consensus sequence, and make one mutation according to the neutral HKY[11] model. The sequence's score is evaluated: if it exceeds the threshold, the mutation is considered fixed and the count of substitutions at that position is incremented, and if not, no increment is made and the sequence reverts back to the original sequence. This mutate-select process is repeated. Assuming that the impact of polymorphism is negligible, removing a given fraction of mutations by selection will reduce the substitution rate by that fraction. Thus, the proportion of accepted over total mutations at each position is evaluated to be the rate of mutation relative to the neutral rate.

We use sum-of-squares as a distance metric to compare each affinity-threshold rate distribution to the empirical distribution, and we considered the best-fitting affinity threshold to be the affinity threshold that generates the distribution with the smallest distance to the empirical relative rates.

3.2. *The affinity-threshold model well describes binding site substitution rates*

The Halpern-Bruno model[12] has been incorporated into effective tools for motif discovery[13] and identification,[14] and it has been shown to well describe yeast binding site relative rates of substitution.[15] These rates are also generated by our model, and so we judged our model's accuracy by comparing its performance to the Halpern-Bruno model's performance (fig. 1). We aligned ChIP-chip bound regions and computed summed position-specific rates of substitution for the aggregate binding sites of the 111 transcription factors that met our conservation requirements. We were able to find a threshold at which the affinity-threshold model better resembled the empirical data than the Halpern-Bruno model did for 42 of the 49 factors with adequate training data (see Methods). The affinity-threshold model well approximates the position-specific substitution rates of most factors.

The best-fitting score threshold for a transcription factor's binding sites may correspond to their minimum non-deleterious affinity for that transcription factor. If this minimum is variable and can be found through our evolutionary analysis, then we should be able to detect that variability robustly. To this end, we used a bootstrap to assess the reliability of our

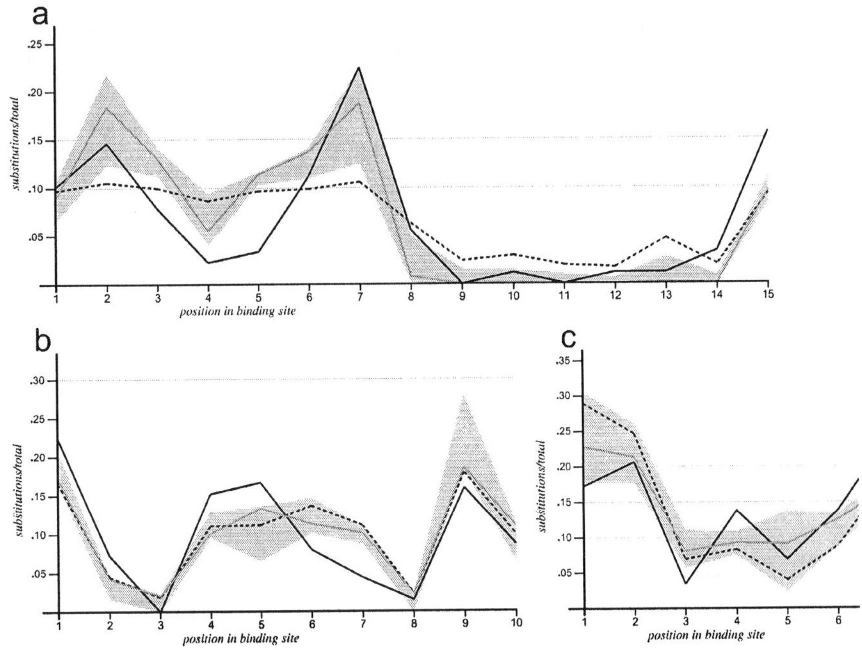

Fig. 1. Position specific rate variation and model predictions for (*a*) Fkh2, (*b*) Fhl1, and (*c*) Aft2: relative rate ($subst/subst_{tot}$) vs position in site. The black line marks the empirical rates, the dashed line marks the Halpern-Bruno predicted rates, and grey line marks the best-fitting affinity-threshold. The grey bar contains the set of rates predicted by all affinity thresholds within the factor's 95% confidence interval

predictions, resampling the the aligned sites. Although most transcription factors had large confidence intervals, they were dispersed over sufficiently wide intervals such that we could form three distinct sets (table 1). We grouped factors with lower bounds greater than 5.9 into a "stringent threshold" set, factors with upper bounds lower than 5.1 into a "lenient threshold" set, and factors with upper bounds lower than 12 and lower bounds greater than -2 into a "medium threshold" set; transcription factors appear to have variable site affinity requirements. We use these sets in all further analysis.

3.3. *The affinity-threshold model predicts extant score distributions for most factors*

If the affinity-threshold model is a reasonable approximation of the evolution of the system, then it should describe other properties of the system beyond the position-specific rate variation of binding sites. One additional prediction of the model is the distribution of binding site scores. For each

Table 1. Affinity threshold confidence intervals and corresponding site prevalence for transcription factors in the stringent (left), medium (middle), and lenient (right) threshold groups

	CI^a	$Prev.^b$		CI^a	$Prev.^b$		CI^a	$Prev.^b$
Reb1p	8.3-11.1	.226-.117	Cin5p	−0.4-8.5	.997-.294	Sut1p	−9.9-4.2	.988-.845
Bas1p	5.8-13.6	.566-.005	Mbp1p	2.7-11.7	.793-.059	Aft2p	−9.8-4.2	.988-.794
Fkh2p	8.1-15.2	.497-.003	Fhl1p	4.2-11.3	.702-.048	Phd1p	−9.8-5.1	.998-.867
Cbf1p	6.2-12.0	.219-.028	Gcn4p	4.0-10.6	.682-.080	Ace2p	−9.9-−0.8	.999-.999
Abf1p	11.0-12.9	.108-.075	Swi6p	3.8-9.9	.854-.166	Yap6p	−9.9-4.2	.993-.909
Sum1p	6.2-14.5	.484-.009	Ste12p	1.0-6.5	.997-.705	Adr1p	−9.5-2.3	.991-.856
Tye7p	8.6-11.3	.183-.037	Nrg1p	−1.3-7.0	.968-.388	Hap5p	−9.4-−2.1	.993-.993
Mcm1p	8.7-19.5	.133-.002				Mot3p	−2.9-5.1	.996-.595
Hap4p	11.0-14.9	.059-.003						

Note: [a] 95% confidence interval, log base two scores
[b] Prevalence: first and second quantities are the fraction of all promoters containing a site meeting the lower and upper bounds of the CI, respectively

factor in the groups determined above we sampled the Markov chain and computed the mean binding site score under the affinity-threshold model. We compared this to the average maximum score for that transcription factor in ChIP-chip bound regions (fig. 2). Although it had a downward bias, the affinity-threshold model predicted the extant distribution of stringent- and medium- threshold transcription factor binding sites. However, it fared worse with the lenient-threshold binding sites, suggesting that the evolution of these sites may not operate within the simplifying bounds of the model, i.e. perhaps their evolution is governed by a more complex fitness landscape instead of our stepwise plateau. Nevertheless, average maximum scores in bound regions for these factors are still found commonly in the genome.

3.4. *Stringent- and lenient-threshold binding sites have distinct patterns of local evolution*

The lenient set of transcription factors allows for binding sites that would be found often by chance in the genome. If this lenient affinity is truly sufficient, these transcription factors may rely on other bound proteins to separate desired from undesired binding sites. In contrast, sites meeting the affinity threshold for stringent-threshold transcription factors should be high-occupancy sites without a need for additional information due to their strong predicted affinity.

To investigate this hypothesis, we counted the average number of different transcription factors bound at each promoter for each of the factors used in the Harbison et al ChIP-chip experiments. Let "lenient-group sites"

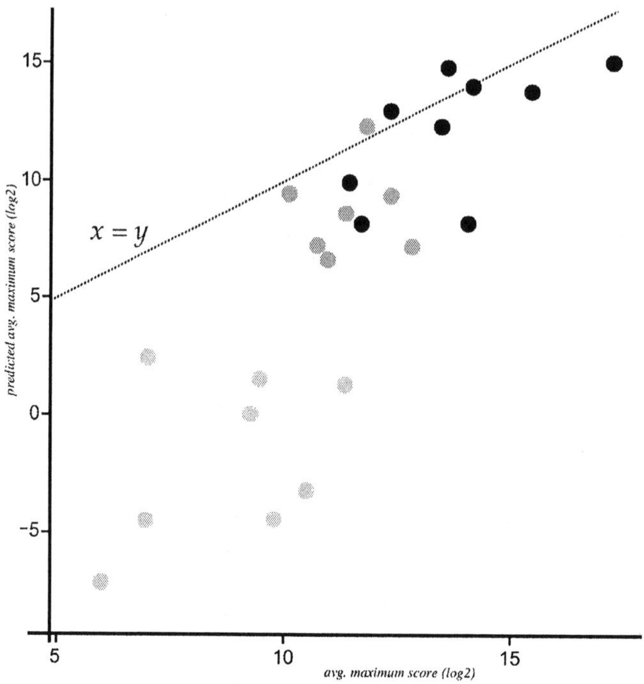

Fig. 2. Predicted average score at best-fitting affinity threshold vs. average maximum score in ChIP-chip bound regions (log base two scores). Stringent-, medium-, and lenient-threshold transcription factors presented as black, dark grey, and light grey dots, respectively.

refer to sites bound by lenient-threshold transcription factors (e.g. Sut1p, table 1), and let "medium-group" and "stringent group" sites be defined similarly. As expected, the stringent and lenient groups were separated, the lenient group promoters having just under three more unique bound factors per promoter for each of three binding significance cutoffs. However, the medium and lenient groups were not well separated.

We used the variation in local substitution patterns to determine whether medium and lenient group factors could be distinguished by an enrichment of local binding events. While medium and lenient group sites have similar numbers of different transcription factors bound to promoters that they also bind, lenient group sites will have a higher density of other binding sites immediately surrounding theirs if recruitment by other proteins is necessary for their function. This density should be reflected in the local pattern of evolution, as the sequence will be comparatively restrained.

We calculated rates of substitution surrounding the binding sites of

Table 2. Average number of binding sites per promoter, grouped by best-fit affinity threshold and ChIP-chip binding p-value

Group	$p < X$		
	.005	.001	.0001
Stringent	7.78	4.74	3.33
Medium	10.30	7.09	5.13
Lenient	10.73	7.59	6.25

stringent-, medium-, and lenient-threshold transcription factors. All transcription factors in each set were pooled and the rate of substitution was calculated and summed by distance to the transcription factor edge. All three sets have a reduced rate of substitution at the position adjacent to the binding site (fig. 3a), suggesting that some of these weight matrices do not describe the entire factor. Lenient group sites have a depressed rate of substitution relative to the areas surrounding the medium and stringent group sites (fig. 3b, $p \cong 0, \chi^2 = 160.8$, 1df), consistent with a hypothesis of increased local binding. In contrast, the regions surrounding stringent group sites are marked by a shoulder of increased substitution rate (fig. 3a). This shoulder suggests a model in which high-affinity sites sterically inhibit transcription factors from binding to adjacent regions, preventing them from being used as regulatory material. The stringent and lenient group sites are distinguished by their expected patterns of local substitution rate variation.

Transcription factors may best interact if they are on the same side of the DNA,[16–18] suggesting that binding sites of interacting factors should be phased at approximately 10.4 base pairs to match the periodicity of the double helix, although this will vary according to the particular nature of interaction between the two proteins. If binding sites coordinated in this manner, the substitution rate should match this periodicity. We evaluated the fit of a model that allowed for a 10.4 base pair periodicity in the rate, although the noted variability between interacting factors will reduce the quality of this match. We fit the twenty base region ten bases from the edge of the transcription factor, allowing for two turns of the DNA while avoiding possible occluding effects of the original bound factor. The regions local to lenient group sites fit this model significantly better than they fit a uniform rate model (fig. 3c, $p = .0053, \chi^2 = 10.53$, 2df), while the regions surrounding medium and stringent group sites did not.

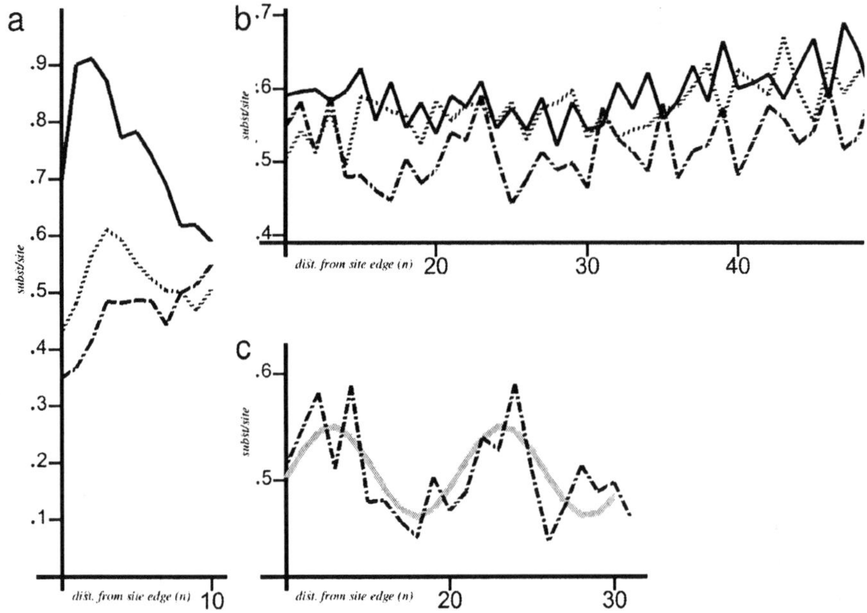

Fig. 3. Local rate of substitution (*subst/site*) vs distance to binding site edge (bp). The solid, dotted, and dot-dashed lines mark the local rates surrounding stringent-, medium-, and lenient-affinity group transcription factor binding sites. In (c), the grey line marks the predicted periodic rate of evolution near lenient-affinity group sites

4. Conclusion

We developed a simple model of binding site evolution to investigate the possibility of differences in transcriptions factors' requirements for binding site affinity. Unlike other models of binding site evolution, the affinity-threshold model is geared toward understanding the transcription factor itself rather than its binding sites. The model was used to create three groups of transcription factors with stringent, lenient, and intermediate requirements for binding site affinity, and these groups were supported by the extant distribution of binding sites and their distinctive patterns of localized substitution rate. We note that some factors appear to evolve and exist at thresholds that poorly distinguish their binding sites from background sequence, perhaps making consideration of context essential for their accurate identification.

5. Methods

5.1. *Rate of binding site evolution*

We downloaded the *S. cerevisiae* sequences used in the Harbison et al[19] study and used bi-directional best FASTA[20] hits ($p < 1e^{-5}$) to find the orthologous subsequences in *S. paradoxus*, *S. mikatae*, *S. kudriavzevii*, and *S. bayanus* contigs available at SGD.[21] We aligned the sequences using Mlagan.[22]

We obtained ChIP-chip binding data from Harbison et al, using all available conditions for each factor. We used a binding p-value cutoff of .001 to determine binding, but the analysis was fairly robust to using different cutoffs: we also calculated rates of evolution for of transcription factor binding sites for binding p-values of .005 and .0001 and observed similar groups, although some stringent-threshold factors were lowered to the medium-threshold group using the former data set. We downloaded weight matrices for 124 factors,[9] and we used Patser[23] to designate the highest-scoring subsequence(s) within each bound locus to be the subsequence responsible for binding. This choice precludes the inclusion of many functional weak sites, but we wished to minimize the impact of non-functional sites. Alignment errors, binding site turnover, and changes in cis-regulation all will introduce neutral sequence evolution into the model training data, biasing our choice of threshold downward. In particular, Borneman et al[24] highlighted rapid changes in binding for two transcription factors across three yeast species. We hoped to minimize the impact of such by imposing minimal criteria for conservation: we discarded alignments with gaps and alignments containing a sequence with a score beneath zero. We used maximum parsimony for all determinations of substitution rate. Although progress has been made towards determining the neutral mutation processes in *S. cerevisiae* intergenic sequence,[25] we wished to avoid remaining uncertainties and so in all cases we compared relative rates within the binding site instead of absolute rates. We did not further analyze transcription factors for which we were unable to train on at least two mutations per position. We calculated the Halpern-Bruno rates according to the method described in Moses et al.[15]

5.2. *Simulation of the affinity-threshold model*

We simulated the affinity-threshold model for a wide range of thresholds for each of the 124 weight matrices described by MacIsaac et al. We calculated position-specific substitution rates for score thresholds between -10 and the position weight matrix's maximum in increments of 0.1. This process starts with the consensus sequence and is run for eighteen million iterations. We determined 95% bootstrap confidence intervals of the best-fitting threshold by finding the best-fitting affinity threshold for each of 10,000 resamples of the aligned binding sites. Software will be available from http://rana.lbl.gov/~rlusk/PSB2008/.

5.3. *Predicted equilibrium distribution of scores*

We sampled every 20,000th sequence generated by the Markov chain for the best-fitting affinity threshold model for each transcription factor in the three groups. We compared the mean score of these sequences with the mean maximum score of the sequences meeting a $p < .001$ ChIP-chip binding cutoff.

5.4. *Periodicity testing*

We evaluated two nested models against the $\pm 10 - 30$ base pair region surrounding each binding site. The first supposed a uniform rate α across the region to determine k_p Poisson-distributed mutation events at each position p, and the second added a periodicity of 10.4 to this rate with magnitude β and phase γ. t_p is the number of gapless alignment columns at that position. The maximum likelihood parameters were discovered by direct search.

$$L(k \mid \alpha, \beta, \gamma; t) = \prod_{p=10}^{30} \frac{e^{-f(\alpha,\beta,\gamma)t_p} \left[f(\alpha, \beta, \gamma) t_p \right]^{k_p}}{k_p!}$$

$$f(\alpha, \beta, \gamma) = \left[1 + \beta \sin(2\pi \frac{p - \gamma}{10.4}) \right] \alpha$$

Significance was determined using a likelihood ratio test with β either allowed to fluctuate between zero and one or held to zero.

This work was supported by a National Institutes Grant R01-HG002779 to MBE. RWL was supported by a NSF graduate research fellowship. This work was supported by the Director, Office of Science, Office of Basic Energy Sciences, and the Assistant Secretary for Energy Efficiency and Renewable Energy, Office of Building Technology, State, and Community Programs, of the U.S. Department of Energy under Contract No. DE-AC02-05CH11231.

References

1. M. Levine and R. Tjian, *Nature* **424**, 147(July 2003).

2. T. I. Lee and R. A. Young, *Annu Rev Genet* **34**, 77 (2000).
3. M. L. Bulyk, *Genome Biol* **5** (2003).
4. E. Sharon and E. Segal, A feature-based approach to modeling protein-dna interactions, in *RECOMB 2007*, eds. T. Speed and H. Huang (Springer-Verlag, Berlin Heidelberg).
5. G. D. Stormo, *Bioinformatics* **16**, 16(January 2000).
6. O. G. Berg and P. H. von Hippel, *J Mol Biol* **193**, 723(February 1987).
7. J. M. Heumann, A. S. Lapedes and G. D. Stormo, *Proc Int Conf Intell Syst Mol Biol* **2**, 188 (1994).
8. E. Segal, Y. Barash, I. Simon, N. Friedman and D. Koller, From promoter sequence to expression: a probabilistic framework, in *RECOMB 2002*, eds. S. Istrail, M. S. Waterman and A. G. Clark
9. K. D. Macisaac, T. Wang, B. D. Gordon, D. K. Gifford, G. D. Stormo and E. Fraenkel, *BMC Bioinformatics* **7**(March 2006).
10. A. Tanay, *Genome Res* (June 2006).
11. M. Hasegawa, H. Kishino and T. Yano, *J Mol Evol* **22**, 160 (1985).
12. A. L. Halpern and W. J. Bruno, *Mol Biol Evol* **15**, 910(July 1998).
13. A. M. Moses, D. Y. Chiang, D. A. Pollard, V. N. Iyer and M. B. Eisen, *Genome Biol* **5** (2004).
14. A. M. Moses, D. Y. Chiang and M. B. Eisen, *Pac Symp Biocomput* , 324 (2004).
15. A. M. Moses, D. Y. Chiang, M. Kellis, E. S. Lander and M. B. Eisen, *BMC Evol Biol* **3**(August 2003).
16. J. Boros, F. L. Lim, Z. Darieva, A. Pic-Taylor, R. Harman, B. A. Morgan and A. D. Sharrocks, *Nucleic Acids Res* **31**, 2279(May 2003).
17. C. Mao, N. G. Carlson and J. W. Little, *J Mol Biol* **235**, 532(January 1994).
18. I. Ioshikhes, E. N. Trifonov and M. Q. Zhang, *Proc Natl Acad Sci U S A* **96**, 2891(March 1999).
19. C. T. Harbison, B. D. Gordon, T. I. Lee, N. J. Rinaldi, K. D. Macisaac, T. W. Danford, N. M. Hannett, J.-B. Tagne, D. B. Reynolds, J. Yoo, E. G. Jennings, J. Zeitlinger, D. K. Pokholok, M. Kellis, A. P. Rolfe, K. T. Takusagawa, E. S. Lander, D. K. Gifford, E. Fraenkel and R. A. Young, *Nature* **431**, 99 (2004).
20. D. J. Lipman and W. R. Pearson, *Science* **227**, 1435(March 1985).
21. J. M. Cherry, C. Adler, C. Ball, S. A. Chervitz, S. S. Dwight, E. T. Hester, Y. Jia, G. Juvik, T. Roe, M. Schroeder, S. Weng and D. Botstein, *Nucleic Acids Res* **26**, 73(January 1998).
22. M. Brudno, C. B. Do, G. M. Cooper, M. F. Kim, E. Davydov, E. D. Green, A. Sidow and S. a. Batzoglou, *Genome Res* **13**, 721(April 2003).
23. G. Hertz and G. Stormo, *Bioinformatics* **15**, 563(July 1999).
24. A. R. Borneman, T. A. Gianoulis, Z. D. Zhang, H. Yu, J. Rozowsky, M. R. Seringhaus, L. Y. Wang, M. Gerstein and M. Snyder, *Science* **317**, 815(August 2007).
25. C. S. Chin, J. H. Chuang and H. Li, *Genome Res* **15**, 205(February 2005).

STRIKING SIMILARITIES IN DIVERSE TELOMERASE PROTEINS REVEALED BY COMBINING STRUCTURE PREDICTION AND MACHINE LEARNING APPROACHES

JAE-HYUNG LEE[1,2†], MICHAEL HAMILTON[5], COLIN GLEESON[6], CORNELIA CARAGEA[3,4], PETER ZABACK[1,2], JEFFRY D. SANDER[1,2], XUE LI[1], FEIHONG WU[1,3,4], MICHAEL TERRIBILINI[1,2], VASANT HONAVAR[1,3,4], DRENA DOBBS[1,2,4]

[1]Bioinformatics & Computational Biology Program, L.H. Baker Center for Bioinformatics & Biological Statistics, [2]Dept. of Genetics, Development & Cell Biology, [3]Dept. of Computer Science, [4]Artificial Intelligence Research Lab & Center for Computational Intelligence, Learning & Discovery, Iowa State University, Ames, IA, 50010, USA

[5]Dept. of Computer Science, Colorado State University, Fort Collins, CO 80523, USA

[6]Dept. of Biological Sciences, Univ. of Illinois, Chicago, IL, 60607, USA

Telomerase is a ribonucleoprotein enzyme that adds telomeric DNA repeat sequences to the ends of linear chromosomes. The enzyme plays pivotal roles in cellular senescence and aging, and because it provides a telomere maintenance mechanism for ~90% of human cancers, it is a promising target for cancer therapy. Despite its importance, a high-resolution structure of the telomerase enzyme has been elusive, although a crystal structure of an N-terminal domain (TEN) of the telomerase reverse transcriptase subunit (TERT) from *Tetrahymena* has been reported. In this study, we used a comparative strategy, in which sequence-based machine learning approaches were integrated with computational structural modeling, to explore the potential conservation of structural and functional features of TERT in phylogenetically diverse species. We generated structural models of the N-terminal domains from human and yeast TERT using a combination of threading and homology modeling with the *Tetrahymena* TEN structure as a template. Comparative analysis of predicted and experimentally verified DNA and RNA binding residues, in the context of these structures, revealed significant similarities in nucleic acid binding surfaces of *Tetrahymena* and human TEN domains. In addition, the combined evidence from machine learning and structural modeling identified several specific amino acids that are likely to play a role in binding DNA or RNA, but for which no experimental evidence is currently available.

1. Introduction

In most eukaryotes, a remarkable ribonucleoprotein enzyme, telomerase, is responsible for the synthesis and maintenance of telomeres, the ends of linear chromosomes [1, 2, 3]. Many exciting discoveries have been made in telomerase biology since 1984, when the enzyme was first identified in the ciliate,

† Corresponding author

Tetrahymena thermophila, by Greider and Blackburn [4]. Recently, pivotal roles for telomerase in signaling pathways that regulate cancer, stress response, apoptosis and aging have been demonstrated [5, 6, 7, 8].

Two essential roles of telomeres are protecting or "capping" chromosome ends and facilitating their complete replication (reviewed in 1, 2, 3). Typically, telomeres consist of arrays of simple DNA sequence repeats, ranging from ~50 copies of 5'-TTGGGG-3' in *Tetrahymena,* to ~1000 copies of 5'-TTAGGG-3' in humans and other vertebrates. The sequence of telomeric repeats is specified by an RNA template (TER), which varies in length from ~160 nts in ciliates to ~1500 nts in vertebrates, and is an essential component of the catalytically active form of telomerase [2, 5]. Human telomerase is composed of hTER and two bound proteins, the telomerase reverse transcriptase component (hTERT) and dyskerin [9]. The regulation of telomerase activity involves interactions with a variety of other cellular proteins, many of which are essential for telomere homeostasis [8, 10].

Telomerase is a promising target for cancer therapy because it is generally present in very low levels in normal somatic cells, but it is highly active in many human malignancies [11]. Telomerase targeting strategies have included short interfering RNA (siRNA) knockdown of endogenous hTER and a combination of siRNA and expression of mutant forms of the hTER RNA, which become incorporated into the enzyme and inhibit proliferation in variety of different human cancer cell lines [11].

Despite its obvious clinical importance, currently there are no experimentally determined structures for the telomerase ribonucleoprotein complex or for telomerase complexes bound to telomeric DNA substrates, presumably because these are multisubunit structures. The telomerase reverse transcriptase component, TERT, is generally thought to consist of four functional domains (see Figure 1): the essential N-terminal (TEN) domain, an RNA-binding domain (TRBD), reverse transcriptase (RT), and a C-terminal extension (TEC). Recently, a crystal structure of the essential N-terminal domain of TERT from *Tetrahymena* has been reported [12] and appears to represent a novel protein fold. Several conserved sequence motifs have been identified within the TEN domain on the basis of multiple sequence alignments and mutagenesis experiments [13, 14]. In addition, experiments directed at mapping DNA and RNA binding sites within TERTs from several organisms have identified specific amino acids that appear to contact either the DNA template or the RNA component [reviewed in 3]. In human telomerase, the TEN domain binds both DNA, specifically interacting with telomeric DNA substrates, and RNA, apparently binding in a non-sequence specific manner [12].

A.

B.

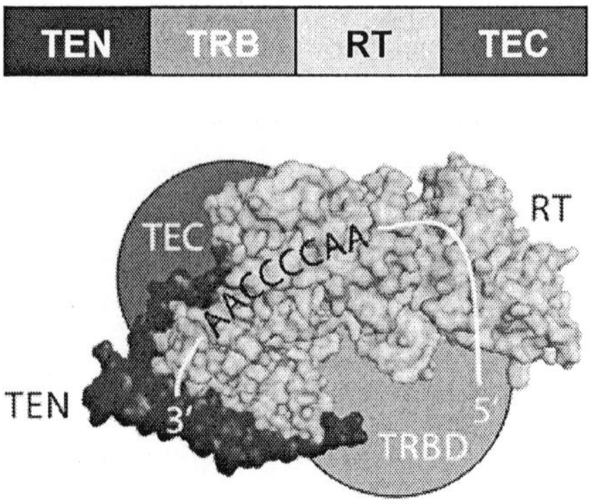

Figure 1. TERT domain architecture. A) The telomerase reverse transcriptase (TERT) comprises 4 functional domains: essential N-terminal (TEN) domain, RNA-binding domain (TRBD), reverse transcriptase (RT), and C-terminal extension (TEC). **B)** Cartoon illustrating TERT domain organization, and the RNA template (TER). The TEN domain is *Tetrahymena* structure (PDB ID: 2B2A), and RT domain is from HIV-RT (PDB ID: 3HVT). Figure modeled after Collins, 2006 [2].

Although vertebrate TEN domain sequences share a high degree of sequence similarity, the TEN domains from more diverse species share very little sequence similarity (<30% identity), suggesting that a homology modeling approach to predicting the structure of the human TEN domain would be difficult. However, an alignment of the N-terminal sequences of TERTs from organisms ranging from human to *T. thermophila* to *S. cerevisiae*, revealed several highly conserved residues distributed throughout the N-terminal domain, suggesting that TEN domains from diverse organisms may share similar architectures [12]. Based on this suggestion, we set out to test the hypothesis that the N-terminal domains of TERTs in diverse organisms not only share a similar overall three-dimensional fold, but may also have phylogenetically conserved DNA and RNA binding surfaces. We used a strategy in which comparative protein structural modeling approaches were integrated with sequence-based machine learning approaches for predicting DNA or RNA binding residues.

2. Datasets, Materials and Methods

2.1 Datasets

RNA-protein interface dataset

A dataset of protein–RNA interfaces was extracted from structures of known protein–RNA complexes in the Protein Data Bank (PDB) [15] solved by X-ray crystallography. Proteins with >30% sequence identity or structures with resolution worse than 3.5 Å were removed using PISCES [16]. The resulting dataset, RB147 [36], contains 147 non-redundant polypeptide chains. RNA-binding residues were identified according to a distance-based cutoff definition: an RNA-binding residue is an amino acid containing at least one atom within 5 Å of any atom in the bound RNA. RB147 contains a total of 6157 RNA-binding residues and 26,167 non-binding residues. The RB147 dataset [36] is larger than the RB109 dataset used in our previous studies [17, 18].

DNA-protein interface dataset

A dataset of protein-DNA interfaces was extracted from structures of known protein-DNA complexes in the PDB [15]. Proteins with >30% sequence identity or structures with resolution worse than 3.0 Å and R factor > 0.3 were removed using PISCES [16]. The resulting dataset, DB208, contains 208 polypeptide chains, each at least 40 amino acids in length. DNA-binding residues were identified according to a definition based on reduction in solvent accessible surface area (ASA): an amino acid is a DNA-binding residue if its ASA computed in the protein-DNA complex using NACCESS [19] is less than its ASA in the unbound protein by at least 1 $Å^2$ [20]. DB208 contains a total of 5,721 interface residues and 39,815 non-interface residues. The DB208 dataset is larger than the DB171 dataset used in our previous studies [21].

2.2 Algorithms for predicting interfacial residues

We used sequenced-based Naïve Bayes classifiers [22, 23] for predicting protein-RNA interfaces [17, 18] and protein-DNA interfaces [21]. Briefly, the input to the classifier is a contiguous window of 2n+1 amino acid residues consisting of the target residue and n sequence neighbors to the left and right of the target residue, obtained from the protein sequence using the "sliding window" approach. The output of the classifier is a probability that the target residue is an interface residue given the identity of the 2n+1 amino acids in the input to the classifier. With Naïve Bayes classifiers, it is possible to tradeoff the rate of true positive predictions against the rate of false positive predictions, by using a classification threshold, θ, on the output probability of the classifier. The target residue is predicted to be an interface residue if its probability returned by the classifier is greater than θ, and a non-interface residue otherwise. The length of the window was set to 21 in the experiments described here.

We used the implementation of the Naive Bayes classifier available in WEKA, an open source machine learning package [23] for training classifiers used to predict interface residues in this study. The performance of the protein-RNA interface predictor trained on RB147 dataset (***RNABindR***, *http://bindr.gdcb.iastate.edu/RNABindR/*), and estimated using leave-one-out sequence-based cross-validation, is documented in [36]. The performance of protein-DNA interface predictor trained on the DB208 dataset (***DNABindR***, *http://cild.iastate.edu/DNABindR*) and estimated using 10-fold sequence-based cross-validation, is comparable to that of the previously published protein-DNA interface predictor, which was trained on the DB171 dataset [21]. The RNA interface predictions on TEN domains were obtained by using Naïve Bayes classifiers trained on the RB147 dataset (high specificity setting of ***RNAbindR***). The DNA interface predictions were obtained by ***DNABindR*** (θ=0.168) trained on the DB208 dataset.

2.3 Structural modeling of telomerase TEN domains in human and yeast

The N-terminal domains from human telomerase (GENBANK NP_937986) and yeast telomerase (GENBANK NP_013422) sequences, were threaded onto the *T. thermophila* telomerase N-terminal domain (TEN) structure (PDB: 2b2a chain A) using FUGUE [24]. The output alignments were used for generating 3D coordinates for the N-terminal domains of human and yeast telomerase by MODELLER [25]. Among 15 generated models, the highest ranking model was chosen and refined using SCWRL [26] to reposition side-chains. Energy minimization was performed by 400 steps of steepest descent using the GROMOS96 force field [27] with a 9Å non-bonded cutoff in the Deep View/Swiss PDB-viewer [28]. One human TEN model was based on the *Tetrahymena* TEN structure in the PDB: 2b2aA, N-terminal domain of tTERT. For a second model, several templates were selected using PSI-BLAST [29] and the Swiss-Model HMM template library [30] to detect remote homologs of hTERT. The chosen templates were portions of the following PDB structures: 1imhC, Tonicity-responsive enhancer binding protein (TONEBP)-DNA complex; 1jfiB, Negative Cofactor 2-TATA box binding protein-DNA complex (NC2-TBP-DNA); 2dyrM, bovine heart cytochrome C oxidase; 1b1uA, bifunctional inhibitor of Trypsin and Alpha-amylase from Ragi seeds; 2b2aA, N-terminal domain of tTERT. The templates were aligned and models were generated using the procedure described above. All generated structures were evaluated using the ANOLEA server [34].

2.4 Experimental identification of RNA and DNA binding residues

Experimentally determined DNA and RNA binding sites in hTERT and tTERT were collected by mining relevant literature. Point mutations that affect RNA binding have not been reported, but Moriarty et al. showed that deletions at positions 30-39 and 110-119 in hTERT result in reduced RNA and DNA association, respectively [31, 32]. Conserved primer grip regions have been mapped in the TEN and RT domains of hTERT, between amino acids 137-141 and 930-934 [33]. Alanine substitutions in the C-terminal region of TEN at positions Q168, F178, and W187 have been shown to substantially decrease tTERT association with DNA [12].

3. Results

3.1 Rationale

Computational and bioinformatic analyses can provide valuable insight into protein sequence-structure-function relationships, especially when the structure of a protein or complex is difficult to solve using experimental approaches. Surprisingly, despite the fascinating structural and regulatory complexity of telomerase, its pivotal role in cellular signal pathways, and its critical interactions with DNA, RNA and protein partners, very few studies have exploited bioinformatic or computational structural biology approaches to investigate the structure and function of telomerase. In this work, we use a combination of comparative structural modeling and sequence-based machine learning methods to test the hypothesis that the N-terminal domains of TERTs in diverse organisms share a similar overall architecture and conserved DNA and RNA binding surfaces.

3.2 Sequence-based prediction of RNA and DNA binding sites in human and Tetrahymena TERT

Conserved domains within the telomerase reverse transcriptase protein of human (hTERT) and *Tetrahymena* (tTERT) are illustrated in Figure 2. In previous work, we used a sequence-based machine learning approach to predict RNA binding residues in TERT sequences and showed that our predictions compared favorably with available experimental data [18]. Results of these previously published predictions are included in Figure 2 for comparison with DNA binding residues predicted in the current study (see Materials and Methods). The predicted DNA and RNA binding regions in hTERT and tTERT are indicated by boxes under the middle sections of Figures 2A and B, respectively. The lower portion of each figure shows specific examples, with boxed amino acids representing short deletions (in hTERT) or alanine-substitution mutations (in tTERT), that have been shown to compromise or abolish DNA binding. Note that for hTERT, the predictions either overlap or surround the amino acids implicated by deletion (Figure 2A). For tTERT, two of three experimentally-identified DNA binding residues lie within the predicted DNA binding region (Figure 2B).

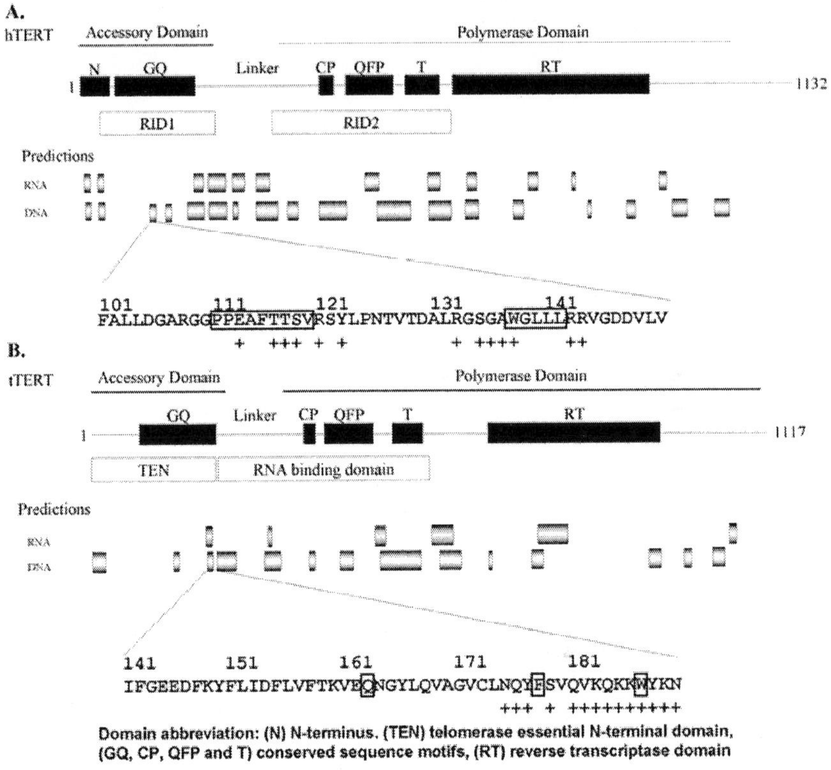

Figure 2. Predicted interface residues and conserved domains for telomerase reverse transcriptase (TERT). Mapped functional domains and conserved motifs of TERT are shown above shaded boxes representing clusters of predicted RNA and DNA interface residues. Predicted interface residues are indicated by a + below the amino acid sequence. **A)** Human telomerase reverse transcriptase (hTERT). In the sequence shown, boxed amino acids 110-119 and 137-141, correspond to the template anchor site and a putative primer grip, implicated in forming the hTERT-DNA active complex [31, 32, 34]. **B)** *Tetrahymena* telomerase reverse transcriptase (tTERT). The amino acid sequence shown represents the C-terminal end of the TEN domain. Alanine mutations at positions Q168, F178 and W187 have been shown to significantly reduce hTERT-DNA association. Predicted interactions spanning amino acids 181-190 are located in a highly flexible, disordered region [12].

A.

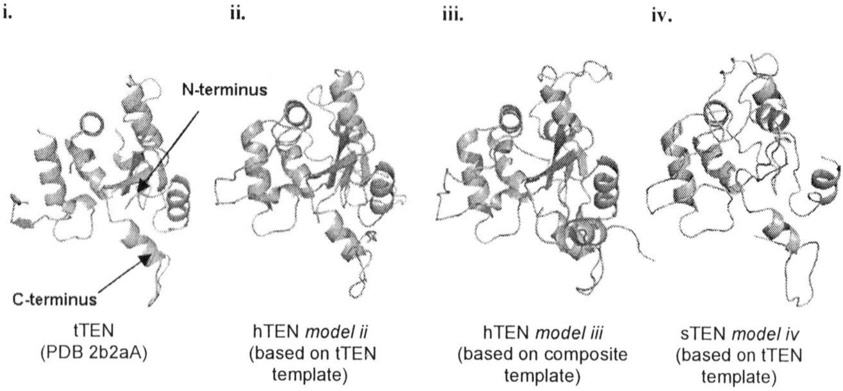

B.

```
T. thermophila   ----MQKINNINNNKQMLTRKEDLLTVLKQISALKYVSN--LYEFLLATEKIVQTSELDT
H. sapiens       ----MPRAPRCRAVRSLLRSHYREVLPLATFVRRLGPQG---WRLVQRGDPAAFRALVAQ
S. cerevisiae    --------------MKILFEFIQDKLDIDLQTNSTYKEN------LKCGHFNGLDEILTT

T. thermophila   QFQEFLTTTII--ASEQNLVENYKQKYN-----QPNFSQLTIKQVID------DSIILLG
H. sapiens       CLVCVPWD-----ARPPPAAPSFRQVSC-----LKELVARVLQRLCE---RGAKNVLAFG
S. cerevisiae    CFALPNSR-------KIALPCLPGDLSH-----KAVIDHCIIYLLTG---ELYNNVLTFG

T. thermophila   NKQNY--VQQIGTTTIGFYVEYENINLSRQTLYSSNFRNLLNIFGEEDFKYFLIDFLVFT
H. sapiens       FALLDGARGGPPEAFTTSVRSYLPNTVTDALRGSGAWGLLLRRVGDDVLVHLLARCALFV
S. cerevisiae    YKIAR------NEDVNNSLFCHSAN-VNVTLLKGAAWKMFHSLVGTYAFVDLLINYTVIQ

T. thermophila   KVEQNGYLQVAGVCLNQYFSVQVKQKKWYKNN----
H. sapiens       LVAPSCAYQVCGPPLYQLGAATQARPPPHASGPRRR
S. cerevisiae    FNG-QFFTQIVGNRCNEPHLPPKWVQRSSSSSAT--
```

Figure 3. Comparison of TEN domain structures and sequences and in *Tetrahymena*, human and yeast, *S. cerevisiae*. **A)** Comparison of *Tetrahymena* TEN domain structure determined by X-ray crystallography with modeled structures of TEN domains from other species. **i)** *T. thermophila*, experimentally-determined structure, PDB ID: 2b2aA [12]; **ii)** human structural model, based on threading using the *T. thermophila* 2b2aA structure as template; **iii)** human structural model, based on threading using a composite of several different structures as template; **iv)** yeast, *S. cerevisiae*, structural model, based on threading using the *T. thermophila* 2b2aA structure as template. **B)** Multiple sequence alignment of telomerase TEN domains from *T. thermophila*, *H. sapiens*, and *S. cerevisiae* [12]. Amino acids conserved in all 3 species in the multiple sequence alignment are highlighted.

3.3 Structural modeling of N-terminal domain of TERT from human and yeast

Our initial attempts to generate structural models of the human and yeast TEN domains by submitting their sequences to several web-based homology modeling servers were unsuccessful, due to failure of the servers to identify appropriate homology modeling templates (the pairwise sequence identity between TEN domains of hTERT and tTERT is < 20%). However, the results of multiple sequence alignment (Figure 3B) and predicted secondary structure similarities (data not shown), led us to try threading, using the FUGUE server (see Materials and Methods). The *Tetrahymena* TEN domain structure (PDB ID 2b2aA) was identified as the highest scoring structural template for both the human and yeast TEN domain sequences (hTERT: certain, with 99% confidence; sTERT: likely, with 95% confidence). Based on the alignments generated by FUGUE, we generated all-atom models and performed energy minimization to generate the final models illustrated in Figure 3A (see Materials and Methods for details). Two different models for the human TEN domain, *model ii*, based on the *Tetrahymena* TEN template, and *model iii*, based on a composite template from several different structures, were very similar to one another as well as to *model iv*, for the yeast TEN domain, despite their highly divergent amino acid sequences. Table 1 shows the root mean square deviation (RMSD) values calculated for comparison of the *Tetrahymena* TEN domain structure (determined by X-ray crystallography [12]) with the hTEN and sTEN modeled structures, using TOPOFIT [35] for structural alignment.

Aligned Structures	RMSD (Å)
tTEN *vs* hTEN	1.11
tTEN *vs* sTEN	1.41
sTEN *vs* hTEN	1.39

Table 1. RMSD computed from structural alignments of TEN domain structures: tTEN, *Tetrahymena,* PDB structure, 2b2aA (Fig.3A, *structure i)*; hTEN, human, modeled structure (Fig. 3A, *model ii*); sTEN, yeast, modeled structure (Fig. 3A, *model iv*). Alignments were performed using TOPOFIT [35]

3.4 Analysis of RNA and DNA binding surfaces in human and Tetrahymena TEN domains

To compare RNA and DNA binding surfaces in human and *Tetrahymena* TEN domains, we examined both our predicted nucleic acid binding sites and available experimental data in the context of the experimentally determined structure of *Tetrahymena* TEN domain [12] and modeled structure of the human TEN domain (*model ii*, Figure 3A). Examples of these analyses are illustrated in Figures 4 and 5. The predicted RNA binding residues in hTEN overlap with several RNA binding sites implicated by deletion experiments (Figure 4A, compare left and right models). Furthermore, additional putative RNA binding residues on the "back" side of the hTEN model (Figure 4B, left, in oval) co-localize with an experimentally defined RNA binding site mapped onto the tTEN crystal structure (Figure 4B, right, in oval).

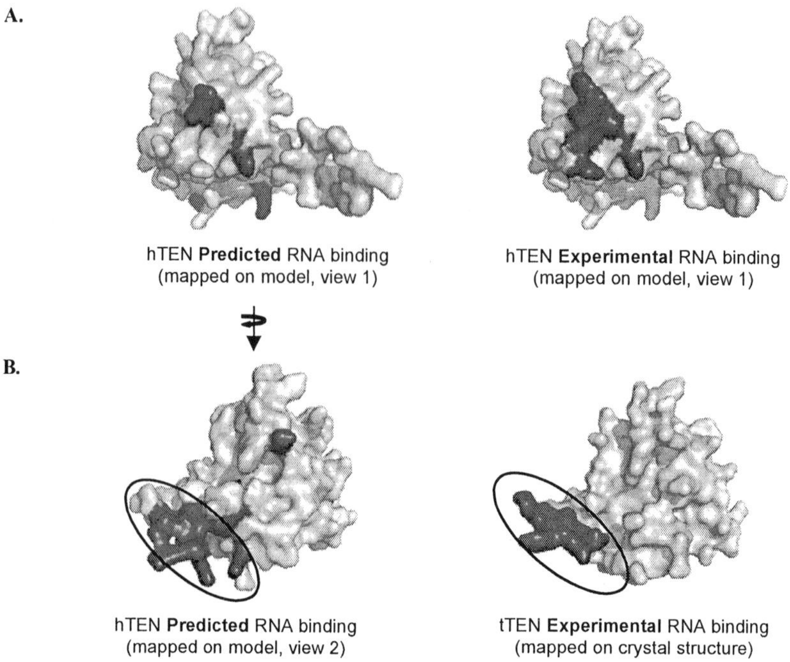

Figure 4. Comparison of predicted and experimentally determined RNA binding surfaces in TEN domains. A) Sequence-based RNA binding site predictions mapped onto the hTERT TEN domain *model ii* (left) overlap with experimentally determined RNA binding residues (right); Black residues are predicted (left) or actual (right) RNA binding residues. **B)** Another patch of predicted RNA binding residues in the hTEN model (left, in oval) co-localizes with an experimentally verified RNA binding region in tTEN (right). Figures 4 and 5 were generated using PyMol (http://pymol.sourceforge.net/).

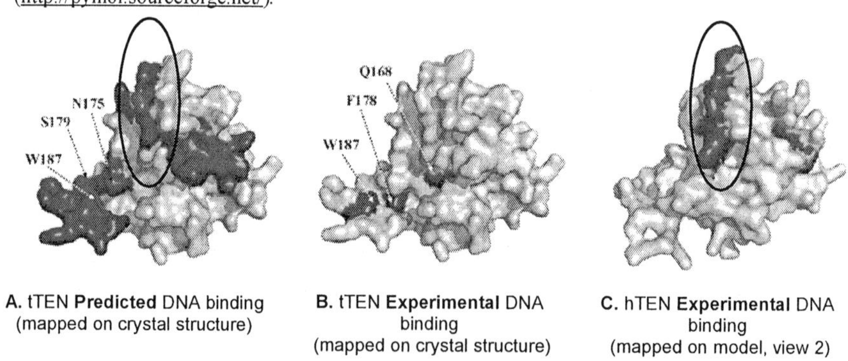

Figure 5. Comparison of predicted and experimentally determined DNA binding surfaces in TEN domains. A) Residues predicted to interact with DNA (black), mapped onto tTEN, PDB 2b2aA. Predicted binding sites encompass residues shown in **B)** which illustrates the only 3 experimentally defined DNA binding residues in tTEN (see Fig. 2B). Note that additional predicted DNA binding residues in **A** (in oval) are consistent with **C)**, which shows experimentally validated DNA binding residues in the human protein mapped onto our modeled structure of hTEN.

Only three DNA binding residues in the TEN domain of tTERT have been experimentally identified: Q168, F178, and W187 (Figure 5B). Several additional putative DNA binding residues are predicted by our machine learning classifiers (Figure 5A). Some of these predicted residues in tTEN (in oval) co-localize with experimentally defined DNA binding residues in the human protein, when viewed in the context of our modeled structure of the hTEN domain (Figure 5C).

Taken together, these results support our hypothesis that TEN domains in diverse organisms have similar three dimensional structures and conserved nucleic acid binding surfaces. Further, they identify additional putative interface residues that could be targeted in experiment studies.

4. Summary and Discussion

Telomerase is one of several clinically important regulatory proteins for which it has been difficult to obtain high resolution structural information. The recent experimental determination of the structure of the N-terminal domain of tTERT, the telomerase reverse transcriptase component from *Tetrahymena*, suggests that at least partial structural information for human telomerase may soon become available. It seems unlikely, however, that experimental elucidation of the structure of the multisubunit RNP complex corresponding to the catalytically active form of telomerase will occur in the near future. Thus, the integrative strategy proposed here, in which structural information gleaned from comparative modeling is combined with machine learning predictions of functional residues, can be expected to provide valuable insights into the sequence and structural correlates of function for telomerase and other "recalcitrant" proteins. We are currently pursuing several avenues for improving the reliability of machine learning predictions, including the use of different sequence representations and additional sources of input information (e.g., structure and phylogenetic information, when available) and more sophisticated machine learning algorithms. We are also pursuing additional approaches for protein structure prediction, including ab initio and fold recognition methods capable of incorporating predicted protein-protein contacts as constraints. Given the large number of proteins with which telomerase interacts and the essential roles of telomerase in cellular signaling, aging, cancer, and other human diseases, this should continue to be rich and challenging area of research.

5. Acknowledgements

This research was supported in part by NIH GM 066387, NIH-NSF BSSI 0608769, NSF IGERT 0504304 and by the ISU Center for Integrated Animal Genomics. We thank Fadi Towfic for critical comments on the manuscript and members of our groups for helpful discussions.

References

1. E. H. Blackburn, *FEBS Letters* **579**, 859 (2005).
2. K. Collins, *Nat. Rev. Mol. Cell. Biol.* **7**, 484 (2006).
3. C. Autexier and N. F. Lue, *Annu. Rev. Biochem.* **75**, 493 (2006).
4. C. W. Greider and E. H. Blackburn, *Cell* **43**, 405 (1985).
5. E. H. Blackburn, *Mol. Cancer. Res.* **3**, 477 (2005).
6. J. W. Shay and W. E. Wright, *J. Pathol.* **211**, 114 (2007).
7. M. A. Blasco, *Nat. Rev. Genet.* **8**, 299 (2007).
8. T. de Lange, Genes. Dev. **19**, 2100 (2005).
9. S. B. Cohen, M. E. Graham, G. O. Lovrecz, et al., *Science* **315**, 1850 (2007).
10. N. Hug and J. Lingner, *Chromosoma* **115**, 413 (2006).
11. A. Goldkorn and E. H. Blackburn, *Cancer Res.* **66**, 5763 (2006).
12. S. A. Jacobs, E. R. Podell, T. R. Cech, *Nat.Struct.Mol.Biol.* **13**, 218 (2006).
13. K. L. Friedman and T. R. Cech, *Genes Dev.* **13**, 2863 (1999).
14. J. Xia, Y. Peng, I. S. Mian, et al., *Mol. Cell. Biol.* **20**, 5196 (2000).
15. H.M. Berman, J. Westbrook, Z. Feng, et al., *Nucleic Acids.Res.* **28**, 235 (2000).
16. G. Wang and R. L. Dunbrack, Jr., *Bioinformatics* **19**, 1589 (2003).
17. M. Terribilini, J. H. Lee, C. Yan, et al., *Pac. Symp. Biocomput.*, 415 (2006).
18. M. Terribilini, J. H. Lee, C. Yan, et al., *RNA* **12**, 1450 (2006).
19. S. J. Hubbard, S. F. Campbell, J.M. Thornton, *J. Mol. Biol.* **220**, 507 (1991).
20. S. Jones and J. M. Thornton, *Proc. Natl. Acad. Sci.* **93**, 13 (1996).
21. C. Yan, M. Terribilini, F. Wu, et al., *BMC Bioinformatics* **7**, 262 (2006).
22. T. Mitchell, *Machine Learning* (McGraw-Hill, 1997).
23. I.H. Witten, E. Frank, *Data Mining: Practical Machine Learning Tools and Techniques* (Morgan Kaufmann, 2005).
24. J. Shi, T. L. Blundell, and K. Mizuguchi, *J. Mol. Biol.* **310**, 243 (2001).
25. R. Sanchez and A. Sali, *Proteins* **Suppl 1**, 50 (1997).
26. A. A. Canutescu, A. A. Shelenkov, and R. L. Dunbrack, Jr., *Protein Sci.* **12**, 2001 (2003).
27. W. R. P. Scott, P. H. Hunenberger, I. G. Tironi, et al., *J. Phys. Chem.* **A 103**, 3596 (1999).
28. N. Guex and M. C. Peitsch, *Electrophoresis* **18**, 2714 (1997).
29. S. F. Altschul, T. L. Madden, A. A. Schaffer, et al., *Nucleic Acids. Res.* **25**, 3389 (1997).
30. J. Kopp and T. Schwede, *Nucleic Acids. Res.* 32, D230 (2004).
31. T. J. Moriarty, S. Huard, S. Dupuis, et al., *Mol. Cell. Biol.* **22**, 1253 (2002).
32. T.J. Moriarty, R.J. Ward, M.A. Taboski, et al., *Mol.Biol.Cell.* **16**, 3152 (2005).
33. H. D. Wyatt, D. A. Lobb, and T. L. Beattie, *Mol. Cell. Biol.* **27**, 3226 (2007).
34. F. Melo and E. Feytmans, *J. Mol. Biol.* **277**, 1141 (1998).
35. V. A. Ilyin, A. Abyzov, and C. M. Leslin, *Protein Sci.* **13**, 1865 (2004).
36. M. Terribilini, J. D. Sander, J. H. Lee, et al., *NAR* **35**, W578-W584 (2007).

TILING MICROARRAY DATA ANALYSIS METHODS AND ALGORITHMS

SRINKA GHOSH AND ANTONIO PICCOLBONI

Affymetrix, Inc.
6550 Vallejo st.
Emeryville, CA 94608

The complete sequencing of the human genome and several other genomes for model organisms and other scientifically or technologically important species has opened what has been dubbed the *post-genomic era.* Notwithstanding the continuing and fruitful sequencing projects, this phase has been marked by a strong emphasis on genome function. The promise that the sequence, once revealed, would pave the way to understanding a variety of other aspects of biology has not been fully realized. For instance, the effort to experimentally characterize the structure of proteins is more vigorous then ever and conformation prediction from sequence information alone, despite progress, remains a challenge. Even coming up with a complete gene list for a newly sequenced genome is still a challenge and there is evidence that the transcribed fraction of the genome has been underestimated. Large collaborative efforts, such as the ENCODE project, have been launched to throw an array of experimental technologies at the problem of functional characterization of the genome – including but definitely not limited to more sequencing and in depth comparative genomics. One such technology is the *tiling microarray* (TM). A variation on the now widespread DNA microarray, the TM contains probes that correspond to regularly spaced position on a target genome, irrespective of their annotation as transcripts, promoters or any other functional determination. Therefore, they are made possible by genome sequencing efforts and complement them as high throughput technologies for the characterization of a variety of functional aspects of the genome. In combination with diverse assays, they have been applied to tasks such as transcript mapping and copy number variation and DNA replication analysis. In particular, the combination of TMs with chromatin immunoprecipitation techniques has enabled the high throughput study of protein-DNA binding and chromatin state. With an increasing number of TM-based datasets available to the scientific community, there is a considerable need for improved algorithms and software for their analysis and processing, and this session sought to provide a forum for investigators in the field to present and discuss the most recent advances.

We accepted three papers for this session. In the first, Kuan, Chun and Keles report on some progress in the analysis of chromatin immunoprecipitation TM data. They observe that a correlation structure exists in this type of data and that this aspect has not received enough attention in the literature. They formulate a model that takes this correlation into account and develop a fast detection algorithm based on this model. They support the usefulness of their approach with simulations and a case study and finally provide an open source implementation. In the second, Zeller, Henz, Laubinger, Weigel and Rätsch focus their attention on the application of TM to the characterization of transcription. They offer two related but distinct contributions: one is a normalization method that reduces within-transcript variability, while enhancing the signal separation between exonic and intronic regions; the second is a segmentation method that extends previous work on the unspliced transcript identification problem to the more challenging spliced case. Finally, Danford, Rolfe and Gifford turn the attention away from data analysis to data processing, storage and retrieval. They present a database design for TM data that can handle the results of a variety of experiments and processing methods, can manage multiple species and genome releases and provides convenient graphical presentation for the data, all built on top of a modular architecture amenable to customizations and extensions. They also present a system to formulate and record relationship between different chromatin immunoprecipitation events and provide a reference implementation.

CMARRT: A TOOL FOR THE ANALYSIS OF CHIP-CHIP DATA FROM TILING ARRAYS BY INCORPORATING THE CORRELATION STRUCTURE

PEI FEN KUAN[1], HYONHO CHUN[1], SÜNDÜZ KELEŞ[1,2*]

[1] *Department of Statistics,*
[2] *Department of Biostatistics and Medical Informatics,*
1300 University Avenue, University of Wisconsin, Madison, WI 53706.
**E-mail: keles@stat.wisc.edu*

Whole genome tiling arrays at a user specified resolution are becoming a versatile tool in genomics. Chromatin immunoprecipitation on microarrays (ChIP-chip) is a powerful application of these arrays. Although there is an increasing number of methods for analyzing ChIP-chip data, perhaps the most simple and commonly used one, due to its computational efficiency, is testing with a moving average statistic. Current moving average methods assume exchangeability of the measurements within an array. They are not tailored to deal with the issues due to array designs such as overlapping probes that result in correlated measurements. We investigate the correlation structure of data from such arrays and propose an extension of the moving average testing via a robust and rapid method called `CMARRT`. We illustrate the pitfalls of ignoring the correlation structure in simulations and a case study. Our approach is implemented as an `R` package called `CMARRT` and can be used with any tiling array platform.

Keywords: ChIP-chip, moving average, autocorrelation, false discovery rate.

1. Background

Whole genome tiling arrays utilize array-based hybridization to scan the entire genome of an organism at a user specified resolution. Among their applications are ChIP-chip experiments for studying protein-DNA interactions. These experiments produce massive amounts of data and require rapid and robust analysis methods. Some of the commonly used methods are ChIPOTle,[1] Mpeak,[2] TileMap,[3] HMMTiling,[4] MAT[5] and TileHGMM.[6] Although these algorithms have been shown to be useful, they don't address the issues due to array designs. The most obvious issue is the correlation of the measurements from probes mapping to consecutive genomic locations.[15] The basis for such a correlation structure is due to both overlapping probe design and fragmentation of the DNA sample to be hybridized on the array. There are several hidden Markov model (HMM) approaches to address the dependence among probes but the current implementations are limited to first order Markov dependence.[4] Generalizations to higher orders increase the computational complexity immensely. We investigate the correlation structure of data from complex tiling array designs and propose an extension of the moving average approaches[1,7] that carefully addresses the correlation structure. Our approach is based on estimating the variance of the moving average statistic by a detailed examination of the correlation structure and is applicable with any array platform. We illustrate the pitfalls of ignoring the correlation structure and provide several simulations and a case study illustrating the power of our approach `CMARRT` (Correlation, Moving Average, Robust and Rapid method on Tiling array).

2. Methods

Let $Y_1, ..., Y_N$ denote measurements on the N probes of a tiling path. Y_i could be an average log base 2 ratio of the two channels or (regularized) paired t-statistic for arrays with two channels (e.g., Nimblegen) and a (regularized) two sample t-statistic for single channel arrays (Affymetrix) at the i-th probe. These wide range of definitions of Y make our approach suitable for experiments with both single and multiple replicates per probe.

A common test statistic for analyzing ChIP-chip data is a moving average of Y_i's over a fixed number of probes or fixed genomic distance.[1,3,7] The parameter w_i will be used to define a window size of $2w_i + 1$, i.e., w_i probes to the right and left of the i-th probe. In the case of moving average across a fixed number of probes for tiling arrays with constant probe length and resolution, the window size w_i is calculated by $L \times (2w_i + 1) - 2w_i \times O = FL$, where L is the probe length, O is the overlap between two probes and FL is the average fragment size. Our framework also covers tiling arrays with non-constant resolution. In this case, w_i will be different for each genomic interval and corresponds to the number of probes within a fixed genomic distance. For simplicity in presentation, we will utilize window size of fixed number of probes. We assume that the data has been properly normalized by potentially taking into account the sequence features,[8] and that $E[Y] = \mu$ and $\text{var}(Y) = \sigma^2$. Consider the following moving average statistic

$$T_i = \frac{1}{2w_i + 1} \sum_{j=i-w_i}^{i+w_i} Y_i. \tag{1}$$

Then, standard variance calculation leads to

$$\mathrm{var}(T_i) = \frac{1}{(2w_i+1)^2}\left((2w_i+1)\sigma^2 + \sum_{j=i-w_i}^{i+w_i}\sum_{k\neq j}\mathrm{cov}(Y_j, Y_k)\right). \tag{2}$$

The standardized moving average statistic is given by

$$S_i = \frac{T_i}{\sqrt{\mathrm{var}(T_i)}}. \tag{3}$$

Standard practice of using moving average statistics relies on (1) estimating σ^2 based on the observations that represent lower half of the unbound distribution; (2) ignoring the covariance term in equation (2); (3) and obtaining a null distribution under the hypothesis of no binding at probe i. In particular, ChIPOTle considers a permutation scheme where the probes are shuffled and the empirical distribution of the test statistic over several shufflings is used as an estimate of the null distribution. As an alternative, a Gaussian approximation is utilized assuming that Y_i's are independent and identically distributed as normal random variables under the null distribution. As discussed by the authors of ChIPOTle, both approaches assume the exchangeability of the probes under the null hypothesis. Exchangeability implies that the correlation within any subset of the probes is the same. However, empirical autocorrelation plots from tiling arrays often exhibit evidence against this (Fig. 1). In particular, in the case of overlapping designs, a correlation structure is expected by design. When the spacing among the probes is large, correlation diminishes as expected (the right panel of Fig. 1), and this was the case for the dataset on which ChIPOTle was developed.

We illustrate the problem with ignoring the correlation structure on a ChIP-chip dataset from an E-coli RNA Polymerase II experiment utilizing a Nimblegen isothermal array (Landick Lab, Department of Bacteriology, UW-Madison). The probe lengths vary between 45 and 71 bp, tiled at a 22 bp resolution. Approximately half of the probes are of length 45 bp. We compute the standardized moving average statistic S_i (assuming $\mathrm{cov}(Y_j, Y_k) \neq 0$) and S_i^* (assuming independence of Y_i's). A method of estimating $\mathrm{cov}(Y_j, Y_k)$ is described in the next section. The p-values for each S_i and S_i^* are obtained from the standard Gaussian distribution under the null hypothesis. We expect the quantiles of S_i and S_i^* for unbound probes to fall along a 45° reference line against the quantiles from the standard Gaussian distribution, whereas the quantiles for bound probes to deviate from this reference line. As evident in Fig. 2, if the correlation structure is ignored, the distribution of S_i^*'s for unbound probes deviates from the standard

Gaussian distribution. Since the data is obtained from a RNA Polymerase II experiment, we expect a larger number of points, corresponding to promoters, to deviate from the reference line. An additional diagnostic tool is the histogram of the p-values. If the underlying distributions for S_i and S_i^* are correctly specified, the p-values obtained should be a mixture of uniform distribution between 0 and 1 and a non-uniform distribution concentrated near 0. The histograms of the p-values (Fig. 2) again illustrate that the distribution for T_i is misspecified.

2.1. *Estimating the correlation structure*

Although it is desirable to develop a structured statistical model that captures the correlations, developing such a model is both theoretically and computationally challenging due to the complex, heterogeneous data generated by tiling array experiments. We propose a fast empirical method that estimates the correlation structure based on sample autocorrelation function. The covariance $\mathrm{cov}(Y_j, Y_{j+k})$ can be estimated from the sample autocorrelation $\hat{\rho}(k)$ and sample variance $\hat{\sigma}^2$,[10]

$$\hat{\rho}(k) = \frac{\sum_{t=1}^{T-k}(Y_t - \bar{Y})(Y_{t+k} - \bar{Y})}{\sum_{t=1}^{T}(Y_t - \bar{Y})^2}, \quad \widehat{\mathrm{cov}}(Y_j, Y_{j+k}) = \hat{\rho}(k)\hat{\sigma}^2. \tag{4}$$

The following strategy is used in `CMARRT` for estimating the correlation structure. The top $M\%$ of outlying probes which roughly correspond to bound probes are excluded in the estimation of $\hat{\rho}(k)$. For the remaining probes, the sample autocorrelation at lag k ($\hat{\rho}_j(k)$) is computed for each segment j consisting of at least N consecutive probes. Genomic regions flanking a large gap or repeat masked regions will be considered as two separate segments. For any lag k, we let $\hat{\rho}(k)$ to be the average of $\hat{\rho}_j(k)$ over j. Here, N can be considered as a tuning parameter and our initial experiments with ENCODE datasets suggest that $N = 500$ works well in practice based on the diagnostic plots discussed in Section 1. M is an anti-conservative preliminary estimate of the percentage of bound probes which can be obtained under the assumption of independence among probes (usually $\sim 1-5\%$, depending on the type of ChIP-chip experiment).

3. Simulation studies

In this section, we investigate the performance of `CMARRT`, the conventional normal approximation approach under the independence assumption (Indep) and the HMM option in TileMap under various scenarios where we know the true bound regions in terms of sensitivity and specificity while controlling FDR at various levels used in practice.

Simulation I: Autoregressive model. We consider the following model

$$Y_i = N_i + R_i, \quad N_i = \sum_{k=1}^{p} \alpha_{i-k} N_{i-k} + \epsilon_i, \tag{5}$$

where N_i is the autoregressive background component and R_i is the real signal. We generate 100,000 N_i from AR(p) to represent the background component under the assumption of $\text{cor}(N_i, N_{i+k}) = \rho^{0.4(k-1)+1}$ and randomly choose 500 peak start sites. We let the size of a peak to be 10 probes, so that $\sim 5\%$ of the probes belong to bound regions. To design scenarios similar to what we have observed in practice, we also allow for ~ 3 outliers within a bound region. The data is simulated from various p (AR order), ρ ($\text{cor}(N_i, N_{i+k})$) and σ ($\text{var}(N_i)$) for the background component, and strength c for the real signal.

Simulation II: Hidden Markov model. In this scenario, the data is simulated from hidden markov models (HMMs)[12] with explicit state duration distribution to introduce direct dependencies at the probe level observations. Let the duration HMM densities be $p_{S_i}(d_i)$ ~Geometric(p_{S_i}). The transition probabilities (a_{ij}) and the parameters p_{S_i} in the duration HMM densities are chosen such that $\sim 5\%$ of the probes belong to bound regions. We consider the joint observation density $f_{N_i}(Y_1, Y_2, ...Y_{d_1}) \sim MVN(0, \Sigma_N)$ for the unbound regions and $f_{B_i}(Y_1, Y_2, ...Y_{d_1}) \sim MVN(\mu, \Sigma_B), \mu > 0$ for the bound regions, where MVN denotes the multivariate normal distribution. The parameters μ, Σ_N and Σ_B are chosen such that generated data resembles observed ChIP-chip data exhibiting correlations at the observation level.

Each simulation scenario is repeated 50 times. A probe is declared as bound if its adjusted p-value[11] is smaller than a pre-specified FDR level α when analyzing with `CMARRT` and Indep. For TileMap, we use the direct posterior probability approach[13] to control the FDR.

3.1. *Results of simulations I and II*

In Fig. 3, we summarize the sensitivity at the peak level and the specificity at various FDR thresholds from Simulation I for `CMARRT`, Indep, and TileMap. `CMARRT` is able to identify most of the bound regions at FDR of 0.05 and above while TileMap tends to be more conservative in declaring bound regions as shown in the sensitivity plots. Although Indep has the highest sensitivity, it also has a high proportion of false positives. The specificity of Indep is significantly lower compared to `CMARRT`, even under the case of low correlation among the probes. Similar results are obtained in Simulation II under the duration HMM (Fig. 4). The left panels show the sensitivity and specificity for the case of smaller peaks with an average peak size of 10 probes while the right panels are for the case of larger peaks of size 20 probes on average. These results illustrate the superior performance of `CMARRT` in terms of both sensitivity and specificity even when the data is generated from a complex model. The heuristic way of estimating the correlation structure in `CMARRT` is able to reduce the false positives (specificity) significantly, but not at the expense of increasing false negatives (sensitivity). On the other hand, ignoring the correlation structure results in a higher proportion of false positives. Additionally, the HMM option in TileMap is more conservative than the moving average approach when the FDR is controlled at the same level.

4. Case study: ZNF217 ChIP-chip data

We provide an illustration of `CMARRT` with a ZNF217 ChIP-chip data tiling the ENCODE regions (available from Gene Expression Omnibus (`http://www.ncbi.nlm.nih.gov/geo/`)[14] with accession number GSE6624). The ENCODE regions were tiled at a density of one 50-mer every 38 bp, leading to $\sim$ 380,000 50-mer probes on the array. We analyze two different replicates of this dataset separately and compare the analysis on these single replicates. In Krig et al.,[14] the bound regions were identified with the Tamalpais Peaks program,[9] which requires a bound region to have at least 6 consecutive probes in the top 2% of the log base 2 ratios. This criteria tends to be too stringent and fails to identify bound regions which contain a few outlier probes with log base 2 ratios below the top 2% threshold and may result in a higher level of false negatives. In the top right panel of Fig. 5, we show one potential peak missed by the Tamalpais Peaks program. In such cases, the sliding window approach is more powerful for finding peaks. Moreover, this method also assumes the observations are independent. As evident in the left panel of Fig. 1, observations from nearby probes in this tiling array are correlated. As shown in Fig 5, the histograms of p-values for the unbound probes under the independence assumption deviates from the expected distribution in both replicates. Similar problem is present in the normal quantile-quantile plots (online supp. mat.) when the correlation structure is ignored.

As in Krig et al.,[14] we require the number of consecutive probes in each bound region to be at least 6. A set of peaks is obtained for each replicate at a given FDR control. We assess the extent of overlaps between the set of peaks in these two replicates. The results are summarized in Table 1. All the methods identified more peaks in replicate 1 than replicate 2. Therefore, using the peaks from rep 1 as reference, the common peaks are defined as the percentage of overlapping peaks in replicate 2. For all FDR thresholds (except 0.01), CMARRT has the highest value of common peaks, followed by Indep and TileMap, which illustrates the consistency of the peaks identified by CMARRT.

As an independent validation, we determine the location of bound regions relative to the transcription start site (TSS) of the nearest gene using GENECODE genes from UCSC Genome Browser as in Krig et al.[14] (Table 1). For a given FDR control, the percentage of peaks located within $\pm 2kb$, $\pm 10kb$ and $\pm 100kb$ of the TSS is the highest in CMARRT, followed by Indep and TileMap. As expected, these numbers decrease as we increase the FDR threshold for all the three methods. These results illustrate the power of CMARRT in detecting biologically more plausible bound regions of ZNF217.

5. Discussion

We have investigated and illustrated the pitfalls of ignoring the correlation structure due to tiling array design in ChIP-chip data analysis. We proposed an extension of the moving average approaches in CMARRT to address this issue. CMARRT is a robust and fast algorithm that can be used with any tiling platform and any number of replicates. Both the simulation results and the case study illustrate that CMARRT is able to reduce false positives significantly but not at the expense of increasing false negatives, thereby giving a more confident set of peaks. We have recently became aware of the work of Bourgon[15] who carefully studies the correlation structure in ChIP-chip arrays and proposes a fixed order autoregressive moving average model (ARMA(1, 1)) and we are in the process of comparing CMARRT with this approach.

CMARRT is developed using the Gaussian approximation approach and the diagnostic plots illustrated can be utilized to detect whether a given dataset violates this assumption. One possible relaxation of this assumption is a constrained permutation approach that aims to conserve the correlation structure among the probes under the null distribution. Implementation of such an approach efficiently is a challenging future research direction.

Acknowledgements

We thank Professor Robert Landick for providing the E-coli ChIP-chip data for our analysis. Supplementary materials are available at `http://www.stat.wisc.edu/~keles/CMARRT.sm.pdf`. This research has been supported in part by a PhARMA Foundation Research Starter Grant (P.K. and S.K.) and NIH grants 1-R01-HG03747-01 (S.K.) and 4-R37-GM038660-20 (H.C.).

References

1. M.J.Buck, A.B. Nobel and J.D. Lieb (2005), ChIPOTle: a user-friendly tool for the analysis of ChIP-chip data, *Genome Biol.* **6**(11).
2. T.H. Kim, L.O. Barrera, M. Zheng, C. Qu, M.A. Singer, T.A. Richmand, Y. Wu, R.D. Green and B. Ren (2005), A high-resolution map of active promoters in the human genome, *Nature* **436**:876-880.
3. H. Ji and W.H. Wong (2005), TileMap: create chromosomal map of tiling array hybridizations, *Bioinformatics* **21**(18):3629-3636.
4. W Li and C.A. Meyer and X.S. Liu(2005), A hidden Markov model for analyzing ChIP-chip experiments on genome tiling arrays and its application to p53 binding sequences, *Bioinformatics* **21**Suppl 1:i274-i282.
5. W.E. Johnson, W. Li, C.A. Meyer, R. Gottardo, J.S. Carroll, M. Brown and X.S. Liu (2006), MAT: Model-based Analysis of Tiling-arrays for ChIP-chip, *Proc Natl Acad Sci USA* **103**:12457-12462.
6. S. Keles (2006), Mixture modeling for genome-wide localization of transcription factors, *Biometrics*, **63**(1):10-21.
7. S. Keles, M. J. van der Laan, S. Dudoit and S.E. Cawley (2006), Multiple Testing Methods for ChIP-Chip High Density Oligonucleotide Array Data, *J. of Comp. Bio.* **13**(3):579-613.
8. T.E. Royce, J.S. Rozowsky and M.B. Gerstein (2007), Assessing the need for sequence-based normalization in tiling microarray experiments, *Bioinformatics*.
9. M. Bieda, X. Xu, M.A. Singer, R. Green and P.J. Farnham (2007), Unbiased location analysis of E2F1-binding sites suggests a widespread role for E2F1 in the human genome, *Genome*.
10. G.P. Box and G.M. Jenkins (1976), *Time series analysis forecasting and control*, Holden-Day.
11. Y. Benjamini and Y. Hochberg (1995), Controlling the false discovery rate: a practical and powerful approach to multiple testing, *JRSS-B* **57**:289-300.
12. L.R. Rabiner (1989), A Tutorial on Hidden Markov Models and Selected Applications in Speech Recognition, *Proceedings of the IEEE* **77**(2):257-286.
13. M.A. Newton, A. Noueiry, D. Sarkar and P. Ahlquist (2004), Detecting differential gene expression with a semiparametric hierarchical mixture method, *Biostatistics* **5**:155-176.
14. S.R. Krig, V.X. Jin, M.C. Bieda, H. O'Geen, P. Yaswen, R. Green and P.J.

Farnham (2007), Identification of genes directly regulated by the oncogene ZNF217 using ChIP-chip assays, *J. Biol. Chem.* **282**(13):9703-9712.
15. R.W. Bourgon (2006), Chromatin immunoprecipitation and high-density tiling microarrays: a generative model, methods for analysis and methodology assessment in the absence of a "gold standard". Ph.D. Thesis, UC Berkeley.

Table 1. Distance of ZNF217-binding sites relative to TSS.

FDR=0.01	CMARRT	Indep	TileMap
Common peaks	0.803(791/935)	0.819(1423/1736)	0.718(799/1113)
% of peaks within ±2kb	0.334	0.278	0.136
% of peaks within ±10kb	0.619	0.565	0.442
% of peaks within ±100kb	0.911	0.903	0.824
FDR=0.05	CMARRT	Indep	TileMap
Common peaks	0.806(1023/1269)	0.790(1796/2272)	0.714(978/1370)
% of peaks within ±2kb	0.321	0.267	0.134
% of peaks within ±10kb	0.589	0.565	0.431
% of peaks within ±100kb	0.903	0.900	0.826
FDR=0.10	CMARRT	Indep	TileMap
Common peaks	0.805(1209/1491)	0.779(2096/2689)	0.703(1071/1524)
% of peaks within ±2kb	0.300	0.265	0.135
% of peaks within ±10kb	0.579	0.561	0.428
% of peaks within ±100kb	0.904	0.894	0.821
FDR=0.15	CMARRT	Indep	TileMap
Common peaks	0.794(1333/1678)	0.763(2301/3051)	0.701(1171/1671)
% of peaks within ±2kb	0.284	0.259	0.136
% of peaks within ±10kb	0.564	0.552	0.434
% of peaks within ±100kb	0.899	0.890	0.827

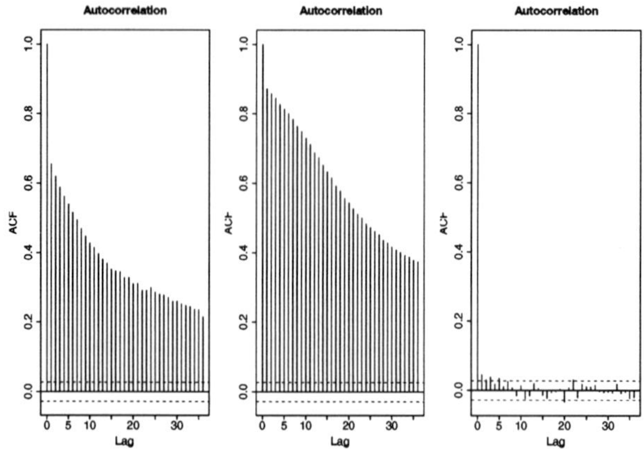

Fig. 1. *Example autocorrelation plots from ChIP-chip data.* The left, middle and right panels are from the data in Krig et al.,[14] Landick Lab and Kim et al.[2] respectively. The autocorrelation plots for Krig et al.[14] and Landick Lab clearly show the presence of correlations among probes. The autocorrelation plot for Kim et al.[2] shows that the correlation structure diminish with increasing spacing between probes. The data from Krig et al.[14] and Landick Lab are from tiling arrays with overlappping probes, whereas the design in Kim et al.[2] have subtantial spacing between probes (i.e., probe length = 50 bp and resolution = 100 bp).

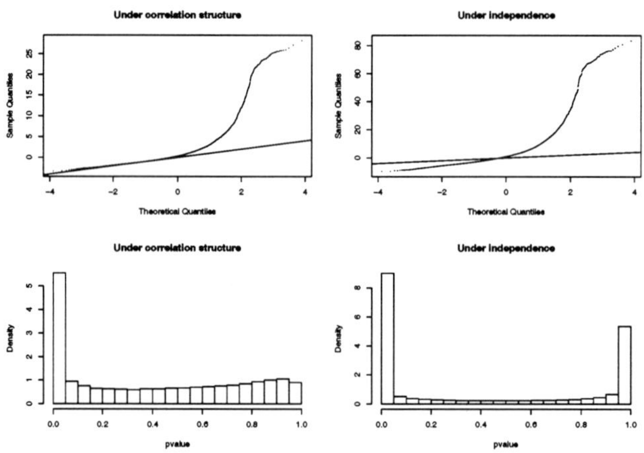

Fig. 2. *Normal quantile-quantile plots (qqplot) and histograms of p-values.* The left panels show the qqplot of S_i and distribution of p-values under correlation structure. The top right panel shows that if the correlation structure is ignored, the distribution of S_i^*'s for unbound probes deviates from the standard Gaussian distribution. The bottom right panel shows that if the correlation structure is ignored, the distribution of p-values for unbound probes deviates from the uniform distribution for larger p-values.

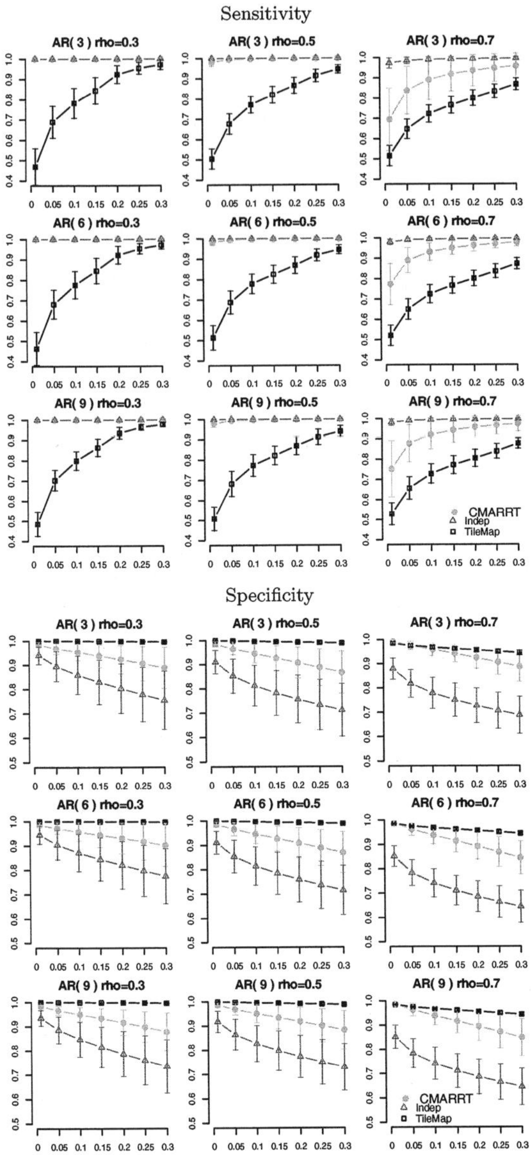

Fig. 3. *Sensitivity at peak level (top figure) and specificity (bottom figure) at various FDR control (x-axis).* The background N is generated from various autoregressive models with $\mathrm{sd}(N_i)=0.3$, $Y_i = N_i + 1.5$, $p = \{3, 6, 9\}$ and $\rho = \{0.3, 0.5, 0.7\}$. Vertical lines are error bars. **CMARRT** is able to identify most of the bound regions at FDR of 0.05 and above. TileMap tends to be more conservative in declaring bound regions. Although Indep gives the highest sensitivity, it also has the highest proportion of false positives. The specificity for **CMARRT** is significantly higher than the Indep approach.

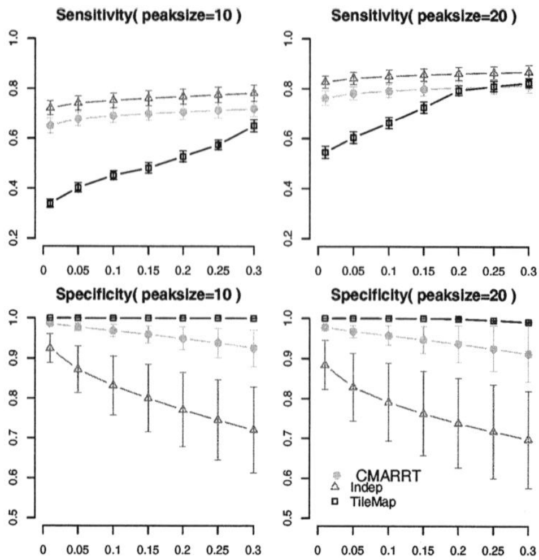

Fig. 4. *Sensitivity and specificity at various FDR control (x-axis).* The left panels are the results under duration HMM simulation with average peak size of 10 probes. The right panels correspond to using average peak size of 20 probes. TileMap tends to be more conservative and has the lowest sensitivity and highest specificity. CMARRT is able to achieve a balance between sensitivity and specificity at each FDR threshold. Indep tends to identify many false positives.

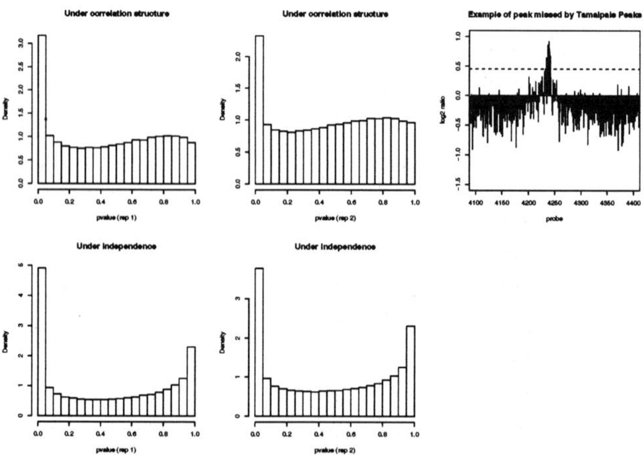

Fig. 5. *Histograms of p-values for replicates 1 and 2 and an example of peak missed by the Tamalpais Peaks program.* The distributions of the probes for unbound regions deviates from uniform distribution when the correlation structure is not taken into account (bottom panels). The dotted line in top right panel is the 98-th percentile of the log base 2 ratios. Tamalpais Peaks requires a peak to have at least 6 probes in a row to be in the top 2 %.

TRANSCRIPT NORMALIZATION AND SEGMENTATION OF TILING ARRAY DATA

GEORG ZELLER

Friedrich Miescher Laboratory of the Max Planck Society &
Max Planck Institute for Developmental Biology, Dept. for Molecular Biology
Spemannstr. 35 & 39, 72076 Tübingen, Germany
E-mail: Georg.Zeller@tuebingen.mpg.de

STEFAN R. HENZ, SASCHA LAUBINGER & DETLEF WEIGEL

Max Planck Institute for Developmental Biology, Dept. for Molecular Biology
Spemannstr. 35, 72076 Tübingen, Germany
E-mail: {Stefan.Henz,Sascha.Laubinger,Detlef.Weigel}@tuebingen.mpg.de

GUNNAR RÄTSCH

Friedrich Miescher Laboratory of the Max Planck Society
Spemannstr. 39, 72076 Tübingen, Germany
E-mail: Gunnar.Raetsch@tuebingen.mpg.de

For the analysis of transcriptional tiling arrays we have developed two methods based on state-of-the-art machine learning algorithms. First, we present a novel transcript normalization technique to alleviate the effect of oligonucleotide probe sequences on hybridization intensity. It is specifically designed to decrease the variability observed for individual probes complementary to the same transcript. Applying this normalization technique to *Arabidopsis* tiling arrays, we are able to reduce sequence biases and also significantly improve separation in signal intensity between exonic and intronic/intergenic probes. Our second contribution is a method for transcript mapping. It extends an algorithm proposed for yeast tiling arrays to the more challenging task of spliced transcript identification. When evaluated on raw versus normalized intensities our method achieves highest prediction accuracy when segmentation is performed on transcript-normalized tiling array data.

Datasets, software and the appendix are available for download at `http://www.fml.mpg.de/raetsch/projects/PSBTiling`

1. Introduction

Tiling arrays on which oligonucleotide probes are spotted at high density have made it feasible to study whole genomes in an unbiased and cost-effective way. They have been used for experiments as diverse as transcriptome analysis, ChIP-chip and DNA sequence variation detection.[4,6,7,8,9,11].

The analysis of tiling array data, however, is not straightforward since intensity measurements are known to be influenced by many factors. In order to allow direct comparisons between arrays potentially hybridized under slightly different experimental conditions, the measurements are typically first normalized as a whole, e.g. by array quantile normalization.[3] Another major reason for variability in hybridization intensity are divergent sequence properties of oligonucleotide probes that have not been optimized due to constraints on tiling array design. In this work we compare a newly developed *transcript normalization* technique for the removal of sequence-specific effects to the recently proposed *sequence quantile normalization.*[16] Our approach particularly aims at reducing the variability around mRNA transcript levels which are ideally assumed to be constant across all exon probes of the same transcript. We have therefore developed a regression model that estimates the deviation between the observed intensities of individual probes and the *transcript intensity* taking probe sequences as input. Such a normalization is expected to be beneficial particularly for transcript mapping approaches attempting to segment the genome into transcriptional units of approximately constant hybridization intensity.

The monitoring of known genes and especially the identification of novel transcripts with whole-genome tiling arrays has received increasing attention over the last years. For the analysis of *S. cerevisiae* tiling arrays, Huber et al.[11] proposed a method that segments the yeast chromosomes such that the sum of squared differences of signal intensities to their mean within a given segment is minimized. To solve this mathematical problem, known as Structural Change Model Segmentation (SCM), they adapted the dynamic programming algorithm proposed by Bai and Perron.[2] While this relatively simple approach has been successfully applied to yeast tiling array data, the segmentation problem is considerably more challenging for the genomes of higher eukaryotes that are capable of (alternative) splicing. Here, gene density is typically lower, exon segments are much shorter and interrupted by potentially very long intron sequences. A more sophisticated model, called GenRate, has been proposed by Frey et al.[9] It explicitly models coregulated units (CoRegs) such as exons of the same gene exhibiting the same expression level. However, the generative model for sequences of hybridization measurements that constitutes the core of their method is based on several assumptions on the structure of a transcript and the distribution of hybridization measurements (e.g. Gaussian distribution of intensity differences from a designated reference probe, geometrically distributed distance of the reference probe from the transcript start, etc.).

Building on this work we propose a novel method that is able to accurately recognize transcripts from tiling array measurements. Our approach is based on a discriminative learning technique closely related to Hidden Markov (HM) Support Vector Machines (SVMs)[1] which combine the advantages of HM models[5] for label sequence learning with those of the discriminative SVM framework. A precursor method can be seen as a reformulation of the SCM method modeling interruptions of active regions (exons) with inactive regions (introns). For this model we still assume Gaussian noise for the deviation of exon probe intensities from their average. Since this assumption is typically not satisfied, we augment the method with more flexible *scoring functions* replacing the squared error terms. Their shapes are estimated from data in order to optimally segment the sequence of intensity measurements. As a supervised learning approach, our algorithm is trained on hybridization intensities together with segmentations determined from known mRNA transcripts.

2. Normalization of Transcriptional Tiling Arrays

2.1. *Array Data and Preprocessing*

We analyzed data from *A. thaliana* tiling arrays manufactured by Affymetrix. For hybridization, total RNA of 21 day-old inflorescences was amplified using oligo-dT-T7 primers. Resulting RNA was converted into double-stranded cDNA, fragmented, labeled and hybridized to Affymetrix *TilingR* arrays following standard protocols (see Appendix A for details).

In a first normalization step, measurements affected by artifacts already apparent from the scanned image of the array were removed using a software called *Harshlight.*[19] To facilitate inter-array comparisons *quantile-normalization* was applied, which involves computing the mean over the empirical intensity distributions of all considered arrays. This mean distribution is then re-assigned to each of the arrays, thus effectively removing differences in intensity distribution between arrays.[3] All intensity measurements were $\log_2$ transformed for the subsequent normalization steps.

2.2. *Sequence Quantile Normalization (SQN)*

Sequence quantile normalization (SQN) has been proposed as an extension of the above described quantile-normalization to remove probe sequence effects.[16] For each 25-mer probe having nucleotide $j \in A, C, G, T$ at position $k = 1, \ldots, 25$ the rank $r_{i,j,k}$ of its intensity y_i among all other probes with the same nucleotide at position k is calculated and normalized by the number of such probes $C_{j,k}$. These position-wise contributions are then averaged: $\hat{S}_i = \frac{1}{25} \sum_{k=1}^{25} \frac{r_{i,j,k}}{C_{j,k}}$. Since the sequence bias is not uniform across positions and summands are not independent, the multivariate regression problem is solved iteratively; in each step the above average is computed and afterwards intensities y_i are replaced by $\hat{S}_i$ which is repeated until convergence.[16]

As a side effect, intensities are substituted by relative ranks that are uniformly distributed between zero and one. In order to obtain normalized intensity values comparable to the original measurements from the array, we modified the averaging as follows. Intensity distributions were approximated by piece-wise linear functions $g_k(r_{i,j,k}) \approx y_i$. In our case, g is parametrized by 200 supporting points with uniformly spaced x-values s_x between zero and one. The corresponding y-values s_y are estimated by linear interpolation between y_m and y_n having ranks $r_{m,j,k} = \max_{m'}\{r_{m',j,k} \mid r_{m',j,k}/C_{j,k} \leq s_x\}$ and $r_{n,j,k} = \min_{n'}\{r_{n',j,k} \mid r_{n',j,k}/C_{j,k} \geq s_x\}$, respectively. Instead of averaging relative ranks, we then calculated the mean $\hat{g} = \frac{1}{25}\sum_{k=1}^{25} g_k$ of the supporting points s_y. From this averaged $\hat{g}$ we reconstructed the normalized intensities by linear interpolation between the supporting points of $\hat{g}$.

2.3. *Transcript normalization techniques*

Ideally, one would expect constant hybridization intensity for all probes measuring the same transcript. Similarly, the background signal of probes in untranscribed or intronic regions of the genome would ideally be constant. However in practice, this is generally not the case (see e.g. Royce et al.[15] for a discussion).

Here we propose a method to reduce within-gene variability caused by probe sequence effects. In a first step we estimate constant transcript and background intensities $\overline{y}_i$ based on the TAIR7 genome annotation,[14] in the following simply referred to as *transcript intensities*: If a probe i is annotated as exon, $\overline{y}_i$ is the median of the intensities y_i of probes in exons of the same gene. Similarly for intron probes, we compute $\overline{y}_i$ as the median over intronic probes of the same gene and for intergenic regions $\overline{y}_i$ is the median of all probes mapped to regions annotated as intergenic. (Probes that were mapped to intron / exon boundaries, more than one splice-form or overlapping genes are excluded from training and evaluation.)

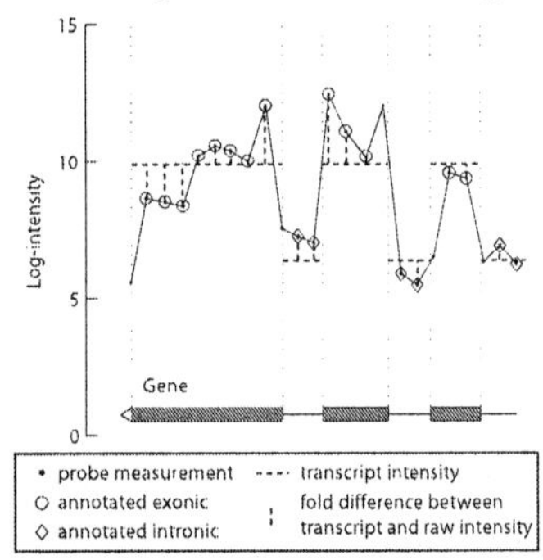

Figure 1.: Illustration of raw and transcript intensities for part of a transcript

Assuming that the concentration of mRNA hybridized to all exon probes of a gene is constant, the differences between the raw intensities and the transcript intensities $\hat{y}_i := y_i - \overline{y}_i$ are mainly due to probe sequence-specific effects (ignoring cross-hybridization, experimental artifacts and thermodynamic noise). Furthermore, it is conceivable that probe effects also depend on the mRNA concentration, and hence the differences $\hat{y}_i$ may also depend on the transcript intensities $\overline{y}_i$ of the exons of the gene. Since it is not *a priori* clear how this dependence should be modeled, it appears reasonable to non-parametrically model the difference by a function of the form $f(\boldsymbol{x}_i, \overline{y}_i) \approx y_i - \overline{y}_i$ that depends both on sequence features $\boldsymbol{x}_i$ of the probe as well as its transcript intensity. However, in order to use this correction, one would have to know in advance whether a certain probe is exonic, intronic or intergenic which is not generally the case. We therefore propose to estimate the function depending not on the transcript intensity, but instead on the raw intensities, i.e. $f(\boldsymbol{x}_i, y_i) \approx y_i - \overline{y}_i$.

Given the large amounts of available data for estimating $f(\boldsymbol{x}, y)$, we can discretize the parameter y into Q quantiles and estimate Q independent functions $f_q(\boldsymbol{x})$. Then $f(\boldsymbol{x}, y)$ is given by

$$f(\boldsymbol{x}, y) = \begin{cases} f_1(\boldsymbol{x}) & \text{for } y \in (-\infty, y_1) \\ \quad\cdots & \quad\cdots \\ f_i(\boldsymbol{x}) & \text{for } y \in [y_i, y_{i+1}) \\ \quad\cdots & \quad\cdots \\ f_Q(\boldsymbol{x}) & \text{for } y \in [y_Q, \infty) \end{cases}$$

As input $\boldsymbol{x}_i$ to the regression function f_q the sequence $\boldsymbol{s}_i$ of probe i was provided together with additional features derived from the sequence: sequence entropy $-\sum_{i=1}^{4} f_i \times \log(f_i)$, where f_i is the frequency of the nucleotide $i \in \{A, C, G, T\}$ in the probe sequence and GC content. Furthermore, two hairpin scores were used: One is the maximum number of base pairs over all possible hairpin structures that a probe can form, the other one is equal to the maximum number of consecutive base pairs over all possible hairpin structures (similarly used for intensity modelling in Zhan et al.[22]).

Based on these sequence features, we considered two methods for learning the functions f_q based on Q sets of n training examples $(\boldsymbol{x}_i^q, \hat{y}_i^q)$, where $\hat{y}_i = y_i - \overline{y}_i$, $i = 1, \ldots, N$ and $q = 1, \ldots, Q$:

Support Vector Regression (SVR) For regression, we applied Support Vector Machines[17] with a kernel function $k(\boldsymbol{x}, \boldsymbol{x}')$ that computes the "similarity" of two examples $\boldsymbol{x}$ and $\boldsymbol{x}'$. Here we used a sum of the Weighted Degree (WD) kernel[13,12] and a linear kernel. The WD kernel has been developed to model sequence properties taking the occurrence and position of substrings up to a certain length d into account.[13] We considered substrings up to order $d = 3$ and allowed a shift of 1 bp between positions of the substring,[12] which can be efficiently dealt with using string indexing data structures.[18] The linear kernel computed the scalar product of the sequence-derived features described above. We used the freely available implementations from the *Shogun toolbox*.[18]

Ridge regression (RR) For every training example we explicitly generated a feature vector from the sequence $\boldsymbol{s}$ having an entry for every possible mono-, di- and tri-nucleotide at every position in the probe (one if present at a position, zero otherwise; similar to the implicit representation in the WD kernel). The resulting feature vector was augmented with the sequence derived features to form $\boldsymbol{x}_i$. In training, the λ-regularized quadratic error is minimized:[10] $\min \lambda||w||^2 + \sum_{i=1}^{n}(\boldsymbol{w}^T\boldsymbol{x}_i - \hat{y}_i)^2$ with $\boldsymbol{w} = \left(\lambda I + \sum_{i=1}^{n} \boldsymbol{x}_i\boldsymbol{x}_i^T\right)^{-1} \sum_{i=1}^{n} \hat{y}_i\boldsymbol{x}_i$ being its solution. Then $f_q(\boldsymbol{x}) = \boldsymbol{w}_q^T\boldsymbol{x}$ is the resulting regression estimate.

Ridge regression is straightforward to implement in any programming language supporting matrix operations and linear equation solvers. In terms of computation time it is much less demanding than both SVR and SQN.

3. Transcript Identification

In this section we describe a novel segmentation algorithm for transcriptional tiling array data. It is based on ideas similarly presented before,[9,11] but uses a different strategy for learning and inference (cf. Section 1).

The goal is to characterize each probe as either intergenic (not transcribed) or as part of a transcriptional unit (either exon or intron). Instead of predicting the label of a probe (intergenic, exonic or intronic) directly, we learn to associate a state with each probe given its hybridization measurements and the local context. From the state sequence we can easily infer the label sequence (see Figure 2).

For learning we first defined the target state sequence, i.e. the "truth" that we attempted to approximate. It was generated from known transcripts and hybridization measurements. We then applied HMSVMs[1] for label sequence learning to build a discriminative model capable of predicting the state and hence the label sequence given the hybridization measurements alone.

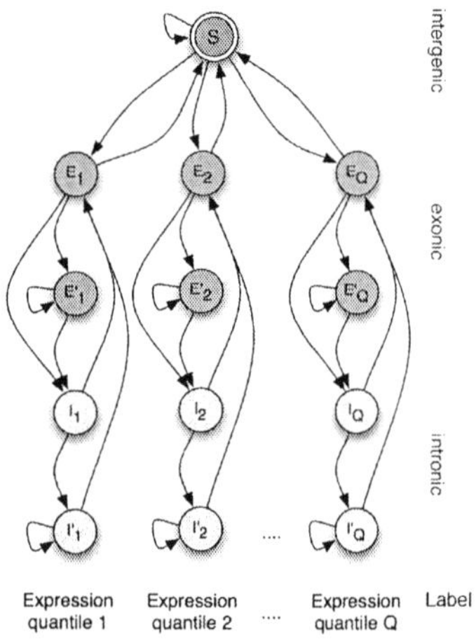

Figure 2.: State model with a subset of states for each expression quantile (columns). The label corresponding to each state is indicated on the right.

State Model The simplest version of the state model had only three states: intergenic, exonic & intronic. It was extended in two ways: (a) by introducing an intron/exon start state that allowed modeling of the start and the continuation of exons & introns separately and (b) by repeating the exon and intron states for each expression quantile which allowed us to model discretized expression levels separately (see below). The resulting state model is outlined in Figure 2. Finally, to compensate for the 3' intensity bias described in Appendix E, we also allow transitions from the exon states of one level to the ones of the next higher or lower level.

Generation of Labelings For genomic regions with known transcripts we considered the sense direction of up to 1 kbp flanking intergenic regions while maintaining a distance of at least 100 bp to the next annotated gene. Within this region we assigned one of the following labels to every probe: *intergenic*, *exonic*, *intronic* and *boundary*. In a second step we subdivided genes according to the median hybridization intensity of all exonic probes into one of $Q = 20$ expression quantiles. For each probe a state was determined from its label and expression quantile. (The boundary probes were excluded in evaluation.)

Parametrization and Learning Algorithm Our goal was to learn a function $f : \mathbb{R}^\star \to \Sigma^\star$ predicting a state sequence $\boldsymbol{\sigma} \in \Sigma^\star$ given a sequence of hybridization measurements $\boldsymbol{\chi} \in \mathbb{R}^\star$, both of equal length T. This was done indirectly via a $\boldsymbol{\theta}$-parametrized discriminant function $F_{\boldsymbol{\theta}} : \mathbb{R}^\star \times \Sigma^\star \to \mathbb{R}$ that assigned a real-valued score to a pair of observation and state sequence.[1,20] Knowing $F_{\boldsymbol{\theta}}$ allowed to determine the maximally scoring state sequence by dynamic programming,[5] i.e. $f(\boldsymbol{\chi}) = \underset{\boldsymbol{\sigma} \in \mathcal{S}^\star}{\operatorname{argmax}} F_{\boldsymbol{\theta}}(\boldsymbol{\chi}, \boldsymbol{\sigma})$.

For each state $\tau \in \Sigma$, we employed a *scoring function* $g_\tau : \mathbb{R} \to \mathbb{R}$. $F_{\boldsymbol{\theta}}$ was then obtained as the sum of the individual scoring contributions and the *transition scores* given by $\phi : \Sigma \times \Sigma \to \mathbb{R}$:

$$F_{\boldsymbol{\theta}}(\boldsymbol{\chi}, \boldsymbol{\sigma}) = \sum_{t=1}^{T} \sum_{\tau \in \Sigma} [[\sigma_t = \tau]] \; g_\tau(\chi_t) + \phi(\sigma_{t-1}, \sigma_t)$$

where $[[.]]$ denotes the indicator function. We modeled the scoring functions g_τ as piecewise linear functions[13] (PLiF) with $L = 20$ supporting points $s_1, \ldots, s_L$. Together with the transition scores ϕ, the y-values at the supporting points $\theta_{\tau,l} =: g_\tau(s_l)$ constituted the parametrization of the model, collectively denoted by $\boldsymbol{\theta}$.

During discriminative training a large margin of separation between the score of the correct path and *any* other wrong path was enforced. (For details on the optimization problem see Appendix C and Altun et al.[1])

4. Results and Discussion

4.1. *Probe Normalization*

The *A. thaliana* genome was partitioned into ≈300 regions while avoiding splits in annotated genes. Mapping perfect match (PM) probes to genome locations resulted in ≈10000 probes per region. We randomly chose 40% of these regions for training, 20% for hyper-parameter tuning and the remaining 40% as a test set for performance assessment. The test regions were further used for the segmentation experiments in Section 4.3.

Removal of Sequence Effects Figure 3 shows that hybridization intensity is strongly correlated with the GC content of the probe causing more than 4-fold changes in median intensity. This sequence effect was reduced by all normalization methods. However, Figure 3 also indicates that the effect is (in part) explained by GC-richness of coding regions.[21] Position-specific sequence effects were further investigated with so-called quantile plots.[16] The strongest reduction of first-order sequence effects was achieved with SQN, although positional sequence effects were reduced by all normalization methods (see Appendix D).

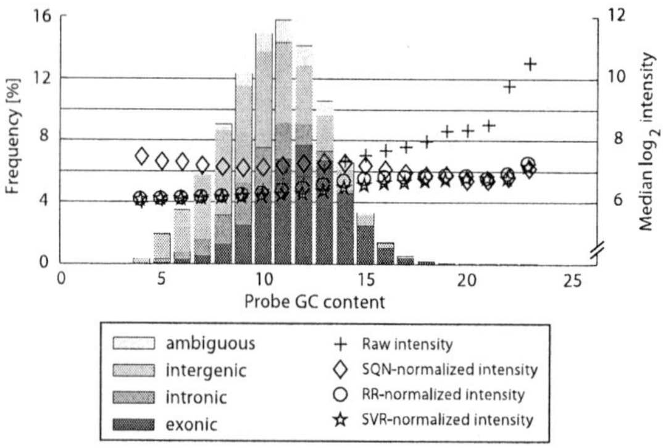

Figure 3. Median hybridization intensity depends on GC content of oligonucleotide probes. The histogram obtained by partitioning probes according to their GC content is shown as bar plots. In each bin the frequency of exonic, intronic and intergenic probes is indicated by different gray-scales, and the median log-intensity is shown before and after the application of normalization methods (see inset).

Reduction of Transcript Intensity Variability For the assessment of transcript variability, i.e. the deviation of individual probe intensities y_i from the constant transcript or background intensity $\overline{y}_i$, we introduced two metrics, T_1 and T_2. Both relate the variability of normalized intensities $y_i - f(\boldsymbol{x}_i, y_i)$ to the variability of raw intensities, and values smaller than 1 indicate a reduction. We defined $T_1 := \frac{\sum_i |y_i - f(\boldsymbol{x}_i, y_i) - \overline{y}_i|}{\sum_i |y_i - \overline{y}_i|}$ as the normalized absolute transcript variability and $T_2 := \frac{\sum_i (y_i - f(\boldsymbol{x}_i, y_i) - \overline{y}_i)^2}{\sum_i (y_i - \overline{y}_i)^2}$ as the normalized squared transcript variability. SVR minimizes the so-called ϵ-insensitive loss closely related to the absolute error, while Ridge regression minimizes the squared loss. Therefore, we expected and observed smaller T_1 values for SVR and smaller T_2 values for RR (see Figure 4). With both methods transcript variability was reduced to approximately half the values of raw intensities. For SQN we observed both T_1 and T_2 greater than 1 indicating increased transcript

Method	T_1	T_2
SQN	1.83	3.16
SVR	0.54	0.47
RR	0.58	0.44

Figure 4.: Within-gene variability after normalization.

variability. One may argue that SQN is therefore not well-suited as a preprocessing routine for transcript mapping (see also Figures 5 and 6). However, as SQN does not directly attempt to reduce transcript variability, this comparison should been interpreted with caution.

4.2. *Exon Probe Identification*

In a simple approach to identify transcribed exonic regions we used a threshold model on the hybridization measurements. Probes with intensities above the threshold were classified as exonic and below the threshold as untranscribed or intronic. We compared the resulting classification of probes with the TAIR7 annotation.[14] For every threshold we calculated precision and recall, defined as the proportion of probes mapped to exons among all probes having intensities greater than the threshold and the proportion of probes with intensities greater than the threshold among all probes that are annotated as exonic, respectively. Thresholding was applied to raw intensity values as well as the normalized intensities from SQN, SVR and RR. The resulting precision-recall curves (PRCs) are displayed in Figure 5 A. We observed that the two transcript normalization methods, SVR and RR, consistently improved exon probe identification compared to raw intensities. For SQN the recognition deteriorated. However, when probes were sub-samples prior to thresholding and evaluation such that the set of exonic probes had the same GC-content the background set (as reported in Royce et al.[16]), the performance of SQN recovered, but was still below SVR and RR (cf. Figure 5 B). Note that the sub-sampling strategy changes the distributions and can not easily be applied to identify exon probes in the whole genome.

In a second experiment we only considered the transcribed regions of the genes in the test regions (exons and introns). We now allowed a threshold to be chosen separately for each gene. Note that this problem is much easier compared to a single global threshold. However, this approach cannot be directly applied when the transcript boundaries are not already known. For each gene we estimated the Receiver-Operator-Characteristic (ROC) curve separately and averaged them over all genes.[a] In Figure 6 we display the area under the averaged ROC curves for genes in different transcript intensity quantiles. As expected, exons could be more accurately identified in highly expressed transcripts. Again, we observed a superior performance of the transcript normalization techniques.

[a]We considered ROC curves instead of PRCs, since the class sizes varied among genes making PRCs incomparable.

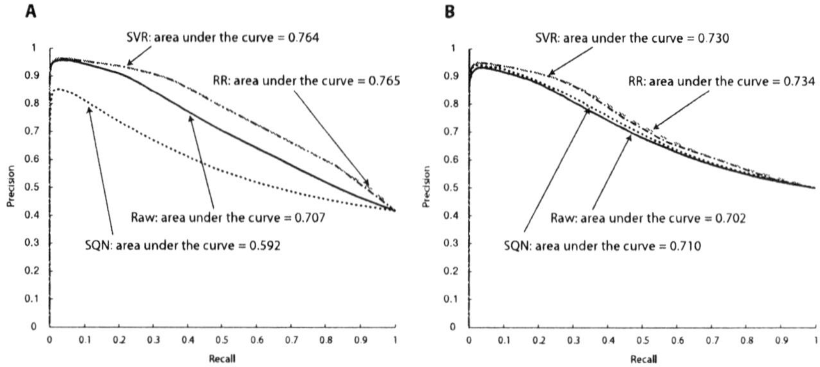

Figure 5. Separation in intensity between probes mapped to known exons and probes in regions annotated as untranscribed or intronic improved after normalization with SVR as well as after normalization with RR. **A** By varying the cutoff value, we calculated the precision-recall curve from all probes in the test regions. **B** Prior to thresholding and precision-recall estimation, probes were sub-sampled to obtain the same GC-content among exonic and intergenic / intronic probes.

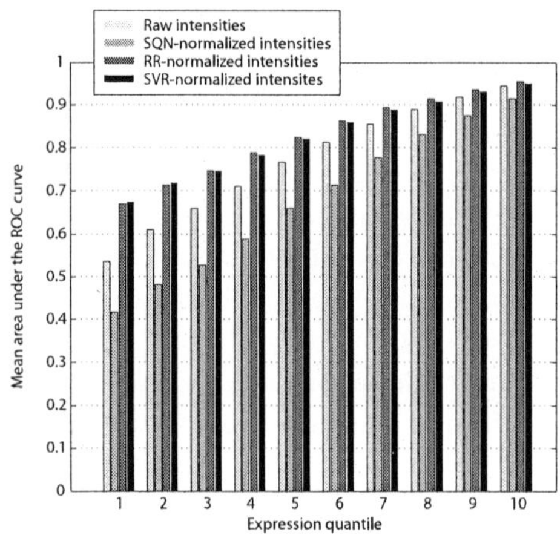

Figure 6. Separation in intensity between intron and exon probes broken down by expression quantiles and normalization methods. Expression values were calculated based on the median intensity of probes annotated as exonic. For each gene the area under the ROC curve (auROC) was obtained by local thresholding and for each expression quantile, auROC values were averaged over all genes in that quantile.

4.3. *Identification of Transcripts*

In a final experiment we show a proof of concept for our transcript identification algorithm. For this we considered genomic regions (from the test set described in Section 2) with known transcripts including 1 kbp of their flanking intergenic regions. We truncated intergenic regions at the boundaries of adjacent, known transcripts. For training, we took 100 randomly chosen regions, containing a single gene each, 500 such regions for model selection and 500 other regions for evaluation. We compared our method with the two simple thresholding approaches described in the previous section. In the first one we used a global

	Global threshold	Local threshold	HMSVMs
Raw intensities	70.4%	79.3%	77.1%
Sequence quantile normalization	65.5%	75.3%	70.9%
Support vector regression	73.5%	82.1%	82.9%
Ridge regression	73.9%	82.1%	82.5%

Figure 7. Accuracy of transcript identification in test regions with exactly one gene. Accuracy is defined as the sum of true positive and true negative exon probes over the total number of probes in a gene.

threshold which could be realistically applied for exon probe identification. In the second one an individual threshold was chosen for each gene to maximize classification accuracy. Note that this method has an advantage in the comparison because the threshold is determined based on expression levels of (unknown) test genes to be identified. Moreover, it cannot be straightforwardly applied to genome-wide detection of exon probes. As input we provided raw as well as normalized hybridization intensities discussed in Section 2 to our segmentation and the two thresholding methods. This resulted in a mapping of probes to exons, introns or intergenic regions. The accuracies of these predictions are summarized in Figure 7. In this comparison our segmentation method was considerably better than global thresholding, and even slightly better than the locally optimal threshold when trancript-normalized intensities were give as input. Moreover, we re-confirmed the findings of the previous section that transcript normalization significantly improved discrimination between exonic and untranscribed / intronic regions not only for thresholding on a per-probe basis, but in particular for a considerably more complex segmentation algorithm.

References

1. Y. Altun, I. Tsochantaridis, and T. Hofmann. Hidden Markov Support Vector Machines. In *Proc. 20th Int. Conf. Mach. Learn.*, pages 3–10, 2003.
2. J. Bai and P. Perron. Computation and analysis of multiple structural change models. *J. Appl. Econom.*, 18:1–22, 2003.
3. B.M. Bolstad, R.A Irizarry, M. Astrand, and T.P. Speed. A comparison of normalization methods for high density oligonucleotide array data based on variance and bias. *Bioinformatics*, 19(2):185–193, 2003.
4. R.M. Clark, G. Schweikert, C. Toomajian, S. Ossowski, G. Zeller, P. Shinn, N. Warthmann, T.T. Hu, G. Fu, D. Hinds, H. Chen, K. Frazer, D. Huson, B. Schölkopf, M. Nordborg, G. Rätsch, J. Ecker, and D. Weigel. Common sequence polymorphisms shaping genetic diversity in Arabidopsis thaliana. *Science*, 317(5836), July 2007.
5. R. Durbin, S. Eddy, A. Krogh, and G. Mitchison. *Biological Sequence Analysis: Probabilistic models of protein and nucleic acids*. Cambridge University Press, 7th edition, 1998.

6. J.S. Caroll et al. Chromosome-wide mapping of estrogen receptor binding. *Cell*, 122:33–43, 2005.
7. L. David et al. A high-resolution map of transcription in the yeast genome. *Proc. Natl. Acad. Sci. USA*, 103:5320–5325, 2006.
8. P. Bertone et al. Global identification of human transcribed sequences with genome tiling arrays. *Science*, 306:2242–2246, 2004.
9. B.J. Frey, Q.D. Morris, and T.R. Hughes. Genrate: A generative model that reveals novel transcripts in genome-tiling microarray data. *Journal of Computational Biology*, 13(2):200–214, 2006.
10. A.E. Hoerl and R.W. Kennard. Ridge regression: Biased estimation for nonorthogonal problems. *Technometrics*, 12(3):55–67, 1970.
11. W. Huber, J. Toedling, and L. M. Steinmetz. Transcript mapping with high-density oligonucleotide tiling arrays. *Bioinformatics*, 22(6):1963–1970, 2006.
12. G. Rätsch, S. Sonnenburg, and B. Schölkopf. RASE: recognition of alternatively spliced exons in *C. elegans*. *Bioinformatics*, 21:i369–i377, 2005.
13. G. Rätsch, S. Sonnenburg, J. Srinivasan, H. Witte, K.-R. Müller, R.J. Sommer, and B. Schölkopf. Improving the Caenorhabditis elegans genome annotation using machine learning. *PLoS Computational Biology*, 3(2):e20, 2007.
14. S.Y. Rhee, W. Beavis, T.Z. Berardini, G. Chen, D. Dixon, A. Doyle, M. Garcia-Hernandez, E. Huala, G. Lander, M. Montoya, N. Miller, L.A. Mueller, S. Mundodi, L. Reiser, J. Tacklind, D.C. Weems, Y. Wu, I. Xu, D. Yoo, J. Yoon, and P. Zhang. The Arabidopsis information resource (TAIR): a model organism database providing a centralized, curated gateway to Arabidopsis biology, research materials and community. *Nucl. Acids Res.*, 31(1):224–8, 2003.
15. T.E. Royce, J.S. Rozowsky, P. Bertone, M. Samanta, V. Stolc, S. Weissman, M. Snyder, and M. Gerstein. Issues in the analysis of oligonucleotide tiling microarrays for transcript mapping. *Trends in Genetics*, 21(8):466–475, 2005.
16. T.E. Royce, J.S. Rozowsky, and M.B. Gerstein. Assessing the need for sequence-based normalization in tiling microarray experiments. *Bioinformatics*, 23(8):988–997, 2007.
17. B. Schölkopf and A.J. Smola. *Learning with Kernels*. MIT Press, 2002.
18. S. Sonnenburg, G. Rïsch, C. Schäfer, and B. Schölkopf. Large scale multiple kernel learning. *Journal of Machine Learning Research*, 7:1531–1565, 2006.
19. M. Suarez-Farinas, M. Pellegrino, K. Wittkowski, and M. Magnasco. Harshlight: a "corrective make-up" program for microarray chips. *BMC Bioinformatics*, 6(1):294, 2005.
20. B. Taskar, C. Guestrin, and D. Koller. Max margin markov networks. In *Advances in Neural Information Processing Systems 13*, 2003.
21. The Arabidopsis Genome Initiative. Analysis of the genome sequence of the flowering plant Arabidopsis thaliana. *Nature*, 408(6814):796–815, 2000.
22. Y. Zhan and D. Kulp. Model-P: A Basecalling Method for Resequencing Microarrays of Diploid Samples. *Bioinformatics*, 21(suppl_2):ii182–189, 2005.

GSE: A COMPREHENSIVE DATABASE SYSTEM FOR THE REPRESENTATION, RETRIEVAL, AND ANALYSIS OF MICROARRAY DATA

TIMOTHY DANFORD, ALEX ROLFE, AND DAVID GIFFORD

MIT Computer Science and Artificial Intelligence Laboratory
32-G538
77 Massachusetts Ave
Cambridge, MA, 02139

We present GSE, the Genomic Spatial Event database, a system to store, retrieve, and analyze all types of high-throughput microarray data. GSE handles expression datasets, ChIP-chip data, genomic annotations, functional annotations, the results of our previously published Joint Binding Deconvolution algorithm for ChIP-chip, and precomputed scans for binding events. GSE can manage data associated with multiple species; it can also simultaneously handle data associated with multiple 'builds' of the genome from a single species. The GSE system is built upon a middle software layer for representing streams of biological data; we outline this layer, called GSEBricks, and show how it is used to build an interactive visualization application for ChIP-chip data. The visualizer software is written in Java and communicates with the GSE database system over the network. We also present a system to formulate and record *binding hypotheses*- simple descriptions of the relationships that may hold between different ChIP-chip experiments. We provide a reference software implementation for the GSE system.

1. Introduction

1.1. *Large-Scale Data Storage in Bioinformatics*

The data storage and computational requirements for high-throughput genomics experiments have grown exponentially over the last several years. Some methods simultaneously collect hundreds-of-thousands, or even millions, of data points. Microarrays contain several orders of magnitude more probes than just a few years ago. Short read sequencing produces 'raw' datasets requiring over a terabyte of computer disk storage[11]. Combine these with massive genome annotation datasets, cross-species sequence alignments mapped on a per-base level, thousands of publicly-available microarray expression experiments, and growing databases of sequence motif information – and you have a wealth of experimental results (and large scale analyses) available to the investigator on a scale unimagined just a few years ago.

Successful analysis of high-throughput genome-wide experimental data requires careful thought on the organization and storage of numerous dataset types. However, the ability to effectively store and query large datasets has often lagged behind the sophistication of the analysis techniques that are developed for that data. Many publicly available analysis packages were developed to work in smaller systems, such as yeast [19]. Flat files are sufficient for simple organisms, but for large datasets they will not fit into main memory and cannot provide the random access necessary for a browsing visualizer.

Modern relational databases provide storage and query capabilities for these vertebrate-sized datasets. Built to hold hundreds of gigabytes to terabytes of data, they provide easy access through a well-developed query language (SQL), network accessibility, query optimizations, and facilities for easily backing up or mirroring data across multiple sites.

Most bioinformatics tools that have taken advantage of database technology, however, are web applications. Often these tools are the front-end interfaces to institutional efforts that gather publicly-available data or are community resources for particular model organisms or experimental protocols. Efforts like UCSC's genome browser and its backing database[12], or the systems of GenBank[2], SGD[6], FlyBase[4], and many others, are all examples of web interfaces to sophisticated database systems for the storage, search, and retrieval of species-based or experiment-based data.

1.2. *A Desktop Analysis Client and a Networked Database Server*

The system that we describe here bridges the gap between the web applications that exist for large datasets and the analysis tools that work on smaller datasets. GSE consists of back-end tools for importing data and running batch analyses as well as visualization software for interactive browsing and analysis of ChIP-chip data.

The visualization software, distributed as a Java application, communicates over the network with the same database system as the as the middle-layer and analysis tools. Our visualization and analysis software is written in Java and are distributed as desktop applications. This lets us combine much of the flexibility of a web-application interface (lightweight, no flat files to install, and can run on any major operating system) with the power of not being confined to a browser environment. Our system can also connect to datastreams from multiple databases simultaneously, and can use other system resources normally unavailable to a browser application.

This paper describes the platform that we have developed for the storage of ChIP-chip and other microarray experiments in a relational database. It then presents our system for intepreting ChIP-chip data to identify binding events using our previously published "Joint Binding Deconvolution" (JBD) algorithm[17]. Finally, we show how we can build a system for the dynamic and automatic analysis of ChIP-chip binding calls between different factors and across experimental conditions.

2. A Database System for ChIP-chip Data

The core of our system is a database schema to represent ChIP-chip data and associated metadata in a manner independent of specific genomic coordinates and of the specific array platform.

2.1. *Common Metadata*

Figure 1 shows the common metadata that all subcomponents of GSE share. We define species, genome builds, and experimental metadata that may be shared by ChIP-chip experiments, expression experiments, and ChIP-seq experiments. We represent factors (e.g. an antibody or RNA extraction protocol), cell-types (tissue identifier or cell line name), and conditions as entries in separate tables.

2.2. *Coordinate Independent ChIP-chip Representation*

In our terminology, an experiment aggregates ChIP-chip datasets which all share the same factor, condition, and cell-type as defined in the common metadata tables. Each replicate of an experiment corresponds to a single hybridization performed against a particular micorarray design. In Section 4, we will outline a system for building biological hypotheses out of these descriptive metadata objects.

GSE stores probes separately from their genomic coordinates as shown in Figure 2. Microarray observations are indexed by probe identifier and experiment identifier. The key data retrieval query joins the probe observerations and probe genomic coordinates based on probe identifier and filters the results by experiment identifier (or more typically a set of experiment

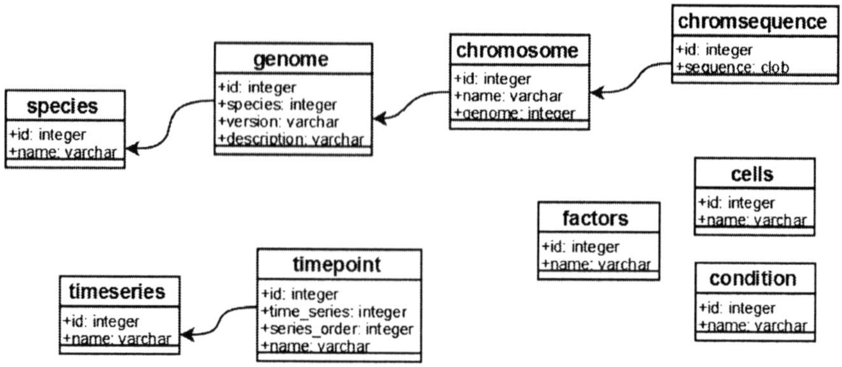

Figure 1. The Genomic Spatial Event database's common metadata defines species, genome assemblies, and terms to describe experiments. *Cells* enumerates the known tissue or cell types. *Conditions* defines the conditions or treatments from which the cells were taken. *Factors* describes antibodies in ChIP-chip experiments or RNA extraction protocols (eg total RNA or polyA RNA) for expression experiments.

identifiers corresponding to replicates of a biological experiment) and genomic coordinate. To add a new genome assembly to the system, we remap each probe to the new coordinate space once and all of the data is then available against that assembly. Since updating to a new genome assembly is a relative quick operation regardless of how many datasets have been loaded, users can always take advantage of the latest genome annotations.

GSE's database system also allows multiple runs of the same biological experiment on different array platforms or designs to be so combined. Some of our analysis methods can cope with the uneven data densities that arise from this combination, and we are able to gather more statistical power from our models when they can do so.

2.3. *Discovering Binding Events from ChIP-chip Data*

Modern, high-resolution tiling microarray data allows detailed analyses that can determine binding event locations accurate to tens of bases. Older low-resolution ChIP-chip microarrays included just one or two probes per gene[9,10]. Traditional analysis applied a simple error model to each probe to produce a bound/not bound call for each gene rather than measurements associated with genomic coordinates[22]. Our Joint Binding Deconvolution (JBD) exploits the dozens or hundreds of probes that cover each gene an intergenic region on modern microarrays with a complex statistical model

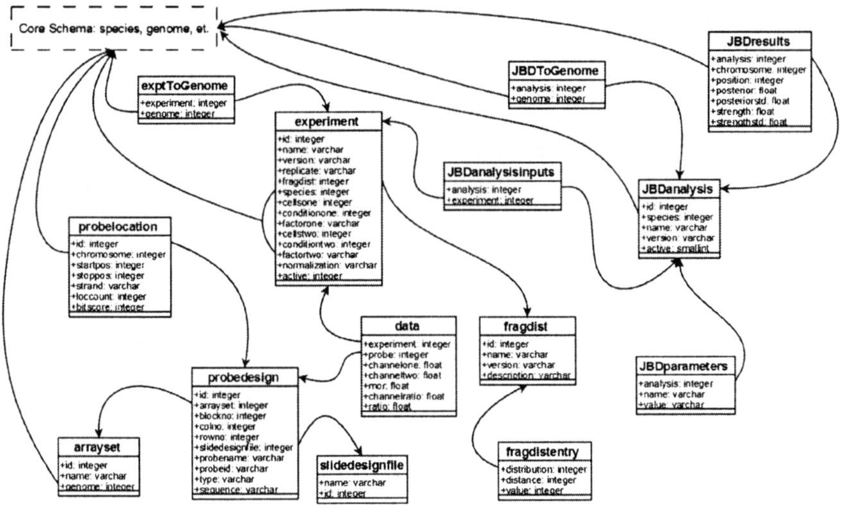

Figure 2. The ChIP-chip schema stores microarray designs, raw microarray observations, and the resulting analyses. We store probe designs as information about a single spot on a microarray. Probes are grouped by slide and by slide sets (arrayset). Genomic coordinates for each probe reside in a separate table to allow easy remapping of probes to new genome assemblies.

that incorporates the results of multiple probes at once and accounts for the possibility of multiple closely-spaced binding events.

JBD produces a probability of binding at any desired resolution (e.g. a per-base probability that a transcription factor bound that location). Figure 2 shows the tables that store the JBD output and figure 3 shows a genomic segment with ChIP-Chip data and JBD results. Unlike the raw probe observations, JBD output refers to a specific genome assembly since the spatial arrangement of the probe observations is a key input. GSE's schema also records which experiments led to which JBD analysis.

2.4. *Prior Work and Performance*

We modeled portions of GSE after several pre-existing analysis and data-handling systems. The core design of an analysis system supported by a relational database was made after experience with the GeneXPress package and its descendant, Genomica[19]. We modeled portions of the GSEBricks system, our modular component analysis system, after the Broad Institute's GenePattern software [18]. There are also several widely-used standards for microarray data storage and annotation databases that we were aware of

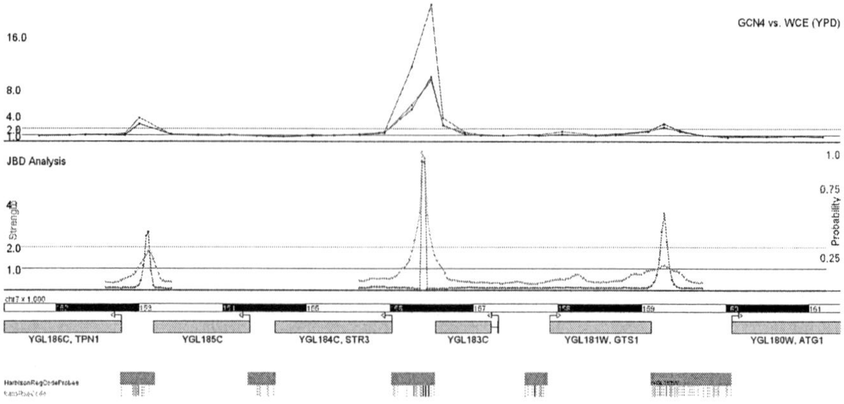

Figure 3. A screenshot from the GSE Visualizer. The top track represents 'raw' high-resolution GCN4 data in yeast, and the bottom track shows two lines for the two output variables of the JBD algorithm. At the bottom are a genomic scale, a representation of gene annotations, and a custom painting of the probes and motifs from the Harbison et. al. Regulatory Code dataset.

when designing our system. For instance, the MIAME standard for microarray information [7] is well-known format and specification for microarray data – however, we made the decision to store significantly less metadata about our ChIP-chip experiments than MIAME requires, since much of it is not immediately useful for biological analysis and it made it harder for our biological collaborators to enter new data into the system. We are also familiar with the DAS system[5], and GSE benefited from close discussions with one of DAS's co-creators, during its design and early implementation. However GSE solves a different problem than DAS, as it is mainly focused on providing a concentrated resource for (often-unpublished) data accumulation and an analysis platform for a small to mid-sized group of researchers.

Measuring the exact performance of a distributed system such as ours is difficult. The system consists of multiple servers running on several heterogeneous platforms, with as many as twenty or thirty regular users. Performance statistics are affected by system load, network latency conditions, and even the complexity of the data itself (the JBD algorithm's runtime is data-dependent, taking longer when the data is more "interesting"). Our group currently runs two database servers, one Oracle and one MySQL, and our computational needs are served by 16 rack-mounted machines with dual 2.2GHz AMD Opteron processors and 4 GB of memory each. We currently store approximately 338 GB of total biological data, which includes 1460 ChIP-chip experiments, 1115 separate results of the JBD algorithm, and over 240 million probe observations. Given this amount of data, and users scattered among at least eight collaborating groups, we are still able to serve up megabase visualizations of most ChIP-chip experiments in a matter of seconds, and to scan single experiments for binding events in times on the order of about 1-2 minutes.

3. GSEBricks: A Modular Library for Biological Data Analysis

GSE's visualization and GUI analysis tools depend on a library of modular analysis and data-retrieval components collectively titled 'GSEBricks'. This system provides a uniform interface to disparate kinds of data: ChIP-chip data, JBD analyses, binding scans, genome annotations, microarray expression data, functional annotations, sequence alignment, orthology information, and sequence motif instances. GSEBricks' components use Java's Iterator interface such that a series of components can be easily connected into analysis pipelines.

A GSEBricks module is written by extending one of three Java interfaces: `Mapper`, `Filter`, or `Expander`. All of these interfaces have an 'execute' method, with a single `Object` argument which is type-parameterized in Java 5. The `Mapper` and `Filter` execute methods have an `Object` (also parameterized) as a return value. `Mapper` produces `Objects` in a one-to-one relationship with its input, while a `Filter` may occasionally return 'null' (that is, no value). The `Expander` execute method, on the other hand, returns an `Iterator` each time it is called (although the `Iterator` may be empty).

3.1. *Ease of Integration and Extensibility*

Each GSEBricks datastream is represented by an `Iterator` object and datastreams are composed using modules which 'glue' existing `Iterators` into new streams. Because we extend the Java `Iterator` interface, the learning curve for GSEBricks is gentle even for novice Java programmers. At the same time, its paradigm of building 'Iterators out of Iterators' lends itself to a Lisp-like method of functional composition, which naturally appeals to many programmers familiar with that language.

Because our analysis components implement common interfaces (eg, `Iterator<Gene>` or `Iterator<BindingEvent>`), it is easy to simply plug them into visualization or analysis software. Furthermore, the modular design lends itself to modular extensions. We have been able to quickly extend our visualizer to handle and display data such as dynamically rescanned motifs (on a base-by-base level within the visualized region), automatic creation of 'meta-genes'[21] (averaged displays of ChIP-chip data from interactively-selected region sets), and the display of mapped reads from ChIP-PET experiments[13].

The final advantage of GSEBricks is the extensibility of the GSEBricks system itself. By modifying the code we use to glue the Iterators together, we can replace sequential-style list-processing analysis programs with networks of asynchronously-communicating modules that share data over the network while exploiting the parallel processing capabilities of a pre-defined set of available machines.

3.2. *GSEBricks Interface*

Figure 4 shows a screenshot from our interface to the GSEBricks system. Users can graphically arrange visual components, each corresponding to an underlying GSEBricks class, into structures that represent the flow of computation. This extension also allows non-sequential computational flows – trees, or other non-simply connected structures – to be assembled and computed. The interface uses a dynamic type system to ensure that the workflow connects components in a typesafe manner.

Workflows which can be laid out and run with the graphical interface can also be programmed directly using their native Java interfaces. The second half of Figure 4 gives an example of a code-snippet that performs the same operation using the native GSEBricks components in Java.

4. Representing and Storing ChIP-chip Binding Hypotheses

The final element of the GSE database is a system to store not just raw experimental data but also a representation of a scientist's beliefs about that data.

Investigators often wish to discover the "regulatory networks" of binding that describe transcriptional control in a particular condition or cell type. For a single experiment, the network is simply a set of genes located near high-confidence binding sites[3,16,14]. With multiple experiments, each set of gene targets (the 'network') is characterized by the binding profiles of multiple factors simultaneously. If the investigator is interested in the

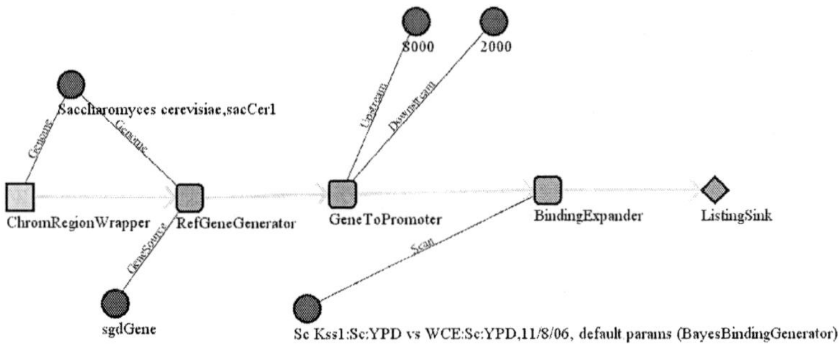


```
BindingScanLoader loader = new BindingScanLoader();
Genome sacCer1 = Organism.findGenome("sacCer1");

ChromRegionWrapper chroms = new ChromRegionWrapper(sacCer1);
Iterator chromItr = chroms.execute();

RefGeneGenerator rgg = new RefGeneGenerator(sacCer1, "sgdGene");
Iterator geneItr = new ExpanderIterator(rgg, chromItr);

GeneToPromoter g2p = new GeneToPromoter(8000, 2000);
Iterator promItr = new MapperIterator(g2p, geneItr);

BindingScan kss1 = loader.loadScan(sacCer1, kss1_id);
BindingExpander exp = new BindingExpander(loader, kss1);
Iterator bindingItr = new ExpanderIterator(exp, promItr);

while(bindingItr.hasNext()) {
    System.out.println(bindingItr.next());
}
```

Figure 4. A GSEBricks pipeline to count the genes in a genome. Each box represents a component that maps objects of some input type to a set of output objects. The circles represent constants that parameterize the behavior of the pipeline. The code on the right replicates the same pipeline using Java components.

behavior of those regulating factors, she will need to summarize the behaviors of the regulators across multiple sets of genes[14]. Once a biologist has outlined what she thinks is the "regulatory network" of a collection of factors, she is faced with the problem of formalizing those conclusions in a way that is useful to other scientists, or even to herself at some distant time in the future.

GSE gives the user a simple language to express relationships between different ChIP-chip experiments whose binding events have been precalculated and saved. GSE also provides the user with a schema for storing those hypotheses in the database and for automatically checking those hypotheses against new and unexamined experiments. In this way, we can think of the Hypothesis system as a kind of basic "lab notebook" for the analysis of ChIP-chip binding data.

Our hypotheses, H, have a simple grammar: $F := \{\text{factors}\}$ and $H := F|H \rightarrow H$. We can treat a hypothesis h as a predicate on the set of distinct genomic coordinates, G. If $h = \texttt{F}$, then $h(x)$ if and only if a binding event of $\texttt{F}$ is located at x. We can also relax this condition to include binding "within a certain distance" from one factor to another. The $\rightarrow$ of our hypothesis language is material implication from logic. If $h = \texttt{H1} \rightarrow \texttt{H2}$, then $h(x)$ holds if and only if either $\texttt{H2}(x)$ or $\neg\texttt{H1}(x)$.

We will evaluate hypotheses in reverse- instead of asking how much the data supports a particular hypothesis, we search for examples that contradict the hypothesis. In other words, we treat different (and distant) genomic coordinates as independent witnesses to the validity of a particular hypothesis and we ask how many locations seem to invalidate the hypothesis. The approach is computationally simple because the logical structure of our language will make it easy to quickly evaluate a fixed set of hypotheses against wide regions of genome which have been assayed with large numbers of binding experiments. We will also be able to easily leverage the high-throughput nature of our experiments, which might slow more complex algorithms to an unusable speed. Our approach is also useful because it gives the user a way to systematically enumerate and test the set of exceptions to a hypothesis.

In Table 1, we show the automatic results generated by our Hypothesis system when compared against the Harbison yeast regulatory code dataset[8]. For three factors we report the top ten ranked hypotheses about genes regulated by Fkh2, Rap1, and Ste12. Each column is followed by the number of 'inconsistent' probes that were found by the Hypothesis system. The results are not given a probabilistic interpretation, or even a description beyond just their ranked lists. It is, however, reassuring that such a simple analysis can easily recover most of the known related or interacting factors for these three simple cases[15,20,1].

5. Conclusion

We have described GSE, a system to represent microarray data and metadata in a relational database, and described a software system which reads and presents that data in a modular, extensible way. A reference imple-

	FKH2	#errors	RAP1	#errors	STE12	#errors
0	→ FKH1	82	→ FHL1	131	→ DIG1	63
1	→ NDD1	86	→ GAT3	195	→ TEC1	98
2	→ SWI6	112	→ YAP5	199	→ NDD1	114
3	→ SWI4	114	→ PDR1	201	→ SWI6	115
4	→ MBP1	116	→ SMP1	205	→ MCM1	116

mentation of this system will be available through the Gifford Lab group website, `http://cgs.csail.mit.edu`. This implementation includes an interactive, Java application for visualization and analysis that uses this modular system to browse and view ChIP-chip experiments and genome annotation data.

We have outlined our opinion that the automatic discovery of regulatory relationships from databases like GSE can only occur when the database itself stores hypotheses about the data. We have sketched a rudimentary hypothesis system which can automatically read simple hypotheses from the GSE database and check them in a non-probabilistic way against pre-computed binding event scans.

In the near future, we will extend our system to handle new kinds of large-scale ChIP-based data. Specifically, we are developing a schema and a set of GSEBricks components to efficiently handle the multi-terabyte datasets we expect to receive from new ChIPSeq machines[11].

References

1. Ziv Bar-Joseph and Georg et al. Gerber. Computational discovery of gene modules and regulatory networks. *Nature Biotechnology*, 21:1337–1342, October 2003.
2. DA Benson, I Karsch-Mizrachi, DJ Lipman, J Ostell, and DL Wheeler. Genbank. *Nucleic Acids Research*, 35:21–25, January 2007.
3. LA Boyer, TI Lee, MF Cole, SE Johnstone, SS Levine, JP Zucker, MG Guenther, RM Kumar, HL Murray, RG Jenner, DK Gifford, DA Melton, R Jaenisch, and RA Young. Core transcriptional regulatory circuitry in human embryonic stem cells. *Cell*, 122(6):947–956, September 2005.
4. MA Crosby, JL Goodman, VB Strelets, P Zhang, WM Gelbart, and Flybase Consortium. Flybase: genomes by the dozen. *Nucleic Acids Research*, 35:486–491, 2007.
5. R. Dowell, R. Jokerst, A. Day, S. Eddy, and L. Stein. The distributed annotation system. *BMC Bioinformatics*, 2, Oct 2001. 10.1186/1471-2105-2-7.
6. SS et. al. Dwight. Saccharomyces genome database: underlying principles and organisation. *Brief Bioinformatics*, 5(1):9–22, Mar 2004.
7. Brazma et. al. Minimum information about a microarray experiment (mi-

ame)[mdash]toward standards for microarray data. *Nature Genetics*, 29:365–371, Dec 2001. 10.1038/ng1201-365.
8. Harbison et al. Transcriptional regulatory code of a eukaryotic genome. *Nature*, 431:99–104, September 2004.
9. Lee et al. Transcriptional regulatory networks in *s*accharomyces cerevisiae. *Science*, 298:799–804, October 2002.
10. Ren et al. Genome-wide location and function of dna binding proteins. *Science*, 290:2306–2309, December 2000.
11. David S. Johnson, Ali Mortazavi, Richard M. Myers, and Barbara Wold. Genome-wide mapping of in vivo protein-dna interactions. *Science*, 316(5830):1497–1502, 2007.
12. D Karolchik, R Baertsch, M Diekhans, TS Furey, A Hinrichs, YT Lu, KM Roskin, M Schwartz, CW Sugnet, DJ Thomas, RJ Weber, D Haussler, and WJ Kent. The ucsc genome browser database. *Nucleic Acids Research*, 31(1):51–54, 2003.
13. YH et. al Loh. The oct4 and nanog transcription network regulates pluripotency in mouse embryonic stem cells. *Nature Genetics*, 38:431–440, March 2006.
14. DT Odom, RD Dowell, ES Jacobsen, W Gordon, TW Danford, KD MacIsaac, PA Rolfe, CM Conboy, DK Gifford, and E Fraenkel. Tissue-specific transcriptional regulation has diverged significantly between human and mouse. *Nature Genetics*, 39:730–732, 2007.
15. Yitzhak Pilpel, Priya Sudarsanam, and George M. Church. Identifying regulatory networks by combinatorial analysis of promoter elements. *Nature Genetics*, 29:153–159, September 2001.
16. Dmitry K. Pokholok, Julia Zeitlinger, Nancy M. Hannett, David B. Reynolds, and Richard A. Young. Activated signal transduction kinases frequently occupy target genomes. *Science*, 313:533–536, July 2006.
17. Yuan Qi, Alex Rolfe, Kenzie MacIsaac, Georg Gerber, Dmitry Pokholok, Julia Zeitlinger, Timothy Danford, Robin Dowell, Ernest Fraenkel, Tommi Jaakkola, Richard Young, and David Gifford. High-resolution computational models of genome binding events. *Nature Biotechnology*, 24(8):963–970, August 2006.
18. M. Reich, Liefeld T., J. Gould, J. Lerner, P. Tamayo, and J.P. Mesirov. Genepattern 2.0. *Nature Genetics*, pages 500–501, 2006.
19. E. Segal, R. Yelensky, A. Kaushal, T. Pham, A. Regev, D. Koller, and N. Friedman. Genexpress: A visualization and statistical analysis tool for gene expression and sequence data. 2004.
20. Priya Sudarsanam, Yitzhak Pilpel, and George M. Church. Genome-wide cooccurrence of promoter elements reveals a cis-regulatory cassette of rrna transcription motifs in *s*accharomyces cerevisiae. *Genome Research*, 12(11):1723–1731, November 2002.
21. F.C. Wardle and D.T. et al Odom. Zebrafish promoter microarrays identify actively transcribed embryonic genes. *Genome Biology*, 7(R71), August 2006.
22. L Weng, H Dai, Y Zhan, Y He, S Stepaniants, and D Bassett. Rosetta error model for gene expression analysis. *Bioinformatics*, 22(9):1111–1121, 2006.

TRANSLATING BIOLOGY: TEXT MINING TOOLS THAT WORK

K. BRETONNEL COHEN
HONG YU
PHILIP E. BOURNE
LYNETTE HIRSCHMAN

1. Introduction

This year is the culmination of two series of sessions on natural language processing and text mining at the Pacific Symposium on Biocomputing. The first series of sessions, held in 2001, 2002, and 2003, coincided with a period in the history of biomedical text mining in which much of the ongoing research in the field focussed on named entity recognition and relation extraction. The second series of sessions began in 2006. In the first two years of this series, the sessions focussed on tasks that required mapping to or between grounded entities in databases (2006) and on cutting-edge problems in the field (2007). The goal of this final session of the second series was to assess where the past several years' worth of work have gotten us, what sorts of deployed systems they have resulted in, how well they have managed to integrate genomic databases and the biomedical literature, and how usable they are. To this end, we solicited papers that addressed the following sorts of questions:

- What is the actual utility of text mining in the work flows of the various communities of potential users—model organism database curators, bedside clinicians, biologists utilizing high-throughput experimental assays, hospital billing departments?
- How usable are biomedical text mining applications? How does the application fit into the workflow of a complex bioinformatics pipeline? What kind of training does a bioscientist require to be able to use an application?
- Is it possible to build portable text mining systems? Can systems be adapted to specific domains and specific tasks without the assistance of an experienced language processing specialist?

- How robust and reliable are biomedical text mining applications? What are the best ways to assess robustness and reliability? Are the standard evaluation paradigms of the natural language processing world—intrinsic evaluation against a gold standard, post-hoc judging of outputs by trained judges, extrinsic evaluation in the context of some other task—the best evaluation paradigms for biomedical text mining, or even sufficient evaluation paradigms?

2. The session

29 submissions were received. Each paper received at least three reviews by members of a program committee composed of biomedical language processing specialists from North America, Europe, and Asia. Nine papers were accepted. All four of the broad questions were addressed by at least one paper. We review all nine papers briefly here.

2.1. *Utility*

A number of papers addressed the issue of utility. Alex et al.[1] experimented with a variety of forms of automated curator assistance, measuring curation time and assessing curator attitudes by questionnaire, and found that text mining techniques can reduce curation times by as much as one third. Caporaso et al.[3] examined potential roles for text-based and alignment-based methods of annotating mutations in a database curation workflow. They found that text mining techniques can provide a quality assurance mechanism for genomic databases. Roberts and Hayes[9] analyzed a large collection of information requests from an understudied population—commercial drug developers—and found that various families of text mining solutions can play a role in meeting the information needs of this group. Wang et al. [11] evaluated a variety of algorithms for performing gene normalization, and found that there are complex interactions between performance on a gold standard, improvement in curator efficiency, portability, and the demands of different kinds of curation tasks.

2.2. *Usability*

Divoli et al.[4] applied a user-centered design methodology to investigate questions about the kinds of information that users want to see displayed in interfaces for performing biomedical literature searches. Among other findings, they report that users showed interest in having gene synonyms displayed as part of the search interface, and that they would like to see extracted information about genes, such chemicals and drugs with which they are associated, displayed as part of the results.

2.3. *Portability*

Leaman and Gonzalez[8] focused on portability of gene mention detection techniques across different semantic classes of named entities and across corpora. Wang et al.[11] took portability issues into account in their study of the effects of various gene normalization algorithms on curator efficiency. The challenge of building systems that can be ported to new domains without the assistance of a text mining specialist remains untackled.

2.4. *Robustness and reliability*

A number of authors looked at issues related to the adequacy of traditional text mining evaluation paradigms, either directly or indirectly. Caporaso et al.[3] examined the correspondence between system performance on intrinsic and extrinsic evaluations, and found that high performance on a corpus does not necessarily predict performance on an actual annotation task well, due in part to the necessity of access to full-text journal articles for database curation. Kano et al.[7] explored the role of well-engineered integration platforms in building complex language processing systems from independent components, and showed that a well-designed platform can be used to determine the optimum set of components to combine for a specific relation extraction task. Wang et al.[11] found that the best-performing algorithms for gene normalization as determined by intrinsic evaluation against a gold-standard data set is not necessarily the most effective algorithm for accelerating curation time.

2.5. *Other topics*

Dudley and Butte[5] explored the use of natural language processing techniques to solve a fundamental problem in translational medicine: distinguishing data subsets that deal with disease-related experimental conditions from those that deal with normal controls. Finally, Brady and Shatkay[2] demonstrated that text mining can be used to apply subcellular localization prediction to almost any protein, even in the absence of published data about it.

3. Conclusions

Some of the most influential and frequently-cited papers in what might be called the "genomic era" of biomedical language processing were presented at PSB. Fukuda et al.'s early and oft-cited paper on named entity recognition for the gene mention problem[6] appeared at PSB in 1998; more recently, Schwartz and Hearst's algorithm for identifying abbreviation definitions in biomedical text[10] rapidly became one of the most frequently used components of biomedical text mining systems after being presented at PSB in 2003. The years since the first PSB text mining sessions have seen phenomenal growth in the amount of work on biomedical text mining, several deployed systems, and an expansion of the range of research in the field from the foundational tasks of named entity recognition and binary relation extraction to cutting-edge work on a wide range of language processing problems. The work presented in this year's session suggests that we are just beginning to tap the potential of text mining to contribute to the work of computational bioscience.

Acknowledgments

K. Bretonnel Cohen's participation in this work was funded by NIH grants R01-LM008111 and R01-LM009254 to Lawrence Hunter. Hong Yu's participation was supported by a Research Committee Award, a Research Growth Initiative grant, and an MiTAG award from the University of Wisconsin, as well as NIH grant R01-LM009836-01A1.

References

1. Beatrice Alex, Claire Grover, Barry Haddow, Mijail Kabadjov, Ewan Klein, Michael Matthews, Stuart Roebuck, Richard Tobin, and Xinglong Wang. Assisted curation: Does text mining really help? In *Pacific Symposium on Biocomputing 2008*, 2008.
2. Scott Brady and Hagit Shatkay. EpiLoc: A (working) text-based system for predicting protein subcellular location. In *Pacific Symposium on Biocomputing 2008*, 2008.
3. J. Gregory Caporaso, Nita Deshpande, J. Lynn Fink, Philip E. Bourne, K. Bretonnel Cohen, and Lawrence Hunter. Intrinsic evaluation of text mining tools may not predict performance on realistic tasks. In *Pacific Symposium on Biocomputing 2008*, 2008.
4. Anna Divoli, Marti A. Hearst, and Michael A. Wooldridge. Evidence for showing gene/protein name suggestions in bioscience literature search interfaces. In *Pacific Symposium on Biocomputing 2008*, 2008.

5. Joel Dudley and Atul J. Butte. Enabling integrative genomic analysis of high-impact human diseases through text mining. In *Pacific Symposium on Biocomputing 2008*, 2008.
6. K. Fukuda, A. Tamura, T. Tsunoda, and T. Takagi. Toward information extraction: identifying protein names from biological papers. In *Pacific Symposium on Biocomputing*, pages 707–718, Human Genome Center, University of Tokyo, Japan. ichiro@ims.u-tokyo.ac.jp, 1998.
7. Yoshinobu Kano, Ngan Nguyen, Rune Sætre, Kazuhiro Yoshida, Yusuke Miyao, Yoshimasa Tsuruoka, Yuichiro Matsubayashi, Sophia Ananiadou, and Jun'ichi Tsujii. Filling the gaps between tools and users: A tool comparator, using protein-protein interaction as an example. In *Pacific Symposium on Biocomputing 2008*, 2008.
8. Robert Leaman and Graciela Gonzalez. BANNER: An executable survey of advances in biomedical named entity recognition. In *Pacific Symposium on Biocomputing 2008*, 2008.
9. Phoebe M. Roberts and William S. Hayes. Information needs and the role of text mining in drug development. In *Pacific Symposium on Biocomputing 2008*, 2008.
10. A.S. Schwartz and M.A. Hearst. A simple algorithm for identifying abbreviation definitions in biomedical text. In *Pacific Symposium on Biocomputing*, volume 8, pages 451–462, 2003.
11. Xinglong Wang and Michael Matthews. Comparing usability of matching techniques for normalising biomedical named entities. In *Pacific Symposium on Biocomputing 2008*, 2008.

ASSISTED CURATION: DOES TEXT MINING REALLY HELP?

BEATRICE ALEX, CLAIRE GROVER, BARRY HADDOW, MIJAIL KABADJOV, EWAN KLEIN, MICHAEL MATTHEWS, STUART ROEBUCK, RICHARD TOBIN, AND XINGLONG WANG

School of Informatics
University of Edinburgh
EH8 9LW, UK
E-mail for correspondence: balex@inf.ed.ac.uk

Although text mining shows considerable promise as a tool for supporting the curation of biomedical text, there is little concrete evidence as to its effectiveness. We report on three experiments measuring the extent to which curation can be speeded up with assistance from Natural Language Processing (NLP), together with subjective feedback from curators on the usability of a curation tool that integrates NLP hypotheses for protein-protein interactions (PPIs). In our curation scenario, we found that a maximum speed-up of 1/3 in curation time can be expected if NLP output is perfectly accurate. The preference of one curator for consistent NLP output and output with high recall needs to be confirmed in a larger study with several curators.

1. Introduction

Curating biomedical literature into relational databases is a laborious task requiring considerable expertise, and it is proposed that text mining should make the task easier and less time-consuming [1, 2, 3]. However, to date, most research in this area has focused on developing objective performance metrics for comparing different text mining systems (see [4] for a recent example). In this paper, we describe initial feedback from the use of text mining within a commercial curation effort, and report on experiments to evaluate how well our NLP system helps curators in their task.

This paper is organised as follows. We review related work in Section 2. In Section 3, we introduce the concept of assisted curation and describe the different aspects involved in this process. Section 4 provides an overview of the components of our text mining system, the TXM (text mining) NLP pipeline, and describes the annotated corpus used to train and evaluate this system. In Section 5, we describe and discuss the results of three different curation experiments which attempt to test the effectiveness of various versions of the NLP pipeline in assisting curation. Discussion and conclusions follow in Section 6.

2. Related Work

Despite the recent surge in the development of information extraction (IE) systems for automatic curation of biomedical data spurred on by the BioCreAtIvE II competition [5], there is a lack of user studies that extrinsically evaluate the usefulness of IE as a way to assist curation. Donaldson et al. [6] reported an estimated 70% reduction in curation time of yeast-protein interactions when using the PreBIND/Textomy IE system, designed to recognise abstracts containing protein interactions. This estimate is limited to the document selection component of PreBind and does not include time savings due to automatic extraction and normalization of named entities (NEs) and relations. Karamanis et al. [7] studied the functionality and usefulness of their curation tool, ensuring that integrating NLP output does not impede curators in their work. In three curation experiments with one curator, they found evidence that improving their curation tool and integrating NLP speeds up curation compared to using a tool prototype with which the curator was not experienced at the start of the experiment. Karamanis et al. [7] mainly focus on tool functionality and presentational issues. They did not analyse the aspects of the NLP output that were useful to curators, how it affected their work, or how the NLP pipeline can be tuned to simplify the curator's job. Recently, Hearst et al. [8] reported on a pilot usability study showing positive reactions to figure display and caption search for bioscience journal search interfaces.

Regarding non-biomedical-related applications, Kristjansson et al. [9] describe an interactive IE tool with constraint propagation to reduce human effort in address form filling. They show that highlighting contact details in unstructured text, pre-populating form fields, and interactive error correction by the user reduces the cognitive load on users when entering address details into a database. This reduction is reflected in the expected number of user actions, which is determined based on the number of clicks to enter all fields. They also integrated confidence values to inform the user about the reliability of extracted information.

3. Assisted Curation

The curation task that we will discuss in this paper requires curators to identify examples of protein-protein interactions (PPIs) in biomedical literature. The initial step involves retrieving a set of papers that match criteria for the curation domain. After an initial step of further filtering the papers into promising candidates for curation, curators proceed on a paper-by-paper basis. Using an inhouse editing and verification tool (henceforth referred to as the 'Editor'), the curators are able to read through an electronic version of the paper and enter retrieved information into a template which will then be used to add a record to a relational database.

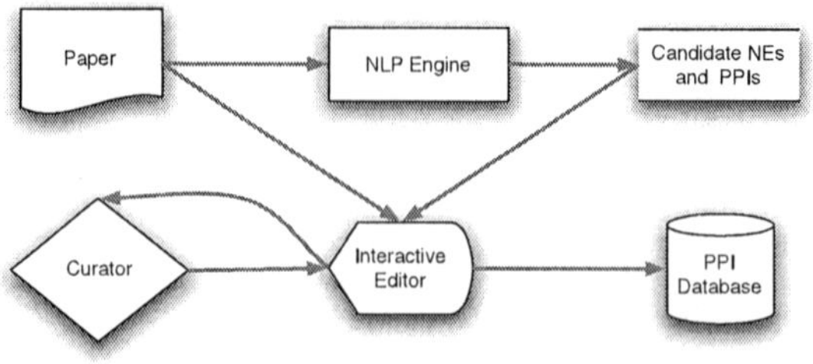

Figure 1. Information Flow in the Curation Process

Curation is a laborious task which requires considerable expertise. The curator spends a significant amount of time on reading through a paper and trying to locate material that might contain curatable facts. Can NLP help the curator work more efficiently? Our basic assumption, which is commonly held [1], is that IE techniques are likely to be effective in identifying relevant entities and relations. More specifically, we assume that NLP can propose *candidate* PPIs; if the curators restrict their attention to these candidates, then the time required to explore the paper can be reduced. Notice that we are not proposing that NLP should *replace* human curators—given the current state of the art, only expert humans can assure that the captured data is of sufficiently high quality to be entered into databases.

Our curation scenario is illustrated in Figure 1. The source paper undergoes processing by the NLP engine. The result is a set of normalised NEs and candidate PPIs. The original paper and the NLP output are fed into the interactive Editor, which then displays a view to the curator. The curator makes a decision about which information to enter into the Editor, which is then communicated to a backend database.

In one sense, we can see this scenario as one in which the software provides *decision support* to the human. Although in broad terms the decision is about what facts, if any, to curate, this can be broken down into smaller subtasks. Given a sentence S, (i) do the terms in S name proteins? If so, (ii) which proteins do they name? And (iii), given two protein mentions, do the proteins stand in an interaction relation? These decision subtasks correspond to three components of the NLP engine: (i) Named Entity Recognition, (ii) Term Identification, and (iii) Relation Extraction. We will examine each of these in turn shortly, but first, we want to consider further the kind of choices that need to be made in examining the usability of NLP for curation. A crucial observation is that the NLP output is bound to be imperfect. How can the curator make use of an *unreliable* assistant?

First, there are **interface design** issues—what information is displayed to the curator, in what form, and what kind of manipulations can the curator carry out?

Second, what is the **division of labour** between the human and the software? For example, there might be some decisions which are relatively cheap for the curator to make, such as deciding what species is associated with a protein mention, and which can then help the software in providing a more focused set of candidates for term identification.

Third, what are the optimal **functional characteristics** of the NLP engine, given that complete reliability is not currently attainable? For example, should the NLP try to improve recall over precision, or *vice versa*?

Although the first and second dimensions are clearly important, in this paper we will focus on the third, namely the functional characteristics of our system.

4. TXM Pipeline

The NLP output displayed in the interactive curation Editor is produced by the TXM pipeline, an IE pipeline that is being developed for use in biomedical IE tasks. The particular version of the pipeline used in the experiments described here focuses on extracting proteins, their interactions, and other entities which are used to enrich the interactions with extra information of biomedical interest. Proteins are also normalised (i.e., mapped to identifiers in an appropriate database) using the term identification (TI) component of the pipeline. In this section a brief description of the pipeline, and the corpus used to develop and test it, will be given, with more implementation details provided by appropriate references.

Corpus In order to use machine learning approaches for named entity recognition (NER) and relation extraction (RE), and for evaluating the pipeline components, an annotated corpus was produced using a team of domain experts. Since the annotations contain information about proteins and their interactions, it is referred to as the enriched protein-protein interaction (EPPI) corpus. The corpus consists of 217 full-text papers selected from PubMed and PubMedCentral as containing experimentally proven PPIs. The papers, retrieved in XML or HTML, were converted to an internal XML format. Nine types of entities (Complex, CellLine, DrugCompound, ExperimentalMethod, Fusion, Fragment, Modification, Mutant, and Protein) were annotated, as well as PPI relations and FRAG relations (which link Fragments or Mutants to their parent proteins). Furthermore, proteins were normalised to their RefSeq[a] identifier and PPIs were enriched with properties and attributes. The properties added to the PPIs are IsProven, IsDirect and IsPositive and the possible attributes are CellLine, DrugTreatment, ExperimentalMethod or ModificationType. More details on properties and attributes can be found in Haddow

[a]http://www.ncbi.nlm.nih.gov/RefSeq/index.html

and Matthews [10]. The inter-annotator agreement (IAA), measured on a sample of doubly and triply annotated papers, amounts to an overall micro-averaged F1-score[b] of 84.9 for NEs, 88.4 for normalisations, 64.8 for PPI relations, 87.1 for properties and 59.6 for attributes. The EPPI corpus ($\simeq$2m tokens) is divided into three sections, TRAIN (66%), DEVTEST (17%), and TEST (17%).

Pre-processing A set of pre-processing steps in the pipeline was implemented using the LT-XML2 tools [11]. The pre-processing performs sentence boundary detection and tokenization, adds useful linguistic markup such as chunks, part-of-speech tags, lemmas, verb stems, and abbreviation information, and also attaches NCBI taxonomy identifiers to any species-related terms.

Named Entity Recognition The NER component is based on the C&C tagger, a Maximum Entropy Markov Model (MEMM) tagger developed by Curran and Clark [12], and augmented with extra features and gazetteers tailored to the domain and described fully in Alex et al. [13]. The C&C tagger allows for the adjustment of the entity decision threshold through the `prior` file, which has the effect of varying the precision-recall balance in the output of the component. This `prior` file was modified to produce the high precision and high recall models used in the assisted curation experiment described in Section 5.3.

Term Identification The TI component uses a rule-based fuzzy matcher to produce a set of candidate identifiers for each recognized protein. Species are assigned to proteins using a machine learning based tagger trained on contextual and species word features [14]. The species information and a set of heuristics are used to choose the most probable identifiers from the set of candidates proposed by the matcher. The evaluation metric for the TI system is *bag accuracy*. This means that if the system produces multiple identifiers for an entity mention, it is counted as a hit as long as one of the identifiers is correct. The rationale is that since a TI system that outputs one identifier is not accurate enough, generating a bag of choices increases chances of finding the correct one. This can assist curators as the right identifier can be chosen from a bag (see [15] for more details).

Relation Extraction Intra-sentential PPI and FRAG relations are both extracted using the system described in Nielsen [16], with inter-sentential FRAG relations addressed using a maximum entropy model trained on features derived from the entities, their context, and other entities in the vicinity. Enriching the relations with properties and attributes is implemented using a mixture of machine learning and rule-based methods described in Haddow and Matthews [10].

[b] Micro-averaged F1-score means that each example is given equal weight in the evaluation.

Component Performance The performance of the IE components of the pipeline (NER, TI, and RE) is measured using precision, recall, and F1-score (except TI– see above), by testing each component in isolation and comparing its output to the annotated data. For example, RE is tested using the annotated (gold) entities as its input, rather than the output of NER, in order that NER errors not affect the score for RE. Table 1 shows the performance of each component when tested on DEVTEST, where the machine learning components are trained on TRAIN.

Table 1. Performance of pipeline components, tested in isolation on DEVTEST and trained on TRAIN.

Component	TP	FP	FN	Precision	Recall	F1
NER (micro-average)	19,925	5,964	7,755	76.96	71.98	74.39
RE (PPI)	1,208	1,173	1,080	50.73	52.80	51.75
RE (FRAG)	1,699	963	1,466	63.82	53.68	58.31
RE (properties micro-average)	3,041	567	579	84.28	84.01	84.14
RE (attributes micro-average)	483	822	327	37.01	59.63	45.67
Component	TP	FP	FN	Precision	Recall	Bag Acc.
TI (micro-average)	9,078	91,396	2,843	9.04	76.15	76.15

5. Curation Experiments

We conducted three curation experiments with and without assistance from the output of the NLP pipeline or gold standard annotations (GSA). In all of the experiments, curators were asked to curate several documents according to internal guidelines. Each paper is assigned a curation ID for which curators create several records corresponding to the curatable information in the document. Curators always use an interactive Editor which allows them to see the document on screen and enter the curatable information into record forms. All curators are experienced in using the interactive curation Editor, but not necessarily familiar with assisted curation. After completing the curation for each paper, they were asked to fill in a questionnaire.

5.1. *Manual versus Assisted Curation*

In the first experiment, 4 curators curated 4 papers in 3 different conditions:

- **MANUAL:** without assistance
- **GSA-assisted:** with integrated gold standard annotations
- **NLP-assisted:** with integrated NLP pipeline output

Each curator processed a paper only once, in one specific condition, without being informed about the type of assistance (GSA or NLP), if any. This experiment

Table 2. Total number of records curated in each condition and average curation speed per record.

		Time per record	
Condition	Records	Average	StDev
MANUAL	121	312s	327s
GSA	170	205s	52s
NLP	141	243s	36s

Table 3. Average questionnaire scores. Scores ranged from (1) for strongly agree to (5) for strongly disagree.

Statement	GSA	NLP
NLP was helpful in curating this documents	2.75	3.25
NLP speeded up the curation of this paper	3.75	3.75
NE annotations were useful for curation	2.50	3.00
Normalizations of NEs were useful for curation	2.75	2.75
PPIs were useful for curation	3.50	3.25

aims to answer the following questions: Does the NLP output which is currently integrated in the interactive Editor accelerate curation? Secondly, do human gold standard annotations assist curators in their work—i.e., how helpful would NLP be to a curator if it performed as well as a human annotator?

Table 2 shows that for all four papers, the fewest records (121) were curated during manual curation, 20 more records (+16.5%) were curated given NLP assistance, and 49 more records (+40.5%) with GSA assistance. This indicates that providing NLP output helps curators to spot more information. Ongoing work involves a senior curator assessing each curated record in terms of quality and coverage. This will provide evidence for whether this additional information is also curatable, i.e. how the NLP output affects curation accuracy, and also give an idea of inter-curator agreement for different conditions. As each curator curated in all three conditions but never curated the same paper twice, inter-document and inter-curator variability must be considered. Therefore, we present curation speed per condition as the average speed of curating a record. Manual curation is most time-consuming, followed by NLP-assisted curation (22% faster), followed by GSA-assisted curation (34% faster). Assisted curation clearly speeds up the work of a curator, and a maximum reduction of $1/3$ in manual curation time can be expected if the NLP pipeline performed with perfect accuracy.

In the questionnaire, curators rated GSA assistance slightly more positively than NLP assistance (see Table 3). However, they were not convinced of either condition speeding up their work, even though the time measurements show otherwise. Considering that they were not familiar with assisted curation prior to the experiment, a certain effect of learning should be allowed for. Moreover, they

Table 4. Total number of records curated in each consistency condition and average curation speed per record.

	Time per record	
Condition	Average	StDev
CONSISTENCY1	128s	43s
CONSISTENCY2	92s	22s

may have had relatively high expectations of the NLP output. In fact, individual feedback in the questionnaire shows that NLP assistance was useful for some papers and some curators, but not others. Further feedback in the questionnaire includes aspects of visualization (e.g. PDF conversion errors) and interface design (e.g. inadequate display of information linked to NE normalizations) in the interactive Editor. Regarding the NLP output, curators also requested more accurate identification of PPI candidates, e.g. in coordinations like "A and B interact with C and D", and more consistency in the NLP output.

5.2. *NLP Consistency*

The NLP pipeline extracts information based on context features and may, for example, recognize a string as a protein in one part of the document but as a drug/compound in another, or assign different species to the same protein mentioned multiple times in the document. While this inconsistency may not be erroneous, the curators' feedback is that consistency would be preferred. To test this hypothesis, and to determine whether consistent NLP output helps to speed up curation, we conducted a second experiment. One curator was asked to curate 10 papers containing NLP output made consistent in two ways. In 5 papers, all NEs recognized by the pipeline were propagated throughout the document (CONSISTENCY1). In the other 5 papers, only the most frequent NE recognized for a particular surface form is propagated, while less frequent ones are removed (CONSISTENCY2). In both conditions, the most frequent protein identifier bag determined by the TI component is propagated for each surface form, and ePPIs are extracted as usual. Subsequent to completing the questionnaire, the curator viewed a second version of the paper in which consistency in the NLP output was not forced, and filled in a second questionnaire regarding the comparison of both versions.

Table 4 shows that the curator managed to curate 28% faster given the second type of consistency. However, examining the answers to the questionnaire listed in Table 5, it appears that the curator actually considerably preferred the first type of consistency, where all NEs that were recognized by the NER component are propagated throughout the paper. While this speed-up in curation may be attrac-

Table 5. Average questionnaire scores. Scores ranged from (1) for strongly agree to (5) for strongly disagree. In questionnaire 2, consistent (CONSISTENCY1/2) NLP output (A) is compared to baseline NLP (B).

Statement	CONSISTENCY1	CONSISTENCY2
Questionnaire 1		
NLP output was helpful for curation	1.6	2.6
NLP output speeded up curation	1.8	3.2
NEs were useful for curation	1.4	4.0
Normalizations of NEs were useful for curation	3.2	4.0
PPIs were useful for curation	3.6	4.2
Questionnaire 2		
A was more useful for curation than B would have been	2.6	4.0
A speeded up the curation process more than B would have	3.0	4.0
A appeared more accurate than B	2.6	4.2
A missed important information compared to B	4.4	1.8
A contained too much information compared to B	3.6	4.6

tive from a commercial perspective, this experiment illustrates how important it is to get feedback from users who may well reject a technology altogether if they are not happy working with it.

5.3. *Optimizing for Precision or Recall*

Currently, all pipeline components are optimized for F1-score, resulting in a relative balance between the correctness and coverage of extracted information, i.e. precision and recall. In previous curation rounds, curators felt they could not completely trust the NLP output, as some of the information displayed was incorrect. The final curation experiment tests whether optimizing the NLP pipeline for F1 is ideal in assisted curation, or whether a system that is more correct but misses some curatable information (high precision) or one that extracts most of the curatable information along with many non-curatable or incorrect facts (high recall) would be preferred. In this experiment, only the NE component was adapted to increase its precision or recall. This is done by changing the threshold in the C&C `prior` file to modify tag probabilities assigned by the C&C tagger.[c] The intrinsic evaluation scores of the NER component optimized either for F1, precision, or recall are listed in Table 6.

In the experiment, one curator processed 10 papers in random order containing NLP output, 5 with high recall NER and 5 with high precision. Note that to simplify

[c]Internal and external features were not optimized for precision or recall. This could be done to increase effects even more. The TI and RE components were also not modified for this experiment.

Table 6. Optimized F1-score versus high precision (P) and high recall (R) NER, along with corresponding counts of true positives (TP), false positives (FP), and false negatives (FN).

Setting	TP	FP	FN	P	R	F1
High F1	20,091	6,085	7,589	76.75	72.58	**74.61**
High P	11,836	1,511	15,844	**88.68**	42.76	57.70
High R	21,880	20,653	5,800	51.44	**79.05**	62.32

the experiment the curator did not normalise entities in this curation round. Subsequent to completing the questionnaire, the curator viewed a second version of the paper with NLP output based on optimized F1-score NER and filled in a second questionnaire regarding the comparison of both versions. The results in Table 7 show that the curator rated all aspects of the high recall NER condition more positively than of the high precision NER condition. Moreover, the curator tended to prefer NLP output with optimised F1 NER over that containing high precision NER, and NLP output containing high recall NER over that with high F1 NER. Although the number of curated papers is small, this curator seems to prefer NLP output that captures more curatable information but is overall less accurate. The curator noted that since her curation style involves skim-reading, the NLP output helped her to spot information that she otherwise would have missed. The results of this experiment could therefore be explained simply by curation style. Another curator with a more meticulous reading style may actually prefer more precise and trustworthy information extracted by the NLP pipeline. Clearly, the last curation experiment needs to be repeated using several curators, curating a larger set of papers, and providing additional timing information per curated record. In general, it would be useful to develop a system that will allow curators to filter information presented onscreen dynamically, possibly based on confidence values, as integrated in the tool described by Kristjansson et al. [9].

6. Discussion and Conclusions

This paper has focused on optimizing functional characteristics of an NLP pipeline for assisted curation, given that current text mining techniques for biomedical IE are not completely reliable. Starting with the hypothesis that assisted curation can support the task of a curator, we found that a maximum reduction of $1/3$ in curation time can be expected if NLP output is perfectly accurate. This shows that biomedical text mining can assist in curation. Moreover, NLP assistance led to the curation of more records, although the validity of this additional information still needs to be confirmed by a senior curator.

In extrinsic evaluation of the NLP pipeline in curation, we have tested several optimizations of the output in order to determine the type of assistance that is

Table 7. Average questionnaire scores. Scores ranged from (1) for strongly agree to (5) for strongly disagree. In questionnaire 2, optimized precision/recall (HighP/HighR) NER output (A) is compared to optimized F1 NER output (B).

Statement	HighP NER	HighR NER
Questionnaire 1		
NLP output was helpful for curation	3.0	2.2
NLP output speeded up curation	3.4	2.4
NEs were useful for curation	3.0	2.0
PPIs were useful for curation	3.2	2.5
Questionnaire 2		
A was more useful for curation than B would have been	4.2	2.6
A speeded up the curation process more than B would have	4.2	3.0
A appeared more accurate than B	4.4	2.8
A missed important information compared to B	1.4	3.2
A contained too much information compared to B	4.8	3.8

preferred by curators. We found that the curator prefers consistency, with all NEs propagated throughout the document, even though this preference is not reflected in the average time measurements for curating a record. When comparing curation with NLP output containing high recall or high precision NE predictions, the curator clearly preferred the former. While this result illustrates that optimizing an IE system for F1-score does not necessarily result in optimal performance in assisted curation, this experiment must be repeated with several curators in view of different curation styles.

Overall, we learnt that measuring curation in terms of curation time is not sufficient to capture the usefulness of NLP output for assisted curation. As recognized by Karamanis et al. [7], it is difficult to measure a curator's performance as one quantitative metric. The average time to curate a record, alone, is clearly not sufficient for capturing all factors involved the curation process. It is important to work closely with the user of a curation system in order to identify helpful and hindering aspects of such technology. In future work, we will conduct further curation experiments to determine the merit of high recall and high precision NLP output for the curation task. We will also invest some time in implementing confidence values of extracted information into the interactive Editor.

Acknowledgements

This work was carried out as part of an ITI Life Sciences Scotland (`http://www.itilifesciences.com`) research programme with Cognia EU (`http://www.cognia.com`) and the University of Edinburgh. The authors are very grateful to the curators at Cognia EU who participated in the experiments. The inhouse curation tool used for this work is the subject of International Patent Application No. PCT/GB2007/001170.

References

1. A. S. Yeh, L. Hirschman, and A. Morgan. Evaluation of text data mining for database curation: Lessons learned from the KDD challenge cup. *Bioinformatics*, 19(Suppl 1): i331–339, 2003.
2. D. Rebholz-Schuhmann, H. Kirsch, and F. Couto. Facts from text - is text mining ready to deliver? *PLoS Biology*, 3(2), 2005.
3. H. Xu, D. Krupke, J. Blake, and C. Friedman. A natural language processing (NLP) tool to assist in the curation of the laboratory mouse tumor biology database. *Proceedings of the AMIA 2006 Annual Symposium*, page 1150, 2006.
4. L. Hirschman, M. Krallinger, and A. Valencia, editors. *Second BioCreative Challenge Evaluation Workshop.* Fundación CNIO Carlos III, Madrid, Spain, 2007.
5. M. Krallinger, F. Leitner, and A. Valencia. Assessment of the second BioCreative PPI task: Automatic extraction of protein-protein interactions. In *Proceedings of the Second BioCreative Challenge Evaluation Workshop*, pages 41–54, Madrid, Spain, 2007.
6. I. Donaldson, J. Martin, B. de Bruijn, C. Wolting, V. Lay, B. Tuekam, S. Zhang, B. Baskin, G.D. Bader, K. Michalickova, T. Pawson, and C.W.V. Hogue. PreBIND and Textomy - mining the biomedical literature for protein-protein interactions using a support vector machine. *BMC Bioinformatics*, 4(11), 2003.
7. N. Karamanis, I. Lewin, R. Seal, R. Drysdale, and E. Briscoe. Integrating natural language processing with FlyBase curation. In *Proceedings of PSB 2007*, pages 245–256, Maui, Hawaii, 2007.
8. M. A. Hearst, A. Divoli, J. Ye, and M. A. Wooldridge. Exploring the efficacy of caption search for bioscience journal search interfaces. In *Proceedings of BioNLP 2007*, pages 73–80, Prague, Czech Republic, 2007.
9. T. T. Kristjansson, A. Culotta, P. Viola, and A. McCallum. Interactive information extraction with constrained conditional random fields. In Deborah L. McGuinness and George Ferguson, editors, *Proceedings of AAAI 2004*, pages 412–418, San Jose, US, 2004.
10. Barry Haddow and Michael Matthews. The extraction of enriched protein-protein interactions from biomedical text. In *Proceedings of BioNLP*, pages 145–152, Prague, Czech Republic, 2007.
11. C. Grover and R. Tobin. Rule-based chunking and reusability. In *Proceedings of LREC 2006*, pages 873–878, Genoa, Italy, 2006.
12. J. Curran and S. Clark. Language independent NER using a maximum entropy tagger. In *Proceedings of CoNLL-2003*, pages 164–167, Edmonton, Canada, 2003.
13. B. Alex, B. Haddow, and C. Grover. Recognising nested named entities in biomedical text. In *Proceedings of BioNLP 2007*, pages 65–72, Prague, Czech Republic, 2007.
14. X. Wang. Rule-based protein term identification with help from automatic species tagging. In *Proceedings of CICLING 2007*, pages 288–298, Mexico City, Mexico, 2007.
15. X. Wang and M. Matthews. Comparing usability of matching techniques for normalising biomedical named entities. In *Proceedings of PSB 2008*, 2008.
16. L. A. Nielsen. Extracting protein-protein interactions using simple contextual features. In *Proceedings of BioNLP 2006*, pages 120–121, New York, US, 2006.

EVIDENCE FOR SHOWING GENE/PROTEIN NAME SUGGESTIONS IN BIOSCIENCE LITERATURE SEARCH INTERFACES

ANNA DIVOLI, MARTI A. HEARST, MICHAEL A. WOOLDRIDGE

School of Information, UC Berkeley

{divoli,hearst,mikew}@.ischool.berkeley.edu

This paper reports on the results of two questionnaires asking biologists about the incorporation of text-extracted entity information, specifically gene and protein names, into bioscience literature search user interfaces. Among the findings are that study participants want to see gene/protein metadata in combination with organism information; that a significant proportion would like to see gene names grouped by type (synonym, homolog, etc.), and that most participants want to see information that the system is confident about immediately, and see less certain information after taking additional action. These results inform future interface designs.

1. Introduction

Bioinformaticians have developed numerous algorithms for extracting entity and relation information from the bioscience literature, and have developed some very interesting user interfaces for showing this information. However, little research has been done on the *usability* of these systems and how to best incorporate such information into literature search and text mining interfaces. As part of an on-going project to build a highly usable literature search tool for bioscience researchers, we are carefully investigating what kinds of biological information searchers want to see, as well as how they want to see this information presented. We are interested in supporting biologists whose main tasks are biological (as opposed to database curators and bioinformaticians doing text mining) and who presumably do not want to spend a lot of time searching.

We use methods from the field of Human Computer Interaction (HCI) for the careful design of search interfaces. We have already used these methods to develop a novel bioliterature search user interface whose focus is allowing users to search over and view figures and captions[5] (see http://biosearch.berkeley.edu). That interface is based on the observation that many researchers, when assessing a research article, first look at the title, abstract, and figures.

In this paper, we investigate whether or not bioscience literature searchers wish to see related term suggestions, in particular, gene and protein names, in response to their queries.[a] This is one step in a larger investigation in which we plan to assess presentation of other results of text analysis, such as the entities corresponding to diseases, pathways, gene interactions, localization information, function information, and so on.

When it comes to presenting users with the output of text mining programs, the interface designer is faced with an embarrassment of riches. There are many choices of entity and relationship information that can be displayed to the searcher. However, search user interface research suggests that users are quickly overwhelmed when presented with too many options and too much information.

Therefore, our approach is to assess the usability of one feature at a time, see how participants respond, and then test out other features. We focus on gene names here because of their prominent role in the queries devised for the TREC Genomics track[6], and because of their focus in text mining efforts, as seen in the BioCreative text analysis competitions[7]. Thus, this paper assesses one way in which the output of text mining can be useful for bioscience software tools.

In the remainder of this paper, we first describe the user-centered design process and then discuss related work. We then report on the results of two questionnaires. The first asked participants a number of questions about how they search the bioscience literature, including questions about their use of gene names. Among the findings were that participants did indeed want to see suggestions of gene names as part of their search experience. The second questionnaire, building on these results, asked participants to assess several designs for presenting gene names in a search user interface. Finally, we conclude the paper with plans for acting on the results of this study.

2. The User-Centered Design Process

We are following the method of *user-centered design*, which is standard practice in the field of Human-Computer Interaction (HCI)[11]. This method focuses on making decisions about the design of a user interface based on feedback obtained from target users of the system, rather than coding first and evaluating later. First a *needs assessment* is performed in which the designers investigate who the users are, what their goals are, and what tasks they have to complete in order to achieve those goals. The next stage is a *task analysis* in which the designers characterize which steps the users need to take to complete their tasks, decide which user goals they will attempt to support, and then create scenarios which exemplify these tasks being executed by the target user population.

---

[a] For the remainder of the paper, we will use the term *gene name* to refer to both gene and protein names.

Once the target user goals and tasks have been determined, design is done in a tight evaluation cycle consisting of mocking up prototypes, obtaining reactions from potential users, and revising the designs based on those reactions. This sequence of activities often needs to be repeated several times before a satisfactory design has been achieved. This is often referred to as "discount" usability testing, since useful results can be obtained with only a few participants. After a design is testing well in informal studies, formal experiments comparing different designs and measuring for statistically significant differences can be conducted.

This iterative procedure is necessary because interface design is still more of an art than a science. There are usually several good solutions within the interface design space, and the task of the designers is to navigate through the design space until reaching some local "optimum." The iterative process allows study participants to help the designers make decisions about which paths to explore in that space. Experienced designers often know how to start close to a good solution; less experienced designers need to do more work. Designing for an entirely novel interaction paradigm often requires more iteration and experimentation.

3. Research on Term Suggestions Usability

An important class of query reformulation aids is automatically suggested term refinements and expansions. Spelling correction suggestions are query reformulation aids, but the phrase *term expansion* is usually applied to tools that suggest alternative wordings.

Usability studies are generally positive as to the efficacy of term suggestions when users are not required to make relevance judgements and do not have to choose among too many terms. Those that produce negative results seem to stem from problems with the presentation interface[2]. Interfaces that allow users to reformulate their query by selecting a single term (usually via a hyperlink) seem to fare better. Anick[1] describes the results of a large-scale investigation of the effects of incorporating related term suggestions into a major web search engine. The term suggestion tool, called Prisma, was placed within the Altavista search engine's results page. The number of feedback terms was limited to 12 to conserve space in the display and minimize cognitive load. In a large web-based study, 16% of users applied the Prisma feedback mechanism at least once on any given day. However, effectiveness when measured in the occurrence of search results clicks did not differ between the baseline and the Prisma groups.

In a more recent study, Jansen et al.[9] analyzed 1.5M queries from a log taken in 2005 from the Dogpile.com metasearch engine. The interface for this engine shows suggested additional terms in a box on the righthand side under the heading

"Are you looking for?" Jansen et al. found that 8.4% of all queries were generated by the reformulation assistant provided by Dogpile. Thus, there is evidence that searchers use such term reformulations, although the benefits are as yet unproven.

4. Current Bioliterature Search Interfaces

There are a number of innovative interfaces for analyzing the results of text analysis. The iHOP system[8] converts the contents of PubMed abstracts into a network of information about genes and interactions, displaying sentences extracted from abstracts and annotated with entity information. The ChiliBot[3] system also shows extracted information in the form of relationships between genes, proteins, and keywords. TextPresso[10] uses an ontology to search over the full text of a collection of articles about *C. elegans*, extracting out sentences that contain entities and relations of interest. These systems have not been assessed in terms of usability of their interface or their features.

The GoPubMed system[4] shows a wealth of information in search results over PubMed. Most prominent is a hierarchical display of a wide range of categories from the Gene Ontology and MeSH associated with the article. Users may sort search results by navigating in this hierarchy and selecting categories. This interface is compelling, but it is not clear which kinds of information are most useful to show, whether a hierarchy is the best way to show metadata information for grouping search results, and whether or not this is too much information to show. The goal of this paper is to make a start at determining which kinds of information searchers want to see, and how they want to select it.

5. First Questionnaire: Biological Information Preferences

Both studies were administered in the form of an online questionnaire. For the first study, we recruited biosciences researchers from 7 research institutions via email lists and personal contacts. The 38 participants were all from academic institutions (22 graduate students, 6 postdoctoral researchers, 5 faculty, and 5 others), and had a wide range of specialties, including systems biology, bioinformatics, genomics, biochemistry, cellular and evolutionary biology, microbiology, physiology and ecology.

Figure 1 shows the percentage of time each participant uses computers for their work. A surprising 37% say they use computers for 80-100% of the time they are working, although only 6 participants listed bioinformatics as one of their fields. Participants were for the most part heavy users of literature search; 84% said they search biomedical literature either daily or weekly.

We asked participants which existing literature search tools they use, and for

When you are doing your work, approximately what percentage of the time involves your using a computer?

0-20 %	2	5%
20-40 %	7	18%
40-60 %	8	21%
60-80 %	7	18%
80-100 %	14	37%
Total	38	100%

How often do you search the biomedical literature?

Every day	18	47%
Every week	14	37%
Every month	3	8%
Rarely	3	8%
Never	0	0%
Total	38	100%

What proportion of your searches include gene/protein names?

I don't use gene/protein names in my queries	5	13%
less than 10% of my searches	5	13%
10-25% of my searches	8	21%
25-50% of my searches	6	16%
50-75% of my searches	11	29%
75-100% of my searches	3	8%
Total	38	100%

Figure 1. Statistics on computer use, search frequency, and percentage of queries that include gene names.

what percent of their searches. 12 participants (32%) said they use PubMed 80% of the time or more; on average it was used 50% of the time. Google Scholar was used on average 25% of the time; all but 3 participants used it at least some of the time. 6 participants used Ovid at least 5% of the time. The other popular search engine mentioned was the ISI Web of Science, which 9 participants used; 2 said they used it more than 90% of the time. Also mentioned were BIOSIS (3 mentions), Connotea (1), PubMedCentral (1), Google web search (1), and bloglines (1).

Figure 1 shows the responses to a question on what proportion of searches include gene names. 37% of the participants use gene names in 50-100% of their queries. Five participants do not use gene names in their queries; one of these people noted that they use literature search in order to discover relevant genes.

Next, participants answered two detailed questions about what kinds of information they would like to see associated with the gene name from their query. Table 1 shows the averaged scores for responses to the question "When you search for genes/proteins, what type of related gene/protein names would you like a system to suggest?" Participants selected choices from a Likert scale which spanned from 1 ("strongly do not want to see this") to 5 ("extremely important to see this information"), with 3 indicating "do not mind seeing this." (These results are for 33 participants, because the 5 participants who said they do not use gene names in their search were made to automatically skip these questions.) The table below also shows the number of participants who assigned either a 1 or a 2 score, indicating that they do not want to see this kind of information.

Table 1. Averaged scores for responses to the question "When you search for genes/proteins, what type of related gene/protein names would you like a system to suggest?" 1 is "strongly disagree," 5 is "strongly agree."

Related Information Type	Avg. rating	# (%) selecting 1 or 2
Gene's synonyms	4.4	2 (5%)
Gene's synonyms refined by organism	4.0	2 (5%)
Gene's homologs	3.7	5 (13%)
Genes from the same family: parents	3.4	7 (18%)
Genes from the same family: children	3.6	4 (10%)
Genes from the same family: siblings	3.2	9 (24%)

The next question, "When you search for genes/proteins what other related information would you like a system to return?" used the same rating scale as above. The results are shown in Table 2.

Table 2. Averaged scores for responses to the question "When you search for genes/proteins what other related information would you like a system to return?" using same rating scale as above.

Related Information Type	Avg. rating	# (%) selecting 1 or 2
Genes this gene interacts with	3.7	4 (10%)
Diseases this gene is associated with	3.4	6 (16%)
Chemicals/drugs this gene is associated with	3.2	8 (21%)
Localization information for this gene	3.7	3 (8%)

When asked for additional information of interest, people suggested: pathways (suggested 4 times), experimental modification, promoter information, lists of organisms for which the gene is sequenced, ability to limit searches to a taxonomic group, protein motifs, hypothesized or known functions, downstream effects and link to a model organism page.

The results of this questionnaire suggest that not only are many biologists heavy users of literature search, but gene names figure prominently in a significant proportion of their searches. Furthermore, there is interest in seeing information associated with gene names. Not surprisingly, the more directly related the information is to the gene, the more participants viewed it favorably. 22 participants said they thought gene synonyms would be extremely useful (i.e., rated this choice with a score of 5). However, as the third columns of the tables show, a notable minority of participants expressed opposition to showing the additional information. In a box asking for general comments, two participants noted that for some kinds of searches, expansion information would be useful, but for others the extra information would be in the way. One participant suggested offering these options at the start of the search as a link to follow optionally. These responses reflect a common view among users of search systems: they do not want to see a cluttered display. This is further warning that one should proceed with caution when adding information to a search user interface.

6. Second Questionnaire: Gene/Protein Name Expansion Preferences

6.1. *The Evaluated Designs*

To reproduce what users would see in a Web search interface, four designs were constructed using HTML and CSS, building upon the design used for our group's search engine. To constrain the participants' evaluation of the designs and to focus them on a specific aspect of the interface, static screenshots of just the relevant portion of the search interface were used in the testing. Example interactions with the interface were conveyed using "before" and "after" screenshots of the designs. Limiting the testing to static screenshots decreased the development time required to set up the tests, since we did not need to anticipate the myriad potential interactions between the testers and a live interface. Figures 2–4 show the screenshots seen by the participants for Designs 1–4. Participants were told they were seeing what happened after they clicked on the indicated link, but not what happens to the search results after the new search is executed.

Design 1, which served as the baseline for comparison with the other designs, showed a standard search engine interface with a text box and submit button in the page header. The gene term "*RAD23*" was used as the example search term, with a results summary showing three results returned.

Design 2 added a horizontal box between the search box and the text summary. The box listed possible expansion terms for the original "*RAD23*" query

Design 1

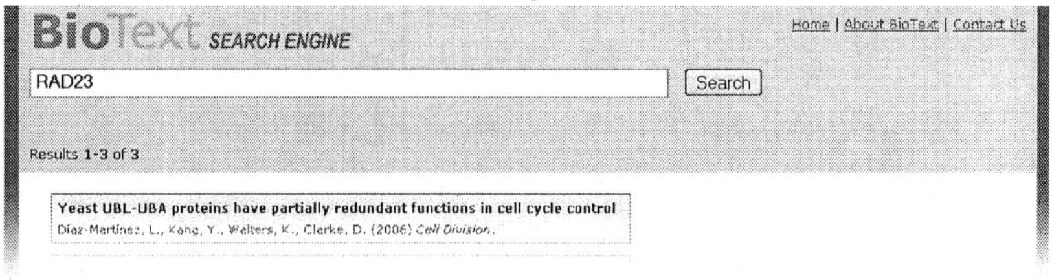

Design 2

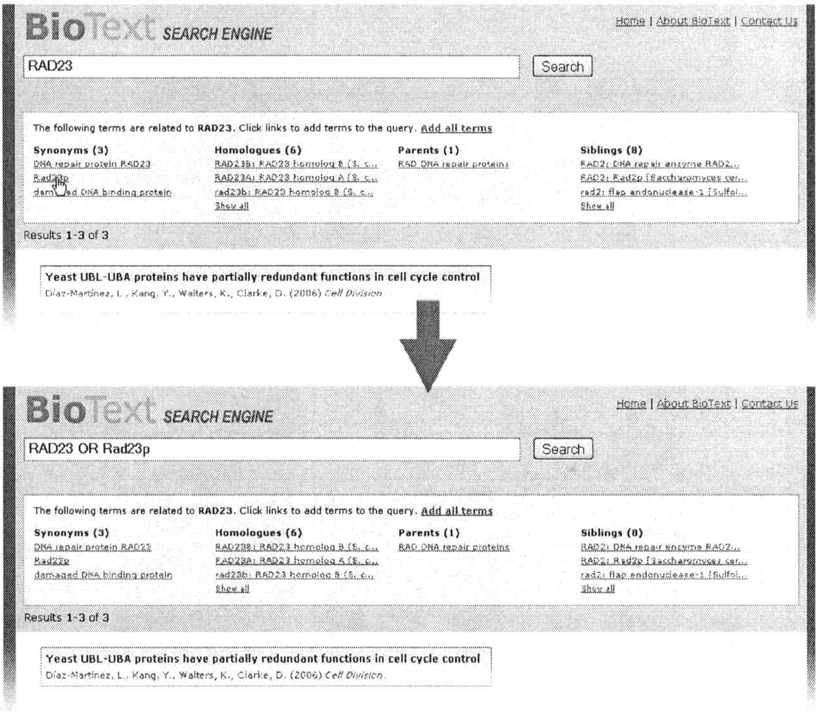

Figure 2. Designs 1 and 2 shown to participants in the second questionnaire.

organized under four categories: synonyms, homologs, parents, and siblings. All the terms were hyperlinked. The "after" screenshot showed the result of clicking a hyperlinked term, which added that term to the query in the text box using an

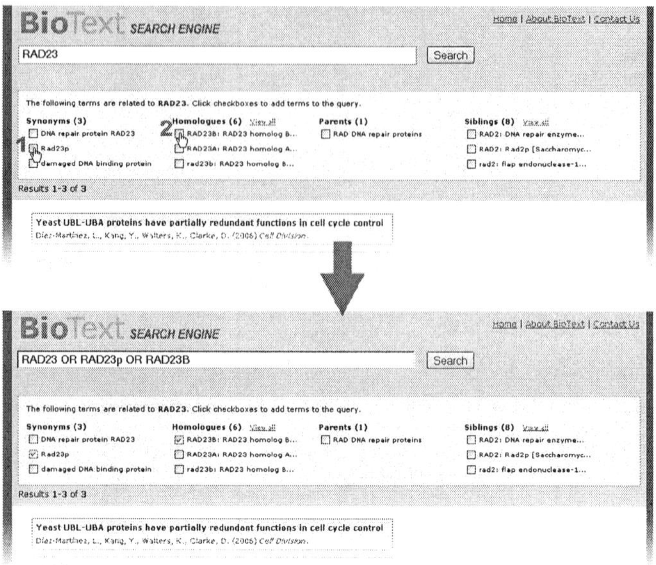

Figure 3. Design 3 shown to participants in the second questionnaire.

OR operator.

Design 3 had a similar layout except that instead of having hyperlinked expansion terms, each expansion term was paired with a checkbox. The terms were organized beneath the same four categories. The "after" screenshot showed that by clicking a checkbox, a user could add the term to the original query.

Design 4 showed a box of plain text expansion terms that were neither hyperlinked nor paired with checkboxes. In this design, each category term had an "Add all to query" link next to it for adding all of a category's terms at once. The "after" screenshot showed the result of clicking a hyperlink, with multiple terms ORed to the original query.

6.2. *Results*

Nineteen people completed the questionnaire. Nine of those who filled out the first questionnaire and who indicated that (a) they were interested in seeing gene/protein names in search results and (b) they were willing to be contacted for a second questionnaire participated in this followup study. Ten additional participants were recruited by emailing colleagues and asking them to forward the

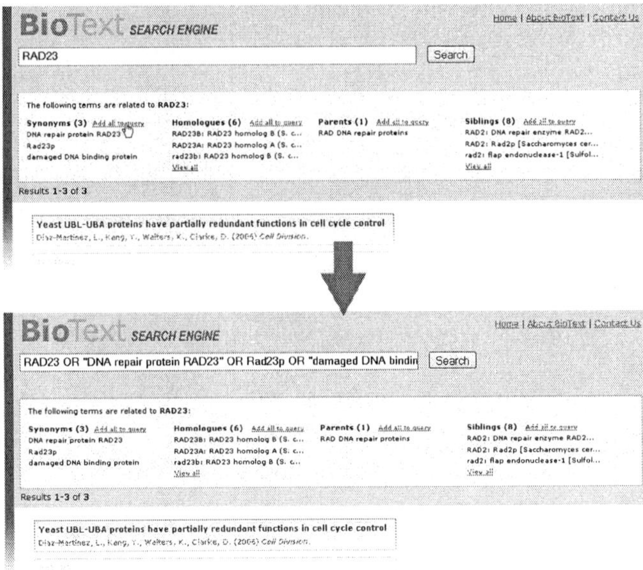

Figure 4. Design 4 shown to participants in the second questionnaire.

request to biologists. Thus, the results are biased towards people who are interested in search interfaces and their improvement.

Again, participants were from several academic institutions (4 graduate students, 7 postdoctoral researchers, 3 faculty, and 5 other researchers). Their areas of interest/specialization included molecular toxicology, evolutionary genomics, chromosome biology, plant reproductive biology, cell signaling networks, and computational biology more generally. The distribution of usage of genes in searches was similar to that of the first questionnaire.

One question asked the participants to rank-order the designs. There was a clear preference for the expansion terms over the baseline, which was the lowest ranked for 15 out of 19 participants. Table 3 shows the results, with Design 3 most favored, followed by Designs 4 and 2, which were similarly ranked.

In the next phase of questions, one participant indicated they would not like to see gene names, and so automatically skipped the questions. Of the remaining 18 participants, when asked to indicate a preference for clicking on hyperlinks versus checkboxes for adding gene names to the query, 10 participants (56%) selected checkboxes and 6 (33%) selected hyperlinks (one suggested a "select all"

Table 3. Design Preferences.

	# participants who rated Design 1st or 2nd	% participants who rated Design 1st or 2nd	Avg. rating (1=low, 4=high)
Design 3	15	79%	3.3
Design 4	10	53%	2.6
Design 2	9	47%	2.5
Design 1	0	0%	1.6

option above each group for the checkboxes). When asked to indicate whether or not they would like to see the organisms associated with each gene name, 16 out of 18 participants said they would like the organism information to be directly visible, showing the organism either alongside (11) or grouping the gene names by organism (5). Two were undecided.

When asked how gene names should be organized in the display, 9 preferred them to be grouped under type (synonyms, homologs, etc). The other participants were split between preferences for showing the information grouped by organism name, grouped by more generic taxonomic information, or not grouped but shown alphabetically or by frequency of occurrence in the collection.

Participants were also asked if they prefer to select each gene individually (2), whole groups of gene names with one click (3), or to have the option to chose either individual names or whole groups with one click (13).

Finally, they were asked if they prefer the system to suggest only names that it is highly confident are related (8), include names that it is less confident about (0), or include names that it is less confident about under a "show more" link (8). In the open comments field, one participant stated that the system should allow the user to choose among these, and another wrote something we could not interpret. These attitudes echo the finding that high-scoring systems in the TREC genomics track[6] often used principled gene name expansion.

7. Conclusions and Future Work

This study addresses the results of the first steps of user-centered design for development of a literature search interface for biologists. Our needs assessment has revealed a strong desire for the search system to suggest information closely related to gene names, and some interest in less closely related information as well. Our task analysis has revealed that most participants want to see organism names in conjunction with gene names, a majority of participants prefer to see term suggestions grouped by type, and participants are split in preference between single-click hyperlink interaction and checkbox-style interaction. The last point suggests that we experiment with hybrid designs in which only hyperlinks are used, but an additional new hyperlink allows for selecting all items in a group. Another hybrid to evaluate would have checkboxes for the individual terms and a link that immediately adds all terms in the group and executes the query.

The second questionnaire did not ask participants to choose between seeing information related to genes and other kinds of metadata such as disease names. Adding additional information will require a delicate balancing act between usefulness and clutter. Another design idea would allow users to collapse and expand term suggestions of different types; we intend to test that as well.

Armed with these results, we have reason to be confident that the designs will be found usable. Our next steps will be to implement prototypes of these designs, ask participants to perform queries, and contrast the different interaction styles.

Acknowledgements: We thank the survey participants for their contributions to this work. This research was supported in part by NSF DBI-0317510.

References

1. P. Anick. Using terminological feedback for web search refinement: a log-based study. *Proceedings of SIGIR 2003*, pages 88–95, 2003.
2. P. Bruza, R. McArthur, and S. Dennis. Interactive Internet search: keyword, directory and query reformulation mechanisms compared. *Proceedings of SIGIR 2000*, pages 280–287, 2000.
3. H. Chen and B.M. Sharp. Content-rich biological network constructed by mining PubMed abstracts. *BMC Bioinformatics*, 5(147), 2004.
4. A. Doms and M. Schroeder. GoPubMed: exploring PubMed with the Gene Ontology. *Nucleic Acids Research*, 33(1):W783–W786, 2005.
5. M.A. Hearst, A. Divoli, J. Ye, and M.A. Wooldridge. Exploring the efficacy of caption search for bioscience journal search interfaces. *Biological, translational, and clinical language processing*, pages 73–80, 2007.
6. W. Hersh, A. Cohen, J. Yang, R.T. Bhupatiraju, P. Roberts, and M. Hearst. TREC 2005 Genomics Track Overview. *the Fourteenth Text Retrieval Conference*, 2005.
7. L. Hirschman, A. Yeh, C. Blaschke, and A. Valencia. Overview of BioCreAtIvE: critical assessment of information extraction for biology. *BMC Bioinformatics*, 6(S1), 2005.
8. R. Hoffmann and A. Valencia. A gene network for navigating the literature. *Nature Genetics*, 36(7):664, 2004.
9. B.J. Jansen, A. Spink, and S. Koshman. Web searcher interaction with the Dogpile.com metasearch engine. *Journal of the American Society for Information Science and Technology*, 58(5):744–755, 2007.
10. H.M. Müller, E.E. Kenny, and P.W. Sternberg. Textpresso: an ontology-based information retrieval and extraction system for biological literature. *PLoS Biol*, 2(11):e309, 2004.
11. B. Shneiderman and C. Plaisant. *Designing the user interface: strategies for effective human-computer interaction, 4/E*. Addison-Wesley, Reading, MA, 2004.

ENABLING INTEGRATIVE GENOMIC ANALYSIS OF HIGH-IMPACT HUMAN DISEASES THROUGH TEXT MINING

JOEL DUDLEY AND ATUL J. BUTTE

Stanford Medical Informatics, Departments of Medicine and Pediatrics

Stanford University School of Medicine

Stanford, CA 94305-5479, USA

Our limited ability to perform large-scale translational discovery and analysis of disease characterizations from public genomic data repositories remains a major bottleneck in efforts to translate genomics experiments to medicine. Through comprehensive, integrative genomic analysis of all available human disease characterizations we gain crucial insight into the molecular phenomena underlying pathogenesis as well as intra- and inter-disease differentiation. Such knowledge is crucial in the development of improved clinical diagnostics and the identification of molecular targets for novel therapeutics. In this study we build on our previous work to realize the next important step in large-scale translational discovery and analysis, which is to automatically identify those genomic experiments in which a disease state is compared to a normal control state. We present an automated text mining method that employs Natural Language Processing (NLP) techniques to automatically identify disease-related experiments in the NCBI Gene Expression Omnibus (GEO) that include measurements for both disease and normal control states. In this manner, we find that 62% of disease-related experiments contain sample subsets that can be automatically identified as normal controls. Furthermore, we calculate that the identified experiments characterize diseases that contribute to 30% of all human disease-related mortality in the United States. This work demonstrates that we now have the necessary tools and methods to initiate large-scale translational bioinformatics inquiry across the broad spectrum of high-impact human disease.

1. Introduction

1.1. *The Role of Text Mining in Translational Bioinformatics*

As the pace at which genomic data is generated continues to accelerate, propelled by technological advances and declining per-experiment costs, our ability to utilize these data to address long-standing problems in clinical medicine continues to lag behind[1]. It is only through the correction of this disparity that we can overcome one of the major obstacles in translating fundamental discoveries from genomic experiments into the world of medicine for the benefit of public health and society[2, 3].

Owing to its capabilities as a high-bandwidth molecular quantification and diagnostic platform, the RNA expression detection microarray has emerged as a premier tool for characterizing human disease[4-6] and developing novel diagnostics[7, 8]. Fortunately, the data generated by microarray experiments is routinely warehoused in a number of public repositories, providing opportunities to address an unprecedented depth and breadth of data for translational research. These repositories include the NCBI Gene Expression Omnibus (GEO)[9], ArrayExpress at EBI[10], and the Stanford Microarray Database[11]. GEO is the largest among these repositories, offering 157,850 samples (microarrays) from 6,062 experiments as of this writing. Given GEO's exponential growth, it is unlikely to lose this position of predominance for the foreseeable future. In light of these characteristics, it is clear that GEO stands as a model public genomic data repository against which novel bioinformatics methods for large-scale translational discovery may be rigorously designed, evaluated and applied.

We recently described a method for the automated discovery of disease-related experiments within GEO using Medical Subject Heading (MeSH) annotations derived from associated PUBMED identifiers[12]. This represented an important first step in enabling large-scale translational discovery by providing an automated means through which an entire body of publicly available genomic data can be mined comprehensively for human disease characterizations. It also demonstrated the utility of applying text mining methods in translational research, as well as their potential role in realizing a fully automated pipeline for translational bioinformatics discovery and analysis of the human "diseasome". The ultimate goal of such an effort is to comprehensively analyze the whole of disease-related experiments for the purpose of developing novel therapeutics and improved clinical protocols and diagnostics. If such a pipeline were realized, we would be able to ask an entirely new class of questions about the nature of human disease, e.g., "Which genes are significantly differentially expressed across all known autoimmune diseases?"

In order to uncover the many putative links between gene expression and human disease, we must first be able to compare the global gene expression of a disease state with that of a comparable disease-free, or normal control state. Given the sheer volume of experiments available in repositories like GEO, there is a need to develop automated tools and techniques to enable the identification of such states on a large-scale.

1.2. *Objective and Approach*

In this study we seek to develop a robust text mining method to automatically identify disease-related GEO experiments that contain samples for both disease and normal control states. To accomplish this, we utilize an upper-level representation of an experiment in GEO known as a GEO DataSet (GDS), in which samples are organized into biologically informative collections known as subsets. These subsets are defined by GEO curators who group samples from a particular experiment according to the experimental axis under examination (e.g.

disease state or *agent*). Each subset is annotated with a brief, free-text description used to further elucidate the nature of the subset (e.g. *disease-free* or *placebo*). The pertinent attributes and relationships of the GEO GDS are illustrated in Figure 1. The definition of GDS has not kept pace with the addition of experiments (GSE), and as of this writing there are 1,936 GDS defined in GEO representing 32% of the total GSE.

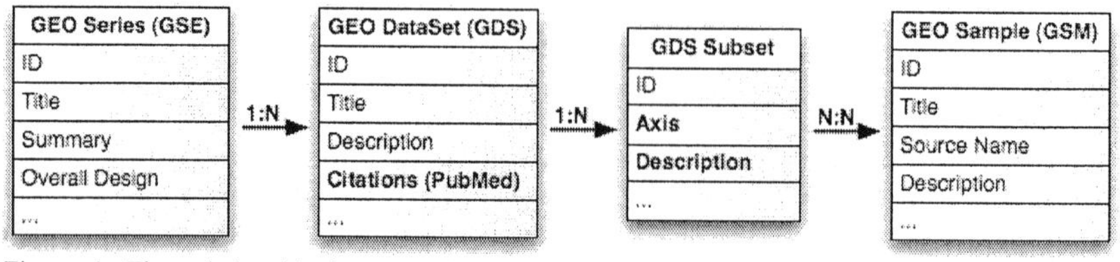

Figure 1. The relationship between GEO Samples, GEO DataSets, GDS Subsets, and GEO Samples is illustrated. The attributes utilized by the proposed method are shown in bold. The label over the arrows indicates the cardinality of the relationship.

We propose that these subset text attributes can be evaluated to determine if a particular subset is representative of either a disease state or a normal control. While the vocabulary used to denote the experimental axes for a subset is principally controlled, currently comprised of twenty-four distinct terms, their utilization within a GDS and their application to sample collections is left completely to curator discretion. Furthermore we find that the content of the descriptions associated with each subset is free-text, constrained by no declared or discernable convention or controlled vocabulary. An example of these subset annotations is shown in figure 2. It is not possible to elucidate control subsets from the experimental axis annotation alone, as these annotations aim to classify the experimental variable being measured (e.g. *cell type* or *development stage*), rather than to describe the context of measurement instances. Thus we are faced with the difficult problem of elucidating the context of each subset based on the free-text descriptions associated with each subset.

Fortunately, simple frequency analysis reveals that a small number of terms commonly used to describe a normal control state are found in the associated subset descriptions for disease-related GDS in high frequency. As shown in Figure 3, the distribution of subset description phrases follows a Zipf-like distribution, with the common used control terms *control*, *normal*, and *wild type* representing the most frequently used phrases by-experiment and by-samples across all disease-related GDS subset descriptions. Thus, it is reasonable to suggest that the problem of large-scale

4 assigned subsets			
Samples		Type	Description
☑ (6)	☑	disease state	type 2 diabetes
☑ (6)		disease state	non-diabetic
☑ (6)	☑	age	8 week
☑ (6)		age	16 week

Figure 2. Example GDS subset designations for GDS402 taken from the GEO website.

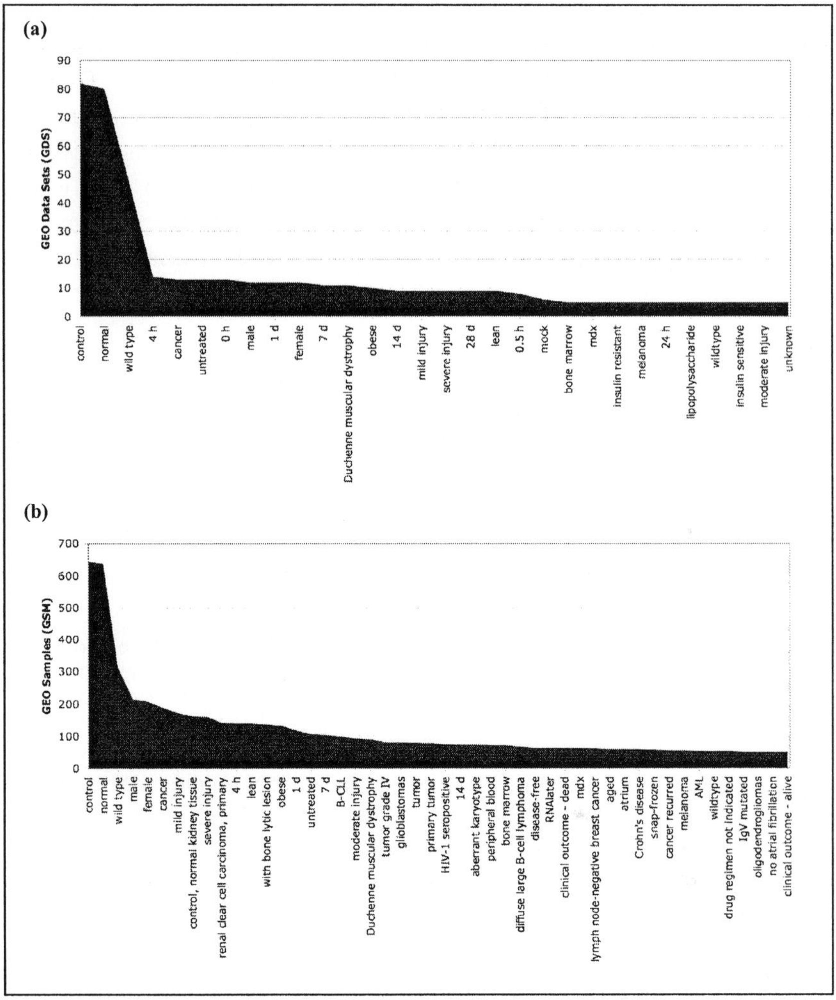

Figure 3. Distribution of GDS subset annotation phrases for all disease-related GDS. The distributions are filtered to terms annotating > 5 GDS and > 50 GSM for display purposes. The distribution shows that the **(a)** majority of disease-related GDS contain subsets annotated with a small set of common control phrases, **(b)** representing a major proportion of samples.

normal control detection within GEO is tractable by the fact that a simple pattern matching approach using three common normal control phrases will identify controls in a majority of experiments representing a majority of samples. However this technique alone is insufficient as many control subsets for unique disease characterizations are found in the "long-tail" of the frequency distributions. In some cases common control terms are found within the subset description, but they do not represent a disease-free state (e.g. *skin cancer control*). In other cases a control subset is annotated using a disease negation scheme (e.g. *diabetes-free*). In such cases the application of a simple pattern matching technique would result in either a false positive or a false negative respectively.

To manage such complex cases we make use of the Unified Medical Language System (UMLS) Metathesaurus[13] to identify terms representing a human disease. With disease terms identified, it is possible to infer control subsets that are implied rather than explicit, for example the negation of a disease term implies a normal control, and avoids incorrectly identifying control subsets that are annotated in a contradictory manner (e.g. *normal skin cancer*).

1.3. *Evaluating the Impact of Translational Text Mining*

The impact of any exercise in *translational* text mining cannot be fully assessed without a clear quantitative evaluation of the clinical impact and overall benefit to human health. For it is through such clinical imperatives that translational bioinformatics is distinguished. It is tempting to measure the clinical impact of the proposed method by way of the total number of unique diseases for which a disease vs. normal control state was identified, however not every human disease carries the same clinical impact.

Therefore in addition to traditional performance measures, we propose to measure translational impact along the axis of human disease-related mortality. In this context, impact is based on the coverage of disease characterizations over the total disease-related human mortality, quantified by the number of deaths for which a disease is responsible. This impact measure is intuitive, because it is reasonable to assume that the diseases causing the greatest number of deaths are the diseases that have the greatest impact on clinical practice.

2. Methods

2.1. *Identifying Disease-Related Experiments*

Similar to our previously described method[12], the disease-related experiments were identified using a MeSH-based mapping approach. We used a February 15th, 2007 snapshot of the Gene Expression Omnibus (GEO)[9] which was parsed into a normalized structure and stored in a relational database.

For the 1,231 GEO DataSets (GDS) experiments associated with a PUBMED identifier, we downloaded the corresponding MEDLINE record and extracted the MeSH using the BioRuby toolkit (http://www.bioruby.org). The extracted MeSH terms were stored in a relational database along with the associated GDS identifier, resulting in 20,654 distinct mappings. These mappings were joined with the UMLS (2007AA release) Concept Names and

Sources (MRCONSO) and Semantic Types (MRSTY) tables to identify GDS associated with MeSH terms having any of the semantic types among *Injury or Poisoning* (T037), *Pathologic Function* (T046), *Disease or Syndrome* (T047), *Mental or Behavioral Dysfunction* (T048), *Experimental Model of Disease* (T050), or *Neoplastic Process* (T191) as disease-related GDS.

2.2. *Control Subset Detection*

For each disease-related GDS we obtained data for the associated subsets using the aforementioned relational snapshot of GEO. The subsets of each disease-related GDS were enumerated and their descriptions evaluated to elucidate control subsets. As previously mentioned, a sizeable proportion of disease-related GDS (41%) have subsets annotated with the common control terms *control*, *normal* and *wild type* or some slight variation thereof. These common control terms were assembled into a set, and any subset with a description annotation comprised of a single term from this set was identified as a normal control subset. Subset descriptions were also transformed into stemmed, word case, spacing and hyphenation variants using porter stemming and regular expressions to detect control term variants (e.g. *controlled* becomes *control*, *wild-type* becomes *wild type*), which represented an additional 14% of disease-related GDS. If any such variant of a common control term was matched in a subset annotation, then the subset was identified as a normal control.

Curiously, a small proportion of disease-related GDS (3%) did not have any subsets defined. It is not clear as to why this was the case. It could be that these GDS are incompletely curated, and subset definitions will be applied in later releases of GEO. Consequently these GDS were removed from consideration.

Subset descriptions not containing common control terms were evaluated using more sophisticated techniques to account for negation and lexical variation.

2.3. *Handling Negation*

We find that GDS subsets are frequently annotated using a negation scheme in which a subset representative of a disease state will be annotated with a UMLS disease concept and the control will be expressed as the negation of that disease concept (e.g. *diabetic* vs. *non-diabetic*). Therefore the identification of control subsets was expanded to include subsets that are annotated using this disease-negation pattern.

The detection of negations in natural languages is non-trivial[14], however there are several properties of GDS subset labels that increase the tractability of the problem. GDS subset descriptions are typically terse (average of 10.7

characters per description), and therefore the word distance between the negation signal and the concept is negligible. This aids negation detection by minimizing a common source of error in tokenizing negation parsers[15], and eliminates the need to engage more complex Natural Language Processing (NLP) approaches, such as parse tree based negation classification[16], to link negation symbols to disjoint disease concepts. Given these properties we chose to identify negation-based control subsets using a modified version of the NegEx algorithm[17]. The NegEx algorithm is a regular-expression based algorithm for the detection of the explicit negation of terms indexed by UMLS. NegEx has been shown to have 78% sensitivity and 84.5 % positive predictive value when detecting negations in medical discharge summaries[17]. It is expected that NegEx will perform better in the detection of negation-based control subsets, as complex syntactic structures, which are not present in terse subset labels, were a major source of error in detecting negations in verbose discharge summaries. Additionally, we constrained the NegEx algorithm to detect negation for UMLS-mapped terms exhibiting any of the five aforementioned disease-related semantic types rather than the broader fourteen semantic type categories used by the unmodified algorithm.

We found that in some cases, a subset description will exhibit the negation of a valid disease term, but does in fact lead to a false positive since the negation is also a valid disease state (i.e. *non-Hodgkins Lymphoma)*. To correct for this case, we first query UMLS to ensure that description does not represent a disease state.

2.4. *Handling Lexical Variations*

In some cases the description for a control subset was expressed in a manner that is lexically inconsistent with the terms used to describe the disease state. For example, GDS887 defines the following subset labels for the *disease state* axis: *type 1 diabetes*, *type 2 diabetes*, and *non-diabetic*. In order to automatically link the subset labeled *non-diabetic* as the negated control of the subset labeled *type 1 diabetes*, we must derive that these lexically incompatible labels are in fact semantically related.

Lexical variations were automatically reconciled using the Normalized Word Index table (MRXNW_ENG) in UMLS. The Normalized Word Index contains tokenized, uninflected forms of UMLS terms, derived either algorithmically or through the SPECIALIST lexicon. Using this table we find that the terms *type 1 diabetes* and *diabetic* share a common association with at least one Concept Unique Identifier (CUI) (C0011854). Therefore we can infer that the subset labeled *non-diabetic* is in fact a valid negated control of the subset labeled *type 1 diabetes*.

2.5. *Performance Evaluation*

To evaluate performance we used an expert human reviewer as a "gold standard", and divided control subsets into two distinct groups. The first group, Group A, represents control subsets identified using common control terms, and the second group, Group B, represents control subsets that did not contain common control terms, and therefore were evaluated using the negation-based approach. We randomly sampled positively and negatively identified control subsets from both groups and calculated True Positive (TP), False Positive (FP), True Negative (TN), and False Negative (FN) counts after each subset from the random samples was evaluated by the expert human evaluator, who positively or negatively identified control subsets. From these counts we calculated sensitivity = TP/TP+FN, specificity = TN/TN+FP, Positive Predictive Value (PPV) = TP/TP+FP, Negative Predictive Value (NPV) = TN/TN+FN, and F1 score = 2·(PPV·(TP/TP+FN))/PPV+(TP/TP+FN). These values were also computed across both groups to provide an overall evaluation of performance for the proposed method.

2.6. *Evaluating Clinical Impact from Mortality Data*

U.S. mortality data from 1999 to 2004 were obtained from the Centers for Disease Control and Prevention (CDC) using the Wide-ranging Online Data for Epidemiologic Research (WONDER) system (http://wonder.cdc.gov). Causes of Death were specified using International Classification of Disease (ICD) codes (10th edition). These codes were mapped to their corresponding MeSH using the MRCONSO table in UMLS. We acknowledge that many ICD10 codes have no direct mapping to MeSH in UMLS, with only ~15% of ICD10 directly linked to MeSH terms. Computational translation between UMLS source vocabularies is an active area of research, with several promising approaches emerging[19, 20]. However it is beyond the scope of this paper to participate in this budding area of research. Therefore we only map ICD10 codes to MeSH terms when they are directly related under the same concept identifier (CUI) in UMLS to provide a minimum estimate of impact. The number of deaths mapped to disease-related GDS in this manner was used to calculate the total disease-related mortality impact.

3. Results

In mapping GDS to MeSH terms, we find that 1,231 (78%) of the 1,588 GDS in our GEO database snapshot were associated with a PUBMED identifier. From the resulting 20,654 GDS to MeSH mappings, we find that 513 GDS are associated with MeSH terms having at least one of the six semantic types considered to be disease-related (T037, T046, T047, T048, T050 or T191).

In detecting common normal control phrases in subset annotations, we find that control subsets are identified in 56% of disease-related GDS. Using the negation and lexical variation compensation techniques, we are able to identify control subsets in an additional 33 GDS, resulting in the automated identification of control subsets in a total of 62% of disease-related GDS. This results in a set of 13,840 samples spanning 141 unique disease-related concepts.

We manually inspected the 38% of disease-related GDS for which normal control subsets could not be identified, and found that they fell into a handful of general categories. A number of GDS experiments were designed to characterize or differentiate among disease subtypes (e.g. expression profiling across different cancer cell lines), and therefore contain no true control subsets. Others annotated subsets using proprietary identifiers for cell lines and animal strains. The latter accounts for a major source of sensitivity dampening in evaluating control subsets. Detailed performance metrics are illustrated in Table 1.

Table 1. Performance Evaluation of Control Detection.

	Sensitivity	Specificity	PPV	NPV	F1
Group A (common control terms) (n=100)	0.979	1.000	1.000	0.980	0.989
Group B (negation-based controls) (n=100)	0.428	0.983	0.937	0.750	0.588
Combined (Group A+Group B) (n=200)	0.750	0.911	0.984	0.840	0.851

We were successful in mapping 2,019 ICD10 codes to MeSH terms, covering 18% of the ICD10 codes represented in the mortality data, and 42% of the total mortality. Using MeSH headings, we were able to map 42% the disease-related GDS with normal controls to ICD10 codes. These ICD10 codes mapped to 77 unique ICD10 codes in the mortality data representing 4,219,703 combined deaths over 5 years, or 30% of the total human disease-related mortality in the United States in the same period. Note that this is a minimum estimation given the limited mapping between ICD10 and MeSH in UMLS.

4. Discussion

Given the current pace of growth experienced by international genomic data repositories, it may be only six years before researchers have access to more than a million microarray samples. Yet, even with less than half that amount available today, it has not been possible to link any significant portion of these genomic measurements to the broad molecular characteristics underlying the broad spectrum of human disease. Here we describe a method that enables the creation of such links, and lays the groundwork for the development of a robust translational bioinformatics pipeline that can be applied to both current and forthcoming volumes of public genomic data.

Through this method we find that we can automatically identify normal control subsets in GDS representing 141 unique disease states and conditions. While cancers make up a significant proportion of the associated diseases, afflictions such as Alzheimer's disease, heart disease, diabetes and other diseases having a major impact on human mortality are also represented.

The techniques developed for the identification of negated control subsets and the reconciliation of lexical variations will become increasingly important as GEO continues its exponential growth. Even if the percentage of disease-related GDS experiments containing non-obvious control subset designations remains the same (17%) or even slightly less, these techniques could enable the automated translational analysis of thousands of disease-related microarray samples.

We have now proven that it is not only possible, but also completely tractable to apply these methods to our current public data collections in an attempt to characterize the broad spectrum of high-impact human disease. Despite the fact that we were only able to identify control subsets in 20% of the total GDS found in GEO, and ultimately only 6% of the total experiments contained within GEO, we were able to associate these GDS experiments with diseases contributing to 30% of the total human mortality in the United States.

The next critical step is to develop a means by which those experiments without associated PUBMED identifiers can be automatically evaluated to identify additional disease-related experiments. In addition, these techniques must be further generalized so that they can be applied to additional public repositories containing data from microarrays and other genome-scale measures.

We acknowledge that while this study provides a successful proof of concept and demonstration of utility, it does not provide a finished product. Therefore the method will not be made available as a public resource, however it will enable the creation of more biologically relevant downstream resources.

Conclusion

Using GEO as a model public data repository, we have developed text mining techniques that enable completely new types and scales of translational research.

As these techniques are applied to new and expanding public data repositories, by means of translational bioinformatics, we will be given the opportunity to discover the fundamental molecular principals and dynamics that underlie the whole of high-impact human disease. It is from this vantage that we will begin to realize the novel diagnostics and therapeutics long-promised in this post-genomic era.

Acknowledgments

The authors would like to thank Alex Morgan for providing critical feedback on an early draft of the manuscript, and Alex Skrenchuck for HPC support. The work was supported by grants from the Lucile Packard Foundation for Children's Health, National Library of Medicine (K22 LM008261), National Institute of General Medical Sciences (R01 GM079719), National Human Genome Research Institute (P50 HG003389), Howard Hughes Medical Institute, and the Pharmaceutical Research and Manufacturers of America Foundation.

References

1. C. A. Ball, G. Sherlock and A. Brazma, Funding high-throughput data sharing. *Nature biotechnology* **22**, 1179-83 (2004)
2. E. A. Zerhouni, Translational and clinical science--time for a new vision. *N Engl J Med* **353**, 1621-3 (2005)
3. M. Chee, R. Yang, E. Hubbell, A. Berno, X. Huang, D. Stern, J. Winkler, D. Lockhart, M. Morris and S. Fodor, Accessing genetic information with high-density DNA arrays. *Science* **274**, 610-4 (1996)
4. S. Calvo, M. Jain, X. Xie, S. A. Sheth, B. Chang, O. A. Goldberger, A. Spinazzola, M. Zeviani, S. A. Carr and V. K. Mootha, Systematic identification of human mitochondrial disease genes through integrative genomics. *Nat Genet* **38**, 576-82 (2006)
5. K. Mirnics and J. Pevsner, Progress in the use of microarray technology to study the neurobiology of disease. *Nat Neurosci* **7**, 434-9 (2004)
6. E. E. Schadt, J. Lamb, X. Yang, J. Zhu, S. Edwards, D. Guhathakurta, S. K. Sieberts, S. Monks, M. Reitman, C. Zhang, P. Y. Lum, A. Leonardson, R. Thieringer, J. M. Metzger, L. Yang, J. Castle, H. Zhu, S. F. Kash, T. A. Drake, A. Sachs and A. J. Lusis, An integrative genomics approach to infer causal associations between gene expression and disease. *Nat Genet* **37**, 710-7 (2005)
7. A. M. Glas, A. Floore, L. J. Delahaye, A. T. Witteveen, R. C. Pover, N. Bakx, J. S. Lahti-Domenici, T. J. Bruinsma, M. O. Warmoes, R. Bernards, L. F. Wessels and L. J. Van't Veer, Converting a breast cancer microarray

signature into a high-throughput diagnostic test. *BMC Genomics* **7**, 278 (2006)

8. G. J. Gordon, R. V. Jensen, L. L. Hsiao, S. R. Gullans, J. E. Blumenstock, S. Ramaswamy, W. G. Richards, D. J. Sugarbaker and R. Bueno, Translation of microarray data into clinically relevant cancer diagnostic tests using gene expression ratios in lung cancer and mesothelioma. *Cancer Res* **62**, 4963-7 (2002)
9. T. Barrett, T. O. Suzek, D. B. Troup, S. E. Wilhite, W. C. Ngau, P. Ledoux, D. Rudnev, A. E. Lash, W. Fujibuchi and R. Edgar, NCBI GEO: mining millions of expression profiles--database and tools. *Nucleic Acids Res* **33**, D562-6 (2005)
10. A. Brazma, H. Parkinson, U. Sarkans, M. Shojatalab, J. Vilo, N. Abeygunawardena, E. Holloway, M. Kapushesky, P. Kemmeren, G. G. Lara, A. Oezcimen, P. Rocca-Serra and S. A. Sansone, ArrayExpress--a public repository for microarray gene expression data at the EBI. *Nucleic Acids Res* **31**, 68-71 (2003)
11. G. Sherlock, T. Hernandez-Boussard, A. Kasarskis, G. Binkley, J. C. Matese, S. S. Dwight, M. Kaloper, S. Weng, H. Jin, C. A. Ball, M. B. Eisen, P. T. Spellman, P. O. Brown, D. Botstein and J. M. Cherry, The Stanford Microarray Database. *Nucleic Acids Res* **29**, 152-5 (2001)
12. A. J. Butte and R. Chen, Finding disease-related genomic experiments within an international repository: first steps in translational bioinformatics. *AMIA Annu Symp Proc* 106-10 (2006)
13. D. A. Lindberg, B. L. Humphreys and A. T. McCray, The Unified Medical Language System. *Methods of information in medicine* **32**, 281-91 (1993)
14. R. M. April and M. E. Caroline, The ambiguity of negation in natural language queries to information retrieval systems. *J. Am. Soc. Inf. Sci.* **49**, 686-692 (1998)
15. P. G. Mutalik, A. Deshpande and P. M. Nadkarni, Use of general-purpose negation detection to augment concept indexing of medical documents: a quantitative study using the UMLS. *J Am Med Inform Assoc* **8**, 598-609 (2001)
16. Y. Huang and H. J. Lowe, A novel hybrid approach to automated negation detection in clinical radiology reports. *J Am Med Inform Assoc* **14**, 304-11 (2007)
17. W. W. Chapman, W. Bridewell, P. Hanbury, G. F. Cooper and B. G. Buchanan, A simple algorithm for identifying negated findings and diseases in discharge summaries. *Journal of biomedical informatics* **34**, 301-10 (2001)
18. O. Bodenreider, S. J. Nelson, W. T. Hole and H. F. Chang, Beyond synonymy: exploiting the UMLS semantics in mapping vocabularies. *AMIA Annu Symp Proc* 815-9 (1998)
19. C. Patel and J. Cimino, Mining Cross-Terminology Links in the UMLS. *AMIA Annu Symp Proc* 624-8 (2006)

INFORMATION NEEDS AND THE ROLE OF TEXT MINING IN DRUG DEVELOPMENT

PHOEBE M. ROBERTS, WILLIAM S. HAYES
Library and Literature Informatics, Biogen Idec, Inc. Cambridge MA USA

Drug development generates information needs from groups throughout a company. Knowing where to look for high-quality information is essential for minimizing costs and remaining competitive. Using 1131 research requests that came to our library between 2001 and 2007, we show that drugs, diseases, and genes/proteins are the most frequently searched subjects, and journal articles, patents, and competitive intelligence literature are the most frequently consulted textual resources.

1. Introduction

Academic research and pharmaceutical research share some common objectives, but there are important differences that influence publishing trends and information needs. Both groups rely heavily on peer-reviewed publications as a source of high-quality information used to formulate hypotheses, design experiments, and interpret results. To remain competitive, both groups must stay abreast of recent developments in order to make informed decisions. Effective search and retrieval is essential for finding high-quality information, which often benefits from integration and visualization due to the sheer volume of information that is available.

Unlike academic biomedical research, where publishing peer-reviewed articles is tied closely to funding, for-profit biomedical research groups are under different constraints. In the competitive marketplace, publishing information can alert competitors to developmental advances. Release of public information, however, is not always avoidable. Drug developers must file applications and data packages with drug approval authorities whose guidelines differ from country to country. Portions of drug application packages are freely available as unstructured text. In addition, drug developers are beholden to patent-granting authorities, filing patents to protect intellectual property and any profits that result from it. This makes legal literature a rich source of early-stage drug discovery information [1]. Publicly traded companies are required by the Securities Exchange Commission to disclose changes in their drug pipeline that have a potential financial impact, all of which are publicly available through the EDGAR database (http://www.sec.gov/edgar.shtml). Conversely, there are times when companies want to make advances known. Publicly traded companies hoping to boost stock prices, or private companies hoping to raise financing, use press releases, industry analyst conferences, and major scientific meetings attended by prescribing physicians to announce advances in their drug pipeline. It is critical to track all of these information resources to stay abreast of competition and spot potential collaborators, and the value of this information is reflected in the success of commercial "competitive intelligence" databases that integrate information in a structured searchable format [2].

Text mining is often raised as an antidote to the exponential expansion of published literature [3, 4]. Instead of relying on one or two keywords to find abstracts and full-text papers, text mining allows more powerful relevance ranking using classification and clustering techniques or class-based searching using entity tagging. Entity extraction adds additional value by structuring unstructured text and generating lists of like items that can be visualized in other ways, allowing the forest to emerge from the trees.

If one were to examine real user information needs, what kinds of questions would benefit from text mining applications? Studies of internet search, and biomedical literature search in particular, indicate that queries tend to be made up of only one or two keywords [5, 6]. Surprisingly, only 1.6% of PubMed queries used the Boolean OR operator [6]. Does this indicate that broadening searches is not important, or does it reflect a lack of familiarity with advanced search capabilities?

One way to understand the potential role of text mining in drug development research is to examine real end-user information needs instead of the terms used to conduct the searches. We describe here classes of queries submitted to the Library and Literature Informatics group at Biogen Idec, a large biotechnology company. The results highlight the entities of greatest value to drug development, and they place in context the utility of peer-reviewed literature versus other information resources.

2. Methods and Results

2.1 Coding Drug Company Research Requests by Subject and Resource

Biogen Idec is the third largest biotechnology company in the world, with strong franchises in multiple sclerosis (MS) and oncology. Historically, Biogen Idec has specialized in developing therapeutic antibodies and biologics, two of which have achieved "blockbuster" status (sales of over a billion dollars a year). The Biogen Idec Library and Literature Informatics group receives requests for research assistance for all aspects of drug development, including research, development, manufacturing, marketing, sales, and post-launch safety. The Library has cataloged 1131 research requests and their results since 2001. This database contains requests for research assistance only. Other Library functions, such as journal article requests or book orders, are not included.

Because of the competitive nature of drug development and the proprietary nature of the research requests, actual user needs will not be explicitly stated here. Instead, we sought a simple classification scheme that would allow us to unambiguously classify queries while maintaining enough information to be valuable to the information retrieval community, even in the absence of user queries. Taxonomies to classify queries have been described for questions asked by clinicians, resulting in an elaborate taxonomy of 64 question types [7]. To simplify our taxonomy, we chose to create controlled vocabularies that captured the main subject(s) of the request (Table 1). Subjects were selected based on their prevalence in the research questions, and questions were coded with as many subjects as applied. Also noted was the resource (e.g. patents, competitive intelligence resources, or journal articles) that was either specified by the requestor or deemed by the information professional to be the best resource for the question (Table 2).

To evaluate the terminologies and their consistent use, both authors (who annotated the full query set) independently coded approximately one-tenth (n=100) of the queries with the controlled vocabulary Subject, Resource, and Text Mining terms shown in Tables 1, 2, and 5 (results are shown in the last column of each table). Interannotator agreement was calculated as the ratio of matches between annotated requests and all requests annotated positively for a specific controlled vocabulary term by either annotator.

Table 1. Information Need Subject Classes

Subject	Description	# requests	Interannotator Agreement (# of Matches)
Drug	Substance administered to humans or animals to reduce or cure disease	355	.82 (46)
Disease	Human disorder or animal model of human disorder. Includes adverse drug reactions.	310	.78 (47)
Gene (includes Protein)	Biological substance that can be mapped to a discrete genetic locus. May be target of a drug.	297	.65 (20)
Company	Institution, public or private, industrial or academic	192	.59 (26)
Methods	Protocols for conducting scientific experiments or administering treatment	120	.47 (9)
Author	Individual who publishes or patents information	89	.70 (7)
Geography	A country or region	64	.62 (5)
Sales/Pricing	Income from or cost of a marketed drug	57	.54 (7)

General	Topics that do not map to subjects above	148	.44 (4)

Table 2. Information Resource Classes

Resource	Description	# requests	Interannotator Agreement (# of Matches)
Journal articles	Scientific literature from biomedical journals	389	.70 (32)
Competitive intelligence resources	Databases and trade publications that draw on company websites, SEC filings, scientific meetings and press releases for information about drugs in development	344	.77 (27)
Patents	Legal literature from worldwide patent agencies	217	.84 (16)
News sources	Newspapers and magazines (not specific to the pharmaceutical industry)	74	.71 (5)
Health statistics resources	Incidence and prevalence of diseases	59	.44 (4)
Other	Sources that do not map to information resources above	123	.29 (4)

Frequently occurring representative queries based on actual user needs are shown in Table 3, illustrating how the controlled vocabularies were applied to categorize query types. Note that the Subject terms were applied to both the input and output of the research request, i.e. the subject of the question, as well as the desired answer. When subject classes were not explicitly stated in the query, they were inferred during query coding based on implicit reference to the subject type. For example, the question, "What's in Phase II for arthritis?" mentions disease as a subject, and drug is inferred. Company information and the gene or gene product targeted by the drug are also provided in the interest of completeness. In our experience, providing drug information in the absence of manufacturer (Company) and mechanism of action (Gene) prompts follow-up requests for that information. Furthermore, by limiting subjects only to those explicitly stated would understate the frequency at which relationships between entities are of interest (see Section 2.2, Table 4). Including subjects from the question and the answer regardless of whether they are explicitly stated impacted interannotator agreement for the Company and Gene subjects, which were most frequently inferred (data not shown).

Table 3. Representative Queries

Representative Query	Subject Terms	Resource Terms	# results
What drugs are in development to treat multiple sclerosis?	company, disease, drug, gene	Competitive intelligence	138
What companies have drugs in Phase II to treat multiple sclerosis, and what	company, disease,	Competitive intelligence	52

are the drugs?	drug, gene		
What patents have been published about TNF-alpha?	company, gene	Patents	49
In what tissues is TNF-alpha expressed?	gene	Journal articles	6
What protocols have been patented for producing large quantities of therapeutic antibodies? By what companies?	methods, company	Patents	4

2.2 Query Analysis

Requests were classified as "navigational" (directed toward a specific piece of information) or "informational" (collecting data about a topic) [8]. Typical navigational queries included information about a patent family, sales figures for a drug, or a recent news article about the pharmaceutical industry. Navigational queries made up 20.2% (228/1131) of research requests. This is lower than the 25.6% mark noted for PubMed queries [6], and it may reflect differences in query analysis methodology, or in how users employ the services of PubMed versus a corporate library. Interannotator agreements for "navigational" and "informational" queries were .37 (10) and .79 (70) respectively.

Questions about drugs, diseases and genes made up the largest class of search requests, representing 31.4% (355/1131), 27.4% (310/1131) and 26.2% (297/1131) of all queries, respectively (Table 1). The first two classes are not surprising when the corporate mission is to create drugs to treat diseases. Gene-based queries are also to be expected, considering that genes and proteins are the targets of drugs, and they provide the key to understanding origins of disease and the mechanism of therapeutic action. Consistent with how authors refer to genes and proteins in the literature [9], Biogen Idec employees favored the long names or synonyms of genes rather than using the official gene symbol the vast majority of the time (data not shown).

Journal articles were the most frequently requested resource type, followed by Competitive Intelligence resources, Patent resources and, to a lesser extent, News. Most competitive intelligence questions could be answered by using commercial databases such as Pharmaprojects (http://www.pharmaprojects.com) or the Investigational Drugs Database (IDdb; http://www.iddb.com), which periodically survey corporate websites, press releases, major conferences, and Securities and Exchange Commission reports (complete listing at http://www.iddb.com/cds/faqs_info_sources.htm) (data not shown). Competitive Intelligence databases also include selected information from journal articles and patents, blurring the lines between our Resource definitions

(Table 2), but they do not constitute enough of the database content to impact our results.

To determine if query topics vary by resource, search subjects from journal articles, competitive intelligence resources, and patents were examined individually (Figure 1). Gene and protein names are common search terms across different resource types, and they are the preferred search subjects in the patent literature. Disease and drug searches are directed primarily to the scientific literature and pipeline databases. Company and Institution queries are largely confined to the competitive intelligence literature, and methods searches are limited to journal articles.

Figure 1. Query Subject by Resource

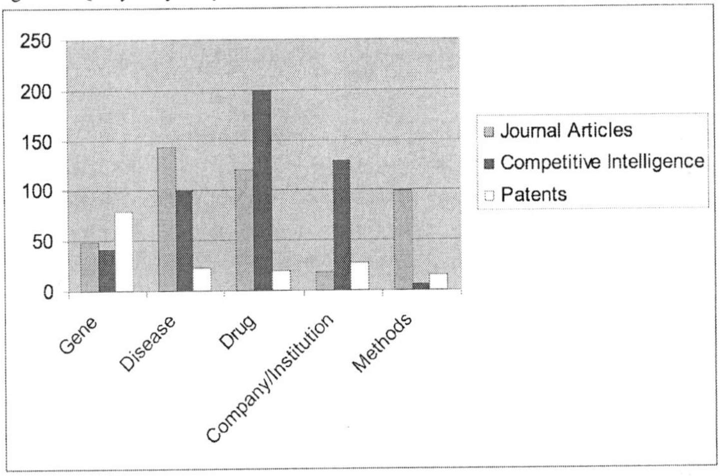

Compound queries, in which more than one subject is represented in the question and/or answer, represented 36.2% (409/1139) of research questions, four examples of which are shown in Table 3. These questions demonstrate the importance of identifying relationships among entity types. Questions requesting information from multiple resources occur in 6.4% (73/1131) of requests. These require answers that involve some degree of data integration, whether it is combining unstructured text from news and journal articles, or merging structured data with unstructured text. This figure is a gross underestimation of data integration requirements, as most journal article, competitive intelligence and patent searches generate results from more than one database [10]. Merging results into a unique set involves extensive post-processing to remove duplicate records, map controlled vocabularies from each database, and apply a uniform format to records from disparate databases.

Table 4. Frequency of Pair-wise Subject Combinations

	Drug	Disease	Gene	Company	Methods	Author	Geography
Drug							
Disease	138						
Gene	22	29					
Company	52	13	13				
Methods	10	10	6	4			
Author	8	12	11	17	0		
Geography	22	23	5	11	1	5	
Sales	41	18	3	7	1	1	14

2.3 Where Does Text Mining Fit In?

Cohen and Hersh define text mining first by distinguishing it from information retrieval, text summarization and natural language processing, then by sub-dividing it into named entity recognition (NER), text classification, synonym and abbreviation extraction, relationship extraction and hypothesis generation [3]. Synonym and abbreviation extraction can be grouped with NER if one assumes that synonyms and abbreviations for each entity are part of the entity extraction process. Similarly, relationship extraction is dependent on NER as a means of identifying which entity classes are related. If the extraction techniques are grouped with NER, that leaves three criteria with which to evaluate the Biogen Idec Library research requests for text mining: extraction, text classification, and hypothesis generation.

A research request was classified as being an Extraction request if the question asked for specific facts ("what are annual sales in Japan?" or "what is the incidence of disease x?"), versus asking for a general search ("please search the patent literature", "I need general information about this disease"). Text Classification was used to describe requests for which large positive training corpora exist. Theoretically, classification can include automated techniques such as unsupervised clustering, which can be applied to all the research requests. Our objective with this category was to quantify the frequency of requests for queries that are executed weekly or monthly over a period of several years, and for which positive training data exist, thereby justifying the effort of building a classifier. A prominent example is product safety literature. The FDA mandates periodic comprehensive literature searches for reports of marketed products in the literature (21 CFR 314.80[1]), which generates a positive training set of documents that can be used to build a classifier. Hypothesis Generation was not used to code the queries, as discussion between the

[1] http://www.accessdata.fda.gov/scripts/cdrh/cfdocs/cfcfr/CFRSearch.cfm?fr=314.80

annotators did not result in a viable protocol for annotation into this proposed category.

Out of the 1131 queries, 304 (26.9%) were classified as Extraction (286/304) or Classification (18/304). Search requests not coded as Extraction (73.1%) typically were at the general search level, suggesting that requesters were conducting a broad search, they wanted context around the facts they were looking for, or they were unaware that entity extraction tools are available.

We examined the queries coded as Extraction further to determine if individual Subjects or Resources were over-represented. The majority of extraction research requests called upon Competitive Intelligence (189/286) or Statistics (53/286) resources (data not shown). Interestingly, the answers for these requests were available in proprietary databases such as IDdb, Adis R&D, and others. Extraction questions not answered using databases were spread across subject categories, with journal articles as the primary Resource type (63/286 queries; data not shown).

Table 5. Text Mining Techniques

Technique	Description	# Requests	Interjudge Agreement
Extraction	Named entity recognition, synonym and abbreviation extraction and relationship extraction	286	.60 (31)
Classification	Text Classification – supervised machine learning	18	.50 (2) * *[n=229]

3. Discussion

3.1. *Impact of Assistance on Research Requests*

Information needs have been studied by examining query logs of search engines and inferring the intended need based on query terms and user sessions [5, 6]. Other studies have gathered information needs directly from clinicians [7] or academic and industry researchers [11]. Our study differs in that the information needs represent questions that require professional assistance, i.e. end-users were not able to find results on their own or they could not find results efficiently. This may be influenced by the query subject; gene and protein names are notoriously difficult to use as search terms due to complicated nomenclature and ambiguity [9]. Drugs also undergo name changes as they traverse the developmental pipeline [12]. Diseases are represented in myriad ways as observed in the Medical Subject Headings terminology. In the absence of a sophisticated indexing and query translation system like the one behind PubMed (http://www.pubmed.org), the low frequency of Boolean OR operator use [6] suggests end-users are missing relevant results, prompting them to seek assistance. Variations in search engine algorithms, database design, and content may also place a naïve end-user at a disadvantage. Even though Competitive Intelligence and Patent end-user tools are available at Biogen Idec, the high frequency of requests for assistance suggests that they are too complex for the casual user to efficiently obtain information.

3.2. *Research Request Subjects and Resources: Why Are Questions Asked?*

A frequently cited application of text mining is database curation; e.g. the extraction of gene names, protein-protein interactions, expression data, and subcellular localization. The predominant subjects in the Biogen Idec research requests overlap with entity types frequently studied in text mining research, notably genes and diseases. Our results support the selection of tasks in text mining challenges such as BioCreAtIvE and the TREC Genomics track as representing real information needs, especially named entity recognition of gene and protein names. Genes were the only subject type of interest across resource types (Figure 1), which may reflect the need to understand gene function throughout the drug development process. Selection of a protein as a drug target requires understanding what it does (a journal article search) and who else is working on it (competitive intelligence and patent searches). As named entity recognition of gene names improves, our results suggest that testing algorithms against multiple text sources is a worthwhile endeavor.

Genes were the primary search subject of patent literature, which was unexpected considering that patents are a significant source of drug development information, especially small molecules and their chemical synthesis [1, 13]. The dearth of patent drug searches in our results is due to chemical structure searches being performed by groups outside the Library who do not need our assistance.

Information about drugs is the most common request subject (Table 1). The high cost of drug development makes awareness of research with comparable compounds essential for maximizing efficacy and minimizing unintended adverse effects. Although named entity recognition of chemical compounds has received some attention in the text mining literature [14], to our knowledge, a broader approach to identify any substance with therapeutic benefit has not. In particular, therapeutics for a specific disease (138/378; Table 4) or against a class of targets (represented by drug-gene compound queries, 22/378, Table 4) are of sufficiently high interest to drive Biogen Idec employees to seek assistance.

Searches about companies or institutions were enhanced in the competitive intelligence literature (Figure 1). One reason for this phenomenon may be the ease with which institution searches can be performed against databases that house journal articles and patents. The second reason reflects the fundamental *raison d'etre* of competitive intelligence literature: to find out what other companies are doing.

3.3. *Existing Databases and Entity Extraction*

The Biogen Idec Library does not typically receive requests to interpret results from transcript profiling or proteomics experiments. There are a number of public and proprietary databases that address these needs, providing extracted entities and relationships among them based on the published literature. Numerous public and proprietary databases permit high-throughput analysis of gene lists and extraction of relationships between genes and diseases, expression patterns, or Gene Ontology terms.

Similarly, in the competitive intelligence space, so-called "pipeline databases" allow users to search by and export lists of drugs, indications (i.e. diseases treatable by drugs), companies, and developmental stages [15]. The success of these databases highlights the importance of entity extraction as a means of managing the vast amount of information available. Furthermore, our quantification supports the need for these resources. Literature and competitive intelligence queries are well-served by existing databases. Patent literature, however, is underserved in this regard. The high incidence of patent gene queries illustrates the need for a reliable and comprehensive resource with extracted information about genes or proteins and their patented use. To some extent, GeneSeq and GeneIT perform this task by isolating nucleotide and amino acid sequences, but not all patents about specific targets contain sequences.

3.4. *Requests in the Future*

The Library tends to receive queries that can be answered, consistent with results from analyzing questions asked by clinicians [7]. To add qualitatively new query types to the ones currently serviced requires training and awareness. New queries resulting in new deliverables often require changing customer behavior to take advantage of new capabilities. An example is inferential analysis, which uses indirect relationships to generate or validate hypotheses. Examples of inferential analysis have been described in the literature [16, 17], but demand for this technique has not surfaced in research requests to our library. The Biogen Idec customer base is increasingly aware of inferential analysis as the tools to service those requests are being deployed and the customer base learns what qualitatively new requests will result in answers.

Acknowledgments

The authors thank Suzanne Szak, Pam Gollis and Lulu Chen for critical reading of the manuscript.

References

1. Grandjean, N., et al., *Competitive intelligence and patent analysis in drug discovery: Mining the competitive knowledge bases and patents.* Drug Discovery Today: Technologies, 2005. **2**(3): p. 211-215.
2. Carlucci, S., A. Page, and D. Finegold, *The role of competitive intelligence in biotech startups (Reprinted from Building a Business section of the Bioentrepreneur web portal).* Nat Biotechnol, 2005. **23**(5): p. 525-527.
3. Cohen, A.M. and W.R. Hersh, *A survey of current work in biomedical text mining.* Brief Bioinform, 2005. **6**(1): p. 57-71.
4. Scherf, M., A. Epple, and T. Werner, *The next generation of literature analysis: integration of genomic analysis into text mining.* Brief Bioinform, 2005. **6**(3): p. 287-97.
5. Chau, M., X. Fang, and O.R.L. Sheng, *Analysis of the query logs of a web site search engine.* J Am Soc Inf Sci Technol, 2005. **56**(13): p. 1363-1376.
6. Herskovic, J.R., et al., *A day in the life of PubMed: Analysis of a typical day's query log.* J Am Med Inf Assoc, 2007. **14**(2): p. 212-220.
7. Ely, J.W., et al., *A taxonomy of generic clinical questions: classification study.* British Medical Journal, 2000. **321**(7258): p. 429-32.
8. Broder, A., *A taxonomy of web search.* SIGIR Forum, 2002. **36**: p. 3-10.
9. Chen, L., H. Liu, and C. Friedman, *Gene name ambiguity of eukaryotic nomenclatures.* Bioinformatics, 2005. **21**(2): p. 248-56.
10. Biarez, O., et al., *Comparison and evaluation of nine bibliographic databases concerning adverse drug reactions.* DICP, 1991. **25**(10): p. 1062-5.

11. Stevens, R., et al., *A classification of tasks in bioinformatics.* Bioinformatics, 2001. **17**(2): p. 180-8.
12. Snow, B., *Drug nomenclature and its relationship to scientific communication*, in *Drug Information: A Guide to Current Resources*, B. Snow, Editor. 1999, Medical Library Association and The Scarecrow Press, Inc.: Lanham, Maryland and London, England. p. 7-19.
13. Simmons, E.S., *Prior art searching in the preparation of pharmaceutical patent applications.* Drug Discov Today, 1998. **3**(2): p. 52-60.
14. Mika, S. and B. Rost, *Protein names precisely peeled off free text.* Bioinformatics, 2004. **20 Suppl 1**: p. i241-7.
15. Mullen, A., M. Blunck, and K.E. Moller, *Comparison of some major information resources in pharmaceutical competitor tracking.* Drug Discov Today, 1997. **2**(5): p. 179-186.
16. Wren, J.D., et al., *Knowledge discovery by automated identification and ranking of implicit relationships.* Bioinformatics, 2004. **20**(3): p. 389-98.
17. Swanson, D.R., *Medical literature as a potential source of new knowledge.* Bull Med Libr Assoc, 1990. **78**(1): p. 29-37.

EPILOC: A (WORKING) TEXT-BASED SYSTEM FOR PREDICTING PROTEIN SUBCELLULAR LOCATION

SCOTT BRADY AND HAGIT SHATKAY

School of Computing, Queen's University
Kingston, Ontario, Canada K7L 3N6

Motivation: Predicting the subcellular location of proteins is an active research area, as a protein's location within the cell provides meaningful cues about its function. Several previous experiments in utilizing text for protein subcellular location prediction varied in methods, applicability and performance level. In an earlier work we have used a preliminary text classification system and focused on the integration of text features into a sequence-based classifier to improve location prediction performance.

Results: Here the focus shifts to the text-based component itself. We introduce *EpiLoc*, a comprehensive text-based localization system. We provide an in-depth study of text-feature selection, and study several new ways to associate text with proteins, so that text-based location prediction can be performed for practically any protein. We show that EpiLoc's performance is comparable to (and may even exceed) that of state-of-the-art sequence-based systems. EpiLoc is available at: *http://epiloc.cs.queensu.ca*.

1. Introduction

Knowing the location of proteins within the cell is an important step toward understanding their function and their role in biological processes. Several experimental methods, such as those based on green fluorescent proteins or on immunolocalization, can identify the location of proteins. Such methods are accurate, but slow and labour-intensive, and are only effective for proteins that can be readily expressed and produced within the cell.

Given the large number of proteins about which little is known, and that many of these proteins may not even be expressed under regular conditions – it is important to be able to computationally infer protein location based on readily available data (e.g. amino acid sequence). Once effective information is computationally elucidated outside the lab, well-targeted lab experiments can be judiciously performed. For well over a decade many computational location-prediction methods were suggested and used, typically relying on features derived from sequence data[7,9,12,13].

Another type of information that can assist in location prediction is derived from text. One option is to explicitly extract location statements from the literature[6]. While this approach offers a way to access pre-existing knowledge, it does not support prediction. An alternative *predictive* approach is to employ classifiers using text-features that are derived from literature discussing the proteins. These features may not state the location, but their relative frequency in the text associated with a certain protein is often correlated with the protein's location. Examples of this approach include work by Nair and Rost[11] and by

Stapley *et al*[17]. They represent proteins using text-features taken from annotations[11] or from PubMed abstracts in which the protein's name occurs[17], and train classifiers to distinguish among proteins from different locations. The main limitations of this earlier work are: a) It was not shown to meet or improve upon the performance of state-of-the-art systems. b) The systems depended on an explicit source of text; in its absence many proteins cannot be localized.

In an earlier work[8,16] we studied the integration of text features into a sequence-based classifier[9], showing significant improvement over state-of-the-art location prediction systems. The text component was a preliminary one, and was not studied in detail. Here we provide an in-depth study and description of a new and complete text-based system, *EpiLoc*. We compare several text-feature selection methods, and extensively compare the performance of this system to other location prediction systems. Moreover, we introduce several alternative ways to associate text with proteins, making the system applicable to practically any protein, even when text is not available from the preferred primary source. Further details about the differences between the preliminary version[8,16] and EpiLoc are given in the complete report of the work[3].

While our work focuses on protein subcellular localization, the ideas and methods, including the study of feature selection and of ways for associating text with biological entities, are applicable to other text-related biological enquiries.

In Section 2 we introduce the methods for associating text with proteins, and the way in which text is used to represent proteins. Section 3 focuses on feature selection methods, while Sections 4 and 5 describe our experiments and results, demonstrating the effectiveness of the proposed methods.

2. Data and Methods

EpiLoc is based on the representation of each protein as an *N*-dimensional vector of weighted text features, $< w_1^p \dots w_N^p >$. Each position in the vector represents a term from the literature associated with the proteins. As not all terms are useful for predicting subcellular location, and to save time and space, feature selection is employed to obtain *N* terms, as discussed in Section 3. Here we describe our primary method for associating text with individual proteins and our term-weighting scheme. We also present three alternative methods that assign text to proteins when the primary method cannot do so.

Primary Text Source: The literature associated with the whole protein dataset is the collection of text related to the individual proteins. For training EpiLoc, text per protein is taken from the set of PubMed abstracts referenced by the protein's Swiss-Prot[2] entry. Abstracts associated with proteins from three or more subcellular locations are excluded, as their terms are unlikely to effectively characterize a single location. Each protein is thus associated with a set of *authoritative* abstracts, as determined by Swiss-Prot curators. As we noted before[16], the abstracts do not typically discuss localization – but rather are authoritative with respect to the protein in general. This choice of text is more specific than that of Stapley *et al.*[17], who used *all* abstracts containing a protein's gene name. Moreover, unlike Nair and Rost[11], who used Swiss-Prot *annotation text* rather than referenced abstracts, our choice is general enough to assign text to the majority of proteins, allowing the method to be broadly applicable.

The text in each abstract is tokenized into a set of terms, consisting of singletons and pairs of consecutive words; a list of standard stop words[a] is removed, and Porter stemming[14] is then applied to all the words in this set. Last, terms occurring in fewer than three abstracts or in over 60% of all abstracts are removed; very rare terms cannot be used to represent the majority of the proteins in a dataset, while overly frequent terms are unlikely to have a discriminative value. The resulting term set typically contains more than 20,000 terms, and is reduced through a feature selection step (see Section 3). The feature-selection process produces a set of *distinguishing terms* for each location, that is, terms that are more likely to be associated with proteins within a certain location than with proteins from other locations. The combined set of all distinguishing terms forms the set of terms that we use to represent proteins, as discussed next.

Term Weighting: Given the set of N distinguishing terms, each protein p, is represented as an N-dimensional weight-vector, where the weight $W_{t_i}^p$ at position i, $(1 \leq i \leq N)$, is the probability of the distinguishing term t_i to appear in the set of abstracts known to be associated with protein p, denoted D_p. This probability is estimated as the total number of occurrences of term t_i in D_p divided by the total number of occurrences of *all* distinguishing terms in D_p. Formally $W_{t_i}^p$ is calculated as: $W_{t_i}^p = (\#\ of\ times\ t_i\ occurs\ in\ D_p)/\Sigma_j(\#\ of\ times\ t_j\ occurs\ in\ D_p)$, where the sum in the denominator is taken over all terms t_j in the set of distinguishing terms T_N.

Once all the proteins in a set have been represented as weighted term vectors, the proteins from each subcellular location are partitioned into training and test sets, and a classifier is trained to assign each protein to its respective location. Our classifier is based on the LIBSVM[5] implementation of support vector machines (SVMs). LIBSVM supports soft, probabilistic categorization for n-class tasks, where each classified item is assigned an n-dimensional vector denoting the item's probability to belong to each of the n classes. Here n is the number of subcellular locations.

Alternative Text Sources: As pointed out by Nair and Rost[11], the text needed to represent a protein is not always readily available. In our case, some proteins

[a] Stop words are terms that occur frequently in text but typically do not bear content, such as prepositions.

may not have PubMed identifiers in their Swiss-Prot entry, and others – newly discovered proteins – may not even have a Swiss-Prot entry. We refer to such proteins as textless, and propose three methods to assign them with text.

HomoLoc – In previous work[16], if a textless protein had a homolog with associated text, we used the text of the homolog to represent the textless protein. HomoLoc extends this idea to consider multiple homologs and re-weight terms accordingly. A BLAST[1] search identifies the set of homologs, and we retain those that share at least 40% sequence identity with the textless protein. (This level of similarity was chosen based on a study by Brenner et al.[4,3]). The retained homologs are then ranked in ascending order according to their E-value, and the set of abstracts associated with the top three homologs are associated with the textless protein. To reflect the degree of homology in the term vector representation, a modified weighting scheme is used where the number of times each term occurs in the abstracts associated with a homolog is multiplied by the percent identity between the homolog and the textless protein. Formally, the modified weight is calculated as:

$$W_{t_i}^{P} = \frac{\sum_{h \in H} (\#\text{ of occurences of } t_i \text{ in } D_h)\cdot(\%\text{ identity of } h)}{\sum_{h \in H} \sum_{t_j \in T_N} (\#\text{ of occurences of } t_j \text{ in } D_h)\cdot(\%\text{ identity of } h)},$$

where h is a homolog, D_h is the set of abstracts associated with h, and a sum is taken over all the homologs in the set of homologs H.

DiaLoc – Proteins are most likely to be textless when they have just recently been sequenced/identified, as little information about them exists in databases such as PubMed or Swiss-Prot. When no close homologs with assigned text are known, HomoLoc cannot be used. The most reliable source of information for such proteins (and the one most likely to be interested in their localization) is the scientist researching the proteins. A user interface (shown in Fig. 2), allows a researcher to type her own short description of the protein based on the current state of knowledge. This description is used as the text associated with the textless protein. DiaLoc is meant to be used as an interactive tool for researchers concerned with individual proteins, and not as a large-scale annotation tool.

PubLoc[b] – Proteins whose Swiss-Prot entries do not contain reference to PubMed may still have PubMed abstracts discussing them. To check if such abstracts exist, the name of the textless protein and its gene are extracted from the Swiss-Prot entry. A query consisting of an *OR*-delimited list of these names is posed to PubMed. The five most recent abstracts returned are used as the protein's text source. This is a simple selection criterion and can be further improved upon.

---

[b] We thank Annette Höglund for suggesting this name.

To select the preferred method for handling textless proteins for large-scale annotation, we compared HomoLoc's and PubLoc's performance on the 614 textless proteins of the MultiLoc dataset (see Section 4). A complete discussion of these experiments is beyond the scope of this paper and is provided elsewhere[3]; we briefly summarize them here. We trained EpiLoc on all the proteins in the MultiLoc dataset that *do* have associated text. We then represented the remaining textless proteins using both PubLoc and HomoLoc, and classified them using the trained system. The overall accuracy obtained (for these 614 proteins) using HomoLoc is 73% for plant and 76% for animal. Using PubLoc the accuracy dropped to 57% and 64%, respectively[c]. As PubLoc is clearly less effective than HomoLoc, it is only applied in cases where neither HomoLoc nor DiaLoc can be used. HomoLoc is thus our method of choice for handling textless proteins, and is further discussed in Section 4.

3. Feature Selection

As stated in Section 2, each protein is represented as a weight-vector defined with respect to a set of *distinguishing terms*. Using a set of selected features can improve performance (even when SVMs are used) and reduces computational time and space. Intuitively, a term t is distinguishing for a location L, if its likelihood to occur in text associated with location L is significantly different from that of occurring in text associated with all other locations. To compare these likelihoods, for each location we assign to each term a score reflecting its probability to occur in the abstracts associated with the location. We formalize this method, referred to as the *Z-Test* method, in Section 3.1, and compare it with several alternatives in Section 3.2.

3.1. *The Z-Test Method*

Let t be a term, p a protein, and L a location. A protein, p, localized to L, is denoted $p \in L$ and has a set of associated abstracts, denoted D_p. The set of all proteins known to be localized to L is denoted P_L. We denote by D_L the set of abstracts associated with location L, (i.e. all abstracts associated with the proteins localized to L). Formally, this set is defined as: $D_L = \cup_{p \in P_L} \{d | d \in D_p\}$, and the number of abstracts in this set is denoted $|D_L|$. The probability of term t to be associated with location L, denoted $Pr(t|L)$, is defined as the conditional probability of t to appear in an abstract d, given that d is associated with location L. This probability is expressed as: $Pr(t|L) = Pr(t \in d | d \in D_L)$. Its maximum likelihood estimate is the proportion of abstracts containing the term t among all abstracts associated with L: $Pr(t|L) \approx$ *(# of abstracts* $d \in D_L$ *such that* $t \in d$*)* $/ |D_L|$. We calculate

[c] We also tested simpler versions of these methods (including the single-homolog method we tried in the past[16]); these were not as effective as the methods presented here[3].

the probability $Pr(t|L)$ for each term t and location L.

Based on the above formulation, a term t is considered *distinguishing* for location L, if and only if its probability to occur in abstracts associated with L, $Pr(t|L)$, is significantly different from its probability to occur in abstracts associated with any other location L', $\Pr(t|L')$. To determine the significance of the difference between the two probabilities, a statistical test is employed that utilizes a Z-score[18]. The test evaluates the difference between two binomial probabilities, $\Pr(t|L)$ and $\Pr(t|L')$, by calculating the following statistics:

$$Z^t_{L,L'} = \frac{\Pr(t|L) - \Pr(t|L')}{\sqrt{\bar{P} \cdot \left(1-\bar{P}\right) \cdot \left(\frac{1}{|D_L|} + \frac{1}{|D_{L'}|}\right)}}, \quad \text{where } \bar{P} = \frac{|D_L| \cdot \Pr(t|L) + |D_{L'}| \cdot \Pr(t|L')}{|D_L| + |D_{L'}|}$$

The higher the absolute value $\left|Z^t_{L,L'}\right|$, the greater is the confidence level that the difference between $\Pr(t|L)$ and $\Pr(t|L')$ is statistically significant. Therefore, we consider a term t as distinguishing for location L if for any other location L', the score $\left|Z^t_{L,L'}\right|$ is greater than a predetermined threshold. Table 1 shows examples of distinguishing terms for several locations; note that the terms do not necessarily state the location, but are merely correlated with it. The precise threshold selected was based on the experiment described next.

3.2. *Feature Selection Comparison*

To determine the effectiveness of the *Z-Test* method, we compare it to four standard feature selection methods: *odds ratio* (*OR*), *Chi-squared* (χ^2), *mutual information* (*MI*), and *information gain* (*IG*)[15]. We also compare it to the *Entropy* method, used by Nair and Rost[11]. Each of the four standard methods attempts to quantify how well a term represents a location by scoring a term t with respect to a location L. The total score for a term is then calculated as a combination of its location-specific scores. Following previous evaluations[15,20], to calculate the total *OR* and the *IG* scores we *sum* the term's scores over all locations, and to calculate the *MI* and χ^2 scores we take the *maximum* score for the term with respect to all locations. The *Entropy* method[11] scores terms with respect to locations, based on the difference between their Shannon information and the maximum attainable information.

To compare among the different feature selection methods we calculated the overall accuracy achieved by classifiers based on each method, on both plant and animal proteins of the MultiLoc dataset. For each of the methods, we used the same text pre-processing and partitioning of the data for five-fold cross-validation. Each of the six methods was evaluated based on its performance over a range of possible number of selected terms (ranging from 500 to 4,000).

Figure 1 shows the overall location prediction accuracy as a function of the number of selected terms for plant proteins. Similar results were obtained for

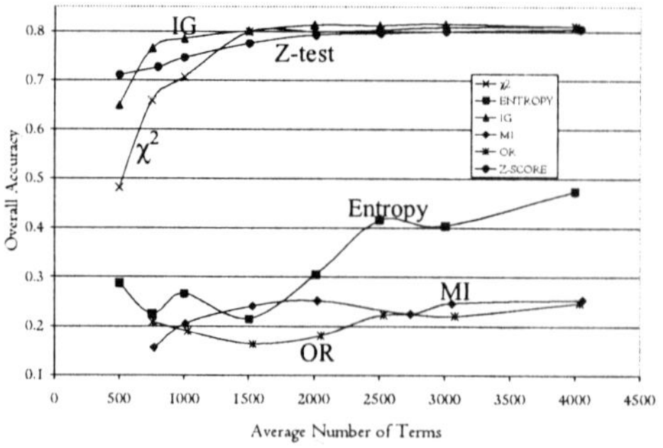

Figure 1. Accuracy of the classifiers (for plant proteins), based on different feature selection methods, as a function of the average number of selected terms (features).

Table 1. Stemmed Distinguishing terms.

Loc.	Example Terms
nu	*bind, base pair, chromatin, DNA*
mi	*acyl coa, cytochrom, electron transport*
go	*acceptor, galactos, golgi, transferase*
Er	*chaperon, disulfid isomeras, endoplasm*

Table 2. The threshold (and confidence level) chosen for each organism and dataset.

Dataset	Organism	Threshold [Confidence]
TargetP	Plant	1.645 [90%]
	Non-Plant	2.576 [99%]
PLOC	Plant	1.150 [75%]
	Animal	1.150 [75%]
MultiLoc	Plant	1.282 [80%]
	Animal	1.645 [90%]

animal proteins[3]. The figure demonstrates that the performance of the *Z-Test, IG*, and χ^2 methods is almost equivalent, and any of them could have been used by our classifier with similar results. We use the *Z-Test* in our experiments as this was our original approach[8,16] and it has a simple statistical interpretation. In contrast, the performance of the *MI, OR*, and *Entropy* methods is not as good. *MI*'s poor performance relative to that of both *IG* and χ^2 was expected, as it has been noted in previous research[20]. The *Entropy* method was originally developed to select features from a relatively small set of potential features compared to the set used here; Nair and Rost used only the *functional keywords in Swiss-Prot annotations* of the proteins, whereas we use a much larger number of potential features. As such, the relatively poor performance of the *Entropy* method shown here is not surprising. Conversely, we expected better results from *OR*. Its poor performance appears to be the result of its preferential selection of terms that occur in the abstracts associated with only a single location, leading to very sparse term vector representations for most proteins (a detailed discussion is provided elsewhere[3]). As mentioned above, we used this experiment as a guide for setting the threshold on the *Z*-score. For each dataset, we place a lower bound of 1.15 on the threshold, and set it to retain about 2,000 terms, as this number attains a balance between a computationally effective feature-space, and classification accuracy. As Figure 1 shows, the accuracy of the top methods does not significantly improve by including over 2,000 features. Table 2 shows the *Z*-score threshold used for each organism in each of the datasets described below.

4. Experimental Setting

EpiLoc was extensively evaluated, and compared to three state-of-the-art prediction systems – TargetP, PLOC, and MultiLoc – using the respective datasets that were used to train and test these systems. HomoLoc's performance is evaluated on the MultiLoc dataset. The datasets and evaluation procedures are described throughout this section.

The following three datasets are used in our comparative study:

TargetP[7] – A total of 3,415 proteins, sorted into four plant (*ch, mi, SP,* and *OT*) and three non-plant (mi, *SP*, and *OT*) locations. The *SP* (Secretory Pathway) class includes proteins from the endoplasmic reticulum (*er*), extracellular space (*ex*), Golgi apparatus (*go*), lysosome (*ly*), plasma membrane (*pm*), and vacuole (*va*); the *OT* (Other) class includes cytoplasmic (cy) and nuclear (nu) proteins.

MultiLoc[9] – The MultiLoc dataset consists of 5,959 proteins extracted from Swiss-Prot release 42.0. Animal, fungal, and plant proteins with annotated subcellular locations were collected and sorted into eleven locations: *ch, cy, er, ex, go, ly, mi, nu, pe, pm,* and *va*. Proteins with a sequence identity greater than 80% were excluded from the dataset, as were any proteins whose subcellular location annotation included the words *by similarity, potential,* or *probable.*

PLOC[13] – This dataset consists of 7,579 proteins with a maximum sequence identity of 80%, extracted from Swiss-Prot release 39.0. In addition to the 11 locations covered by the MultiLoc dataset, proteins from the cytoskeleton (*cs*) are also included. This set is larger than the MultiLoc dataset, due to the inclusion of proteins whose subcellular location line in Swiss-Prot included the words *by similarity, potential,* or *probable.*

Using these three datasets, we compare the performance of EpiLoc to that of TargetP, PLOC, and MultiLoc. Following previous evaluations[7,9,13] we use strict, stratified, five-fold cross-validation. We do not use the same partitions as used to evaluate each of TargetP, PLOC, and MultiLoc, as these partitions include textless proteins, which are not included in the evaluation of the primary EpiLoc method, (the TargetP, PLOC, and MultiLoc datasets contain 292, 1076, and 614 textless proteins, respectively). Therefore, for each dataset we perform five sets of five-fold cross-validation runs to ensure the robustness of the evaluations.

The metrics used here for performance evaluation are those used for evaluating previous systems[7,9,13]. For each dataset, and each location, performance is measured in terms of sensitivity (*Sens*), specificity (*Spec*), and Matthew's Correlation coefficient (*MCC*)[10]. These are formally defined as:

$$Sens = \frac{TP}{TP + FN}, \quad Spec = \frac{TP}{TP + FP}, \text{ and } MCC = \frac{TP \cdot TN - FP \cdot FN}{\sqrt{(TP + FN)\cdot(TP + FP)\cdot(TN + FN)\cdot(TN + FP)}},$$

where *TP, TN, FP,* and *FN* represent the number of true positives, true negatives, false positives, and false negatives, respectively, with respect to a given location. We also measure the *overall accuracy*, $Acc = C/N$, where C is the total number of correctly classified proteins and N is the total number of classified proteins. Finally, we calculate the *average sensitivity*, *Avg*, over all locations.

To evaluate HomoLoc's performance, we conducted an experiment in which the text associated with the proteins in each of the five test subsets used for the cross-validation of MultiLoc was removed. Each protein in each test subset was then assigned the text of its homologs by HomoLoc, without including the text associated with the protein itself.

5. Results and Discussion

Tables 3, 4, and 5 show the results of running EpiLoc on the TargetP, PLOC and MultiLoc datasets, respectively. For comparison, we also list the results reported by the authors of TargetP[7], PLOC[13], and MultiLoc[9] on their corresponding datasets, taken from the respective publications. Table 5 also shows earlier results of applying our basic text-based system[8,16] (denoted here *EarlyText*) to the MultiLoc dataset, demonstrating EpiLoc's improvement relative to the early system. Each table shows the overall accuracy (*Acc*), average sensitivity (*Avg*), and location-specific results. The highest values for each measure appear in bold, and standard deviations (denoted ±) are provided where available.

The results in Tables 3, 4, and 5 clearly indicate that the EpiLoc classifier performs at a level similar to earlier prediction systems. EpiLoc's overall accuracy and average sensitivity slightly exceed those of TargetP (Table 3), while each of the two systems scores higher than the other on some of the location-specific measures. On the MultiLoc dataset (Table 5), EpiLoc's overall accuracy, average sensitivity, and almost all location-specific scores are higher than those of the MultiLoc classifier.

On the PLOC dataset (Table 4) PLOC's overall accuracy is higher than EpiLoc's, while EpiLoc's average sensitivity is much higher than PLOC's. EpiLoc's sensitivity is actually higher for most locations. Whereas PLOC works well primarily on over-represented locations for which a large number of proteins are known (*ex*, *cy*, *pm*, *nu*, all have at least 860 proteins), EpiLoc performs well even for locations with relatively few associated proteins (*pe*, *er*, *ly*, *cs*, *go*, all with at most 125 proteins). These results all demonstrate that EpiLoc's performance is comparable to state-of-the-art prediction systems.

We note that EpiLoc's performance on both the TargetP and the MultiLoc datasets is better than it is on the PLOC set. As the criteria used for selecting proteins for the MultiLoc and TargetP datasets were stricter than those employed for the PLOC dataset (see Section 4), the resulting protein distribution among locations, and thus the distribution of associated text, is quite different among the datasets. As such, a lower *Z*-score threshold, as shown in Table 2, was needed to select a sufficient number of features (only about 1,250 actually chosen) for the PLOC set. As these terms are fewer and less distinguishing, using them to represent the PLOC dataset results in EpiLoc's lower performance.

As stated in Section 4, our evaluation of EpiLoc does not include the textless proteins from each of the three datasets. Consequently, when applied to the

Table 3. Prediction performance of TargetP and EpiLoc on the TargetP dataset, for both plant and non-plant proteins.

Loc.	TargetP	EpiLoc	TargetP	EpiLoc
	Plant (*Sens Spec MCC*)		**Non-Plant** (*Sens Spec MCC*)	
ch	0.85 **0.69 0.72**	**0.92** 0.53 0.68	N/A	
mi	0.82 **0.90** 0.77	**0.89** 0.81 **0.82**	0.89 0.67 0.73	**0.92 0.84 0.86**
SP	**0.91 0.95 0.90**	0.89 0.84 0.80	**0.96 0.92 0.92**	0.93 0.86 0.84
OT	**0.85** 0.78 0.77	0.84 **0.95 0.78**	**0.88 0.97 0.82**	**0.88** 0.95 0.81
Acc	0.853 (±0.035)	**0.862 (±0.004)**	0.900 (±0.007)	**0.901 (±0.006)**
Avg	0.856 (n/a)	**0.883 (±0.001)**	0.907 (n/a)	**0.908 (±0.003)**

Table 4. Prediction performance of PLOC and EpiLoc on the animal proteins of the PLOC dataset. Specificity and MCC values were not available for PLOC, hence only its sensitivity is listed and compared with our sensitivity values.

		PLOC Dataset (Animal)										
	Loc.	*go*	*cs*	*ly*	*er*	*pe*	*mi*	*ex*	*cy*	*pm*	*nu*	**Acc/Avg**
PLOC	(*Sens*)	0.15	0.59	0.62	0.47	0.25	0.57	**0.78**	**0.72**	**0.92**	**0.90**	**0.796 (± 0.009)**/ 0.579 (± 0.021)
EpiLoc	(*Sens*)	**0.76**	**0.84**	**0.89**	**0.72**	**0.85**	**0.79**	0.74	0.53	0.79	0.81	0.743 (±0.002)/ **0.773 (±0.0012)**
	(*Spec*)	0.51	0.32	0.32	0.30	0.55	0.85	0.68	0.63	0.85	0.90	
	(*MCC*)	0.62	0.51	0.53	0.45	0.68	0.80	0.66	0.50	0.78	0.80	

Table 5. Prediction performance of MultiLoc, EarlyText (our basic text-based system used in earlier work[8,16]), EpiLoc and HomoLoc on the animal[d] proteins of the MultiLoc dataset.

	MultiLoc Dataset (Animal)			
Loc.	**MultiLoc**	**EarlyText**	**EpiLoc**	**HomoLoc**
	(*Sens Spec MCC*)			
go	0.71 0.43 0.53	0.86 0.40 0.57	0.88 0.62 0.73	**0.90 0.72 0.80**
ly	0.69 0.36 0.48	0.75 0.32 0.47	**0.86** 0.39 0.57	0.85 **0.49 0.63**
er	0.68 0.56 0.60	0.74 0.48 0.58	0.74 0.59 0.65	**0.77 0.67 0.71**
pe	0.71 0.31 0.44	**0.93** 0.60 0.74	0.90 **0.77 0.82**	0.80 0.69 0.74
mi	**0.88** 0.82 **0.83**	0.80 0.79 0.77	0.82 0.82 0.80	0.79 **0.84** 0.80
ex	0.79 **0.83** 0.77	0.76 0.78 0.72	0.80 0.82 0.77	**0.83 0.83 0.79**
cy	0.67 **0.85** 0.68	0.51 0.77 0.53	0.68 0.79 0.65	**0.72** 0.80 **0.67**
pm	0.73 0.90 0.76	0.80 0.91 0.81	0.85 0.90 0.84	**0.89 0.91 0.87**
nu	0.82 0.73 0.73	0.84 0.71 0.73	0.84 0.81 0.80	**0.87 0.84 0.83**
Acc	0.746 (± 0.01)	0.725 (±0.007)	0.792 (±0.008)	**0.812 (± 0.010)**
Avg	0.741 (± 0.025)	0.775 (±0.015)	0.818 (±0.005)	**0.822 (± 0.005)**

TargetP, PLOC, and MultiLoc datasets, EpiLoc predicts the location of 91.4%, 85.8%, and 89.7% of the proteins, respectively. We note that if HomoLoc (as described in Section 2) is used to assign text to the textless proteins, EpiLoc predicts the location of 100% of the proteins, while maintaining its high accuracy (e.g. overall accuracy of *0.81* on the MultiLoc dataset).

Table 5 shows the performance of HomoLoc on the MultiLoc dataset. HomoLoc's overall accuracy actually exceeds EpiLoc's, and its average sensitivity is at least as high. Moreover, HomoLoc produces many of the highest location-specific results. HomoLoc's improved performance on the MultiLoc

[d] Similar results were obtained for plant and fungus proteins.

dataset is most likely the result of the large amount of text that it associates with each protein. Having more abstracts, originating from the three close homologs, provides a larger sample of representative terms for the protein than the single set of abstracts referenced by the protein's single Swiss-Prot entry.

HomoLoc's performance on the MultiLoc dataset clearly demonstrates its utility for handling textless proteins. These results strongly support the idea that in the absence of curated text for a protein, using the text of its homologs to represent the protein yields a very good prediction.

Finally, we demonstrate by example the use of the DiaLoc method. Its proper evaluation requires a study over a prolonged period of time, in which researchers will use the web-interface to enter text and assess the results. Thus no formal evaluation is given here. Our example is the histone H1, a nuclear protein involved in the structure of DNA. For the "expert" text describing the protein, we use the description of H1 given by Wikipedia[19]. This choice of example is reasonable as it provides the high-level description we expect to obtain from an expert who has some knowledge of the protein, but is still searching for more details. Any word starting with the letters *nucle*, which might be viewed as a hint for a nuclear protein, was removed from the text. The resulting text is the input to the DiaLoc web server (Fig. 2), and the output is a location prediction. DiaLoc correctly assigns H1 to the nucleus with a probability of 0.5661, (a high value within a multinomial distribution over *9* possible locations). Although this example clearly does not test DiaLoc's overall predictive ability, it demonstrates DiaLoc as a working tool. As the prediction engine used by DiaLoc is the same one used by EpiLoc, given the same PubMed abstracts as were used for testing EpiLoc, DiaLoc's performance is the same as EpiLoc's. DiaLoc's strength lies in its ability to serve as an interactive tool for researchers.

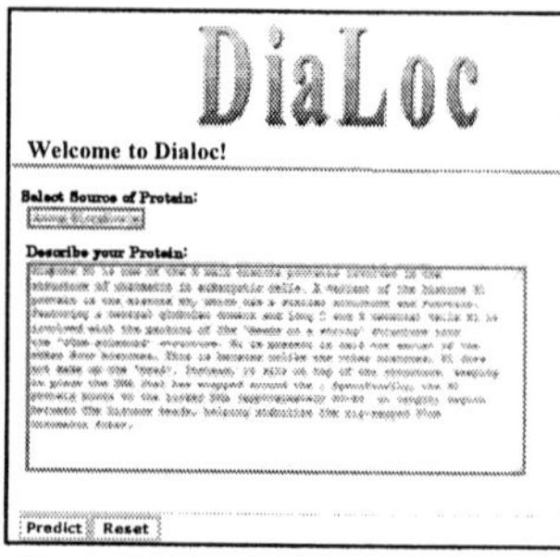

Figure 2. User interface for DiaLoc.

6. Conclusion and Future Directions

The work presented here clearly demonstrates that EpiLoc can predict the subcellular location of proteins as reliably as other state-of-the-art systems. Moreover, we have demonstrated that the HomoLoc method is an effective way to represent proteins for location prediction. By using HomoLoc, PubLoc and DiaLoc, our system can associate text with practically any protein, and predict its location. DiaLoc is expected to be a useful tool for lab scientists, while EpiLoc and HomoLoc are primarily large-scale annotation tools.

In an earlier study[8,16] we showed that the integration of a relatively basic text-based system with the sequence-based MultiLoc system[9] produced a much improved prediction performance with respect to the state-of-the-art. While the work presented here focuses on EpiLoc as a text based system, we expect that its integration with MultiLoc will further improve the overall performance. We plan to study such integration in the near future. Other future directions include a thorough evaluation of DiaLoc, and the extension of EpiLoc to predict sub-subcellular locations of proteins. EpiLoc and DiaLoc are available online at: *http://epiloc.cs.queensu.ca* and *http:// epiloc.cs.queensu.ca/DiaLoc.html.*

Acknowledgments

Many thanks to Oliver Kohlbacher's group at Tübingen, and particularly to Annette Höglund and Torsten Blum, for working with us on the early integration of text-features into their Multiloc system. The research is supported by CFI award #10437 and NSERC Discovery grant #298292-04.

References

1. Altschul SF, *et al. Basic Local Alignment Search Tool.* J. Mol. Biol., **215,** 403–410, 1990.
2. Bairoch A, Apweiler R. *The SWISS-PROT protein sequence database and its supplement in TrEMBL in 2000.* Nucleic Acids Res., **28,** 45–48, 2000.
3. Brady S. *Improved Prediction of Protein Subcellular Location through a Text-based Classifier.* M.Sc. Thesis, Queen's University, *http://www.cs.queensu.ca/~shatkay/papers/ScottBradyThesis.pdf*, 2007.
4. Brenner SB, *et al. Assessing sequence comparison methods with reliable structurally identified distant evolutionary relationships. PNAS*, **95**, 6073-6078, 1998.
5. Chang CC, Lin CJ. *LIBSVM: A library for support vector machines.* 2003. *http://www.csie.ntu.edu.tw/~clin/libsvm/.*
6. Craven M, Kumlien J. *Constructing Biological Knowledge Bases by Extracting Information from Text Sources.* Proc. of the ISMB, 77–86, 1999.
7. Emanuelsson O *et al. Predicting subcellular localization of proteins based on their N-terminal amino acid sequence.* J. Mol. Biol., **300,** 1005–1016, 2000.
8. Höglund A *et al. Significantly Improved Prediction of Subcellular Localization by Integrating Text and Protein Sequence Data.* Proc. of the Pacific Symp. on Biocomput. (PSB), 16–27, 2006.
9. Höglund A *et al. MultiLoc: prediction of protein subcellular localization using N-terminal targeting sequences, sequence motifs and amino acid composition.* Bioinformatics, **22**, 1158–1165, 2006.
10. Matthews, BW. *Comparison of predicted and observed secondary structure of T4 phage lysozyme.* Biochim. Biophys. Acta., **405,** 442–451, 1975.
11. Nair R, Rost B. *Inferring sub-cellular localization through automated lexical analysis.* Bioinformatics, **18,** S78–S86, 2002.
12. Nakai, K and Kanehisa, M. *A knowledge base for predicting protein localization sites in eukaryotic cells.* Genomics, **14**, 897–911, 1992.
13. Park, KJ, Kanehisa, M. *Prediction of protein subcellular locations by support vector machines using compositions of amino acids and amino acid pairs.* Bioinformatics, **19**, 1656–1663, 2003.
14. Porter MF. *An Algorithm for Suffix Stripping* (Reprint). In: Readings in Information Retrieval, Morgan Kaufmann, 1997. *http://www.tartarus.org/~martin/PorterStemmer/.*
15. Sebastiani F. *Machine Learning in Automated Text Categorization.* ACM Computing Surveys, **34,** 1–47, 1999.
16. Shatkay H *et al. SherLoc: High-Accuracy Prediction of Protein Subcellular Localization by integrating Text and Proteins Sequence Data.* Bioinformatics, 23, 1410–1417, 2007.
17. Stapley *et al. Predicting the sub-cellular location of proteins from text using support vector machines.* Proc. of the Pacific Symp. On Biocomputing. (PSB), 374–385, 2004.
18. Walpole RE *et al.* Probability and Statistics for Engineers and Scientists, Prentice-Hall, 235–335, 1998.
19. Wikipedia contributors. *Histone H1.* Wikipedia, The Free Encyclopedia.
20. Yang Y, Pedersen JO. *A Comparative Study on Feature Selection in Text Categorization.* Proc. of International Conference on Machine Learning (ICML), 1997.

FILLING THE GAPS BETWEEN TOOLS AND USERS: A TOOL COMPARATOR, USING PROTEIN-PROTEIN INTERACTION AS AN EXAMPLE

YOSHINOBU KANO[1], NGAN NGUYEN[1], RUNE SÆTRE[1], KAZUHIRO YOSHIDA[1], YUSUKE MIYAO[1], YOSHIMASA TSURUOKA[3], YUICHIRO MATSUBAYASHI[1], SOPHIA ANANIADOU[2,3], JUN'ICHI TSUJII[1,2,3]

[1]*Department of Computer Science, University of Tokyo*
Hongo 7-3-1, Bunkyo-ku, Tokyo 113-0033 Tokyo

[2]*School of Computer Science, University of Manchester*
PO Box 88, Sackville St, MANCHESTER M60 1QD, UK

[3]*NaCTeM (National Centre for Text Mining), Manchester Interdisciplinary Biocentre, University of Manchester, 131 Princess St, MANCHESTER M1 7DN, UK*

Recently, several text mining programs have reached a near-practical level of performance. Some systems are already being used by biologists and database curators. However, it has also been recognized that current Natural Language Processing (NLP) and Text Mining (TM) technology is not easy to deploy, since research groups tend to develop systems that cater specifically to their own requirements. One of the major reasons for the difficulty of deployment of NLP/TM technology is that re-usability and interoperability of software tools are typically not considered during development. While some effort has been invested in making interoperable NLP/TM toolkits, the developers of end-to-end systems still often struggle to reuse NLP/TM tools, and often opt to develop similar programs from scratch instead. This is particularly the case in BioNLP, since the requirements of biologists are so diverse that NLP tools have to be adapted and re-organized in a much more extensive manner than was originally expected. Although generic frameworks like UIMA (Unstructured Information Management Architecture) provide promising ways to solve this problem, the solution that they provide is only partial. In order for truly interoperable toolkits to become a reality, we also need sharable type systems and a developer-friendly environment for software integration that includes functionality for systematic comparisons of available tools, a simple I/O interface, and visualization tools. In this paper, we describe such an environment that was developed based on UIMA, and we show its feasibility through our experience in developing a protein-protein interaction (PPI) extraction system.

1. Introduction

In the biomedical domain, an increasing number of Text Mining (TM) and Natural Language Processing (NLP) tools, including part-of-speech (POS) taggers [1], named entity recognizers (NERs) [10], protein name normalizers [2], syntactic parsers [3,4], and relation or event extractors (ERs) have been developed, and some of them are now ready for biologists and database curators to use for their own purposes [5]. However, it is still very difficult to integrate independently developed tools into an aggregated application that achieves a specific task. The difficulties are caused not only by differences in programming platforms and different input/output data formats, but also by the lack of higher level interoperability among modules developed by different groups.

UIMA, the Unstructured Information Management Architecture [11], was originally developed by IBM. It recently became an open project in OASIS and Apache. It provides a promising framework for tool integration. UIMA has a set of useful functionalities, such as type definitions shared by modules, management of complex objects, linkages between multiple annotations, and the original text, and a GUI for module integration. However, since UIMA only provides a generic framework, it requires a user community to develop their own end-to-end analysis pipelines with a set of actual software modules. A few attempts have already been made to establish platforms for the biomedical domain, including toolkits by the Mayo Clinic [25], the Biomedical Text Mining Group at the University of Colorado School of Medicine [6][26], and Jena University [22], as well as for the general domain, including toolkits by OpenNLP [8], the CMU UIMA component repository [20], and GATE [21] with its UIMA interoperability layer.

However, simply wrapping existing modules for UIMA does not offer a complete solution for flexible tool integration, necessary for practical applications in the biomedical domain. Users, including both the developers and the end-users of TM systems, tend to be confused when choosing appropriate modules for their own tasks from a large collection of tools.

Individual user groups in the biomedical domain have diverse interests. Requirements for NLP/TM modules vary significantly depending on their interests [18]. For example, an NER module developed for a specific user group usually cannot satisfy the needs of another group. Different groups may need different types of entities to be recognized. They may also need to process different types of texts, such as scientific papers, reports, or medical records. Due to this range of needs, significant effort is often required to combine modules that were developed independently for different user groups, even after they are wrapped for UIMA. (Wrapping a tool for UIMA is a process of adding a conversion layer, which wraps the original I/O of the tool in order to communicate with the UIMA framework).

Furthermore, a task in the biomedical domain is composite in nature, from the TM/NLP point of view, and can only be solved by combining several modules. Although the selection of modules affects the performance of the aggregated system, it is difficult to estimate how this selection affects the ultimate performance of the system. Users need careful guidance in the selection of modules to be combined.

In this paper, we discuss our strategy of using comparators and automatic generators of processing streams to facilitate module integration and to guide the selection of modules. Taking the extraction of protein-protein interaction (PPI) as a typical example of a composite task, we illustrate how our platform helps users construct a system for their own needs.

There are several other technical issues that we encountered as UIMA users. For example, the issue of efficiency cannot be ignored, since we want to process a large collection of documents including all of Medline and full papers in a collection of open access journals in BMC (BioMed Central). From the viewpoint of a tool provider, the burden of making an existing module compatible with a specific platform should be minimized. Some of these issues are discussed in this paper.

2. Motivation and Background

2.1. *Goal Oriented Evaluation, Module Selection and Inter-operability*

There are standard evaluation metrics for NLP/TM modules, including precision, recall, and F-measure. For basic tasks such as sentence splitting, POS tagging, and named-entity recognition, these metrics can be estimated using existing gold-standard test sets. However, accuracy measurements based on standard test sets are sometimes deceptive because the accuracy may change significantly in practice, depending on the types of texts and the actual tasks at hand.

For example, in the bioinformatics task of recognizing occurrences of entities of specific types (e.g. cell-lines, cell locations) in text when comprehensive lexicons for those entities are available, an NER system for an open set of entities (e.g. proteins or metabolites) trained using a gold-standard data set may not be the best choice, even if it yields the best performance on a standard test set. Moreover, systems which have similar levels of performance according to standard metrics often behave differently in specific cases. Because these accuracy metrics do not take into account the importance of different types of errors to any particular application, the practical utility of two systems with seemingly similar levels of accuracy may in fact differ significantly. To users and developers alike, a detailed examination of how systems perform (on the text they would like to process) is often more important than standard metrics and test sets. Naturally, far greater importance is placed in measuring the end-to-end performance of a composite system than in measuring the performance of individual components.

In reality, because selection of modules usually affects the performance of the entire system, careful selection of modules that are appropriate for a given task is crucial. This is the main reason for having a collection of interoperable modules. What we need to be able to test is how the ultimate performance will be affected by selection of different modules and what would be the best combination of modules in terms of the performance of the whole aggregated system for the task at hand.

Since the number of possible combinations of component modules is typically large, the evaluation system has to be able to enumerate and execute them semi-automatically. This requires a higher level of interoperability for individual modules than just wrapping them for UIMA.

2.2. UIMA

2.2.1. *CAS and Type System*

The UIMA framework uses the "stand-off annotation" style [16]. The raw text in a document is kept unchanged during the analysis process. When processing is performed on the text, the result is added as new stand-off annotations with references to their positions in the raw text. A *Common Analysis Structure* (CAS) maintains a set of these annotations, which in turn are objects themselves. The annotation objects in a CAS belong to *types* that are defined separately in a hierarchical *Type System*. The features of an *annotation*[*] object have values which are typed as well.

2.2.2. *Components and Component Descriptors*

The analysis process, which includes any sort of processing of the text, is performed by one or more *Annotators*, the smallest processing components in UIMA. Components in UIMA are divided into three types: *Collection Reader*, *Analysis Engine* and *CAS Consumer*. An *Analysis Engine* analyzes a document and creates annotation objects. For example, a named entity recognizer receives a CAS, detects named entities in the text, and adds *annotation* objects of a corresponding *type(s)* (`NamedEntity` in our case) to the received CAS. There are two types of *Analysis Engines*. An *Analysis Engine* with a single *Annotator* is called a *Primitive Analysis Engine*, and an *Analysis Engine* with more than two *Annotators* inside is called an *Aggregate Analysis Engine*. A *Collection*

[*] In the UIMA framework, `Annotation` is a base type which has *begin* and *end* offset values, as a subtype of the root type `TOP`. In this paper we call any objects (any subtype of `TOP`) *annotations*.

Reader reads documents from outside of a UIMA framework and generates CASs, while a *CAS Consumer* does not output CASs.

Every UIMA component (i.e. *Collection Reader*, *Analysis Engine* and *CAS Consumer*) has a *descriptor* XML file, which provides its behavioral information. For example, the *Capability* property in a descriptor file describes what *types* of objects the component may take as input and what *types* of objects it produces as output. The compatibility of their *capabilities* is the pre-requisite for two components to be combined.

It is possible to deploy any UIMA component as a SOAP web service. Therefore, we can combine a remote component on a web service with local component freely inside a UIMA-based system.

3. Integration Platform and Comparators

3.1. *Shared Type System*

Although UIMA provides a useful set of functionalities for an integration platform of NLP/TM tools, users still have to develop the actual platform to use these functionalities effectively. The designer of an integration platform must make several decisions.

Firstly, as a crucial decision, the designer must decide how to use *types* in UIMA. At one extreme, the designer may wrap existing programs without using explicit *types*, putting information into a single String field of a common generic *type*. Since compatibility among modules is already automatically guaranteed, such a design decision would be easy to follow; however, it would not be appropriate if we aim to attain the higher level of inter-operability required for goal-oriented module selection and evaluation.

At the other extreme, the designer may force all modules developed by different groups to accept a unique *type system* which the platform defines. While this would make inter-operability readily attainable, it puts too much of a burden on the individual modules. In the worst case, we may have to re-program all of the tools developed by other groups. Thus, this design is impractical.

Our decision lies in the middle between these two extremes. That is, if necessary, we keep different *type systems* by individual groups as they are. We require, however, that individual *type systems* have to be related through a common, shared *type system* which our platform defines. Such a shared *type system* can bridge modules with different *type systems*, though bridging module may lose some information during the translation process.

Whether such a shared *type system* can be defined or not is dependent on the nature of each problem. For example, a shared *type system* for POS tags in

English can be defined rather easily, since most of POS-related modules, such as POS taggers (their output is a sequence of POSs), shallow parsers (their input is a sequence of words with their POS assignments), etc., more or less follow the well-established types defined by the Penn Treebank [24] tag set for POS types.

Figure 1 shows a part of our shared *type system.* We deliberately define a highly organized type hierarchy, since the structure of a shared common *type system* directly influences the loss of information during the translation process. For instance, it is better to express each POS as a distinct *type*, not as a String feature value, in order to identify each POS uniquely. It is also better to make abstract *types* in hierarchies as much as possible, in order not to lose information during the translation between *type systems*. For example, if a local *type system* has a *type* of general verb but has no *type* of past tense verb, then the shared *type system* should have an abstract *type* (like `Verb`) in order to capture the local *type* information.

Secondly we should consider that the *type system* could be used to compare and/or mix similar tools. *Types* should be defined in a distinct and hierarchical manner; both tokenizers and POS taggers generate a variety of tokens, but their roles are different when we assume a cascaded pipeline. We defined `Token` as a supertype (tokenizer) and `POSToken` (POS tagger) as a subtype of `Token`. Each tool should have an individual *type* to make clear which tool generated which instance; this is necessary because each tool may have a slightly different definition of output *types* even if they are the same sort of tools.

3.2. *General Combinatorial Comparison Generator*

Even if the *type system* is defined in the way previously described, there are still some issues to consider when comparing tools. We illustrate these issues using

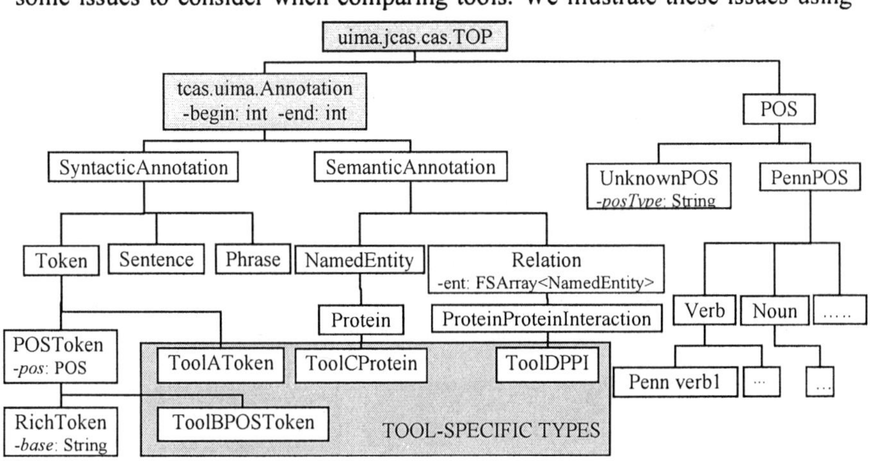

Figure 1 Part of our type system

the PPI *workflow* that we utilized in our experiments.

Figure 2 shows the *workflow* of our whole PPI system conceptually. If we can prepare two or more *Annotators* for some type of the components in the *workflow* (e.g. two sentence detectors and three POS taggers), then we could make combinations of these tools to form a multiplied number of *workflow* patterns (2x3 = 6 patterns). See Table 1 for the details of UIMA components used in our experiments.

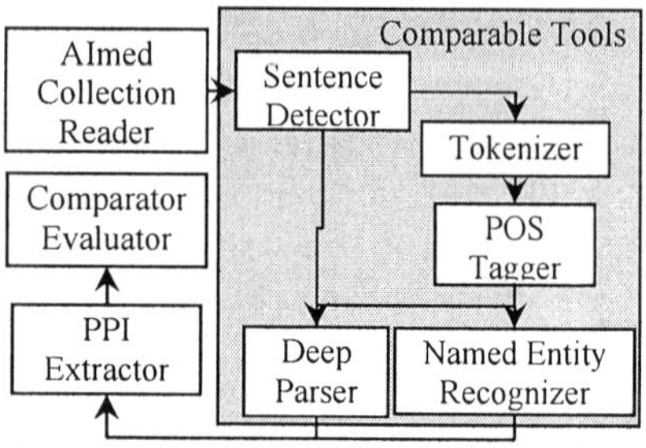

Figure 2. PPI system workflow (conceptual)

We made a pattern expansion mechanism which generates possible *workflow* patterns automatically from a user-defined *comparable workflow*. A *comparable workflow* is a special *workflow* which explicitly specifies which set of *Annotators* should be compared. Then, users just need to group comparable components (e.g. ABNER[†] and MedT-NER as a comparable NER group) without making any modifications to the original UIMA components. This aggregation of comparable *Annotators* is controlled by our *custom workflow controller*.

In some cases, a single tool can play two or more roles (e.g. the GENIA Tagger performs tokenization, POS tagging, and NER; see Figure 4). It may be possible to decompose the original tool into single roles, but in most cases it is difficult and unnatural to decompose such a complex tool. We designed our comparator to detect possible input combinations automatically by the *types* of previously generated *annotations,* and the input *capability* of each posterior *Annotator*. As described in the previous section, *Annotator* should have appropriate *capabilities* with proper *types* in order to permit this detection.

When an *Annotator* requires two or more input *types* (e.g. our PPI extractor

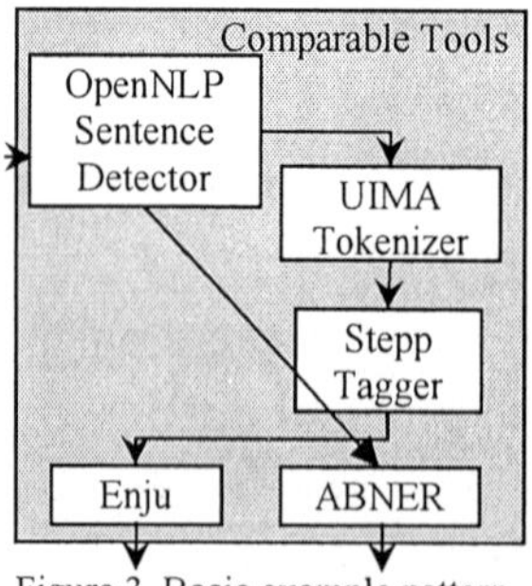

Figure 3. Basic example pattern

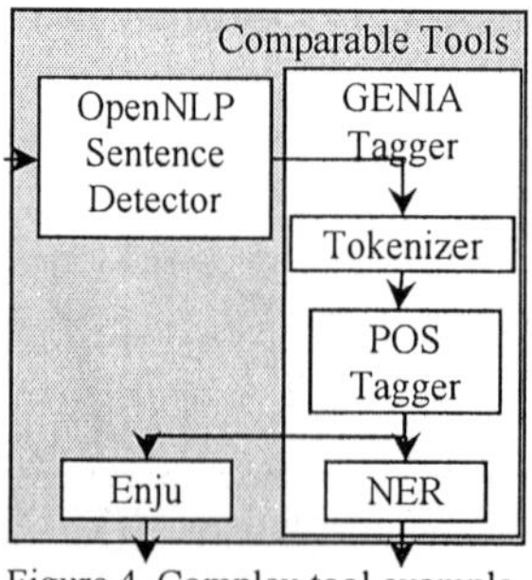

Figure 4. Complex tool example

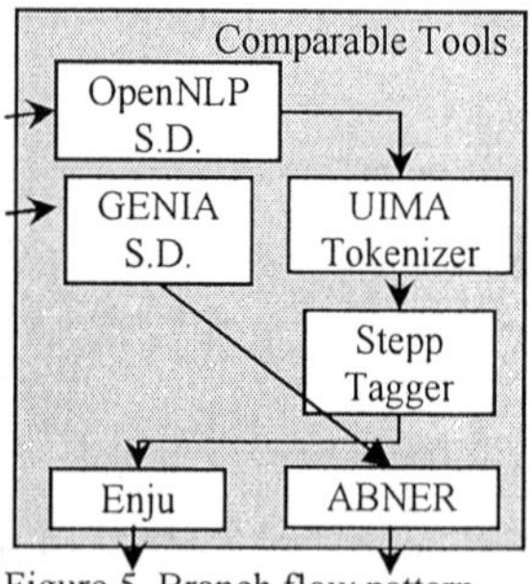

Figure 5. Branch flow pattern

[†] In the example figures, ABNER requires `Sentence` to make the explanation clearer, though ABNER does not require it in actual usage.

requires outputs of a deep parser and a protein NER system), there could be different *Annotators* used in the prior flow (e.g. OpenNLP and GENIA sentence detectors in Figure 5). Thus, our comparator calculates such cases automatically.

Because of limitations of the current Apache UIMA implementation, we originally defined `AnnotationGroup`, each of which holds *annotations* generated by a single *Annotator* in a specific *workflow* pattern. An `AnnotationGroup` has dependency links to the prior `AnnotationGroups`. Because an expanded combinatorial *workflow* is cascaded, `AnnotationGroups` are shared within posterior *Annotators* in order to increase performance.

Although it is efficient to share `AnnotationGroups`, whole combinatorial results are put into a single CAS in this design and a CAS may contain a large number of *annotations*. When web services or network communications are used, a large CAS could be costly with respect to transmission time, and may therefore decrease the performance of the system. In addition it is impossible for normal UIMA components to process such a mixture of combinatorial *annotations*. We made a special adapter component which generates a temporary CAS by the *CAS Multiplier* functions. This temporary CAS contains only a set of required *annotations* for each component in order to avoid these problems.

Table 1. List of UIMA-compliant tools that we used in the experiment.
Legend: Input type(s) required for that tool; Input required optionally; Output type(s)

<table>
<tr><th>Sentence</th><th>Token</th><th>POSToken</th><th>RichToken</th><th>Protein</th><th>Phrase</th><th>PPI</th></tr>
<tr><td colspan="7">GENIA Tagger: Trained on the WSJ, GENIA, and PennBioIE corpora (POS). Uses Maximum Entropy [17], trained on the NLPBA data set [9], derived from the GENIA corpus (NER) [15].</td></tr>
<tr><td colspan="7">Enju: HPSG parser with predicate-argument structures (PAS) as well as phrase structures. Although trained with newswire articles (Penn Treebank [24]), it can compute accurate analyses of biomedical texts due to our method for domain adaptation. The evaluated bio-performance is 86.9 F-score [4].</td></tr>
<tr><td colspan="7">STePP Tagger: Based on probabilistic models, tuned to biomedical text trained by WSJ, GENIA, and PennBioIE corpora, with state-of-the-art performance (97.3% on the standard WSJ test set).</td></tr>
<tr><td colspan="7">MedT-NER: Statistical recognizer trained on the JNLPBA [9] data. NEs are normalized to Uniprot entries to provide a high-quality link into existing databases. When a name is ambiguous between several Uniprot entries, a MaxEnt classifier is used to rank the candidate IDs.</td></tr>
<tr><td colspan="7">ABNER: From the University of Wisconsin [10], wrapped by the Center for Computational Pharmacology at the University of Colorado.</td></tr>
<tr><td colspan="7">Akane++: A new version of the AKANE system [12], trained by SVM^{light}-TK [14,19] and the AImed Corpus, achieving state-of-the-art F-measures for PPI using 10-fold cross validation (F=52- 69).</td></tr>
<tr><td colspan="7">Annotation Comparator and Evaluator: Compares annotations using the type system hierarchy to decide which annotations can be compared; generates statistical results and visualization.</td></tr>
<tr><td colspan="7">UIMA Examples: Provided in the Apache UIMA example. Sentence Splitter and Tokenizer.</td></tr>
<tr><td colspan="7">OpenNLP Tools: Part of the OpenNLP project [8], included in the Apache UIMA example.</td></tr>
<tr><td colspan="7">AImed Corpus: 225 Medline abstracts with proteins and protein-protein interactions annotated [13].</td></tr>
</table>

3.3. *User- and Developer-Friendly Utilities*

For the end-user utilities, our comparator provides a filtering function and visualization of the results, in addition to providing statistical results.

Web services are a better option when a specific runtime environment or rich computational resources are required, when a tool cannot be distributed due to licensing issues, or when it is necessary to save the time needed for module initialization. We deployed most of our components as SOAP web services so that users can launch our entire *workflow* from any environment.

We also made a single-click-to-launch system based on the Java Web Start technology. Users need not follow any explicit installation process or settings, if their machines already have Java installed.

Although Apache UIMA provides its Java APIs and C++ enhancement kit with rich functionality, it is cumbersome for developers to make their existing tools UIMA-compliant. For developers, we provide a simpler I/O interface that does not depend on any specific programming languages, so that the developers do not need to learn anything about Java or UIMA when they need to wrap existing tools into UIMA. Wrapper developers should only have to make stand-off annotations, using specified *type* and feature names, via the standard I/O streams. Our Java adapter then automatically performs all tasks to wrap the tools.

4. Experiments and Results

We have performed experiments using our PPI extraction system as an example. The PPI system (Figure 2) is similar to our BioCreative PPI system [7]. It differs in that we have decomposed the original system into seven different components.

4.1. *Combinatorial Comparison*

As summarized in Table 1, we have several comparable components and AImed as gold standard data. In this case, possible combination *workflow* patterns are 36 for `PosToken`, 589 for `ProteinProteinInteraction`, etc.

Table 2. Screenshot of a POS combinatorial comparison. Values are precision/recall in "labeled (unlabeled)" pairs, and total numbers of instances are shown.

.POSToken .POS	.Sentence	.SimpleToken	gold	OpenNLP Genia OpenNLP 3105	Stepp Genia OpenNLF 1693
Stepp	Genia	UIMA	1661	40/75 (40/75)	62/63 (62/6
Stepp	AIMED	UIMA	1661	40/75 (40/75)	62/63 (62/6
GENIA	Genia	UIMA	1661	39/73 (39/73)	53/54 (53/5
GENIA	AIMED	UIMA	1661	39/73 (39/73)	53/54 (53/5
Stepp	AIMED	GENIA	1690	31/57 (31/57)	63/63 (63/6
Stepp	AIMED	OpenNLP	1692	31/57 (31/57)	63/63 (63/6

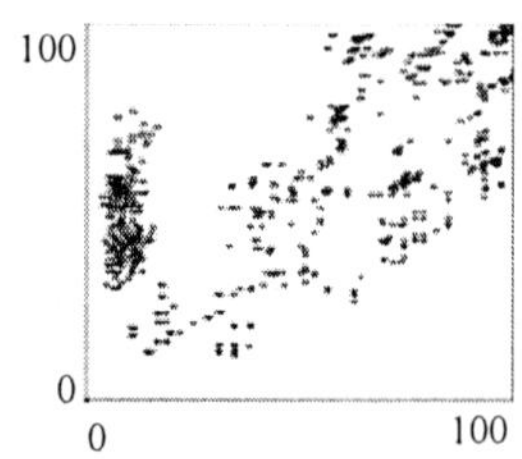

Figure 6. NER comparison distribution of precisions (x-axis) and recalls (y-axis).

Table 2 and Figure 6 show a part of the comparison result screenshots between these patterns on 20 articles from the AImed corpus. In Table 2, *labeled* scores represent complete matches of every feature of *annotations*, while unlabeled scores ignore primitive fields excluding offsets (e.g. compare offsets but ignore protein IDs). Table 3 shows a part of PPI extraction results from which we discern which combination of tools generate the best result.

When neither of compared results include the gold standard data (AImed in this case), the comparison results show a *similarity* of the tools for this specific task and data, rather than an evaluation. Even if we lack an annotated corpus, it is possible to run tools and compare results in order to understand the characteristics of tools depending on the corpus and the tool combinations.

4.2. *Performance with Multi-threading*

Apache UIMA provides an option to enable multi-threading of a *workflow* or multi-deployment of components without modifying UIMA components. We have tested multi-threading performance and the result suggests that we can increase the overall performance easily by using a parallel architecture. Because CPU architectures are evolving rapidly towards multi-cores in order to increase global performance, the capability of UIMA to support multi-threading promises considerable advantages, despite the wrapper overheads or web service communication overheads.

Table 3. PPI Evaluated on AImed, with 5631 protein pairs. (1068 true interactions). DEP means our dependency parser. Values are percentages from 10-fold cross-validation on abstracts. "pairwise" is the widely used 10-fold cross-validation on protein pairs. Refer to [23] for details.

Features	Prec	Recall	F1
DEP	67.6	26.3	37.1
WORDS	55.7	29.2	37.8
PAS (Enju)	72.0	28.7	41.0
DEP+WORDS	59.9	39.3	46.9
PAS+DEP	68.9	37.8	48.6
PAS+WORDS	61.3	40.7	48.6
ALL	64.3	44.1	52.0
ALL (pairwise)	78.1	62.7	69.5

5. Conclusion and Future Work

Although UIMA provides a general framework with much functionality, we still need to fill the gaps between what is already provided and what the users need for their specific tasks. Biomedical tasks typically consist of many components, and it is necessary to show which sets of tools are most suitable for each specific task and data. In this paper, we provided an answer to this problem using extraction of protein-protein interaction as an example task.

With any set of UIMA components that have *types* designed in the way described in this paper, our general combinatorial comparator generates possible combinations of tools for a specific *workflow* and compares/evaluates the results.

We are preparing to make a portion of the components and services described in this paper available publicly (http://www-tsujii.is.s.u-tokyo.ac.jp/uima/).

The system shows which combination of components yields the best score, and also succeeds in generating comparative results. This helps users to grasp the characteristics of and differences between the tools, which cannot be easily observed just by the widely used F-score metric.

Future directions for this work include combining the output of several modules of the same kind (such as NER systems) to obtain better results, collecting other tools developed by other groups using bridging type systems, making machine learning tools UIMA-compliant, and making grid computing available with UIMA *workflows* to increase overall performance.

Acknowledgments

We wish to thank Dr. Lawrence Hunter's text mining group at the Center for Computational Pharmacology for discussing with us and making their tools available for this research. This work was partially supported by NaCTeM (the UK National Centre for Text Mining), Grant-in-Aid for Specially Promoted Research (MEXT, Japan) and Genome Network Project (MEXT, Japan). NaCTeM is jointly funded by JISC/BBSRC/EPSRC.

References

1. Y. Tsuruoka, Y. Tateishi, J. D. Kim, T. Ohta, J. Tsujii and S. Ananiadou, *Developing a Robust Part-of-Speech Tagger for Biomedical Text.* Volos: In the Advances in Informatics. LNCS 3746: pp. 382-392 (2005).
2. N. Okazaki and S. Ananiadou, *Building an abbreviation dictionary using a term recognition approach.* Bioinformatics, pp. 22(24):3089-3095 (2006).
3. S. Pyysalo, T. Salakoski, S. Aubin and A. Nazarenko, *Lexical adaptation of Link grammar to the biomedical sublanguage: a comparative evaluation of three approaches.* BMC Bioinformatics. Suppl 3:S2 (2006).
4. T. Hara, Y. Miyao and J. Tsujii, *Evaluating Impact of Re-training a Lexical Disambiguation Model on Domain Adaptation of an HPSG Parser.* In the Proceedings of IWPT 2007. Prague, Czech Republic, June 2007.
5. L. Hirschman, M. Krallinger and A. Valencia, *Proc. of Second BioCreative Challenge Evaluation Workshop.* Madrid: Centro Nacional de Investigaciones Oncologicas (2007).
6. H. L. Johnson, W. A. Baumgartner, M. Krallinger, K. B. Cohen and L. Hunter. *Corpus refactoring: a feasibility study.* J Biomed Discov Collab pp. 2:4 (2007).
7. R. Sætre, K. Yoshida, A. Yakushiji, Y. Miyao, Y. Matsubayashi and T. Ohta, *AKANE System: Protein-Protein Interaction Pairs in the BioCreAtIvE2 Challenge,* PPI-IPS subtask (2006).

8. J. Baldrige and T. Morton, *OpenNLP*. http://opennlp.sourceforge.net/
9. J. D. Kim, T. Ohta, Y. Tsuruoka, Y. Tateishi and N. Collier, *Introduction to the bio-entity recognition task at JNLPBA*. Geneva, Switzerland. JNLPBA-04. pp. 70–75 (2004).
10. B. Settles, *ABNER: an open source tool for automatically tagging genes, proteins, and other entity names in text*. Wisconsin: Bioinformatics. pp. 21(14):3191-2 (2005).
11. A. Lally and D. Ferrucci, *Building an Example Application with the Unstructured Information Management Architecture*, IBM Systems Journal 43, No. 3, pp. 455–475 (2004).
12. A. Yakushiji, *Relation Information Extraction Using Deep Syntactic Analysis,* PhD. thesis. University of Tokyo (2006).
13. R. C. Bunescu and R. J. Mooney, *Subsequence kernels for relation extraction*. NIPS (2005).
14. T. Joachims, *Making large-Scale SVM Learning Practical*. In B. Schölkopf and C. Burges and A. Smola (ed.), Advances in Kernel Methods - Support Vector Learning, MIT-Press (1999).
15. J. D. Kim, T. Ohta, Y. Tateishi, and J. Tsujii, *GENIA corpus - a semantically annotated corpus for bio-textmining*. Bioinformatics. pp. 19(suppl. 1):i180–i182 (2003).
16. D. Ferrucci et al. *Towards an Interoperability Standard for Text and Multi-Modal Analytics*. IBM Research Report, RC24122 (2006).
17. A. L. Berger, S. D. Pietra, and V. J. D. Pietra, *A maximum entropy approach to natural language*. Comp. Ling., pp. 22(1):39–71 (1996).
18. S. Ananiadou, D. B. Kell and J. Tsujii, *Text mining and its potential applications in systems biology*. Trends Biotechnol, Vol. 24 (2006).
19. A. Moschitti, *Making tree kernels practical for natural language learning*. Trento, Italy. In Proc. EACL-2006.
20. The Carnegie Mellon University, *UIMA component repository*. http://uima.lti.cs.cmu.edu/
21. H. Cunningham, D. Maynard, K. Bontcheva and V. Tablan. *GATE: an Architecture for Development of Robust HLT*. In Proc. ACL-2002.
22. The JULIE Lab (the Jena University Language & Information Engineering Lab). http://www.julielab.de/
23. R. Sætre et al. *Syntactic features for protein-protein interaction extraction*. LBM2007, to be submitted.
24. M. P. Marcus, B. Santorini, and M. A. Marcinkiewicz. *Building a large annotated corpus of english: The penn treebank*. Comp. Ling. 19:313--330 (1993).
25. S. Pakhomov, J. Buntrock, P. Duffy. *High Throughput Modularized NLP System for Clinical Text (Interactive Poster)*. ACL 2005; Ann Arbor, MI.
26. W. A. Baumgartner, Z Lu, H. L. Johnson, J. G. Caporaso, J. Paquette, A. Lindemann, E. K. White, O. Medvedeva, L. M. Fox, K. B. Cohen, and L. Hunter. *An integrated approach to concept recognition in biomedical text*. Proc. of the Second BioCreative Challenge Evaluation Workshop (2006).

COMPARING USABILITY OF MATCHING TECHNIQUES FOR NORMALISING BIOMEDICAL NAMED ENTITIES

XINGLONG WANG AND MICHAEL MATTHEWS
School of Informatics, University of Edinburgh
Edinburgh, EH8 9LW, UK
{xwang,mmatsews}@inf.ed.ac.uk

String matching plays an important role in biomedical Term Normalisation, the task of linking mentions of biomedical entities to identifiers in reference databases. This paper evaluates exact, rule-based and various string-similarity-based matching techniques. The matchers are compared in two ways: first, we measure precision and recall against a gold-standard dataset and second, we integrate the matchers into a curation tool and measure gains in curation speed when they were used to assist a curator in normalising protein and tissue entities. The evaluation shows that a rule-based matcher works better on the gold-standard data, while a string-similarity based system and exact string matcher win out on improving curation efficiency.

1. Introduction

Term Normalisation (TN) [1] is the task of grounding a biological term in text to a specific identifier in a reference database. TN is crucial for automated processing of biomedical literature, due to ambiguity in biological nomenclature [2, 3, 4, 5]. For example, a system that extracts protein-protein interactions (PPIs) would ideally collapse interactions involving the same proteins, even though these are named by different word forms in the text. This is particularly important if the PPIs are to be entered into a curated database, which refers to each protein by a canonical unique identifier.

A typical TN system consists of three components: an ontology processor, which expands or prunes the reference ontology; a string matcher, which compares entity mentions in articles against entries in the processed ontology; and finally a filter (or a disambiguator) that removes false positive identifiers using rules or statistical models [6, 7]. The string matcher is arguably the core component: a matcher that searches a database and retrieves entries that exactly match an entity mention can form a simple TN system. The other two components are important but they can be viewed as extras that may help further improve the performance of the matcher. A reasonable assumption is that if a matching system can help improve curation speed, then more complex TN systems should be even more helpful. Indeed, the matching systems described in this paper can be used as stand-alone TN modules, and can also work in conjunction with external ontology processors and filters.

Much work has been carried out on evaluating performance of TN systems on Gold Standard datasets [6, 8]. However, whether such systems are really helpful in speeding up curation has not yet been adequately addressed. This paper focuses on investigating matching techniques and attempts to answer which ones are most helpful in assisting biologists to perform TN curation. We emphasise *assisted*, rather than *automated* curation because, at least in the short term, replacing human curators is not practical [9, 10], particularly on TN tasks that involve multiple types of biological entities across numerous organisms. We believe that designing tools that help improve curation efficiency is more realistic. This paper compares different techniques for implementing matching: exact, rule-based, and string similarity methods. These are tested by measuring recall and precision over a Gold Standard dataset, as well as by measuring the time taken to carry out TN curation when using each of the matching systems. In order to examine whether the matching techniques are portable to new domains, we tested them on two types of entities in the curation experiment — proteins and tissues (of human species).

This paper is organised as follows: Section 2 gives a brief overview of related work. Section 3 summarises the matching algorithms that we studied and compared. Section 4 presents experiments that evaluated the matching techniques on Gold Standard datasets, while Section 5 describes an assisted curation task and discusses how the fuzzy matching systems helped. Section 6 draws conclusions and discusses directions of future work.

2. Related Work

TN is a difficult task because of the pervasive variability of entity mentions in the biomedical literature. Thus, a protein will typically be named by many orthographic variants (e.g., *IL-5* and *IL5*) and by abbreviations (e.g., *IL5* for *Interleukin-5*), etc. The focus of this paper is how fuzzy matching techniques [11] can handle such variability. Two main factors affect performance of fuzzy matching: first, the quality of the lexicon, and second, the matching technique adopted. Assuming the same lexicon is used, there are three classes of matching techniques: those that rely on exact searches, those that search using hand-written rules, and those that compute string-similarity scores.

First, with a well constructed lexicon, **exact matching** can yield good results [12, 13]. Second, **rule-based methods**, which are probably the most widely used matching mechanism for TN, have been reported as performing well. Their underlying rationale is to alter the lexical forms of entity mentions in text with a sequence of rules, and then to return the first matching entry in the lexicon. For example, one of the best TN systems submitted to the recent BioCreAtIvE 2 Gene Normalisation (GN) task [14] exploited rules and background knowledge extensively.[a] The third category is **string-similarity matching** approaches. A large amount of work has been carried out on matching by string similarity in fields such as database record linkage. Cohen et al. [15] provided a good overview on a number of metrics, including edit-distance metrics, fast heuristic string comparators, token-based distance metrics, and hybrid methods. In the BioCreAtIvE 2 GN task, several teams used such techniques, including Edit Distance [16], SoftTFIDF [17] and JaroWinkler [18].

Researchers have compared the matching techniques with respect to performance on Gold Standard datasets. For example, Fundel et al. [12] compared their exact matching approach to a rule-based approximate matching procedure implemented in ProMiner [19] in terms of recall and precision. They concluded that approximate search did not improve the results significantly. Fang et al. [20] compared their rule-based system against six string-distance based matching algorithms. They found that by incorporating approximate string matching, overall performance was slightly improved. However, in most scenarios, approximate matching only improved recall slightly and had a non-trivial detrimental effect upon precision. Results reported by Fang et al. [20] and Fundel et al. [12] were based on measuring precision and recall on Gold Standard datasets which contained species-specific gene entities. However, in practice, curators might need to curate not only genes, but many other types of entities.Section 5 presents our investigation on whether matching techniques can assist curation in a setup more analogous to these real-world situations.

3. Matching Techniques

This section outlines the rule-based and the string similarity-based algorithms that were used in our experiments. Evaluation results from the BioCreAtIvE 2 GN task on human genes seem to indicate that rule-based systems perform better. The weakness of rule-based systems, however, is that they may be less portable to new domains. By contrast, string similarity-based matching is more generic and can be easily deployed to deal with new types of entities in new domains.

---

[a]See Hirschman et al. [6] for an overview of the BioCreAtIvE 1 GN task and Morgan and Hirschman [8] for the BioCreAtIvE II GN task.

3.1. *Rule-based Matching*

For each protein mention, we used the following rules[b] to create an ordered list of possible RefSeq[c] identifiers.

(1) Convert the entity mention to lowercase and look up the synonym in a lowercase version of the RefSeq database.
(2) Normalise the mention[d] (NORM MENTION), and look up the synonym in a normalised version of the RefSeq database (NORM lexicon).
(3) Remove prefixes (*p*, *hs*, *mm*, *m*, *p* and *h*), add and remove suffixes (*p*, 1, 2) from the NORM MENTION and look up result in the NORM lexicon.
(4) Look up the NORM MENTION in a lexicon derived from RefSeq (DERIVED lexicon).[e]
(5) Remove prefixes (*p*, *hs*, *mm*, *m*, *p* and *h*), add and remove suffixes (*p*, 1, 2) from the NORM MENTION, and look up result in the DERIVED lexicon.
(6) Look up the mention in the abbreviation map created using the Schwartz and Hearst [21] abbreviation tagger. If this mention has a corresponding long form or corresponding short form, repeat steps 1 through 5 for the corresponding form.

3.2. *String Similarity Measures*

We considered six string-similarity metrics: *Monge-Elkan*, *Jaro*, *JaroWinkler*, *mJaroWinkler*, *SoftTFIDF* and *mSoftTFIDF*. Monge-Elkan is an affine variant of the Smith-Waterman distance function with particular cost parameters, and scaled to the interval $[0, 1]$. The Jaro metric is based on the number and order of the common characters between two strings. A variant of the Jaro measure due to Winkler uses the length of the longest common prefix of the two strings and rewards strings which have a common prefix. A recent addition to this family is a modified JaroWinkler [18] (mJaroWinkler), which adapts the weighting parameters and takes into account factors such as whether the lengths of the two strings are comparable and whether they end with common suffixes.

We also tested a 'soft' version of the TF-IDF measure [22], in which similar tokens are considered as well as identical ones that appear in both strings. The similarity between tokens are determined by a similarity function, where we used

[b]Some of the rules were developed with reference to previous work [13, 20].

[c]See `http://www.ncbi.nlm.nih.gov/RefSeq/`.

[d]Normalising a string involves converting Greek characters to English (e.g., $\alpha \rightarrow$alpha), converting to lowercase, changing sequential indicators to integer numerals (e.g., *i*, *a*, *alpha*$\rightarrow$1, etc.) and removing all spaces and punctuation. For example, *rab1*, *rab-1*, *rabα*, *rab I* are all normalised to *rab1*.

[e]The lexicon is derived by adding the first and last word of each synonym entry in the RefSeq database to the lexicon and also by adding acronyms for each synonym created by intelligently combining the initial characters of each word in the synonym. The resulting list is pruned to remove common entries.

JaroWinkler for SoftTFIDF and mJaroWinkler for mSoftTFIDF. We deem two tokens similar if they have a similarity score that is greater than or equal to 0.95 [17], according to the corresponding similarity function.

4. Experiments on Gold Standard Datasets

We evaluated the competing matching techniques on a Gold Standard dataset over a TN task defined as follows: given a mention of a protein entity in a biomedical article, search the ontology and assign one or more IDs to this protein mention.

4.1. *Datasets and Ontologies*

We conducted the experiments on a protein-protein interaction (PPI) corpus annotated for the TXM [18, 23] project, which aims at producing NLP-based tools to aid curation of biomedical papers. Various types of entities and PPIs were annotated by domain experts, whereas only the TN annotation on proteins was of interest in the experiments presented in this section.[f] 40% of the papers were doubly annotated and we calculated inter-annotator agreement (IAA) for TN on proteins, which is high at 88.40%.

We constructed the test dataset by extracting all 1,366 unique protein mentions, along with their manually normalised IDs, from the PPI corpus. A lexicon customised for this task was built by extracting all synonyms that are associated with RefSeq IDs that were assigned to the protein mentions in the test dataset. In this way, the lexicon was guaranteed to have an entry for every protein mention and the normalisation problem can be simplified as a string matching task.[g] Note that as our data contains proteins from various model organisms, and thus this TN task is more difficult than the corresponding BioCreAtIvE 1 & 2 GN tasks, which dealt with species-specific genes.

4.2. *Experimental Setup*

We applied the rule-based matching system and six similarity-based algorithms to the protein mentions in the test dataset.[h] A case-insensitive (CI) exact match baseline system was also implemented for comparison purpose.

---

[f] We have an extended version of this dataset in which more entity types are annotated. The curation experiment described in Section 5 used protein and tissue entities in that new dataset.

[g] Although we simplified the setup for efficiency, the comparison was fair because all matching techniques used the same lexicon.

[h] We implemented the string-similarity methods based on the *SecondString* package. See `http://secondstring.sourceforge.net/`

Given a protein mention, a matcher searches the protein lexicon, and returns one match. The exact and rule-based matchers return the first match according to the rules and the similarity-based matchers return the match with the highest confidence score. It is possible that a match maps to multiple identifiers, in which case all identifiers were considered as answers.

In evaluation, for a given protein mention, the ID(s) associated with a match retrieved by a matcher are compared to the manually annotated ID. When a match has multiple IDs, we count it as a hit if one of the IDs is correct. Although this setup simplifies the TN problem and assumes a perfect filter that always successfully removes false positives, it allows us to focus on investigating the matching performance without interference from NER errors or errors caused by ambiguity.

4.3. *Results and Discussion*

We used metrics precision (P), recall (R) and $F1$, for evaluation. Table 1 shows performance of the matchers.

Table 1. Precision (P), recall (R) and $F1$ of fuzzy matching techniques as tested on the PPI corpus. Figures are in percentage.

Matcher	P	R	$F1$	*Matcher*	P	R	$F1$
MongeElkan	51.3	51.3	51.3	mSoftTFIDF	66.0	61.5	63.7
Jaro	59.4	59.3	59.3	**Rule-based**	81.4	57.5	**67.4**
JaroWinkler	61.7	61.6	61.7	mJaroWinkler	61.2	61.1	61.1
SoftTFIDF	66.5	62.2	64.3	Exact Match (CI)	77.2	33.8	47.0

Both the rule-based and the string-similarity based approaches outperformed the exact match baseline, and rule-based system outperformed the string-similarity-based ones. Nevertheless, the SoftTFIDF matcher performed only slightly worse than the winner,[i] and we should note that string-similarity based matchers have the advantage of portability, so that they can be easily adopted to other types of biomedical entities, such as tissues and experimental methods, as long as the appropriate lexicons are available.

Among the similarity-based measures, the two SoftTFIDF-based methods outperformed others. As discussed in [22], two advantages of the SoftTFIDF over other similarity-based approaches are: first, token order is not important so permu-

[i]The rule-based system yields higher recall but lower precision than the similarity-based systems. Tuning the balance between recall and precision may be necessary for different curation tasks. See [23] for more discussion on this issue.

tation of tokens are considered the same, and second, common but uninformative words do not greatly affect similarity.

5. Curation Experiment

We carried out a TN curation experiment where three matching systems were supplied to a curator to assist in normalising a number of tissue and protein entities. A matcher resulting in faster curation is considered to be more helpful.

5.1. *Experimental Setup*

We designed a realistic curation task on TN as follows: a curator was asked to normalise a number of tissue and protein entities that occurred in a set of 78 PubMed articles.[j] Tissues were to be assigned to MeSH[k] IDs and proteins to RefSeq IDs. We selected only human proteins for this experiment, because although species is a major source of ambiguity in biological entities [7], we wanted to focus on investigating how matching techniques affect curation speed in this work.

Figure 1. A screenshot of the curation tool.

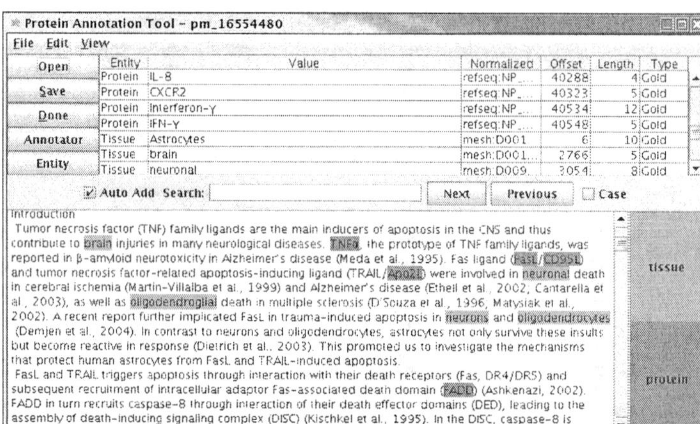

Curation was carried out using an in-house curation tool (as shown in Figure 1). When loaded, the tool displays a full-length article and highlights a number

---

[j]The articles were taken from an extended version of the dataset described in Section 4.1, in which tissues and proteins were already manually marked up and normalised. The normalisations were concealed from the curator and only used after the experiment to assess the quality of the curation.

[k]See http://www.nlm.nih.gov/mesh/MBrowser.html.

of randomly selected protein and tissue entities Only unique entity mentions in each article were highlighted. To make sure that the numbers of entities were distributed evenly in the articles, a maximum of 20 tissues and 20 proteins were highlighted in each article.[l]

We integrated three matching techniques into the curation tool to assist curation: (1) SoftTFIDF, the best performing string-similarity-based matching method in our previous experiment; (2) rule-based matching;[m] and (3) exact matching. The 78 articles were randomly divided into three sets, each of which used a different matching technique, and then the articles were randomly presented to the curator. When an article was loaded into the tool, the normalisations guessed by one of the matchers were also added. When the curator clicked on a highlighted entity mention, a dialogue window would pop up, showing its pre-loaded normalisations, along with a brief description of each ID in order to help the curator select the right ID. The descriptions were extracted from RefSeq and MeSH, consisting of synonyms corresponding to the ID.

The curation tool also provided a search facility. When a matcher misses the correct IDs, the curator can manually query RefSeq and MeSH lexicons. The search facility was rather basic and carried out 'exact' and 'starts with' searches. For example, if a matcher failed to suggest a correct normalisation for protein mention "*α-DG*" and if the curator happened to know that "*DG*" was an acronym for "*dystroglycan*", then she could query the RefSeq lexicon using the term "*alpha-dystroglycan*". We logged the time spent on manual searches, in order to analyse the usefulness of the matching techniques and how they can be further improved. As with the experiments carried out on the Gold Standard dataset, we followed a 'bag' approach, which means that, for each mention, a list of identifiers, instead of a single one, was shown to the curator.[n]

5.2. *Results and Discussion*

Tables 2 and 3[o] show the average curation time that the curator spent on normalising a tissue or a protein with respect to the matching techniques. There are two

[l]Because articles contain different numbers of entities, the total numbers of protein and tissue entities in this experiment are different. See Table 2 and 3 for exact figures.

[m]We used the same system as described in Section 3 for protein normalisation. For tissue normalisation, a rudimentary system was used that first carries out a case-insensitive (CI) match, followed by a CI match after adding and removing an *s* from the tissue mention, and finally adding the MeSH ID for the Cell Line if the mention ends in *cells*.

[n]This is in-line with our evaluation on the gold-standard dataset where a metric of *top n* accuracy was used.

[o]The standard deviations were high due to the fact that some entities are more difficult to normalise than others.

types of normalisation events: 1) a matcher successfully suggested a normalisation and the curator accepted it; and 2) a matcher failed to return a hit, and the curator had to perform manual searches to normalise the entity in question.

Table 2. Time spent on normalising tissues with three matching techniques.

	including manual searches			*excluding manual searches*		
Matcher	*# of entities*	*time(ms)*	*StdDev*	*# of entities*	*time(ms)*	*StdDev*
Exact	283	7,078	8,757	127	2,198	2,268
Rule-based	326	6,639	8,607	172	2,158	1,133
SoftTFIDF	292	**6,044**	7,596	208	2,869	2,463

Table 3. Time spent on normalising proteins with three matching techniques.

	including manual searches			*excluding manual searches*		
Matcher	*# of entities*	*time(ms)*	*StdDev*	*# of entities*	*time(ms)*	*StdDev*
Exact	196	**6,972**	8,859	147	3,714	4,479
Rule-based	129	8,615	12,809	110	6,744	11,030
SoftTFIDF	108	11,218	17,334	88	7,381	9,071

The columns titled "excluding manual searches" and "including manual searches" reflect the two types of events. By examining averaged curation time cost on each, we can see how the matchers helped. For example, from the "excluding manual searches" column in Table 2, we observe that the curator required more time (i.e., 2,869 *ms.*) to find and accept the correct ID from the candidates suggested by SoftTFIDF, whereas the time in the "including manual searches" column shows that overall using SoftTFIDF was faster than the other two matchers. This is because in the majority of cases (208 out of 292), the correct ID was in the list returned by SoftTFIDF, which allowed the curator to avoid performing manual searches and thus saved time. In other words, the curator had to perform time-consuming manual searches more often when assisted by the exact and the rule-based matchers.

Overall, on tissue entities, the curator was faster with help from the SoftTFIDF matcher, whereas on proteins the exact matcher worked better.[p] To explain this, we should clarify that the major elements that can affect curation speed are: 1) the

[p] We performed significance tests on both the protein and tissue data using R. Given that the data is not normally distributed as indicated by the Kolomorov-Smirnov normality test, we used the non-parametric Kruskal-Wallis test which indicates that the differences are significant with $p = .02$ for both data sets.

Table 4. Average bagsizes of tissue and protein entities.

Type	*Matcher*	*Cnt (bagsize>=0)*	*Avg. bagsize*	*Cnt (bagsize=0)*	*Percentage*
	Exact	283	0.43	160	56.5%
Tissue	Rule-based	326	0.66	111	34.0%
	SoftTFIDF	292	5.38	7	2.4%
	Exact	196	0.90	51	26.02%
Protein	Rule-based	129	5.12	14	10.85%
	SoftTFIDF	108	13.97	9	8.50%

performance of the matcher, 2) time cost in eyeballing the IDs suggested, and 3) the time spent on manual searches when the matcher failed.

Therefore, although we evaluated the matchers on a Gold Standard dataset and concluded that the rule-based matcher should work best on normalising protein entities (see Section 4), this does not guarantee that the rule-based matcher will lead to an improvement in curation speed.

The second factor is due to the sizes of the bags. The SoftTFIDF matcher returns smaller sets of IDs for tissues but bigger ones for proteins. Table 4 shows the average bagsizes and the percentage when bagsize is zero, in which case the matcher failed to find any ID. One reason that SoftTFIDF did not help on proteins might be the average bagsize is too big at 13.97, and the curator had to spend time reading the descriptions of all IDs.

As for the third factor, on tissues, 56.5% of the time the exact matcher failed to find any ID and the curator had to perform a manual search; by contrast, the SoftTFIDF matcher almost always returned a list of IDs (97.6%), so very few manual searches were needed.

As mentioned, the articles to curate were presented to the curator in random order, so that the potential influence to performance of normalisation resulting from training curve and fatigue should distribute evenly among the matching techniques and therefore not bias the results. On the other hand, due to limitation in time and resources, we only had one curator to carry out the curation experiment, which may cause the results to be subjective. In the near future, we plan to carry out larger scale curation experiments.

6. Conclusions and Future Work

This paper reports an investigation into the matching algorithms that are key components in TN systems. We found that a rule-based system that performed better in terms of precision and recall, as measured on a Gold Standard dataset, was not the most useful system in improving curation speed, when normalising protein and tissue entities in a setup analogous to a real-world curation scenario. This result highlights concerns that text mining tools achieving better results as measured by traditional metrics might not necessarily be more successful in enhancing curators' efficiency. Therefore, at least for the task of TN, it is critical to measure the usability of text mining tools extrinsically in actual curation exercises. We have learnt that, besides the performance of the matching systems, many other factors are also important. For example, the balance between precision and recall (i.e., presenting more IDs with higher chances to include the correct one, or less IDs where the answer is more likely to be missed), and the backup tool (e.g., the manual search facility in the curation tool) used when the assisting system fails, can both have significant effects on usability. Furthermore, in real-world curation tasks that often involve more than one entity type, approaches with better portability (e.g., string-similarity-based ones) may be preferred. Our results also indicated that it might be a good idea to address different types of entities with different matching techniques.

One direction for future work is to conduct more curation experiments so that the variability between curators can be smoothed (e.g., some curators may prefer seeing more accurate NLP output whereas others may prefer higher recall). Meanwhile, we plan to improve the matching systems by integrating ontology processors and species disambiguators [7].

Acknowledgements

The work reported in this paper was done as part of a joint project with Cognia (`http://www.cognia.com`), supported by the Text Mining Programme of ITI Life Sciences Scotland (`http://www.itilifesciences.com`). We also thank Kirsten Lillie, who carried out curation for our experiment, and Ewan Klein, Barry Haddow, Beatrice Alex and Claire Grover who gave us valuable feedback on this paper.

References

1. M. Krauthammer and G. Nenadic. Term identification in the biomedical literature. *Journal of Biomedical Informatics (Special Issue on Named Entity Recogntion in Biomedicine)*, 37(6):512–526, 2004.
2. L. Hirschman, A. A. Morgan, and A. S. Yeh. Rutabaga by any other name: extracting biological names. *J Biomed Inform*, 35(4):247–259, 2002.
3. O. Tuason, L. Chen, H. Liu, J. A. Blake, and C. Friedman. Biological nomenclature: A source of lexical knowledge and ambiguity. In *Proceedings of PSB*, 2004.
4. L. Chen, H. Liu, and C. Friedman. Gene name ambiguity of eukaryotic nomenclatures. *Bioinformatics*, 21(2):248–256, 2005.
5. K. Fundel and R. Zimmer. Gene and protein nomenclature in public databases. *BMC Bioinformatics*, 7:372, 2006.

6. L. Hirschman, M. Colosimo, A. Morgan, and A. Yeh. Overview of BioCreAtlvE task 1B: normalised gene lists. *BMC Bioinformatics*, 6, 2005.
7. X. Wang. Rule-based protein term identification with help from automatic species tagging. In *Proceedings of CICLING 2007*, pages 288–298, Mexico City, 2007.
8. A. A. Morgan and L. Hirschman. Overview of BioCreative II gene normalisation. In *Proceedings of the BioCreAtlvE II Workshop*, Madrid, 2007.
9. I. Donaldson, J. Martin, B. de Bruijn, C. Wolting, V. Lay, B. Tuekam, S. Zhang, B. Baskin, G. Bader, K. Michalickova, T. Pawson, and C. Hogue. PreBIND and Textomy - mining the biomedical literature for protein-protein interactions using a support vector machine. *BMC Bioinformatics*, 4:11, 2003.
10. Nikiforos Karamanis, Ian Lewin, Ruth Seal, Rachel Drysdale, and Edward Briscoe. Integrating natural language processing with FlyBase curation. In *Proceedings of PSB*, pages 245–256, Maui, Hawaii, 2007.
11. G. Nenadic, S. Ananiadou, and J. McNaught. Enhancing automatic term recognition through term variation. In *Proceedings of Coling*, Geneva, Switzerland, 2004.
12. K. Fundel, D. Güttler, R. Zimmer, and J. Apostolakis. A simple approach for protein name identification: prospects and limits. *BMC Bioinformatics*, 6(Suppl 1):S15, 2005.
13. A. Cohen. Unsupervised gene/protein named entity normalization using automatically extracted dictionaries. In *Proceedings of the ACL-ISMB Workshop on Linking Biological Literature, Ontologies and Databases: Mining Biological Semantics*, 2005.
14. J. Hakenberg, L. Royer, C. Plake, H. Strobelt, and M. Schroeder. Me and my friends: Gene mention normalization with background knowledge. In *Proceedings of the BioCreAtlvE II Workshop 2007*, Madrid, 2007.
15. W. W. Cohen, P. Ravikumar, and S. E. Fienberg. A comparison of string distance metrics for name-matching tasks. In *Proceedings of IIWeb-03 Workshop*, 2003.
16. W. Lau and C. Johnson. Rule-based gene normalisation with a statistical and heuristic confidence measure. In *Proceedings of the BioCreAtlvE II Workshop 2007*, 2007.
17. C. Kuo, Y. Chang, H. Huang, K. Lin, B. Yang, Y. Lin, C. Hsu, and I. Chung. Exploring match scores to boost precision of gene normalisation. In *Proceedings of the BioCreAtlvE II Workshop 2007*, Madrid, 2007.
18. C. Grover, B. Haddow, E. Klein, M. Matthews, L. A. Nielsen, R. Tobin, and X. Wang. Adapting a relation extraction pipeline for the BioCreAtIvE II task. In *Proceedings of the BioCreAtlvE II Workshop 2007*, Madrid, 2007.
19. D. Hanisch, K. Fundel, H-T Mevissen, R Zimmer, and J Fluck. ProMiner: Organism-specific protein name detection using approximate string matching. *BMC Bioinformatics*, 6(Suppl 1):S14, 2005.
20. H. Fang, K. Murphy, Y. Jin, J. Kim, and P. White. Human gene name normalization using text matching with automatically extracted synonym dictionaries. In *Proceedings of the HLT-NAACL BioNLP Workshop*, New York, 2006.
21. A.S. Schwartz and M.A. Hearst. Identifying abbreviation definitions in biomedical text. In *Proceedings of PSB*, 2003.
22. W. W. Cohen and E. Minkov. A graph-search framework for associating gene identifiers with documents. *BMC Bioinformatics*, 7:440, 2006.
23. B. Alex, C. Grover, B. Haddow, M. Kabadjov, E. Klein, M. Matthews, S. Roebuck, R. Tobin, and X. Wang. Assisted curation: does text mining really help? In *The Pacific Symposium on Biocomputing (PSB)*, 2008.

INTRINSIC EVALUATION OF TEXT MINING TOOLS MAY NOT PREDICT PERFORMANCE ON REALISTIC TASKS

J. GREGORY CAPORASO[1], NITA DESHPANDE[2], J. LYNN FINK[3], PHILIP E. BOURNE[3], K. BRETONNEL COHEN[1], AND LAWRENCE HUNTER[1]

[1] *Center for Computational Pharmacology*
University of Colorado Health Sciences Center, Aurora, CO, USA
[2] *PrescientSoft Inc., San Diego, CA, USA*
[3] *Skaggs School of Pharmacy and Pharmaceutical Sciences*
University of California, San Diego, San Diego, CA, USA

Biomedical text mining and other automated techniques are beginning to achieve performance which suggests that they could be applied to aid database curators. However, few studies have evaluated how these systems might work in practice. In this article we focus on the problem of annotating mutations in Protein Data Bank (PDB) entries, and evaluate the relationship between performance of two automated techniques, a text-mining-based approach (MutationFinder) and an alignment-based approach, in intrinsic versus extrinsic evaluations. We find that high performance on gold standard data (an intrinsic evaluation) does not necessarily translate to high performance for database annotation (an extrinsic evaluation). We show that this is in part a result of lack of access to the full text of journal articles, which appears to be critical for comprehensive database annotation by text mining. Additionally, we evaluate the accuracy and completeness of manually annotated mutation data in the PDB, and find that it is far from perfect. We conclude that currently the most cost-effective and reliable approach for database annotation might incorporate manual and automatic annotation methods.

1. Introduction

Biomedical text mining systems have been reaching reasonable levels of performance on gold standard data, and the possibility of applying these systems to automate biological database construction or annotation is becoming practical. These systems are generally evaluated intrinsically—for example, against a gold standard data set with named entities that are tagged by human annotators, judging the system on its ability to replicate the human annotations. Systems are less commonly evaluated extrinsically—i.e., by measuring their contribution to the performance of some task. Intrinsic evaluations of text mining tools are critical to accurately assessing their basic functionality, but they do not necessarily tell us how well a system will perform in practical applications.

Hunter and Cohen (2006) list four text mining systems that are being or have been used to assist in the population of biological databases (LSAT,[2] MuteXt,[3] Textpresso,[4] and PreBIND[5]). Of these four, data on the actual contribution of the tool to the database curation effort is available for only one: the PreBIND system is reported to have reduced the time necessary to perform a representative task by 70%, yielding a 176-person-day time savings. More recently, Karamanis et al. (2007) recorded granular time records for a "paper-by-paper curation" task over three iterations in the design of a curator assistance tool, and noted that curation times decreased as user feedback was incorporated into the design of the tool. In the information retrieval (IR) domain, Hersh et al. (2002) assessed the ability of an IR tool (Ovid) to assist medical and nurse practitioner students in finding answers to clinical questions, and found that performance of the system in intrinsic evaluation did not predict the ability of the system to help users identify answers. Some tasks in the recent BioCreative shared tasks (particularly the GO code assignment task in BioCreative 2004, PPI task in BioCreative 2006, and the GN tasks in both years), and to a lesser extent, of the TREC Genomics track in some years, can be thought of as attempts at extrinsic evaluations of text mining technologies[a]. Camon et al. (2005) gives an insightful analysis of the shortcomings of the specific text mining systems that participated in the BioCreative 2004 GO code assignment task. We are not aware of any work that directly assesses the ability of an automated technique to recreate a large, manually curated data set, although the importance of such evaluations has been noted.[8]

There has recently been much interest in the problem of automatically identifying point mutations in text.[3,9–15] Briefly, comprehensive and accurate databases of mutations that have been observed or engineered in specific biological sequences are often extremely valuable to researchers interested in those sequences, but because the requisite information is generally dispersed throughout the primary literature, manually compiling these databases requires many expert hours[b]. To address this issue, we have developed MutationFinder,[9] an open source, high performance system for identifying descriptions of point mutations in text. We performed an in-depth intrinsic evaluation of MutationFinder on blind, human-annotated test data. For extracting mentions of point mutations from MEDLINE abstracts, the most difficult task it was evaluated on, it achieved 98.4% precision and 81.9% recall.

The availability of this system allows us to ask subsequent questions. First, how effective are manual biological database annotation techniques in terms of accuracy and coverage; and second, does the performance of an automated annotation technique in intrinsic evaluation predict the performance of that system in an extrinsic evaluation? The first question

[a]In fact, some of the earliest work on information extraction in the modern era of BioNLP, such as Craven and Kumlein (1999) and Blaschke et al. (1999), can be thought of as having extrinsic, rather than intrinsic, evaluation.

[b]We present the problem of identifying mutations in text, our approach to addressing it, and a review of the approaches taken by other groups in Caporaso et al. (2007b).

addresses the issue of whether replacement or augmentation of manual database annotation methods with automatic methods is worth exploring, while the second addresses whether the most commonly performed evaluations of automatic techniques translate into information regarding their applicability to real-world tasks.

To address these questions, we compare and evaluate three approaches for annotating mutations in Protein Data Bank (PDB)[17] entries[c]: manual annotation, which is how mutations are currently annotated in the PDB; and two automatic approaches—text-mining-based annotation using MutationFinder, and alignment-based annotation—which we are exploring as possibilities to replace or augment manual annotation. (The PDB is the central repository for 3D protein and nucleic acid structure data, and one of the most highly accessed biomedical databases.) In the following section we present our methods to address these questions and the results of our analyses. We identify problems with all of the annotation approaches, automatic and manual, and conclude with ideas for how to best move forward with database annotation to produce the best data at the lowest cost.

2. Methods and Results

In this section we describe the methods and results of three experiments. First, we evaluate the accuracy and comprehensiveness of the manual mutation annotations in the Protein Data Bank. Then, we extrinsically evaluate our two automated techniques by comparing their results with the manually deposited mutation data in the PDB. Finally, we compare MutationFinder's performance when run over abstracts and full text to address the hypothesis that MutationFinder's low recall in extrinsic evaluation is a result of the lack of information in article abstracts. Unless otherwise noted, all evaluations use a snapshot of the PDB containing the 38,664 PDB entries released through 31 December 2006. All data files used in these analyses are available via `http://mutationfinder.sourceforge.net`.

2.1. Evaluation of manual annotations

When a structural biologist submits a structure to the PDB, they are asked to provide a list of any mutations in the structure. Compiling this information over all PDB entries yields a collection of manually annotated mutations associated with PDB entries, and this mapping between PDB entries and mutations forms the basis of our analyses. We evaluate the accuracy of these annotations by comparing mutation field data with sequence data associated with the same entries. We evaluate the completeness of this data by looking for PDB entries which appear to describe mutant structures but do not contain data in their mutation fields.

[c]Entries in the PDB are composed of atomic cartesian coordinates defining the molecular structure, and metadata, including primary sequence(s) of the molecule(s) in the structure, mutations, primary citation, structure determination method, etc.

2.1.1. Manual mutation annotations

Manual mutation annotations were compiled from the *mutation field* associated with PDB entries. The mutation field is a free-text field in a web-based form that is filled in by researchers during the structure deposition process. The depositor is expected to provide a list of mutations present in the structure (e.g., 'Ala42Gly, Leu66Thr'[d]), but the information provided is not always this descriptive. For example, many mutations fields contain indecipherable information, or simply the word *yes.* In cases where the depositor does not provide any information in the mutation field (as is often the case), differences identified by comparison with an aligned sequence are suggested to the author by a PDB annotator. The author can accept or decline these suggestions.

2.1.2. Normalization of manual mutation annotations

Because the mutation field takes free text input, automated analysis requires normalization of the data. This was done by applying MutationFinder to each non-empty mutation field. Point mutations identified by MutationFinder in a mutation field were normalized. To evaluate this normalization procedure, a non-author biologist manually reviewed a random subset (n=400) of non-empty mutation fields and the normalized mutations output by MutationFinder. Precision of the normalization procedure was 100.0%. Recall was 88.9%. This high performance is not surprising, since the task was relatively simple. It suggests that normalizing mutation fields with MutationFinder is acceptable. 10,504 point mutations in 5971 PDB records were compiled by this approach. This data set is referred to as the *manually deposited mutation annotations.*

2.1.3. Accuracy of manually deposited mutation annotations

To assess the accuracy of the manually deposited mutation annotations, each mutation was validated against the sequence data associated with same PDB entry. (This process is similar to that employed by the MuteXt[3] system.) If a mutation could not be validated against the sequence data, that entry was considered to be inaccurately annotated and was reported to PDB. (Note that this discrepancy could indicate an error in the sequence, an error in the mutation annotation, or a mismatch in sequence numbering.)

Validation of mutations against sequence data was performed as follows. Sequences were compiled for all PDB entries. For a given entry, we checked whether the putative mutant residue was present at the annotated sequence position. For example, PDB entry 3CGT is annotated with the mutation `E257A`. The sequence associated with 3CGT was scanned to determine if alanine, the mutant residue, was present at position 257. In this case it was, so the annotation was retained. If alanine were not present at position 257, the

---

[d]This is a common format for describing point mutations, which indicates that alanine at position 42 in the sequence was mutated to glycine, and leucine at position 66 was mutated to threonine.

annotation would have been labelled as invalid. In cases where PDB entries contain multiple sequences (e.g., a protein composed of several polypeptide chains), each sequence was checked for the presence of the mutant residue.

2.1.4. Coverage of manually deposited mutation annotations

To assess the coverage of the manually annotated data, we attempted to identify PDB records of mutant structures that did not contain data in their mutation field. To identify records of mutant structures, we searched PDB entry titles for any of several keywords that suggest mutations (case insensitive search query: `muta* OR substitut* OR varia* OR polymorphi*`). MutationFinder was also applied to search titles for mentions of point mutations. If a title contained a keyword or mutation mention and the entry's mutation field was blank, the entry was labelled as insufficiently annotated. An informal review of the results suggested that this approach was valid.

2.1.5. Results of manual annotation evaluation

40.6% (4260/10504) of mutations mentioned in mutation fields were not present at the specified position in the sequence(s) associated with the same PDB entry. These inconsistencies were present in 2344 PDB entries, indicating that 39.3% of the 5971 PDB entries with MutationFinder-normalizable mutation fields may be inaccurately annotated. As mentioned, these inaccurate annotations could be due to errors in the mutation annotation or the sequence, or mismatches between the position numbers used in the mutation and the sequence. We expect that in the majority of cases the errors arise from mismatches in numbering, as there is generally some confusion in how mutations should be numbered (i.e., based on the sequence in the structure or based on the UniProt reference sequence). PDB entries now contain mappings between the structure and UniProt sequences, and in a future analysis we will use these mappings to determine if any of these apparent errors are instead inconvenient discrepancies which could be avoided automatically.

Additionally, 21.7% (1243/5729) of the PDB entries that contained a mutation keyword or mention in the title were found to contain an empty mutation field. These entries appear to be underannotated. (As a further indication of the scope of the "underannotation problem," note that 12.9% (1024/7953) of the non-empty mutation fields simply contain the word *yes*.) Again, this is likely to be an overestimate of the true number of underannotated PDB entries (due to promiscuity of the search query), but even if we are overestimating by a factor of 10, this is still a problem.

These results suggest that the manually deposited mutation data is far from perfect, and that not just the quantity, but the *quality* of manual database annotation stands to be improved. In the next section, we explore automated techniques for mutation annotation in the PDB to determine if they may provide a means to replace or augment manual annotation. These automated techniques are evaluated against the manually deposited mutation annotations, although we have just shown that it is far from perfect. Performance of the automated techniques is therefore underestimated.

2.2. Automated mutation annotation evaluated extrinsically

In this section two automated mutation annotation techniques are evaluated by assessing their ability to reproduce the manually deposited mutation annotations in the PDB. The first automated method, text mining for mutations using MutationFinder, has been shown to perform very well on blind test data (i.e., in intrinsic evaluation). Our second approach, detecting differences in pre-aligned sequences, is not inherently error-prone, and therefore does not require intrinsic evaluation. We might expect that the near-perfect and perfect abilities of these systems (respectively) to perform the basic function of identifying mutations would suggest that they are capable of compiling mutation databases automatically. Assessing their ability to recreate the manually deposited mutation annotations allows us to evaluate this expectation.

2.2.1. Text-mining-based mutation annotation: MutationFinder

Two sets of MutationFinder mutation annotations were generated—with and without the sequence validation step described in Section 2.1.3. The unvalidated data should have higher recall, but more false positives. To compile the unvalidated MutationFinder annotation set, MutationFinder was applied to primary citation abstracts associated with PDB records. For each record in the PDB, the abstract of a primary citation (when both a primary citation and an abstract were available) was retrieved, and MutationFinder was applied to extract normalized point mutations. 9625 normalized point mutations were associated with 4189 PDB entries by this method, forming the *unvalidated MutationFinder mutation annotations.* To compile the *validated MutationFinder mutation annotations*, we applied sequence validation to the unvalidated MutationFinder mutation annotations. This reduced the results to 2602 normalized mutations in 2061 PDB entries.

2.2.2. Alignment-based mutation annotation

Validated and unvalidated data sets were also compiled using an alignment-based approach. Sequences associated with PDB entries were aligned with UniProt sequences using `bl2seq`. Differences between aligned positions were considered point mutations, and were associated with the corresponding entries. The alignment approach yielded 23,085 normalized point mutations in 9807 entries (the *unvalidated alignment mutation annotations*). Sequence validation[e] reduced this data set to 14,284 normalized mutations

[e]Sequence validation was somewhat redundant in this case, but was included for completeness. Surprisingly, it was not particularly effective here. The positions assigned to mutations in this approach were taken from the aligned UniProt sequence when sequence start positions did not align perfectly, or when the alignment contained gaps. This resulted in different position numbering between the manually- and alignment-produced annotations, and reduced performance with respect to the manual annotations.

in 6653 entries (the *validated alignment mutation annotations*).

2.2.3. Extrinsic evaluation of automated annotation data

To assess the abilities of the MutationFinder- and alignment-based annotation techniques to recreate the manual annotations, mutation annotations generated by each approach were compared with the manually deposited mutation annotations in terms of precision, recall, and F-measure using the `performance.py` script[f]. Two metrics were scored: *mutant entry identification*, which requires that at least one mutation be identified for each mutant PDB entry, and *normalized mutations*, which requires that each manually deposited mutation annotation associated with a PDB entry be identified by the system. *Mutant entry identification* measures a system's ability to identify structures as mutant or non-mutant, while *normalized mutations* measures a system's ability to annotate the structure with specific mutations.

Normalized mutations were judged against the manually deposited mutation annotations, constructed as described in Section 2.1.1. This set contained 10,504 mutations in 5971 PDB records from a total of 38,664 records. As we noted earlier, many non-empty mutation fields do not contain mutations (e.g., when they contain only the word *yes*). However, in the vast majority of cases, a non-empty mutation field indicates the presence of mutations in a structure. We therefore constructed a different data set for scoring mutant entry identification. We generated a *manually curated mutant entry* data set from all PDB entries which contained non-empty mutation fields. This data set contained 7953 entries (out of 38,664 entries in the PDB snapshot).

2.2.4. Extrinsic evaluation results

We assess the utility of the automated techniques (and combinations of both) for identifying mutant PDB entries (*mutant entry identification*, Table 1a) and annotating mutations associated with PDB entries (*normalized mutations*, Table 1b).

On both metrics, the highest precision results from the intersection of the validated MutationFinder mutation annotations (method 2) and the unvalidated alignment mutation annotations (method 3) data, while the highest recall results from the union of these. Generally, method 2 achieves high precision, and method 3 achieves high recall. None of these approaches achieves a respectable F-measure, although as we point out in Section 2.1.5, these performance values are likely to be underestimates due to noise in the manually deposited mutation annotations.

2.3. MutationFinder applied to abstracts versus full-text

Table 1 shows that MutationFinder (with and without validation) achieves very low recall with respect to the manually deposited mutation annotations. We evaluated the hypothesis that this was a result of mutations not

[f] Available in the MutationFinder package at `http://mutationfinder.sourceforge.net`.

Table 1. Six automated methods for identifying mutant PDB entries are assessed against (a) manually curated mutant entry data, and (b) manually deposited mutation annotations. True positives (TP), false positives (FP), false negatives (FN), precision (P), recall (R), and F-meaure (F) are presented. Sequence validation (methods 2 and 4) is described in Section 2.1.3.

	(a) Mutant Entry Id.	TP	FP	FN	P	R	F
1	MutationFinder	2690	1499	5263	0.642	0.338	0.443
2	MutationFinder + validation	1665	396	6288	**0.808**	0.209	0.333
3	Alignment	6079	3728	1874	0.620	**0.764**	0.685
4	Alignment + validation	4104	2549	3849	0.617	0.516	0.562
5	2 *and* 3	1403	275	6550	**0.836**	0.176	0.291
6	2 *or* 3	6258	3816	1695	0.621	**0.787**	0.694

	(b) Normalized Mutations	TP	FP	FN	P	R	F
1	MutationFinder	2575	7050	7929	0.268	0.245	0.256
2	MutationFinder + validation	1803	799	8701	**0.693**	0.172	0.275
3	Alignment	7681	15404	2823	0.333	**0.731**	0.457
4	Alignment + validation	5059	9225	5455	0.354	0.482	0.408
5	2 *and* 3	1584	532	8920	**0.749**	0.151	0.251
6	2 *or* 3	7900	15671	2604	0.335	**0.752**	0.464

being mentioned in article abstracts, but rather only in the article bodies. A PDB snapshot containing the 44,477 PDB records released through 15 May 2007 was used for this analysis.

2.3.1. Compiling and processing full-text articles

PubMed Central was downloaded through 15 May 2007. XML tags and metadata were stripped. All articles were searched for occurrences of a string matching the format of a PDB ID. (IDs are four characters long: a number, a letter, and two letters or numbers, e.g. 3CGT.) If such a string was found, it was compared to a list of valid PDB IDs; if the string matched a valid PDB ID, the article was retained. This returned 837 articles. From this set, articles that were primary citations for PDB structures were selected, resulting in a set of 70 PDB entries (with 13 manually annotated mutations) for which full text was available.

2.3.2. Comparing abstracts versus full text

MutationFinder with sequence validation (as described in Section 2.1.3) was applied to the abstracts and full-text articles, yielding two mutation data sets. The results were compared against the manually annotated mutation data, allowing us to directly assess the contribution of the article bodies to MutationFinder's performance.

2.3.3. Abstract versus full text results

A 10-fold increase in recall was observed when the article body was provided to MutationFinder in addition to the abstract, with no associated degradation of precision (Table 2). While 70 PDB entries with 13 mutations is a small data set, these data strongly suggest that access to full text is critical for automated mutation annotation by text mining tools. When sequence validation was not applied, normalized mutation and mutant entry identifi-

Table 2. MutationFinder with sequence validation was applied to abstracts and full articles (abstract + article body) for 70 PDB entries. Results are compared with manually annotated data. True positives (TP), false positives (FP), false negatives (FN), precision (P), recall (R), and F-meaure (F) are presented, describing each approach's ability to replicate manually curated data.

Metric	Input	TP	FP	FN	P	R	F
Normalized Mutations	Abstracts	1	0	12	1.000	0.077	0.143
	Full text	10	0	3	1.000	0.769	0.870
Mutant Entry Id.	Abstracts	1	0	9	1.000	0.100	0.182
	Full text	7	0	3	1.000	0.700	0.824

cation recall were perfect, but precision was 11.7% and 38.5%, respectively.

3. Conclusions

These experiments address two questions. First, how effective are manual biological database annotation techniques in terms of accuracy and coverage; and second, does the performance of an automated annotation technique in intrinsic evaluation predict the performance of that system in an extrinsic evaluation? We now present our conclusions regarding these questions, and discuss their implications for database curation.

3.1. Reliability of mutation annotation approaches

The manual *and* automatic approaches to annotating mutations appear to yield significant Type I and II errors when analyzed on the PDB as a whole. This suggests that these methods may be insufficient to generate the required quality and quantity of annotation that is necessary to handle the barrage of data in the biomedical sciences.

Manual annotation of PDB entries is error-prone, as illustrated by our sequence-validation of these data described in Section 2.1.5, and does not guarantee complete annotation. (It should be noted that many of the results that are classified as errors in the manually annotated data are likely to be due to sequence numbering discrepanices. Mappings between PDB sequences and UniProt sequences in the PDB can be used to identify these, and in a future analysis these mappings will be used to reevaluate the manually annotated data.)

The automated mutation annotation approaches also appear to have limitations. MutationFinder (with validation against sequence data) performs well, but full text is probably required for any text mining approach to achieve sufficient recall. Conversely, the alignment-based approach is comprehensive, but overzealous. The manual and automatic methods do frequently validate and complement one another (data not shown due to space restrictions)—in parallel, they may provide a means for improving the quality, while reducing the cost (in person-hours), of database annotation.

3.2. MutationFinder: intrinsic versus extrinsic evaluations

In an intrinsic evaluation against blind gold standard data, MutationFinder achieved 97.5% precision and 80.7% recall on normalized mutations extraction, and 99.4% precision and 89.0% recall on document retrieval.[9,10] In our extrinsic evaluation against manually deposited mutation annotations in the PDB, the exact same system achieved 26.8% precision and 24.5% recall for normalized mutation extraction, and 64.2% precision and 33.8% recall for mutant entry identification (the equivalent of document retrieval in this work). While these are likely to be underestimates of the true utility (Section 2.1.5), the large difference in performance cannot be explained completely by the imperfect extrinsic evaluation. The discrepancy appears to be chiefly explained by two factors: introduction of a systematic source of false positives, and missing data. These issues illustrate that accurately and comprehensively pulling desired information from text is just the beginning of deploying a text mining system as a database curation tool.

False positives were systematically introduced when a single article was the primary citation for several PDB entries, and MutationFinder associated all mutations mentioned in the article with all the citing entries[g]. Our sequence validation step addressed this issue, and improved normalized mutation precision by 42.5 percentage points with an associated degradation in recall of 7.4 percentage points.

False negatives were most common when the targeted information was not present in the primary citation abstracts. In our abstract versus full text analysis, we found that processing the full text with MutationFinder plus sequence validation resulted in a nearly 70 percentage point increase in recall, with no precision degradation. These data result from an analysis of only a small subset of the PDB, but they clearly illustrate the importance of full text for high-recall mutation mining.

We conclude that while it is an essential step in building a text mining system, evaluating a system's performance on gold standard data (intrinsic evaluation) is not necessarily indicative of its performance as a database curation tool (extrinsic evaluation). Identifying and gaining access to the most relevant literature, and identifying and responding to sources of systematic error, are central to duplicating the performance observed on a (well-chosen) gold standard data set in an extrinsic evaluation.

3.3. Alignment-based mutation annotation: extrinsic evaluation

Compiling differences in aligned sequences is not inherently error prone, unlike text mining—beyond unit testing to avoid programming errors, no intrinsic evaluation is necessary. However, this method does not perform perfectly for annotating mutations in the PDB, but rather achieves high recall with low precision.

---

[g]For example, PDB entries 1AE3, 1AE2, and 1GKH all share the same primary citation (PMID: 9098886). This abstract mentions five mutations, all of which MutationFinder associates with each of the three PDB entries. Each of the structures contains only one of the five mutations, so four false positives are incurred for each entry. (The other two mutations referred to are in other structures.) Sequence validation eliminated all of these false positives while retaining all of the true positives.

Error analysis suggests that the primary cause of both false positives and false negatives obtained by alignment-based mutation annotation with respect to the manually deposited mutation annotations is differences in sequence position numbering between the PDB sequence and the UniProt sequence. In PDB entry 1ZTJ, for example, the authors annotated mutations S452A, K455A, T493A, and C500S, while sequence comparison identified S75A, K78A, T115A, and C123S. The (almost identical) relative sequence positions and wild-type and mutant residues suggest that these are the same mutations, but the sequence position offset results in four false positives and four false negatives. Utilizing the mappings between PDB sequence positions and UniProt sequence positions in the PDB should help to alleviate these discrepancies in position numbering. This will be explored in future work, and is expected to significantly reduce these types of errors.

False positives additionally occur as a result of slight differences in the sequence of the solved structure and the closest sequence in UniProt. Differences in the sequences are not *necessarily* mutations induced for analysis, and are therefore not annotated as such. For example, sequence comparison identified six mutations in the PDB entry 1QHO, and the primary citation authors acknowledge several of these as sequence 'discrepancies.' False negatives can also occur when a sequence cannot be aligned to a UniProt sequence, and the alignment-based method cannot be applied, or alternatively if inaccurate information was provided by the depositor. For example, PDB entry 1MWT is annotated with the Y23M mutation, but valine is present at position 23 in the associated sequences. In this case the classification as false negative is an artifact of a problematic manual annotation, rather than a statement about the performance of the annotation technique.

4. Discussion

Automatic annotation can not yet replace manual database curation, even for the relatively simple task of annotating mutations in molecular structures. We evaluated manual curation and two automated methods, and showed that all three are unreliable. Genomic data and their reliable annotation are essential to progress in the biomedical sciences. It has been shown empirically that manual annotation cannot keep up with the rate of biological data generation;[20] furthermore, we have shown here that even if manual annotation could keep pace with data generation, it is still error prone.

A reasonable approach to pursue is the incorporation of automated techniques into manual annotation processes. For example, when a scientist deposits a new PDB structure, their primary citation and sequences can be scanned for mutations. The depositor could be presented with suggestions: *In your abstract, you mention an A42G mutation—is this mutation present in your structure?* Additionally, these tools can be applied as quality control steps. Before a mutation annotation is accepted, it could be validated against sequence data. Responses to such prompts could be recorded and used to generate new gold standards that could be used to improve existing or future tools for automating annotation procedures. 'Smart' annotation deposition systems could be the key to improved quality of data in the present and improved automated techniques in the future.

Acknowledgments

The authors would like to acknowledge Sue Brozowski for evaluating the MutationFinder normalization of PDB mutation fields, Jeffrey Haemer, Kristina Williams, and William Baumgartner for proof-reading and helpful discussion, and the four anonymous reviewers for their insightful feedback. Partial funding for this work came from NIH grants R01-LM008111 and R01-LM009254 to LH.

References

1. Hunter, L. and Cohen, K.B., *Mol Cell* 21, 589–594 (2006).
2. Shah, P.K. and Bork, P., *Bioinformatics* 22, 857–865 (2006).
3. Horn, F., Lau, A.L. and Cohen, F.E., *Bioinformatics* 20, 557–568 (2004).
4. Müeller, H., Kenny, E.E. and Sternberg, P.W., *PLoS Biol* 2, e309 (2004).
5. Donaldson, I., Martin, J., de Bruijn, B., Wolting, C., Lay, V., Tuekam, B., Zhang, S., Baskin, B., Bader, G.D., Michalickova, K., Pawson, T. and Hogue, C.W.V., *BMC Bioinformatics* 4, 11 (2003).
6. Hersh, W.R., Crabtree, M.K., Hickam, D.H., Sacherek, L., Friedman, C.P., Tidmarsh, P., Mosbaek, C, and Kraemer, D. *J American Medical Informatics Association* 9, 283–293 (2002).
7. Karamanis, N., Lewin, I., Sealy, R., Drysdaley, R., and Briscoe, E., *Pacific Symposium on Biocomputing* 12, 245–256 (2007).
8. Cohen, A.M. and Hersh, W., *Briefings in Bioinformatics* 6, 57–71 (2005).
9. Caporaso, J.G., Baumgartner Jr., W.A., Randolph, D.A., Cohen, K.B. and Hunter, L., *Bioinformatics* 23, 1862–1865 (2007).
10. Caporaso, J.G., Baumgartner Jr., W.A., Randolph, D.A., Cohen, K.B. and Hunter, L., *J. Bioinf. and Comp. Bio. (accepted, pub. Dec. 2007)*, (2007b).
11. Rebholz-Schuhmann, D., Marcel, S., Albert, S., Tolle, R., Casari, G. and Kirsch, H., *Nucl. Acids Res.* 32, 135–142 (2004).
12. Baker, C.J.O. and Witte, R., *Journal of Information Systems Frontiers* 8, 47–57 (2006).
13. Lee, L.C., Horn, F. and Cohen, F.E., *PLoS Comput Biol* 3, e16 (2007).
14. Bonis, J., Furlong, L.I. and Sanz, F., *Bioinformatics* 22, 2567–2569 (2006).
15. Witte, R., Kepler, T., and Baker, C.J.O., *Int J Bioinformatics Research and Methods* 3 389–413 (2007).
16. Camon, E.B., Barrell, D.G., Dimmer, E.C., Lee, V., Magrane, M., Maslen, J., Binns, D. and Apweiler, R., *BMC Bioinformatics* 6 Suppl 1, S17 (2005).
17. Berman, H.M., Westbrook, J., Feng, Z., Gilliland, G., Bhat, T.N., Weissig, H., Shindyalov, I.N. and Bourne, P.E., *Nucleic Acids Res* 28, 235–242 (2000).
18. Craven, M. and Kumlien, J., *ISMB 1999*, (1999).
19. Blaschke, C., Andrade, M.A., Ouzounis, C. and Valencia, A., *ISMB 1999*, 60–67 (1999).
20. Baumgartner Jr., W.A., Cohen, K.B., Fox, L.M., Acquaah-Mensah, G. and Hunter, L. *Bioinformatics* 23, 14 (2007).

BANNER: AN EXECUTABLE SURVEY OF ADVANCES IN BIOMEDICAL NAMED ENTITY RECOGNITION

ROBERT LEAMAN

Department of Computer Science and Engineering, Arizona State University

GRACIELA GONZALEZ*

Department of Biomedical Informatics, Arizona State University

There has been an increasing amount of research on biomedical named entity recognition, the most basic text extraction problem, resulting in significant progress by different research teams around the world. This has created a need for a freely-available, open source system implementing the advances described in the literature. In this paper we present BANNER, an open-source, executable survey of advances in biomedical named entity recognition, intended to serve as a benchmark for the field. BANNER is implemented in Java as a machine-learning system based on conditional random fields and includes a wide survey of the best techniques recently described in the literature. It is designed to maximize domain independence by not employing brittle semantic features or rule-based processing steps, and achieves significantly better performance than existing baseline systems. It is therefore useful to developers as an extensible NER implementation, to researchers as a standard for comparing innovative techniques, and to biologists requiring the ability to find novel entities in large amounts of text.

BANNER is available for download at http://banner.sourceforge.net.

1. Introduction

With molecular biology rapidly becoming an information-saturated field, building automated extraction tools to handle the large volumes of published literature is becoming more important. This need spawned a great deal of research into named entity recognition (NER), the most basic problem in automatic text extraction. Several challenge evaluations such as BioCreative have demonstrated significant progress [19, 20], with teams from around the world implementing creative solutions to the known challenges in the field such as the unseen word problem and the mention boundary problem. Although there are other open-source NER systems such as ABNER [11] and LingPipe [1] which are freely available and have been extensively used through the years as

* Partially supported by NSF CISE grant 0412000 (BPC supplement)

baseline systems, the advances since the creation of these systems have mostly remained narrated in published papers, and are generally not available as easily deployable code. Thus the field now sees a great need for a freely-available, open-source system implementing these advances for a more accurate reflection of what a baseline system should achieve, allowing researchers to focus on alternative approaches or extensions to the known techniques. In other words, the field needs an updated measuring stick.

We present here BANNER, an open-source biomedical named-entity recognition system implemented using conditional random fields, a machine learning technique. It represents an innovative combination of known advances beyond the existing open-source systems such as ABNER and LingPipe, in a consistent, scalable package that can easily be configured and extended with additional techniques. It is intended as an executable survey of the best techniques described in the literature, and is designed for use directly by biologists, by developers as a building block, or as a point of comparison when experimenting with alternative techniques.

2. Background

Named entity recognition (NER) is the problem of finding references to entities (*mentions*) such as genes, proteins, diseases, drugs, or organisms in natural language text, and labeling them with their location and type. Named entity recognition in the biomedical domain is generally considered to be more difficult than other domains, such as newswire, for several reasons. First, there are millions of entity names in use [19] and new ones are added constantly, implying that neither dictionaries nor training data will be sufficiently comprehensive. Second, the biomedical field is moving too quickly to build a consensus on the name to be used for a given entity [6] or even the exact concept defined by the entity itself [19], while very similar or even identical names and acronyms are used for different concepts [6], all of which results in significant ambiguities. Although there are naming conventions, authors frequently do not follow them and instead prefer to introduce their own abbreviation and use that throughout the paper [2, 19]. Finally, entity names in biomedical text are longer on average than names from other domains, it is generally much easier – for both humans and automated systems – to determine whether an entity name is present than it is to detect its boundaries [7, 19, 20].
Named entity recognition is typically modeled as a *label sequence problem*, which may be defined formally as follows: Given a sequence of input tokens $\boldsymbol{x} = (x_1 \ldots x_n)$, and a set of labels $\boldsymbol{L}$, determine a sequence of labels $\boldsymbol{y} = (y_1, \ldots, y_n)$ such that $y_i \in \boldsymbol{L}$ for $1 \leq i \leq n$. In the case of named entity recognition the labels

incorporate two concepts: the type of the entity (e.g. whether the name refers to a protein or a disease), and the position of the token within the entity. The simplest model for token position is the *IO* model, which indicates whether the token is **I**nside an entity or **O**utside of a mention. While simple, this model cannot differentiate between a single mention containing several words and distinct mentions comprising consecutive words [21]. The next-simplest model used is *IOB* [11], which indicates whether each token is at the **B**eginning of an entity, **I**nside an entity, or **O**utside. This model is capable of differentiating between consecutive entities and has good support in the literature. The most complex model commonly used is *IOBEW*, which indicates whether each token is at the **B**eginning of an entity, **I**nside an entity, at the **E**nd of an entity, a one-**W**ord entity, or **O**utside an entity. While the *IOBEW* model does not provide greater expressive power than the *IOB* model, some authors have found it to provide the machine learning algorithm with greater discriminative power, which may translate into higher accuracy [16]. Example sentences annotated using each label model can be found in table 1.

Table 1. Example sentences labeled using each of the common labeling models, taken from the BioCreative 2 GM training corpus [19].

<table>
<tr><th>Label model</th><th>Example</th></tr>
<tr><td>IO</td><td>Each|O immunoprecipitate|O contained|O a|O complex|O of|O N1|I-GENE (|I-GENE 3eltaic|I-GENE)|I-GENE and|O CBF1|I-GENE .|O</td></tr>
<tr><td>IOB</td><td>TNFalpha|B-GENE and|O IL|B-GENE -|I-GENE 6|I-GENE levels|O were|O determined|O in|O the|O culture|O supernatants|O .|O</td></tr>
<tr><td>IOBEW</td><td>CES4|W-GENE on|O a|O multicopy|O plasmid|O was|O unable|O to|O suppress|O tif1|B-GENE -|I-GENE A79V|E-GENE .|O</td></tr>
</table>

Conditional random fields (CRF) [14] are a machine learning technique that forms the basis for several other notable NER systems including ABNER [11]. The technique can be seen as a way to “capture” the hidden patterns of labels, and “learn” what would be the likely output considering these patterns. Like all supervised machine learning techniques, a CRF-based system must be trained on labeled data. In general, a CRF is modeled as an arbitrary undirected graph, but linear-chain CRFs, their linear form, are used for sequence labeling. In a CRF, each input x_i from the sequence of input tokens $\boldsymbol{x} = (x_1 \dots x_n)$ is a vector of real-valued *features* or descriptive characteristics, for example, the part of speech. As each token is labeled, these features are used in conjunction with the pattern of labels previously assigned (the history) to determine the most likely label for the current token. To achieve tractability, the length of the history used, called the *order*, is limited: a 1st-order CRF uses the last label output, a 2nd-order CRF uses the last two labels, and so on. There are several good introductions to conditional random fields, such as [14] and [18].

As a discriminant model, conditional random fields use conditional probability for inference, meaning that they maximize $p(\boldsymbol{y}|\boldsymbol{x})$ directly, where $\boldsymbol{x}$ is the input sequence and $\boldsymbol{y}$ is the sequence of output labels. This gives them an advantage over generative models such as Hidden Markov Models (HMMs), which maximize the joint probability $p(\boldsymbol{x}, \boldsymbol{y})$, because generative models require the assumption that the input features are independent given the label. Relaxing this assumption allows discriminatively trained models such as CRFs to retain high performance even though the feature set contains highly redundant features such as overlapping n-grams or features irrelevant to the corpus to which it is currently being applied. This, in turn, enables the developer to employ a large set of rich features, by including any arbitrary feature the developer believes may be useful [14]. In addition, tolerating irrelevant features makes the feature set more robust with respect to applications to different corpora, since features irrelevant to one corpus may be quite relevant in another [6].

In contrast, another significant machine learning algorithm – support vector machines (SVMs) – also tolerate interdependent features, but the standard form of SVMs only support binary classification [21]. Allowing a total of only 2 labels implies that they may only recognize one entity type and only employ the *IO* model for label position, which cannot distinguish between adjacent entities.

3. Architecture

The BANNER architecture is a 3-stage pipeline, illustrated in Figure 1. Input is taken one sentence at a time and separated into *tokens*, contiguous units of meaningful text roughly analogous to words. The stream of tokens is converted to *features*, each of which is a name/value pair for use by the machine learning algorithm. The set of features encapsulates all of the information about the token the system believes is relevant to whether or not it belongs to a mention. The stream of features is then *labeled* so that each token is given exactly one label, which is then output.

The tokenization of biomedical text is not trivial and affects what can be considered a mention since generally only whole tokens are labeled in the output [20]. Unfortunately, tokenization details are often not provided in the biomedical named entity recognition literature. BANNER uses a simple tokenization which breaks tokens into either a contiguous block of letters and/or digits or a single punctuation mark. For example, the string "Bub2p-dependent" is split into 3 tokens: "Bub2p", "-", and "dependent". While this simple tokenization generates a greater number of tokens than a more compact representation would, it has the advantage of being highly consistent.

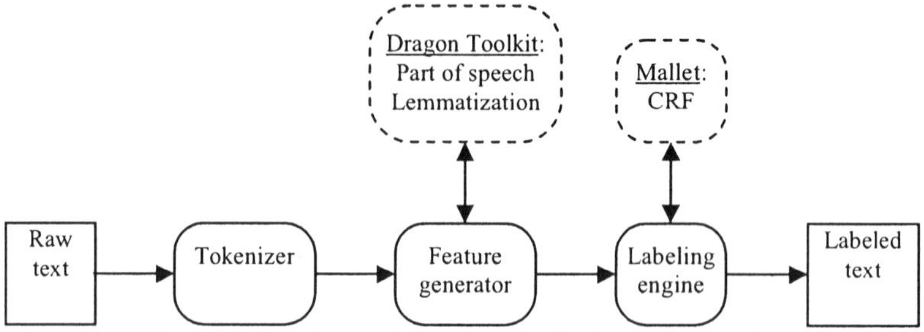

Figure 1. BANNER architecture. Raw sentences are tokenized, converted to features, and labeled. The Dragon toolkit [22] (POS) and Mallet [8] are used for part of the implementation.

BANNER uses the CRF implementation of the latest version of the Mallet toolkit (version 0.4) [8] for both feature generation and labeling using a second order CRF. The set of machine learning features used primarily consist of orthographic, morphological and shallow syntax features and is described in table 2. While many systems use some form of stemming, BANNER instead employs lemmatization [16], which is similar in purpose except that words are converted into their base form instead of simply removing the suffix. Also notable is the numeric normalization feature [15], which replaces the digits in each token with a representative digit (e.g. "0"). Numeric normalization is useful since entity names often occur in series, such as the gene names *Freac1*, *Freac2*, etc. The numeric-normalized value for all these names is *Freac0*, so that forms not seen in the training data have the same representation as forms which are seen. The entire set of features is used in conjunction with a token window of 2 to provide context, that is, the features for each token include the features for the previous two tokens and the following two tokens.

Table 2. The machine learning features used in BANNER (aside from the token itself), primarily based on orthographic, morphological and shallow syntax features.

Feature set description	Notes
The part of speech which the token plays in the sentence	Provided by the Dragon toolkit [22] implementation of the Hepple tagger [5]
The lemma for the word represented by the token, if any	Provided by the Dragon toolkit [22]
A set of regular expression features	Includes variations on capitalization and letter/digit combinations, similar to [9, 11, 15]
2, 3 and 4-character prefixes and suffixes	
2 and 3 character n-grams	Including start-of-token and end-of-token indicators
Word class	Convert upper-case letters to "A", lower-case letters to "a", digits to "0" and other characters to "x" [11]
Numeric normalization	Convert digits to "0" [15]
Roman numerals	
The names of the Greek letters	

There are features discussed in the literature which are not implemented in BANNER, particularly *semantic features* such as a match to a dictionary of names and *deep syntactic features*, such as information derived from a full parse of each sentence. Semantic features generally have a positive impact on overall performance [20] but often have a deleterious effect on recognizing entities not in the dictionary [11, 21]. Moreover, employing a dictionary reduces the flexibility of the system to be adapted to other entity types, since comparable performance will only be achieved after the creation of a comparable dictionary. While such application-specific performance increases are not the purpose of a system such as BANNER, this is an excellent example of an adaptation which researchers may easily perform to improve BANNER's performance for a specific domain.

Deep syntactic features are derived from a full parse of the sentence, which is a noisy and resource-intensive operation with no guarantee that the extra information derived will outweigh the additional errors generated [6]. The use of deep syntactic features in biomedical named entity recognition systems is not currently common, though they have been used successfully. One example is the system submitted by Vlachos to BioCreative 2 [16], where features derived from a full syntactic parse boosted the overall F-score by 0.51.

Unlike many similar-performing systems, BANNER does not employ rule-based post-processing steps. Rules created for one corpus tend to not generalize well to other corpora [6]. Not using such methods therefore enhances the flexibility of the system and simplifies the process of employing it on different corpora or for other entity types [9].

There are, however, two types of general post-processing which have good support in the literature and are sufficiently generic to be applicable to any biomedical text. The first of these is detecting when matching parenthesis, brackets or double quotation marks receive different labels [4]. Since these punctuation marks are always paired, detecting this situation is useful because it clearly demonstrates that the labeling engine has made a mistake. BANNER implements this form of processing by dropping any mention which contains mismatched parenthesis, brackets or double quotation marks.

The second type of generally-applicable post-processing is called abbreviation resolution [21]. Authors of biomedical papers often introduce an abbreviation for an entity by using a format similar to "antilymphocyte globulin (ALG)" or "ALG (antilymphocyte globulin)". This format can be detected with a high degree of accuracy by a simple algorithm [12], which then triggers additional processing to ensure that both mentions are recognized. The implementation of this form of post-processing is left as future work.

Extending BANNER for use in a specialized context or for testing new ideas is straightforward since the majority of the complexity in the implementation resides in the conversion of the data between different formats. For instance, most of the upgrades above the initial implementation (described in the next section) required only a few lines of code. Configuration settings are provided for the common cases, such as changing the order of the CRF model or adding a dictionary of terms.

4. Analysis

BANNER was evaluated with respect to the training corpus for the BioCreative 2 GM task, which contains 15,000 sentences from MEDLINE abstracts and mentions over 18,000 entities. The evaluation was performed by comparing the system output to the human-annotated corpus in terms of the precision (p), recall (r) and their harmonic mean, the F-measure (F). These are based on the number of true positives (TP), false positives (FP) and false negative (FN) returned by the system:

$$p = \frac{TP}{TP+FP} \qquad r = \frac{TP}{TP+FN} \qquad F = \left(\frac{p^{-1}+r^{-1}}{2}\right)^{-1} = \frac{2pr}{p+r}$$

The entities in the BioCreative 2 GM corpus are annotated at the individual character level, and approximately 56% of the mentions have at least one alternate mention annotated, and mentions are considered a true positive if they exactly match either the main annotation or any of the alternates. The evaluation of BANNER was performed using 5x2 cross-validation, which Dietterich shows to be more powerful than the more common 10-fold cross validation [3]. Differences in the performance reported are therefore more likely to be due to a real difference in the performance of the two systems rather than a chance favorable splitting of the data.

The initial implementation of BANNER included only a naïve tokenization which always split tokens at letter/digit boundaries and employed a 1st-order CRF. This implementation was improved by changing the tokenization to not split tokens at the letter/digit boundaries, changing the CRF order to 2, implementing parenthesis post-processing and adding lemmatization, part-of-speech and numeric normalization features. Note that both the initial and final implementations employed the *IOB* label model. In table 3 we present evaluation results for the initial and final implementations, as well as several system variants created by removing a single improvement from the final implementation.

Table 3. Results of evaluating the initial version of the system, the final version, and several system variants created by removing a single improvement from the final implementation.

BANNER System Variant	Precision (%)	Recall (%)	F-Measure
Initial implementation	82.39	76.21	79.18
Final implementation	**85.09**	**79.06**	**81.96**
With IO model instead of IOB	84.71	79.40	81.96
Without numeric normalization	84.56	79.09	81.74
With IOBEW model instead of IOB	85.46	78.15	81.64
Without parenthesis post-processing	84.05	79.27	81.59
Using 1st order CRF instead of 2nd order	84.49	78.72	81.50
With splitting tokens between letters and digits	84.54	78.35	81.33
Without lemmatization	84.44	78.00	81.09
Without part-of-speech tagging	84.02	77.83	80.81

The only system variant which had similar overall performance was the *IO* model, due to an increase in recall. This setting was not retained in the final implementation, however, due to the fact that the *IO* model cannot distinguish between adjacent entities. All other modifications result in decreased overall performance, demonstrating that each of the improvements employed in the final implementation contributes positively to the overall performance.

5. Comparison

We compare the performance of BANNER against the existing freely-available systems in use, we compare its performance against ABNER [11] and LingPipe [1], chosen because they are the most commonly used baseline systems in the literature [17, 19]. The evaluations are performed using 5x2 cross validation using the BioCreative 2 GM task training corpus, and reported in table 4. To demonstrate portability we also perform an evaluation using 5x2 cross validation on the disease mentions of the BioText disease-treatment corpus [10]. These results are reported in table 5. We believe that the relatively low performance of all three systems on the BioText corpus is due to the small size (3655 sentences) and the fact that no alternate mentions are provided.

Table 4. Results of comparing BANNER against existing freely-available software, using 5x2 cross-validation on the BioCreative 2 GM task training corpus.

System	Precision (%)	Recall (%)	F-Measure
BANNER	85.09	79.06	81.96
ABNER [11]	83.21	73.94	78.30
LingPipe [1]	60.34	70.32	64.95

Table 5. Results of comparing BANNER against existing freely-available software, using 5x2 cross-validation on the disease mentions from the BioText disease/treatment corpus [10].

System	Precision (%)	Recall (%)	F-Measure
BANNER	68.89	45.55	54.84
ABNER [11]	66.08	44.86	53.44
LingPipe [1]	55.41	47.50	51.15

Like BANNER, ABNER is also based on conditional random fields; however it uses a 1st-order model and employs a feature set which lacks part-of-speech, lemmatization and numeric normalization features. In addition, it does not employ any form of post-processing, though it does use the same *IOB* label model. ABNER employs a more sophisticated tokenization than BANNER, however this tokenization is incorrect for 5.3% of the mentions in the BioCreative 2 GM task training corpus.

LingPipe is a well-developed commercial platform for various information extraction tasks that has been released free-of-charge for academic use. It is based on a 1st-order Hidden Markov Model with variable-length n-grams as the sole feature set and uses the *IOB* label model for output. It has two primary configuration settings, the maximum length of n-grams to use and whether to use smoothing. For the evaluation we tested all combinations of max n-gram={4...9} and smoothing={*true*, *false*} and found that the difference between the maximum and the minimum performance was only 2.02 F-measure. The results reported here are for the maximum performance, found at max n-gram=7 and smoothing=*true*. Notably, LingPipe requires significantly less training time than either BANNER or ABNER.

The large number of systems (21) which participated in the BioCreative 2 GM task in October of 2006 provides a good basis for comparing BANNER to the state of the art in biomedical named entity recognition. Unfortunately, the official evaluations for these systems used a test corpus that has not yet been made publicly available. The conservative 5x2 cross-validation used for evaluating BANNER still allows a useful direct comparison, however, since BANNER achieves higher performance than the median system in the official BioCreative results, even with a significant handicap against it: the BioCreative systems were able to train on the entire training set (15,000 sentences) while BANNER was only trained on half of the training set (7,500 sentences) because the other half was needed for testing. These results are reported in table 6.

Table 6. Comparison of BANNER to select BioCreative 2 systems [19]. A difference in the F-measure of 1.23 or more is significant and a difference of 0.35 or less is not ($p < 0.05$).

System or author	Rank at BioCreative 2	Precision (%)	Recall (%)	F-Measure
Ando [19]	1	88.48	85.97	87.21
Vlachos [16, 19]	9	86.28	79.66	82.84
BANNER	–	85.09	79.06	81.96
Baumgartner et. al. [19]	11 (median)	85.54	76.83	80.95
NERBio [15, 19]	13	92.67	68.91	79.05

Unlike BANNER, most of the systems submitted to BioCreative 2 were competitive systems employing features or post-processing rules specific to genes [19], a notable exception being the system submitted by Vlachos [16]. The results reported for those systems may therefore not generalize to other entity types or corpora. Moreover the authors are unaware of any of the BioCreative 2 GM systems being publicly available, as of July 2007, except for NERBio [15], which is available for limited manual testing over the Internet*, but not for download.

6. Conclusion & Future Work

We have shown that BANNER, an executable survey of advances in named entity recognition, achieves significantly better performance than existing open-source systems. This is accomplished using features and techniques which are well-supported in the more recent literature. In addition to confirming the value of these techniques and indicating that the field of biomedical named entity recognition is making progress, this work demonstrates that there are sufficient known techniques in the field to achieve good results using known techniques.

We anticipate that this system will be valuable to the biomedical NER community both by providing a benchmark level of performance for comparison and also by providing a platform upon which more advanced techniques can be built. We also anticipate that this work will be immediately useful for information extraction experiments, possibly by including minimal extensions such as a dictionary of names of types of entities to be found.

Future work for BANNER includes several general techniques which have good support in the literature but have not yet been incorporated. For example, authors have noted that part-of-speech systems trained on biomedical text gives superior performance to taggers such as the Hepple tagger which are not specifically intended for biomedical text [6]. We performed one experiment using the Dragon toolkit implementation of the MedPost POS tagger [13], which resulted in slightly improved precision (+0.18%), but significantly lower recall (-1.44%), degrading overall performance by 0.69 F-measure. We plan to test other taggers trained on biomedical text and anticipate achieving a small improvement to the overall performance.

A second technique which has strong support in the literature but is not yet implemented in BANNER is feature induction [7, 9, 15]. Feature induction is the creation of new compound features by forming a conjunction between adjacent singleton features. For example, knowing that the current token contains capital letters, lower-case letters and digits (a singleton pattern probably indicating an acronym) *and* knowing that next token is "gene" is a

* http://140.109.19.166/BioNER

stronger indication that the current token is part of a gene mention than either fact alone. Feature induction employs feature selection during training to automatically discover the most useful conjunctions, since the set of all conjunctions of useful length is prohibitively large. While this significantly increases the amount of time and resources required for training, McDonald & Pereira [9] report an increase in the overall performance of their system by 2% F-measure and we anticipate BANNER would experience a similar improvement.

Acknowledgements

The authors wish to thank Jörg Hackenburg for helpful discussions and for suggesting the BioText corpus. The authors would also like to thank the anonymous reviewers for many useful suggestions.

References

1. Baldwin, B.; and B. Carpenter. LingPipe. *http://www.alias-i.com/lingpipe/*
2. Chen, L.; H. Liu; and C. Friedman. (2005) Gene name ambiguity of eukaryotic nomenclatures. *Bioinformatics* 21, pp 248-255.
3. Dietterich, T. (1998) Approximate Statistical Tests for Comparing Supervised Classification Learning Algorithms. *Neural Computation* 10, pp. 1895-1923.
4. Dingare, S.; et al. (2005) A system for identifying named entities in biomedical text: how results from two evaluations reflect on both the system and the evaluations. *Comparative and Functional Genomics* 6, pp. 77-85.
5. Hepple, M. (2000) Independence and commitment: Assumptions for rapid training and execution of rule-based POS taggers. *Proceedings of the 38th Annual Meeting of the Association for Computational Linguistics (ACL-2000)*, Hong Kong.
6. Leser, U.; and J. Hakenberg. (2005) What makes a gene name? Named entity recognition in the biomedical literature. *Briefings in Bioinformatics* 6, pp. 357-369.
7. McCallum, A. (2003) Efficiently Inducing Features of Conditional Random Fields. *Proceedings of the 19th Annual Conference on Uncertainty in Artificial Intelligence (UAI-03)*, San Francisco, California, pp. 403-441.
8. McCallum, Andrew. (2002) MALLET: A Machine Learning for Language Toolkit. *http://mallet.cs.umass.edu*
9. McDonald, R.; and F. Pereira. (2005) Identifying gene and protein mentions in text using conditional random fields. *BMC Bioinformatics* 6 (Suppl. 1):S6.

10. Rosario, B.; M. A. Hearst. (2004) Classifying Semantic Relations in Bioscience Text. *Proceedings of the 42nd Annual Meeting of the Association for Computational Linguistics (ACL 2004)*
11. Settles, B. (2004) Biomedical Named Entity Recognition Using Conditional Random Fields and Rich Feature Sets. *Proceedings of the COLING 2004 International Joint Workshop on Natural Language Processing in Biomedicine and its Applications (NLPBA/BioNLP).*
12. Schwartz, A.S.; and Hearst, M.A. (2003) A Simple Algorithm for Identifying Abbreviation Definitions in Biomedical Text. *PSB 2003* pp 451-462.
13. Smith, L.; T. Rindflesch; and W.J. Wilbur. (2004) MedPost: a part-of-speech tagger for bioMedical text. *Bioinformatics* 20, pp. 2320-2321.
14. Sutton, C.; and A. McCallum. (2007) An Introduction to Conditional Random Fields for Relational Learning. *Introduction to Statistical Relational Learning*, MIT Press.
15. Tsai, R.; et al. (2006) NERBio: using selected word conjunctions, term normalization, and global patterns to improve biomedical named entity recognition. *BMC Bioinformatics* 7 (Suppl. 5):S11.
16. Vlachos, A. (2007) Tackling the BioCreative 2 gene mention task with Conditional random fields and syntactic parsing. *Proceedings of the Second BioCreative Challenge Workshop* pp. 85-87.
17. Vlachos, A.; C. Gasperin; I. Lewin; and T. Briscoe. (2006) Bootstrapping the recognition and anaphoric linking of named entities in Drosophila articles. *PSB 2006* pp. 100-111.
18. Wallach, H.M. (2004) Conditional Random Fields: An Introduction. University of Pennsylvania CIS Technical Report MS-CIS-04-21.
19. Wilbur, J.; L. Smith; and T. Tanabe. (2007) BioCreative 2. Gene Mention Task. *Proceedings of the Second BioCreative Challenge Workshop* pp. 7-16.
20. Yeh, A.; A. Morgan; M. Colosimo; and L. Hirschman. (2005) BioCreAtIvE Task 1A: gene mention finding evaluation. *BMC Bioinformatics* 6 (Suppl. 1):S2.
21. Zhou, G.; et al. (2005) Recognition of protein/gene names from text using an ensemble of classifiers. *BMC Bioinformatics* 6 (Suppl 1):S7.
22. Zhou, X.; X. Zhang; and X. Hu. (2007) Dragon Toolkit: Incorporating Auto-learned Semantic Knowledge into Large-Scale Text Retrieval and Mining. *Proceedings of the 19th IEEE International Conference on Tools with Artificial Intelligence (ICTAI).*